WORLD *of* MATHEMATICS

WORLD *of* MATHEMATICS

Brigham Narins, *Editor*

Volume 2

M - Z

General Index

GALE GROUP

Detroit
New York
San Francisco
London
Boston
Woodbridge, CT

STAFF

Brigham Narins, *Editor*

Zoran Minderovic, *Associate Editor*

Ellen S. Thackery, *Associate Editor*

Mark Springer, *Editorial Technical Consultant*

Margaret A. Chamberlain, *Permissions Specialist*
Shalice Shah-Caldwell, *Permissions Associate*

Mary Beth Trimper, *Composition Manager*
Evi Seoud, *Assistant Production Manager*
Stacy Melson, *Buyer*

Kenn Zorn, *Product Design Manager*
Michelle DiMercurio, *Senior Art Director*

Barbara Yarrow, *Manager, Imaging and Multimedia Content*
Randy Bassett, *Imaging Supervisor*
Robert Duncan, *Senior Imaging Specialist*

Indexing provided by Julie Shawvan.
Illustrations created by Electronic Illustrators Group, Morgan Hill, California.

ISBN: 0-7876-3652-5 (set)
ISBN: 0-7876-5064-1 (Volume 1)
ISBN: 0-7876-5065-X (Volume 2)

Printed in the United States of America
10 9 8 7 6 5 4 3 2 1

Library of Congress Cataloging-in-Publication Data
World of mathematics / Brigham Narins, editor.
    p. cm.
  Includes bibliographic references and index.
    ISBN 0-7876-3652-5 (set : alk. paper)—ISBN 0-7876-5064-1 (v. 1)—ISBN 0-7876-5065-X (v. 2)
    1. Mathematics-Encyclopedias. I. Narins, Brigham, 1962-
QA5 .W67 2001
510'.3—dc21
                                                00-051408

CONTENTS

INTRODUCTION

The works of mathematics are all around us. Born into the age of technology, we don't think twice about relying on machines whose interiors are mysteries to us, and that depend on mathematics in order to function. From eyeglasses and contact lenses, whose shapes depend on the principles of geometric optics, to CD players, which owe their existence to the ideas of quantum mechanics, the fruits of our mathematical understanding have improved the quality and length of our lives dramatically. One of the oldest and deepest of enigmas is, Why is our physical world governed by the laws of mathematics? The nature of the connection between mathematics and physics is enormously complex and well beyond the scope of this introduction; but that connection is an essential ingredient in our understanding of the world. Nobel physicist Richard Feynman said, "To those who do not know mathematics it is difficult to get across a real feeling as to the beauty, the deepest beauty of nature. If you want to learn about nature, to appreciate nature, it is necessary to understand the language she speaks in."

Most mathematicians, however, did not choose to study mathematics because they wanted to build a better VCR. Instead, many were drawn to mathematics by its aesthetic qualities—its purity, its spare elegance. Beauty in mathematics is hard to define, but mathematicians instantly understand what their colleagues mean when they call a proof "beautiful," and they think of their calling almost as an art. The great mathematician Henri Poincaré said, "A scientist worthy of his name, above all a mathematician, experiences in his work the same impression as an artist; his pleasure is as great and of the same nature." So strong is this feeling among mathematicians that they have an instinctive distrust for awkward or complicated mathematical work. G. H. Hardy wrote in his *Mathematician's Apology,* "The mathematician's patterns, like the painter's or the poet's must be beautiful; the ideas, like the colors or the words must fit together in a harmonious way. Beauty is the first test: there is no permanent place in this world for ugly mathematics."

The great Hungarian mathematician Paul Erdös tried to explain his sense of the aesthetic quality of mathematics by the allegorical idea that God has a Book that contains all the most beautiful, most perfect proofs of mathematical truths. For Erdös, the highest achievement for a mathematician is to find a proof worthy of being in the Book. "You don't have to believe in God, but you should believe in the Book," he said. In *World of Mathematics,* you will read about proofs—such as the proof of the four-color theorem—so large and unwieldy that they are almost impossible to check, and you will see other proofs—such as the proof of the Pythagorean theorem or the irrationality of the square root of 2—so simple and elegant that they perhaps deserve a place in the Book. You will also find an entry in *World of Mathematics* called "Beauty in Mathematics" in which this issue is discussed at greater length.

Erdös's Book metaphor emerges from a question that has stimulated some of the most energetic debates of mathematical philosophy: Is mathematics invented or discovered? Do mathematicians build an edifice of theorems from a foundation of their own design, or do mathematical truths exist in some realm that human beings can only brush against, occasionally being fortunate enough to knock some small piece of the truth within human reach?

This question dates back to almost the earliest days of mathematical thought. In the 6th century B.C., Pythagoras, awed by the unexpected patterns he found in numbers, concluded that mathematical objects have a natural harmony, one that is only partially understood by mathematicians. Later, Plato put forth the idea that mathematical truths have their own existence in an ideal world of essences, one that human beings can only try to discover; since then, this belief has been referred to as Platonism. Equally influential on the other hand was the work of the great geometer Euclid. His *Elements,* written in the 4th century B.C., stripped down the ideas of geometry to a few basic building blocks called axioms, and then constructed a huge structure of theorems on this foundation, following rigorous logical arguments. Euclid's work, which spanned 13 books, was an intellectual triumph that has set the standard for logical thought for more than two millenia. The idea that human beings invent mathematics using sheer logical power has since been termed reductionism.

Towards the end of the 19th century, the advantage in the debate seemed to be with the reductionists. With the development of the principles of computers and algorithms, the reductionists set themselves a new goal: to construct a system of axioms from which all of mathematics could be derived, following the strict laws of logic. Their efforts culminated in a monumental opus by Bertrand Russell and Alfred North Whitehead published between 1910 and 1913, the *Principia Mathematica*; in it, Russell and Whitehead tried to construct a system of axioms and rules of deduction from which every mathematical truth could be proved. They could not, of course, include proofs of every theorem, but the proofs that they did present followed the laws of logic to the smallest detail; they were so careful that it took several hundreds of pages before they got to the definition of the number 1. While this might seem like absurd overkill, to the reductionists it was worth the effort to be certain that their work rested on a solid foundation. This foundation was swept out from under their feet, however, when in 1930 the young mathematician Kurt Gödel proved a theorem that shook the mathematical community to its core. Gödel's theorem says that in any axiomatic system powerful enough to prove significant mathematical theorems, there will be statements that are true but which cannot be proved using the axioms and deductive laws of the system.

Gödel's "Incompleteness" theorem reopened the question, What is mathematical truth? The response of the Platonists to Gödel's theorem is to say that truth is not the same as provability; that mathematical truth exists independent of any particular system of axioms. The reductionists, on the other hand, say that mathematical truth simply does not exist—that the only thing mathematicians can say is whether a statement is provable in a given axiomatic system, not whether it is true or false. The mathematical community continues to be divided on this question.

In the nearly 800 entries in this book, you will find not just ammunition for such philosophical arguments, but also the everyday, bread-and-butter ideas of mathematics—the patterns and puzzles, theorems and proofs, shapes and symmetries that have fascinated both professional and amateur mathematicians over the centuries. Read about prime numbers, or fractals, or the nature of infinity, and form your own opinion about whether mathematics is invented or discovered.

You will also find biographies of the great mathematicians whose lives span the whole of human history, and whose stories range from comical to tragic, from humble to heroic. You will read of duels, suicides, ritual sacrifices, murders, and also quiet lives devoted to reflection and study. Through the centuries, mathematical scholars have come from the aristocracy and from the slums; they have dazzled the world as youthful prodigies or have worked long years in patient obscurity. The scope of their experiences shows above all that mathematics is for everyone—not just the Fields medalist at

Harvard or Princeton, but also the child who looks at the night sky with wonder, or who happens upon an unexpected pattern in numbers, and is struck by its beauty.

Erica Klarreich
Assistant Professor of Mathematics
University of Michigan
May 2000

How to Use the Book

This first edition of *World of Mathematics* has been designed with ready reference in mind.

- **Entries are arranged alphabetically**, rather than by chronology or scientific field.
- **Boldfaced terms** direct reader to related entries.
- **Cross-references** at the end of entries alert the reader to related entries that may not have been specifically mentioned in the body of the text.
- A **Sources Consulted** section lists many worthwhile print and electronic materials encountered in the compilation of this volume. It is there for the inspired reader who wants more information on the people and concepts covered in this work.
- The **Historical Chronology** includes over 500 important events in the history of mathematics and related fields spanning the period from around 35,000 B.C. through 1995.
- A **comprehensive general index** guides the reader to all topics and persons mentioned in the book. Boldface page references refer the reader to the term's full entry. Page numbers in italics refer to illustrations. Page numbers followed by *f* indicate figures.

Advisory Board

In compiling this edition, we have been fortunate in being able to call upon the following people—our panel of advisors—who contributed to the accuracy of the information in this premier edition of *World of Mathematics*, and to them we would like to express sincere appreciation:

Kristin Chatas
Department Chair—Mathematics
Pioneer High School
Ann Arbor, Michigan

Erica Klarreich, Ph.D.
Assistant Professor of Mathematics
University of Michigan

Dana Mackenzie, Ph.D.
Freelance Mathematics and Science Writer
Santa Cruz, California

The editor would like to thank the advisors for all of their good work reviewing and compiling the entry list, for their patience and good will answering my many questions, and for their outstanding writing. You're the best math teachers I never had. Thanks are also due to all the writers who contributed to this book; in particular, William Arthur Atkins, Lewis Bowen, Fran Hodgkins, Elisabeth Morlino, and Stephen R. Robinson. Finally, thanks go to Zoran Markovic of the Institute of Mathematics in Belgrade for his assistance early in the project; and to Julie Shawvan for her indexing and critical review at the end.

Cover

The image on the cover represents the symmetry operation called inversion.

ACKNOWLEDGMENTS

Abacus, drawing by Hans and Cassidy. Gale Group.-Abel, Niels Henrik, screen print. Corbis-Bettmann. Reproduced by permission.-Agnesi, Maria Gaetana, engraving. Corbis-Bettmann. Reproduced by permission.-Aiken, Howard, photograph. Corbis-Bettmann. Reproduced by permission.-Anaxagoras with Pericles, in his palace, engraving. Corbis-Bettmann. Reproduced by permission.-Archimedes, sculpture. Corbis-Bettmann. Reproduced by permission.-Archytas, engraving. Corbis-Bettmann. Reproduced by permission.-Aristotle, photograph. Corbis-Bettmann. Reproduced by permission.-Bernoulli, Daniel, woodcut. Corbis-Bettmann. Reproduced by permission.-Bernoulli, Jacques, engraving by P. Dupin. Corbis-Bettmann. Reproduced by permission.-Bernoulli, Jean I, engraving. Corbis-Bettmann. Reproduced by permission.-Boolean algebra, illustrations by Hans & Cassidy. Gale Group.-Calculation of the shortest distance between two points on a graph, using the Pythagorean theorem, diagram by Hans & Cassidy. Gale Group.-Calculator, Babylonian digits, diagram by Hans & Cassidy. Gale Group.- Calculus, bar graph by Hans & Cassidy. Gale Group.-Calculus, graph approximating line segment, illustration by Hans & Cassidy. Gale Group.-Calderon, Alberto P., Israel, 1989, photograph. Wolf Foundation. Reproduced by permission.-Carnap, Rudolph, 1963, photograph. AP/Wide World Photos. Reproduced by permission.-Carroll, Lewis, photograph. The Granger Collection. Reproduced by permission.-Cartan, Elie Joseph, c. 1936, photograph. AP/Wide World Photos. Reproduced by permission.-Cartesian coordinate plane, graphs by Hans & Cassidy. Gale Group.-Cartesian plane, illustrations by Hans & Cassidy. Gale Group.-Chandrasekhar, Subrahmanyan, with King Carl Gustaf of Sweden, 1983, photograph. AP/Wide World Photos. Reproduced by permission.-Chern, Shiing S., photograph. Wolf Foundation. Reproduced by permission.-Complex numbers figure, acute angle, photograph by Hans & Cassidy. Gale Group.-Conic section, cone, oval, circle, and arch, illustrations by Hans & Cassidy. Gale Group.-Continuity, graphs by Hans & Cassidy. Gale Group.- Coulomb, Charles August de, photograph. Corbis-Bettmann. Reproduced by permission.-Courant, Richard, 1961, photograph. AP/Wide World Photos. Reproduced by permission.-Cramer, Gabriel, 18th century, painting. The Granger Collection, New York. Reproduced by permission.-De Morgan, Augustus, photograph by Ernest Edwards. Corbis-Bettmann. Reproduced by permission.-Determinants, four small boxes, variables in corners, illustration by Hans & Cassidy. Gale Group.-Determinants, graph by Hans & Cassidy. Gale Group.-Determinants, six small boxes, photograph. Hans & Cassidy. Gale Group.-Diagram showing addition on an abacus, drawing by Hans and Cassidy. Gale Group.- Dipolar vortices, collision, photograph by G. Van Heijst and J. Flor. Photo Researchers, Inc. Reproduced by permission.-Dirichlet, Peter Gustav Lejeune, engraving. Archive Photos, Inc. Reproduced by permission.-Durer, Albrecht, self-portrait. AP/Wide World Photos. Reproduced by permission.-Electrons, classical and quantum, diagram by Hans & Cassidy. Gale Group.-"Euclid," painting by Justus van Ghent. Corbis-Bettmann. Reproduced by permission.- Fermat, Pierre, illustration. Corbis-Bettmann. Reproduced by permission.-Fibonacci, Leonardo, engraving. The Granger Collection, New York. Reproduced by permission.-Figurative numbers, illustrations by Hans & Cassidy. Gale Group.-Figurative numbers represented by pentagons, illustration by Hans & Cassidy. Gale Group.- Figurative numbers, three squares in a row, diagram by Hans & Cassidy. Gale Group.-Fractal, three lines, illustration by Hans & Cassidy. Gale Group.-Freedman, Dr. Michael, La Jolla, California, 1982, photograph. AP/Wide World Photos. Reproduced by permission.-Galois, Evariste, drawing. Corbis-Bettmann. Reproduced by permission.-Gauss, Carl Friedrich, drawing. Corbis-Bettmann. Reproduced by permission.-Germain, Sophie, engraving after a life mask. The Granger Collection, New York. Reproduced by permission.-Gödel, Kurt, 1951, photograph. AP/Wide World Photos. Reproduced by permission.-Hadamard, Jacques, 1920, photograph. Reuters/Corbis-Bettmann. Reproduced by permission.-Hermite, Charles, photograph. Corbis-Bettmann. Reproduced by permission.-Hilbert, David, photograph. Corbis-Bettmann. Reproduced by permission.-Hipparchus, engraving. Archive

Photos, Inc. Reproduced by permission.- Hypatia, conte crayon drawing. Corbis-Bettmann. Reproduced by permission.-Hyperbola, circle, graph by Hans & Cassidy. Gale Group.-Hyperbola, directrix, focus, illustration by Hans & Cassidy. Gale Group.-Hyperbola, graphs and planks, photographs by Hans & Cassidy. Gale Group.-Inequality, graph, left half shaded, photograph by Hans & Cassidy. Gale Group.-Inequality, line graph, photograph by Hans & Cassidy. Gale Group.-Jacobi, Karl Gustav, painting. The Granger Collection. Reproduced by permission.-"Karl Weierstrass," etching by Hans Thoma. Corbis-Bettmann. Reproduced by permission.-Keen, Linda, photograph. Reproduced by permission of Linda Keen.- Khayyam, Omar, drawing. Corbis-Bettmann. Reproduced by permission.-Kolmogorov, Andrei, Rome, 1963, photograph. UPI/ Corbis-Bettmann. Reproduced by permission.-Kovalevskaya, Sonya, engraving. Corbis-Bettmann. Reproduced by permission.-Locus figures and line graphs, illustrations by Hans & Cassidy. Gale Group.-Maclaurin, Colin, illustration. Corbis-Bettmann. Reproduced by permission.-Mandelbrot, Benoit, photograph. Copyright © Hank Morgan. Photo Researchers, Inc. Reproduced by permission.-Marquise du Chatelet, 1873, wood engraving. The Granger Collection, New York. Reproduced by permission.-Mersenne, Marin, 18th century, engraving by P. Dupin. The Granger Collection. Reproduced by permission.-Minkowski, Hermann, photograph. Corbis-Bettmann. Reproduced by permission.-Newton, Sir Isaac, engraving. Archive Photos, Inc. Reproduced by permission.-Parabola, cones and graphs, illustrations by Hans & Cassidy. Gale Group.-Peirce, Charles Sanders, photograph. Corbis-Bettmann. Reproduced by permission.-Poincare, M. Henri, photograph. Hulton-Deutsch Collection/Corbis-Bettmann. Reproduced by permission.-Polar coordinates, acute and obtuse angles, diagrams by Hans & Cassidy. Gale Group.-Polya, Dr. George, with his book "How To Solve It," 1978, photograph. AP/Wide World Photos. Reproduced by permission.- Poncelet, Jean Victor, marble sculpture. Corbis-Bettmann. Reproduced by permission.-Pythagoras, engraving. Corbis-Bettmann. Reproduced by permission.-Queneau,

Raymond, photograph by Jerry Bauer. © Jerry Bauer. Reproduced by permission.-Ramanujan, Srinivasa, photograph. The Granger Collection. Reproduced by permission.-Real numbers, line and triangle, illustration by Hans & Cassidy. Gale Group.-Regiomontanus, Johannes, illustration. Corbis-Bettmann. Reproduced by permission.-Riemann, G.F.B., 1868, engraving after a photograph. Corbis-Bettmann. Reproduced by permission.-Rotation (pie slice), graph by Hans & Cassidy. Gale Group.-Set of points on a coordinate plane (right angles), diagram by Hans & Cassidy. Gale Group.-Set theory, difference, photograph by Hans & Cassidy. Gale Group.-Set theory figures, illustrations by Hans & Cassidy. Gale Group.-Snell, Willebrord, painting. Photo Researchers, Inc. Reproduced by permission.-Stevin, Simon, painting. Corbis-Bettmann. Reproduced by permission.-Taylor, Brook, engraving. Archive Photos, Inc. Reproduced by permission.-Texas map, illustration by Hans & Cassidy. Gale Group.-Thales of Miletus, engraving by Ambroise Tardieu. Corbis-Bettmann. Reproduced by permission.-Thales of Miletus, lithograph. Corbis-Bettmann. Reproduced by permission.-Third-order determinants (graph and equation), illustration by Hans & Cassidy. Gale Group.-Translation of a triangle, in one and two directions, diagrams by Hans & Cassidy. Gale Group.-Turing, Alan, finishing second in a 3 mile race at Dorking, England, photograph. The Granger Collection, Ltd. Reproduced by permission.-Two points plotted in a perpendicular coordinate system, graph by Hans & Cassidy. Gale Group.-Two triangles and an acute angle in a rotation, diagram by Hans & Cassidy. Gale Group.-Two-dimensional complex-number plane, graph by Hans & Cassidy. Gale Group.-Von Neumann, John, testifying before the Atomic Energy Commission, photograph. AP/Wide World Photos. Reproduced by permission.-Weyl, Hermann, 1954, photograph. AP/Wide World Photos. Reproduced by permission.-Whitehead, Alfred North, photograph. AP/Wide World Photos. Reproduced by permission.-Wiles, Andrew J. , photograph. Wolf Foundation. Reproduced by permission.-Young, Grace Chishol, 1923, photograph. Reproduced by permission of Sylvia Wiegand.

M

MACLAURIN, COLIN (1698-1746)
Scottish geometer and physicist

Colin Maclaurin was one of Europe's foremost mathematicians during the 1700s. He was the first to provide systematic **proof** of **Isaac Newton**'s theorems. Some of his noted accomplishments include explanations of the properties of conics and the theory of tides. Maclaurin was also a brilliant mathematician in his own right, who solved many problems in **geometry** and applied physics. Besides being an esteemed mathematician, Maclaurin was a creative inventor who loved to devise mechanical appliances. He was skilled in astronomy, mapmaking and sometimes spent his spare time acting as an actuary for insurance companies.

Maclaurin was born in Kilmodan, Scotland, in 1698, the son of a minister named John Maclaurin, a man of great learning. Unfortunately, his father died when Colin was just six years old. When his mother died nine years later, Maclaurin moved in with his uncle, Daniel Maclaurin.

Maclaurin's eldest brother, John, studied for the ministry and became a noted religious expert. Following his brother's lead, Maclaurin studied divinity at the University of Glasgow for a year. While at Glasgow, he met Robert Simson, professor of mathematics, who inspired Maclaurin's interest in geometry, especially the geometry of ancient mathematicians such as **Euclid**.

Maclaurin became interested in Newton's theories early on in his career. In 1715, he presented his thesis "On the Power of Gravity," which demonstrated real distinction, even though he was only in his teens when he defended it and earned a master of arts degree. The thesis also won him an appointment as a professor in mathematics at Marischal College in Aberdeen a year later.

In 1719, Maclaurin made two trips to London where he met some of the renowned scientists of the day, including Newton and Martin Folkes, who later became the president of the Royal Society of London. In 1720, Maclaurin published one of his signature works, *Geometrica Organica, sive descriptio linearum curvarum universalis*, which explained higher plane curves and conics. It proved many of the theories that Newton had proposed as well as solving other important problems in geometry. Maclaurin, for instance, showed that the cubic and the quartic could be represented by rotating these angles around their vertices. Newton had demonstrated similar properties for the **conic sections**.

In 1722, Maclaurin left Marischal to take on the tutoring of the son of Lord Polwarth, a powerful British diplomat. The two traveled through France, where Maclaurin produced another masterpiece *On the Percussion of Bodies*. For this work, the French Académie Royale des Sciences presented him with its prize in 1724.

When Polwarth's son died, Maclaurin hoped to reclaim his teaching position at Marischal, but during his three–year absence, it had been declared vacant and filled. When the chair of professor of mathematics at the University of Edinburgh fell vacant in 1725, he was appointed there. The prodigious Maclaurin then began lecturing on some of his favorite topics: the theories of Euclid, conics, astronomy, **trigonometry**, and Newton's *Principia.*

Maclaurin moved in Scotland's most inner circles. He was a fellow of the Royal Society and the Philosophical Society, where he acted as secretary. In 1733, he married Anne Stewart, the daughter of the solicitor general in Scotland. They had seven children.

In 1740, Maclaurin attracted international notice when he submitted an essay called "On the Tides" for the Académie Royale des Sciences prize. In the essay, Maclaurin explained the theory of tides, based on Newton's *Principia,* and defined the tides of the sea as an ellipsoid revolving around an inner point. Maclaurin shared the prize for his theory of tides with the mathematicians **Leonhard Euler** and **Daniel Bernoulli**. They all provided proof of Newton's theories about the movement of the ocean. The mathematician **Alexis Clairaut** was so taken with Maclaurin's success in explaining tides that he began probing the mystery of the Earth's shape with geometry.

Colin Maclaurin

Maclaurin's *Treatise of Fluxions*, published in 1742, was another of his major works. The treatise was written as a reply to **George Berkeley**'s 1734 publication *The Analyst, A Letter Addressed to an Infidel Mathematician,* in which Berkeley criticized Newton's theory of **fluxions** as "ghosts of departed quantities." Other mathematicians had also decried Newton's methods for their lack of systematic foundation. In his two–volume work, Maclaurin explained Newton's theories in detail and also solved other great quandaries of mathematics. The book contains descriptions of the **infinite series** as well as much praised work on the curves of quickest descent.

Though the *Treatise of Fluxions* was considered noteworthy at the time, it had little influence on international mathematics. The book persuaded British scientists to continue with the geometrical methods of Newton rather than the analytical **calculus** just being devised in other nations in Europe. As a result, Britain was left far behind in the development of mathematics during the late 1700s.

Maclaurin worked ceaselessly to defend Edinburgh during an attack by Jacobites in the rebellion of 1745. While planning and erecting the defenses of the city, however, he became so exhausted that he fell physically ill. When the city surrendered to the Jacobites, Maclaurin fled to England. A year later he returned to Edinburgh, where he died at age 48.

Until his death in 1746, Maclaurin remained the consummate mathematician and defender of Newtonian theories. A few hours before his death he dictated his last writing on Newton's work, which firmly set forth Maclaurin's belief in life after death. His *Treatise on Algebra* and *An Account of Sir Isaac Newton's Philosophical Discoveries*, an incomplete work, were published posthumously.

MACLAURIN'S THEOREM

Maclaurin's **theorem** is a specific form of Taylor's theorem, or a Taylor's **power series** expansion, where $c = 0$ and is a series expansion of a function about **zero**. The basic form of Taylor's theorem is: $\Sigma^{\infty}_{n=0} (f^{(n)}(c)/n!)(x - c)^n$. When the appropriate substitutions are made Maclaurin's theorem is: $f(x) = f(0) + f'(0)x + f''(0)x^2/2! + f^{(3)}(0)x^3/3! +... f^{(n)}(0)x^n/n! +....$ The Taylor's theorem provides a way of determining those values of x for which the Taylor series of a function f converges to $f(x)$.

In 1742 Scottish mathematician **Colin Maclaurin** attempted to put **calculus** on a rigorous geometric basis as well as give many applications of calculus in the work. It was the first logical and systematic exposition of the method of **fluxions** and originated as a reply to Berkeley's attack on Newton's methods of calculus. In this text, among several other monumental ideas, Maclaurin gave a **proof** of the theorem that today holds his name, Maclaurin's theorem, and is a special case of Taylor's theorem. He obtained this theorem by assuming that $f(x)$ can be expanded in a power series form and then, upon differentiation and substituting $x = 0$ in the results, the values of the coefficients of each term can be obtained. He did this but did not investigate the convergency of the series at that time. Although this theorem holds Maclaurin's name it was previously published by another Scottish mathematician, **James Stirling**, in his book *Methodus Differentialis*, published in 1730. It is no doubt but that this theorem is what Maclaurin is best remembered.

MAGIC CUBES

A magic cube is a three-dimensional (or higher dimensional) analogue of a magic **square**. Specifically, an order n magic cube is an n x n x n array of numbers chosen from the set $\{1,2,...,n^3\}$ so that each number appears exactly once and the sum of the numbers in every row, column, file, and diagonal is the same, magic constant. The magic constant must therefore be the sum of all the numbers in the cube divided by the number of rows (columns or files). So, it is $(1 + 2 +... + n^3)/n^2 = n(n^3 +1)/2$.

A perfect magic cube is such that each square parallel to a face of the cube is a magic square. Here is one:

19	497	255	285	432	78	324	162
303	205	451	33	148	370	128	414
336	174	420	66	243	273	31	509
116	402	160	382	463	45	291	193
486	8	266	236	89	443	181	343
218	316	54	472	357	135	393	107
185	347	85	439	262	232	490	12
389	103	361	139	58	476	214	312

134	360	106	396	313	219	469	55
442	92	342	184	5	487	233	267
473	59	309	215	102	392	138	364
229	263	9	491	346	188	438	88
371	145	415	125	208	302	36	450
79	429	163	321	500	18	288	254
48	462	196	290	403	113	383	157
276	242	512	30	175	333	67	417

306	212	478	64	414	367	97	387
14	496	226	260	433	83	349	191
109	399	129	355	466	52	318	224
337	179	445	95	238	272	2	484
199	293	43	457	380	154	408	118
507	25	279	245	72	422	172	330
412	122	376	150	39	453	203	297
168	326	76	426	283	249	503	21

423	69	331	169	28	506	248	278
155	377	119	405	296	198	460	42
252	282	24	502	327	165	427	73
456	38	300	202	123	409	151	373
82	436	190	352	493	15	257	227
366	144	386	100	209	307	61	479
269	239	481	3	178	340	94	448
49	467	221	319	398	112	354	132

381	159	401	115	194	292	46	464
65	419	173	335	510	32	274	244
34	419	173	335	510	32	274	244
286	256	498	20	161	323	77	431
140	362	104	390	311	231	475	57
440	86	348	186	11	489	231	261
471	53	315	217	108	394	136	358
235	265	7	485	344	182	444	90

492	10	264	230	87	437	187	345
216	310	60	474	363	137	391	101
183	341	91	441	268	234	488	6
395	105	359	133	56	470	220	314
29	511	241	275	418	68	334	176
289	195	461	47	158	384	114	404
322	164	430	80	253	287	17	499
126	416	146	372	449	35	301	207

96	446	180	338	483	1	271	237
356	130	400	110	223	317	51	465
259	225	495	13	192	350	84	434
63	477	211	305	388	98	368	142
425	75	325	167	22	504	250	284
149	375	121	411	298	204	454	40
246	280	26	508	329	171	421	71
458	44	294	200	117	407	153	379

201	299	37	455	374	152	410	124
501	23	281	251	74	428	166	328
406	120	378	156	41	459	197	295
170	332	70	424	277	247	505	27
320	222	468	50	131	353	111	397
4	482	240	270	447	93	339	177
99	385	143	435	480	62	308	210
351	189	435	81	228	258	16	494

magic of any order other than 2, 3, 4, 5, 6, or 10. However, it has been proved that there are no perfect magic cubes of order 2, 3, or 4 so only the order 5, 6, and 10 are in question.

Here is a **proof** (reproduced from Martin Gardner's *Time Travel and other Mathematical Bewilderments*) that there are no order 3 perfect cubes. Suppose that such a cube exists. Consider any square parallel to a face of the cube. Let A, B, C be the numbers on the first row and D, E, F be the numbers on the third row. Let X be the middle number. Since the magic constant of any order 3 magic cube is 42, $3X + A + B + C + D + E + F = 3 \times 42$, $A + B + C = 42$, and $D + E + F = 32$.

This implies that $3X = 42$ and $X = 14$. In a pure magic cube, however, no number is repeated twice. So, 14 cannot be the middle number of every cross-section. Thus, a perfect order-3 magic cube is impossible.

A magic tesseract is a 4-dimensional magic cube. Here is one:

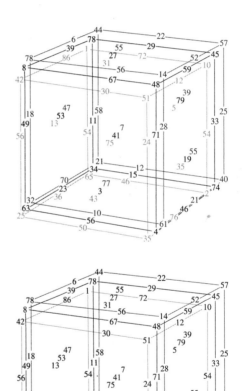

There are 58 magic tesseracts of order 3. Hendricks has constructed a pandiagonal order 4 magic tesseract and a perfect order 16 magic tesseract. He also proved there are no perfect magic tesseracts of order less than 16.

See also Magic pentagrams and hexagrams; Magic squares

If a magic cube is not perfect then it is called semi-perfect. A pandiagonal cube has the property that all of its broken **space** diagonals sum up to the magic constant. In other words, if copies of the cube were stacked on top of each other and to the right and left as well (without rotating), then any diagonal string of length n will have all of its numbers adding up to the magic constant. Here is a pandiagonal magic cube.

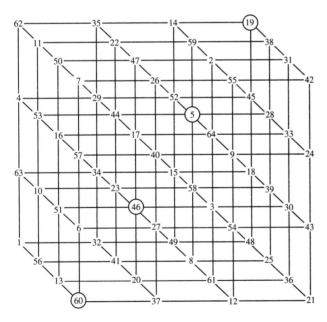

Methods are known for constructing magic cubes of every order other than 2. If A and B are magic cubes and A can be rotated or reflected to look just like B, then A and B are considered to be the same magic cube. Accordingly, there are 4 different order 3 magic cubes. The number of order 4 magic cubes is not known. There are 7680 pandiagonal order-4 magic cubes. Methods are known for constructing perfect

MAGIC PENTAGRAMS AND HEXAGRAMS

A magic hexagram is a hexagram such that each outer point and each crossing point is assigned a number so that different points are assigned different numbers. The sum of the numbers on any line is required to be the same for all lines. In a pure magic hexagram, the set of assigned numbers is the set of consecutive **integers** from one to twelve. Finding magic hexagrams is a popular recreation.

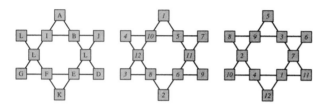

In any pure magic hexagram, the sum of the numbers of the corners of each large **triangle** must be the same. Here is the **proof**. Let M be the magic number. From the diagram follows these six equations:

- 1. B = M - I - L - J
- 2. C = M - B - A - D
- 3. E = M - C - K - J
- 4. F = M - E - D - G
- 5. H = M - F - L - K
- 6. I = M - H - A - G

Substitute (M - I - L - J) for B is equation 2 to get a new equation for C. Substitute the right side of this new equation for C in equation 3 to get a new equation for E. Continue this until you get a new equation for I. This equation happens to be:

- 7. I = I + 2(L + J + K) - 2(A + G + D).

So, L + J + K = A + G + D.

H. E. Dudeney was the first to attempt to enumerate all pure magic hexagrams. His work appeared in 1926 in the book *Modern Puzzles*. But, he missed seven hexagrams. Not counting reflections and rotations, there are 80 different pure magic hexagrams. Twelve of these are such that the outer points sum up to the magic number, also. The complement of a pure magic hexagram is formed in this way: if x is the number assigned to a vertex v, then the complement has 13-x assigned to the same vertex.

In contrast to the hexagram, there are no pure magic pentagrams. That is, the integers from one to ten cannot be assigned to the vertices of a pentagram in such a way that the sum of the numbers on every line is the same.

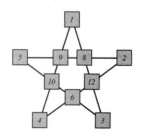

Here is a proof. If a pure magic pentagram exists, then the sum of all five lines would be equal to five times the magic number. But, each point is in two lines, so this is also equal to two times the sum of all the integers from one to ten. Thus the magic number is 22. The numbers other than 1 on the two lines that contain the number 1 must sum up to 42. But 9 + 8 + 7 + 6 + 5 + 4 = 39, so one of the numbers on those lines must be 10. Similarly, the numbers other than 2 on the two lines that contain the number 2 must sum up to 40. So, one of the numbers on those two lines must be 10. Let L1 be the line containing 1 and 10 and let L2 be the line containing 2 and 10. Then the other numbers on L1 are either {3, 8}, {4,7} or {5,6}. Similarly, the other numbers on L2 are either {3,7} or {4,6}. These two lines cannot intersect more than once, so either (case one) L1 contains 1, 10, 3, and 8 and L2 contains 2, 10, 4, and 6 or (case two) L1 contains 1,10, 5, and 6 and L2 contains 2, 10, 3, and 7. Let L3 be the line containing 3 that does not contain 10. In the first case, L3 must contain a number in L2 that is not 10. But L3 cannot contain any number other than 3 of L1. This is impossible. In the second case, the same considerations lead to one possibility: L3 must contain 6, 4, and 9. Let L4 be the line that contains 4 but not 3. This line must contain 8 since 8 is not in L1, L2, or L3 and every number is in two lines. On the other hand, it must contain either 5 or 1 since it must intersect L1 and it must contain either 2 or 7 since it must intersect L2. Neither of these possibilities will allow the numbers on L4 to sum up to 22, however. So, a pure magic pentagram is impossible.

Weakly magic pentagrams are such that the assigned numbers are either the integers one through twelve with 7 and 11 omitted. There are twelve weakly magic pentagrams not counting rotations and reflections. Almost magic pentagrams are such that only four out of the five lines sum up to the same number. It is also required that the numbers are chosen from the set of integers from one to ten so that each integer appears exactly once. There are seven almost magic pentagrams not counting rotations and reflections.

A prime magic pentagram is such that all the assigned numbers are **prime numbers**. Here is the smallest such pentagram:

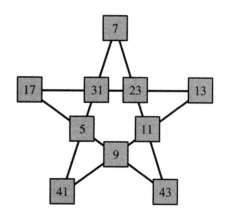

Here is the smallest magic pentagram using consecutive primes:

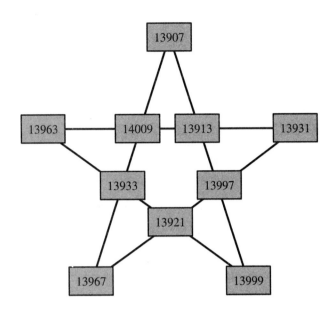

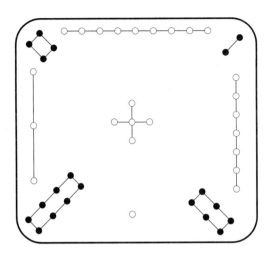

Magic septagrams, octagrams and so on are also possible. There are 72 pure magic septagrams, 112 pure magic octagrams, and more than 2,000 pure magic nonagrams. These amounts are, of course, only up to **rotation** and reflection.

See also Magic cubes; Magic squares

MAGIC SQUARES

8	1	6
3	5	7
4	9	2

16	2	3	13
5	11	10	8
9	7	6	12
4	14	15	1

17	24	1	8	15
23	5	7	14	16
4	6	13	20	22
10	12	19	21	3
11	18	25	2	9

32	29	4	1	24	21
30	31	2	3	22	23
12	9	17	20	28	25
10	11	18	19	26	27
13	16	36	33	5	8
14	15	34	35	6	7

30	39	48	1	10	19	28
38	47	7	9	18	27	29
46	6	8	17	26	35	37
5	14	16	25	34	36	45
13	15	24	33	42	44	4
21	23	32	41	43	3	12
22	31	40	49	2	11	20

64	2	3	61	60	6	7	57
9	55	54	12	13	51	50	16
17	47	46	20	21	43	42	24
40	26	27	37	36	30	31	33
32	34	35	29	28	38	39	25
41	23	22	44	45	19	18	48
49	15	14	52	53	11	10	56
8	58	58	5	4	62	63	1

A magic square is an n x n grid in which numbers have been written in such a way that the sum of the numbers along any row, column, or main diagonal is always the same magic constant. In a pure magic square, the numbers in the grid must be the consecutive **integers** from one to n² with each number used exactly once. The number n is called the order of the square. The unique order 3 magic square was known in China at least by 500 BC where it is called the *lo shu*.

Most literature on magic squares is about how to construct magic squares in general and those with special properties. A magic square in which all broken diagonals add up to the magic constant is called pandiagonal. A magic square in which cells that are situated directly opposite from the center is called associative. The smallest magic squares that are both pandiagonal and associative have order 5. Here is one.

1	15	22	18	9
23	19	6	5	12
10	2	13	24	16
14	21	20	7	3
17	8	4	11	25

If all the numbers in a magic square are squared and the result is a magic square, the square is called bi-magic. Here is one.

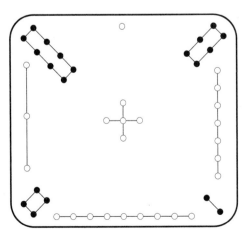

Martin Gardner has offered $100 prize money to the first person who can find an order-3 magic square whose entries are all different perfect squares or prove that none can exist. If all the numbers in a magic square are prime, the square is called prime-magic. Here is an order-3 magic square containing consecutive primes:

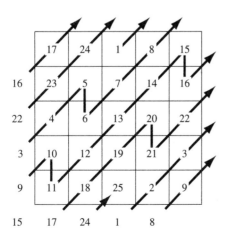

It is standard not to count rotations and reflections when counting magic squares. Accordingly, there is only one magic square of order-1 or order-3. There are none of order-2. All 880 magic squares of order-4 were first enumerated by Bernard Frenicle de Bessy in 1693. The website http://www.pse.che.tohoku.ac.jp/~msuzuki/MagicSquare.4x4. total.html has pictures of all of them. In 1973, Richard Schroeppel developed a computer program to enumerate all order-5 magic squares. After 100 hours, the computer found all 275,305,224 of them. Pinn and Wieczerkowski used Monte Carlo simulation and methods from statistical **mechanics** to estimate that the number of order-6 magic squares is between 1.7766×10^{19} and 1.7743×10^{19}.

There are different methods for constructing magic squares, but all of them depend on whether the order of the square to be constructed is odd, even but not divisible by four, or divisible by four. William Andrews gave the following method for constructing odd-order magic squares in his pioneering book *Magic Squares and Cubes* (1917).

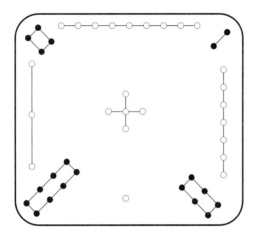

First, imagine that the top edge of the square is attached to the bottom edge so that if one moves up from the top edge, one arrives at the bottom edge. Similarly, imagine the right edge is attached to the left edge. Now, start in the middle top square and write a number 1. Move up one and to the right one and write a number 2. Next, move up one and to the right one and write a number 3. Continue in this manner unless your path is blocked. In the above picture, this first happens after writing the number 5. In this situation, move down one and write the next number. Then continue as before, moving up one and to the right one and writing down the next number. After n^2 steps, you have a magic square.

In recent times, Allen Adler has developed a way to "multiply" magic squares. For example, let A be the 3x3 magic square below and let B be the 4x4 magic square.

8	1	6
3	5	7
4	9	2

1	15	14	4
12	6	7	9
8	10	11	5
13	3	2	16

A*B is defined to be the square shown below:

9	1	6	134	127	132	125	118	123	35	28	33
3	5	7	129	131	133	120	122	124	30	32	34
4	9	2	130	135	128	121	126	119	31	36	29
107	100	105	53	46	51	62	55	60	80	73	78
102	104	106	48	50	52	57	59	61	75	77	79
103	108	101	49	54	47	58	63	56	76	81	74
71	64	69	89	82	87	98	91	96	44	37	42
66	68	70	84	86	88	93	95	97	39	41	43
67	72	65	85	90	83	94	99	92	40	45	38
116	109	114	26	19	24	17	10	15	143	136	141
111	113	115	21	23	25	12	14	16	138	140	142
112	117	110	22	27	20	13	18	11	139	144	137

It is formed in this way. First draw an empty 4x4 square. Find where the number 1 is on B. In this case, it is the top left corner. Place a copy of A in this corner square. Now find where the number 2 is on B. In this square place a copy of A but add 9 to all its numbers. In the square where the number 3 is on B, place a copy of A and add two times 9, or 18, to all its numbers, and so on. At each cell where a number x was on B, a copy of A with 9 times x-1 added to all its numbers is placed. The result is a 12x12 magic square. Adler proved that this operation is associative, that is if A, B, and C are any magic squares then $(A*B)*C = A*(B*C)$. It is not commutative however. In the example, $A*B$ does not equal $B*A$. The 1x1 magic square with a number 1 in its only square is denoted by I. It has the property that $A*I = I*A = A$ for any magic square A. If a magic square $C = A*B$ and neither A nor B is equal to I, then C is called "composite" in analogy with the integers. If P is a magic square that is not composite, it is called prime. If C is any magic square then there are prime magic squares P1, P2..., Pn such that $C = P1*P2*...*Pn$. Adler proved in 1992 that if Q1,..,Qm are primes and $C = Q1*Q2*...*Qm$ also, then m=n and Qi = **Pi** for every I.

In 1998, Kathleen Ollerenshaw and David Brée's *Most-Perfect Pandiagonal Magic Squares* was published. In this work, the authors give a method for constructing all most-perfect pandiagonal magic squares. This type of pandiagonal magic square has the property that every 2x2 subsquare adds up to $2(n^2 - 1)$ and every pair of squares that are n/2 squares apart along a diagonal sum up to $n^2 - 1$. It's been known for a century, that any most-perfect pandiagonal magic square's order is divisible by 4. They prove a formula for the number of most-perfect pandiagonal squares. For order 4 the number is 48. For order 8, it is 368,640. For order 32, there are 6×10^{37}. Their accomplishment marks the first complete classification of any type of magic square with orders greater than 5.

See also Magic cubes; Magic Pentagrams and Hexagrams

MANDELBROT, BENOIT B. (1924–)
Polish-born American geometer

Benoit B. Mandelbrot is a mathematician who conceived, developed, and named the field of fractal **geometry**. This field describes the everyday forms of nature—such as mountains, clouds, and the path traveled by lightning—that do not fit into the world of straight lines, circles, and smooth curves known as Euclidean geometry. Mandelbrot was also the first to recognize fractal geometry's value as a tool for analyzing a variety of physical, social, and biological phenomena.

Mandelbrot was born November 20, 1924, to a Lithuanian Jewish family in Warsaw, Poland. His father, the descendant of a long line of scholars, was a manufacturer and wholesaler of children's clothing. His mother, trained as a doctor and dentist, feared exposing her children to epidemics, so instead of sending her son to school, she arranged for him to be tutored at home by his Uncle Loterman. Mandelbrot and his uncle played chess and read maps; he learned to read, but he claims that he never did learn the whole alphabet. He first attended elementary school in Warsaw. When he was eleven years old, his family moved to France, first to Paris and then to Tulle, in south central France. When Mandelbrot entered secondary school, he was 13 years old instead of the usual 11,

Benoit B. Mandelbrot

but he gradually caught up with his age group. His uncle Szolem Mandelbrojt, a mathematician, was a university professor, and Mandelbrot became acquainted with his uncle's mathematician colleagues. Mandelbrot's teenage years were disrupted by World War II, which rendered his school attendance irregular. From 1942 to 1944, he and his younger brother wandered from place to place. He found work as an apprentice toolmaker for the railroad, and for a time he took care of horses at a château near Lyon. He carried books with him and tried to study on his own.

After the war, at the age of 20, Mandelbrot took the month-long entrance exams for the leading science schools. Although he had not had the usual two years of preparation, he did very well. He had not had much formal training in **algebra** or complicated integrals, but he remembered the geometric shapes corresponding to different integrals. Faced with an analytic problem, he would make a drawing, and this would often lead him to the solution. He enrolled in Ecole Polytechnique. Graduating two years later, he was recommended for a scholarship to study at the California Institute of Technology. In 1948, after two years there, he returned to France with a master's degree in aeronautics and spent a year in the Air Force.

Mandelbrot next found himself in Paris, looking for a topic for his Ph.D. thesis. One day his uncle, rummaging through his wastebasket for something for Mandelbrot to read on the subway, pulled out a book review of *Human Behavior and the Principle of Least Effort,* by George Zipf. The author discussed examples of frequency distributions in the social sciences that did not follow the Gaussian "bell-shaped curve," the so-called normal distribution according to which statistical data tend to cluster around the average, scattering in a regular fashion. Mandelbrot wrote part of his 1952 University of Paris Ph.D. thesis on Zipf's claims about word frequencies; the second half was on statistical **thermodynamics**. Much later, Mandelbrot commented that the book review greatly influenced his early thinking; he saw in Zipf's work flashes of genius, projected in many directions yet nearly overwhelmed by wild notions and extravagance, and he cited Zipf's career as an example of the extraordinary difficulties of doing scientific work that is not limited to one field. At the time, Mandelbrot had read **Norbert Wiener** on **cybernetics** and **John von Neumann** on **game theory**, and he was inspired to follow their example in using mathematical approaches to solve long-standing problems in other fields.

Mandelbrot was invited to the Institute for Advanced Studies at Princeton University for the academic year 1953–54. On returning to Paris, he became an associate at the Institut Henri Poincaré. In 1955, he married Aliette Kagan, who later became a biologist; they had two children, Laurent and Didier. From 1955 to 1957 Mandelbrot taught at the University of Geneva. In 1957 and 1958 he was junior professor of **applied mathematics** at Lille University and taught mathematical **analysis** at Ecole Polytechnique. In 1958 he became a member of the research staff at the IBM Thomas J. Watson Research Center in Yorktown Heights, New York.

In the 1960s Mandelbrot studied stock market and commodity price variations and the mathematical models used to predict prices. A Harvard professor had studied the changes in the price of cotton over many years and had found that the changes in price did not follow the bell-shaped distribution. The variations appeared to be chaotic. Existing statistical models for stock-market prices assumed that the rise and fall was continuous, but Mandelbrot noted that prices may jump or drop suddenly. He showed that a **model** that assumes continuity in prices will turn out to be wrong. He used IBM computers to analyze the data, and he found that the pattern for daily price changes matched the pattern for monthly price changes. Statistically, the choice of time scale made no difference; the patterns were self-similar. Using this concept, he was able to account for a great part of the observed price variations, where earlier statistical techniques had not succeeded.

Shortly thereafter, IBM scientists asked for Mandelbrot's help on a practical problem. In using electric current to send computer signals along wires, they found occasional random mistakes, or "noise." They suspected that some of the noise was being caused by other technicians tinkering with the equipment. Mandelbrot studied the times when the noise occurred. He found long periods of error-free transmission separated by chunks of noise. When he looked at a noisy chunk in detail, he saw that it, in turn, consisted of smaller error-free periods interspersed with smaller noisy chunks. As he continued to examine chunks at smaller and smaller scales,

he found that the chance of the noise occurring was the same, regardless of the level of detail he was looking at. He described the probability distribution of the noise pattern as self-similar, or scaling—that is, at every time scale the ratio of noisy to clean transmission remained the same. The noise was not due to technicians tinkering with screwdrivers; it was spontaneous. In understanding the noise phenomenon, Mandelbrot used as a model the **Cantor set**, an abstract geometric construction of **Georg Cantor**, a 19th-century German mathematician. The model changed the way engineers viewed and addressed the noise problem.

For centuries humankind has tried to predict the water level of rivers like Egypt's Nile in order to prevent floods and crop damage. Engineers have relied on such predictions in building dams and hydroelectric projects. In the 1960s, Mandelbrot studied the records of the Nile River level and found that existing statistical models did not fit the facts. He found long periods of drought along with smaller fluctuations, and he found that the longer a drought period, the more likely the drought was to continue. The resulting picture looked like random noise superimposed on a background of random noise. Mandelbrot made graphs of the river's actual fluctuations. He showed the unlabeled graphs to a noted hydrologist, along with graphs drawn from the existing statistical models and other graphs based on Mandelbrot's statistical theories. The hydrologist dismissed the graphs from the old models as unrealistic, but he could not distinguish Mandelbrot's graphs from the real ones. For Mandelbrot, this experience illustrated the value of using visual representations to gain insight into natural and social phenomena. Other researchers found similar support for Mandelbrot's statistical model when they showed fake stock charts to a stockbroker; the stockbroker rejected some of the fakes as unrealistic, but not Mandelbrot's.

Early in this century, mathematicians and geometers created curves that were infinitely wrinkled and solids that were full of holes. Much later, Mandelbrot found their abstract mathematics useful in models for shapes and phenomena found in nature. He had read an article about the length of coastlines in which Lewis Fry Richardson reported that encyclopedias in Spain and Portugal differed on the length of the border between the two countries; Richardson found similar discrepancies—up to 20 percent—for the border between Belgium and the Netherlands. Mandelbrot took up the question in a paper he called "How Long is the Coast of Britain?" The answer to the question, according to Mandelbrot, depended on the length of the ruler you used. Measuring a rocky shoreline with a foot ruler would produce a longer answer than measuring it with a yardstick. As the scale of measurement becomes smaller, the measured length becomes infinitely large. Mandelbrot also investigated ways of measuring the degree of wiggliness of a curve. He worked with programmers to develop computer programs to draw fake coastlines. By changing a number in the program, he could produce relatively smooth or rough coastlines that resembled New Zealand or those of the Aegean Sea. The number determined the degree of wiggliness and came to be known as the curve's fractal **dimension**.

Fascinated with this approach, Mandelbrot looked at patterns in nature, such as the shapes of clouds and mountains, the meanderings of rivers, the patterns of moon craters, the frequency of heartbeats, the structure of human lungs, and the patterns of blood vessels. He found that many shapes in nature—even those of ferns and broccoli and the holes in Swiss cheese—could be described and replicated on the computer screen using fractal formulas.

In Mandelbrot's reports and research papers during this period, he made clear that his methods were part of a more general approach to irregularity and chaos that was applicable to physics as well. Editors, however, usually preferred a more narrowly technical discussion. But then he was invited to give a talk at the Collège de France in 1973. Rather than selecting one of his many areas of research, he decided to explain how his many different interests fit together. Mandelbrot wanted a name for this new family of geometric shapes, which typically involved statistical irregularities and scaling. Looking through his son's Latin dictionary, he found the adjective *fractus,* meaning "fragmented, irregular," and the verb *frangere,* "to break," and he came up with *fractal.* His lecture aroused considerable interest and was published in expanded form in 1975 in French as *Les Objets Fractals: Forme, Hasard et Dimension.* Revised and expanded versions were published later in English in the United States as *The Fractal Geometry of Nature.* The publication of his book, which Mandelbrot called a manifesto and a casebook, attracted interest from researchers in fields from mathematics and engineering to economics and physiology. Mandelbrot remained at IBM as an IBM fellow but with various concurrent positions and visiting professorships at universities, including Harvard University, Massachusetts Institute of Technology, Yale University, **Albert Einstein** College of Medicine, and the University of California.

Using fractal formulas, computer programmers could produce artificial landscapes that were remarkably realistic. This technology could be used in movies and computer games. Among the first movies to use fractal landscapes were George Lucas's *Return of the Jedi,* for the surface of the Moons of Endor, *Star Trek II: The Wrath of Khan,* and *The Last Starfighter.* Some fractal formulas produced fantastic abstract designs and strange dragon-like shapes. Mathematicians of the early 20th century had done research in this area, but they did not have the advantage of seeing visual representations on a computer screen. The formulas were studied as abstract mathematical objects and, because of their strange properties, were called "pathological."

In the 1970s, Mandelbrot became interested in investigations carried out during World War I by French mathematicians Pierre Fatou and Gaston Julia, the latter having been one of his teachers years before at Polytechnique. Julia had worked with mathematical expressions involving **complex numbers** (those which have as a component the **square** root of negative one). Instead of graphing the solutions of **equations** in the familiar method of **René Descartes,** Julia used a different approach; he fed a number into an equation, calculated the answer, and then fed the answer back into the equation, recy-

cling again and again, noting what was happening to the answer. Mandelbrot used the computer to explore the patterns generated by this approach. For one set, he used a relatively simple calculation in which he took a complex number, squared it, added the original number, squared the result, continuing again and again; he plotted the original number on the graph only if its answers did not run away to **infinity**. The figure generated by this procedure turned out to contain a strange cardioid shape with circles and filaments attached. As Mandelbrot made more detailed calculations, he discovered that the outline of the figure contained tiny copies of the larger elements, as well as strange new shapes resembling fantastic seahorses, flames, and spirals. The figure represented what came to be known as the **Mandelbrot set**. Representations of the Mandelbrot set and the related **sets** studied by Julia, some in psychedelic colors, soon appeared in books and magazines—some even in exhibits of computer art.

Through his work with **fractals** and computer projections of various equations, Mandelbrot had discovered tools that could be used by scientists and engineers for strengthening steel, creating polymers, locating underground oil deposits, building dams, and understanding protein structure, corrosion, acid rain, earthquakes, and hurricanes. Physicists studying dynamical systems and fractal basin boundaries could use Mandelbrot's model to better understand phenomena such as the breaking of materials or the making of decisions. If images could be reduced to fractal codes, the amount of data necessary to transmit or store images could be greatly reduced.

Fractal geometry showed that highly complex shapes could be generated by repeating rather simple instructions, and small changes in the instructions could produce very different shapes. For Mandelbrot, the striking resemblance of some fractal shapes to living organisms raised the possibility that only a limited inventory of genetic coding is needed to obtain the diversity and richness of shapes in plants and animals.

In 1982 Mandelbrot was elected a fellow of the American Academy of Arts and Sciences. In 1985 he received the Barnard Medal for Meritorious Service to Science, awarded every five years by the National Academy of Sciences for a notable discovery or novel application of science beneficial to the human race. In 1986 he received the Franklin Medal for his development of fractal geometry. In 1987 he became a foreign associate of the U. S. Academy of Sciences. In 1988 he received the Harvey Prize and in 1993 the Wolf Prize for physics for having changed our view of nature. He officially retired from IBM in 1993, but he continued to work at Yale and at IBM as a fellow emeritus, preparing a collection of his papers and doing further research in fractals.

MANDELBROT SET

The Mandelbrot set is the most famous object in the branch of mathematics known as **fractals** and **chaos theory**. It was discovered in 1980 by the mathematician **Benoit Mandelbrot**, who pioneered this relatively new field.

•

A picture of the Mandelbrot set is reminiscent of a beetle with two main segments: a large circular bulb adjoining the left end of an even larger cartioid shaped central region. While these two segments visually dominate the Mandelbrot set, the geometric intricacies of its border truly define it.

A glance at that border reveals a number of bulbs of varying sizes attached to the two main regions of the set. Each of these bulbs, no matter the size, has a number of antennae protruding from it. Like the main region of the Mandelbrot set, the border of these antennae are decorated with a series of bulbs, which in turn have antennae protruding from them. Zooming in on those antennae will reveal more bulbs, each of which have antennae protruding from them. Zooming in on these antennae again reveals more bulbs, which contain more antennae, which contain more bulbs, and so on and so forth. Magnifying various regions of the Mandelbrot set's border not only reveals an infinitely repeating series of bulbs and antennae. It also reveals copies of the Mandelbrot set itself within the antennae. Each of these "miniature" Mandelbrot **sets** contain all of the geometric properties and similarities described above, including additional copies of the Mandelbrot set, which in turn contain all the previously discussed properties of the Mandelbrot set. These patterns, which reveal themselves at every scale of magnification, intrigue not only mathematicians, but scientists and artists as well.

The Mandelbrot set is a graph of an algebraic function, much like a **parabola** is a graph. Unlike a parabola, which exists on the Cartesian plane, the Mandelbrot set exists on the complex plane. The horizontal axis in this plane is the x-axis, as is true in the Cartesian plane. The vertical axis in the complex plane, however, is known as the i-axis. Each point in the complex plane represents a complex number. It is these **complex numbers** that produce the Mandelbrot set. In order to understand the **geometry** of this graph, it is necessary to understand the **algebra** behind it.

The fundamental principle underlying the Mandelbrot set is **iteration**, which means to repeat a process over and over again. In mathematical terms, to iterate a function means to substitute a value derived from that function back into the function to derive the next value. This can be repeated indefinitely. This process of iteration is illustrated in the following example using the function $f(x) = x^2 + 1$ and an initial value of $x = 0$.

$f(0) = 0^2 + 1 = 1$
$f(1) = 1^2 + 1 = 2$
$f(2) = 2^2 + 1 = 5$
$f(5) = 5^2 + 1 = 26$
$f(26) = 26^2 + 1 = large$
$f(large) = large^2 + 1 = very large$
etc.

As shown in the example above, the **orbit**, or iteration, of $x^2 + 1$, with an initial value of $x = 0$, tends to **infinity**.

Orbits do not always tend to infinity. This is illustrated in the iteration of the function $f(x) = x^2 + 0$ with an initial value of $x = 0$.

$f(0) = 0^2 + 0 = 0$
$f(0) = 0^2 + 0 = 0$

$f(0) = 0^2 + 0 = 0$
etc.

The orbit of $f(x) = x^2 + 0$, with an initial value of $x = 0$, is a fixed point.

The above examples illustrate an important dichotomy in iterative **functions**: Sometimes orbits of a function go to infinity, sometimes they do not. This dichotomy provides a **definition** of the Mandelbrot set: The Mandelbrot set consists of all c-values in the function $x^2 + c$, where x and c are both complex numbers, for which the orbit of 0 does not go to infinity.

This explains how the graph of the Mandelbrot set is created. A complex number is used as a c-value in the function $x^2 + c$. That function is then iterated using an initial value of $x = 0$. If this iteration does not tend to infinity, then that c-value is in the Mandelbrot set. Its corresponding point on the complex plane is colored black. If the iteration does tend to infinity, then the c-value is not in the Mandelbrot set. Its corresponding point on the complex plane is represented by a color other than black. This explains why the black region of the graph is the actual Mandelbrot set. It also hints at the extraordinary number of calculations required to create the graph of the Mandelbrot set, as this iteration process is applied to thousands of points on the complex plane.

Though the colored regions in the graph of the Mandelbrot set are not actually part of the Mandelbrot set, they do have meaning. They represent how quickly the orbit of 0 escapes to infinity when iterated in the function $x^2 + c$. The points that directly border the Mandelbrot set are colored violet. When used as c-values in the function $x^2 + c$, these points result in an orbit of 0 that tends to infinity very slowly. It may take 40 to 50 iterations for these orbits to explode to infinity. The points farthest away from the Mandelbrot set are colored red. When used as c-values in the function $x^2 + c$, these points result in an orbit of 0 that escapes to infinity very rapidly. These orbits will explode to infinity after only 5 to 10 iterations.

The Mandelbrot set results from a relatively simple algebraic procedure. Yet this procedure produces an intricate geometry that is full of **similarity** and patterns in the midst of apparent chaos. These patterns have provided new understanding of many diverse fields, including ecology, economics, and meteorology.

See also Chaos theory; Fractals

MANIFOLD

A manifold is a curve, surface, or higher-dimensional **space** that has, at every point in that space, a small neighborhood around each point that looks like a ball in the corresponding Euclidean space of the appropriate **dimension**. More precisely, the neighborhood should be topologically equivalent to a ball—in other words, it should be possible to stretch and distort the neighborhood, without tearing or gluing it, so that it is a ball.

The easiest manifolds to visualize are the 2-dimensional manifolds. In dimension 2, a "ball" in Euclidean space is simply a filled-in **circle**, or a disk. So a 2-dimensional manifold (or "2-manifold") is an object for which every point has a small neighborhood that looks like a (distorted) disk—these are what we commonly think of as surfaces. One of the simplest examples of a 2-manifold is the surface of the earth, a **sphere**; every point on the surface of the earth has a neighborhood that looks like a distorted disk. Another example is a **torus**, the surface of a donut. Manifolds do not have to be orientable—the **Möbius strip**, a non-orientable surface formed by gluing the opposite ends of a strip of paper with a 180-degree twist, is a manifold, since each point has a neighborhood that looks like a flat disk. Thus, manifolds can look quite different from the Euclidean space on which they are modeled; it is only locally that they must resemble Euclidean space.

In dimension 1, the Euclidean space is a line, and a "ball" is simply a line segment (this is the proper analogue, since we think of a ball as the **locus** of points within a certain fixed **distance** from a given point). Thus, a 1-dimensional manifold is any "space" that locally looks like a distorted line segment. A circle is a 1-dimensional manifold, but the letter "X" is not; at almost every point of the "X", there is a neighborhood that looks like a line segment, but at the center point, where four line segments come together, there is no neighborhood that looks like a single piece of line segment.

The **correspondence** between a small neighborhood and a piece of Euclidean space is often called a "map". This terminology arises from maps of the surface of the earth, which create correspondences between pieces of the sphere and pieces of the plane. A fruitful way of thinking of a manifold is in terms of its maps, which give concrete ways of visualizing small pieces of the manifold. A given point in the manifold can be contained in many different maps, in just the same way that a point on the surface of the earth can appear in many different maps, since there are many useful ways of **mapping** the earth. We can think of a manifold as being built from its maps, by gluing together pieces of Euclidean space whenever they correspond to the same piece of the manifold. In this way, a manifold (which often is very difficult to visualize) can be described by a collection of pieces of Euclidean space, together with gluing rules. These rules must satisfy some technical "compatibility" requirements in order for the result of the gluing to be a manifold; roughly speaking, these conditions ensure that maps are always glued together over ball-shaped pieces, and that they are glued without any folding or tearing of the pieces.

One of the most important questions in the study of manifolds is: How many different manifolds are there in each dimension? In dimension 1, the only kinds of manifolds are curves (possibly infinitely long) and circles (possibly distorted). In dimension 2, there is a complete classification of the "closed" manifolds, surfaces that have no boundary curves and that fit in a finite region. The orientable, compact 2-manifolds are the sphere, the torus, the double torus (a torus with 2 holes instead of 1), the triple torus, and so on. The non-orientable compact 2-manifolds are the projective plane (a man-ifold formed from a flat circular paper by gluing opposite points on the boundary circle), the **Klein bottle**, which is obtained from the projective plane by attaching a handle, and surfaces formed from the projective plane by adding 2 handles, 3 handles, and so forth. The Möbius strip is not included in this list since it is a surface with boundary, hence is not closed.

In dimensions higher than 2, it becomes much more difficult to classify, or even visualize, manifolds. We are all familiar with one 3-dimensional manifold: the space around us. We generally assume, based on our local picture, that the universe is shaped like 3-dimensional Euclidean space, extending infinitely far in all dimensions. However, physicists entertain the possibility (which at present cannot be tested conclusively) that our universe might be a more complicated 3-manifold. This seems counterintuitive, but it stems from the fact that we can only see a small portion of the universe, and a small piece is not enough to give information about global properties. For example, an ant that lived on the Möbius strip but could only see a very tiny part of it might never realize that it was not the Euclidean plane. In the same way, the universe might have some 3-dimensional "twists" in it that are simply too far away for us to see.

Still higher-dimensional manifolds have proven to be important in describing the physical world. Einstein's theory of relativity is based on the idea of thinking of the universe as a 4-dimensional manifold, in which time is one of the dimensions, and then considering geometric structures on that 4-manifold. More recently, with the advent of string theory, has come the idea that the universe might be an even higher-dimensional manifold, several of whose dimensions are too small for us to perceive. Another important manifold that arises in the study of Hamiltonian dynamics is the phase space, a high-dimensional manifold whose points represent the different possible configurations (positions, velocities, and so forth) of a given collection of physical objects.

See also Topology

MAPPING

A map, or mapping, is a rule, often expressed as an equation, that specifies a particular element of one set for each element of another set. To help understand the notion of map, it is useful to picture the two **sets** schematically, and map one onto the other, by drawing connecting arrows from members of the first set to the appropriate members of the second set. For instance, let the set mapped from be well-known cities in Texas, specifically, let A = {Abilene, Amarillo, Dallas, Del Rio, El Paso, Houston, Lubbock, Pecos, San Antonio}. We will map this onto the set containing **whole numbers** of miles. The rule is that each city maps onto its **distance** from Abilene. The map can be shown as a diagram in which an arrow points from each city to the appropriate distance.

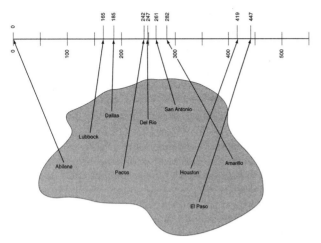

Figure 1.

A relation is a set of ordered pairs for which the first and second elements of each ordered pair are associated or related. A function, in turn, is a relation for which every first element of an ordered pair is associated with one, and only one, second element. Thus, no two ordered pairs of a function have the same first element. However, there may be more than one ordered pair with the same second element. The set, or collection, of all the first elements of the ordered pairs is called the **domain** of the function. The set of all second elements of the ordered pairs is called the **range** of the function. A function is a set, so it can be defined by writing down all the ordered pairs that it contains. This is not always easy, however, because the list may be very lengthy, even infinite (that is, it may go on forever). When the list of ordered pairs is too long to be written down conveniently, or when the rule that associates the first and second elements of each ordered pair is so complicated that it is not easily guessed by looking at the pairs, then it is common practice to define the function by writing down the defining rule. Such a rule is called a map, or mapping, which, as the name suggests, provides directions for superimposing each member of a function's domain onto a corresponding member of its range. In this sense, a map is a function. The words map and function are often used interchangeably. In addition, because each member of the domain is associated with one and only one member of the range, mathematicians also say that a function maps its domain onto its range, and refer to members of the range as values of the function.

The concept of map or mapping is useful in visualizing more abstract **functions**, and helps to remind us that a function is a set of ordered pairs for which a well defined relation exists between the first and second elements of each pair. The concept of map is also useful in defining what is meant by composition of functions. Given three sets A, B, and C, suppose that A is the domain of a function f, and that B is the range of f. Further, suppose that B is also the domain of a second function g, and that C is the range of g. Let the symbol o represent the operation of composition which is defined to be the process of mapping A onto B and then mapping B onto C. The result is equivalent to mapping A directly onto C by a third function, call it h. This is written g o f = h, and read "the composition of f and g equals h."

MARKOV, ANDREI ANDREEVICH (1856-1922)

Russian number theorist

Andrei Andreevich Markov's research covered a number of fields in mathematics, including **number theory**, **differential equations**, and quadrature formulas. He is best known, however, for his work in **probability theory** and his derivation of a powerful predictive tool now known as **Markov chains**. Although Markov himself saw few applications for this tool, Markov chains are now widely used in many fields of modern science.

Markov was born in Ryazan, Russia, on June 14, 1856. His mother was Nadezhda Petrovna, daughter of a government worker, and his father was Andrei Grigorievich Markov, an employee of the state forestry department who also managed a private estate. Markov suffered from poor health as a child, walking with crutches until the age of ten. He was not a particularly good student, although he demonstrated an interest and skill in mathematics at an early age. While still in high school he wrote a paper on the integration of linear differential **equations** that drew the attention of members of the mathematics faculty at the University of St. Petersburg. Markov entered that university in 1874, where he studied with the renowned **Pafnuty L. Chebyshev**. Markov formed a long-term working relationship with Chebyshev, with whom he wrote a number of papers. In one of his earliest papers, Markov reworked one of Chebyshev's theorems, called the **central limit theorem**, and corrected certain errors his teacher had made.

Markov received his bachelor's degree in 1878 for a thesis on differential equations and continuing **fractions**, for which he was also awarded a gold medal. Markov then stayed on at St. Petersburg to work for his master's degree, which was granted in 1880, then for his doctorate, which he received in 1884. His doctoral thesis was also on continuing fractions, and this was a subject that would remain central to much of his career.

Markov began teaching at the University of St. Petersburg in 1880 while still a graduate student. By 1886 he was named extraordinary professor of mathematics and in 1893 full professor, a post he would hold until his retirement in 1905. Markov also held parallel appointments in the St. Petersburg Academy of Sciences during this period. He was elected an adjunct member of the academy in 1886, extraordinary academician in 1890, and an ordinary academician in 1896. During his years at St. Petersburg Markov pursued a rather wide variety of topics in mathematics, including the search for minima in indefinite quadratic **functions**; the evaluation of **limits** for functions, integrals, and derivatives; and the method of moments. It was not until 1907, however, after his retirement from St. Petersburg, that he completed the work in probability theory for which he is now best known.

His most important contribution to probability theory began when he was writing a textbook on probability **calculus**, originally published in 1900. During preparation of the book, Markov encountered a particular type of probabilistic event, in which it is sometimes possible to predict the future status of some collection of random events if certain information is available about the present status of the sequence of events. As

Andrei Andreevich Markov

an example, the movement of gas molecules in a container is random. It can never be predicted exactly what any one molecule will do at any given moment. But, Markov showed, there are certain circumstances under which a later state of molecules in the container can be predicted if certain conditions exist among the molecules now. The sequence of events under which this situation can occur is called a Markov chain.

Markov apparently saw few practical applications for his **theorem**, mostly because of the state of science at that time, when it was believed that natural laws determined most events. Within a decade, however, the nature of science had undergone a dramatic revolution. Phenomena were seen to be the result of the *probable* behavior of fundamental particles and waves, though these actions often could not be proved. Markovian **analysis** became useful for predicting this type of probable behavior, and today Markov's work finds applications in a countless number of ways in the biological and physical sciences and in technology.

Beyond his academic work, Markov was also involved with Russian politics in the years leading up to the Russian revolution. He vigorously supported the liberal movement that swept through the country in the early twentieth century. In 1902, for example, he protested the action of Czar Nicholas II in withholding membership in the St. Petersburg Academy from dissident writer Maxim Gorky. Markov refused govern-

ment honors later offered to him and in 1907 resigned from the academy in protest of the czar's opposition to government reform. During the revolution of 1917 Markov volunteered to teach mathematics without pay at the remote village of Zaraisk. He became ill shortly after his return to St. Petersburg and died there in 1922.

MARKOV CHAINS

A Markov chain is a process, invented by the Russian mathematician **Andrei Markov** (1856-1922), used for predicting future outcomes or "states" of a system based upon the current state of the system. More formally, a Markov chain is a probabilistic dynamical system consisting of a finite number of states in which the probability that the system will be in a given state at time n depends only upon the state of the system at time n-1. The states of the system form the "chain." The movement of the system from one state to the next is called a transition. Every Markov chain has a "transition matrix" which contains the probabilities involved in moving from one state to the next. Also associated with each Markov chain is an initial state vector giving the state of the system at time 0. Beginning with the initial state vector, succeeding states of the Markov chain may be computed by a recursive process. If X_0 is the initial state vector, X_n the state vector at time n, and T the transition **matrix**, then the recursion formula $X_n = X_{n-1}T$ for n=1,2,3_, computes the state vectors for each successive state of the system.

As an example of how the above schema can be put into practice, suppose that the probability that an automated assembly line works correctly is 0.9 if it worked correctly the last time it was used, but only 0.7 if it did not work correctly the last time it was used. Also, suppose that statistical records indicate that the assembly line has worked correctly 80% of the time in the past. We want to predict the long term probabilities that the system works correctly or incorrectly. Our initial state vector would then be [0.8 0.2] where the first entry indicates the probability that the system is currently working correctly and the second entry is the probability that it is not working correctly. The transition matrix for this Markov chain would be a 2 by 2 matrix whose first row is [0.9 0.1], indicating the probabilities that the assembly line will work correctly (0.9) or not work correctly (0.1) given that it did work correctly the last time it was used; and whose second row is [0.7 0.3], indicating the probabilities of working and not working given that it did not work the last time it was used. So we are starting at state 0 which is [0.8 0.2]. To find state 1 of the system, we simply multiply [0.8 0.2] by the transition matrix, which gives [0.86 0.14]. This is interpreted to mean that the probability that the assembly line will work correctly at state 1 is 0.86. To find the probabilities for state 2 we multiply the new state vector [0.86 0.14] by the transition matrix, which gives [0.872 0.128]. Thus at state 2, the probability that the line will work correctly is 0.872. Now we simply continue this recursive procedure of multiplying each new state vector by the transition matrix to obtain the probabilities at the next state and we notice something interesting happening. Here are the state vectors for states 2 through 5: [0.8744

0.1256], [0.87488 0.12512], [0.874976 0.125024], and [0.8749952 0.1250048]. If we continue this process, we notice that at each new state of the system the state vector gets ever closer to [0.875 0.125]. In fact, after 13 iterations of this process, the **calculator** rounds off to [0.875 0.125] and, from that point on, each new **iteration** gives [0.875 0.125]. We interpret this to mean that, in the long run, we can expect our assembly line to work correctly 87.5% of the time. The vector [0.875 0.125] is called a "steady state" vector or an "equilibrium" vector for this Markov chain. A Markov chain which has an equilibrium vector is said to be regular.

Although Markov chains have been around since 1907, their use in applications was limited until the advent of fast digital computers with sufficient power to handle the often high dimensional matrices needed to **model** large systems. Currently Markov chains are used to predict future trends in business, economics, sociology, transportation, marketing, engineering, and the physical and biological sciences.

MASS AND ENERGY

Mass and energy are terms often used in physics calculations. While they were assumed to be interrelated quantities from their inception, it was only with the development of Einstein's theory of relativity that their truly intertwined nature became known, with the famous equation $E = mc^2$.

The concept of mass can be dealt with in more than one way. Inertial mass is the reaction an object or particle has to being accelerated or decelerated. That is, it is the expression of Newton's concept of inertia. Gravitational mass expresses the reaction an object has to being in a gravitational **field**. While there is nothing in theory that says these two quantities must be the same, the most sensitive of physical tests have found that they are.

Energy also comes in several forms. Kinetic energy is the energy a particle has due to its **motion**. Potential energy is due to chemical bonds, gravitational attraction, electromagnetic attraction, and other considerations.

Conservation of mass and energy are separate principles at low speeds. At high speeds (near the speed of **light**) they become one unified principle. Either way, these conservation laws are essential to giving quantitative solutions to physical problems. Energy conservation **equations** exist in algebraic and differential forms, and are used regularly in almost all physics and engineering applications.

See also Albert Einstein; Euler-Lagrange equation; Newton's laws; Relativity

MATH PHOBIA

Math phobia, which is exhibited by many students, is the persistent, illogical, intense fear of not succeeding in math. It is the belief that one is unable to handle the difficulty associated with learning math. Many people incorrectly assume that math phobia and an inability to be successful in mathematics are inherited from one's parents. Several legitimate factors contribute to, and increase the severity of, this perception. For instance, gender and ethnic backgrounds are not determining factors in mathematical competence, but peers' and teachers' attitudes toward gender and ethnicity may increase or decrease one's confidence in mathematical skills. The methods used to teach mathematics skills may affect whether a student feels successful and develops mathematical self-confidence. Finally, family and peer attitudes may positively or negatively influence students' attitudes toward mathematics, which in turn affect their levels of confidence.

Unless someone is diagnosed with a specific learning disability associated with processing numbers or learning intuitive concepts, math phobia is not a permanent condition. Math phobia can be overcome with the patience of an experienced and enthusiastic teacher, parent, coach, or therapist. Once a person gains even minimal amounts of success with mathematical concepts, the anxiety usually abates. Many teachers and other people concerned with the issue agree with Vanessa Stuart who writes: "My own hypothesis is that mathematics is like a sport: 90 percent mental—one's mathematics confidence—and 10 percent physical—one's mathematics competence in performing mathematical skills."

MATHEMATICAL SYMBOLS

Symbols play a key role in mathematics. They indicate sign (positive or negative) and grouping, describe operations and relationships, and represent numbers, shapes, constants, and variables in **equations**. The most basic type of mathematical symbol is a numeral, which is a symbol for a number. Ancient cultures used many different symbols to represent numbers. The numerals used today for the cardinal numbers (1,2,3...) are traced to Hindu and Arabic sources. The Egyptians, Babylonians, and Romans used different symbols, for example, the Hindu-Arabic numeral 10 is represented by X in **Roman notation**.

Although the exact origin of some mathematical symbols is not known, many were derived from the Arabic, Latin, and Greek languages and first appeared in print in Europe during the fifteenth through the seventeenth centuries. For example, the plus sign, +, is thought to originate from the Latin word *et* which means "and" in English. Its mathematical use has been found in manuscripts dating back to the 1400s.

The introduction or popularization of some symbols can be traced to specific mathematicians. **Robert Recorde** is credited with popularizing the use of two parallel lines, =, to indicate an equivalent relationship between mathematical terms. He first used this symbol in print in 1557. The common use of the letters x, y, and z to represent unknown variables in algebraic equations is attributed to **René Descartes** (1596-1650). **Leonhard Euler** (1707-1783) is often credited for the use of **e** for the base of natural **logarithms**, r for the radius of a **circle**, i for an imaginary number, f(x) for a function of x, Σ for summation, and Δ for finite difference.

The latter two symbols, Σ and Δ, are letters from the Greek alphabet. Greek letters are widely used in mathematics, physics, engineering, and other sciences to represent variables and constants. In mathematics, Σ is an example of **additive notation** and indicates that a series is to be added. For example, Σx_i for i=1 to 3 means the summation of x_1, x_2, and x_3 or $x_1 + x_2 + x_3$. Δ represents a difference between values. For example, if T_1=50 and T_2=30 then ΔT=50-30=20. Sometimes upper- and lowercase versions of Greek letters have different meanings. For example, the lowercase letter **pi**, π, represents the ratio of the **circumference** of a circle to its **diameter**. However, an uppercase pi, Π, is a **multiplicative notation** that represents series **multiplication**, e.g., Πx_i for i=1 to 3 means the product of x_1, x_2, and x_3 or $x_1 \times x_2 \times x_3$.

Note that multiplication can also be represented by a raised dot, as in $x_1 \cdot x_2 \cdot x_3$, by an asterisk, as in $x_1 * x_2 * x_3$, using parentheses to show grouping, as in $(x_1)(x_2)(x_3)$, or by juxtaposition of the terms with no symbol, as in $x_1 x_2 x_3$. Throughout history, mathematicians often used different symbols to represent the same concept. For example, Leonhard Euler and **Johann Bernoulli** used p and c, respectively, to represent π at various times in their writings. Certain symbols became popular over time, while others did not. Today there is a general consensus about symbols that are widely used in various branches of mathematics. These symbols are shown below along with their meanings:

General Arithmetic and Algebra		Differential Calculus and Integral Calculus	
Symbol	Meaning	Symbol	Meaning
+	Plus (also indicates positive sign)	$\int x$	Integral of x
−	Minus (also indicates negative sign)	f (x) or Φ(x)	Function of x
±	Plus or minus	dx	Differential of x
× or · or *	Multiplied by (also called times)	dx/dy or x′	First derivative of x with respect to y
/ or ÷	Divided by	$d^n x/dy^n$	n^{th} derivative of x with respect to y
:	Ratio	∂x/∂y	Partial derivative of x with respect to y
=	Equal to	ẋ	Partial derivative of x with respect to time
≠	Does not equal	Δx	Increment of x
≈ or ≐	Approximately to	Γ (n)	Gamma function of n
: : or ∝	Proportional to		
~	Similar to		
≅	Congruent to (also means defines)		
⇒	Implies		
→	Approches		
<	Less than		
<<	Much less than		
>	Greater than		
>>	Much greater than		
≤	Less than or equal to		
≥	Greater than or equal to		
$\sqrt{x}$	Square root of x		
$\sqrt[n]{x}$	n^{th} root of x		
Σ	Summation of		
Π	Product of		
!	Factorial (e.g., 3!=3*2*1)		
|x|	Absolute value of x (e.g., |−3| =3)		
∞	Infinity		
i	Imaginary number		
{ }	Set		
%	Percent		

Trigonometry and Geometry	
Symbol	Meaning
⊥	Perpendicular to
||	Parallel to
∠ or θ	Angle
∟	Right angle

Constants	
Symbol	Meaning
e	Base of natural logarithm system = 2.71828...
π	pi = 3.14159...
ϒ	Euler-Mascheroni constant = 0.577216...

MATHEMATICS AND ART

Although not immediately obvious to the untrained eye, both ancient and contemporary art depend on a great quantity of mathematical concepts. Artists vary scale, perspective, proportion and other ideas from mathematics in order to evoke a particular feeling, convey a particular idea, or represent a particular scene. Some artists, such as Jean Baptiste Camille Corot, use scale and proportion common to human experience in order to create "life-like" works, while other artists, such as Salvador Dali, use purposefully misleading elements of scale and perspective in order to craft surrealist images. Indeed entire schools of thought in art have centered on basic principles of mathematics—Corot is generally classified as part of the Realist movement and Dali generally classified as part of the Surrealist movement. Even colors, or hues, used by artists are related to mathematics—in order to create a desired color, for example, many mediums require artists to combine base, or primary, colors into a new hue. To achieve the same color later, the same proportion of the base colors must be used.

Artists work in both two-dimensional (such as paintings) and three-dimensional (such as sculptures) media. Works created in two-dimensional **space** are contained within a confined area of a plane. Works created in three-dimensional space have a finite mass that encompasses a finite **volume**.

The idea of the line is significant in both science and art; it represents the path of a object moving in space. In art, a line may be straight or serpentine (S-shaped); it may form the distinct outline of an object or simply be inferred as a border between two objects of an piece of art. In either case, the line draws a viewer's eyes through a painting, sculpture, or other type of artwork; it leads the viewer from one object to the next within the composition. Lines have a recognized effect on the human psyche—a thin, delicate line may invoke a sense of fragility, while a diagonal line is generally suggestive of movement, vigor, or instability. When a number of objects in a composition are placed along a particular line, the line is called an axis. A piece of art may have more than one axis contained within it, but if so, generally one axis will dominate. City plans provide excellent examples of the use of lines as axes.

Some art of ancient Egypt portrayed three-dimensional objects—such as animals, boats, and humans—on two-dimensional surfaces, such as temple walls and pottery. Although beautiful for their detail, the objects generally appear "flat"—the three-dimensional qualities of the objects are not captured in the artists' images. In other works, however, artists incorporate perspective—an organization of objects in space in terms of a single point—to evoke a sense of continuity in the composition. As the point toward which the lines converge is approached, objects diminish in size and appear to recede into the **distance**. Leonardo da Vinci's *The Last Supper* provides an excellent example of perspective. In the painting, the lines of perspective converge on Christ's head, projecting the image "into" the wall on which the painting appears. Although da Vinci's painting uses a point in space centrally located on the frame, other artists have used points in space offset from the center, or even beyond the borders of the canvas.

In art, proportion is the mathematical relationship between two or more objects in a composition; it provides an aesthetic means to relate the size of objects relative to each other. Generally speaking, proportion is both intuitive and common to the human experience—as our cognition develops in childhood and beyond, we generally associate an expected size to objects, especially relative to each other. We expect trains to be larger than cars, cars to be larger than bicycles, and bicycles to be larger than roller skates. We also expect objects further away to appear smaller than objects closer to us. When an object appears out of proportion—a woman walking on stilts, for example, whose long legs appear "out of proportion" with her torso—the object attracts our attention.

Artists manipulate the proportion of objects to draw attention to a particular object (or even the relationship between objects) or perhaps evoke a particular sentiment. Figures of authority—kings, deities, emperors—may be portrayed as larger than their subjects in order to evoke a sense of power or perhaps awe; at other times, they may be portrayed as the same relative size of their subjects of evoke a sense of commonality or accessibility. Some artists use skewed proportion to disturb viewers of their works and perhaps focus them on a particular idea. On the other hand, one of the many reasons that Michelangelo's *David* is considered so magnificent is that despite its superhuman size (13' 5"), David's entire figure is well-proportioned relative to itself.

Mathematical principles—especially scale, proportion, and perspective—underlie all forms of art, regardless of medium. Although artists may work with these concepts intuitively instead of making specific measurements, the net effect is the same—using the common human experience with basic mathematics to convey a specific idea.

See also Area; Dimension; Line; Perspective and projection; Plane figures and laminas; Plane geometry; Proportion; Volume

MATHEMATICS AND COMPUTERS

Computers not only operate on principles of mathematics, but are very useful in the often complex and time-consuming tasks mathematicians and scientists are required to perform. The operations and seemingly limitless abilities of computers are all based on numbers, but it is a more basic numbering system than that to which we humans are accustomed. Just as the most common numbering system we use is based on the number of digits most familiar to us (our ten fingers), computers use only two numbers, the **binary number system**, to carry out their work. This system matches the basic operating unit of the computer, its **logic circuits**, which is made up, in a sense, of on-off switches. In computer operations, off is equivalent to 0 and on to 1 in the binary numbering system. Computers are composed of millions or billions of these switches going on and off, or equaling 1 or 0, in ways that create an amazing number of possible outcomes. Each of these individual switches is called a bit, which is short for Binary DigITS. Eight of these bits working together compose a byte. While one bit can only be the number 0 or 1, a byte can represent numbers from 0 to 255. Larger combinations can be used to create larger numbers, such combinations of bits forming the various types of memory, central processing and video display circuits withing a computer. Some of the more commonly known numbers talked about in computer jargon are the kilobyte (k), the megabyte (MB), and the gigabyte (GB).

While binary is the basic numbering system of the computer, programmers work with different combinations of bytes to deal with computers at the most basic levels without the cumbersome methods required to program in binary. They use Octal (base 8) and Hexadecimal (base 16) notation, and of course, complex and very useful programming languages. No matter what the method used to program the computer, the basic operations are reduced to simple 1's and 0's in the end. But using many different combinations of these two numbers manipulated in special ways provides all the operations necessary for everything from computer games to complex scientific calculations.

Actual math operations begin with simple **addition**. In binary, when it comes to positive integer addition, this is quite simple. Using several bits, one can add two numbers, say 2 + 3, which, in binary are represented as 010 + 011=101. **Subtraction**, **negative numbers**, and other operations require more than simple **addition**. It is necessary to assign special **functions** to various parts of the computer circuitry in order for certain bits or combinations of bits to accomplish specific tasks. The simplest example of this is when a negative number is denoted by the use of a fourth bit. If 2 = 010 in binary, -2 would be 1010 with the first 1 meaning "negative."

Fractions, which are calculated by the computer in decimal form, can be created by having bits to the left of a decimal point equal decreasing positive powers of two and to the right of the decimal point increasing negative **powers** of two. Since in binary only two states are possible, numbers larger than two are created by combining bits in a format where the first place left of the decimal can be 0 or 1 (that is, 0 or 2^0), the next place to the left represents 0 or 2 (and $2=2^1$), the third place represents 0 or 4 (2^2) and so forth. To the right of the decimal, these bits would be, first 0 or .5 (2^{-1}), second 0 or .25 (2^{-2}), etc. Given enough bits on either side of the decimal point, any number can be represented in this way which is called fixed point notation. With floating point computer **arithmetic**, some bits are assigned to represent the number, others are **exponents**, and one works as the sign for the number. With such assignments made for the numbers, **logic circuits** are combined to perform the needed operations, including the basic four, plus **Boolean algebra**, comparisons, **iteration**, **factor analysis**, and many more.

See also Boolean algebra; Decimal fractions; Decimal position system

MATHEMATICS AND CULTURE

Throughout human history the use and understanding of mathematics has been so closely related with the rise of cultures that it would not be unreasonable to say that mathematics is actually one of the prerequisites of culture. Nearly all of the well-known ancient cultures—the Sumerians, the Egyptians (see **Egyptian mathematics**), the Babylonians (see **Babylonian mathematics**), Greeks, Romans, and others—founded their civilizations upon mathematics and mathematically related engineering.

It is interesting to note that knowledge of mathematics appears to predate the rise of civilization. A Neolithic site in Japan, dating from approximately 7000 years ago, contains numerous structures that clearly show cedar-trunk post holes set unerringly 4.2 meters apart. Such uniform accuracy suggests skills in both measurement and engineering. Artifacts have been unearthed at several other Neolithic and Paleolithic sites that can be interpreted as counting or tallying devices. And it has been accurately shown that clay tablets from the Fertile Crescent region of Iraq, tablets that may have been the earliest examples of written language, were actually commercial "contracts" used to tally the amount of commodities (grain, cattle, etc.) to be bartered.

Perhaps the most startling fact regarding the cultural **history of mathematics** is that it may not have only predated culture, it may have predated humanity itself. An exhaustive examination of fossil hand-axes, fashioned between 500,000 and 1,000,000 years ago and wielded by Neanderthals and other hominids, show an understanding of the principle of proportion. Although the individual axes vary widely in their actual sizes, the proportion of length to width is almost always the same. This grasp of the idea of proportion was not directly addressed again until Euclid (c. 325 B.C.-c. 265 B.C.) set it down in his *Elements* hundreds of thousands of years later.

See also Babylonian mathematics; Egyptian mathematics; History of mathematics; Mathematics and magic; Rhind papyrus

MATHEMATICS AND LITERATURE

It may not be obvious to the casual observer that the fields of mathematics and literature have anything whatsoever to do with each other. Certainly other fields of human endeavor have proven more appealing to poets and writers, perhaps because they have been more comprehensible. However, mathematicians have inspired work from across the spectrum of literary arts—fiction, poetry, plays, and essays—and in some cases have even contributed to these arts themselves.

Few mathematicians since the Vedic scholars in India have been as concerned with the relationship of verse and mathematics. Their proofs and theorems were often written as poetry, with the structure and meter of the verse reflecting the numerical nature of their subject matter. Omar al-Khayyam was, in his own time, equally famous for his work in verse and for his mathematical reasoning, although he did not combine the two pursuits as thoroughly as the Vedic mathematicians. English mathematicians such as Hamilton, Boole, Sylvester, and Maxwell all wrote poetry as a hobby, although none of them ever enjoyed nearly the critical success al-Khayyam did. Mathematician **Felix Hausdorff** wrote plays, and his colleague Sonya Kovalevskaya did work in theatre, poetry, and short novels, as well as writing the seminal autobiography of a mathematician. However, perhaps the most famous mathematician-author was Charles Dodgson, also known as **Lewis Carroll**. Dodgson's most famous works, *Alice in Wonderland* and *Alice Through the Looking-Glass*, were for children, although he also wrote verse, essays, short stories, and other children's books. Much of the delight of the Alice books comes from their inspired use of mathematics and logic puzzles, an interest which was certainly developed in Dodgson's practice of mathematics.

Other literary works have been more directly mathematical in their nature, have been inspired by mathematical progress, or have dealt with the lives of fictional mathematicians. Shakespeare was focused on the concept of **zero**, which was fairly new to his society; it kept coming up in metaphors and asides in his plays. *Flatland*, a study in tolerance, was also a description of dimensionality in **geometry**. Even picture books such as *The Dot and the Line: A Mathematical Romance* and *Math Curse* have been written to demonstrate mathematical concepts in more palatable ways to the average reader.

In the late 1990s, there were a few movies made about mathematics or mathematicians. However, the theatre has been fairly short on math plays. **Fractals**—a particularly popular and accessible branch of mathematics because so many of them can be visually beautiful—were one of the topics of Tom Stoppard's play *Arcadia*, which chronicled the life of a young mathematical genius. Other works do not abound.

The genre of science fiction has probably used mathematics less than any other technical field as inspiration for its tales. However, some short stories, particularly "A Streetcar Named Mobius" and the works of Greg Egan, do use mathematical themes. Also, Isaac Asimov's classic *Foundation* series is about the life's work of a mathematician who applies his craft to the behavior of human beings. For the most part, though, in science fiction as in other genres, the related fields of physics and computer science have proven more fruitful inspiration for math-related literature than mathematics itself.

MATHEMATICS AND MAGIC

Mathematical applications so permeate the many spheres of human interest that they are often easily overlooked. Occultists, for example, have been keen utilizers of mathematics right down the millennia. Astrology, numerology and sacred **geometry** all utilize mathematics; indeed, practitioners of these and similar pursuits were the discoverers of some of the fundamental mathematical thought that we still follow today. **Pythagoras** (580? B.C to 500? B.C.), for example, was as much interested in magic and the occult as he was geometry—if not more so.

Modern magicians use math, too. Illusionists who perform for entertainment make use of a variety of calculations; everything from the fixing of angles of line-of-sight, for large-scale stage illusions; to mathematical elimination computed mentally in scores of card tricks. Some of the most familiar magic tricks we've seen, like the lady cut in half and variations on the levitation trick, are dependent upon applied geometry.

Math and magic have been working together for so long that their relationship has become nearly invisible; the magicians themselves forget that they are part mathematician. But like so many challenges of intellect that humankind has taken on throughout history, magic can only build upon what has been learned before. And since an awareness of mathematics has been one of the most ancient of human experiences, it cannot but effect nearly every other.

See also History of mathematics; Mathematics and culture

MATHEMATICS AND MUSIC

While many artists are not comfortable with the more complicated work carried out by mathematicians and vice versa, mathematics has been an important part of the art of music for centuries, from the simple counting of the beats in a musical composition to the complex computer **algorithms** used to produce realistic, computer-synthesized music. Musical time signatures are written and spoken of like **fractions**, with the "numerator" telling musicians how many beats to count per measure and the "denominator" showing the type of note (half note, quarter note, etc.), or the length in relation to a whole note, that is assigned to each beat. For example, any time signature in the form 1/x means there is one beat in each measure, 5/x means there are five beats in each measure, etc. A time signature of x/1 means each beat is worth one whole note, or x/4 means each beat is worth one quarter note, and so on. Considering this mathematically, it can be seen that this is a case of very simple **fractions**. In math, 3/2 means the whole is divided into two parts, and that there are three of these parts. In music, this would mean that the whole (the whole note) is divided into two parts or beats (half notes), and there are three

half notes in each measure of the music. Notes themselves are worked with as fractions of the whole note. Each is played for a length of time relative to the whole note which is, in turn, related to real time by the assigned pace of the musical beat. This fraction analogy is helpful in understanding the basics of music, and those who understand fractions can more quickly understand musical notation, but while in math, the fractions 1/2 and 2/4 are equal to one another, in music, 1/2 time and 2/4 time are not.

Another way of using mathematics to understand musical notes it to think of each of the main note durations as a power of two. As mentioned previously, all notes have length values that are some fraction of the whole note. If we think of the whole note as equaling one, then it can be given a value of $2^0 = 1$. Half notes would then have a value of $2^{-1} = 1/2$, quarter notes 2^{-2}, and so on. There are notation conventions for rests, or silence within music that have these same basic values. Composers often wish to use notes of lengths other than these values and there are a few different methods for doing this. One of the most common is the triplet which places three note in the **space** of two. If the original two notes were eighth notes, then each individual note in the triplet is worth one third of a quarter note. This result in the creation of a (1/4 x 1/3) 1/12 note. There are many combinations like this that can be put together to create notes of all different fractional amounts of a whole note. Another method of creating notes of different lengths is the use of the dot. A dot increases a note by one half its original value. In other words, a quarter note in 4/4 time is worth one beat. Adding a dot to the note makes it worth one and a half beats. Therefore, a dotted quarter note is the equivalent of 1/4 x 3/2 = 3/8, or a three-eighths note. This method can be used to make many notes.

Because music is an artistic thing and composers desire more subtlety in creating it than would be allowed by a very few discreet lengths of notes, they add notes of different lengths together to create any desired length. This is called the tie because two notes are "tied together" and played as if they were one note with the length of the two together. The methods mentioned previously are somewhat limited since they may not be able to provide very small steps in lengthening notes. Often, composers may make use of ties, which require musicians to be able to quickly add fractions in their heads. In real performances, this is more a matter of "feel" than actual calculation, however. But to illustrate what is happening simply, two quarter notes can be tied together to make a half note (1/4 + 1/4 = 1/2). This technique becomes more generally useful when tying together notes of different lengths to step outside the basic timings available through the use of individual notes. One could create a note with a length relative to the whole note of 9/16 by tying a 1/2 note and a 1/16 note together (1/2 + 1/16 = 8/16 + 1/16 = 9/16). Using ties, a wide variety of fractional note values can be achieved allowing perhaps infinite variety and versatility in the creation of melodies and harmonies.

Studies have shown that students that are good at math have a tendency to do better at playing or interpreting music than those who are not, and that, conversely, the study and playing of music helps students excel at math. The examples above should make it clear why the former is true. Any student that is adept with fractions will be much more able to think quickly when counting rhythms, or transposing note steps in music, or figuring out note values and time signatures. Given a measure of music with a number of 1/16, or even 1/64 notes, or dotted notes and triplets, someone with strong math abilities should be able to count it out more easily than someone who doesn't. More sophisticated math abilities allow composers to more easily handle frequencies, harmonies, and other aspects of music.

Listening to music also means interpreting it. To interpret music, certain math skills are used. This may not immediately improve math skills, but it can heighten ones awareness of these mathematical concepts. One study showed what has become known as the "Mozart Effect." Students were isolated, with one group listening to the music of Mozart, another left in complete silence, while a third listened to meditation tapes. A short period of time afterward, each group was given a series of skill and academic tests. In almost all of these tests, the Mozart group performed better than the other two groups showing that classical music can not only be pleasant to listen to, but can help in academic achievement.

During the 1970s, another way of using mathematics to create music began to emerge. While the idea of the personal computers was still in its infancy, computer technology began to revolutionize the music industry with the advent of electric keyboards and synthesizers. The most basic of sound-creating instruments are simple audio oscillators which emit tones of one or more pitches. These have the form of **sine** waves. Combinations of these simple waves and electronics that modify them, not only change their pitch at the touch of a button or turn of a knob, but make them display a wide variety of complex patterns that allow large numbers of unique sounds to be created. Early synthesizers not only created new sounds impossible to attempt on conventional musical instruments, but tried to imitate actual instruments as well. This was done by working with combinations of electrical signals that were sent through various devices which converted them to audio that could be output to speakers or put onto recording tape for later playback.

As with so many areas in the world today, advances in computers have brought advances in the creation of synthetic music. It is now possible for complete, full-length compositions to be created electronically with an amazing variety of sounds and a large number of virtual instruments. It is even common to imitate real instruments with such accuracy that even the most discerning ear has difficulty telling that the music is synthesized. These remarkable capabilities are all made possible through the same type of mathematical operations as carried out in computers but with a strictly audio output. As a matter of fact, multimedia computers have musical synthesizers, commonly known as sound cards, inside them that allow voice, music, or other sounds to accompany the visual output displayed on the computer screen.

See also Fractions; Mathematics and art; Mathematics and computers

MATHEMATICS AND PHILOSOPHY

On the surface, there would seem to be little in common between mathematics and philosophy. These two disciplines, however, share a notable history of parallel academic pursuit as well as much nomenclature and concepts that apply equally well to both.

Classical Greek philosophers, particularly **Plato** (428 BC-347 B.C.) and **Aristotle** (384 BC-322 B.C.), saw not only math, but also philosophy, as hard, empirical sciences. Plato's Academy developed analytical methodology for **arithmetic** and plane **geometry** that was comparable to analyses used in philosophy. Platonic philosophers also sought to develop the mathematics of astronomy and musical harmony, which were then understood to be philosophical pursuits, and even see med to have an early grasp of Euclidean geometry. The arithmetic they recognized consisted primarily of methods of counting and measuring, and it was perhaps thanks to their philosophical bent that they spent much time and effort attempting to develop mathematical systems to measure concepts which we now think of as unquantifiable, such as pleasure and pain.

Such pursuits demonstrate, to the modern mind at least, a seemingly inevitable divergence between mathematics and philosophy that results in the gulf between them that we know today. But an open-minded consideration of how the ancient philosophers viewed mathematics, might gain us new insights into similarities that we wouldn't normally recognize as such. The concepts of paradox, **infinity** and even **zero** have profound philosophical implications. Philosophy and mathematics share the processes of postulates and axioms to argue theories. Even the concept of numbers themselves seem philosophical, when one considers that numbers, in of themselves, are unobservable; they are neither physical nor mental, neither moving nor at rest, have no mass or any definable location in **space-time**.

Does this mean that philosophy and mathematics share relevant parallels that should alter how we study them today? Probably not. It does mean, however, that the modern student should always remember that mathematics never exists in a vacuum; that it affects and is effected by a myriad of other disciplines, and that a rich symbiosis of knowledge and learning is the result.

See also History of mathematics; Mathematics and culture; Mathematics and magic

MATHEMATICS AND PHYSICS

Until the end of the nineteenth century, the people who studied mathematics and those who studied physics would probably not have separated themselves out professionally. Both would have called themselves "natural philosophers," and many of them did research work in both fields. The Greeks who did much original work in **geometry** also formulated many theorems about the physical world around them, most memorably in Archimedes' bathtub revelation about the calculation of **vol-**

ume. The Arabic scholars who invented **algebra** also preserved what texts remained of Greek physics. However, it was in the person of **Isaac Newton** that mathematics and physics were most completely united. Newton's physics informed his mathematics. The study of the rate of change in **calculus** was necessary for the Newtonian theory of **mechanics**. While it is still not yet well-determined whether Newton or Leibniz originally formulated calculus, Newton was the one who applied that type of mathematics extensively to the physical universe. This application allowed physics to make a transition from a largely observational science, as it was when astronomy was the viable component of it, to a science with considerable predictive and explanatory power.

Though most of the great nineteenth-century physicists were also mathematicians, mathematics and physics were already separating themselves into separate disciplines. Distinguished mathematicians such as Helmholtz and the great **Gauss** made contributions in the then-new science of electromagnetics as well as formulating much of the theory of **differential equations, non-Euclidean geometries**, and diverse other mathematical fields of interest. By the dawn of the twentieth century, physicists considered mathematics the language in which physics is expressed. It had its own validity, of course, and there were vast fields of arcane mathematics that seemed to have no physical applications. For its part, physics was making fewer contributions to mathematics, although every once in awhile a physicist would find some **theorem** or property useful and would prove it or ask a mathematician to do so. The contributions physics makes to mathematics are subtler, because physics allows for some concrete examples and visualizations of mathematics as it is taught. However, as **quantum mechanics** grows more arcane by the decade, the evolving relationship between mathematics and physics continues to change. The arcane theories of yesterday (**group theory**, for example) are essential to explaining the concrete physical data of tomorrow. While the two disciplines have both grown too large to allow for many people who do work in both, they continue to inform each other and contribute problems and explanations back and forth across the subject boundaries.

See also Mechanics; Isaac Newton; Space-time

MATHEMATICS AND PSYCHOLOGY

Psychology is the scientific study of the relationships between mental processes, emotions, and behavior. Mathematics and psychology are linked in three major ways. First, psychologists study mathematical cognition, i.e., the brain's development, acquisition, and application of mathematical skills. Second, psychologists investigate people's feelings and attitudes regarding mathematics. Third, psychologists use mathematics, particularly **statistics**, as a professional tool to quantify and analyze their scientific findings.

Psychologists working in the field of mathematical cognition study how humans process information, interpret **mathematical symbols**, and develop and use strategies to solve mathematical problems. For example, these skills are particularly important for so-called "word" problems, where written descriptions must be translated into **equations**. Most students consider "word" problems more difficult to solve than other types of math problems. This is because "word" problems require a variety of skills from the brain, including the ability to read and comprehend the meaning and context of the words, the ability to perceive and define the mathematical problem, the ability to assign mathematical symbols to unknown variables, and finally, the ability to apply problem-solving strategies and calculate the correct answer.

Mathematical cognition is a very important field in psychology. It benefits scientists and doctors studying the brain, and it helps educators develop better teaching methods for mathematics. In addition, its study is crucial to the development of "smart" computers, neural networks, fuzzy logic, robots, and artificial intelligence.

Psychologists also study how people feel about mathematics, because a person's feelings about a subject influence their willingness to learn and use it. For example, cultural and gender differences in attitudes about mathematics affect test scores. Another area receiving a lot of attention is called **math phobia** or math anxiety. Math phobia is a fear of mathematics. People with math phobia become so uncomfortable and anxious when confronted with mathematical tasks that they can experience physical symptoms including increased heart rate, nervous stomach, and breathing difficulties that prevent them from concentrating and learning. These feelings have been traced to a variety of sources, including negative experiences in the class room, poor self-image, lack of appreciation for the application of mathematics to "real life," and shyness that prevents asking questions.

The third major link between psychology and mathematics is that psychologists use mathematical and statistical tools to quantify and analyze their research findings. This use is called psychometrics and arises from the application of the **scientific method** in psychology, i.e., a systematic method of data collection, hypothesis development, and experimental testing that can be duplicated and verified by other scientists.

One example of psychometrics is the Intelligence Quotient (IQ) test, a standardized test that measures a person's relative intelligence. An IQ score is a relative measurement; it is compared to a reference IQ of 100 for the average score. IQ scores for a large population are an example of a statistical function called the normal distribution. Normal curves or Gaussian curves are the familiar bell-shaped curves in which measurements are graphed along the x-axis and frequency is graphed along the y-axis. The majority of IQ scores fall in the wide part of the curve near the **mean** value of 100. As scores deviate negatively or positively from 100, they decrease in frequency.

Q methodology is a type of **analysis** used in psychology to measure and quantify the feelings of a group of people regarding a particular subject. For example, a large group of students could be asked the following question: "How do you feel about your school?" A wide variety of answers would be collected ranging from "I hate it" to "I love it" with many opinions in between pointing out good and bad qualities of the school. This entire opinion set is called the concourse. From it, a limited number of opinions (the Q sample) would be selected that represent the spectrum of responses. During the next interview, the students would read the Q sample and rank their level of agreement with each opinion using a scale of -4 to +4, where -4 indicates strong disagreement and +4 indicates strong agreement with the opinion. This process is called Q sorting. The resulting numerical data can be analyzed using statistical **functions** to provide a mathematical description of student opinions about their school.

Common statistical concepts and tools studied and used by psychologists include **correlation**, regression, sampling distributions, probability density functions, and **factor** analysis.

MATRIX

A matrix is a rectangular array of numbers or number-like elements:

$$\begin{pmatrix} 1 & 1 \\ 2 & 0 \end{pmatrix} \quad \begin{pmatrix} a_{11} & a_{12} \\ a_{21} & a_{22} \\ a_{31} & a_{32} \end{pmatrix}$$

In the example on the left, 1 1 and 2 0 are its rows; 1 2 and 1 0, its columns. In the example on the right there are three rows and two columns, making it a 3×2 matrix. When subscripted variables are used to represent the elements, the first subscript names the row, the second, the column: $a_{\text{row, column}}$. For example, a_{21} is in the second row and first column, but a_{12} is in the first row, second column. Except when there is danger of confusion, the subscripts need not be separated by a comma. Some authors enclose a matrix in brackets: other authors use parentheses, as above.

Matrices can also be represented with single letters A, I, or with a single subscripted **variable** $(a_{ij} = b_{ij})$ if and only if $a_{ij} = b_{ij}$ for all i, j which says symbolically that two matrices are equal when their corresponding elements are equal.

Under limited circumstances matrices can be added, subtracted, and multiplied. Two matrices can be added or subtracted only if they are the same size. Then $(a_{ij}) + $ or $- (b_{ij}) = (a_{ij}) = (b_{ij})$ which says that the sum or difference of two matrices is the matrix formed by adding or subtracting the corresponding elements.

$$\begin{pmatrix} 1 & 2 & 1 \\ 0 & 4 & 7 \end{pmatrix} + \begin{pmatrix} -3 & 1 & 4 \\ -1 & 2 & 2 \end{pmatrix} = \begin{pmatrix} -2 & 3 & 5 \\ -1 & 6 & 9 \end{pmatrix}$$

These rules for adding and subtracting matrices give matrix **addition** the same properties as ordinary addition and **subtraction**. It is closed (among matrices of the same size), commutative, and associative. There is an additive identity (the matrix consisting entirely of zeros) and an additive inverse:

$-(a_{ij}) = (-a_{ij})$

This latter **definition** allows one to subtract a matrix by adding its opposite:

A - B = A + (-B)

Multiplication is much trickier. For multiplication to be possible, the matrix on the left must have as many columns as the matrix on the right has rows. That is, one can multiply an m × n matrix by an n × q matrix but not an m × n matrix by an p × q matrix if p is not equal to n. The product of an m × n matrix and an n × q matrix will be an m × q matrix.

Multiplication is best explained with an example:

$$\begin{pmatrix} 1 & 3 \\ 2 & 1 \end{pmatrix} \begin{pmatrix} 5 & 2 & 1 \\ 0 & -1 & 2 \end{pmatrix} = \begin{pmatrix} 5 & -1 & 7 \\ 10 & 3 & 4 \end{pmatrix}$$

The 5 in the product comes from (1) (5) + (3) (0). The -1 comes from (1) (2) + (3) (-1). The 7 comes from (1) (1) + (3) (2). In the second row of the product, 10 = (2) (5) + (2) (0); 3 = (2) (2) + (1) (-1); and 4 = (2) (1) + (1) (2).

Each row in the matrix on the left has been "multiplied" by each column in the matrix on the right. We say "multiplied" because each row on the left is a two-number row, and each column on the right is a two-number column. These numbers have been paired off, multiplied, and added. This kind of "multiplication" is somewhat more complicated than the ordinary sort. Those who are familiar with vectors will recognize this as forming the dot product of each row of the matrix on the left with each column on the right.

Multiplication is associative, but not commutative. That is (AB)C = A(BC) but, in general, AB not equal BA.

In the example above, multiplication is not even possible if the 2 × 3 matrix is placed on the left.

There is a multiplicative identity, I. It is a **square** matrix of an appropriate size. It has 1's down the main diagonal and 0's elsewhere.

$$\begin{pmatrix} 1 & 0 \\ 0 & 1 \end{pmatrix} \begin{pmatrix} 5 & 2 & 1 \\ 0 & -1 & 2 \end{pmatrix} = \begin{pmatrix} 5 & 2 & 1 \\ 0 & -1 & 2 \end{pmatrix}$$

or

$$\begin{pmatrix} 5 & 2 & 1 \\ 0 & -1 & 2 \end{pmatrix} \begin{pmatrix} 1 & 0 & 0 \\ 0 & 1 & 0 \\ 0 & 0 & 1 \end{pmatrix} = \begin{pmatrix} 5 & 2 & 1 \\ 0 & -1 & 2 \end{pmatrix}$$

A matrix may or may not have a multiplicative inverse, which is a matrix A^{-1} such that $A^{-1}A = I$
Since

$$\begin{pmatrix} 1 & -1 \\ 1 & 2 \end{pmatrix} \begin{pmatrix} \dfrac{2}{3} & \dfrac{1}{3} \\ -\dfrac{1}{3} & \dfrac{1}{3} \end{pmatrix} = \begin{pmatrix} 1 & 0 \\ 0 & 1 \end{pmatrix}$$

the two matrices on the left side of the equation are multiplicative inverses of each other.

An example of a matrix which does not have an inverse is

$$\begin{pmatrix} 1 & 1 \\ 2 & 2 \end{pmatrix}$$

This can be seen by trying to solve the matrix equation

$$\begin{pmatrix} 1 & 1 \\ 2 & 2 \end{pmatrix}\begin{pmatrix} a & b \\ c & d \end{pmatrix} = \begin{pmatrix} 1 & 0 \\ 0 & 1 \end{pmatrix}$$

Using the row-by-column rule for multiplying gives a + c = 1 and 2a + 2c = 0, which is impossible.

Typically one limits the concept of an inverse to matrices which are square. Without this limitation a matrix such as

$$\begin{pmatrix} 1 & 0 & 1 \\ 1 & 1 & 2 \end{pmatrix}$$

would have no left inverse at all and an infinitude of right inverses. Working only with square matrices, it is possible to show that a matrix and its inverse commute, that is, that any left inverse is also a right inverse. It is also possible to show that any inverse is unique.

Matrices are used in many ways. The following examples show three of those ways.

A matrix can be used to solve systems of linear **equations**. If

$$A = \begin{pmatrix} 1 & -1 \\ 1 & 2 \end{pmatrix} \quad X = \begin{pmatrix} x \\ y \end{pmatrix} \quad \text{and } B = \begin{pmatrix} 9 \\ 3 \end{pmatrix}$$

then the matrix equation AX = B represents the system x - y = 9, x + 2y = 3. If one multiplies both sides of the matrix equation by the inverse of A (computed above) $A^{-1}AX = A^{-1}B$ then $X = A^{-1}B$.

Writing these matrices in expanded form

$$\begin{pmatrix} x \\ y \end{pmatrix} = \begin{pmatrix} \frac{2}{3} & \frac{1}{3} \\ -\frac{1}{3} & \frac{1}{3} \end{pmatrix}\begin{pmatrix} 9 \\ 3 \end{pmatrix}$$

and multiplying

$$\begin{pmatrix} x \\ y \end{pmatrix} = \begin{pmatrix} 7 \\ -2 \end{pmatrix}$$

or x = 7, y = -2.

For such a small system of equations, using matrices is rather inefficient. For systems with a large number of unknowns and equations, using matrices is very efficient, especially if one turns the work over to a computer. Computers love matrices.

Two-by-two matrices can be used to represent **complex numbers**:

$$a + bi \longleftrightarrow \begin{pmatrix} a & b \\ -b & a \end{pmatrix}$$

They behave like complex numbers, and they sneak around the sometimes disturbing property

$i^2 = -1$: $\begin{pmatrix} 0 & 1 \\ -1 & 0 \end{pmatrix}\begin{pmatrix} 0 & 1 \\ -1 & 0 \end{pmatrix} = \begin{pmatrix} -1 & 0 \\ 0 & -1 \end{pmatrix}$

Matrices can be used for enciphering messages. If the message were "OUT OF WATER," it would first be converted to numbers using a = 1, b = 2, etc. to become 15 21 20 15 6 23 1 20 5 18. These numbers would then be broken into pairs, and each pair, treated as a 2 × 1 matrix, would then be multiplied by a secret enciphering matrix:

$$\begin{pmatrix} 5 & 2 \\ 7 & 3 \end{pmatrix}\begin{pmatrix} 15 \\ 21 \end{pmatrix} = \begin{pmatrix} 117 \\ 168 \end{pmatrix} = \begin{pmatrix} 13 \\ 12 \end{pmatrix}$$

where 117 and 168 are reduced to numbers 26 or below by subtracting 26 as many times as needed. When this is done for the entire message, the numbers are converted back to letters, ML..., and the enciphered message is sent.

The recipient goes through the same steps, but uses a secret deciphering matrix:

$$\begin{pmatrix} 3 & 24 \\ 19 & 5 \end{pmatrix}\begin{pmatrix} 13 \\ 12 \end{pmatrix} = \begin{pmatrix} 327 \\ 307 \end{pmatrix} = \begin{pmatrix} 15 \\ 21 \end{pmatrix}$$

which can be converted back to "OU...." This works because the product of the enciphering and the deciphering matrices is, after reducing the numbers by subtracting 26s, the identity matrix:

$$\begin{pmatrix} 5 & 2 \\ 7 & 3 \end{pmatrix}\begin{pmatrix} 3 & 24 \\ 19 & 5 \end{pmatrix} = \begin{pmatrix} 53 & 130 \\ 78 & 183 \end{pmatrix} = \begin{pmatrix} 1 & 0 \\ 0 & 1 \end{pmatrix}$$

Multiplying the message first by the enciphering matrix, then by the deciphering is equivalent to multiplying it by the identity matrix. Therefore the original message is restored.

A two-by-two enciphering matrix doesn't conceal the message very well. A skilled crytanalyst could crack a long message or series of short ones very easily. (This one, by itself, would be too short for the cryptanalyst to do any of the statistical analyses needed for cracking it.) If the enciphering and deciphering matrices were bigger, say ten-by-ten, the encipherment would be pretty secure.

MATRIX INVERSE

If an n x n **matrix** A has a nonzero determinant, then there is a matrix called "A inverse," denoted by A^{-1} such that $AA^{-1} = A^{-1}A = $ Id. Here Id denotes the identity matrix, that is the matrix that has ones on its diagonal entries and zeros everywhere else. If there is matrix B, say, that has the property that BA = AB = Id then $B = A^{-1}$. Also, the determinant of A is nonzero because the determinant of a product of matrices is equal to the product of the matrices' **determinants**. Hence det(A)det(B) = det (Id) = 1. So, det(A) = 1/det(A^{-1}).

The entries of A^{-1} can be found with **Cramer's rule**. Since $AA^{-1} = $ Id, if the jth column of A^{-1} is denoted by A^{-1}_j then the jth column of Id is equal to AA^{-1}_j. By Cramer's rule, the kth entry of the jth column of A is equal the determinant of $A_k(e_j)$

divided by the determinant of A. Here e_j denotes the jth column of Id. The determinant of $A_k(e_j)$ is equal to $(-1)^{j+k}$ times the determinant of the matrix A(j,k) which is obtained from A by removing the jth row and the kth column. The number equal to $(-1)^{j+k}$ times the det A(j,k) is called the (j,k)-cofactor of A.

Here is a method for finding the determinant of the matrix A. First pick a column or a row. If the chosen column or row has a lot of zeros, the determinant will be easier to calculate. Suppose that the jth row is the chosen one. Then the determinant is equal to the sum from k = 1 to n of the (j,k)-entry of A times C(j,k). This **definition** of determinant does not depend on the row or column chosen.

This fact can be proven using elementary matrices (for definitions see the article **Matrix multiplication**). A is equal to a product of the form $BE_1...E_m$ in which E_i is an elementary matrix for each i and B has all of its nondiagonal entries equal to **zero** and all of its diagonal entries equal to either one or zero. The **proof** that A equals this uses Gaussian elimination (see the article on **Systems of Linear Equations**). Gaussian elimination implies that any matrix, A in particular, can be multiplied by a sequence of elementary matrices to result in an upper triangular matrix. That is a matrix with all of its (i,j) entries equal to zero whenever i > j. But, Gaussian elimination can be continued by exchanging the words "row" and "column." After this continuation, the result is a matrix B that has the required form. It is easy to prove that for the matrix B, its determinant does not depend on the row or column chosen for cofactor expansion. By **induction**, we assume that there is a number N ≥ 0 such that if m is ≤ N then any matrix M of the form $M = BE_1...E_m$ has the property that its determinant does not depend on the particular row or column chosen for cofactor expansion. It is easy to show that for any elementary matrix E, the product ME also has this property. By induction, therefore, it is true that the determinant of any matrix does not depend on the particular row or column chosen.

See also Cramer's rule; Determinant; Linear algebra; Matrix; Matrix multiplication; Systems of linear equations

MATRIX MULTIPLICATION

The formula for **matrix multiplication** is as follows. Let A be an n x m matrix and let B be an m x k matrix. The product AB is a n x k matrix whose (i,j)-entry is equal to the sum over an index p that ranges from 1 to m of the products of the (i,p)-entry of A with the (p,j)-entry of B. Another way to say this is that the (i,j)-entry is equal to the dot product (also called the inner product) of the ith row of A with the jth column of B. Another way is to multiply by columns. Denote the ith column of B by B_i. The ith column of AB is AB_i.

Matrix multiplication is not always commutative. For example, |1 0| | 1 1 | | 1 1 | | 1 1 | |1 0| |1 -1| |0 -1| x | 0 1 | = | 0 -1 |. But | 0 1 | x |0 -1| = |0 -1|.

Matrices often represent linear maps on **vector spaces**. Multiplication of matrices, in this case, represents composition of maps. For example, suppose f is a linear map from R^n to

R^mm and g is a linear map from R^m to R^k. Then, by using the standard coordinates, f can be represented by a m x n matrix F, and g can be represented by a k x m matrix G. The map equal to g composed with f is represented by the product GF.

An elementary matrix is an n x n matrix of one of the following types. A matrix of the first type has all nondiagonal entries equal to **zero**, all but one diagonal entry equal to one, and all diagonal entries are nonzero. A matrix of the second type is equal to the identity matrix with two of its rows exchanged. A matrix of the third type is equal to the identity matrix with two of its columns exchanged. A matrix of the fourth type is equal to the identity matrix except that one nondiagonal entry is not equal to zero. An important fact is that every invertible matrix is the product of elementary matrices. The entry on **Systems of linear equations** in this volume explains Gaussian elimination. This process shows how to multiply a matrix A, say, by elementary matrices so that the result is an upper triangular matrix (which means all (i,j)-entries in which i > j are zero). It is also possible to continue this process exchanging the words "row" and "column" so that the result is the a matrix with a nondiagonal entries equal to zero and all other entries equal to either one or zero. Since all elementary matrices are invertible and since the product of invertible matrices is invertible, the result of Gaussian elimination on an invertible matrix must be an invertible matrix. But since the result has nonzero entries on all of its diagonal entries, its determinant is equal to the product of its diagonal entries. Its determinant is nonzero if and only if it is invertible. So, if A is invertible, then the result must be the identity matrix. This proves that A^{-1} is the product of elementary matrices. But since this is true, $A = (A^{-1})^{-1}$ is also the product of elementary matrices.

Another important fact is that the determinant of a product AB of n x n matrices is equal to the product of their **determinants**. To see this, first note that if either A or B have a zero determinant, then one of them is not one-to-one (see **Correspondence**) and so the product cannot be one-to-one and onto. Thus it cannot be invertible and so it has zero determinant. On the other hand, if both A and B have nonzero determinants, then they can both be written as the product of elementary matrices. By **induction** on the number of elementary matrix factors, it suffices to prove that the determinant of a product of a matrix with nonzero determinant with an elementary matrix is equal to the product of their determinants. This is an easy, straightforward calculation.

The n x n identity matrix is the matrix that has its (i,i)-entry equal to one for all i and all other entries equal to zero. It is commonly denoted by Id. Its name comes from the fact that if A is any n x n matrix, then A x Id = Id x A = A.

Matrix multiplication is associative. Therefore, the set of all n x n invertible matrices forms a group, called the rank n general linear group. It can be represented as a subset of R^{n2} by the map that sends any matrix A to the coordinates given by its entries. By this representation, the general linear group is a Lie group. Most **Lie groups** are subgroups of a general linear group. Lie **group theory** is an important and active research area for mathematical physicists.

See also Determinants; Group theory; Lie groups; Matrix; Matrix inverse; Orthogonal coordinate system; Systems of linear equations; Vector spaces

MAUPERTUIS, PIERRE LOUIS MOREAU DE (1698-1759)

French physicist and author

Pierre Louis Moreau de Maupertuis created the least action principle, which he claimed, among other things, proved the existence of God. Something of a Renaissance man, he also participated in a famous scientific expedition to Lapland to measure the Earth, studied biology, and introduced his homeland to the principles of **Isaac Newton**. A difficult and cantankerous character, Maupertuis was in later years the target of ridicule despite his many scientific accomplishments.

Maupertuis was born in St.-Malo, France on September 28, 1698. His doting mother is said to have spoiled him until he was unable to accept criticism of any sort and unwilling to do anything other than what he wanted. Maupertuis attended local private schools as a boy and at age 16 traveled to Paris for more advanced studies, although he mainly disliked the traditional classical classes he was offered. In 1717 he began studying music, but quickly found that he preferred mathematics instead. By 1723 he had been elected to the illustrious French Academy of Sciences, where he did mathematical and biological research and taught mathematics.

Maupertuis made a trip to London in 1728 that would change his life forever. There he was introduced to the world of Newtonian **mechanics**, which was a radical change from the Cartesian principles had grown up with. A rapid convert to Newton's views of the universe—especially concerning the shape of the Earth and gravity—Maupertuis returned to France as the foremost proponent of these revolutionary views. In 1732 he published his first work on Newtonian physics. The paper was so persuasive that by the following year Maupertuis was considered the leading European expert on Newton.

For this reason, Maupertuis was asked in 1736 to lead an expedition to Lapland to test Newton's prediction that the Earth was not perfectly round, as scientists then believed, but rather flattens toward both of the poles. His method was based on measuring the length of a degree along the meridian of longitude, which, if Newton were correct, would reveal that the degree of longitude would be longer in the planet's far northern reaches than near the equator. It was not until a similar expedition to Peru returned to France in 1738 that Newton's hypothesis was proved correct. Germany's Frederick the Great (Frederick II of Prussia) was so impressed with Maupertuis's work that he asked the scientist to join the Berlin Academy of Sciences. Meanwhile, Maupertuis had published his *Elements of Geography.*

Soon after the War of Austrian Succession began, Maupertuis joined Frederick II in Silesia, but was captured when his headstrong horse bolted with him behind enemy lines. Feared dead, Maupertuis eventually showed up safe in Vienna only to become the butt of jokes for some time—a situation that did not agree with his sensitive, quick-tempered nature. Nevertheless, Maupertuis's scientific career continued to prosper, and he was chosen as a member of the Academie Francaise in 1743.

In 1745 Maupertuis published *Venus physique,* an argument against the period's widely accepted biological theory that all embryos are preformed and that an "essence" from one of the parents could affect the preformed fetus in the other parent. Later that year, he accepted Frederick's invitation to move to Berlin, where he soon married and became president of the Academy of Sciences in 1746.

Also in 1746, Maupertuis published "The Laws of Movement and of Rest," which contained his first discussion of the least action principle. Simply put, the principle states that nature choose the path of least resistance for moving **light** rays, physical bodies, etc. (Later, mathematician **William Rowan Hamilton** would refine and extend Maupertuis's work in this area, while the principle itself would become a major contribution to quantum physics and the idea of homeostatis in biological terms.) Maupertuis used the least action principle in an effort to prove the objective existence of God and also hoped that it might be a unifying theory for all laws of the universe. He reportedly regarded the least action principle as his own most important work. In 1750, he published *An Essay on Cosmology,* which contained further discussion of the principle.

As Academy of Sciences president, Maupertuis had a volatile and uneasy tenure, although he attracted many noteworthy scientists to the institution. However, he managed to keep up his work and in 1751 published *The System of Nature,* one of his most significant works. A study of the occurrence of polydactyly (having too many fingers and toes) in a local family, the work was the first rigorously scientific analysis of transmission of a dominant trait in humans. Based on his findings, Maupertuis proposed a theory of fetal formation and heredity that was far ahead of its time.

In 1752 a longstanding conflict between Maupertuis and the philosopher and writer Voltaire, a former friend, reached a climax when Voltaire published *Micromegas,* which made fun of the scientist and his work. A sudden laughingstock, Maupertuis suffered a break in health and left Berlin to recuperate in the town of his birth, resigning from the Berlin Academy of Science in 1753. He returned to Berlin in 1754, left again for France in 1756, and finally stopped in Switzerland on his way back to Germany in 1758. There he visited his old friend **Johann Bernoulli**, and died on July 27, 1759.

MAXWELL, JAMES CLERK (1831-1879)

Scottish physicist

James Clerk Maxwell created groundbreaking work in an impressive number of fields from a very young age. He is considered one of the founders of the kinetic theory of gases, and his work in electrodynamics laid the groundwork for the theories of **Albert Einstein**.

James Clerk Maxwell was the son of James Clerk (Maxwell), who added the name Maxwell in order to hold on to family property, and Frances Cay Maxwell. He was born June 13, 1831, in Edinburgh, Scotland, and spent most of his childhood on the family's estate near Dalbeatie, in the Galloway region. In 1841 he was enrolled at the Edinburgh Academy.

James published his first scientific paper at age 15. It was a description of a new way of drawing ovals, and it was presented to the Edinburgh Royal Society in March, 1846.

This first paper was the beginning of a wide-ranging but sadly short-lived career in science. Maxwell's work included research on **thermodynamics**, Saturn's rings, color vision, geometrical optics, photo- and visco-elasticity, servomechanisms, relaxation processes, and reciprocal diagrams in engineering processes.

In 1847 he enrolled at the University of Edinburgh, where he continued his studies of optics and branched out into elastic solids (such as gelatin). He remained at the university until 1850, when he entered Cambridge University as an undergraduate, first enrolling at Peterhouse College but moving to Trinity College. In 1854, he graduated, receiving numerous honors, and was named a fellow at Trinity College, teaching optics and hydrostatics. In 1856, he left Trinity to become a professor of natural philosophy at Marischal College in Aberdeen, Scotland.

There, he met Katherine Mary Dewar, whom he would marry in 1858. They had no children.

During his time at Marischal, he became interested in the question of the state of Saturn's rings—were they solid, fluid, or something else? Pierre Simon de Laplace, a French mathematician, had done calculations that showed that the rings could not be solid, or they would be unstable. Maxwell's calculations showed that the rings were probably made up of small solid particles. Not only did this work earn him the Adams Prize, but it also allowed him to develop statistical methods that he would later use in his kinetic theory of gases.

At about this time, Marischal College joined with King's College to create the University of Aberdeen, and Maxwell lost his position. However, he moved to King's College in London, where he became professor of physics and astronomy.

Despite his heavy teaching load there, he continued his research. He presented a device for mixing colors of the spectrum to the British Association in 1860. (He had long been interested in color, and discovered that people who are color-blind lack certain color receptors in their eyes.) But he devoted most of his time to studying **electromagnetism**.

In 1870 the chancellor of Cambridge decided to build a physics laboratory for Cambridge. The Cavendish Laboratory was completed in 1874, and Maxwell became its first director, and chair of experimental physics, in 1871. At the laboratory Maxwell led investigations into the nature of electricity, including electrical resistance and establishing the ohm as a unit of measurement for electricity.

Maxwell had long been interested in electricity and magnetism, starting in 1856 when he investigated Michael

James Clerk Maxwell

Faraday's theory of lines of **force**. But Maxwell went beyond Faraday.

He devised four **equations**, now called **Maxwell's equations**, that summarized the nature of the electric and magnetic fields. He showed that this electromagnetic **field** moved through **space** at the speed of **light**, and concluded that light was really a type of electromagnetic field. He also theorized that light was not the only electromagnetic field of this sort; fields made of longer waves existed, he believed. He was proved right in 1887, when Heinrich Hertz discovered the existence of radio waves.

Maxwell's work was both groundbreaking and difficult to understand. But a bright young Dutchman, **Hendrik Antoon Lorentz**, built on Maxwell's work and laid the foundations of theoretical physics that would culminate in Albert Einstein's theory of relativity.

Maxwell explained his theory in his *Treatise on Electricity and Magnetism*, a two-volume tome that appeared in 1873.

Another of Maxwell's great contributions to science was his work on the kinetic theory of gases. In 1865 he and his wife Katherine (who was her husband's able assistant from his early days of color study) measured gas viscosity at various temperatures and pressures. This work showed that perceiving gas molecules simply a randomly bouncing balls was not sufficient. Maxwell produced a formula regarding the distribution of velocities among gas molecules at a uniform pressure. Describing physical processes as formulas and **functions** was something new. He was able to apply his statistical method to various properties of gases, include head conduction and viscosity, and the **motion** of molecules.

Besides his work on electricity and magnetism, Maxwell also published *Theory of Heat* (1871) and *Matter and Motion* (1874).

In 1877, Maxwell began to suffer stomach pains, and in 1879 was diagnosed with stomach cancer. He died November 5, 1879, in Cambridge, only 49 years old.

MAXWELL'S EQUATIONS

Maxwell's **equations** are the four equations formulated by **James Clerk Maxwell**, a British physicist, that explain all classical electromagnetic phenomena. They are a set of **differential equations** that describe and predict the behavior of electromagnetic waves in dielectrics, free **space** and at the boundary of conductors and dielectric materials. Maxwell's equations unify the laws of **Gauss**, Ampere, and Faraday in concise statements concerning the fundamentals of electricity and magnetism. They embody a high level of mathematical sophistication and are usually found in advanced courses. The laws are usually written in the form of equations in the absence of magnetic or polarizable media.

The first of Maxwell's equations is a form of Gauss' law for electricity. Written in differential form it is: $\nabla \cdot E = \rho/\varepsilon_0$, where E is the electric **field** vector, ρ is the density of electric charge and ε_0 is the permittivity of a vacuum. Basically it says that the electric flux out of any closed surface is proportional to the total charge enclosed within this surface. In its integral form this law is useful in calculating electric fields around charged objects.

Maxwell's second equation is like the first equation but applicable to magnetic fields. It is written as: $\nabla \cdot B = 0$, where B is the magnetic field vector or magnetic flux density. This law is a form of Gauss' law for magnetism. This equation tells us that the magnetic flux out of any closed surface is **zero**, hence telling about the sources of magnetic fields. For a magnetic dipole, the magnetic flux directed outward from the north pole will equal the flux directed inward from the south pole. Although the poles are only imaginary in this case they help in orienting the system. The net flux will always be zero for dipole sources. A magnetic monopole source would give a nonzero net magnetic flux but this equation indicates that there are no magnetic monopoles since it would violate this equation.

The third of Maxwell's equations concerns electric fields in a circuit. This equation is a derivation of Faraday's law of **induction** and is written: $\nabla \times E = -\partial B/\partial t$. This equation says that the electric field around a closed loop or circuit is equal to the negative of the rate of change of the magnetic field over the area enclosed by the loop. It is equal to the elec-tromotive **force** or generated voltage in the circuit. This equation forms the basis for the operation of electric generators, inductors, and transformers.

The last of Maxwell's equations is another form of Ampere's law concerning the relationship between the magnetic field and electric current. It is written as: $\nabla \times B = \partial E/\partial t + J$, where J is the current density vector. This equation is applicable in the situation of a static electric field. In this case the magnetic field around a closed loop or circuit is proportional to the electric current flowing through the circuit. It is possible to use this equation to calculate the magnetic field in simple situations.

The first two of Maxwell's equations do not involve a time parameter and can be considered static equations. These equations taken alone seem to indicate that the electric field and magnetic field are similar to the gravitational field in that they are instantaneous forces at a **distance**. Although this static nature is indicated by the first two equations it is untrue as is indicated by the last two equations. The latter two of Maxwell's equations indicate that the magnetic and electric fields are dynamically linked together. The fields are not only **functions** of each other but they also involve time. The magnetic and electric fields depend not only on the distribution of charge at a given instant of time but also on the movement of charge and on the rates of change of the fields themselves at particular instants of time.

MEAN

The **arithmetic** mean, or as it is sometimes referred to, the average of a data set, is the sum of all data values divided by the number of items in the set. A mean is considered to be a measure of central tendency—it describes a "typical" value to represent the data set. It describes or gives a summary of a whole distribution of events or measurements. The information given by measures of central tendency can be used for description or comparison.

An example of finding an arithmetic mean using a particular data set is as follows: in the set 1, 2, 5, 10, 12, 14, 16, 17, 20 the mean is calculated as $1 + 2 + 5 + 10 + 12 + 14 + 16 + 17 + 20 \div 9 = 10.8$. It is very likely that the mean will be a number not directly represented in the data set. However, every number in the data set affects the mean. If an "outlier" (a data value very unlike the rest) is present, the mean can misrepresent the data set. The symbols used for mean are x or μ. The arithmetic mean is the most common measure used for central tendency—the others being **median** and **mode**.

The geometric mean, m, of two numbers (x and y) is the **square** root of their product. $x \div m = m \div y$ or $m = \sqrt{xy}$. The geometric mean is also referred to as the geometric proportional. It is used in **geometry** to find the lengths of sides in similar triangles using the **altitude**.

See also Median; Mode

MEAN-VALUE THEOREM

First presented in a variant form by the French mathematician **Michel Rolle** (1652-1719) in an obscure book, the mean-value **theorem** is one of the fundamental principles of the discipline of **calculus**. Stated mathematically, the mean-value theorem asserts that if a function $f(x)$ is both continuous and differentiable over the closed **interval** $[a, b]$ (that is, the interval includes the endpoints a and b), then there exists at least one number c (and perhaps more) such that the first derivative of the function (symbolized by $f'(x)$), evaluated at c, is equal to the difference of the function evaluated at b and a divided by the difference of b and a. Given the conditions of **continuity** and a closed interval, the mean-value theorem can be written (in symbolic form) $f'(c) = [f(b)—f(a)] / (b—a)$. If the quantity $[f(b)—f(a)] / (b—a)$ is considered the average, or **mean**, rate of change of the function f over the given interval $[a, b]$ and $f'(c)$ is considered an instantaneous change, there must exist at least one interior point c at which the instantaneous rate of change equals the average rate of change.

Relationship to Rolle's Theorem

Published in 1691, **Rolle's theorem** is a simplified version of the mean-value theorem. Rolle's theorem asserts that between any two "zeros" a and b of a continuous and differentiable function $f(x)$ (that is, $f(a) = f(b) = 0$), there exists at least one point c (and perhaps more) such that $f'(c) = 0$ in the interval $[a, b]$. The value c is called a critical point. If $f(x)$ is not equal to a constant over this interval, Rolle's theorem implies that there exists at least one (and perhaps more) local maximum *or* local minimum in the interval at which the **slope** of the function changes sign, such as $f' > 0$ changing to $f' < 0$ or $f' < 0$ changing to $f' > 0$.

Integral Form of the Mean-Value Theorem

The form of the mean-value theorem given above is its differential case, and as one would expect, a form of the mean-value theorem also exists for the integral case. This case involves the **definite integral** over the interval $[a, b]$, and states that at some point c, the continuous function $f(x)$ is equal to 1 / $(b—a)$ multiplied by the integral from a to b of $f(x)dx$. Written symbolically, if f is continuous on $[a, b]$, then there exists some point c such that $f(c) = 1 /(b—a) * \text{int}(f(x)dx)$.

Like the differential case, the integral case of the mean-value theorem over a closed interval $[a, b]$ also has a relationship to an average value of the function. It turns out that if f is integrable over the interval $[a, b]$, then the average, or mean, value of the function in this interval is exactly 1 / $(b—a) * \text{int}(f(x)dx)$. Any points satisfying the condition that $f(c)$ is equal to this quantity are *mean values*. It should be noted that if f is integrable, the quantity $1 /(b—a) * \text{int}(f(x)dx)$ is in itself an expression of the average change of the function $F(x)$, where $F'(x) = f(x)$. This means that $f(c)$ is actually a point on the interval $[a, b]$ where the instantaneous rate of change of the function $F(x)$ is equal to its average rate of change over the same interval.

See also Calculus; Continuity; Derivatives and Differentials; Definite integral; Differential calculus; Division; Extrema and critical points; Functions; Instantaneous events; Instantaneous velocity; Integral calculus; Interval; Mathematical symbols; mean; Multiplication; Necessary condition; Rolle, Michel; Rolle's Theorem; Slope; Sums and differences; zero

MEASURABLE AND NONMEASURABLE

Around the turn of the century, mathematicians such as **Emile Borel** and **Henri Lebesgue** were looking for a precise **definition** of the measure of a subset of the real n-dimensional **space** R^n. Four considerations bounded their search. First, the measure of any "normal" object such as a line segment in R^1, a **square** in R^2, or a cube in R^3 should be equal to its length, **area**, or **volume** respectively. Second, if A and B are disjoint subsets then the measure of A union B should be equal to the measure of A plus the measure of B. Third, if A and B are congruent **sets** then their measures should be equal. Fourth, it should be possibly to measure "most" sets or at least all "nice" ones. In 1924, **Stefan Banach** and **Alfred Tarski** dramatically demonstrated that not all subsets could be measurable if the first three considerations held true. They showed that a ball is equal to the union of six disjoint pieces that can be reassembled into two balls, both of the same volume. Thus the pieces must be non-measurable.

A fifth consideration aided the development of measure theory; if A is a subset of B, then the measure of A should be less than or equal to that of B. Since n-dimensional rectangles have to be measurable by the first consideration, any subset that can be approximated by disjoint unions of rectangles should be measurable too. The measure of an open set (see the article on **topology** for definitions) is defined as the supremum over all measures of disjoint unions of rectangles contained within the open set. The measure of a compact set is the infimum over all measures of open sets that contain the compact set. The inner measure of any subset is the supremum over all measures of compact sets contained in the subset. The outer measure is the infimum over all measures of open sets that contain the subset. A subset S is measurable if its inner measure equals its outer measure. In this case, the measure of S is to defined to be its inner (or outer) measure.

Lebesgue showed that the set of all measurable subsets is a sigma-algebra. This means that the empty set and the whole space are measurable, complements of measurable sets are measurable, and **denumerable** unions of measurable sets are measurable. He also proved that the measure of a denumerable disjoint union of measurable sets is equal to the infinite sum of the measures of the sets. Suppose that X and Y are measurable and that Y is contained in X. Then, the measure of X - Y is equal to the measure of X minus the measure of Y. Nowadays, an (abstract) measure on a space is defined to be a sigma **algebra** of measurable sets and a function from the sigma algebra that satisfies all the above properties. This generality enables the basic ideas of measure theory to be applied to a variety of objects such as groups, manifolds, and function spaces. The principle use of measure theory, however, is Lebesgue integration. Lebesgue integration has the virtue that it can be applied to noncontinuous yet "measurable" **functions**. A function f is measurable if for every open set V is the **range** of f, f⁻¹ (V) is measurable.

Here is a non-measurable set. Let E be a subset of the **real numbers** with the following two properties. First, for any real number x there is a y in E such that x - y is rational. Second, if x and y are both in E and not equal then x - y is irrational. The existence of a set with these properties can be proved with **Zorn's lemma**. Since any number is arbitrarily close to a rational number, the set E can be chosen so that it is contained in the **interval** [0,1]. Suppose for a contradiction that E is measurable. For any rational number r, E + r is the set of all numbers of the form x + r where x is an element of E. So, E + r is disjoint from E and congruent to E. The union U, of all sets E + r for all rational r in the interval [-1,1] is contained in [-1,2]. Since U is a denumerable union of measurable sets, it is measurable. U contains [0,1] since any y in [0,1] can be written as e + r for some e in E and r in R. But since e is in [0,1], r must be in [-1,1]. So y is in U. So, the measure of U is in between the measure of [0,1], which is 1, and the measure of [-1,2], which is 3. On the other hand, its measure is equal to the sum of the measures of E + r for r in [-1, 1]. This is because if r and s are distinct **rational numbers** then E + r is disjoint from E + s. Since there are infinitely many rational numbers in [-1, 1], the measure of U must be **infinity** times the measure of E. This number is either **zero** or infinity depending on whether the measure of E is zero or positive. But since U contains [0,1] and is contained in [-1, 2], the measure of U must be between 1 and 3. This contradiction shows that E is not measurable.

Measure theory relies on Cantor's theory of infinite sets. Cantor's theory was controversial. In fact, it was referred to as a "disease of which later generations will cure itself of" by Henri Poincaré. On the other hand, measure theory has been so successful that most mathematicians accepted Cantor's theory in order to validate measure theory. Today, both theories are generally accepted and used without reservation.

See also Area; Borel, Emile; Cantor set; Lebesgue integration; Rational numbers; Real numbers; Topology; Volume

MECHANICS

Mechanics is the branch of physics that studies the **motion** of particles or large bodies. Currently, mechanics can be divided into two main branches: classical and **quantum mechanics**. The two branches take different mathematical approaches reflecting the different philosophies, though some of the tools overlap. Further, each type of mechanics can be done using more than one mathematical route to get to the same answer. In mechanics, the "answer" comes in the solution of the **equations** of motion, which describe the motion of the particle with respect to time.

Classical mechanics uses traditional **calculus** in either of its formulations. It assumes that the behavior of particles is deterministic. The method of solution usually taught in elementary physics is the solution of the **force** equation. Using Newton's F = dp/dt, where F is the vector sum of the forces on the particle and p is the momentum vector, a second-order differential equation is obtained, which is then solved to determine the position over time. The Lagrangian method for solving mechanics problems deals with mechanics in an energy formulation, making the quantities calculated scalars. The Hamiltonian method even further reduces complications by presenting a first-order scalar differential equation. The classical formulation of mechanics is valid when the object in question is sufficiently large, or as an average result of quantum mechanical situations.

Quantum mechanics, on the other hand, is the science of the very small levels of matter and their behavior. Unlike the macroscopic situations we observe with the naked eye, quantum mechanical situations behave probabilistically. That is, even if all initial conditions are known, only probabilities may be stated about the particle's further behavior. However, quantum mechanics provides several alternatives for calculating this probable behavior. They are mathematically equivalent but often involve quite different procedures in arriving at the results. The three main options in quantum mechanics are Schrödinger's integral formulation, Heisenberg's **matrix** formulation, and Feynman's sum-over histories. While the consequences of this theory are often counterintuitive, they have been proven to work on the physical level, and approximations demonstrate that quantum behavior will "average out" to classical behavior.

While much of mechanics is already solved, it is still necessary and useful for students of physics to see how things are moving and what they can expect of basic force-influenced motion in experiments on other topics. There are also elements of mathematics that tie in with current "hot" topics in mechanics, notably **chaos theory**. In the real world, chaotic equations often express the situation most accurately, complicating the physics and making it more interesting to physicists.

See also Isaac Newton

MEDIAN

The median of a data set is the value above which half of the data lie and below which half the data lie. It is considered to be the "middle" number of the data set. If the data set is composed of an odd number of entries, the median is the exact middle number. If the data set is composed of an even number of entries, the median is the average of the two middle numbers. The median may not be a number in the data set. The median is influenced very little by extreme data values. When computing the median, the data list must be ordered from lowest number to highest number.

The following data set is odd: 1, 2, 5, 7, 9. The median here is 5, because 5 is the absolute middle number (there are two numbers on either side of it). The following data set is even: 1, 2, 5, 7, 9, 10. Here the median is 6; 5 + 7 (the two numbers in the middle) ÷ 2 = 6.

The median of a data set is a measure of central tendency, like the **mean** and the **mode**. It describes the distribution or gives a summary of events that take place in a data set. It can be used for comparison purposes.

The median of a **triangle** is a segment from a vertex in the triangle to the **midpoint** of the opposite side. Every triangle has three possible medians—one from each vertex. The centroid is the place inside the triangle where all three medians intersect. This is considered to be the balance point of the triangle.

See also Mean; Mode

MENAECHMUS (CA. 380 B.C.-CA. 320 B.C.)
Greek philosopher, author, and teacher

A student of **Eudoxus**, Menaechmus is said by some sources to have been a mathematics tutor to Alexander the Great. He is most famous for his discovery of **conic sections**.

Menaechmus was born in about 380 B.C. in Alopeconnesus, Asia Minor, in what is now Turkey. Little is known of the ancient mathematician's personal life. However, his alleged response to Alexander's request for a shortcut to learning **geometry** suggests something of a sense of humor as well as a reverence for his chosen field of study. He is said to have told Alexander, "O king, for traveling through the country there are private roads and royal roads, but in geometry there is one road for all." Some sources suggest that **Aristotle** might have introduced the two men to each other.

It seems likely that Menaechmus took over from Eudoxus as head of an illustrious mathematics school in Cyzicus, a Turkish city. Other references indicate that as a teacher Menaechmus would have stringently focused on the technology and philosophy of mathematics. For instance, ancient writings record the development of his theory that all **propositions** are **theorems** and that there are two types of theorems: those that seek a concrete answer and those that seek something's properties (e.g., what it is, what group it belongs to, how it has changed, its relationship to another, etc.).

Records also show that Menaechmus was involved in mathematical astronomy. He is said to have introduced the concept of "counteracting" and "deferent" concentric spheres as part of the era's widely held explanation for planetary movement. According to Menaechmus, one **sphere** would bear a heavenly body and another would correct its **motion**, thus explaining the planets' seemingly irregular paths.

Although these contributions were important in their day, Menaechmus's discovery of **conic sections** is mainly why he is still remembered in modern times. The breakthrough while he was looking for a way to duplicate a cube—specifically, how to find two **mean** proportionals between two straight lines. Indeed, Menaechmus proved that one could obtain the two means by intersecting a **hyperbola** and a

parabola. For his trouble, Menaechmus become the brunt of anger. According to writings of the period, **Plato** was infuriated that Menaechmus and others used mechanical devices in their efforts to double the cube, saying that resorting to such means debased geometry as the highest human achievement.

It was actually more important in terms of mathematical discovery that Menaechmus was the first to show that hyperbolas, parabolas, and **ellipses** can be obtained by cutting a cone in a plane that is not parallel to the base of the cone. Most sources assert that Menaechmus did not coin the terms "parabola" and "hyperbola," although some disagree.

MENGER, KARL (1902-1985)
Austrian-American professor

One of the foremost mathematicians of the twentieth century, Karl Menger was especially recognized for his work in curve and **dimension** theory.

Menger was born on January 13, 1902 in Vienna, Austria, the son of eminent economist Carl Menger and novelist Hermione Andermann. From 1913 to 1920 he attended the Doblinger Gymnasium (a classical college-preparatory school), where in his last year he decided that he wanted a career in literature. Menger seems to have formulated this goal in the face of discouragement from his teachers, however, who thought he had far more talent in **mathematics and physics**. One of his teachers at the *gymnasium* described Menger as "not quite normal," megalomaniacal, and "a strange fellow."

Menger entered the University of Vienna later in 1920 and began studying physics. A year later his academic interests shifted to mathematics when he heard a lecture in which the professor mentioned that at the time there was no satisfactory **definition** of a curve. Menger was intrigued by this statement and immediately went to work on the problem. Within a few days, he had found a solution and presented it to the stunned lecturer. It was apparently this incident that led to Menger's further work in the area of curves and dimension.

Severe tuberculosis slowed Menger down considerably during the rest of his undergraduate career. In 1921 doctors sent him to a sanatorium in the Austrian Alps to recover. He remained there until 1923, but during that time did much of his fundamental mathematics work. Menger finally returned to the University of Vienna, receiving his doctorate in mathematics in 1924.

A year later, he moved to Amsterdam to begin research work with **Egbertus Jan Luitzen Brouwer**. Although the two did not get along particularly well, Menger did more valuable work there on curve and dimension theory as well as gaining a deeper understanding of logic and the foundations of mathematics. Meanwhile, he served as a docent in mathematics for nearby university students. Menger earned his teaching credential in 1926 and returned to the University of Vienna as chair of the **geometry** department.

Menger's prestige as a mathematician received a major boost in 1927 when he was chosen as a member of the elite Vienna Circle, a group of some three dozen scientists from wide-ranging disciplines. He remained at the University of Vienna for the next decade, except for the time he spent in 1930-1931 as a visiting professor at Rice and Harvard universities. He decided to resign from Vienna in 1937 and accepted a position at the University of Notre Dame in Indiana to avoid the worsening political situation in his homeland.

Once he had settled into his new position in the United States, Menger established the Mathematical Colloquium as a rival to the Vienna Circle. However, World War II soon began to affect the United States as well, and the group never gained the momentum it needed to make it as influential as the older body. Menger's passion for mathematics continued unabated, however, and he found his interests broadening to include probabilistic and **hyperbolic geometry** as well as the **algebra** of **functions**. Also, much of his work at the university was centered on teaching **calculus** to Navy cadets as part of the war effort.

Menger remained at Notre Dame until 1946, when he joined the new Illinois Institute of Technology (IIT) in Chicago. Because of his work during the war, which had convinced him that there must be a better way to teach basic mathematics, he expended great effort during the 1950s and 1960s to bring about changes in mathematics education. He wrote a textbook on calculus to this end.

After becoming an emeritus professor at IIT in 1971, Menger dedicated more time to leisurely pursuits. He enjoyed trying new foods, taking long walks, and listening to music. Menger and his wife, an actuarial student, had married in 1934 and had four children together. He died in his sleep on October 5, 1985 in Highland Park, Illinois.

MERSENNE, MARIN (1588-1648)

French number theorist and writer

Had it not been for the tireless efforts of French Minimite friar Marin Mersenne, communication describing the discoveries in science would have never been dispersed to the far corners of the mathematical and scientific worlds during the 17th century. During this critical period in the **history of mathematics** there were no scientific publications, bulletins, or newsletters. Instead, information was disseminated via scientific discussion circles such as Accademia dei Lineei (host to **Galileo**) and Accademia del Cimento in Italy, the Invisible College in England, and by written correspondence. Fortunately, Mersenne had a particular penchant for correspondence and a personal interest in the advancement of mathematical knowledge. It was said that to notify Mersenne of a discovery was the same as informing all of Europe.

Mersenne was born on September 8, 1588, near Oize, Sarthe, France, and later entered into service for the Roman Catholic Church as a devoted teacher in 1611. He attended school with **René Descartes** and the two men remained friends throughout their lives. It was Mersenne who defended Descartes' philosophy against his critics in the Church and it was to him that Descartes wrote first whenever he developed a new theory in mathematics or philosophy. It was also

through Mersenne that Descartes gained fame in the circles of European intellectuals. Mersenne was also a staunch defender of **Galileo**, assisting him in translations of some of his mechanical works. And it was Mersenne, Galileo's representative in France, who circulated Galileo's question of the path of falling objects on a rotating Earth, leading Descartes to put forth the equiangular or logarithmic **spiral** $r = e^{a\Theta}$ as the possible path. This is just one of many instances in which Mersenne acted as correspondent and intermediary for the mathematical breakthroughs of the 17th century.

Very often Mersenne was given privileged previews of works that were later lauded as masterpieces; Descartes' *Le monde* is one example. The treatise was given to Mersenne as a New Year's gift in 1634 while all the rest of Paris, aware of its creation, had to wait until after Descartes' death to read it.

In addition to his responsibilities as a teacher of philosophy and theology at Nevers and Paris, Mersenne also conducted weekly scientific discussions, from which the French Académie Royale des Sciences was established in 1699 in Paris. While credit for starting two separate discussion groups at around the same time is usually given to Descartes and **Blaise Pascal**, it was actually Mersenne who originated the gathering of great minds to exchange information. One proof of this is that the discussions led by "Father Mersenne" were already in place at the time Pascal began to join them when he was only 14 years old. Another is that Descartes spent much of his time living in Holland and outside of Paris.

Mersenne's writings was not exclusive to acting as the broker for the mathematical world, however. He wrote extensively on physics, **mechanics**, navigation, **geometry**, and philosophy. A portion of his life's work was also devoted to editing the compositions of **Euclid, Apollonius, Archimedes**, Theodosius, and Menelaus, as well as other ancient Greek mathematicians.

Much is owed to Mersenne for his work as correspondent within the mathematical world, but the reason he is still known today can be attributed to his "Mersenne numbers." Although Mersenne has been called "the famous amateur of science and mathematics," his contribution did much to launch the discoveries of those who succeeded him. Mersenne published his *Cogitata Physico–Mathematica* in 1644 and it is for this paper that he is best known. In the paper, he asserted that a particular formula could be used for finding **prime numbers**, positive numbers which are only divisible by one and the number itself. In the paper Mersenne gave no reasons for why he believed his number theories were correct and later the formulas were proved to be incomplete. Many of the large numbers he alleged to be prime were not. Even though Mersenne was not successful in his attempt to create an ironclad formula, his conjectures were the stimulus and basis for later research into the theory of numbers and the search for large prime numbers (called Mersenne primes).

Mersenne spent his life as a staunch supporter of experimentation and was indirectly responsible for the invention of the pendulum clock. He suggested to **Christiaan Huygens** that he experiment with timing objects rolling down a slanted sur-

Marin Mersenne

face by using a pendulum. Galileo had previously noted the characteristic timekeeping property of the pendulum but through Mersenne's suggestion it was Huygens who developed the general application of the pendulum as a time controller in clocks in 1656.

Cogitata Physico–Mathematica also contained explanations of some of Mersenne's experiments in the field of physics. These experiments were, no doubt, stimulated by his correspondence with Galileo, Huygens, **Pierre de Fermat**, and other men of science, but it appears that Mersenne was better at chronicling and appending the work of others than he was at immortalizing his own. The last of Mersenne's papers were published in 1644, containing a condensation of mathematics. Mersenne died in Paris in 1668 at the age of 60.

MERSENNE NUMBERS

Mersenne numbers are numbers of the form 2^p-1 where p is a prime number. If the Mersenne number 2^p-1 itself is prime then it is called a Mersenne prime. The first few are $2^2-1=3$, $2^3-1 = 7$, $2^5-1= 31$, $2^7-1 = 127$ are all **prime numbers** but the next one, $2^{11}-1 = 2047$ is divisible by 23 and therefore is not prime.

Although Mersenne primes have been considered since antiquity because of their relation with **perfect numbers**, they are named after the French clergyman, Father **Marin Mersenne**, who, in 1644, made a list (with some errors) of the Mersenne primes with exponent p at most 257.

Most Mersenne numbers are not prime and, to date, 38 Mersenne primes have been discovered out of the more than 400,000 Mersenne numbers tested. It is believed, but not known for sure, that there exists infinitely many Mersenne primes. The largest known Mersenne prime is $2^{6972593}-1$, a number of more than two million digits, which was discovered by N. Hajratwala, G. Woltman and S. Kurowski who are all amateur mathematicians from the United States. It was found as part of the Great International Mersenne prime search, which is a worldwide effort to use idle time on home computers to find Mersenne primes and has led to advances in distributed computing. They use an implementation by G. Woltman of a test devised by the French mathematician E. Lucas and improved by the American mathematician D. N. Lehmer specifically to test the primality of Mersenne numbers.

See also Fermat numbers; Perfect numbers

METAMATHEMATICS

Metamathematics, sometimes called metalogic, is the branch of logic concerning the combination and application of **mathematical symbols**. The primary goal of metamathematics is to determine the nature of mathematical reasoning. It consists of basic principles, mainly concerned with proofs of consistency that attempt to formulate mathematical theories. Metamathematics today is mainly used as a mechanism of proving mathematical theories in an automated way.

Metamathematics arose during the attempts in the late 1800s to mechanize the verification of mathematical proofs. At that time precise mathematical logic coexisted on equal terms with vague intuitions and it was becoming apparent that a better, more firm basis for mathematics be developed. From the late 19th century, studies in formal logic attempted to develop a complete, consistent formulation of mathematics such that proposals could be formally stated and proved or disproved using a limited number of symbols that had well defined meanings. Theorems, formulated to follow strict rules of **inference**, emerged from axioms like branches of a tree. The axioms were the primordial seeds from which all other things evolved. It was thought that if a systematic way of mathematics was formulated then all need for thought or judgment would be eliminated. The only requirements were that the axioms were true statements and the rules of inference were **truth** preserving. This, it was thought, would lead to mathematics in which falsehoods simply could not be present. Although statements were formally written using standard numerals, **arithmetic** signs, parentheses and so on, it was thought that this notation was not a necessary feature. The statements could be built out of any arbitrary set of symbols as long as they were consistent and defined. This way of looking at mathematics was novel and named metamathematics. The difficulties in this task started to become realized in 1925 when Whitehead and Russell published *Principia Mathematica*. Hundreds of pages of symbols were required before the simple statement $1 + 1 = 2$ could be deduced. In 1931 the realization of the impossible feat of developing such a formulation was fully realized as Gödel's **proof** of his incompleteness **theorem** showed that nearly a century of efforts by the world's greatest mathematicians was doomed to failure. He determined that by thinking of theorems as patterns of symbols that it is possible for a statement in a formal system not only to talk about itself but to also deny its own theoremhood. Thus Gödel showed that this formal system would not capture all true statements of mathematics. All formal systems are incomplete because they are able to express statements that say of themselves that they are unprovable. It is not math itself that is incomplete but any formal system that attempts to capture all the truths of mathematics in a finite set of axioms and rules that is incomplete. Gödel was 25 when he developed the incompleteness theorem. Although the proof of the theorem removed any possibility of mechanizing the verification of mathematical proofs in a formal system it also did something else. The article in which the incompleteness theorem was published invented the theory of recursive **functions**, which today is the basis of a powerful theory of computing.

METHODS OF APPROXIMATION

Ideally, every function would be easy to deal with in all mathematical situations, abstract and applied. Unfortunately, this is not always the case. It sometimes becomes convenient to have an approximation to the function, usually a polynomial or some other "well-behaved" series of functional terms. There are several ways to formulate an approximation when this is desirable.

The simplest method of approximation is the least accurate one: estimation. With a certain amount of mathematical intuition, sometimes it is possible to guess a curve (or some related quantity of interest) that is fairly close to the actual desired curve. However, this method is only recommended as a starting point for more exact approximations, since it is an educated guess in the most literal sense of that phrase.

A Taylor series approximation relies upon the derivatives of the function being approximated, and it requires a central point about which to base the approximation. With an infinite number of k values starting at **zero**, each term in the approximation is $(x - x_0)^k f^{(k)}(x_0)/k!$. The point x_0 is often chosen to be zero itself, making for a simplified series form. Another desirable choice of x_0 is such that many of the derivatives will disappear at that point—all of the odd or even derivatives, for example, or all derivatives greater than a certain number. This polynomial series is exact as long as it is infinite; its error rate depends largely on how many terms can be used.

A Legendre polynomial makes it possible to interpolate data between known points into a function. The polynomial in its simplest form is written as $y = y_0 (x - x_1)/(x_0 - x_1) + y_1 (x - x_0)/(x_1 - x_0)$. This will provide a result between the two selected points x_0 and x_1, with their functional values y_0 and y_1. This method may be used for data that do not have a known function associated with them, or for **functions** that are used to calculate the y-values, and will provide a polynomial **interpolation** either way. This method has the advantage over the Taylor series approximation in that it uses more than one point in its approximation. However, it is by no means exact. The method may be improved by changing to quadratic or cubic terms, in which case the **polynomials** each gain a term in both numerator and denominator and the number of points used goes up.

The problem with Legendre polynomials of successively higher accuracy is that each polynomial must be calculated separately. With Newton polynomials, once one polynomial of degree N has been obtained, the next, N + 1, is simply the Nth polynomial plus the term $a_N (x - x_0)(x - x_1)...(x - x_{N-1})$. Each a-term is a fit coefficient obtained from the parameters known about the data or function. The Chebyshev polynomials use the same type of techniques as Legendre and Newton polynomials, but they use a limit **definition** to determine which coefficients of each group are most appropriate for the function over some specified **interval**, to maximize the accuracy of each method.

The above approximations are the ones most commonly used by human beings directly. There are several other approximations which are sometimes used in computer programs for various reasons. Pade approximations use a "rational approach," meaning that they feature two calculated polynomials of similar types to the above series, dividing one by the other for the net approximation. The largest restriction on these approximations is that the function and all its derivatives must vanish at zero in order for the polynomial to be manageable. Spline functions are not very accurate except on the order of many repetitions; an approximation by spline functions interpolates a chosen degree of polynomial (usually linear or quadratic) between each set of two adjacent points. This choppy method is sometimes useful for obtaining numerical results (especially when computer fitting can make the **distance** between any two adjacent selected points fairly small), but it does not allow for very much mathematical manipulation of the approximation and is of limited use on that basis.

Of course, there is nothing magical about polynomials as far as approximation goes; they are merely used because of their simplicity and their ease in mathematical use. They are also fairly easy to envision for even the most novice mathematician. However, any **Hilbert space**—that is, any set of orthogonal polynomials—may be used to obtain an exact series representation of a function over an infinite sum. The second most common set of functions used for approximation is the **trigonometric functions**, for Fourier approximation. This approximation, using sines and cosines, is favored because it gives approximations over a well-defined and physically useful periodicity.

Approximation is not a skill which is emphasized in some formal portions of mathematics, but it is essential to any branch of **applied mathematics**. Computer science uses approximate calculations for almost any command, and physics and engineering are almost as dependent upon approximation and interpolation in their experimental phases.

METHODS OF INTEGRATION

Integration is one of the basic operations of **calculus**. However, its results are not always straightforward; there are some integrals whose solution is not apparent by inspection. Because this is the case, and because closed form and numerical solutions are valuable in many cases, mathematicians have developed several different methods of integration.

In many cases, the integral may be evaluated at all points by direct anti-differentiation. The rules of differentiation are then followed backwards to find a closed form solution of the integral in question. When a **definite integral** is desired in this case, the anti-differentiated form is simply evaluated at the **limits**.

In some cases, an apparently difficult integral can be vastly simplified by "u-substitution." In u-substitution, the integral is shifted from the original x **variable** to a new variable, u, with the limits (if any) changed appropriately and appropriate attention given to the differential of u and how it differs from the differential of x. For example, a common substitution is u = cos x, du = -sin x dx. When the integral has been carried out in simplified form, the original expression is resubstituted for the u value.

Another integration "trick" is integration by parts. Integration by parts is the reverse of the product rule for differentiation, where d/dx (uv) = u (dv/dx) + v (du/dx). Then if the integral is of the form u dv, it's equal to uv minus the integral of v du. If v is a simpler function to integrate than u, this can be a very helpful way of solving the integral. This technique is often used with **fractions**, where the numerator and denominator are regrouped into equivalent terms.

There are several numerical methods of solving an integral when closed form solutions (as described above) are not feasible but a quantitative answer is desired. The trapezoid or trapezoidal rule is the simplest **numerical integration** method. The desired area of integration is divided into a selected number of subdivisions. Each of these forms a trapezoid, with the sides of the region being the parallel sides, the axis acting as the base of the trapezoid, and the line connecting the endpoints of the curve on that **interval** forming the top of the trapezoid. The error in this method may be reduced, as with most numerical integration techniques, by using smaller and smaller subdivisions within the same area. The equivalent result will come of calculating the areas of rectangles whose top was drawn through the **midpoint** of the function on each interval.

Simpson's rule uses the same type of regional subdivisions as the trapezoid rule, but instead of taking a straight line fit across the function, the top of the interval is approximated to a quadratic equation. The area is then calculated as though the integral was of a **parabola** instead of over the function in question. Boole's rule uses the same type of quadratic fit but then employs a recursive relationship to make the quadrature more precise. While this is still not an exact calculation of the integral, the error rate can be far reduced.

Rather than selecting values at predetermined, even intervals as the other numerical techniques do, Gauss-Legendre quadrature evaluates the integral based on random points throughout the interval. A table of abscissa and weights allows the random selection to give a fair approximation of the average height of the function throughout the interval (a better and better approximation as more points are selected).

The wide variety among these techniques allows for the person evaluating the integral to select which method is most appropriate to the function and interval and to the application at hand. The advent of computer technology has made the numerical methods more common and more feasible with high precision levels, but each method, exact or approximate, has advantages that keep it in the list of good methods for integration.

METRIC SYSTEM

The metric system of measurement is an internationally agreed-upon set of units for expressing the amounts of various quantities such as length, mass, time, temperature, and so on.

Whenever we measure something, from the weight of a sack of potatoes to the **distance** to the moon, we must express the result as a number of specific units: for example, pounds and miles in the English system of measurement (although even England no longer uses that system), or kilograms and kilometers in the metric system. As of 1994, every nation in the world has adopted the metric system, with only four exceptions: the United States, Brunei, Burma and Yemen.

The metric system that is in common use around the world is only a portion of the broader International System of Units, a comprehensive set of measuring units for almost every measurable physical quantity from the ordinary, such as time and distance, to the highly technical, such as the properties of energy, electricity, and radiation. The International System of Units grew out of the 9th General [International] Conference on Weights and Measures, held in 1948. The 11th General Conference on Weights and Measures, held in 1960, refined the system and adopted the French name *Système International d'Unités*, abbreviated as SI.

Because of its convenience and consistency, scientists have used the metric system of units for more than 200 years. Originally, the metric system was based on only three fundamental units: the meter for length, the kilogram for mass, and the second for time. Today, there are more than 50 officially recognized SI units for various scientific quantities.

Measuring units in folklore and history

In the biblical story of Noah, the ark was supposed to be 300 cubits long and 30 cubits high. Like all early units of size, the cubit was based on the always-handy human body, and was most likely the length of a man's forearm from elbow to fingertip. You could measure a board, for example, by laying your forearm down successively along its length. In the Middle Ages, the inch is reputed to have been the length of a medieval king's first thumb joint. The yard was once defined as the distance between the nose of England's King Henry I and the tip of his outstretched middle finger. The origin of the foot as a unit of measurement is obvious.

In Renaissance Italy, **Leonardo da Vinci** used what he called a *braccio*, or arm, in laying out his works. It was equal to two *palmi*, or palms. But arms and palms, of course, will differ. In Florence, the engineers used a *braccio* that was 23 inches long, while the surveyors' *braccio* averaged only 21.7 inches. The foot, or *piede*, was about 17 inches in Milan, but only about 12 inches in Rome.

Eventually, ancient "rules of thumb" gave way to more carefully defined units. The metric system was adopted in France in 1799 and the British Imperial System of units was established in 1824. In 1893, the English units used in the United States were redefined in terms of their metric equivalents: the yard was defined as 0.9144 meter, and so on. But English units continue to be used in the United States to this day, even though the Omnibus Trade and Competitiveness Act of 1988 stated that "it is the declared policy of the United States ... to designate the metric system of measurement as the preferred system of weights and measures for United States trade and commerce."

English vs. metric units

Why do scientists and everybody else in the world except the United States and three tiny, non-industrialized nations believe that the metric system is superior to the English system? There are four main reasons.

(1) English units are based on silly standards. When that medieval king's thumb became regrettably unavailable for further consultation, the standard for the inch was changed to the length of three grains of barley, placed side by side—not much of an improvement. Metric units, on the other hand, are based on nature, not on the whims of humans.

(2) The standards behind the English units aren't reproducible. Arms, hands, and grains of barley will obviously vary in size; the size of a 3-foot yard depends on whose feet are in question. But metric units are based on standards that are precisely reproducible, time after time.

(3) There are simply too many English units. We have buckets, butts, chains, cords, drams, ells, fathoms, firkins, gills, grains, hands, knots, leagues, three different kinds of miles, four kinds of ounces, and five kinds of tons, to name just a few. There are literally hundreds more. For measuring **volume** or bulk alone, the English system (now more accurately called the American system) uses ounces, pints, quarts, gallons, barrels and bushels, among many others. In the met-

ric system, on the other hand, there is only one basic unit for each type of quantity.

(4) Any measuring unit, in whatever system, will be too big for some applications and too large for others, so we must have a variety of sizes. People would not appreciate having their waist measurements in miles or their weights in tons. That's why we have inches and pounds. The problem, though, is that in the American system the conversion factors between various-sized units—12 inches per foot, 3 feet per yard, 1760 yards per mile—have no rhyme or reason to them. They're completely arbitrary. Metric units, on the other hand, have conversion factors that are all **powers** of ten. That is, the metric system is a decimal system, just like dollars and cents. In fact, our entire system of numbers is decimal, based on tens, not threes or twelves. Therefore, converting a unit from one size to another in the metric system is just a matter of moving the decimal point.

The metric units

The SI starts by defining seven basic units: one each for length, mass, time, electric current, temperature, amount of substance, and luminous intensity. ("Amount of substance" refers to the number of elementary particles in a sample of matter. Luminous intensity has to do with the brightness of a **light** source.) But only four of these seven basic quantities are in everyday use by nonscientists: length, mass, time, and temperature. Their defined SI units are the meter for length, the kilogram for mass, the second for time and the degree Celsius for temperature. (The other three basic units are the ampere for electric current, the mole for amount of substance and the candela for luminous intensity.) Almost all other units can be derived from the basic seven. For example, **area** is a product of two lengths: meters squared, or **square** meters. Velocity or speed is a combination of a length and a time: kilometers per hour.

The meter was originally defined in terms of the earth's size; it was supposed to be one ten-millionth of the distance from the equator to the North Pole, going straight through Paris. But because the earth is subject to geological movements, this distance can't be depended upon to remain the same forever. The modern meter, therefore, is defined in terms of how far light will travel in a given amount of time when traveling at—naturally—the speed of light. The speed of light in a vacuum is considered to be a fundamental constant of nature that will never change, no matter how the continents drift. The standard meter turns out to be 39.3701 inches.

The kilogram is the metric unit of mass, not weight. Mass is the fundamental measure of the amount of matter in an object. The mass of a baseball won't change if you hit it from the earth to the moon, but it will weigh less—have less weight—when it lands on the moon because the moon's smaller gravitational **force** is pulling it down less strongly. Astronauts can be weightless in **space**, but they can lose mass only by dieting. As long as we don't leave the earth, though, we can speak loosely about mass and weight as if they were the same thing. So you can feel free to "weigh" yourself (not "mass" yourself) in kilograms. Unfortunately, no absolutely unchangeable standard of mass has yet been found to standardize the kilogram on. The kilogram is therefore defined as the mass of a certain bar of platinum-iridium alloy that has been kept (very carefully) since 1889 at the International Bureau of Weights and Measures in Sèvres, France. The kilogram turns out to be 2.2046 pounds.

The metric unit of time is the same old second that we've always used, except that it is now defined in a super-accurate way. It no longer depends on the wobbly **rotation** of our eccentric old planet (1/86,400th of a day), because Mother Earth is slowing down; her days keep getting a little longer as she grows older. So the second is now defined in terms of the vibrations of a certain kind of atom known as cesium-133. One second is defined as the amount of time it takes for a cesium-133 atom to vibrate in a particular way 9,192,631,770 times. This may sound like a strange definition, but it is a superbly accurate way of fixing the standard size of the second, because the vibrations of atoms depend only on the nature of the atoms themselves, and cesium atoms will presumably continue to behave exactly like cesium atoms forever. The exact number of cesium vibrations was chosen to come out as close as possible to what was previously the most accurate value of the second.

The metric unit of temperature is the degree Celsius (°C), which replaces the English system's degree Fahrenheit (°F). In the scientists' SI, the fundamental unit of temperature is actually the Kelvin (K). But the Kelvin and the degree Celsius are exactly the same size: 1.8 times as large as the degree Fahrenheit. You can't convert between Celsius and Fahrenheit simply by multiplying or dividing by 1.8, however, because the scales start at different places. That is, their zero-degree marks have been set at different temperatures.

Bigger and smaller metric units

Because the meter (1.0936 yards) is much too big for measuring an atom and much too small for measuring the distance between two cities, we need a variety of smaller and larger units of length. But instead of inventing different-sized units with completely different names, as the English-American system does, we can create a metric unit of almost any desired size by attaching a prefix to the name of the unit. For example, since kilo- is a Greek form meaning a thousand, a kilometer (kil-OM-et-er) is a thousand meters. Similarly, a kilogram is a thousand grams; a gigagram is a billion grams or 10^9 grams; a nanosecond is one billionth of a second or 10^{-9} second.

Minutes are permitted to remain in the metric system for convenience or for historical reasons, even though they don't conform strictly to the rules. The minute, hour, and day, for example, are so customary that they're still defined in the metric system as 60 seconds, 60 minutes, and 24 hours—not as multiples of ten. For volume, the most common metric unit is not the cubic meter, which is generally too big to be useful in commerce, but the liter, which is one thousandth of a cubic meter. For even smaller volumes, the milliliter, one thousandth of a liter, is commonly used. And for large masses, the metric ton is often used instead of the kilogram. A metric ton (often spelled tonne in other countries) is 1,000 kilograms. Because a kilogram is about 2.2 pounds, a metric ton is about 2,200

pounds: ten percent heavier than an American ton of 2,000 pounds. Another often-used, nonstandard metric unit is the hectare for land area. A hectare is 10,000 square meters and is equivalent to 0.4047 acre.

Converting between English and metric units

The problem of changing over a highly industrialized nation such as the United States to a new system of measurements is a substantial one. Once the metric system is in general use in the United States, its simplicity and convenience will be enjoyed, but the transition period, when both systems are in use, can be difficult. Nevertheless, it will be easier than it seems. While the complete SI is intimidating because it covers every conceivable kind of scientific measurement over an enormous range of magnitudes, there are only a small number of units and prefixes that are used in everyday life.

MIDPOINT

The midpoint is defined as a position midway between two extreme points. In mathematics it is the point on a line segment or curvilinear arc that divides it into two parts of equal length. In a right **triangle** the midpoint of the **hypotenuse** is equidistant from the three vertices of the triangle.

There are several ways to find the midpoint of a particular figure. Two of the most widely used are the lens method and the Mascheroni construction. The lens method is used to determine the midpoint on a line segment. First a lens using circular arcs is constructed around the line segment so that the line connecting the cusps of the lens is perpendicular to the ends of the line segment. Where the line connecting the cusps of the lens intersects the line segment is the midpoint of that line segment. The Mascheroni construction is more complex but allows one to determine the midpoint of a line segment using only a compass. In about 1797 Mascheroni proved that all constructions possible with a compass and straightedge are possible with a movable compass alone. This construction involves drawing a series of circles starting with the endpoints of the line segment as the centers of the first two circles. Drawing a series of seven circles with varying centers and radii eventually produces the midpoint of the initial line as an intersection between the last two drawn circles. It is a complex construction but allows one to determine the midpoint of the line segment with only a movable compass.

Archimedes developed a **theorem** that relates the midpoint of an arc on a **circle** to line segments drawn within the circle. Given the circle below let M be the midpoint of the arc AMB. If point C is chosen at random and point D is chosen such that the line segment MD is perpendicular to AC then the length of AD is equal to the sum of the lengths of DC and BC. AD = DC + BC: this is known as Archimedes' midpoint theorem and can serve a variety of uses relating the midpoint of the arc on a circle to specific line segments within.

MILANKOVICH, MILUTIN (1879-1958)

Serbian professor and research scientist

Milutin Milankovich is best remembered for formulating the orbital or astronomical variation theory of climatic change in the 1930s. The Milankovich theory, as it is known, proposes that as the Earth travels through **space**, three distinct cyclic movements combine to cause variations in the amount of sunlight that falls on the planet. These variations, according to the theory, are what produce changes in the ebb and flow of ice fields. Most of the scientist's best efforts were aimed at reconstructing Earth's and the other planets' past climates.

Milankovich was born on May 28, 1879 in the town of Dalj, near Osijek, Croatia (then Austria-Hungary). After receiving a degree in 1902 from the School of Civil Engineering for a thesis on building a bridge using reinforced concrete, Milankovich finished his doctorate in 1904 at the Institute of Technology. The following year, he joined a prestigious civil engineering firm in Vienna and began working on projects to build bridges, viaducts, dams, and aqueducts using reinforced concrete.

In 1909, Milankovich left the professional world for good when he accepted a position as chair of the Applied Mathematics Department at the University of Belgrade. He continued to investigate applications for reinforced concrete, but from that time, Milankovich dedicated most of his energy to fundamental research. Beginning in about 1912, he turned his attention to solar climates and how temperatures affect the planets.

World War I began in 1914. Milankovich, who had just married at the time, became a prisoner of war soon afterward. He was interned at Nezsider and then in Budapest. In Budapest, however, his captors allowed him to continue his climatic research at the Hungarian Academy of Sciences Library. The war ended in 1918, and by 1920 Milankovich had published his first monograph on climatology, which he called "A Mathematical Theory of Thermic Phenomena Caused by Solar Radiation."

The paper brought Milankovich a significant amount of attention from the academic world, especially for his description of a "curve of isolation" at the Earth's surface. Although admired, this theory was not widely accepted until 1924, when other meteorologists discussed the curve more extensively in a textbook on geological climates. His fame renewed, in 1927 Milankovich was invited to contribute to two handbooks on climatology and geophysics.

Continuing his research into the 1930s, Milankovich developed the theory that the key to decoding the characteristics of past climates was the amount of solar radiation that the Earth received ("insolation"). This always varies by latitude, but according to Milankovich it also depends on three factors: the ellipticity of the Earth's **orbit**, which changes over time from virtually circular to more elliptical; precessional changes over about 21,000 years that determine which hemisphere receives more sunlight; and the subtly changing tilt of the planet's axis over thousands of years. He used this theory to explain the advance and retreat of ice over periods of 10,000-100,000 years, although he could not explain what caused the ice ages to happen in the first place.

In 1941, on the eve of World War II, Milankovich published his *Canon of Insolation of the Earth and Its Application to the Problem of the Ice Ages.* However, it would not be until the late 1960s and early 1970s, after sampling of deep-sea sediments and advanced climate **modeling**, that scientists would accept Milankovich's views as correct. His Northern Hemisphere summer radiation curves for the last 650,000 years, on the other hand, are not considered valid today.

Milankovich was also interested in the history of scientific development and wrote two books on the topic: *Through the Realm of Science* and *Through Space and Centuries.* He wrote an autobiography, *Recollection, Experiences, and Vision,* although this has not been translated into English. Milankovich died in Belgrade on December 12, 1958.

MINKOWSKI, HERMANN (1864-1909)

Russian-born German analyst

In spite of a relatively short career, Hermann Minkowski played an important role in the development of modern mathematics. His work formed the basis for modern functional **analysis**, and he did much to expand the knowledge of quadratic forms. Minkowski also developed the mathematical theory known as the **geometry** of numbers and laid the mathematical foundation for **Albert Einstein**'s theory of relativity by pioneering the notion of a four-dimensional **space-time** continuum.

Minkowski was born in Alexotas, Russia, on June 22, 1864, to German parents. The family returned to their native Germany in 1872, to the city of Königsberg, where Minkowski spent the rest of his childhood and also attended university. His brother, Oskar Minkowski, became famous as the physiologist who discovered the link between diabetes and the pancreas.

Even as a student at the University of Königsberg, Minkowski demonstrated a rare mathematical talent. In 1881, the Paris Académie Royale des Sciences offered a prize, the Grand Prix des Sciences Mathématiques, for a **proof** describing the number of representations of an integer as a sum of five squares of integers—a proof that, unbeknownst to the Académie, the British mathematician H. J. Smith had already outlined in 1867. Minkowski produced the proof independently, while Smith filled in the details of his outline and submitted it. In 1883, both Smith and Minkowski received the prize. At that time, the 19–year–old Minkowski was two years away from receiving his doctorate from the University of Königsberg. The work contained in his 140–page solution was, in fact, considered a better formulation than Smith's because Minkowski used more natural and more general definitions in arriving at his proof.

While he was a university student, Minkowski began a lifelong friendship with fellow student **David Hilbert,** who would eventually edit Minkowski's collected works. After receiving his doctorate from the University of Königsberg in

1885, Minkowski taught at the University of Bonn until 1894. Returning to teach at the University of Königsberg for two years, he then taught at the University of Zurich until 1902. One of his closest colleagues at Zurich was a former teacher, A. Hurwitz, who is best known for his **theorem** on the composition of quadratic forms.

Throughout his life, Minkowski worked on the **arithmetic** of quadratic forms, particularly in *n* variables. According to mathematician Jean Dieudonné, who profiled him for the *Dictionary of Scientific Biography,* Minkowski made two important contributions to this field. One was a characterization of **equivalence** of quadratic forms with rational coefficients, under a linear **transformation** with rational coefficients. The other, published in 1905, completed **Charles Hermite**'s theory of reduction for positive definite *n*-ary quadratic forms with real coefficients by finding a unique reduction form for each equivalence class. Pursuing the results of his 1905 paper, Minkowski developed a more geometric style of work, which led to what Dieudonné called his "most original achievement"—the geometry of numbers.

In 1889, Minkowski had introduced the geometrical concept of **volume** into his work on ternary quadratic forms. This technique involved centering ellipses on lattice points in the plane and looking at the areas of the ellipses as the lattice points increased in number. The limiting case as the number of non-overlapping ellipses in the plane approaches **infinity** gave Minkowski an estimate of the minimal solution of a specified quadratic equation in two variables. Using this type of geometrical technique, he was able to prove various theorems about numbers without performing any numerical calculations, a feat that was praised by Hilbert as "a pearl of Minkowski's creative art," reported Harris Hancock in the introduction to his book *Development of the Minkowski Geometry of Numbers.*

Minkowski generalized the technique to ellipsoids and other convex shapes (such as cylinders and polyhedrons) in three dimensions. This work led to investigations in packing efficiency (how to fill up **space** most densely with given shapes), a topic that has applications in chemistry, biology, and other sciences. Generalizing further to various types of convex objects in *n*-dimensional space, he produced numerous results in **number theory** through geometry. As Dieudonné wrote, "Long before the modern conception of a metric space was invented, Minkowski realized that a symmetric convex body in an *n*-dimensional space defines a new notion of 'distance' on that space and, hence, a corresponding 'geometry'"—a development that laid the foundation for modern functional analysis.

Hancock wrote of Minkowski, "His grasp of geometrical concepts seemed almost superhuman." However, he also noted that Minkowski's publications and notes were often incomplete and poorly explained. In part this was simply Minkowski's style, although it was complicated by his early death that brought an abrupt end to his works in progress. He died at the age of 44 from a ruptured appendix in Göttingen, Germany, on January 12, 1909. Hancock wrote his book to

Hermann Minkowski

"reconstruct and clarify much that Minkowski would have done, had he lived."

At Hilbert's urging, the University of Göttingen created a new professorship for Minkowski in 1902. It was during his tenure at Göttingen that Minkowski turned his attention to relativity theory. Einstein, who published his initial work in relativity in 1905, had taken nine classes from Minkowski in Zurich, more than he had taken from anyone else (even the physics professor). The two men had no particular liking for one another. Lewis Pyenson wrote in *Archive for History of Exact Sciences* that "By the time he graduated in 1900 ... Einstein had become indifferent to Minkowski's approach to mathematics and physics," and "Minkowski later thought that his own interpretation of the principle of relativity was superior to Einstein's because of Einstein's limited mathematical competence."

In 1905, Minkowski participated in an electron theory seminar that discussed the current theories of electrodynamics. With this background, he studied the competing theories of subatomic particles proposed by Einstein and **Hendrik Lorentz**. Minkowski was the first to realize that both theories led to the necessity of visualizing space as a four-dimensional, non-Euclidean, space-time continuum. Dieudonné wrote that Minkowski "gave a precise definition and initiated the mathematical study [of this four-space]; it became the frame of all later developments of the theory and led Einstein to his bolder conception of generalized relativity."

Rather than a mathematical adjustment of Einstein's theory, Minkowski developed his ideas as an alternate theory. As soon as Minkowski's first publication on relativity appeared, Pyenson wrote, "Einstein turned to the only part of Minkowski's paper that contained a physical prediction" and showed how it failed to account for a known phenomenon. The rivalry between the two men was carried on at a level that few recognized; the differences between their derivations were subtle. Ultimately, Einstein used Minkowski's ideas to develop his general theory of relativity, which was published seven years after Minkowski's death.

MINKOWSKI SPACE-TIME

Minkowski **space-time** is a single, four-dimensional continuum composed of the three coordinates defining position and a fourth **dimension** of time. It is a concept developed by German mathematical physicist **Hermann Minkowski** in 1907, and it provides the framework for all mathematical work in relativity, including Albert Einstein's general theory of relativity. Minkowski space-time is also often referred to as Minkowski **space** or the Minkowski universe. Although Minkowski space-time is used predominately in special relativity, it can be applied to other studies involving the coupling of time to spatial vectors. In order to be able to use Minkowski space-time in this manner, this type of coupling must show that the time **transformation** of a process is not independent of the spatial components, that the two are strictly interdependent.

In an attempt to understand work previously done by **Hendrik Lorentz** and Einstein, Hermann Minkowski developed a four-dimensional space-time continuum in 1907. Until the development of Minkowski space-time, the three-dimensional coordinate system describing position and the other dimension, time, were considered as independent entities in Newtonian physics. Minkowski realized that the preliminary work on relativity could best be understood in a non-Euclidean space that involved a combination of these two separate systems. He coupled these two systems together into a four-dimensional space-time continuum and employed it in his own treatments of a four-dimensional study of electrodynamics. He noticed that the **invariant interval** between two events has some of the properties of the **distance** in Euclidean **geometry** and formulated it as the **square** root of a sum and difference of squares of intervals of both space and time. Using this idea, events localized with regards to both space and time are the analogues of points in three-dimensional geometry. This means that in Minkowski space-time, the time in the history of one particle resembles the arc length of a curve in three-dimensional space. Minkowski space-time does not add anything to the original physical concepts of relativity, but its contribution to the conceptual development of relativity is immeasurable.

See also Relativity, general; Relativity, special; Space-time

MITTAG-LEFFLER, MAGNUS GÔSTA (GUSTAF) (1846-1927)
Swedish mathematician and editor

Founder of the prestigious international mathematical journal *Acta Mathematica* and a fixture in the Scandinavian school of mathematics, Magnus Gôsta Mittag-Leffler is also remembered for his work on the analytic representation of a one-valued function, which became the basis of the **theorem** that bears his name.

The eldest son of a school principal, Mittag-Leffler was born in Stockholm, Sweden on March 16, 1846. His parents recognized the boy's mathematical talents early on. Mittag-Leffler trained as an actuary at the University of Uppsala, where he received his doctorate for a thesis on analytic function theory in 1872 and immediately accepted an offer to work as a lecturer. The period from 1873 to 1876 was a formative one for him, as he studied in Paris and Berlin with such illustrious mathematicians as **Charles Hermite**, **Jules-Henri Poincare**, and **Karl Thodor Wilhem Weierstrass**.

In 1877 Mittag-Leffler finished his teaching preparation (habilitation) with a paper on the theory of elliptic function, after which he was appointed mathematics professor at the University of Helsinki. Four years later, he returned to Stockholm to take up the same position at the newly established university there, also serving as rector on two occasions. A successful businessman and husband of a wealthy Finnish woman, Mittag-Leffler spent most of his free time at his two homes in Tallberg and Djursholm, a suburb of Stockholm where he housed his impressive mathematical library. Even then, the library was acknowledged as the world's finest.

With financial help from his wife and Sweden's King Oscar II, Mittag-Leffler launched *Acta Mathematica* in 1882. His motivation was reportedly to maintain a record of the modern **history of mathematics**; he would remain the journal's editor-in-chief for the rest of his life. In 1884 he combined all of his work on one-valued **functions** into a single paper that is still considered his masterpiece, publishing it in *Acta Mathematica*. In the paper he proposed a series of general topological ideas on infinite point **sets** based on the new Cantor **set theory**.

In Copenhagen in 1925, Mittag-Leffler lectured to an international congress of mathematicians for the last time. One of the attendees, English mathematician **Godfrey Harold Hardy**, remembered that "the whole audience rose and stood as [Mittag-Leffler] entered the room.... it was an entirely spontaneous expression of the universal feeling that to him, more than any other single man, the great advance in the status of Scandinavian mathematics during the last 50 years was due."

Mittag-Leffler died in Stockholm on July 7, 1927. The Mittag-Leffler Mathematical Institute, which he and his wife founded in 1916, became part of the Swedish Royal Academy of Sciences in 1919. His library remains one of the school's major assets.

Möbius, August (Augustus) Ferdinand (1790-1868)

German mathematician

Known principally for his discovery of the **Möbius strip**, which made him a pioneer in the field of **topology**, August Ferdinand Möbius also did important work in theoretical astronomy and **analytic geometry**.

Möbius, born in Schulpforta, Germany on November 17, 1790, was the only child of a dance teacher and his wife. Until he was 13, Möbius received his education at home. Already noted for his talent in mathematics, he began attending regular school in Schulpforta in 1803 and in 1809 started classes at Leipzig University. At first believing that he wanted to study law (perhaps due to his family's influence), Möbius quickly returned to mathematics as his dominant academic interest. Astronomy and physics were close seconds, however, and he took many classes in both.

In 1813 Möbius left Leipzig for the academic mecca of Göttingen, having been chosen for a fellowship. There he spent half a year studying theoretical astronomy with **Johann Carl Friedrich Gauss**. The following year Möbius received his doctorate and returned to Leipzig to serve as junior astronomy professor. He was never to be known as a good lecturer, however, and sometimes had to resort to offering free courses to convince students to attend. In 1816, having somehow avoided an attempt to draft him into the Prussian Army, he also became an observer at the school's observatory. From 1818 to 1821 Möbius presided over the renovation and refurbishment of the outdated facility.

Despite several offers from other universities for better positions with more pay, Möbius declined to leave Leipzig University. Over the next decade, he devoted himself to astronomy, publishing several important works on occultation phenomena, the path of Halley's comet, and the basic laws of astronomy. *The Principles of Astronomy* came out in 1836. His most influential book on astronomy, however, was *The Elements of Celestial Mechanics* (1843), in which he managed to discuss his subject in mathematical terms, but without using higher mathematics. This meant that even the layperson could understand the work, although professional astronomers also found it useful.

Möbius's dedication to astronomy complemented, rather than competed with, his talent for mathematics. His work in the former yielded insight into the latter, and vice versa. In 1827 he published *The Calculus of Centers of Gravity,* which introduced homogeneous coordinates into analytic **geometry** and discussed projective geometry. It was in this work that he first mentioned the configuration now known as the Möbius strip.

During the 1830s, Möbius spent most of his time writing about statics, a branch of **mechanics**. His *Handbook on Statics,* which appeared in 1837, gave the **area** a geometric treatment, which led to the study of systems of lines in **space** and the null system of planes and points.

Despite his many academic accomplishments, Möbius did not receive a full professorship in astronomy and higher mechanics at Leipzig University until 1844. In 1848, the school made him directory of the observatory.

It was not until the last decade of his life that Möbius did the work for which he would become most famous. In 1858 he entered a contest put on by the Paris Academy of Sciences for research on the geometrical theory of polyhedrons. Möbius published his results in two papers, the last of which described his discovery of what came to be known as the Möbius strip. A rectangular, flat strip with a half twist, its ends connect to create a continuous, single-edged loop. Theoretically, it is a two-dimensional surface with only one side, but it can be constructed in three dimensions. There is some question as to whether another mathematician (Johann Listing) also discovered the configuration at about the same time.

Möbius died in Leipzig on September 26, 1868. He had married in 1820 and had three children.

Möbius strip

A Möbius strip is a one-sided surface formed by gluing together two opposite edges of a strip of paper after twisting one of the edges 180 degrees. The Möbius strip was named after the German mathematician and astronomer **August Ferdinand Möbius**, one of the pioneers of the mathematical field known as **topology**, which concerns the properties of shape that do not change when an object is stretched or bent without being torn. Möbius investigated the curious properties of the Möbius strip in 1858, in connection with a question about the **geometry** of polyhedra posed by the Paris Academy. Most historians agree, however, that Möbius was not the first to discover the strip named after him: by coincidence, the German mathematician Johann Benedict Listing also discovered the Möbius strip in 1858, edging out Möbius by a few months.

In spite of the fact that the Möbius strip is formed from a piece of paper that has two sides, the Möbius strip has only one side. If you start drawing a line down the middle of the strip on one side, and continue drawing the line until you come back to where you started, you will find that you have drawn the line on both "sides" of the strip, so that in fact the strip has only one side. Because of this property, Möbius strips have found a use in industry as conveyor belts, since when an ordinary loop is used, one side of the belt eventually gets very worn, while the other side does not get worn at all. The Möbius strip also has only one edge, as can be seen by drawing a line along one of the edges and continuing the line until it returns to its starting point; by that time, the line has run along the entire edge of the strip.

The Möbius strip has several other surprising attributes. If you make a cut all along the center loop of the strip, the result will not be two thinner Möbius strips, but a single long strip with two twists in it. Furthermore, if you cut the Möbius strip into thirds lengthwise, you will end up neither with three short loops nor with one long loop; rather, you will have one short Möbius strip intertwined with a longer loop that has two twists in it.

The Möbius strip is the simplest example of what is known as a non-orientable surface, one on which it is impossible to form a consistent notion of the difference between, say, a right hand and a left hand. Suppose that you draw a right hand on the Möbius strip, using an ink marker that soaks through the paper to the other side. Imagine that you start drawing copies of the right hand along the strip, being careful always to draw a right hand. By the time you get back to the starting place you have gone through a twist and are drawing on the reverse side of the paper, so that it faces the opposite way from the original hand: if you view the new hand from the same vantage point as the original hand, the new hand will look like a left hand. Thus there is no way to define the notion of a right or left hand, so the Möbius strip is non-orientable. What's more, mathematicians have proven that every non-orientable surface contains at least one Möbius strip, so that Möbius strips are the essential building blocks for all non-orientable surfaces.

See also Manifold; Topology

MODAL LOGIC

Modal logic is an extension of ordinary propositional logic. Its purpose is to **model** our reasoning about statements that do not correspond to actual facts in the world. In general, the goal of logic is to formalize the structure of human reasoning. With a standardized abstract language of thought we hope to communicate facts about the world efficiently and accurately. Propositional logic and its extensions, first and second order logic, are meant to resolve questions about how the world *is*, e.g., "is the sum of the angles of a **triangle** 180 degrees?"

However, we do not always contemplate the way the world is. Often we think and argue about how the world *might* or *should* be. Could the South have won the Civil War? Should we feed the starving of the world? The answers to these questions are not statements about how the world actually is; nevertheless these questions are interesting and relevant. Modal logic allows us to formally discuss these non-actual statements.

In the strictest sense, modal logic refers to the distinction between different *modes* of **truth**, *necessity* and *possibility*. We often believe that some facts are necessarily true, others are contingently true, that is, they could easily have been false under slightly different circumstances. One way to look at this distinction is by considering the mental image of **possible worlds** introduced by philosopher and mathematician Gottfried Wilhelm Leibniz. We live in the actual world, the world that is real. But every time we choose between alternative courses of action, we can envision ourselves having made an alternate choice that would have led to a different chain of events. The world would have been different if you had chosen to eat leftovers this morning instead of pancakes: You might have had indigestion now, and nothing for lunch this afternoon. The actual world differs minutely from the imaginary "leftovers for breakfast world."

In modal logic, we consider a set of possible worlds that includes the actual world and any number of other imaginary worlds. We say that a statement p is necessarily true if it is true in all possible worlds. No matter which world in our set we look at, p is always true in that world. We say that a statement q is possibly true if there exists some world in our set of possible worlds in which q is true. It need not be the actual world. In order to articulate these notions we enrich the standard language of logic (**propositions** and the operators "not," "and," "or," and "if-then") with the modal operators "it is necessary that" and "it is possible that."

Suppose we have the proposition "Marcus eats meat." In ordinary logic we can only ask if this statement is true or false. In modal logic we can construct the more complex propositions: "Marcus necessarily eats meat," or "It is possible that Marcus eats meat." We can easily imagine a world in which Marcus is a vegetarian, so "Marcus necessarily eats meat" is a false statement. Consider however the proposition "Vegetarians do not eat meat." This statement is true in all worlds since it is a simple matter of **definition**. In modal terms, the statement "Vegetarians necessarily do not eat meat" is true. All logical **tautologies**, mathematical theorems, and **arithmetic** are thought to be necessary truths.

Note that necessity and possibility can be defined in terms of each other. To say that p is necessarily true is equivalent to saying that it is not possible that p be false. If p is true in all possible worlds, then there is no world in which p is not true. Hence the statement "it is possible that not p" is false. Likewise, to say that q is possible is equivalent to saying that it is not necessarily true that q is false.

Another area in which the central ideas of modal logic can be brought to bear is in *deontic* logic (moral reasoning). We can distinguish between obligatory, permitted, and forbidden actions. Instead of the modal operators "it is necessary that" and "it is possible that" we would have the operators "it is permitted that," "it is obligatory that," and "it is forbidden that." Other areas in which modal logic is used include temporal logic (operators "it is the case that," "it will always be the case that," and "it will never be the case that") and the logic of knowledge (operator "X knows that").

Like any other system of logic or mathematics, the theory of modal logic must be built on a set of axioms from which all theorems are deduced. In addition to the standard axioms of propositional logic, modal logic requires several axioms that define the meanings and relationships of the modal operators. Different interpretations of modality (logic of necessity or deontic logic) require different axioms. For instance, one **axiom** of the logic of necessity is "if it is necessarily true that p, then p obtains in the actual world." In deontic logic this is not the case. "If p is obligatory then p obtains" is demonstrably false. Paying taxes is obligatory, but some people evade their civic duty. There are many different axioms of modal logic, and each combination of them generates its own characteristic system of modal logic. Each individual modal system may often work well for some applications but not for others.

Although modal logic was first formalized in the early part of this century by American philosopher C. I. Lewis, the concept of modality has been present in logic for many centuries. **Aristotle** distinguished between necessary truth and contingent possibility. Anselm of Canterbury produced a **proof** for the existence of God in the 11th century that is essentially a modal argument and can easily be translated into modern modal terminology. Today modal logic is fertile ground for research in logic and philosophy and has found many applications in computer science.

MODE

The mode of a data set is the entry or entries that occur most frequently in the set. If a data set has two different values occurring most frequently, it is said to be bimodal. For the data set 1, 2, 4, 8, 8, 10, 10, 10, 12, 14, 15, 15, the mode is 10. For the data set: 1, 2, 4, 8, 8, 8, 10, 10, 10, 12, 14, 15, 15, the modes are 8 and 10. If all data values are equally represented, the data set will have no mode. Unlike the **mean**, the mode is always a member of the data set and it is not subject to "outliers" (data values that are extremes, very unlike the other members). The mode is a type of descriptive statistic that gives summary information about a data set. It is a measure of central tendency in that it describes or gives a summary or "average" of a whole set of events or measurements. The other two measures of central tendency are mean and **median**. The information given by measures of central tendency can be used for description or comparison.

See also Mean; Median

MODEL

A mathematical model is an equation or system of **equations** and/or **inequalities** used to describe data from real world situations and to make predictions from that data. The model may be as simple as a linear function or as complicated as a system of hundreds or even thousands of higher degree equations programmed into a powerful computer. Whatever the complexity of the model, the goal of the modeler is to create a mathematical system which comes as close as possible to capturing both the quantitative and qualitative properties of the real world system being modeled. Thus if an economist collects data on price and demand and observes that her data seems to be quite linear, she will undoubtedly attempt to find a linear function that is a good fit for the collected data. Her desire is to be able to make predictions about future demand based on future prices or vice-versa. If her assumption that this relationship will continue to remain linear is correct, then her simple linear model should do a good job at predicting future events. On the other hand, she might be interested in **modeling** the entire economy of a country. In this case, she would undoubtedly need to construct a far more sophisticated model consisting of many equations with many variables to account for the vast array of forces that influence the workings of a national economy. Such a complicated model would undoubtedly be programmed into a computer and many different simulations of the economy could be run on the computer based upon changing various critical assumptions about different components of the economy.

Some mathematical models give predictions that are consistently so accurate that scientists use them repeatedly with near certain confidence. One of the most famous examples of a successful mathematical model is the system of **differential equations** developed by Sir **Isaac Newton** (1642-1727) to describe the **motion** of the planets about the sun. Newton's laws of planetary motion have been used not only to accurately predict motion in the solar system, but to provide modern-day scientists and engineers a basis for sending satellites into **space** and controlling their motion. Perhaps the greatest engineering achievement of the 20th century was sending humans to the moon and returning them safely to earth. Newton's mathematical model was the foundation for this tremendous feat. On the other hand some real world systems have more or less defied attempts to model them with such a high degree of certainty as Newton's model. A notorious example is the weather. When meteorologists try to predict where a hurricane will strike land, they are able to state their predictions only within a very broad range of probabilities. The problem is that there are so many potential variables involved in weather systems that it has been impossible to construct a sophisticated enough mathematical model to take all these variables into account. Some mathematicians and meteorologists have suggested that the weather will always be impossible to predict with a high degree of accuracy because weather systems fall into a class of dynamical systems which are called chaotic. A chaotic system is one which exhibits extreme sensitivity to its initial conditions, meaning that any small error in an initial value put into a mathematical model will lead to very great errors in the final predictions of the model. Nevertheless, with computing power increasing exponentially every year, scientists continue to build more and more sophisticated mathematical models to predict the outcome of such varied systems as the weather, the economy, the stock market, and more.

It is important to understand that with very complicated systems, the best model is not always the first one proposed. Often many models are suggested, tested, and rejected before a correct model is adopted. Even then, the modelers are open to the possibility that new and unexpected conditions may arise that will necessitate modifications to the model. In fact, the modeling of sophisticated systems is very often a trial and error process, an attempt to build the best model by successive approximations to reality. One might think that a mathematician trying to capture the essence of a very complicated system would, from the beginning, attempt to account for every **variable** that she could possibly imagine would have an impact on the system. This is not typically the case. Usually one begins with a very simple model and only increases its complexity as the evidence warrants. It is a golden rule among mathematicians that, when it comes to models, simpler is better as long as the simpler model is able to account for the essential behavior of the system being modeled. This is not only an esthetic desire, but a practical one as well. A simpler model means a simpler mathematical **analysis** will be required to explain the system.

See also Modeling

MODEL THEORY

Model theory is a branch of mathematical logic concerned with the study of formal theories viewed as mathematical structures or objects. Those mathematical structures are studied by examining the first order sentences that describe those structures and the **sets** in those structures that are defined by first order formulas. Model theory is that part of mathematics that demonstrates how to apply mathematical logic to the study of structures in mathematics. Those structures investigated in model theory can be expressed in a formal language that usually consists of first order statements.

Model theory's fundamental tools are interpretations. Mathematical **truth** is like all truth in that it is relative to specific situations and depends upon how and where it is interpreted. This is because of the language used to express mathematical ideas rather than to mathematics itself. The theory itself is generally one consisting of interpretations of axiomatic **set theory**. Axiomatic set theory is a specific version of set theory, a branch of mathematics closely associated to logic, consisting of axioms that are taken as un-interpreted as opposed to being formalizations of pre-existing truths. Mathematical structures that obey axioms in a system are considered the models of that system. The second order axioms of **analysis** are known to have **real numbers** as their model. Nonstandard analysis is a form of model theory in which the axioms are weakened to include only the first order axioms. Model theory itself utilizes the full power of set theory.

Model theory also involves investigations concerned with the expressive power of formal languages in the sense of what they can say about specific mathematical structures. Model theory is the ultimate abstraction in the sense of its methods of analysis yet it has immediate applications to practical mathematics. An alternative to model theory is the view that **deduction** depends on formal rules of **inference** like the rules of logical **calculus**. The differences between model theory and deduction are similar to the differences in logic between proof-theoretic methods that are based on truth tables and proof-theoretic methods that are based on formal rules. The two main functions model theory attempts to perform are explaining the relationship between language and experience and specifying the idea of logical consequence. Classical model theory can be thought of as the act of dealing with static relationships among individuals.

The origins of model theory can be traced back to the continuum hypothesis proposed by Cantor in 1878. In this hypothesis he put forth the idea that every infinite subset of the continuum is either countable or has the **cardinality** of the continuum. This hypothesis was part of logic and more specif-

ically a part of set theory. **David Hilbert** understood the importance of this hypothesis and in the early 1900s sought to propose methods to more fully understand and prove the hypothesis. In 1902 **Ernst Zermelo** adopted Hilbert's ideas and published his first work on set theory. In 1905 he began to attempt to axiomatic set theory and in 1908 published his axiomatic system although he failed to prove consistency. In 1922 Skolem and Fraenkel, working independently, proceeded to improve Zermelo's **axiom** system and in doing so created the most commonly used system for axiomatic set theory to date. Later, Skolem extended Löwenheim's work from 1915 and formulated a nonstandard model of **arithmetic** in 1919. This was the first introduction of model theory but in a weaker form. Malcev's first publications in about 1933 were on model theory and these ideas later appeared in Robinson's pioneering work on model theory, for which he received his Ph.D. in 1949, and nonstandard analysis, which he introduced in 1961. Tarski continued the advancement of model theory by developing a semantic method to more thoroughly study formal scientific languages. After attending a course taught by Tarski in 1946, Lyndon began work on model theory and eventually led to the development of **group theory**. Mostowski also began working on the formulation and understanding of model theory at about the same time under the direction of Tarski. In 1975 a collection of works were published called *Model theory and algebra* as a memorial tribute to Robinson who had died a year earlier. Robinson made significant advances in model theory and is attributed with inventing nonstandard analysis, a weakened form of model theory. More recently there has been a fruitful interplay between model theory and **algebra**. From this there have been methods developed in a pure model-theoretic setting that have been applied in such areas as group theory, **number theory**, and **algebraic geometry**. Model theory has also enjoyed an extensive interaction with the field of infinite permutation groups. Research continues that not only employs model theory but contributes to its advancement as well.

MODELING

Mathematical modeling is the process of constructing a mathematical system designed to describe, analyze, and predict future outcomes of real world phenomena. Mathematical models are commonly used in physics, chemistry, biology, business, economics, finance, sociology, anthropology, and many other areas as well. Practitioners in these disciplines believe that their areas of study will be most seriously regarded if they have a solid mathematical foundation under them. A mathematical **model** may be as simple as a linear function or as complicated as a system of thousands of **equations** in thousands of variables which requires a powerful computer to run simulations based upon the model. However simple or complex the model may be, the steps in the modeling process generally follow a process similar to the following steps. First, a scientist makes observations and collects data from some real world system that she wishes to understand. Second, the scientist focuses on what she believes to be the essential elements of the real-world system and strips away from the modeling process anything that seems to be extraneous to the workings of the system. Third, based upon an initial analysis of the data, she hypothesizes a mathematical model—an equation or inequality or some system of these. Fourth, she tests the model by doing computations to see if the model generates values which are reasonably close to the data collected from the real-world system. If so, she may either accept the model or, more likely, attempt to make some adjustments which will give even closer **correspondence** to the real-world system. If the original model gives results which are far away from the data values, then more dramatic adjustments will need to be made, possibly even bringing into the model terms or equations which account for some of the processes originally believed to be extraneous. In either case, the modified model is tested again to see how well it simulates the real-world data. This process of modifying and testing is repeated until the scientist is satisfied that she has a model system which generates results close enough to the real system data that it can be successfully used for making predictions of future events or of events in the far past before any data had been collected.

Consider an example of the modeling process described in the preceding paragraph. Suppose an economist wishes to study the dynamics of price, supply, and demand for a certain commodity in an economy. First, he would collect data either from his own or someone else's research. Ideally, this data would be based on a fairly large sample of several years duration and would consist of recorded values of price, supply, and demand for the commodity. Suppose that after looking at some graphical representations of the data, the economist decides to use a system of **linear equations** to model this data. He tests the model and finds that it is not so good at simulating the dynamics of the real-world system. He makes modifications, perhaps changing one or more of the equations from linear to non-linear, with the hope that these modifications will allow the model to give results that are in closer correspondence with the data. He continues to make modifications in successive stages of the model until his simulations based on the model are satisfactory. The process of developing mathematical models in this way is both creative and inventive, involving both science and art in addition to mathematics. Very often a mathematical model can be viewed as a "work in progress" as the modeler continues to tweak it to get better and better correspondence with data actually collected from the real world.

With the ascent of increasingly more powerful computers, mathematicians have been able to create remarkably sophisticated models, but some of the most important breakthroughs in mathematical modeling came long before computers were available and involved relatively simple models. Sir **Isaac Newton** instituted a revolution in science with his laws of gravitation and planetary **motion** based upon a simple but powerful mathematical model involving just two **differential equations** and their ramifications. Newton's remarkable model formed the basis of 20th-century **space** exploration including the first landing of a man on the moon in 1969 and the numerous communications satellites currently orbiting the earth above us. In fact, Newton's model was the accepted explanation for all motion in the universe until the beginning of the 20th century when **Albert Einstein** proposed a more sophisticated model, called relativity theory, to account for certain inadequacies found in Newton's model when studying motion at near light-speed. Other famous mathematical models include James Clerk Maxwell's differential equations of **electromagnetism**, the equations of quantum **mechanics**, the laws of population growth and radioactive decay, Mandelbrot's fractal **geometry**, and the so-called "grand unification theories" of modern nuclear physics and cosmology. Several recent Nobel prizes in economics have been awarded to economists who constructed complex mathematical models of some portion of the national economy. Wherever there is a natural system to be studied and understood, one is likely to find a mathematical modeler on the scene constructing a model which, she or he hopes, will capture the essential features of that system.

See also Model

MODULAR ARITHMETIC

Modular **arithmetic** is a generalization of odd and even. We say that two **integers**, x and y, are congruent modulo a third integer n if and only if x - y is divisible by n. For example, 5 and 8 are congruent modulo 3 because 5 - 8 = -3 is divisible by 3. However, 4 and 5 are not congruent modulo 3. A number is even if it is congruent to **zero** modulo two. It is odd if it is congruent to one modulo two. The notation "x ≡ y mod(n)" means "x is congruent to y modulo n." The notation gcd(m, n) means the greatest common divisor (or **factor**) of m and n. Here are the basic facts about modular arithmetic:

- if x ≡ y mod(n) and z ≡ w mod(n), then x + z ≡ y + w mod(n) and xz ≡ yw mod(n).
- if xm ≡ ym mod (n) and gcd(m, n) = 1 then x ≡ y mod(n).

Here is the **proof** of the first fact. Since x ≡ y mod(n) and z ≡ w mod(n) there are integers j and k such that x - y = nj and z - w = nk. Then x + z - (y + w) = (x - y) + (z - w) = nj + nk = n(j + k). Thus x + z ≡ y + w mod(n). Also xz = (y + nj)(w + nk) = yw + n(jw + ky + jkn). Thus xz ≡ yw mod(n).

Here is the proof of the second fact. Since xm ≡ ym mod(n) there is an integer k such that xm - ym = kn. Since gcd(m, n) = 1, and m divides kn, m must divide k. So k/m is an integer and x - y = (k/m)n. So x ≡ y mod(n). On the other hand, if gcd(m, n) is not one then x does not have to be congruent to y. For example, suppose m = 2, n = 4, x = 1, and y = 3. Then, 2 is congruent to 6 modulo 4 but 1 is not congruent to 3 modulo 4.

Modular arithmetic is a fundamental tool used by mathematicians in practically every field but most commonly, it used by number theorists. Modern cryptography uses modular arithmetic extensively. Here is how public key cryptosystems work. First there is a method, known to everyone, for encoding messages into numbers or sequences of numbers. If, say, Jane wants to receive secret messages, then she publishes two numbers, e and N, say. If Bob wants to send Jane a message, then he first writes his message and his computer translates it into a number M. Then his computer raises M to the power of e and sends the remainder after dividing by N to Jane's computer. In other words, Bob's computer sends M^e modulo N. Jane has a number, that she keeps secret, called d. Her computer raises the number it received from Bob to the power of d and takes the remainder modulo N. The trick is that e, d, and N are specially chosen so that if M is any number, then $M^{ed} \equiv$ M mod(N). So, now Jane's computer can translate M back to Bob's message. This system works for two main reasons. First, a computer can quickly calculate M^e and M^{ed} modulo N. Second, it is currently very hard (even for a computer) to figure out what d is if you only know what e and N are and if they are large (100 digits or more). Most cryptosystems today use some variation of this method.

Here is an example illustrating how easy it is to calculate M^e modulo N. Suppose M is 7, e is 1000 and N is 10. 7^{1000} = $(7^2)^{500}$ = $(49)^{500} \equiv 9^{500}$ mod(10). $9^{500} = (9^2)^{250} = (81)^{250} \equiv 1^{250}$ mod (10). $1^{250} = 1$. So the answer is, $7^{1000} \equiv 1$ mod(10).

Gauss' favorite **number theory theorem**, the law of quadratic reciprocity, uses modular arithmetic. We say that x is a quadratic residue modulo N if $x \equiv y^2$ mod(N) for some integer y. The **Legendre symbol** (p/q) is defined to equal 1 if p is a quadratic residue modulo q, and it is equal to -1 otherwise. Gauss' "aureum theorema" is the following: If p and q are distinct odd primes, then $(p/q)(q/p) = (-1)^{(p-1)(q-1)/4}$. Although Euler stated this theorem in 1783, Gauss was the first to prove it in 1796. Since Gauss, more than fifty different proofs have been found.

Another theorem from number theory states that if p is an odd prime then there are integers x and y such that $p = x^2 + y^2$ if and only if p is congruent to 1 modulo 4. This is called the genus theorem.

See also Legendre symbol; Number theory

MODUS PONENS

Modus ponens (*MP*) is one of the simplest rules of **inference**. The term is Latin, and means "way of affirming."

Given a set of true **propositions**, we use rules of inference to find out what other propositions must also be true. For instance, when we assume the **truth** of **Euclid's axioms**, we can infer the truth of other geometrical propositions such as the **Pythagorean Theorem**.

MP tells us that if a conditional proposition (a proposition of the form "if A, then B"), and its antecedent (the statement preceded by "if" in the conditional proposition, i.e., A) are true, we can deduce the truth of its consequent (the statement followed by "then" in the conditional, i.e., B).

MP can be illustrated by a simple example. Consider the statements: "If Timothy is a cat, then Timothy is an animal,"

and "Timothy is a cat." By the rule of *modus ponens*, we can infer that Timothy is an animal.

Systems of logic (there are many different logics), like other areas of mathematics, are constructed by choosing a minimal number of simple, fundamental axioms whose truth is considered beyond question. However, given only axioms, one cannot derive any other true statements unless one also accepts the validity of a rule of inference as axiomatic. Using such a rule one can then derive other propositions from the axioms, and if the rule is followed accurately, these new statements are called theorems. Other rules of inference can then be derived and considered theorems of that logic. *Modus ponens* is fairly easy to understand, and seems intuitively to be "common sense." Therefore, when logicians construct systems of logic, they often choose *MP* as their first, unquestionable rule of inference, using it and their postulated axioms to derive theorems as well as other rules of inference such as transitivity or *modus tollens*.

MODUS TOLLENS

Modus tollens is a rule of logical **inference**. We use rules of inference when we argue from a set of premises to a conclusion, for example when we prove that the **Pythagorean Theorem** follows from **Euclid's Axioms**. Roughly, *modus tollens* means "way of removing," alluding to the fact that one of the relevant premises is negated or "removed."

Faced with **propositions** of certain forms, a rule of inference tells you how to logically combine them to derive valid conclusions. In particular, *modus tollens* tells you what propositions you can derive from a **conditional proposition** and the **negation of its consequent**. A conditional is a proposition of the form "if A then B." A conditional is made up of two parts, the **antecedent**, the part preceded by if, i.e., "A," and the **consequent**, the part followed by "then," i.e., "B." The negation of the consequent would be "not-B." By *modus tollens*, if a conditional is true and the negation of its consequent is true, then the negation of the antecedent must be true as well.

Modus tollens can be best illustrated by an example. Consider the conditional "If Timothy is a cat, then Timothy is an animal." Suppose you know that Timothy is not an animal. Then obviously, Timothy cannot possibly be a cat.

An informal **proof** would be to assume for the sake of argument that Timothy is a cat after all (this strategy is known as proof by contradiction). Then by the conditional, since Timothy is a cat, he must also be an animal. But we had already ruled out his membership in the animal kingdom! This is a contradiction. We can conclude that our assumption must be false: Timothy is not a cat. Essentially, *modus tollens* is a rule that we use as a shortcut for this pattern of reasoning.

One very common mistake is to confuse and scramble *modus tollens* and *modus ponens*. Quite often, one encounters an argument that starts with a conditional and its consequent, and concludes that the antecedent must be true. This is a grave error that can be avoided by keeping in mind another Timothy example. Suppose we have again the conditional "If Timothy is a cat, then Timothy is an animal." Suppose also that we know that the consequent is true: "Timothy is an animal." Absent any other information, it would be unreasonable to conclude that Timothy is a cat. Timothy could be anything: a dog, a starfish, or a hippopotamus. The conclusion "Timothy is a cat" is invalid as long as no further evidence is brought to bear. This erroneous pattern of reasoning is sometimes referred to as *modus moron*: "the way of the fool."

Systems of logic (there are many different logics), like other areas of mathematics, are constructed by choosing a minimal number of simple, fundamental axioms whose **truth** is considered beyond question. However, given only axioms, one cannot derive any other true statements unless one also accepts the validity of a rule of inference as axiomatic. Using such a rule one can then derive other propositions from the axioms, and if the rule is followed accurately, these new statements are called theorems. Other rules of inference can then be derived and considered theorems of that logic.

In principal, when logicians construct systems of logic, they could choose *modus tollens* as their first, unquestionable rule of inference, using it and their postulated axioms to derive theorems as well as other rules of inference such as transitivity or *modus ponens*. Usually, however, logicians use rules that are more immediately obvious such as *modus ponens* or proof by contradiction, and then derive *modus tollens*.

See also Modus Ponens

MONGE, GASPARD (1746-1818)

French geometer and educator

Considered the founder of descriptive and differential **geometry**, Gaspard Monge was one of the most famous mathematicians of his day and renowned for his application of geometry to problems of construction. A man of many interests, Monge also worked in chemistry and physics, as well as education, training a generation of mathematicians and paving the way for the extension of his theories to projective geometry. A fervent republican, Monge was a supporter of the French Revolution, and was for a time minister of the navy. He helped to establish the metric system and was a founder of the École Polytechnique, which he directed in its early years. Made a count in 1808, Monge is sometimes known as Comte de Peluse. His major work, *Geometrie descriptive*, leads directly to the methods employed in modern mechanical drawing known as orthographic **projection**.

Monge was born on May 9, 1746, in Beaune, France, to Jacques Monge and Jeanne Rousseaux. The elder Monge was a knife grinder and peddler with a belief in the power of education. Young Monge was considered the genius of the family and excelled at the college of the Orations in Beaune which he attended until 1762. From 1762 to 1764, he attended college in Lyons where, at age 16, he was made a physics instructor. On vacation in Beaune in 1762, Monge spent his free time by completing a map of the town, developing both the means of such large-scale projection as well as the surveying equipment

necessary for its completion. This work caught the eye of an officer of engineers who recommended Monge for a place at the military school at Mezieres, even though as the son of a commoner, he would never be eligible for an officer's commission.

Monge made an early name for himself by devising a plan for gun emplacements in a proposed fortress. He managed to substitute a geometrical process for such a calculation, avoiding the cumbersome **arithmetic** techniques then in use, and it was this breakthrough which in part led to his contributions in descriptive geometry. Initially, Monge's plan was ignored, as the officer in charge felt there was some trickery involved, though when finally his plan was examined, it was found to be of such value that he was pledged to secrecy. For the next fifteen years, Monge's method of descriptive geometry—representing three–dimensional objects in two dimensions—was kept under wraps as a military secret. Monge was made a professor at the École Royale du Genie in Mezieres, and from 1768 to 1783 taught both physics and mathematics, developing the field of geometry in service to the solution of construction and mechanical problems: everything from fortifications to scaffolding and general architecture.

Monge married Catherine Huart in 1777, and the couple eventually had three daughters. With his election to the French Académie Royale des Sciences in 1780, he began to spend more time in Paris, dividing his time between the capital city and Mezieres. He participated in research sponsored by the Académie and presented several papers there. In 1783, Monge was named examiner of naval cadets, a position which made it impossible for him to continue his professorship in Mezieres. For the next nine years he was busy with tours of inspection of the naval schools and his academic duties in Paris. The most fruitful years of research were behind him.

At the outbreak of revolution in 1789, Monge was one of the best-known French scientists. His support of the Revolution, however, was kept relatively quiet until 1792 with his membership in several revolutionary clubs. With the fall of the monarchy and the takeover of the Legislative Assembly, Monge was appointed minister of the navy, a position he held for a scant eight months. His politics were considered too moderate for the men in charge, and thereafter Monge's political activity was at a minimum. He did, however, continue to work for the democratic goals of the Revolution, even after the suppression of the Académie in 1793. Working with the Temporary Commission on Weights and Measures, Monge helped to develop the **metric system**. Lecturing at the École Normale in Paris in 1794, he was first allowed to teach his methods of descriptive geometry in public. Heeding an appeal for scientific men to come to the aid of French industry in support of the war effort, Monge even supervised foundries and wrote a factory handbook. Important for the future course of mathematics, Monge was influential in preparing the way for the École Polytechnique, which opened in 1795. His lectures in **infinitesimal** geometry held there were eventually printed first as *Feuilles d'analyse appliquee a la geometrie*, and later as *Application de l'analyse a la geometrie*. This text estab-

Gaspard Monge

lished the algebraic principles of three–dimensional geometry and helped to revolutionize engineering design.

In 1796, Monge left for what would be a protracted absence from France, first on a mission to Italy, and thereafter in service to Napoleon Bonaparte in Egypt, where he was assigned various technical and scientific tasks, including the establishment of the Institut d'Egypt in Cairo. With the defeat of Napoleon by the British fleet, Monge escaped back to France. During these years, he had continued his analytical researches, adding chapters to his *Application de l'analyse a la geometrie,* making it in effect a textbook of differential geometry with his introduction of the idea of lines of curvature in **space**.

Back in Paris, Monge resumed his duties at the École Polytechnique, and with the advent of the new century, honors and awards started coming his way. He was named a grand officer of the Legion of Honor in 1804, president of the Senate in 1806, and made a count in 1808. His duties as a senator accordingly took away from his work at the École Polytechnique, yet he was still able to see his monumental *Application de l'analyse a la geometrie* published in 1807. With the ultimate defeat of Napoleon and his abdication, Monge fled France for a time. Stripped of his honors and professional position by the restored Bourbon monarchy, Monge returned to Paris in 1816, and lived out the last two years of

his life reviled by many for his part in the Bonaparte regime. At his death in 1818, and despite government censure, many in the scientific community, both his students and colleagues, paid Monge respect by defying a government ban and placing a wreath on his grave the day after his funeral.

Monge is remembered primarily for his development of descriptive geometry, which he developed from 1766 to 1775, and its practical applications in the fields of construction and architecture. Though much of this was a codification of what others had done before, Monge made a system of such projective techniques. Such techniques paved the path ultimately for the development of projective geometry. Monge also pioneered techniques in analytic and infinitesimal geometry, a particular favorite of his, researching the properties of surfaces and of space curves. In particular, Monge is known for the application of the **calculus** to the examination of the curvature of surfaces. Additionally, Monge made important progress in the field of **differential equations**.

Less well known is Monge's work in **mechanics**, chemistry, and technology. In addition to books on the theory of machines, Monge also worked in caloric theory, acoustics, and optics. Perhaps his most famous work in chemistry dealt with the composition of water. In all, Monge's accomplishments in mathematical research and technology have won him a lasting position in the canon of mathematical greats. He was, as E. T. Bell described him in *Men of Mathematics,* "a born geometer and engineer with an unsurpassed gift for visualizing complicated space–relations."

MONTE CARLO METHOD

The Monte Carlo method, used to investigate a wide variety of problems, is a **stochastic** technique based on the **statistics** of random events applied to large numbers. Monte Carlo methods use **random numbers** and probability statistics to explain a variety of phenomena. Although there are several types of Monte Carlo methods, the simplest is the hit-and-miss integration method and is the one most often employed. Monte Carlo methods require only experience and this differentiates it from other stochastic methods because it requires no prior knowledge of the environment's dynamics, yet can still attain optimal behavior. This method is useful for obtaining solutions to problems that are too complicated to solve analytically. There are two types of Monte Carlo methods: the simple Monte Carlo method and the sophisticated Monte Carlo method. The first, simple Monte Carlo method, is a direct **modeling** of a random process. The second type, sophisticated Monte Carlo method, recasts deterministic problems in probabilistic terms.

By averaging sample returns the Monte Carlo methods provide ways to solve the reinforcement learning problem. Because of this Monte Carlo methods are described for episodic tasks only. That is, experience is divided into episodes and all episodes eventually terminate no matter what outcome may be. So rather than be incremental in a step-by-step fashion Monte Carlo methods require that the action is incremental in an episode-by-episode sense. It is a method based on determining results based on averaging complete returns rather than learning from partial returns. The Monte Carlo method is one of the most powerful techniques available in terms of the application to such a wide range of problems.

All Monte Carlo methods are comprised of several different major components. The physical or mathematical system must be described by a set of probability distribution **functions**. These functions enable the investigator to determine if a positive or negative result will occur for a particular input. A source of random numbers uniformly distributed on the **interval** of interest is required. A sampling rule that describes a prescription for sampling from the probability distribution functions on the interval of interest must be given. The outcome, whether it is positive or negative, must be accumulated and tallied. The variance as a function of the number of trials must be determined and a variance reduction technique must be available to reduce the computational time for the Monte Carlo simulation. Lastly, algorithms that allow Monte Carlo methods to be implemented on advanced computer systems can be helpful in analyzing the data.

As an example we will use the popular hit-and-miss integration method to calculate the value of π. If we draw a **circle** in a **square** and shade inside of the circle and imagine that we are throwing darts randomly at the figure that sometimes the darts hit inside of the square and sometimes inside of the circle. It should be apparent that of the total number of darts that hit within the square, the number of darts that hit within the shaded circle will be proportional to the **area** of the circle so that: (# of darts hitting shaded circle)/(# of darts hitting inside square) = (area of circle)/(area of square). From **geometry** we know that: (# of darts hitting shaded circle)/(# of darts hitting inside square) = $(\pi r^2)/r^2) = \pi$. If each dart that is thrown lands inside of the square then the ratio of dart hits in the shaded circle to throws will be equal to the value of π. In reality it actually takes a very large number of throws to get a decent value of π but there are methods to reduce the number of attempts and get a better estimate.

The Monte Carlo method is named after the city in the Monaco principality known for gambling. More precisely S. Ulam named it in 1946 after a roulette wheel, which is a simple random number generator. S. Ulam had a relative that had a propensity to gamble and so he named the method in the relative's honor. The well-formulated methods date back to the second world war when it was first employed during the Manhattan Project, where very complex **equations** had to be solved that could not be approached using traditional methods. There are, however, a number of isolated instances on much earlier occasions where similar methods were employed. In a paper of 1873, A. Hall describes an experimental determination of π which used methods similar to Monte Carlo methods. In 1899 Lord Rayleigh showed that without absorbing barriers a one-dimensional random walk could provide an approximate solution to a parabolic differential equation. Later, in 1931, Kolmogorov demonstrated a relationship between Markov stochastic processes and certain intergro-differential equa-

tions. Although crude Monte Carlo methods were employed in studies that predate that of the development of the atomic bomb the systematic development of Monte Carlo ideas emerged in 1948 when a group of scientists obtained Monte Carlo estimates for the eigenvalues of the Schrödinger equation. Today Monte Carlo methods are used in virtually all areas of science and engineering with problems too difficult to be addressed by other methods.

MOORE, ROBERT LEE (1882-1974)

American mathematician

Author of the influential 1932 book *The Foundations of Point Set Topology,* Robert Lee Moore was a pioneer in that area of mathematics. He was a lifelong mathematics teacher and professor, as well as a fixture in the American Mathematical Society (AMS).

Moore was born in Dallas, Texas on November 14, 1882. He began his college career at the University of Texas in 1898 and received a bachelor of sciences in 1901, after which he was appointed a fellow at the school. During the 1902-1903 school year, Moore taught high school mathematics in the town of Marshall, Texas, but later in 1903 he returned his attention to his academic career and entered the University of Chicago. He emerged with a PhD in 1905, having written his thesis on **sets** of metrical hypotheses for **geometry**.

From 1905 to 1906, Moore worked at the University of Tennessee as an assistant professor in the mathematics department, then moving to New Jersey's Princeton University to serve as an instructor. He remained at Princeton until 1908, when he accepted a position as instructor at Northwestern University. In 1911 the restless Moore relocated once again, this time to the University of Pennsylvania. He began there as an instructor, but in 1916 was promoted to assistant professor of mathematics. He stayed at that university until 1920. In the meantime, Moore published a paper entitled "On A Set of Postulates Which Suffice to Define a Number-Plane." Published in *Transactions of the American Mathematical Society* in 1915, Moore's work received positive attention among academic mathematicians.

Finally, in 1920 Moore came home again to Texas and his alma mater, taking up the position of assistant professor at the Austin campus. He received a promotion to full professor in 1923, working in that capacity until 1927. Beginning in 1929, however, Moore exchanged his professorship for the posts of Colloquium lecturer and editor of *Colloquium Publications,* which recorded the proceedings, findings, and communications of the AMS.

Moore spent only one year as Colloquium lecturer, then becoming visiting lecturer for the group from 1931 to 1932. In 1932 he published his *Foundations,* which was based on that year of lecturing. His main work with the AMS was as editor of *Colloquium Publications* from 1929 to 1936 and as its editor-in-chief from 1930 to 1933. Moore was also president of the AMS from 1936 to 1938.

Returning to the University of Texas in 1937 as distinguished professor of pure mathematics, Moore would spend an extended period (until 1953) in that post. The school appointed him professor of astronomy and professor of mathematics in 1953. He taught astronomy until 1959 and mathematics until 1970, when the school gave Moore emeritus status despite his protests; he had a strong desire to continue active teaching despite his advanced age.

Moore kept hours in his university office until shortly before his death in Austin, Texas on October 4, 1974.

See also Topology

MORGENSTERN, OSKAR (1902-1977)

German-American economist

Perhaps best known for his coauthorship with **John von Neumann** of *The Theory of Games and Economic Behavior,* Oskar Morgenstern was a longtime professor and prolific author of many other influential books on economics and **game theory**.

Morgenstern was born in Gôlitz, Silesia, Germany on January 24, 1902. He received his higher education at the University of Vienna, earning a doctorate in political science there in 1925. Morgenstern spent the following four years on an American fellowship, at the end of which he published *Economic Forecasting* (1928). He worked as a lecturer and professor in economics at his alma mater from 1929 to 1938, publishing his *Frontiers of Economic Policy* in 1934.

In 1938, Morgenstern traveled to the United States to assume a post as visiting lecturer in economics at Princeton University in New Jersey. He would remain in the country for the rest of his life, becoming a naturalized citizen in 1944. Morgenstern soon received a promotion to full professor at Princeton, where he began working with von Neumann. Their joint book, one of the first and most important on the complex science of game theory, appeared in 1944. A mathematical way to examine strategies, game theory concentrates on maximizing gains and minimizing losses within certain prescribed limitations. This book applied von Neumann's 1928 theory of games of strategy to competitive business situations.

Morgenstern continued to write on his own throughout the 1950s, publishing *On the Accuracy of Economic Observations* in 1950, *Prolegomena to a Theory of Organization* in 1951, and *The Question of National Defense and International Transactions and Business Cycles* in 1959. In 1960 he helped found Vienna's Institute for Advanced Studies and three years later published *Game Theory and Economic Science.*

After retiring from Princeton in 1970, Morgenstern accepted a professorship in economics at New York University, where he would remain until 1977. This decade was Morgenstern's busiest as an author; he cowrote *The Predictability of Stock Market Prices* (1970), *Long-Term Predictions of Power: Political, Economic, and Military*

Forecasting (1973), and *Mathematical Theory of Expanding and Contracting Economies* (1976).

Also in 1976, Morgenstern wrote a revealing paper, "Collaborating with von Neumann," in which he discussed his work with the man who built one of the world's first computers. New York University appointed Morgenstern its distinguished professor of game theory and mathematical economics in 1977. He died soon thereafter, on July 26, 1977 in Princeton. The economist had married in 1948.

MOTION

The process by which something moves from one position to another is referred to as motion; that is, a changing position involving time, velocity, and acceleration. Motions can be classified as linear or translational (motion along a straight line), rotational (motion about some axis), or curvilinear (a combination of linear and rotational). A detailed description of all aspects of motion is called kinematics and is a fundamental part of **mechanics**.

The kinematical description of motion really began with **Galileo**. From observations Galileo introduced two concepts: velocity as the time rate of change of position and acceleration as the time rate of change of velocity. With velocity, acceleration, time and **distance** traveled (change of position) the complete kinematical description of motion was possible. Four algebraic **equations** resulted, each involving three variables and an initial position or velocity.

The position of an object must be given, or implied, relative to a frame of reference and its motion is then described relative to this frame. Within this frame, position, change of position, velocity, and acceleration require a magnitude (how much) and a direction, both being equally important for a complete description. Physical concepts having this nature are called vectors in contrast to scalar concepts which require only a magnitude for their description (for example, time and mass). Saying the mall is a 5 mi (8 km) drive may be true but doesn't guarantee one will find the mall. However, specifying 5 mi (8 km) north would give the mall's precise location. Magnitude and direction are equally important.

In circular motion velocity is always parallel to the direction of motion and perpendicular to the radius of motion. The acceleration required to change the velocity's direction, called centripetal acceleration, is always perpendicular to the velocity and toward the center of motion. To change the velocity's magnitude an acceleration is required in the direction of the velocity. Hence, acceleration is required to change both magnitude and direction of velocity and are in different directions. This is applicable to curvilinear motion in general.

MULTIPLE INTEGRALS

Multiple integrals are a set of integrals that are taken over more than one **variable**. In general the more integrals that are taken the higher the **dimension** of **space** that is involved in the area's shape. For instance if two integrals comprise a multiple integral then the outcome corresponds to an **area**, three integrals correspond to a **volume**. A multiple integral generally appears as: $\iiint f(x, y, z)\, dx\, dy\, dz$. To compute a multiple integral one begins from the most inner integral and evaluates that one first, then proceeds to the next inner integral and so on. In the example above the most inner integral would treat x as the variable whilst y and z are taken as constants. The order in which one carries out computations can be varied but care must be exercised to correctly transform the **limits** to correspond to the proper integral. Just as with a single integral, a multiple integral is defined as the limit of a Riemann sum.

Multiple integrals arose when mathematicians confronted **functions** that depend on more than one variable. It was determined that in order to integrate with respect to two different variables that one simply integrated with respect to one of the variables first, then integrated the result with respect to the other variable. While integrating with respect to one variable the other variables are treated as constants. **Guido Fubini**, an Italian mathematician, concentrated his studies on expression of surface integrals in terms of two simple integrations. He is the author of Fubini's **theorem** which establishes a connection between multiple integrals and repeated ones, integrals that are several times over a single variable.

Multiple integrals can be integrated via a summation method or **numerical integration** methods. A double integral can be thought of as kind of a double sum. Integrating a function with respect to multiple variables involves integrating over a **range** of one variable and then summing the result of this integration over a range of the other variable. In this case, with two variables, the result is an area in a plane rather than just a section of one axis, which corresponds to a single integral. The trapezoidal approximation as well as other quadrature methods, such as Simpson's rule, can be applied to multiple integrals just as to single integrals. This is accomplished in the same way, i.e. first the numerical method is applied to the innermost integral and after evaluation it is applied to successive integrals working outward.

On occasions it is sometimes useful to use cylindrical and spherical coordinates when calculating multiple integrals. These substitutions often make evaluation of the results much simpler. In triple integrals in order to convert to cylindrical coordinates substitutions for x, y and z must be made: $x = r\cos\Theta$, $y = r\sin\Theta$, $z = z$.. After such substitutions have been made the volume element is given by: $dV = r\,dz\,dr\,d\Theta$. For converting a triple integral to spherical polar coordinates the following substitutions need to be made: $x = s\,\sin\Theta\cos\Phi$, $y = s\,\sin\Theta\sin\Phi$, $z = s\,\cos\Theta$ and the volume element is given by: $dV = s^2\,\sin\Theta\,ds\,d\Theta\,d\Phi$. These substitutions often make the evaluation of triple integrals generated by **rotation** of a curve around an axis or line much simpler.

Multiple integrals are found in several areas of science and engineering. They are used to determine the area of a two-dimensional region, volume, mass of two-dimensional plates, **force** on a two-dimensional plate, the average of a function, the **center of mass**, the moment of inertia, and surface area. Chemistry and physics readily employ multiple integrals in quantum **mechanics** as well as to determine optical cross sections.

MULTIPLICATION

Multiplication is a fundamental operation of **arithmetic** whose use stretches far back into antiquity. Ancient civilizations found the need to "tally up" the quantities of various goods that were gathered together to be stored, sold, or bartered. Hence the need arose for arithmetic. The most basic and earliest arithmetic operation used would have been for the combining of numbers (i.e., **addition**). Given large numbers of transactions, it should not be too difficult to imagine how early mankind developed a notation to indicate the repetitive addition of numbers, thereby devising multiplication. That is to say, multiplication was (and is) a kind of shorthand notation for the operation of addition. For instance, the multiplication of 6 by 5 indicates the successive addition of "6" five times (6 + 6 + 6 + 6 + 6); the resulting number "30" is called the product of "6 times 5." Over the course of time, multiplication tables and other techniques were devised to aid in finding the product of numbers. (Today, children still refer to multiplication cards or tables in order to memorize the products of various numbers.) The word multiply is from the Middle English "multiplien," derived from the Old French "multiplier," which in turn was based on the Latin "multiplicare." The first known appearance of the word multiply was in 1390 in the *Confessio ameantis n* by John Gower.

Various symbols are used to indicate the multiplication of two numbers. For numbers "a" and "b," "a × b," "a · b," "(a)(b)," and "ab" are equivalent ways of expressing the multiplication of "a times b." However, "×" is most commonly used in arithmetic expressions, while the other three multiplicative notations typically appear in higher mathematics, such as in algebraic expressions. In the expression "a × b," the numbers a and b are both called factors.

As mentioned above, multiplication is a fundamental operation of arithmetic (altogether there are four basic arithmetic operations: addition, **subtraction**, multiplication, and **division**). By calling multiplication an operation, it is meant that any two numbers multiplied together results in a third unique number. As pointed out earlier, the process of multiplication stretches back into antiquity. It would have originally been used for the counting numbers (i.e., the **whole numbers** {1, 2, 3, 4,...}). Over time, the use of multiplication was expanded to included **fractions**, **negative numbers**, **irrational numbers**, etc., on up to the so-called real number system. For all these number systems the following rules are valid for the operation of multiplication. (The laws are given in terms of **real numbers** "a, b, and c," but the laws are also valid for the number systems discussed previously, like the **rational numbers**. Also, a bracketed operation, like "(b × c)," is to be carried out before an unbracketed operation):

- Law 1: "a × b" is a unique real number, closure law;
- Law 2: "a × b = b × a," commutative law;

- Law 3: "a × (b × c) = (a × b) × c," associative law;
- Law 4: "a × 1 = a," identity law;
- Law 5: for every number "a" (except **zero**) there exists a corresponding and unique number "1 / a" such that "a × (1 / a) = 1," multiplicative inverse law; and
- Law 6: "a × (b + c) = (a × b) + (a × c)," distributive law of multiplication.

It should be noted that law 5 above does not apply for the counting or natural numbers nor for the set of **integers**, since by definition those number systems do not encompass reciprocals. By using the multiplicative laws above one can derive additional rules, or theorems, governing multiplication of the real numbers. For example, the rule of "equality for numbers" states that if "a = b" and "c = d," then "a × c = b × d" is valid. The (above) multiplicative laws and the "laws of equality" demonstrate the validity of this statement:

- Step 1: "a × c" is a real number, closure law;
- Step 2: "a × c = a × c," reflexive rule of equality (i.e., any number equals itself);
- Step 3: "a × c = b × c," substitution rule of equality (i.e., "b" substituted for "a");
- Step 4: "a × c = b × d," substitution rule of equality (i.e., "d" substituted for "c") [QED].

Multiplication of fractions is encountered in the study of arithmetic. Briefly, the product of two fractions is equal to the product of the two numerators over the product of the two denominators, and the resulting fraction is reduced to lowest terms, as in the following example: "(2 / 3) × (5 / 6) = (2 × 5) / (3 × 6) = 10 / 18 = 5 / 9."

Multiplication is also defined as an operation on **complex numbers**, vectors, **functions**, matrices, tensors, and so forth. For these mathematical objects the laws of multiplication given previously for the real numbers, rational numbers, etc., may or may not hold. By way of highlighting some of the deviations from the ordinary rules of multiplication, the multiplication of matrices is described below.

The multiplication of matrices is valid only if the order of **matrix** A is $(i \times j)$ and the order of matrix B is $(j \times k)$; that is multiplication is only valid if the column (j) of matrix A is equal in order to the row (j) of matrix B. In addition, the commutative law of multiplication is not valid for **matrix multiplication**. The product of matrix multiplication is "C = A × B" and the order of the resulting matrix C is of order $(i \times k)$ and its elements are $c_{ik} = (a_{ij}b_{jk})$, where j is summed over for all possible values of i and k.

The study of multiplication is taken to a deeper level in the branch of mathematics known as abstract **algebra**. Beginning in the nineteenth century, in an attempt to further the theory of solutions to algebraic **equations**, certain patterns or rules (like the multiplicative laws stated previously) were related to the operations of addition and multiplication as applied to the real numbers. These investigations eventually led to applications in disparate fields of mathematics (like algebra and **topology**), and became important in their own right. For example, it turns out that the non-zero rational, real and complex numbers form what are called infinite Abelian groups under the operation of multiplication. Similarly, the dual arithmetic operations of multiplication and addition form what are called fields upon various **sets**, such as the set of real numbers.

See also Addition; Algebra; Arithmetic; Complex numbers; Division; Products and quotients; Fractions; Field; Functions; Irrational numbers; Matrix; Matrix multiplication; Multiplication principle; Multiplication; Multiplicative notation; Negative numbers; Numbers and numerals; Rational numbers; Real numbers; Set theory; Subtraction; Topology; Vector analysis; Whole numbers

MULTIPLICATIVE NOTATION

Multiplication is one of the four basic operations of **arithmetic** (the others being **addition**, **subtraction**, and **division**). Multiplication operates on the set of **real numbers** such that for any real numbers multiplied together, a unique real number is determined. **Multiplicative notation** is defined as the system of symbols used in the operation of multiplication to represent numbers and actions. For the expression "2 × 3" the natural numbers "2" and "3" are multiplied together to produce a unique natural number, namely "6." In the expression "2 × 3" the operation of multiplication is denoted by "×," while the numbers "2" and "3" are called the "multiplicand" and "multiplier," respectively. Since multiplication is commutative on the real numbers (that is, the multiplicand and the multiplier can be interchanged) they are, also, both called factors.

Various symbols are used to indicate the multiplication of two numbers. The symbols "×" (lying cross), "·" (raised dot), and "*" (asterisk) are all used; so for numbers "a" and "b," "a × b," "a · b" and "a * b" are equivalent ways of expressing the multiplication of "a" times "b." Sometimes multiplication is represented with only adjacent parenthesis (i.e., by writing "(a)(b)," or simply "ab," to denote the multiplication of numbers a and b).

The use of particular symbols in mathematics is not arbitrary. Rather, the symbols in use today can be traced back along a definite historical path. In the next few paragraphs a brief background is given on the multiplicative symbols described above.

English mathematician William Oughtred (1574-1660), who gave free private lessons to pupils interested in mathematics, used the symbol "×" for times (multiplication). In 1631 he used the lying cross "×" as a symbol for multiplication in his book *Clavis Mathematicae* (Key to Mathematics). The symbol "×" appeared earlier in 1618 in an anonymous appendix to Edward Wright's translation of John Napier's *Descriptio*. However, Oughtred is believed to have written this appendix. The symbols were introduced as a way to make writing faster and easier, and to take up less written **space** for the new printing process of the times.

German mathematician **Gottfried Wilhelm von Leibniz** (1646-1716) advocated the raised dot, stating that the "×" looked too much like the unknown **variable** "x." In 1631

English mathematician Thomas Harriot (1560-1621) (posthumously) used the raised dot "·" in his book *Analyticae Praxis ad Aequationes Algebraicas Resolvendas* and in 1655 Thomas Gibson used the raised dot in *Syntaxis mathematica*. However, it is felt that neither author meant for these dots to represent multiplication. On the other hand, several research papers indicated that Harriot occasionally used the dot to denote multiplication, but admitted that its use was not accepted until Leibniz adopted it. Leibniz also used the cap symbol "∩" for multiplication. Today the cap symbol is used to indicate intersection in **set theory**. The asterisk "*" was used by Johann Rahn (1622-1676) in his 1659 book *Teutsche Algebra*.

To place variables side by side (sometimes called juxtaposition) in order to indicate multiplication was first found in a manuscript found buried near the village of Bakhshali, India. It is believed that the manuscript was written between the eighth and tenth centuries. Multiplication by juxtaposition was also found in fifteenth-century manuscripts, specifically by al-Qalasadi. In 1544 Michael Stifel (1487 or 1486-1567) used the notation in his *Arithmetica integra* and repeated the use in 1553. In 1637 the French mathematician **René Descartes** (1596-1650) also used juxtaposition.

The term **powers** is used with reference to the multiplication of equal terms. The nth power of "a," denoted as "a^n," is the product of "n" factors of "a." That is to say: "$a^2 = a \cdot a$" and "$a^5 = a \cdot a \cdot a \cdot a \cdot a$," and so forth. In the expression "a^n," "a" is called the basis and "n" the exponent. Moreover, when dealing with powers, both "a" and "n" customarily have integer values only. The physical sciences as well as mathematics make frequent use of powers. For instance, the **volume** (V) of a cube having sides of length "x" may be expressed as "$V = x \cdot x \cdot x$," or more compactly as the third power of x, "$V = x^3$." If x = 5 centimeters, then the cube's volume is the third power of 5, or "$V = 5^3 = 125$ cubic centimeters." As powers grow larger, the need for this compact notation becomes even more apparent.

The **factorial** of a positive integer "n" is another multiplicative notation. Factorial, represented as "n!," is the product of all positive **integers** from 1 to n. It is generically written out as "$n! = 1 \cdot 2 \cdot 3 \cdot ... \cdot n$," where by convention, "$0! = 1$." For example, "$5! = 1 \cdot 2 \cdot 3 \cdot 4 \cdot 5 = 120$."

Besides its use with respect to the real numbers, multiplicative notation is used to denote the products of **complex numbers**, vectors, **functions**, matrices, and tensors. The symbols described previously to denote multiplication of real numbers (i.e., lying cross, raised dot, asterisk, brackets, and juxtaposition) are also used to indicate multiplication of functions, complex numbers, etc. However, different symbols may denote different types of multiplication. A good illustration is vector multiplication. The lying cross, "×," is used to denote the "cross product" of two vectors, resulting in another vector (ie., $f \times g = h$). In contrast the dot symbol, "·," indicates the "dot product" of two vectors (ie., $f \cdot g = h$), which is a scalar quantity (i.e., a real number).

It is perhaps worth noting that multiplicative notation is encountered in both set theory and abstract **algebra**. These two branches of mathematics were created and extensively developed within the last two centuries, and have proven to be of tremendous use in unifying, heretofore, seemingly disparate mathematical disciplines. In set theory one encounters Cartesian products, denoted "$A \times B$," which is the set of all ordered pairs of **sets** A and B. Multiplicative notation is also used in the definitions of groups and fields, which are central concepts in abstract algebra.

See also Algebra; Addition; Arithmetic; Complex numbers; René Descartes; Division; Factorial; Functions; Integers; Gottfried Wilhelm von Leibniz; Matrix; Multiplication principle; Multiplication; Numbers and numerals; Powers; Products and quotients; Real numbers; Scalar and vector components; Set theory; Subtraction; Vector analysis

N

NAPIER, JOHN (1550-1617)

Scottish inventor of logarithms

John Napier is best remembered as the inventor of the first system of **logarithms**. A logarithm is the power to which a number, called the base, must be raised to produce a given number. Napier's work with logarithms was described in two Latin treatises, *Mirifici logarithmorum canonis descriptio* (1614; *A Description of the Wonderful Canon of Logarithms*) and *Mirifici logarithmorum canonis constructio* (1619; *The Construction of the Wonderful Canon of Logarithms*). This work was immediately recognized by other mathematicians as a great advance. Three hundred years later, in an essay marking the publication anniversary of the *Descriptio*, P. Hume Brown wrote that Napier's most notable achievement "has given him a high and permanent position in the history of European culture."

Napier, the eighth laird of Merchiston, was born in 1550 at Merchiston Castle near Edinburgh, Scotland. He came from a line of influential noblemen and statesmen. His father, Sir Archibald Napier, was a prominent public figure who was allied with the Protestant cause. Among other roles, Sir Archibald served for more than 30 years as Master of the Mint. Sir Archibald's first wife and Napier's mother, Janet Bothwell, was the daughter of an Edinburgh burgess.

Napier entered the University of St. Andrews at age 13, which was typical of wellborn boys of the time. However, he was a dropout, staying at the university for only a short while. It is likely that Napier then traveled to the European continent to continue his studies, although little is known of this period. By 1571, however, he had returned to Scotland and was living at Gartnes, where he built a castle. In 1572 Napier married Elizabeth Stirling. The couple had two children, Archibald and Joanne, before Elizabeth's death in 1579. Napier later married Agnes Chisholm, with whom he had ten children. These included a son named Robert, who eventually served as his father's literary executor. The family stayed in Gartnes until

1608, when Napier inherited Merchiston Castle upon Sir Archibald's death.

Napier lived an active life as a Scottish landowner. He was an amateur scientist who never held a professional post. Nevertheless, his varied accomplishments earned him the nickname of "Marvelous Merchiston." He experimented with fertilizers to improve his land, and he invented a hydraulic screw and revolving axle that could be used to remove water from flooded coal pits. Not surprisingly, Napier was also intrigued by the religious and political controversies of his day. A staunch Protestant, he published *A Plaine Discovery of the Whole Revelation of St. John* in 1593. This book, a virulent attack on Catholicism that concluded the Pope was the Antichrist, was read widely and translated into several languages.

Napier's preoccupation with defending his faith and country prompted him to design various weapons. These included burning mirrors for setting enemy ships on fire, an underwater craft, and a tank-like vehicle. He invested much time and money in these projects, even building prototypes in some cases. In one instance, Napier was said to have built and tested an experimental rapid-fire gun that he claimed would have killed 30,000 Turks without the loss of a single Christian.

Astronomy was another of Napier's passions, and it was this pursuit that led to his greatest discovery. Napier's astronomy research required him to do a number of tedious calculations involving trigonometric **functions**. Over the course of more than two decades, he gradually developed and refined new ideas for speeding such calculations through the use of logarithms. The *Descriptio* briefly explained his invention and presented the first logarithmic tables. The *Constructio*, published after his death, described how the tables had been computed.

The *Descriptio* attracted the attention of English mathematician Henry Briggs, who traveled to Edinburgh to discuss the new tables. Napier and Briggs worked on such improvements as using the base 10. The result was Briggs's development of a standard form of logarithmic table that remained in common use until the advent of calculators and computers. Thanks to his efforts, logarithms were quickly adopted by mathematicians throughout Europe.

Napier made other advances in spherical **trigonometry**. The so-called "Napier's analogies" were formulas for solving spherical triangles. "Napier's rules of circular parts" were ingenious rules for stating the interrelationships of the parts of a right spherical **triangle**. In addition, Napier pioneered the use of the decimal point to separate the whole number part of a number from its fractional part. As one of the first to use **decimal fractions**, he helped to popularize them.

Napier's contributions to mathematics did not end there, however. His concern with simplifying calculations led him to invent mechanical aids for doing **arithmetic**, described in *Rabdologia*(1617). Among these aids were rods with numbers marked off on them, often known as "Napier's bones" because they were usually made of bone or ivory. Using the rods, multiplication became a process of reading the appropriate figures and making minor adjustments. In addition, Napier invented other types of rods for extracting **square roots** and cube roots.

It is not clear when Napier first began to dabble in mathematics. However, some early writings, dealing mainly with arithmetic and **algebra**, seem to date from the period just after his first marriage. These writings were collected and transcribed after Napier's death by his son Robert. They were first published in 1839 by descendant Mark Napier under the title *De arte logistica*, revealing the keenness of Napier's mathematical ruminations. Among other subjects, it appears that he investigated imaginary roots of **equations**.

During his lifetime, Napier was reputed to be not only a mathematician, but a magician as well. It was rumored that he possessed supernatural powers, and that he owned a black rooster as a spiritual familiar. Given his wide-ranging interests, Napier was almost never idle. A combination of overwork and gout finally led to his death at Merchiston on April 4, 1617. His burial place is uncertain, but it is probably at the old church of St. Cuthbert's parish in Edinburgh.

In 1914, the 300th anniversary of the publication of the *Descriptio* was commemorated by the Royal Society of Edinburgh. In his inaugural address, Lord Moulton lauded Napier as one who "stands prominent among that small band of thinkers who by their discoveries have substantially increased the powers of the human mind as a practical agent." In 1964 Napier University, named for the mathematician, was founded in Edinburgh. Among its campuses is one at Merchiston, which houses courses in science, technology, and design.

N-BODY GRAVITATIONAL PROBLEM

The *n*-body gravitational problem is the mathematical formulation of a question that has fascinated and puzzled astronomers for millennia: How can we predict the **motion** of the moon and the planets?

According to **Isaac Newton**'s theory of gravitation, any two objects (or "bodies") attract each other by a gravitational

force inversely proportional to the **square** of their **distance** from each other. Furthermore, Newton's second law of motion states that the acceleration of each body is proportional to the sum of the forces on it from each of the other bodies. Thus, if a certain number (*n*) of bodies are released at a given time, the acceleration of each one can be computed precisely from the position of the others. Their motion is completely determined, for all time, by their initial positions and velocities.

Newton's theory would appear to be the end of the *n*-body gravitational problem, but in fact it was just the beginning. One reason is that Newton's formulation was highly idealized: It assumed all the objects are "point masses." (That is, they occupy only a single point in space.) Thus it omitted tidal effects, which can be quite important in the actual solar system. Newton also, of course, ignored the effects of relativity theory, which were discovered by **Albert Einstein** over 200 years later.

But these are physical objections. There were also pressing mathematical problems, namely that the theory gave a differential equation that the motion of the *n* bodies must obey, but did not say how to solve that differential equation. The reputation of Newton's theory was cemented by the fact that he did succeed in solving the equation *exactly* for the case of two bodies: The two bodies travel in ellipses. Philosophically, this means that the orbits are stable for all time. It also means the two bodies can never collide unless they *start out* on a collision course, which can be considered "infinitely unlikely." (Imagine trying to fire two bullets so that they collide in midair.)

Newton was unable to perform the same feat for three bodies, even the three bodies of most consequence to astronomers: the earth, the moon, and the sun. Even today, only a very few special configurations of three bodies can be solved exactly, most notably a rotating equilateral **triangle** with the sun at one vertex, the earth at another, and a very small object, such as a satellite, at the third vertex, called a Lagrange point. Such a configuration has actually been used for some **space** missions.

Lacking exact solutions, mathematicians have made much more progress on the qualitative, or philosophical, questions related to the *n*-body problem: Is the motion stable? Is it predictable, in the sense that small errors in observation of a planet will only lead to small errors in the prediction of its motion? How often do collisions occur? Are other "singularities" possible, such as the ejection of a planet to an infinite distance away in a finite amount of time?

At the risk of oversimplifying a beautiful and deep theory, here are the major results on these questions, along with the names of the mathematicians who proved them. A very small perturbation of an exactly solvable system (such as the perturbation of one planet's motion by another) is likely—but not certain—to be stable for all time (Kolmogorov, Arnold, Moser). Collisions continue to be "infinitely unlikely" (Saari); however, bear in mind here that we are talking about systems of point masses, since collisions *do* occur in the real solar system. Ejection singularities are not possible for the three-body problem, but are possible for five or more bodies

(Painlevé, Xia, Gerver). Perhaps most importantly, the solutions to the *n*-body problem are usually *not* predictable over the long term but are chaotic (Poincaré, Birkhoff, Smale). The last **theorem**, in particular, explains why it is futile to hope for an exact general solution of the three (or more)-body problem. We can only settle for short-term approximations by computer, such as those NASA uses when planning space missions, or long-term solutions in special cases, such as the Lagrange configuration.

See also Chaos theory; Differential equations; Kepler's laws

NECESSARY CONDITION

The **conditional** is one of the most common types of logical and mathematical **propositions**. The generic conditional takes the form: "If *p* is true, then *q* is true, too." The first part of the conditional (the "if-clause") is known as the **antecedent**, the second part as the **consequent**. The term conditional is used because the antecedent describes a condition for the consequent. Under the condition that *p* is the case, *q* is the case, also.

We can distinguish between two types of conditions: **necessary** and **sufficient**. The idea is fairly simple: A necessary condition is a condition without which the consequent cannot possibly be true. A sufficient condition guarantees by itself, without the support of any other conditions, that the consequent is true.

The principle can be illustrated with easy examples. Suppose we know an animal, Suzie, and we want to figure out if Suzie is a snail. The fact that she is an animal is a necessary condition for her being a snail. If she weren't an animal, she could not be a snail. Suppose on the other hand that we know from the start that Suzie is a snail. Then we immediately know that Suzie is an animal. Her being a snail is a sufficient to make her an animal as well; we do not need to find out any other information about Suzie to judge if she is an animal.

A necessary condition need not be sufficient. By itself, the fact that Suzie is an animal does not tell us that she is a snail—we need to know more about her to draw that conclusion. Likewise, a sufficient condition need not be necessary. If we knew that Suzie was not a snail after all, she still could be an animal. She could be an aardvark or a muskrat or a gnu. But a condition can be both necessary and sufficient, as is the case for definitions. Being unmarried is both a necessary and sufficient condition for being a bachelor.

In a simple conditional, "if *p* then *q*," the antecedent, *p*, is a sufficient condition for *q*. That is what the conditional means: the fact that *p* is the case guarantees that *q* is the case. Conversely, *q* is a necessary condition for *p*. Why is this so? Suppose *q* is false. If *p* were true, then by the conditional *q* would be true, too. But this would contradict the hypothesis that *q* is false. Therefore, *p* cannot be true. In other words, if *q* is false, then *p* is false, too. That means *q* is a necessary condition for *p*.

NEGATIVE

Negative is a term in mathematics that usually means "opposite." An electron's charge is called negative not because it is "below" but because it is opposite that of a proton. A surface with negative curvature bulges in from the point of view of someone on one side of the surface but bulges out from the point of view of someone on the other side. A line with negative **slope** is downhill for someone moving to the right but uphill for someone moving to the left.

The term negative is most commonly applied to numbers. When negative is an adjective applied to a number or integer, the reference is to the opposite of a positive number. As a noun, negative is the opposite of any given number. Thus, -4, -3/5, and $-\sqrt{2}$ are all negative numbers, but the negative of -4 is +4. The **integers**, for example, are often defined as the natural numbers plus their negatives plus **zero**. Sometimes the word opposite is used to mean the same as the noun negative.

Technically, negative numbers are the opposites with respect to **addition**. If *a* is a positive number then -*a* is a negative number because: a + (-a) = 0.

Allowing numbers and other mathematical elements to be negative as well as positive greatly expands the generality and usefulness of the mathematical systems of which they are a part. For example, if one owes a credit card company $150 and mistakenly sends $160 in payment, the company automatically subtracts the payment from the balance due, leaving-$10 as the balance due. It does not have to set up a separate column in its ledger or on its statements. A balance due of -$10 is mathematically equivalent to a credit of $10.

When the Fahrenheit temperature scale was developed, the starting point was chosen to be the coldest temperature which, at that time, could be achieved in the laboratory. This was the temperature of a mixture of equal weights of ice and salt. Because the scale could be extended downward through the use of negative numbers, it could be used to measure temperatures all the way down to absolute zero.

The idea of negative numbers is readily grasped, even by young children. They usually do not raise objection to extending a number line beyond zero. They play games that can leave a player "in the hole." Nevertheless, for centuries European mathematicians resisted using negative numbers. If solving an equation led to a negative root, it would be dismissed as without meaning.

In other parts of the world, however, negative numbers were used. The Chinese used two abaci, a black one for positive numbers and a red one for negative numbers, as early as two thousand years ago. **Brahmagupta**, the Indian mathematician who lived in the seventh century, not only acknowledged negative **roots** of quadratic **equations**, he gave rules for multiplying various combinations of positive and negative numbers. It was several centuries before European mathematicians became aware of the work of Brahmagupta and others and began to treat negative numbers as meaningful.

Negative numbers can be symbolized in several ways. The most common is to use a minus sign in front of the number. Occasionally the minus sign is placed behind the number, or the number is enclosed in parentheses. Children, playing a game, will often draw a **circle** around a number which is "in the hole." When a minus sign appears in front of a letter representing a number, as in -x, the number may be positive or negative depending on the value of x itself. To guarantee that a number is positive, one can put absolute value signs around it, for example |-x|. The absolute value sign can also guarantee a negative value, which is -|x|.

NEGLIGIBLE TERMS

Elements that comprise a sequence, series, or equation are known as terms, of which there may be a finite or infinite number. Negligible terms are those terms in a sequence, series, or equation that can be largely ignored, primarily because they hold such small values as to not affect the end result in a meaningful way. Thus, they are termed *negligible*, or unimportant, in relation to the task at hand. Use of the concept of negligible terms is found whenever one uses numerical methods to approximate a solution. It is applied frequently in applied and computational mathematics.

Most commonly, negligible terms are discarded when one is attempting to approximate a value within a specified error. The specified error may represent either a finite number—such as approximating **pi** within 1/1000th of its actual value—or an order of magnitude, such as using a Taylor series to approximate sin x around a particular point to $O(x^5)$, that is, approximating the value of the function out to the fifth-power term. In each case, the maximum allowable error represents a bound; the exact value of the function lies somewhere in the closed **interval** that is bounded by sum of the values of all *previous* terms, plus and minus the value of the *last* term to be considered. Obviously, higher-order terms can be discarded only when the method for obtaining the solution converges. If the methodology applied to obtain a solution does *not* converge, the higher-order terms are *not* negligible and therefore cannot be ignored or discarded.

The concept of negligible terms is extremely useful in math, physics, engineering, and even business. Its use is widespread. For example, the concept enables us to use computers and calculators to make approximations of functions—such as natural and base 10 **logarithms** (ln x/log x), exponentials (e^x), trigonometric (sin x) and others—with great speed and accuracy. In fact, the average **calculator** is unlikely to maintain a table of values for all possible natural logarithms (such as ln 2.7132) in its memory. Maintaining such a table would simply require too much memory storage, making it too bulky to carry and too expensive to produce. Instead, a calculator will be programmed with an **algorithm**, possibly based on Taylor's series or iterative process, to calculate the natural logarithm of any value specified by the user. This calculation will be performed to a level of precision specified by the manufacturer and thus provides a means of comparison between various makes and models. Once the desired level of **convergence** (and thus precision) is achieved, the algorithm is stopped, and

the approximate value is given to the user. The value, then, is not calculated exactly, but a useful approximation is made.

Most students encounter the concept of negligible terms in early mathematics education when they are taught to round decimal numbers to a particular decimal place. The decimal approximation of any rational or irrational number, for example, can be considered a linear combination of **decimal fractions**. The decimal approximation of pi, 3.14159..., for example, is a linear combination of the decimal **fractions** $3*1$ + 1/10 + 4/100 + 1/1000 + 5/10,000 + 9/100,000.... When estimating pi to two decimal places, only the first four decimal fractions listed above are used, resulting in 3.14. All other terms of the decimal fraction expansion do not affect the final value and are ignored; they are negligible terms. (Note that 1/1000 is *not* negligible because it determines whether pi is approximated as 3.14 or 3.15). If pi were approximated to an additional decimal place, the first *five* decimal fractions listed would be used; the decimal fraction 9/100,000 and all subsequent terms are negligible and therefore ignored. The concept of ignoring negligible terms can be similarly applied any algorithmic process, such as truncating numbers or calculating **square roots**. It is a powerful concept that makes many computations practical.

See also Applied mathematics; Convergence; Equations; Fibonacci and related series; Finite sequences and series; Geometric series; Infinite series; Mathematics and computers; Numerical analysis; Power series; Sequences and series; Taylor's and Maclaurin's series

NEWTON, ISAAC (1643-1727)

English mathematician and physicist

Isaac Newton was one of the leading mathematical and scientific geniuses of the 17th and 18th centuries, best known for his far-reaching discoveries in mathematics, physics, and optics. Among Newton's many achievements was his invention of **calculus** (the German philosopher and mathematician **Gottfried Wilhelm Leibniz** would independently invent it a few years later), the formulation of the three laws of **motion** and the universal theory of gravitation, and he proved that sunlight is the combination of many colors. Newton also served as master of the Royal Mint in London and president of the Royal Society of London for many years.

Newton was born on January 4, 1643. He was so small and sickly at birth that two women who were sent into town to get supplies for him delayed coming back because they thought he would not live more than a few hours. His father, who died a few months before his son was born, was also named Isaac. His mother, Hannah Ayscough, remarried three years after her son's birth to Reverend Barnabas Smith, and they had three children.

After his mother left to live with Reverend Smith, Newton stayed behind on the family farm with his maternal grandmother. His was a solitary childhood, fostering a habit that persisted for the rest of his life. When Newton was 12, he went to the grammar school at Grantham. There he studied Latin, and learned it so well that later he could write it as fluently as English. Early in his stay at Grantham, Newton was involved in a prophetic incident. A boy kicked him in the stomach on the way to school one day. Newton challenged him to a fight after school, and even though he was much smaller physically, Newton won the fight, and he rubbed his opponent's nose on the church wall. Later in his life, Newton was known to be easily provoked to retaliation, and he was usually very determined, nasty, and successful against his opponents.

When not in school, Newton spent his time at Grantham making sundials, drawings, and wooden models. He constructed a **model** windmill which included a mouse on a treadmill to supply power. He created a four–wheeled cart for himself, which he powered by turning a crank he had installed. Sometimes Newton spent so much time on these inventions that he fell behind at school. When that happened, he applied himself to his studies and regained his high academic standing in his class.

When Newton was almost 17 years old, his mother called him home to learn to manage the family farm. He was a disaster as a farmer. Newton was more interested in building models and reading, and he had no patience for watching the livestock or anything to do with farming. His schoolmaster at Grantham and his uncle William Ayscough had noticed his brilliance and successfully persuaded his mother to let him finish grammar school.

In June 1661, Newton was admitted to Cambridge University. He found the traditional Aristotelian curriculum at Cambridge so unrewarding that he never finished any of the assigned books, but he did read prodigiously about philosophy, science, and mathematics. Newton had learned little about mathematics at Grantham, but at Cambridge he devoured books by **Euclid, René Descartes, Galileo**, and **Johannes Kepler** and took copious notes. When Newton was working on an interesting problem, he completely forgot about eating or sleeping. He worked so hard on **mathematics and physics** that within a year he began to record original insights in his notebooks. In 1665, Newton received his bachelor's degree. In the summer of 1665, the plague came to Cambridge, and the university shut down until the spring of 1667. Newton went home and spent his next two years laying the foundations for calculus, which he called the "fluxional method."

Calculus is a mathematical tool that allows people to solve problems about the **limits** and rates of changes in dynamic relationships among such factors as time, velocity, position, **force**, **distance**, and so on. If one **factor** is a function of the other, (i.e., dependent upon it—the faster you go, the more distance you cover), then information about an unknown factor can be determined if the other factor is known. Since many factors in human experience are related, calculus is crucial in many specialities, for example, civil engineering (e.g., what is the maximum stress a building can tolerate if it is made of x materials and is subjected to hurricane winds of y miles per hour?), celestial navigation (e.g., where will a satellite be at a certain time if it is traveling at x speed at y latitude?), and economics (e.g., what will a corporation's profit be if its costs

go up at rate x, its sales go up at rate y, and inflation goes down at rate z?).

No one knew that Newton had developed calculus at home in 1665 and 1666. In 1669, however, he made some of his work known to a few people, including **Isaac Barrow**, the Lucasian professor of mathematics at Cambridge. Barrow was impressed. Newton received his master's degree in 1668, and when Barrow resigned his professorship in 1669, Newton took his place. The Lucasian professor gave a series of lectures every term, but Newton's lectures were so difficult that few or no students attended them. When no one attended, he would lecture for fifteen minutes to the empty room and leave.

Newton's work in 1669 and the early 1770s was on optics and a theory of colors. To aid in his studies, Newton built a reflecting telescope which magnified 40 times. The Royal Society was so impressed by its quality that they elected him a member in January 1672. In February 1672, the Royal Society published an article by Newton on his theory of colors. **Robert Hooke**, another scientist, challenged Newton in a very superior tone. Newton's response to Hooke was vicious. Newton had thought about colors for years, had conducted many experiments, and knew his subject well. He had no patience for Hooke's tone or ignorance. Newton's disputes with Hooke and others made him withdraw into silence, but while he was silent, Leibniz was publishing his own work on calculus.

In 1684, Newton was stirred to more work on mathematics and physics by **Edmund Halley**. In 1687, Newton published his *Philosophiae Naturalis Principia Mathematica*, which formulated the three laws of motion and the universal principle of gravitation. It remains today one of the most famous works of western science.

Newton's most famous quarrel was with Gottfried Wilhelm Leibniz over who first invented calculus. In 1684, Leibniz published a version of calculus which had a superior system of notation, making it easier to use. When Newton's *Principia* and other works by English mathematicians were published after Leibniz's calculus, mathematicians throughout Europe assumed that Newton had taken his method from Leibniz.

Newton, however, had invented calculus in 1665 and 1666, but did not publish his work for years after Leibniz did. Leibniz received letters from Newton in 1671 and 1676 discussing mathematics, and he also saw one of Newton's unpublished manuscripts in London in 1676 and took many notes on it. The issue of who first invented calculus came to a head in 1699 when Leibniz was indirectly accused of plagiarizing Newton. There followed years of accusations on both sides, which diminished after Leibniz died in 1716. Many scholars today believe that Leibniz did not plagiarize Newton and that both men discovered calculus separately, though Newton discovered it earlier. Leibniz's version, however, was easier to use and consequently caught on faster than Newton's.

Tired of academia, Newton accepted an offer to become warden of the Royal Mint in 1696. He was so well organized that he improved the Mint's operations enormously. Newton was promoted to Master of the Mint in 1700 and became wealthy because he earned a commission on the coins minted.

Newton was elected president of the Royal Society from 1703 until his death. In 1705, Newton was knighted by Queen Anne. In the years before his death, he suffered from a number of physical ailments. Newton died in London on March 20, 1727.

NEWTON'S LAWS OF MOTION

Earthly and heavenly motions were of great interest to Newton. Applying an acute sense for asking the right questions with reasoning, Newton formulated three laws which allowed a complete analysis (mathematical) of dynamics, relating all aspects of **motion** to basic causes, **force** and mass. So great was Newton's work that it is referred to as the first revolution in physics.

First law of motion

Galileo's observation that without friction a body would tend to move forever challenged Aristotle's notion that the natural state of motion on Earth was one of rest. **Galileo** deduced that it was a property of matter to maintain its state of motion, a property he called inertia. Newton, grasping the meaning of inertia and recognizing that **Aristotle's** reference to what keeps a body in motion (outside influence) really should have been what changes a body's state of motion, set forth a first law of motion which states: A body at rest remains at rest or a body in constant motion remains in constant motion along a straight line unless acted on by an external influence, called force.

Examples of the first law

(1.) Why use seat belts? Riding in a car you and the car have the same motion. When the brakes are applied, the brakes stop the car. What stops you? Eventually the steering wheel, the dashboard, or the window unless they are replaced by a seat belt, which stops your body. When the accelerator is depressed with the car in gear the motor turns the wheels and the car moves forward. What moves you forward? As the car moves forward the seatback comes forward, contacts, you and pushes you forward.

(2.) While you are riding in the front passenger seat of a car, the driver suddenly turns left. What about you? You continue to move in a straight line until the door to your right, turning left, eventually runs into you. In the car it may appear to you that you slid outward and hit the door.

Second law of motion

The first law of motion concentrates on a state of constant motion but adds unless an outside influence, force, acts on it. Force produces a change in the state of motion (velocity describes a body's motion); that is, an acceleration. Newton found that the greater a body's mass the greater the force required to overcome its inertia and mass is taken as a quanti-

tative measure of a body's inertia. He also found that applying equal force to two different masses, the ratio of their accelerations was inversely proportional to the ratio of their masses. Newton's Second Law of Motion is thus stated: A net force acting on a body produces an acceleration; the acceleration is inversely proportional to its mass and directly proportional to the net force and in the same direction.

This law can be put mathematically $F = ma$ where F is the net force, m is the mass, and a the acceleration. The second law is a cause-effect relationship. The net force acting on a body is determined from all forces acting and the resultant acceleration calculated (assuming a known mass). From the acceleration, velocity and **distance** traveled can be determined for any time.

Applications of the second law

(1.) Objects, when released, fall to the ground due to the earth's attraction. Newton's Universal Law of Gravity gave the force of attraction between two masses, m and M, as $F = GmM/R^2$ where G is the gravitational constant and R is the distance between mass centers. This force, weight, produces gravitational acceleration g, thus weight = GmM/R^2 = mg(2nd Law) giving g = GM/R^2. This relationship holds universally. For all objects at the earth's surface, g =32 ft/sec/sec or 9.0 m/sec/sec downward and on Jupiter 84 ft/sec/sec. Since the dropped object's mass does not appear, g is the same for all objects. Falling objects have their velocity changed downward at the rate of 32 ft/sec each second on earth. Falling from rest, at the end of one second the velocity is 32 ft/sec, after 2 seconds 64 ft/sec, after 3 seconds 96 ft/sec, etc.

For objects thrown upward, gravitational acceleration is still 32 ft/sec/sec downward. A ball thrown upward with an initial velocity of 80 ft/sec has a velocity after one second of 80-32= 48 ft/sec, after two seconds 48-32= 16 ft/sec, and after three seconds 16-32= -16 ft/sec (now downward), etc. At 2.5 seconds the ball had a **zero** velocity and after another 2.5 seconds it hits the ground with a velocity of 80 ft/sec downward. The up and down motion is symmetrical.

(2.) Friction, a force acting between two bodies in contact, is parallel to the surface and opposite the motion (or tendency to move). By the second law, giving a mass of one kilogram(kg) an acceleration of 1 m/sec/sec requires a force of one Newton(N). However, if friction were 3 N, a force of 4 N must be applied to give the same acceleration. The net force is 4N (applied by someone) minus 3N (friction) or 1N.

Free fall, example (1), assumed no friction. If there were atmospheric friction it would be directed upward since friction always opposes the motion. Air friction is proportional to the velocity; as the velocity increases the friction force (upward) becomes larger. The net force (weight minus friction) and the acceleration are less than due to gravity alone. Therefore, the velocity increases less rapidly, becoming constant when the friction force equals the weight of the falling object (net force=0). This velocity is called the terminal velocity. A greater weight requires a longer time for air friction to equal the weight, resulting in a larger terminal velocity.

(3.) A contemporary and friend of Newton, Halley, observed a comet in 1682 and suspected others had observed it many times before. Using Newton's new **mechanics** (laws of motion and Universal Law of Gravity) Halley calculated that the comet would reappear at Christmas, 1758. Although Halley was dead, the comet reappeared at that time and became known as Halley's comet. This was a great triumph for Newtonian mechanics.

Using Newton's Universal Law of Gravity (see example 1) in the second law results in a general solution (requiring **calculus**) in which details of the paths of motion (velocity, acceleration, period) are given in terms of G, M, and distance of separation.

While these results agreed with planetary motion known at the time there was now an explanation for differences in motions. These solutions were equally valid for applying to any system's body: earth's moon, Jupiter's moons, galaxies, truly universal.

(4.) Much of Newton's work involved rotational motion, particularly circular motion. The velocity's direction constantly changes, requiring a centripetal acceleration. This centripetal acceleration requires a net force, the centripetal force, acting toward the center of motion. Centripetal acceleration is given by $a(central) = v^2/R$ where v is the velocity's magnitude and R is the radius of the motion. Hence, the centripetal force $F(central) = mv^2/R$, where m is the mass. These relationships hold for any case of circular motion and furnish the basis for "thrills" experienced on many amusement park rides such as ferris wheels, loop-the-loops, merry-go-rounds, and any other means for changing your direction rather suddenly. Some particular examples follow.

(a.) Newton asked himself why the moon did not fall to the earth like other objects. Falling with the same acceleration of gravity as bodies at the earth's surface, it would have hit the earth. With essentially uniform circular motion about the earth, the moon's centripetal acceleration and force must be due to earth's gravity. With gravitational force providing centripetal force, the centripetal acceleration is $a(central) = GM/R^2$ (acceleration of gravity in example 1 above). Since the moon is about 60 times further from the earth's center than the earth's surface, the acceleration of gravity of the moon is about .009 ft/sec/sec. In one second the moon would fall about .06 inch but while doing this it is also moving away from the earth with the result that at the end of one second the moon is at the same distance from the earth, R.

(b.) With the gravitational force responsible for centripetal acceleration, equating the two acceleration expressions given above gives the magnitude of the velocity as $v^2 = GM/R$ with the same symbol meanings. For the moon in (a) its velocity would be about 2,250 MPH. This relationship can be universally applied.

Long ago it was recognized that this analysis could be applied to artificial "moons" or satellites. If a satellite could be made to encircle the earth at about 200 mi it must be given a tangential velocity of about 18,000 mph and it would encircle the earth every 90 seconds; astronauts have done this many times since 1956, 300 years after Newton gave the means for predicting the necessary velocities.

It was asked: what velocity and height must a satellite have so that it remains stationary above the same point on the earth's surface; that is, have the same **rotation** period, one day, as the earth. Three such satellites, placed 120 degrees apart around the earth, could make instantaneous communication with all points on the earth's surface possible. From the fact that period squared is proportional to the cube of the radius and the above periods of the moon and satellite and the moon's distance, it is found that the communication satellite would have to be located 26,000 mi from earth's center or 22,000 mi above the surface. Its velocity must be about 6,800 mph. Many such satellites are now in **space** around the earth.

(c.) When a car rounds a curve what keeps it on the road? Going around a curve requires a centripetal force to furnish the centripetal acceleration, changing its direction. If there is not, the car continues in a straight line (first law) moving outward relative to the road. Friction, opposing outward motion, would be inward and the inward acceleration a(cent) $= friction/mass = v^2/R$. Each radius has a predictable velocity for which the car can make the curve. A caution must be added: when it is raining, friction is reduced and a lower velocity is needed to make the curve safely.

Third law of motion or law of action-reaction

Newton questioned the interacting force an outside agent exerted on another to change its state of motion. He concluded that this interaction was mutual so that when you exert a force on something you get the feeling the other is exerting a force on you. Newton's Third Law of Motion states: When one body exerts a force on a second body, the second body exerts an equal and opposite force on the first body.

In the second law, only forces exerted on a body are important in determining its acceleration. The third law speaks about a pair of forces equal in magnitude and opposite in direction which are exerted on and by two different bodies. This law is useful in determining forces acting on an object by knowing forces it exerts. For example, a book sitting on a table has a net force of zero. Therefore, an upward force equal to its weight must be exerted by the table on the book. According to the third law the book exerts an equal force downward on the table. When two objects are in contact they exert equal and opposite forces on each other and these forces are perpendicular to the contacting surface.

Examples of the third law

(1.) What enables us to walk? To move forward parallel to the floor we must push backward on the floor with one foot. By the third law, the floor pushes forward, moving us forward. Then the process is repeated with the other foot, etc. This cannot occur unless there is friction between the foot and floor and on a frictionless surface we would not be able to walk.

(2.) How can airplanes fly at high altitudes and space crafts be propelled? High **altitude** airplanes utilize jet engines; that is, engines burn fuel at high temperatures and expel it backward. In expelling the burnt fuel a force is exerted backward on it and it exerts an equal forward force on the plane. The same analysis applies to space crafts.

(3.) A father takes his eight-year-old daughter to skate. The father and the girl stand at rest facing each other. The daughter pushes the father backwards. What happens? Whatever force the daughter exerts on her father he exerts in the opposite direction equally on her. Since the father has a larger mass his acceleration will be less than the daughter's. With the larger acceleration the daughter will move faster and travel farther in a given time.

NEYMAN, JERZY (1894-1981)

Moldavian-American statistician

The founder of modern theoretical **statistics**, Jerzy Neyman was also an integral factor in the development of the statistical theory of hypothesis testing. His works has come to be widely applicable in the fields of medical diagnosis, astronomy, genetics, meteorology, and high-tech agriculture. Neyman is particularly remembered for combining a theory and its practical applications in his work, and for making advances in the statistical methods of confidence intervals and survey sampling.

Neyman was born Jerzy Splawa-Neyman on April 16, 1894 in Bendery, Moldavia (formerly Bessarabia, now Moldova), which then was part of Russia. He began classes at the Ukraine's Kharkov State University in 1912, ultimately receiving a master's degree from the school in 1916 for his work on Lebesgue **integers**. The following year, the Institute of Technology in Kharkov hired him as a lecturer.

Neyman worked at the Institute until 1921, when he accepted a post as statistician at the Institute of Agriculture in Bydgoszcz, Poland. After two years there, he moved again, this time to Warsaw, Poland to work as a lecturer in mathematics and statistics at the College of Agriculture. It was there that he received his Ph.D. in 1924. After attending lectures on a fellowship during 1925-1927 by **Jacques Hadamard** and **Henri Lebesgue** in Paris and London, Neyman returned to Warsaw and renewed his interest in statistics.

As soon as he got back to Poland, Neyman launched an effort to establish a biometrics (or biostatistics) laboratory in Warsaw. It took him a year, but by 1928, he had succeeded in creating the lab at the Nencki Institute for Experimental Biology. He joined the University of Warsaw faculty in 1928, when he also began collaborating with E. S. Pearson on such influential papers as "On The Problem of The Most Efficient Tests of Statistical Hypotheses" (1933) and "The Testing of Statistical Hypotheses in Relation to Probabilities A Priori" (1933). Meanwhile, however, Neyman was having an increasingly difficult time keeping his laboratory open and his researchers busy because of the Polish government's inefficiency and general political upheaval in the region. By 1932, he complained, "I simply cannot work, the crisis and the struggle for existence takes all my time and energy."

In 1934, Neyman went to the University of London to fill a temporary post as special lecturer in statistics, but when

the job became permanent the following year, he stayed on. Neyman remained on the University of London staff until 1938, when he emigrated to the United States. He found work at the University of California at Berkeley as a mathematics professor, but by 1941 he had created and been made director of a new statistical laboratory, as well as professor of statistics. Neyman would soon make the lab one of the world's top centers for the study and application of mathematical statistics. He also oversaw a series of lectures on probability and statistics that drew participants from all over the world.

During World War II, the U.S. Navy made good use of Neyman's statistical expertise, putting him to work on the National Defense Research Committee Project from 1942 to 1945. After the war, he spent 1946 as a visiting professor at New York's Columbia University, but aside from that brief foray, he remained at UC Berkeley. In 1955, he became chairperson of the school's new Department of Statistics and from 1958 to 1959 he was resident professor at UC's Miller Institute of Basic Research Science. In 1968 Neyman received the National Medal of Science.

Neyman remained at UC Berkeley until his death on August 5, 1981 in Oakland, California. He had married in 1920 and had one child.

NICHOLAS OF ORESME (CA. 1320-1382)

French scientist, clergyman, translator

The French clergyman, scientist, economist, and translator Nicholas of Oresme is best known for his treatise on money, *De moneta,* and for his services to King Charles V of France.

Nicholas of Oresme was born at Allemagne in Normandy. Little is known of his early years, except that he studied theology. He attended the College of Navarre of the University of Paris in 1348 and served as master of that college from 1356 to 1361. By 1370 he had become royal chaplain to King Charles V, and he had probably been Charles's tutor during the reign of Charles's father, King John II.

Nicholas wrote on a great variety of scientific subjects, but he is best known for his economic theory, his translations of the works of **Aristotle**, and his opposition to astrology. Charles V, a patron of the early Renaissance in France, collected a library of several thousand volumes. He commissioned Nicholas to translate, from Latin into French, Aristotle's *De caelo (On the Heavens), Ethics, Politics,* and *Economics.* The influence of the king can also perhaps be seen in Nicholas's chief interest, economics. Under the pressure of the economic disruption caused by the Hundred Years War, Charles V reorganized royal finances into a system that was preserved until 1789. In these circumstances, Nicholas wrote his treatise *De moneta (On Money)* between 1355 and 1360. In it Nicholas maintained that money is the property of the community, not of the ruler, and that, therefore, the ruler has an obligation to preserve the purity of coinage and may not debase it. *De moneta* is not always a realistic reflection of late medieval economy, but it became very popular in the 17th century.

Astrology was a fad of Nicholas's times, and he wrote in both French and Latin against the notion that the future can be predicted from a study of the stars. For example, borrowing his title and purpose from Cicero, Nicholas wrote *De divinatione* (*On Divination*) in order to attack dream interpreters and horoscopes. In general, Nicholas argued that supposedly magical events can be explained by natural causes. To support his arguments, he studied astronomy; and although he accepted the Ptolemaic system, in which the universe was believed to revolve around the earth, he granted that terrestrial **motion** cannot be disproved. Nicholas encouraged the study of nature and the use of reason in examining the Christian faith, remarking that "Everything contained in the Gospels is highly reasonable."

Nicholas became Bishop of Lisieux in 1377, and he died there on July 11, 1382.

NICOMACHUS OF GERASA (CA. 60-CA. 100)
Greek arithmetician, musical theorist, and philosopher

Born in Gerasa, Palestine (now Jerash, Jordan), Nicomachus is mainly known for his work in **arithmetic**. He has been credited with founding Greek arithmetic as a separate mathematical discipline, and is also considered a notable representative of neo–Pythagorean philosophy.

Before discussing Nicomachus' work in arithmetic, it is important to understand, as Carl B. Boyer explains, that the Greek term *arithmetike*, although derived from the word *arithmos*, which means number, did not refer to actual calculation, but rather denoted theoretical, speculative thinking about numbers. Nicomachus' ideas about numbers is pretty much defined by **Pythagoras,** or, more precisely, the neo–Pythagorean tradition that he represented. In his *Introduction to Arithmetic*, Nicomachus discusses various kinds of numbers (odd, even, prime, composite, figurate, perfect), but also considers numbers as divine entities with apparently anthropomorphic qualities, such as, for instance, goodness. However, his discussion of odd and even numbers, which for Pythagoreans raised fundamental ontological questions, also contains practical aspects, including the following **theorem**. By grouping the odd **integers** in the following manner: 1; 3 + 5; 7 + 9 + 11; 13 + 15 + 17 + 19, Nicomachus discovered that the sums were equal to the cubed integers. For example, $1^3 = 1$; $2^3 = 3 + 5$; $3^3 = 7 + 9 + 11$; $4^3 = 13 + 15 + 17 + 19$, etc.

Nicomachus' world view is a fascinating blend of **mathematics and philosophy**, incorporating the main currents of Greek scientific, idealistic, and mystical thought. "In his system," Frederick Copleston has written, "the Ideas existed before the formation of the world (**Plato**), and the Ideas are numbers (Plato again). But the Number–Ideas did not exist in a transcendental world of their own: rather, they were Ideas in the Divine Mind, and so patterns or archetypes according to which the things of this world were formed (cf. Philo the Jew, Middle Platonism and Neo–Platonism)."

In his *Manual of Harmonics*, Nicomachus discusses musical scales, notes, and intervals. Following the Pythagorean tradition as well as his interest in numerical **ratios**, he defines notes and intervals on the basis of numerical ratios. However, he also seems to have accepted the idea, developed by the noted music theorist Aristoxenus, that in music, for example, in determining whether an **interval** is consonant, auditive perception plays a crucial role. It is interesting that Nicomachus attempts to reconcile two opposing traditions in music theory.

NOETHER, EMMY (1882-1935)
German-born American algebraist and educator

Emmy Noether's innovative approach to modern abstract **algebra** not only produced significant results, but it inspired highly productive work by students and colleagues who emulated her technique. Dismissed from her university position at the beginning of Nazi rule in Germany—for she was both Jewish and female—Noether emigrated to the United States, where she taught at several universities and colleges. In a letter published in the *New York Times*, **Albert Einstein** eulogized her as "the most significant creative mathematical genius thus far produced since the higher education of women began."

Noether was born on March 23, 1882, in Erlangen, Germany. Her first name was Amalie, but she was known by her middle name of Emmy. Her mother, Ida Amalia Kaufmann Noether, came from a wealthy family in Cologne. Her father, **Max Noether,** was an eminent mathematics professor at the University of Erlangen who worked on the theory of algebraic **functions**. Two of her three brothers became scientists—Fritz was a mathematician and Alfred earned a doctorate in chemistry.

Noether was educated as a typical German girl of her era. Besides learning to cook and do household chores, she took piano lessons and enjoyed going to dances. Since girls were not eligible to enroll in the *gymnasium* (college preparatory school), Noether attended the Städtischen Höheren Töchterschule, where she studied **arithmetic** and languages. In 1900, Noether passed the Bavarian state examinations, becoming certified to teach French and English at female-only institutions.

Rather than seeking a language teaching position, Noether decided to undertake university studies. Since she had not graduated from a *gymnasium*, she first had to pass an entrance examination. At the University of Erlangen, Noether obtained her instructors' permission to audit courses during 1900 and 1902, and in 1903 she passed the matriculation exam. Noether attended the University of Göttingen for a semester and heard such notable mathematicians as **Hermann Minkowski, Felix Klein,** and **David Hilbert**. Noether enrolled at the University of Erlangen when women were accepted in 1904. At Erlangen, she studied with Paul Gordan, a mathematics professor who was also a family friend. Noether completed her dissertation, entitled "On Complete Systems of

Invariants for Ternary Biquadratic Forms," and was awarded her Ph.D., *summa cum laude,* in 1908.

Throughout her career, Noether faced severe discrimination because of her gender, but she studied and worked in whatever capacity she was allowed. An inheritance from her father apparently allowed her to work for little or no pay when necessary. From 1908 until 1915, Noether did research at the Mathematical Institute of Erlangen, where she served as dissertation adviser for two students. She occasionally delivered lectures for her father, who suffered lingering effects of a childhood case of polio. During this period, Noether began to work with Ernst Fischer, an algebraist who directed her toward the broad theoretical style characteristic of Hilbert.

Klein and Hilbert invited Noether to join them at the Mathematical Institute in Göttingen. They were working on the mathematics of the newly announced general theory of relativity, and they believed Noether's expertise would be helpful. **Albert Einstein** later wrote an article for the 1955 *Grolier Encyclopedia,* in which he characterized the theory of relativity by the basic question: "How must the laws of nature be constituted so that they are valid in the same form relative to ... an arbitrary **transformation** of **space** and time?" It was precisely this type of invariance under transformation on which Noether focused her mathematical research.

In 1918, Noether proved two theorems that formed a cornerstone for general relativity. They were very general statements, though rigorous and succinct, that mathematically validated certain relationships suspected by physicists of the time. One, now known as Noether's **theorem**, established the **equivalence** between an invariance property and a conservation law. The other involved the relationship between an invariance and the existence of certain integrals of the **equations** of **motion**.

While Noether was proving these profound and useful results, she was working without pay at Göttingen University, where women were not admitted to the faculty. Hilbert, in particular, attempted to obtain a position for her but could not persuade the historians and philosophers on the faculty to vote in a woman's favor. He arranged for her to teach, however, by announcing classes in mathematical physics under his name and letting her lecture in his place. By 1919, regulations were eased, and she was designated a *Privatdozent* (a licensed lecturer who could receive fees from students but not from the university). In 1922, Noether was given the title of unofficial associate professor, an honorary position bearing no responsibilities or income. She was, however, hired as an adjunct teacher and paid a modest salary.

A keen mind and infectious enthusiasm for mathematical investigations made Noether an effective teacher, although her classroom technique was nontraditional. Rather than simply lecturing, she conducted discussion sessions in which she and her students would jointly explore some topic. She loved to spend free time with her students, especially taking long walks while discussing their personal interests or pursuing some mathematical topic. She would sometimes become so engrossed in the conversation that her students would have to remind her to watch for traffic.

Brilliant mathematicians often make their greatest contributions early in their careers; Noether was one of the notable exceptions to that rule. She began producing her most powerful and creative work at about the age of 40. Her change in style started with a 1920 paper on noncommutative fields (systems in which an operation such as **multiplication** yields a different answer for a x b than for b x a). During the years that followed, Noether developed a very abstract and generalized approach to the axiomatic development of algebra. As Weyl attested, "she originated above all a new and epoch-making style of thinking in algebra."

Noether's 1921 paper on the theory of ideals in rings is considered to contain her most important results. It extended the work of **Julius Dedekind** on solutions of polynomials and laid the foundation for modern abstract algebra. Rather than working with specific operations on **sets** of numbers, this branch of mathematics looks at general properties of operations. Because of its generality, abstract algebra represents a unifying thread connecting such theoretical fields as logic and **number theory** with **applied mathematics** useful in chemistry and physics.

During the winter of 1928-1929, Noether lectured as a visiting professor at the University of Moscow and the Communist Academy. She held a similar position during the summer of 1930 at Frankfurt. Noether was the main speaker at one of the section meetings of a conference of the International Mathematical Congress in 1928 in Bologna. In 1932, she addressed the Congress' general session during its conference in Zurich. Noether enjoyed learning about other cultures when she traveled.

When **Hermann Weyl** joined the Göttingen faculty in 1930, he tried unsuccessfully to obtain a better position for Noether because (as he later said in a eulogy delivered at Bryn Mawr and published in *Scripta Mathematica*) "I was ashamed to occupy such a preferred position beside her whom I knew to be my superior as a mathematician in many respects." In 1932, Noether again was honored by those outside her own university when she was named co-winner of the Alfred Ackermann-Teubner Memorial Prize for the Advancement of Mathematical Knowledge.

During the early 1930s, many considered the flourishing Mathematical Institute in Göttingen the premier mathematical center in the world, and its most active element was Noether's group of algebraists. On Noether's fiftieth birthday, the algebraists held a celebration, and her colleague Helmut Hasse dedicated in her honor a paper validating one of her ideas on noncommutative algebra. This successful and congenial environment ended in 1933 when the Nazi Party seized absolute power in Germany. Within months, anti-Semitic policies became law throughout the country, and on April 7, 1933, Noether was formally notified that she could no longer teach at the university. She was a dedicated pacifist, and Weyl later recalled, "[Noether's] courage, her frankness, her unconcern about her own fate, her conciliatory spirit were, in the midst of all the hatred and meanness, despair and sorrow surrounding us, a moral solace."

For a time, Noether continued to meet informally with students and colleagues, inviting groups to her apartment. By summer, the Emergency Committee to Aid Displaced German Scholars was entering into an agreement with Bryn Mawr, a women's college in Pennsylvania, to offer her a professorship. The Rockefeller Foundation provided a grant matching the Committee's appropriation, thus funding Noether's first year's salary.

In the fall of 1933, Noether was supervising four graduate students at Bryn Mawr. Starting in February 1934, she also delivered weekly lectures at the Institute for Advanced Study at Princeton University. She bore no malice toward Germany, and maintained friendly ties with her former colleagues. With her characteristic curiosity and good nature, she settled into her new home in America. Her English was adequate for conversation and teaching, although she occasionally lapsed into German when concentrating on technical material.

During the summer of 1934, Noether visited Göttingen to arrange shipment of her possessions to the United States. She returned to Bryn Mawr in the early fall, having received a two-year renewal on her teaching grant. In the spring of 1935, Noether underwent surgery to remove a uterine tumor. The operation was a success, but four days later, she suddenly developed a very high fever and lost consciousness. She died on April 14, apparently from a postoperative infection. Her ashes were buried near the library on the Bryn Mawr campus.

Over the course of her career, Noether supervised a dozen graduate students, wrote 45 technical publications, and inspired countless other research results through her habit of suggesting topics of investigation to students and colleagues. After World War II, the University of Erlangen attempted to show Noether the honor she had deserved during her lifetime. In 1958, the institution presented a conference commemorating the 50th anniversary of Noether's doctorate, and in 1982—the 100th anniversary year of Noether's birth—the university dedicated a memorial plaque to her in its Mathematics Institute. That same year, the Emmy Noether Gymnasium, a coeducational school emphasizing mathematics, the natural sciences, and modern languages, opened in Erlangen.

Non-denumerably infinite sets

The German mathematician **Georg Cantor** (1845-1918), who laid the foundation of modern **set theory**, utilized the idea of "one-to-one correspondence" as a way of "counting" the members of a set. Thus he defined a finite set as either the empty set or a set whose members can be placed in one-to-one **correspondence** with the set {1, 2, 3,..., n} where n is some positive integer. An infinite set, then, was defined as a set that was not finite. Many of Cantor's most interesting and sometimes provocative results concerned his theory of infinite **sets**. These he broke into two types, the denumerably infinite and the non-denumerably infinite sets. The denumerably (or countably) infinite sets are sets whose members can be placed in one-to-one correspondence with the infinite set {1, 2, 3,...} of positive **integers**. The even integers {2, 4, 6,...} are denu-

merably infinite since we can correspond 2 with 1, 4 with 2, 6 with 3,..., and so on. On the other hand, non-demumerably (or uncountably) infinite sets were those which could not be placed in one-to-one correspondence with the positive integers. Cantor produced a clever argument to show that, for example, the **real numbers** in the **interval** [0,1] were non-denumerably infinite. He used an indirect **proof**, assuming that a one-to-one correspondence between the real numbers in [0,1] and the positive integers could be made and then showing that this assumption led to a contradiction. Since any real number in [0,1] can be written as an infinite decimal, Cantor set up the following "alleged" one-to-one correspondence: 1 corresponds to $0.a_{11}a_{12}a_{13}...$, 2 corresponds to $0.a_{21}a_{22}a_{23}...$, 3 corresponds to $0.a_{31}a_{32}a_{33}...$, and, in general, n corresponds to $0.a_{n1}a_{n2}a_{n3}...$, and so on. Then Cantor showed how to come up with a real number in [0,1] that had been left out of this correspondence. His number had the form $0.b_1b_2b_3... b_n...$; where a_{11} is not equal to b_1, a_{22} is not equal to b_2, a_{33} is not equal to b_3, and, in general, a_{nn} is not equal to b_n. Since this number cannot be equal to any of the numbers given in the correspondence, it follows that such a correspondence cannot be made.

In a sense, Cantor was saying that the interval [0,1] has "more" elements than the set of positive integers; but "more" can be a very tricky and misleading concept when dealing with infinite sets. For example, the set {1, 2, 3,... } "seems" to have "more" elements than the set {2, 4, 6,...}; in fact "maybe" twice as many. Yet we showed above that these two sets can be placed in one-to-one correspondence. To get around this seeming paradox, Cantor defined the concept of **cardinality** for sets. The empty set would have cardinality 0, and a finite set with n elements would have a cardinality of n. Thus for a finite set, the cardinality is just the number of elements in the set. So any two finite sets with the same number of elements have the same cardinality, and clearly any two finite sets with the same cardinality can be placed in one-to-one correspondence with each other. But what about infinite sets? Since we cannot meaningfully talk about the "number" of elements in an infinite set, Cantor simply carried over the one-to-one correspondence idea and said that any two infinite sets which can be placed in one-to-one correspondence with each other have the same cardinality. He also introduced the symbol $\aleph_0$, read "aleph-null" for the cardinality of the positive integers. ($\aleph$ is the first letter of the Hebrew alphabet.) Therefore, {2, 4, 6,...} has cardinality $\aleph_0$, as does any other set that can be placed in one-to-one correspondence with the positive integers. Sets with cardinality $\aleph_0$ are the **denumerable** (countable) sets. To the cardinality of the set [0,1] Cantor gave the symbol C, for "continuum" since any interval of real numbers has no "gaps" or "holes." He also showed that the set of all real numbers can be placed in one-to-one correspondence with the interval [0,1], and, therefore, also has cardinality C. In fact, any interval of real numbers has cardinality C, as does any **square** and its interior, as does the entire Cartesian plane. Thus all of these sets are non-denumerably infinite, whereas, the sets with cardinality $\aleph_0$ are denumerably infinite.

Cantor defined an order relation for cardinality as follows: The cardinality of a set T is greater than the cardinaltiy of a set S if and only if S can be placed in one-to-one correspondence with a subset of T but T cannot be placed in one-to-one correspondence with a subset of S. Thus the cardinality C is greater than the cardinality $\aleph_0$. It is natural to ask if there are sets with cardinality greater than C. To answer this question, Cantor considered the "power" set of a given set, i.e., the set of all subsets of S. For example, the set {1, 2, 3} has the following subsets: the empty set, {1}, {2}, {3}, {1,2}, {1,3}, {2,3}, {1,2,3}. (The empty set is a subset of every set and a set is always considered to be a subset of itself.) Note that the power set of {1, 2, 3} has $8 = 2^3$ elements. In general a finite set with n elements will have a power set with 2^n elements, hence the term "power" set. Clearly, the cardinality of the power set of any finite set will always be greater than the cardinality of the set itself, and Cantor was able to show that this is also true for infinite sets. Therefore, we may conclude that the cardinality of the power set of any set with cardinality C will be greater than C. This cardinality is sometimes designated by 2^C. In this way, Cantor was able to generate an infinite number of so-called "transfinite" cardinal numbers. One of the most famous questions of **set theory** is whether or not there is a transfinite **cardinal number** between $\aleph_0$ and C. Cantor thought not, and his statement of this fact became known as "the continuum hypothesis." In 1938, the great mathematical logician Kurt Gödel (1906-1978) showed that no contradiction would arise if the continuum hypothesis were added to the axioms of conventional **Zermelo-Fraenkel set theory**; but in 1963, **Paul Cohen** (1934-) showed that no contradiction would arise if the negation of the continuum hypothesis were added to those axioms. These two results taken together showed that, within conventional Zermelo-Fraenkel set theory, the continuum hypothesis is undecidable.

See also Cardinal number; Cardinality; Continuum hypothesis

NON-EUCLIDEAN GEOMETRY

Non-Euclidean **geometry** refers to certain types of geometry which differ from plane and **solid geometry** which dominated the realm of mathematics for several centuries. There are other types of geometry which do not assume all of Euclid's postulates such as **hyperbolic geometry**, elliptic geometry, spherical geometry, descriptive geometry, differential geometry, geometric **algebra**, and multidimensional geometry. These geometries deal with more complex components of curves in **space** rather than the simple plane or solids used as the foundation for Euclid's geometry. The first five postulates of Euclidean geometry will be listed in order to better understand the changes that are made to make it non-Euclidean.

1.) A straight line can be drawn from any point to any point.

2.) A finite straight line can be produced continuously in a straight line.

3.) A **circle** may be described with any point as center and any **distance** as a radius.

4.) All right angles are equal to one another.

5.) If a transversal falls on two lines in such a way that the interior angles on one side of the transversal are less than two right angles, then the lines meet on the side on which the angles are less than two right angles.

A consistent logical system for which one of these postulates is modified in an essential way is non-Euclidean geometry. Although there are different types of non-Euclidean geometry which do not use all of the postulates or make alterations of one or more of the postulates of Euclidean geometry, hyperbolic and elliptic are usually most closely associated with the term non-Euclidean Geometry.

Hyperbolic geometry is based on changing Euclid's **parallel postulate**, which is also referred to as Euclid's Fifth **postulate**, the last of the five postulates of Euclidian geometry. Euclid's parallel postulate may also be stated as one and only one parallel to a given line goes through a given point not on the line.

Elliptic geometry uses a modification of Postulate II. Postulate II allows for lines of infinite length, which are denied in Elliptic geometry, where only finite lines are assumed.

The history of non-Euclidean geometry

Euclid was thought to have instructed in Alexandria after Alexander the Great established centers of learning in the city around 300 B.C. Euclid was the mathematician who collected all of the definitions, postulates, and theorems that were available at that time, along with some of his insights and developments, and placed them in a logical order and completed what we now know as Euclid's *Elements*.

The influence of Greek geometry on the mathematics communities of the world was profound for in Greek geometry was contained the ideals of deductive thinking with its definitions, corollaries, and theorems which could establish beyond any reasonable doubt the **truth** or falseness of **propositions**. For an estimated 22 centuries, Euclidean geometry held its weight.

Despite the general acceptance of Euclidean geometry, there appeared to be a problem with the parallel postulate as to whether or not it really was a postulate or that it could be deduced from other definitions, propositions, or axioms. The history of these attempts to prove the parallel postulate lasted for nearly 20 centuries, and after numerous failures, gave rise to the establishment of non-Euclidean geometry and the independence of the parallel postulate.

Several Greek scientists and mathematicians considered the parallel postulate after the appearance of Euclid's *Elements*, around 300 B.C. Aristotle's treatment of the parallel postulate was lost. However, it was the Arab scholars who appeared to have obtained some information on the last text and reported that Aristotle's treatment was different from that of Euclid since his **definition** depended on the distance between parallel lines. Proclus and Ptolemy also published some attempts to prove the parallel postulate.

Omar Khayyam provided extensive coverage on the **proof** of the parallel postulate or theory of parallels in his discussions on the difficulties of making valid proofs from

Euclid's definitions and theorems. During the 13th century Husam al-Din al-Salar wrote a text on the parallel postulate in an attempt to improve on the development by Omar Khayyam.

The 18th century produced more sophisticated proofs and although not correct, produced developments that were later used in non-Euclidean geometry. The Italian mathematician, **Girolamo Saccheri**, in one of his proofs considered non-Euclidian concepts by making use of the acute-angle hypothesis on the intersection of two straight lines.

The attempt to solve this problem was made also by Farkas Bolyai, the father of Johann Bolyai, one of the founders of non-Euclidean geometry, but his proof was also invalid. It is interesting to note that Johann's father cautioned his son not to get involved with the proof of the parallel postulate because of its complexity.

The founders of non-Euclidean geometry

The writings of **Gauss** showed that he too, first considered the usual attempts at trying to prove the parallel postulate. However, a few decades later, in his unpublished reports in his correspondence with fellow mathematicians such as W. Bolyai, Olbers, Schumacker, Gerling, Tartinus, and Bessel showed that Gauss was working on the rudiments of non-Euclidian geometry, the name he attributed to his mathematics of parallels. Gauss shared his thoughts on this topic and asked them not to disclose this information but Gauss never published them. It has been proposed by historians that Gauss was concerned that these concepts were too radical for acceptance by mathematicians at that time. And if this was the case, it probably was correct since the two founders of non-Euclidean geometry, Bolyai and Lobachevsky, received very little acceptance until after their deaths.

It was at the University of Kazan, in the Russian province of Kazakhstan, that **Nicolai Ivanovitch Lobachevsky** made his contributions in non-Euclidean geometry. In his early days at the university, he did try to find a proof of the parallel postulate, but later changed direction. As early as 1826, he made use of the hypothesis of the acute **angle** already developed by Saccheri and Lambert in his lecture noting that two parallels to a given point can be drawn from a point where the sum of the angles of the **triangle** is less than two right angles. His works *On the Imaginary Geometry, New Principles of Geometry, With a Complete Theory of Parallels, Applications of the Imaginary Geometry to Certain Integrals and Geometrical Researches* are on the theory of parallel lines. He later completed his work in one French and two German publications. Lobachevsky developed his *PanGeometry* on the 28 propositions of Euclidean geometry and the negation of the parallel postulate. He developed the concepts for non-Euclidean geometry by introducing two new figures—the Horocycle and the Horoscope. Using these two concepts and some **transformation** formulas, he developed his new geometry.

Although Lobachevsky continued throughout his career improving the development of non-Euclidean geometry, Johann Bolyai, the other mathematician given credit for its development apparently only spent slightly over a decade in his mathematical considerations. As indicated previously, Johann's father suggested that he not waste his time working on the complex problems of the parallel postulate. However, Johann and his friend Carl Szasz worked on the theory of parallels while students at the Royal College for Engineers at Vienna from 1817-1822. In 1823 Bolyai discovered the formula for the transformation which connected the angle of parallelism to the corresponding line. He continued with his development and sent his manuscript to his father who published it in 1832. The article was entitled "The Science of Absolute Space" in the Appendix of his father's book. Prior to its publication, Johann's father had sent the paper to Gauss for his consideration. It is reported that the paper originally sent in 1831 to Gauss was lost. Three months after the publication, the article was sent again to Gauss and in 1832 his father received his reply. Gauss indicated that he was impressed by the work but noted that he had been working on the same problem with similar results and was pleasantly surprised to have the development completed by his friend's son. Johann was deeply suspicious of this reply and apparently suspected Gauss of trying to take credit for his work. However, in this instance there was no problem, since Gauss had no publications on the topic and could not claim priority but Johann continued to be suspicious. After the publication of his work, Johann did very little significant mathematical research. And even though he was interested in having his work published before Lobachevsky when he heard of Lobachevsky's contributions, he never completed the necessary research to report to the mathematical journals.

The most important conclusions of Bolyai's research in non-Euclidean geometry were the following: 1.) The definition of parallels and their properties independent of the Euclidian postulate. 2.) The circle and the **sphere** of infinite radius. The geometry of the sphere of infinite radius is identical with ordinary **plane geometry**. 3.) Spherical **trigonometry** is independent of Euclid's Postulate. Direct demonstration of the formula. 4.) Plane trigonometry in non-Euclidean geometry. Applications to the calculation of areas and volumes. 5.) Problems which can be solved by elementary methods. **Squaring the circle**, on the hypothesis that the Fifth Postulate is false.

A later development following that of Bolyai's and Lobachevsky's hyperbolic non-Euclidean geometry was that of elliptic non-Euclidian geometry. The rudiments of elliptic non-Euclidean geometry were developed by Georg Friedrich **Bernhard Riemann**. His introduction to his foundations of spherical geometry apparently was used as the basis for his elliptic Geometry which made use of the postulate that the sum of the angles of a triangle in space are greater than 180°. Based on the foundations that Riemann had introduced, Klein was able to further develop elliptic non-Euclidean geometry and was actually the mathematician who defined this new field as elliptic non-Euclidian geometry. Klein's Erlanger Program made a significant contribution in providing major distinguishing features among parabolic (Euclidean geometry), hyperbolic, and elliptic geometries.

NORMAL CURVE

In **statistics**, the term "normal curve" refers to a specific member of an entire **family of curves** that share a number of important characteristics. First, they are curves that **model** probability distributions for random **variables**. Second, they are all symmetric about the vertical line with equation of the form x = μ, where μ is the **mean** of the probability distribution. Third, all normal curves are shaped like the profile of a bell and are, therefore, sometimes called "bell curves." Fourth, the edges of the bell are infinite in length and approach the horizontal axis (the line y = 0) asymptotically from above. Fifth, since the curves represent probability distributions, the total area under each normal curve is 1. Finally, members of the family of normal curves have equations of the form $y = (1/(\sigma\sqrt{(2\pi)}))e^{((-1/2)(x-\mu)/\sigma)^2}$, where μ is the mean of the given distribution and σ is the **standard deviation**. These formulas are derived in advanced courses in mathematical statistics, but the typical researcher uses statistical tables generated from these formulas without needing to understand the derivations.

Normal curves are important in statistics because many kinds of natural phenomena and human-generated experiments produce data sets that can be very accurately modeled by them. For example, heights, weights, sizes of heads, arm lengths, shoe sizes, and other such physical measurements, tend to be normally distributed for random samples of certain groups of people and animals. The scores on standardized achievement tests for nationwide samples of students also tend to be normally distributed. The great German mathematician **Carl Friedrich Gauss** observed that the distribution of errors in certain astronomical measurements could be modeled using a normal curve. For this reason, normal curves are often called "Gaussian" curves to honor Gauss's pioneering work in this area of statistics.

The most important and useful member of the family of normal curves is the one having a mean of 0 and a standard deviation of 1. This curve is called the "standard normal curve" and has the equation $y = (1/\sqrt{(2\pi)})e^{(-1/2)x^2}$. All other normal curves may be mapped onto the standard normal by using mathematical **transformations**. When a normal distribution is transformed in this way, it is said to have been "standardized." The importance of standardization is that a single probability distribution table, called the standard normal table, may then be used to calculate probabilities for any normal distribution. Probabilities are represented by areas of regions under a normal curve. Since the normal curve is symmetric, exactly 50% of the population will fall below the mean and 50% of the population will fall above the mean. For the standard normal distribution, this means that 50% of the area under the curve lies in the region to the right of 0 and 50% lies below 0. Also, it can be shown that approximately 68% of the area under the standard normal curve will lie between –1 and 1. Similarly, about 95% of the area will fall between –2 and 2 and about 99.7% will lie between –3 and 3. To illustrate the process of standardizing a set of data, consider the case of an achievement test given to students across the nation resulting in a nor-

mal distribution of scores with a mean of 500 and a standard deviation of 100. Since the mean of this data is 500 units to the right of the origin, we translate this data 500 units to the left so that the center of the distribution is now 0, which is the mean of the standard normal distribution. There is still a problem, though. The standard deviation of the original data is 100, not 1. Therefore, we must also scale all the data values by a factor of 1/100 (or, equivalently, divide each one by 100) in order to complete the **mapping** of the data onto the standard normal curve. The composite of these two transformations may be represented by the mathematical expression (x – 500)/100, where x represents any score on the test. Thus a score of 600 on the test would correspond to (600 – 500)/100 = 1 on the standard normal table. The table gives areas under the standard normal curve. It tells us that approximately 14% of the area lies to the right of 1. We can interpret this to mean that approximately 14% of the students who took this test scored 600 or above.

Using the method described in the preceding paragraph, researchers, statisticians, scientists, and other individuals use areas under the standard normal curve to determine the percentage of a population that falls within a particular interval. Typically, a *sample* of the desired population is collected. A normal distribution, based on the sample mean and sample standard deviation, may then be used to estimate the information desired about the population represented by the sample. The science of creating representative samples of a desired population in order to model it is a major underpinning of the science of statistics.

See also Central limit theorem; Johann Carl Friedrich Gauss; Mean; Standard deviation; Statistics

NORMAL NUMBERS

A real number is said to be *simply normal to the base b* if every possible digit appears in its fractional part with a frequency of occurrence of *1/b*, where *b* is the base of the expansion. By this we understand that if a number α is expressed to the base *b* (as, for example, we regularly express numbers to the base 10) and if for each digit, d (0 ≤ d ≤ b–1), we count the occurrence of *d* in the first *N* positions after the decimal point, divide this count by *N* and then let *N* increase without bound; and if the ratio of this count approaches the limit *1/b* we say that the number is *simply normal to the base b*.

In, say, the expansion of π (which has been expanded, at last count, to 5.4 billion digits), we would naturally expect any particular digit, say 3, to appear no more nor less than any of the other nine digits, that is, one in 10 times. And this is what would happen if it turned out that π were normal. Each digit should have an equal distribution or equal likelihood of occurrence.

Symbolically, taking A(d,b,N) to mean the number of occurrences of digit *d* when our number is expressed to the base *b* in the first *N* digits, then lim A(d,b,N)/N = 1/b is our standard for simple normality base b, where we take the limit as N grows toward **infinity**.

If in **addition**, α is simply normal to base b^n, for all positive **integers** n, then we say that α is *entirely normal to the base b*.

And if, moreover, α is entirely normal to every base b greater than 1, we say that α is *absolutely normal*, or sometimes just *normal*.

It can be shown that a number simply normal to the base b^n is also simply normal to base b; however, the converse is not true. For example, the rational number 0.0123456789... with this block of numbers repeated indefinitely is simply normal to the base 10, as any of the digits 0 through 9 appear with a frequency of 1/10; but it is not normal to the base 10^{10}, since it misses $10^{10}.- 1$ digits. That is, 0123456789 is a single digit base 10^{10}, and it is the only one that appears.

What is signal about normal numbers is *Borel's normal number theorem*, which asserts that *almost all* numbers are normal. The phrase "almost all" must be understood in the context of Lebesgue measure.

We say that a subset of the real line has *Lebesgue measure zero* if it can be covered by countably many intervals whose total length is less than any preselected positive quantity, however small. Borel's normal number **theorem** is then equivalent to the claim that the set of numbers that are not normal has Lebesgue measure **zero**. Thus, for example, on the **interval** [0,1], the set of normal numbers will have measure 1 and its compliment; the set of numbers that are not normal will have measure 0.

It may also be demonstrated that the Borel normal number theorem is equivalent, in **probability theory**, to the *strong law of large numbers*. In this setting we exploit the concept of probability measure, which is nothing other than Lebesgue measure restricted to the [0,1] interval, so that events are identified with points or **sets** of points in [0,1], and their probability of occurrence is their Lebesgue measure (roughly, their total length), which is not less than 0 nor greater than 1.

All this may be clarified by considering the celebrated **Cantor set**, which is an example of a set that fails to be normal.

We begin by considering the [0,1] interval and represent the numbers between 0 and 1 in ternary expansion, that is their expansion to the base 3. The digits are all either 0, 1, or 2, and in the first position 0 will represent all numbers in the first third of the unit interval, 1 the numbers in the middle third of the unit interval, and 2 the final third of the interval. Each of these three intervals may again be subdivided into thirds and the second digit of the expansion, 0, 1, or 2 represents the first, second, or third trisection of each of the three initial divisions, and so on.

The Cantor set, then, may be represented by the set of numbers in [0,1] whose ternary expansion does not contain a 1. Geometrically, such representations designate, in the first stage, corresponding to the first digit, the first and third subinterval of [0,1] with the middle third deleted. The second digit, which is either 0 or 2, corresponds to the first and last

third of each of the two thirds of [0,1] whose first digit is 0 or 2, and so on.

The Cantor set is not normal, for the digit 1 not only fails to occur with a frequency of 1/3, but fails to appear at all. According to Borel's normal number theorem, the Cantor set must have measure 0.

Probabilistically, we may interpret the Cantor set as representing random events in which one of the three possible outcomes never occurs. The probability of such a "Cantor-like" event occurring is zero.

The Cantor set is of interest because it possesses a variety of unusual properties, among them is the fact that it is an uncountable set whose measure is zero. This, though, is a property that it shares with the compliment of normal numbers, which is also uncountable.

Another property of the Cantor set is that it is possible to construct on the interval [0,1] a function that is singular, meaning that it has zero derivative on all but a set of measure zero, namely on all but the Cantor set, yet the function is nonconstant. Analogously, it is possible to construct on [0,1] a singular function, strictly increasing, with zero derivative on all but a set of measure 0, namely on all but the compliment of the set of normal numbers.

NUMBER THEORY

Number theory is a broad and diverse part of mathematics that developed from the study of the **integers**. In this entry we describe two kinds of problems that have stimulated the development of number theory.

A very old problem in number theory is to determine the integer solutions to **equations** or systems of simultaneous equation. Equations to be solved in integers are called *Diophantine equations* after **Diophantus of Alexandria**. Diophantus probably lived about 250 A.D. He is known for his work *Arithmetica* in which he posed and solved various types of equations in integers. This part of number theory has strong connections to **algebraic geometry** and is sometimes referred to as *Diophantine geometry*. In some cases it is possible to completely determine all integer solutions to a particular Diophantine equation. For example, it is possible to describe all of the triples of integers (x, y, z) that satisfy the equation $x^2 + y^2 = z^2$. To do so, first observe that it suffices to determine all solutions such that x, y, and z are positive integers and have no common prime divisor. Then all positive solutions having no common prime divisor have the form

$$x = 2uv, \quad y = u^2 - v^2, \quad z = u^2 + v^2,$$

where u and v are relatively prime integers, $v < u$, and uv is odd. A further example of a Diophantine equation for which all solutions can be explicitly determined is *Pell's equation: $x^2 - dy^2 = 1$*, where d is a positive integer and d is not divisible by the **square** of a prime. In this case the solutions depend on the continued fraction expansion for the number $\sqrt{d}$.

There are several important **Diophantine equations** that are known to have no solutions or only trivial solutions. Perhaps the best known example is the equation $x^n + y^n = z^n$

where $n \geq 3$. In this case the trivial solutions are those for which one of the variables is 0. In 1995, **Andrew Wiles**, building on the work of several other mathematicians, proved that there are no nontrivial solutions. This result had long been known, but inappropriately, as "Fermat's last theorem," because Fermat claimed to have found a **proof**. On the other hand, in 1772 Euler conjectured that the Diophantine equation

$$w^4 + x^4 + y^4 = z^4$$

has only trivial solutions in which two of the variables are 0. However, in 1988, N. Elkies found the example

$$2682440^4 + 15365639^4 + 18796760^4 = 20615673^4$$

and thereby disproved Euler's conjecture.

Another old subject of investigation in number theory is the distribution of **prime numbers** among the positive integers. We recall that a positive integer $p > 1$ is called a *prime number*, or simply a prime, if the only positive integers that divide p are 1 and p, (see the entry on the **Sieve of Eratosthenes**). Thus the first few prime numbers are

2, 3, 5, 7, 11, 13, 17, 19, 23, 29, 31, 37, 41,....

Prime numbers are important in number theory because every positive integer $n > 1$ can be written as a product of prime numbers. From this result it is easy to prove that there are infinitely many primes. Suppose that in fact there are only N distinct prime numbers, and then write $p_1, p_2, ... p_N$ for the complete list of all primes. Let $Q = (p_1)(p_2)... (p_N) + 1$ and observe that Q is not divisible by any of the primes p_1, $p_2, ... p_N$. Since Q can be written as a product of prime numbers there must be additional primes not accounted for in the finite list $p_1, p_2, ... p_N$. This contradicts the assumption that there are only N distinct prime numbers and so verifies that the set of primes is infinite. The argument given here was first stated by Euclid.

An interesting proof that there are infinitely many primes was given by Euler in 1737. Euler showed that the **infinite series**

$$\sum_p \frac{1}{p}$$

diverges, where the sum is over all prime number p. And in 1837, G. L. Dirichlet proved that if $q \geq 2$, and if a is an integer relatively prime to q, then the infinite series

$$\sum_{p \equiv a \bmod q} \frac{1}{p}$$

diverges. In particular, this shows that there are infinitely many prime numbers congruent to a modulo q whenever a and q are relatively prime.

Another way to formulate questions about prime numbers is to define $\pi(x)$ to be the number of prime numbers less than or equal to the positive real number x. Then our previous remarks establish that $\pi(x) \to \infty$ as $x \to \infty$. This raises an important question that has long been of interest in number theory: is it possible to give a reasonably good estimate, or

even an exact formula, for the function $\pi(x)$? One of the major results in number theory that begins to shed some light on this question is the **prime number theorem**. This was proved in 1896 by **J. Hadamard** and, independently, by **C. J. de la Vallèe Poussin** using properties of the Riemann **zeta-function**. The prime number **theorem** asserts that

$$\lim_{x \to \infty} \frac{\pi(x) \log x}{x} = 1$$

Thus for large values of x the function $x/\log x$ is a reasonably good approximation to $\pi(x)$. However, this is a little unsatisfactory because it does not give us any information about how close the approximation is. Many mathematicians have worked to improve the statement of the prime number theorem and today we know that an even better approximation to $\pi(x)$ is given by the function

$$\mathrm{li}(x) = \int_2^x \frac{1}{\log t} \, dt.$$

A stronger form of the prime number theorem, which can be obtained from a deeper **analysis** or the Riemann zeta-function, states that there exist constants $A > 0$ and $c > 0$ such that

$$|\pi(x) - \mathrm{li}(x)| \leq Ax \exp(-c\sqrt{\log x})$$

for all **real numbers** $x \geq 2$.

NUMBERS AND NUMERALS

Numbers and numerals are closely connected, but distinct, concepts. Numerals represent and symbolize numbers. Numbers themselves are abstract concepts that may or may not correlate to phenomena in the physical world.

Numerals are any symbols used to denote a number. Ancient societies began the practice of keeping written records thousands of years ago. Records indicating the quantities of things stored or sold would obviously have been of great use to the merchant class. A convenient and standardized system of symbols was, therefore, needed to meet such a requirement. Such a system is called a "numeral system." A numeral system is defined as a particular set of symbols and associated rules used to represent numbers. In past ages a variety of numeral systems have been employed, including "simple grouping systems" (e.g., Roman numerals), "multiplicative grouping systems" (e.g., Chinese-Japanese), and "ciphered systems" (e.g., Greek alphabetic). Modern numeral systems are place-value systems, meaning that the value of the symbol depends upon its position in the representation. For example, the "binary number system" is a positional system wherein any natural number is represented using only the symbols "0" and "1," and is used extensively in electronic computing. But by far the most familiar numeral system is the Hindu-Arabic numeral system, which is the decimal positional numeral system based on the symbols "0, 1, 2, 3, 4, 5, 6, 7, 8, 9" and the **powers** of ten. In this system any rational number is expressed using the base symbols given above and then as the sum of products involving powers of ten. For instance, "342.1" denotes the summation: $(3 \times 10^2) + (4 \times 10^1) + (2 \times 10^0) + (1 \times 10^{-1})$.

In the broadest sense, a number is a unit of a mathematical system whose elements are subject to particular rules within the system. The bulk of modern mathematics can be made to rest upon the real number system, which in turn may be deduced from the whole or natural numbers. Through the use of **set theory** and logical methods, a rigorous mathematical construction of the infinite set of **whole numbers** can be attained, and through the whole numbers other number systems may be derived including the real and **complex numbers**.

The following list (and adjoining flow chart) describes various number systems:

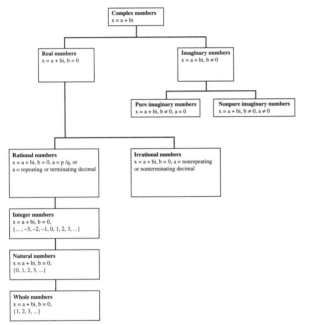

- Complex numbers consist of the set {x | x is a real or imaginary number} in the form x = a + bi, where a and b are **real numbers** and i = $(-1)^{1/2}$ is an imaginary number (e.g., x = 4 + 5i, x = 4, or x = 5i).
- Real numbers consist of the set {x | x is a rational or irrational number} in the form x = a + bi, where b = 0. (e.g., x = 4).
- **Imaginary numbers** consist of the set {x | x is a pure or nonpure number} in the form x = a + bi, where b ≠ 0. (e.g., x = 5i or x = 4 + 5i).
- Pure imaginary numbers consist of numbers in the form x = a + bi, where b ≠ 0, a = 0. (e.g., x = 5i).
- Nonpure imaginary numbers consist of numbers in the form x = a + bi, where b ≠ 0, a ≠ 0. (e.g., x = 4 + 5i).
- **Irrational numbers** consist of the set {x | x has a nonrepeating or nonterminating decimal expansion} in the form x = a + bi, where b = 0 (e.g., x = π).
- **Rational numbers** consist of the set {p / q | p and q are **integers**, a = p / q and q ≠ 0} or {x | x has a repeating

or terminating decimal expansion} where x = a + bi, b = 0 (e.g., x = 4 / 1 or x = 4 / 3 = 1.333...).

- Integer numbers consist of the set {..., -3, -2, -1, 0, 1, 2, 3,...} where x = a + bi, b = 0.
- Natural numbers consist of the set {0, 1, 2, 3,...} where x = a + bi, b = 0.
- Whole (counting) numbers consist of the set {1, 2, 3,...} where x = a + bi, b = 0.

The simplest numbers are the whole numbers that consist of the set {1, 2, 3,...}. At the most fundamental level, a whole number is related to the quantity of elements composing a group or set. The sum and product of whole numbers is always a whole number. The difference and quotient may not result in a whole number. In order to make the **subtraction** of any whole numbers from itself a valid operation, it is necessary to introduce **zero** (0). This creates the natural numbers, consisting of the set {0, 1, 2, 3,...}. To make the subtraction of whole numbers always valid it is necessary to introduce negative integers. This creates the set of integer numbers denoted by {..., -3, -2, -1, 0, 1, 2, 3,...}. To make **division** of whole numbers always valid it is necessary to introduce the positive **fractions**. Because the difference of two positive fractions is not always a positive fraction, it is necessary to introduce the negative fractions. The positive and negative integers and fractions, and the number zero, comprise the rational numbers. The sum, difference, product, or quotient of two rational numbers is always a rational number (except for division by zero, which is undefined).

The ancient Greeks, through geometric methods, were able to construct **polygons** possessing incommensurable sides (i.e., line segments whose lengths cannot be expressed as a rational number). Hence, the existence of irrational numbers was demonstrated. When expressed in ordinary decimal format, irrational numbers, like the **square** root of 2 ($2^{1/2}$), have decimal expansions which are nonrepeating or nonterminating. The totality of the rational and irrational numbers makes up the real numbers.

Any real number multiplied by itself is 0 or positive. Therefore, "$x^2 = -1$" has no real number solution. The creation of so-called "imaginary numbers" allows a solution to such an algebraic equation. All numbers of the form "x = a + bi," where a and b are real numbers and "i = $(-1)^{1/2}$" is an imaginary number, belong to complex numbers.

See also Addition; Binary number system; Complex numbers; Decimal fractions; Decimal position system; Division; Fractions; Imaginary numbers; Integers; Irrational numbers; Multiplication; Negative numbers; Number theory; Products and quotients; Rational numbers; Real numbers; Roman notation; Subtraction; Sums and differences; Whole numbers; Zero

NUMERICAL ANALYSIS

Unlike most fields of mathematics, numerical **analysis** is not about proving theorems or finding generalized solutions to problems. Instead, numerical analysis is the field that focuses on finding a usable solution to a specific question about a known function. Some of the most common problems in which numerical analysis is used are integrating specific regions of a curve and finding the zeroes of a function. Instead of finding the desired answer exactly, numerical analysis gives a way to find the answer within a specified margin of error.

There are many popular techniques in numerical analysis. All involve the knowledge of the function itself, which yields the value at all points. This value can be used to approximate **distance** from the axis or other useful approximation quantities in the process of calculating the desired quantity. For some results there will be numerous techniques for numerical solution, all of which converge upon the same result, but at different speeds and with different degrees of certainty. For example, for **numerical integration**, one may have occasion to use the trapezoid rule, Simpson's rule, Boole's rule, or Gauss-Legendre quadrature. The trapezoid (or trapezoidal) rule divides the **area** of interest into a selected number of trapezoids and calculates the area of each trapezoid, then sums the areas to get the total integration area. (This could also be expressed by rectangles with heights at the **midpoint** of each interval.) Simpson's rule takes the same type of subdivided regions and approximates the area of each region using a quadratic polynomial fit instead of a straight line across the top forming a trapezoid. Boole's rule uses a recursive relationship of the type of quadratic fits used above in Simpson's rule. Gauss-Legendre quadrature uses uneven choices for values of the function and tables of abscissa and weights to find the average values of the function across the **interval**, from which the area is calculated. Each of these rules has an error rate and several methods of improving the accuracy of the procedure. This is only one example of the many problems for which numerical analysis can give multiple paths to an approximately correct solution. Of course factors such as interval size and placement will come to bear on this particular example, as will the errors inherent in the techniques themselves, so that even the poorest technique can be made fairly feasible with technical improvements.

Other useful applications of numerical analysis are to find polynomial interpolations and extrapolations of **functions**, solutions to matrices or linear systems of **equations**, curve fitting (including least-squares fit), numerical differentiation, functional optimization, solution of eigenvalue and eigenvector equations, and solutions of ordinary and partial **differential equations**.

While numerical analysis techniques have been known for hundreds of years, they have become much more widely used since the advent of computers and electronic calculators. First, many of the simple computations performed by these devices are approximations within a certain **range** of error. The number of decimal places available in some of the computations will provide the source of that error. Second, the availability of computer programs to do more complex calculations allows many more iterations on an approximation, bringing it ever closer to an exact answer and thus rendering it more useful. The number of available calculations per minute means that a problem that would take hours to approx-

imate closely by hand would take less than a minute to approximate using a computer program. Because of this capability, numerical analysis is more widely used today than it ever has been before. With the knowledge of a structured programming language, numerical analysis can be a very useful tool in scientific or engineering applications where exact solutions would prove to be impossible or much more difficult.

See also Applied mathematics; George Boole; Calculator; Ferdinand Georg Frobenius; Johann Carl Friederich Gauss; Infinitesimal; Adrien-Marie Legendre; Mathematics and computers; Isaac Newton; Recursive functions; Thomas Simpson

NUMERICAL INTEGRATION

Numerical integration involves calculation of an integral by numerical techniques. An integral is a mathematical representation of an **area** or generalization of an area and is one of the fundamental objects of **calculus**. The word quadrature can mean the numerical computation of an integral containing only one **variable**. Gaussian quadratures and Newton-Cotes formulas are methods of numerical integration. Cubature means the numerical computation of a multiple integral, a set of integrals taken over multiple variables, and includes Monte Carlo integration.

The main difference between the two classifications of numerical integration methods, Gaussian quadratures and Newton-Cotes formulas, is the way a curve is divided into parts for analysis. Gaussian quadratures provide flexibility and efficiency for known smooth **functions** but Newton-Cotes formulas are the most widely used numerical integration methods. All Gaussian quadratures involve evaluating an integral by the summation of the product of the function evaluated at optimal points and a weighting function related to that point plus an error function: $\int_a^b f(x)dx = \Sigma^n_{k=1} w(x_k)f(x_k) + R_n(x)$, where x_k are the points at which the function is evaluated, $w(x)$ is the weighting function for each point, and R_n is the error function. Gaussian quadratures allow one to pick the optimal abscissas as which to evaluate the function. The fundamental **theorem** of Gaussian quadratures shows that the optimal abscissas are the **roots** of the orthogonal polynomial for the same **interval** and weighting function. Gaussian quadratures have a higher degree of accuracy than numerical integrations carried out using Newton-Cotes formulas. This is because Gaussian quadratures fit all **polynomials** up to degree $2n$ exactly. Some commonly used Gaussian quadratures are the Gauss-Legendre formula and the Gauss-Chebyshev formula, which are used for closed, definite integrals, the Gauss-Hermite formula, which is used for integrals that have $-\infty$ and ∞ as the **limits** of integration, and the Gauss-Laguerre formula, which is used for integrals on the interval $[0, \infty)$.

The Newton-Cotes formulas are numerical integration methods involve dividing up the interval over which integration is to occur into equal parts. Then polynomials that approximate the function are substituted for the function. The polynomials are calculated at the nodal points and summed with weighting coefficients to find the solution to the original integral. These methods are not as accurate as using the Gaussian quadrature methods but with computers being readily available calculating integrals this way is the most widely used numerical integration technique. Newton-Cotes formulas include the trapezoidal rule, which uses linear **equations**, Simpson's rule, which uses parabolic equations, Simpson's 3/8 rule, which uses **cubic equations**, and Bode's rule, which uses fourth degree polynomial equations. The Romberg integration is an extension of the trapezoidal rule that uses refinements to remove some error terms. There are many other Newton-Cotes formulas employed to numerically integrate functions, these are just a few.

Numerical integration methods can be traced back to about 260 B.C. when **Archimedes of Syracuse**, an Italian mathematician, perfected methods of numerical integration. He employed an early form of integration called the method of exhaustion. This method allows calculation of an area by calculating the areas of a sequence of **polygons** that approximate the original area. Archimedes gave an accurate approximation of π using this method. He published a book, *Quadrature of the parabola*, in which he determines the area of a segment of a **parabola** cut off by any chord. Enhancements in methods of numerical integration were achieved by several mathematicians over the next century but Ibrahim, an Arab mathematician, made a special contribution when he introduced a method of integration that was more general than that of Archimedes in the 10th century A.D.. **Thomas Simpson**, an English mathematician, is also another important contributor to methods of numerical integration. He introduced a Newton-Cotes formula that bears his name in the 18th century. Since then many other methods have been introduced and employed to determine areas by numerical integration.

O

ORBIT

An orbit is the path followed by a celestial body moving in a gravitational **field**. When a single object, such as a planet, is moving freely in a gravitational field of a massive body, such as a star, the orbit is in the shape of a conic section, that is, elliptical, parabolic, or hyperbolic. Most orbits are elliptical.

The exact path and position of an object in **space** can be determined by taking into account seven orbital elements. These elements deal with the mathematical relationships between the two bodies. To determine the orbit of a celestial body, it must be observed and precise measurements taken at least three times. However, at least 20 precise observations, covering at least one full revolution, are needed for accurate orbital elements to be determined. If two bodies that move in elliptical orbits around their common **center of mass** (for example, the Sun and Jupiter) were alone in an otherwise empty universe, we would expect that their orbits would remain constant. However, the solar system consists of the Sun, eight major planets, and an enormous number of much smaller bodies all orbiting around the solar system's center of mass. The masses of these objects all influence the orbits of each other in small and large ways.

Perturbation theory

The Sun's gravitational attraction is the main **force** acting on each planet, but there are much weaker gravitational forces between the planets, which produce perturbations of their elliptical orbits; these make small changes in a planet's orbital elements with time. The planets which perturb the Earth's orbit most are Venus, Jupiter, and Saturn. These planets and the sun also perturb the moon's orbit around the Earth—Moon system's center of mass. The use of mathematical series for the orbital elements as **functions** of time can accurately describe perturbations of the orbits of solar system bodies for limited time intervals. For longer intervals, the series must be recalculated.

Today, astronomers use high-speed computers to figure orbits in multiple body systems such as the solar system. The computers can be programmed to make allowances for the important perturbations on all the orbits of the member bodies. Such calculations have now been made for the Sun and the major planets over time intervals of up to several tens of millions of years.

As accurately as these calculations can be made, however, the behavior of celestial bodies over long periods of time cannot always be determined. For example, the perturbation method has so far been unable to determine the stability either of the orbits of individual bodies or of the solar system as a whole for the estimated age of the solar system. Studies of the evolution of the Earth-Moon system indicate that the Moon's orbit may become unstable, which will make it possible for the Moon to escape into an independent orbit around the Sun. Recent astronomers have also used the theory of chaos to explain irregular orbits.

The orbits of artificial satellites of the Earth or other bodies with atmospheres whose orbits come close to their surfaces are very complicated. The orbits of these satellites are influenced by atmospheric drag, which tends to bring the satellite down into the lower atmosphere, where it is either vaporized by atmospheric friction or falls to the planet's surface. In addition, the shape of Earth and many other bodies is not perfectly spherical. The bulge that forms at the equator, due to the planet's spinning **motion**, causes a stronger gravitational attraction. When the satellite passes by the equator, it may be slowed enough to pull it to earth.

Types of orbits

A synchronous orbit around a celestial body is a nearly circular orbit in which the body's period of revolution equals its **rotation** period. This way, the same hemisphere of the satellite is always facing the object of its orbit. This orbit is called a geosynchronous orbit for the Earth where, with its sidereal rotation period of 23 hours 56 minutes 4 seconds, the geosynchronous orbit is 21,480 miles (35,800 km) above the equator on the Earth's surface. A satellite in a synchronous orbit will seem to remain fixed above the same place on the body's equator. But perturbations will cause synchronous satellites to drift away from this fixed place above the body's equator. Thus, frequent corrections to their orbits are needed to keep geosynchronous satellites in their assigned places. They are very useful for communications and making global meteorological observations. Hence, the vicinity of the geosynchronous orbit is now crowded with artificial satellites.

The Space Age has greatly increased the importance of hyperbolic orbits. The orbits of spacecraft flybys past planets, their satellites, and other solar system bodies are hyperbolae. Other recent flybys have been made past Halley's Comet in March 1986 by three spacecraft, and past the asteroids 951 Gaspra in October 1991 and 243 Ida in August 1993; both flybys were made by the Galileo spacecraft enroute to Jupiter. Although accurate masses could not be found for these small bodies from the hyperbolic flyby orbits, all of them were extensively imaged.

Orbits of double and multiple stars

The orbits of double stars, where the sizes of the orbits have been determined, provide the only information we have about the masses of stars other than the Sun. Close double stars will become decidedly non-spherical because of tidal distortion and/or rapid rotation, which produces effects analogous to those described above for close artificial planetary satellites. Also, such stars often have gas streaming from their tidal and equatorial bulges, which can transfer mass from one star to the other, or can even eject it completely out of the system. Such effects are suspected to be present in close double stars where their period of revolution is found to be changing.

Multiple stars with three (triple) or more (multiple) members have very complicated orbits for their member stars, and require many perturbing effects to be considered. The investigation of the orbits of double and multiple stars is important for solving many problems in astrophysics, stellar structure, and stellar evolution.

ORDINAL NUMBER

The number 8 can be used in three ways: to tell "how many," to tell "where" in a ranking, and to name someone or something. The girl with the number 8 on her baseball uniform, who is 8th in the batting order, playing on a team that scores 8 runs, is using the same number in each of these ways. When she is 8th in the batting order, she is using the number as an ordinal number. An ordinal number is one which is used to indicate where in an ordered list someone or something occurs. A number used to tell how many is a "cardinal number." A number which is used to name something is neither a cardinal nor an ordinal number.

The ordinal name of a number differs somewhat from the cardinal name. In most instances the cardinal name can be converted into the ordinal name by adding "th." Thus the **car-**

dinal number one thousand becomes the ordinal number one thousandth; four becomes fourth; and so on. In the case of 1, 2, and 3, however completely different names are used: the cardinal 1 is the ordinal "first" and the ordinal symbol "1st"; 2 is second and 2nd (or 2d); 3 is third and 3rd (or 3d); 4 is fourth and 4th; 20 is twentieth and 20th; 21 is twenty-first and 21st; 100 is one hundreth and 100th; 101 is one hundred first and 101st.

The clear distinction between cardinal and ordinal forms of 1 and 2 arises from the way in which the events or things they describe differ. A runner who comes in first comes in ahead of anyone else, and that is what is most notable about the event, not that one runner has crossed the line. Likewise, someone coming in second "follows," and that, too, is something which can be noted without consciously counting the two runners. By the time the third runner comes along, counting becomes a helpful if not necessary aid in determining his or her position. The **similarity** between "three" and "third" (and the Latin roots from which they come) reflects this. Beyond 3, counting is almost essential, and the cardinal and ordinal forms are almost the same.

The ordinal name of a number is used in some instances where no ranking is intended or where the ranking is vestigial. The names given to the denominators of common fraction are ordinal names although they signify the number of uniform parts into which each unit is cut. Thus three-fifths indicates that each unit has been divided into five equal parts, and the fraction represents three of them. However, when the fraction is written with numerals, 3/5, both numerator and denominator are written in the cardinal form.

On the other hand there are times when the cardinal form of a number is used in an ordinal sense. In counting a group of objects one is putting them into one-to-one **correspondence** with the numbers 1, 2, 3,... in order. That is why the counting process works. Nevertheless one counts, "one, two, three,...," not, "first, second, third...." Mathematicians who work with **infinity** have extended the concept of ordinal number to apply to certain classes of infinite numbers as well.

ORTHOGONAL COORDINATE SYSTEM

A coordinate system is a way of naming the points of n-dimensional **space** with n-tuples of **real numbers** so that no two points have the same name. For example, a line can be assigned coordinates be naming one of the points 0 (also called the origin) and deciding which side of 0 is positive and which is negative. Any point that is on the positive side is assigned the number equal to its **distance** from 0. A number on the left side is assigned the number equal to minus its distance from 0. These coordinates are called the Cartesian coordinates of the line. Cartesian coordinates can also be extended to the plane and n-dimensional space for any positive integer n. In polar coordinates, each point (x, y) in the plane is given the coordinates (r,a) in which r is the distance (0, 0) to (x, y) and a is the **angle** between the ray containing (0, 0) and (1, 0) and the ray containing (0, 0) and (x, y). In other words, r = √(x² + y²) and a = arctan (y/x). In cylindrical coordinates, each point (x, y, z) of three-dimensional space is assigned the coordinates (r, a, z) in which, as in polar coordinates, $r = \sqrt{(x^2 + y^2)}$ and a = arctan (y/x). In spherical coordinates, each point (x, y, z) is assigned the point (r, a, b) in which r is the distance from (0, 0, 0) to (x, y, z), a is, again, the angle between the ray containing (0, 0) and (1, 0) and the ray containing (0, 0) and (x, y), and b is the angle between the ray containing (0, 0, 0) and (0, 0, 1) and the ray containing (0, 0, 0) and (x, y, z). In each of these examples, there are circles, cylinders, or spheres defined the equation r = constant. In some sense, these coordinates are based upon these objects. Coordinate systems can be based on ellipsoids, paraboloids, torii, hyperboloids, or other surfaces.

Orthogonality is a generalization of perpendicularity that uses tangent spaces. Two intersecting curves in the plane are orthogonal if, near the point of intersection, the two curves together almost form a plus sign. To be precise requires some differential **calculus**. Suppose the two curves are given by differentiable **functions** f and g from the real numbers to the plane such that f(0) = g(0) is the point of intersection. Then the lines consisting of all points of the form f'(0)t + f(0) and g'(0)t + g(0) for all real numbers t, are called the tangent lines to f and g (at the point of intersection) respectively. Here f' and g' are the derivatives of f and g. If these tangent lines intersect at right angles, then the curves are said to be orthogonal.

Two planes in three-dimensional space are orthogonal if their union is congruent to the union of the xy plane with the xz plane. In other words, if we look at one plane from its side so that it appears as a line, the other plane appears as a line, too and these two lines are perpendicular.

Two (differentiable) surfaces are orthogonal if in a small neighborhood of any point of intersection, the intersection is "looks" like the intersection of two orthogonal planes. If P is a point on a surface and f is a differentiable curve from the real numbers to the surface, then the tangent line to f at the point P is called a tangent line to S at P. The tangent plane of the surface at P is the union of all such tangent lines. Two surfaces are said to be orthogonal if at every point of intersection, their tangent planes are orthogonal. Orthogonality can be defined more generally for (n-1)-dimensional manifolds in n-dimensional space in a similar manner.

An orthogonal coordinate system for three-dimensional space is a triplet of functions (u_1, u_2, u_3), say, that satisfies the following properties.

- (1) Each u_i maps three-dimensional space to the real line.
- (2) The function that sends (x, y, z) to $(u_1(x, y, z), u_2(x, y, z), u_3(x, y, z))$ is a one-to-one, differentiable function.
- (3) For every point (x,y,z), the three surfaces defined by the **equations** $u_1 = x$, $u_2 = y$, and $u_3 = z$ are orthogonal. Equivalently, grad u_1, grad u_2, and grad u_3 are orthogonal vector fields.

Orthogonal coordinate systems for n-dimensional spaces in which n is not three are defined in a similar manner.

Cylindrical coordinates, for example, are given by $u_1(x, y, z) = \sqrt{(x^2 + y^2 + z^2)}$, $u_2(x, y, z) = \arctan(y/x)$, $u_3(x, y, z) = z$. The surfaces defined by the equations mentioned in (3) are just

the **cylinder** of radius r that is perpendicular the xy-plane, the plane that contains the z-axis and the point (x, y, z), and the plane parallel to the xy-plane at height z.

Different coordinate systems are used depending on the problem one is trying to solve. For example, objects with rotational **symmetry** are often easier to analyze if they are given in cylindrical or spherical coordinates whereas objects built from cubes are easiest to analyze with Cartesian coordinates.

See also Cartesian coordinates; Manifolds; Polar coordinates; Shifting orthogonal coordinates; Surfaces and solids of revolution; Vector analysis

OSCILLATIONS

An oscillation is a particular kind of **motion** in which an object repeats the same movement over and over. It is easy to see that a child on a swing and the pendulum on a grandfather clock both oscillate when they move back and forth along an arc. A small weight hanging from a rubber band or a spring can also oscillate if pulled slightly to start its motion, but this repeated motion is now linear (along a straight line). On a larger scale, you can notice oscillations when bungee jumpers fall to the end of their cords, are pulled back up, fall again, etc. Actually, oscillations are all around us, even in the pages of this book.

Anything, no matter how large or small, can oscillate if there is some point where the object is in stable equilibrium. Stable equilibrium means that an object always wants to return to that position. Suppose you placed a marble at the exact center inside a very smooth bowl. If you tap the marble slightly to move it a small **distance**, it rolls back towards the center, overshoots, rolls back, overshoots, etc. The marble is oscillating as it continues to return to center of the bowl, its point of stable equilibrium. If you think of the marble and the bowl as a "unit," you can see that the "unit" stays together even though the marble is oscillating (unless you tap the marble so hard that it flies out of the bowl). This is the reason for using the term stable.

On the other hand, what if you turned the bowl over and tried the same experiment by placing the marble on top at the center. You might succeed in balancing the marble for a short time, but eventually you will touch the table or a breeze will move the marble a small amount and it will fall. When this happens, the "unit" of marble and bowl comes apart and no oscillation can happen. In this case, the center of the bowl would be a point of unstable equilibrium, since you can balance the marble there, but the marble cannot return to that point when disturbed to keep the "unit" from disintegrating.

For the motion of a child on a swing, the bottom of the arc (when the swing hangs straight down) is the point of stable equilibrium. The point of stable equilibrium for a weight on the rubber band is the location at which the weight would hang if it was very slowly lowered. In either case, an oscillation occurs when the object (child or weight) is moved away from stable equilibrium. If we pull the swing back some distance the child will move toward the bottom of the arc. At the instant the swing is at the point of stable equilibrium, the child is moving the fastest since as the swing proceeds up the arc on the other side, it slows down. The higher the swing was when the motion was started, the faster the child moves at the bottom. The swing overshoots stable equilibrium and the child rises to the same distance on this side of the bottom as on the starting side. For a brief instant the swing will stop before the swing begins to retrace its path, traveling in the other direction.

This simple example demonstrates several properties shared by all oscillations: 1) The point of stable equilibrium is the center of the oscillating motion since the object moves the same distance on either side. That distance is called the amplitude of the oscillation. 2) At either end of the motion, the object stops briefly (slowest location) while the fastest location is when the object is just passing through the point of stable equilibrium. 3) The energy that an object has when it is oscillating is related to the amplitude. The larger the amplitude, the larger the energy.

Oscillations also have two very specific properties regarding time. Every oscillation takes a certain amount of time before the motion begins to repeat itself. Since the motion repeats, we really only need to worry about what happens in one cycle, or repetition of the oscillation. If we pick any point in the motion and follow the object until it has returned to that same point ready to repeat, then the oscillation has completed one cycle. The amount of time that it took to complete one cycle is called the period, and every cycle will take the same amount of time. Suppose for the child on a swing we pick the point at the bottom of the arc. When the swing moves through

that position, we start a timer. The child will swing up, stop, swing back down through the bottom (but traveling in the other direction), swing up, stop, and swing back through our point. Now the child has returned to our starting point and the motion is about to repeat, so we stop our timer. The curious thing is that even when the amplitude is changed, the period stays the same. This is because even though the child moves faster when pulled higher to start the motion, the swing also has farther to travel to complete a cycle.

The other time property is called the frequency, which tells how often the motion repeats. This really gives the same information as the period since if it takes 0.5 second for 1 cycle, then the frequency will be (1 cycle)/(0.5 second) = 2 cycles per second. Often a unit called the Hertz (Hz) is used to represent a cycle per second. The cycles per second should sound familiar because the magnitude, or amplitude, of the electrical current in most households oscillates at 60 cycles per second. The time properties of an oscillation are very important since they control how best to add energy to the motion.

The child on a swing, weight on a spring, and bungee jumper on an elastic cord are all types of oscillations which we can see with our eyes. However, if you kick a rock (disturb it) and it does not disintegrate, then the atoms within the rock must be in stable equilibrium. The atoms within the rock are therefore capable of oscillating. Small oscillations are actually occurring all the time in every seemingly solid substance, including the rock and this page. We cannot see this motion, but we do feel it. The larger the amplitude of those small oscillations of the atoms, the hotter the object, and that is something we can detect directly.

P

P VERSUS NP PROBLEM

The premier unsolved question in the study of algorithms, the "P versus NP" problem, asks whether there is really any such thing as a "hard" problem for a computer. The security of financial transactions on the Internet depends upon it. All encryption algorithms start from the basic assumption that decoding a message is "easy" for someone with the right decoding key, but "hard" for someone who is trying to steal the message. Yet no one has ever proven that this assumption is correct. In the unlikely event that it is not—that is, if mathematicians ever discover a way to crack "hard" problems (formally known as "NP-complete" problems)—the viability of commerce on the Internet would be seriously threatened.

Factorization is a classical example of an apparently hard problem. Most of us can calculate 127 ö 83 (which equals 10,541) much faster than we could solve the inverse problem (i.e., find the factors of 10,541). In general, when multiplying two n-digit numbers by hand, we compute roughly n^2 products (each digit of the first number times each digit of the second) and roughly $2n$ sums (one for each digit of the final answer). So it takes about $n^2 + 2n$ steps to find the answer. Since $n^2 + 2n$ is a polynomial, the **multiplication algorithm** is said to take "polynomial time" with respect to the length of its input.

Now compare this to the number of steps it takes to find the factors of a $2n$-digit number, X. We could try a trial-and-error approach: simply divide X by every number until we either find one that divides into it exactly, or until we have tried every number with n digits or fewer. We might get lucky and find a small **factor**, such as 2 or 3 or 7—but in the worst case, we might need to go through every number up to 10^n (because that is how many numbers have n digits or fewer). Hence, in a worst-case scenario, this algorithm takes at least 10^n steps. Because 10^n is an exponential function, the factorization algorithm is said to take "exponential time" with respect to the length of its input.

Exponential-time algorithms are vastly inferior to polynomial-time algorithms. If $n = 100$, for example, the above arguments suggest that it would take about 10,200 steps to multiply two n-digit numbers and get a $2n$-digit number. But the reverse process, factoring a $2n$-digit number into two n-digit numbers, would require at least 10,000,000,000,... (1 followed by 100 zeros) steps. Not surprisingly, then, there is virtually no limitation on the size of the numbers that computers can multiply; but at present they cannot even come close to factoring 200-digit numbers. The first and most popular "public-key" cryptosystem, the Rivest-Shamir-Adleman (RSA) algorithm, is based on precisely this contrast between the relatively easy problem of multiplication and the intractable inverse problem of factorization.

Yet, remarkably, no one has ever been able to prove with that this contrast is inevitable. Some day, someone might find a polynomial-time algorithm to factor numbers. (In fact, Peter Shor of AT&T Labs discovered such an algorithm that works on a quantum computer, taking advantage of the fact that quantum systems can exist in a superposition of many distinct states. However, the technical problems make it questionable whether a working quantum computer will ever be built. Until then, the solution time of algorithms is generally stated for conventional computers.)

Problems that can be solved by an algorithm in polynomial time are said to belong to class P. A broader class is NP: the problems whose solutions can be checked in polynomial time. (This class, as well as the "P versus NP" problem, was first named by Stephen Cook of the University of Toronto in 1971, although he defined the class NP in a more complicated way.) The hardest problems, in a sense, are NP-complete. These are problems which, if one could solve them quickly, would provide an "oracle" for solving any other NP problem. Of the problems discussed so far,

- Multiplication is in class P, therefore also in class NP;
- Factorization is in class NP but is not known to be in P. It is also not known to be NP-complete.

The million-dollar question is: Are the classes P and NP really different? If they are the same, then there are no hard problems, because everything that can be checked in polynomial time can also be solved in polynomial time. As mentioned before, this would be a disaster for e-businesses and a delight for hackers. On the other hand, if P is different from NP, then hard problems do exist. That is what most mathematicians believe, and what most Internet security experts are counting on.

Because the factorization problem is not known to be NP-complete, it is conceivable that it could turn out to be "easy" without affecting the status of the "P versus NP" problem as a whole. On the other hand, if a polynomial-time solution is found for even one NP-complete problem, such as the "traveling salesman problem" of graph theory, then the theoretical distinction between easy and hard problems will essentially cease to exist.

At present, mathematicians seem to be a long way from either proving either that P and NP are different or that they are the same. In 2000, this was named as one of the seven "Millennium Prize Problems" by the Clay Mathematics Institute. Anyone who publishes a solution that is accepted by the mathematical community will receive a one-million dollar prize.

PACIOLI, LUCA BARTOLOMES (CA. 1445-CA. 1517)

Italian scholar and author

Luca Bartolomes Pacioli is widely regarded as the developer of the double-entry accounting system, which revolutionized the way people kept track of money and is still the dominant accounting method in use today. He wrote extensively on the subject and helped to make double entry the standard accounting practice throughout the world.

Pacioli was born in Sansepolcro, Italy in about 1445. For some reason not clear today, the boy grew up with the nearby Befolci family. It is said that the great artist Piero de la Francesca fostered and encouraged Pacioli's love of mathematics from an early age, perhaps giving him part of his formal education. As a young man, Pacioli served as an apprentice to a wealthy Venetian merchant, also living with him and tutoring his three sons while learning mathematics at a nearby church school.

After writing his first work on **arithmetic** in 1470, Pacioli stayed briefly in Rome at the home of architect **Leone Alberti**, who served as Pacioli's patron and friend. Sometime in the next several years, however, Pacioli was ordained as a friar in the Franciscan order and completed theological studies at an Italian monastery. Some sources say that Alberti's death prompted this sudden, life-changing action. Pacioli then began wandering throughout Italy, teaching arithmetic in some of the larger cities before settling at the University of Perugia from 1477 to 1480. There he wrote a treatise on mathematics in 1478.

Beginning in 1481, Pacioli started another period of itinerant teaching, working for brief periods in Zara (in what is now Croatia), Naples, and Rome, during which he completed the equivalent of what today would be a doctoral degree. In 1489 he returned to Sansepolcro and spent a few years there writing his largest work, *Summa de Arithmetica, Geometrica, Proportioni et Proportionalita (Everything about Arithmetic, Geometry, and Proportion).* He took the manuscript to Venice, where it was printed on Guttenberg's expensive new metal type machine, and published it in 1494. Although the book had 36 chapters on bookkeeping, the topic represented only a fraction of what the book covered, since it was intended as a compendium of existing mathematical knowledge. Thus, it did not contain a great deal of original work, but rather summarized the contributions of many important mathematicians. Pacioli openly acknowledged his use of their material, although he would encounter much criticism for his liberal use of others' work.

Pacioli's bookkeeping chapters, collectively entitled "De computis et scripturis" ("Of Reckonings and Writings"), discussed in detail the double entry system, known then as "the method of Venice," that would change accounting forever. Although Benedetto Cotrugli is actually thought to have invented the system, it was Pacioli who developed and codified it in his huge work, making it accessible to anyone who could use it by writing the treatise in laymen's terms. The first 16 chapters describe the basic layout of books and accounts. The Pacioli system comprises three parts: the memorandum, for recording transactions as they occurred; the journal, a private account book for the merchant; and the ledger, which features a double page with debits on the left and credits on the right. The remaining 20 chapters concern merchant-related issues such as deposits and withdrawals, drafts, barter transactions, expense disbursements, and closing and balancing the books. The only real differences between Pacioli's double entry accounting and that in use today are related to the needs of larger-scale businesses.

In 1496, Pacioli accepted an invitation to teach mathematics in Milan at the royal court. There he met **Leonardo da Vinci**, with whom he soon became close friends. In fact, when the French Army marched into Milan in 1499, the men fled together to Venice. In 1500, Pacioli began teaching **geometry** at the University of Pisa, remaining there until 1506. Meanwhile, Pacioli and da Vinci had remained close friends; Da Vinci even illustrated the cover and some of the figures of Pacioli's 1509 trilogy *Divina proportione (The Golden Ratio).*

Pacioli returned to the University of Perugia as a lecturer in 1510. He taught again in Rome in 1514, but soon returned to Sansepolcro, where he died in about 1517.

PAPPUS OF ALEXANDRIA (CA. 300-CA. 350)

Greek mathematician

Pappus of Alexandria was a late Greek geometer whose **theorem**s provided a foundation for modern projective **geometry**. Virtually nothing is known about his life. He wrote his major work, *Synagoge,* or the *Mathematical Collection,* as a guide to Greek geometry. This collection of mathematical writings in eight books is thought to have been written around 340, although some historians believe that Pappus had completed the work by 325. Pappus' work discusses the theorems of more than thirty different mathematicians of antiquity, including **Archimedes**, **Euclid**, **Apollonius**, and **Ptolemy**.

Synagoge is important for two reasons. First, it was a type of concordance to the study of Greek geometry with historical interpretations and amendments to existing theorems. The initial purpose of the text was to allow the student to use the *Synagoge* as a supplement while the original works are read. He generalized the postulates and theorems of his predecessors, as with the **Pythagorean theorem** found in Euclid's *Elements*.

Secondly, *Synagoge* was composed in an era that witnessed the gradual decay of the classical world—it could thus be construed as an attempt on Pappus' part to preserve the mathematical traditions of the Greco-Roman world. The text chronicles many of the works of ancient mathematicians and philosophers. Although the first book and part of the second are lost, it is a highly valuable historical record of Greek mathematics.

See also Geometry

PARABOLA

Geometrically, a parabola is the set of all points in the plane which are equally distant from a fixed point, called the focus, and a fixed line, called the directrix. We will say more about this geometric **definition** below. Algebraically, a parabola is the graph, in the x-y coordinate system, of an equation of the form $y=ax^2+bx+c$. Such **equations** define **functions** called "quadratic" functions. The simplest of all **quadratic functions** is defined by $y=x^{*2}$. This function's graph is the parabola with its lowest point, the vertex, at the origin and which is symmetric to the y-axis. It passes through the set of points of the form (a,a^2); in other words, the second coordinate of any point on this parabola is the **square** of the first coordinate. All other parabolas can be generated from the $y=x^2$ parabola with some combination of translations, reflections, dilations, and/or rotations. For this reason, we sometimes call the $y=x^2$ parabola the "parent" of all other parabolas. Since all other parabolas "inherit" the basic characteristics of the $y=x^2$ parabola, the mathematician can learn most of what she needs to know about parabolas by studying the parent in depth.

Using the geometric definition, given in the first sentence of this article, let the focus have coordinates $(0,p)$ and the directrix have equation $y=-p$. One point on this parabola must then be at $(0,0)$ since that point is halfway between $(0,p)$ and $(0,-p)$, and this latter point is on the line $y=-p$, the directrix. In fact, $(0,0)$ is the vertex of this parabola. If (x,y) is a **variable** point on the parabola, then the **distance** formula from **algebra** gives $(x-0)^2+(y-p)^2=(y+p)^2$. Expanding the binomials in this equation and "simplifying," we arrive at the equation $x^2=4py$, or $y=(1/(4p))x^2$. Now in the special case where the focus is at $(0,1/4)$ and the directrix is $y=-1/4$, we have the "parent" parabola $y=x^2$. In fact, any parabola whose vertex is at the origin may be mapped onto the $y=x^2$ parabola and vice versa by a **transformation** called a dilation or size transformation. A dilation is a transformation which maps points of the form (x,y) onto points of the form (kx,ky), where k is a nonzero real number called the dilation **factor** or magnitude. In particular, the $y=x^2$ parabola can be mapped onto a parabola with equation $y=ax^2$ by the dilation which maps (x,y) onto $((1/a)x,(1/a)y)$. Dilations belong to a class of transformations called **similarity** transformations, which means that dilations map geometric figures onto other geometric figures that are similar to the original figures. In fact, the transformational definition of similarity is: Two figures are similar if and only if one can be mapped onto the other by a similarity transformation. This gives us the rather remarkable fact that all parabolas with vertex at the origin are similar to the $y=x^2$ and, by transitivity, all parabolas centered at the origin are similar to one another. In other words, they all have the same shape in the same sense that all circles have the same shape, or all similar triangles have the same shape. This is hard for the beginning student to understand because when one graphs, for example, $y=x^2$ and $y=2x^2$ on the same axis system, the $y=2x^2$ parabola looks "thinner" or is closer to the y-axis for any specific value of x than is the $y=x^2$ parabola. But if the graphs are accurately plotted on two separate transparencies, with the scale for the $y=2x^2$ parabola adjusted by a factor of ½, then if one transparency is placed on top of the other so that the origin and axes coincide, the portion of the $y=2x^2$ parabola on its transparency will exactly cover the portion of the $y=x^2$ parabola on its transparency. The story does not end here, however. As mentioned above, every parabola with vertex at the origin can be mapped onto any other parabola in the plane by translations, reflections, and rotations. These are also similarity transformations and this means that all parabolas are similar.

The mathematical properties of parabolas make them excellent models for physical objects in which a focusing component is essential. It can be shown that parallel lines drawn on the inside of any parabola are reflected from the curve of the parabola to its focus. Thus, many telescopes are designed using parabolic reflectors with the **light** collection instrument located precisely at the focus of the parabola. Satellite television receivers use this same focal property to gather television signals from satellites in stationary **orbit** around the earth. Searchlights that require concentrated beams of light use this property in reverse. The light source is located at the focus of a parabolic reflector so that when the light is turned on, it bounces off the sides of the reflector and is directed outward as beam of parallel light rays. Parabolas also **model** the **motion** of a body in free fall towards the surface of the earth. Isaac Newton's theories of gravitation and motion lead to an equation of the parabolic form $y=-16t^2$ for a body falling towards the earth with negligible air resistance. In this case, time is measured in seconds and distance is measured in feet. Newton also showed that the path of a projectile launched from a point on the earth's surface will follow a parabolic path until it hits the ground at the end of its flight. Parabolas are also used in the design of bridges and other structures involving arches.

See also Quadratic functions

PARALLEL POSTULATE

The parallel **postulate** was Euclid's famous, and sometimes infamous, fifth postulate. Euclid's five postulates, or axioms, as they are sometimes called, appear in his masterpiece the *Elements*, one of the most influential mathematics books ever written. Euclid's program in the *Elements* was to build a deductive **geometry** starting with the minimum number of postulates necessary. Since postulates are essentially assumptions stated without **proof** and since a deductive system is primarily about proving theorems, mathematicians from Euclid (c. 300 BC) to the present have desired to assume as little as necessary and prove as much as possible. Euclid thought that five, but no more than five, postulates were necessary to build up his deductive geometry. The first four of Euclid's postulates caused no controversy at all. They were four simple assumptions that nobody could doubt; and those are the properties, simplicity and certainty, for which one looks in a postulate or **axiom**. The first four of Euclid's five postulates are stated here for reference:

1. A straight line segment can be drawn joining any two points.

2. Any straight line segment can be extended indefinitely in a straight line.

3. Given any straight line segment, a **circle** can be drawn having that segment as its radius and one endpoint of the segment as its center.

4. All right angles are congruent.

These four postulates have very simple statements and are about as "self-evident" as statements get. Unfortunately, the same cannot be said of the fifth postulate:

5. If two lines are drawn which intersect a third in such a way that the sum of the inner angles on one side is less than two right angles, then the two lines inevitably must intersect each other on that side if extended far enough.

Note that in this form, postulate 5 is neither simple nor obvious. This fact was to account for much consternation, as well as much good mathematics, during the next two millennia. The fifth postulate looks so out of place among the other four, that generations of mathematicians were troubled by its inclusion as a postulate in the *Elements*. Several attempts were made to come up with a simpler, but equivalent, form of the postulate. The earliest on record was a version by the Greek philosopher and mathematician Proclus (410-485) which said, "If a line intersects one of two parallel lines, it must intersect the other line also." This is certainly a simpler statement than Euclid's and probably more "obvious," but, as we shall see, the use of words such as "obvious" and "self-evident" can sometimes depend upon subtleties of which we may be unaware. Another very simple statement that is equivalent to Euclid's fifth postulate is this, called the Equidistance Postulate: "Parallel lines are everywhere equidistant." Perhaps the most famous statement that is equivalent to the parallel postulate is due to John Playfair (1748-1819) and is called Playfair's Axiom: "Given a line and a point not on that line, there is exactly one line which passes through the given point and is parallel to the given line." This is perhaps a more complicated statement than the equidistance pos-

tulate, but, historically, Playfair's Axiom is the one most often used in discussions of the parallel postulate. Therefore, we will stay with that tradition in the remainder of this entry.

Even with the simpler versions of Euclid's fifth postulate in place, many mathematicians remained uneasy about its presence with the other four postulates. A suspicion arose that the parallel postulate was not independent of the other four as it should be if it is truly needed as a postulate. If the parallel postulate were not independent of the other four, then it should be possible to prove it as a **theorem** by using some combination of the first four postulates. This became the program of a number of mathematicians of the seventeenth and eighteenth centuries, most notably the Italian mathematician Geralamo Saccheri (1667-1733). Saccheri, with great gusto, set out to "vindicate Euclid of every blemish" by showing that the parallel postulate could be proved using just the first four of Euclid's postulates. This would show that the fifth "postulate" was really a theorem and did not need to be assumed. Alas, Saccheri's "proof" at the end of a very long treatise was found to be fallacious. However, the enterprise was not a total waste of Saccheri's effort, for in the first part of his book, which was without error, he developed a geometry completely independent of the parallel postulate. This made Saccheri, unwittingly, the inventor of what is now called absolute geometry and has been a very fruitful area of mathematical research.

Following Saccheri's "accidental" development of absolute geometry, a host of eighteenth and nineteenth century mathematicians turned their attention to the question of what would happen if the parallel postulate were replaced by an axiom that was contradictory to it. This could take one of two directions. The first, proposed by the Russian mathematician **Nikolai Lobachevsky** (1792-1856), replaces the parallel postulate with this: "Given a line and a point not on that line, there are at least two lines which pass through the given point and are parallel to the given line." At first this seems counter-intuitive because we are used to thinking in Euclidean terms, i.e. the geometry of a "flat" plane. Of course, this is precisely the geometry which results from assuming the parallel postulate; but with the parallel postulate replaced by Lobachevsy's, a different geometry arises, which, nevertheless, is just as internally consistent as Euclidean geometry. The great French mathematician Henri Poncaré (1854-1912) produced a **model** of a geometry which satisfied Lobachevsky's axiom. In this model, unfamiliar things happen. For example, the sum of the measures of the angles of a **triangle** is always less than 180 degrees, whereas in Euclidean geometry, this sum is always equal to 180 degrees. Nevertheless, Poincaré's model showed that **Lobachevskian geometry**, also called **hyperbolic geometry**, is just as consistent as Euclidean geometry. The second direction one may take in denying the parallel postulate is to replace it with this axiom: "Given a line and a point not on that line, there are no lines which pass through the given point and are parallel to the given line." This was the direction taken by the great German mathematician **Bernhard Riemann** (1826-1866). Note that Riemann's replacement of the parallel postulate is equivalent to saying that no parallel lines exist in this geometry. At first this may seem absurd, but consider the surface of a **sphere** in which "lines" are defined to be the great circles, e.g., lines of longitude on the surface of the earth. Then all lines intersect at the north and south poles, meaning that none are parallel. In this "Riemannian" or "spherical" geometry, the sum of the angles of any triangle is always greater than 180 degrees and some triangles have two right angles. Yet this geometry is also just as consistent as Euclidean geometry. In fact, **Riemannian geometry** is the geometry of Albert Einstein's Relativity Theory, which changed the way physicists view the universe. Lobachevskian geometry and Riemannian geometry are examples of **non-Euclidean geometries**. Their developments are among the great results of nineteenth century mathematics and both are the result of denying the "obvious" parallel postulate of Euclid.

See also Euclid; Euclidean Axioms; Euclidean geometry

PARALLELOGRAM RULE

The parallelogram rule, also called the parallelogram law, provides a straightforward means to perform vector **addition** with two vectors in two-dimensional **space**. It can also be extended to 3-dimensional or even *n*-dimensional space, but is most easily understood in the two-dimensional **model**. Simply stated, the parallelogram law states that the sum of two two-dimensional vectors *P* and *Q* can be determined by:

- drawing the vector *P*, then
- drawing the vector *Q*, taking care to place the tail of vector *Q* at the head of vector *P*, and finally
- drawing a vector *S* from the free tail of vector *P* to the free head of vector *Q*

This third vector *S*, which extends from the free tail vector *P* to the free head of vector *Q*, represents the sum of the two vectors. If the process is repeated from the same starting point—but initiated with vector *Q* instead of vector *P*, a parallelogram will result. The sum vector *S* becomes the long diagonal of the parallelogram. Interestingly enough, the difference vector *D*, which is equal to *P* - *Q*, is the short diagonal of the parallelogram.

Basics of Vector Addition

Suppose two vectors *P* and *Q* exist in two-dimensional space such that $P = (x_1, y_1)$ and $Q = (x_2, y_2)$. Under vector addition, the sum *S* of these two vectors is simply $S = P + Q = (x_1, y_1) + Q = (x_2, y_2) = Q = (x_1 + x_2, y_1 + y_2)$. If $P = (3, 5)$ and $Q = (1, -1)$, for example, then $S = (4, 4)$. The concept of vector addition can be easily extended to *n*-dimensional vectors. Suppose, for example, the *n*-dimensional vector *A*, $(x_1, x_2, x_3,..., x_{n-1}, x_n)$ is added to the *n*-dimensional vector *B*, $(y_1, y_2, y_3,..., y_{n-1}, y_n)$. The result is simply the one-on-one addition of each pair of associated vector elements. That is, $A + B = (x_1, x_2, x_3,..., x_{n-1}, x_n) + (y_1, y_2, y_3,..., y_{n-1}, y_n) = (x_1 + y_1, x_2 + y_2, x_3 + y_3,..., x_{n-1} + y_{n-1}, x_n + y_n)$. As the two-dimensional example illustrated, the values $x_1, x_2,..., x_n$ in the vector may be positive, negative, or even **zero**. Note that since $(x_1 + y_1, x_2 + y_2,..., x_n + y_n) = (y_1 + x_1, y_2 + x_2,..., y_n + x_n)$, vector addition is commutative, and thus $A + B = B + A$.

Basics of Vector Subtraction

Vector **subtraction** is much like vector addition. If one wishes to find the difference between two vectors *A* and *B*, for example, it is the equivalent of adding *A* to the negative of *B* (that is, each element of the vector *B* is multiplied by -1). Thus $A - B$ is simply $(x_1 - y_1, x_2 - y_2,..., x_n - y_n)$.

Relationship of Complex Numbers to Vectors

Complex numbers, which consist of both a real number component and an imaginary number component, can be considered vectors. If a complex number is usually written in the form $a + bi$, it can written in vector form (a, b). Adding or subtracting two complex numbers then follows the same form as vector addition. Specifically, $(a + bi) + (c + di) = (a, b) + (c, d) = (a + c, b + d)$, and $(a + bi) - (c + di) = (a, b) - (c, d) = (a - c, b - d)$.

See also Addition; Commutative property; Complex numbers; dimension; Imaginary numbers; Negative numbers; Real numbers; Scalar and vector components; Space; Sums and differences; Vector algebra; Vector analysis; Vector spaces; Zero

PARAMETRIC EQUATIONS

Parametric **equations** are those that relate typical *x* and *y* values to another **variable** or arbitrarily chosen constant. Such equations are widely found in studies dealing with **motion** as a function of time. There are scalar and vector parametric equations. In vector parametric equations the arbitrary variable is called a parameter and it is related to a vector via an ordered pair. Scalar parametric equations are more commonly referred to as just parametric equations. They are the set of equations that relates the arbitrarily chosen constant or variable to the traditional Cartesian coordinates *x* and *y*. For example let us consider a cycloid that is formed as a point on a **circle** traces out a curve as the circle rolls along a line. Although one can represent the cycloid as the graph of a function it is also possible to represent the curve *C* using parametric equations. In this case the point *(x,y)* on *C* can be expressed as **functions** of time *t*: $x = f(t)$ and $y = g(t)$, where *f(t)* and *g(t)* are continuous functions on an **interval** such that *C* consists of all points *(x,y)*. In this case the equations noted above are called parametric equations of *C* and we say that *C* is parametrized by those equations with *t* being a parameter of *C*.

The vector parametric equation $r = tv + b$ can be interpreted as giving the position *r* at time *t* of a particle that is at *b* when $t = 0$ and moves with a constant vector velocity *v*. This interpretation yields a graph of the equation that is called the trajectory of the particle. These equations are only applicable to vector situations.

The more commonly used type of parametric equation is the scalar parametric equation, which from here onwards will be referred to simply as parametric equations. Relating parametric equations to the corresponding Cartesian equation is often times a simple task. Take for example the parametric equations $x = sin\Theta$, $y = cos2\Theta$. They can easily be converted to the corresponding Cartesian form: $y = cos2\Theta = 1 - 2sin^2\Theta = 1 - 2x^2$. Although the Cartesian equation can be obtained from parametric equations by eliminating the parameter from the set of parametric equations it is often not equivalent to the original set of parametric equations. This is because the coordinates of points that are not given by the parametric equations may satisfy the Cartesian equation. In order to obtain equivalent parametric and Cartesian equations any restrictions on the domain that may be implicit in the parametric form of the equation but do not appear in the Cartesian form must also be stated explicitly. In the previous example it is clear that for $x = sin\Theta$ that $-1 \leq x \leq 1$. As Θ takes on all real values in the parametric equations it is clear that the curve traces and retraces the arc of $y = 1 - 2x^2$ that is bounded by $(-1, -1)$ and $(1, -1)$. The Cartesian equation should therefore be stated with the restriction $-1 \leq x \leq 1$.

PARETO, VILFREDO FREDERICO DAMASO (1848-1923)

Italian economist, sociologist, and author

Vilfredo Pareto's application of mathematics to economic analysis and sociology did much to advance the new fields' standing as legitimate sciences. Such developments as the Pareto optimum and his curves of indifference are analytical tools still used by some modern sociologists. He is still regarded today as the founder of mathematical economics—a title granted to him early in his own lifetime.

One of three children of an exiled Italian nobleman who supported his family through a successful career as a hydrological engineer, Pareto was born in Paris, France on July 15, 1848. He enjoyed an upper-middle-class upbringing, receiving a quality education in local private schools. Pareto completed a degree in engineering in 1869 at the Polytechnic Institute of Turin, finishing first in his class.

In 1871, he returned to a unified Italy with his parents and, like his father, found employment as a civil engineer with the government-run railroad. He was transferred to Florence in 1872, but two years later accepted a position as supervisor of an industrial firm that operated mining and ironworking companies in the Arno Valley. In this position, he was required to travel to England and Scotland, where he witnessed the effects of and admired the British government's free-market policies.

Evincing an increasing interest in politics, Pareto made a bid for a parliamentary seat in 1881. It was not successful, but did signify the beginning of a new period in his life. Although he continued to work as an engineer and manager throughout the 1880s, Pareto began writing many political commentaries and editorials. When he and his first wife, whom he married in 1889, retired to the countryside later that year, his writing became prolific. The government rebuffed his offer to put on a free lecture series on political economy, so Pareto began accepting speaking engagements. This finally attracted the attention of hostile bureaucrats, who sent police

and (allegedly) paid thugs to stop his controversial speeches on economics. These mainly consisted of arguments in favor of free trade, open competition, and reduced government involvement in the economy.

In 1891, Pareto published his first papers in which he put complex economic theories into concise forms using mathematical formulas. Although the government continued to regard Pareto as a threat, he began to receive positive attention from the world of academia. It was because of this that he received an appointment as professor of political economy at Switzerland's University of Lausanne in 1893. There he quickly came to be known as "the father of mathematical economics." He published his first major work, *A Course in Political Economics,* in 1896. This contained a discussion of his controversial "law of income distribution," in which Pareto used a sophisticated mathematical technique to show that the distribution of wealth and income in society is not random, but rather follows a definite pattern that applies to all cultures the world over.

In 1898 there came another big change in Pareto's life. This took the form of a large inheritance that allowed him to buy a house in the Swiss countryside on Lake Geneva and concentrate on his writing. Although his wife left him in 1901, causing him considerable distress, most of Pareto's best work occurred during this period, including what many regard as his masterpiece, *A Manual of Political Economy* (1906). This publication developed his economic theory, expanded on his analysis of "ophelimity" (the power to satisfy), and established the foundation of modern welfare economics. The last was based on the Pareto optimum, which states that the best allocation of resources has not been achieved if it remains possible to make one person feel better about his economic situation while keeping others at the same level of satisfaction. The book also introduced curves of indifference, which became popular analytical tools for sociologists beginning in the 1930s.

It was in the early 1900s that Pareto began writing about sociology, a relatively new discipline that he felt could unify the discrete fields of economics and politics. His most influential book on this topic was *A Treatise on General Sociology,* published in English in 1916 as *Mind and Society*. Pareto considered this to be his greatest work. Here he introduced one of his most controversial ideas: the "circulation of elites." In this theory, Pareto argued that, as water will always seeks its own level, people will always sink or rise to the level of social status where they naturally belong, no matter into what sort of circumstances they were born. Based on what he called "the superiority of the elite," Pareto argued, low-born people with superior abilities would inevitably rise to the higher echelons of society while the converse was also true, whether through complacency or inferiority. He became the brunt of accusations of fascism for this belief, however.

Pareto had completed most of his important work by 1914 following years of self-imposed exile at his Swiss estate. He published *Facts and Theories* in 1920 and *The Transformation of Democracy* in 1921. Shortly after his first wife left, he fell deeply in love with another woman, Jane Regis, who became his companion and lover for the rest of his life. Pareto gradually withdrew even more from the mainstream of life, entertaining only close friends occasionally and spending many hours with his 18 Angora cats. A longtime sufferer of insomnia, he was an avid and voracious reader. Pareto and Regis married in the last months of his life after his first wife finally granted him a divorce. At about that time, he received a nomination as senator of the Italian Kingdom. Pareto died of heart disease at his Lake Geneva home in Celigny on August 19, 1923.

PARTIAL DIFFERENTIATION

Differentiation is the mathematical act of taking the derivative of a function. Partial differentiation is the mathematical act of taking the derivative of a function that depends on more than one **variable** with respect to one or more of those variables. If a differential equation contains partial derivatives with respect to more than one **independent variable** then it is called a partial differential equation (PDE). Partial differentiation is a pivotal technique used in chemistry, physics, and engineering. Partial **differential equations** are in general very difficult to solve but their importance in applications warrants their use. Many partial differential **equations** describe situations in which a property is dependent upon not only time but also on position. Such a situation would be heat flow through a thin wire which involves position in the wire as well as time.

Partial differentiation involves a process by which the derivatives of a function containing multiple independent variables are found by considering all but the variable of interest as fixed during differentiation. Partial differentiation corresponds to the same thing as ordinary differentiation in that it represents an **infinitesimal** change in the function with respect to a given parameter. The difference is that partial differentiation is performed on an equation with more than one independent variable whereas ordinary differentiation is performed on an equation with only one independent variable. The partial derivative is usually denoted as: $\partial f/\partial x$ or f_x. These denote the partial derivative of the function f, that is comprised of more independent variables than just x, with respect to x. If a partial derivative is of second order or greater with respect to two or more different variables then it is called a mixed partial derivative. For example if f is a function of x, y, z ($f(x, y, z)$) and the partial derivative with respect to x is taken followed by the partial derivative with respect to y then it is called a mixed partial derivative and is denoted as: f_{xy} or $\partial^2 f/\partial x \partial y$. For such **functions** whose partial derivatives exist and are continuous, for nice functions, the mixed partial derivatives are equal no matter which differentiation is performed first: $f_{xy} = f_{yx}$.

In general partial differential equations are more difficult to solve using analytical methods than are ordinary differential equations. This is because often times they are more complex because of the multiple variables involved. There are several methods that have been developed over the years specifically designed to solve partial differential equations. Some of these methods include the Bäcklund **transformation**, characteristic partial differential equation, Green's function, Lagrange multiplier method, integral transform, Lax pair, and separation of variables. The Lagrange multiplier method was probably the first of these methods formally devised to solve partial differential equations. Joseph Louis Lagrange, an Italian mathematician, published this method involving multipliers to investigate the **motion** of a particle in **space** that is constrained to move on a surface defined by an equation involving three, independent variables. This method was published in Lagrange's book, *Mecanique analytique*, in 1778 and is currently used to maximize or minimize a function that is subject to a constraint. It can be employed in a variety of situations such as minimizing the fuel required for a spacecraft to reach its desired trajectory as well as maximizing the productivity of a commercial enterprise limited by the availability of financial, natural, and personnel resources. As well as these methods numerical methods can be applied to solve partial differential equations.

Although partial differential equations are in general difficult to solve, second-order partial differential equations are often easily solved via analytical solutions. They can be classified as elliptic, hyperbolic, or parabolic on the basis of a particular **matrix** or the discriminate of that matrix. Each class has a solution that is quite different from the other classes. Second-order partial differential equations that fall into the elliptic class produce stationary and energy-minimizing solutions such as Laplace's equation and Poisson's equation. Those that are classified as hyperbolic yield a propagating disturbance such as the wave equation. The last of the classes, the parabolic equations, produce a smooth-spreading flow of an initial disturbance such as the heat conduction equation and other diffusion equations.

PASCAL, BLAISE (1623-1662)

French geometer, probabilist, physicist, inventor, and philosopher

Although Blaise Pascal can be seen in retrospect as an important scientist in his time, he was a controversial figure to his contemporaries. There is no doubt that Pascal was of a superior intelligence, but he was a modest man, embarrassed by his own genius. The reason for his contemporaries' doubt was perhaps in part due to the fact that much of his work was not published in his lifetime, which limited how his accomplishments could be viewed. Still, Pascal gave the world, mathematical and otherwise, many things: he opened up new forms of **calculus**, projective **geometry**, **probability theory**, and he designed and manufactured the first calculating machine run by cogs and wheels.

Pascal was born in Clermont (now known as Clermont-Ferrand), Auvergne, France, on June 19, 1623. He was the son of a mathematician and civil servant Ètienne and Antoinette (nee Bégon) Pascal. His mother died when Pascal was three. With his two sisters, Gilberte (Madame Périer after marrying Florin Périer in 1641) and Jacqueline, Pascal was educated at home, primarily by his father. Pascal was part of a very tightly knit family, and he was especially close to his sisters. Pascal's sister Jacqueline was a child literary prodigy, and, despite being sickly all his life, Pascal was also a gifted child. In 1631, the family moved to Paris, where Pascal began his mathematical education when he was 10 or 11. His father insisted that his education start with the study of ancient languages, before learning geometry.

At age 13, Pascal and his father began attending discussions in Paris with a group of scientists and mathematicians, such as **René Descartes,** called the Académie Parisienne. By the time he was 16 years old, Pascal had already done a significant amount of his mathematical groundwork. He continued his studies in Rouen where the family moved in 1639, when his father was appointed to the tax office there. Pascal continued to go to Paris occasionally while living in Rouen, and it was on one of these trips that he presented one of his important mathematical discoveries.

Published in 1640 as a pamphlet, *Essai sur les coniques* was a vital step in the development of projective geometry and it contained what came to be known as Pascal's mystic hexagram. Pascal began writing this treatise on **conic sections** to clarify the 1639 publication of **Gérard Desargues'** *Brouillon project d'une atteinte aux evenements des rencontres du cone avec un plan* ("Experimental project aiming to describe what happens when the cone comes into contact with a plane"). Desargues had written his book in a manner that was very difficult to understand, even for other mathematicians.

But as Pascal began to work with the **propositions** that Desargues made, he went beyond what Desargues accomplished. Pascal developed his own **theorem** which he used to deduce some 400 propositions as corollaries. His theorem, describing a figure known as Pascal's mystic hexagram, states that the three points of intersection of the pairs of opposite sides of a hexagon inscribed in a conic are collinear. When he presented his findings to the Académie Parisienne, Descartes could not believe a 16–year–old boy had written this work. Only part of the manuscript was published in the 1640 essay, but a whole manuscript did exist at one time.

Soon after this pamphlet was published, in 1641, Pascal's health began to decline. He suffered from headaches (perhaps caused by a deformed skull), insomnia, and indigestion, but he continued his work. To help with his father's lengthy tax work in Rouen, Pascal worked on what became the first manufactured **calculator** from 1642 to 1644. This machine could automatically add and subtract, using cogged wheels to do the calculations. The invention was patented and Pascal received a monopoly by a royal decree dated May 22, 1649. He wanted to manufacture these machines as a full scale business enterprise but it proved too costly. The basic principle behind Pascal's calculator was still used in this century before the electronic age.

The year 1646 was key in Pascal's life. He became part of an anti–Jesuit Catholic sect called Jansenism, which believed in predestination and that divine grace was the only way to achieve salvation. He persuaded his family to join him, and the influence of Jansenism played a dominant role in the rest of his life.

Pascal also began doing work in physics, conducting experiments in atmospheric and barometric pressure, and vacuums. Pascal used the theories of **Evangelista Torricelli** as a starting point for his work. In Pascal's experiments, he had his brothers–in–law climb Puy de Dôme with tubes filled with different liquids to test his theories. The results were not just more information about atmospheric pressure, for Pascal

invented the syringe and the hydraulic press based on them. More importantly, he delineated what came to be known as Pascal's principle, which says that pressure will be transmitted equally throughout a confined fluid at rest, regardless of where the pressure is applied.

Pascal published some of his results in 1647 under the title *Experiences nouvelles touchant le vide*, and in 1648 as *Récit de la grande experience sur l'equilibre des liqueurs*. He completed the work already done on hydrostatics theory, bringing together the **mechanics** of both fluids and rigid bodies. His whole treatise on this subject was not published until a year after his death.

The Pascal family returned to Paris in 1647. His father died there in 1650, and his sister Jacqueline entered the Jansenism Convent at Port-Royale. Pascal himself had a profound religious experience four years later on the night of November 23, 1654. That night, Pascal was nearly killed in a riding accident. A few months later, Pascal left the secular world to live in the Port-Royal Convent. He did a little more work in mathematics and science, but primarily published religious philosophy.

Before Pascal's religious experience, earlier in 1654, he and **Pierre de Fermat** began writing each other about problems on dice and other games of chance. This **correspondence** laid the foundation for the mathematical theory of probability.

During his work on probability, Pascal made a comprehensive study of the **arithmetic triangle**. Although this triangle of numbers was more than 600 years old, Pascal used it so ingeniously in his probability studies that it became known as **Pascal's triangle**. His work on the binomial coefficients that make up the triangle helped to lead **Isaac Newton** to his discovery of the general **binomial theorem**.

Pascal's last mathematical work was on the cycloid, the curve traced by the **motion** of a fixed point on the **circumference** of a **circle** rolling along a straight line. This curve was known as far back as the early 16th century, but in 1658, over the course of eight intensive days of effort, Pascal solved many of the remaining problems about the geometry of the cycloid. The "theory of indivisibles," a forerunner of **integral calculus**, allowed him to find the **area** and center of gravity of any segment of the cycloid. He also computed the **volume** and surface area of the solid of revolution formed by rotating the curve around a straight line. As was customary in those days, once having found these solutions, Pascal proposed a challenge to other mathematicians to solve a set of problems about the cycloid and offered two prizes. Neither prize was ever awarded and Pascal eventually published his own solutions.

The fruition of Pascal's correspondence with Fermat came in 1658, when he was trying to forget the pain of a toothache. Pascal came up with solutions to problems related to the curve cycloid, also known as roulette. He solved the problems using what became known as Pascal's arithmetic triangle (also known as the triangle of numbers) to calculate probability. His results were published in 1658 as *Lettre circulaire relative a la cycloïde*. This work played a major role in the development of calculus, both differential and integral.

With this framework, areas and volumes could be calculated, and **infinitesimal** problems could be solved.

Though Pascal had been sickly all his life, his health became much worse later in 1658. His last project was developing a public transportation system of carriages in Paris in the first part of 1662. He did not live to see the system running. He died in his sister Gilberte's home on August 19, 1662, probably of a malignant stomach ulcer. Before his death, he may have parted company with his Jansenist friends.

Pascal was often underestimated in his time, and the bulk of his work was published posthumously. For example, *Traité de la pesanteur de la masse de l'air* was published in 1663, and *Traité du triangle arithmétique* in 1665, although these are only two of many.

PASCAL'S TRIANGLE

Pascal's triangle is a well-known set of numbers aligned in the shape of a pyramid. The numbers represent the binomial coefficients. Binomial coefficients represent the number of subsets of a given size. The numbers in Pascal's triangle are also the coefficients of the expansion of $(a+b)^n$, $(a+b)$ raised to the n-th power. So for n equals to three, the expansion is $(a+b) \times (a+b) \times (a+b)$ which equals $(a^2 + 2ab + b^2) \times (a+b)$ which equals $(a^3 + 3ab^2 + 3ba^2 + b^3)$. The coefficients are 1,3,3,1. These are listed in the third row of Pascal's triangle.

Pascal's triangle was also known as the Figurate Triangle, the Combinatorial Triangle, and the Binomial Triangle. The triangle was first given the name, "Pascal's triangle," by a mathematician named Montmort in 1708. Montmort wrote the numbers in the form below known as the combinatorial triangle.

1 1 1 1 1 1 2 3 4 1 3 6 1 4 1

The combination of numbers that form Pascal's triangle were well known before Pascal, but he was the first one to organize all the information together in his treatise, "The Arithmetical Triangle." The numbers originally arose from Hindu studies of **combinatorics** and binomial numbers, and the Greek's study of **figurate numbers**. The Chinese also wrote about the binomial numbers in "Precious Mirror of the Four Elements" in 1303. The figurate numbers were known over 500 years before Christ. There are **square** and triangular figurate numbers. The first four of each are shown below.

The Triangular numbers:

1 3 6 10

The Square numbers:

1 4 9 16

Additional square and triangular numbers are formed by increasing the size of each respectively. Actually, figurate numbers can be formed from any polygon. Another set of figurate numbers could be formed using the pentagon, a polygon with five sides. The figurate numbers were studied heavily to learn about counting numbers and arrangements. For example, if a woman was asked to determine which of two sacks of gold coins was worth more, she would probably have to count the coins. To count the coins, the best approach would be to stack the coins into short stacks of a given number. Then the number of stacks could be counted. Counting numbers, looking at the patterns and studying the ways objects could be arranged led to the numbers in Pascal's triangle. The study of combining or arranging objects by various rules to create new arrangements of objects is called combinatorics, an important branch of mathematics. Pascal's triangle in its current form is shown below. It is the same as the above combinatorial triangle rotated 45 degrees clockwise.

$$
\begin{array}{lccccccccccc}
(A+B)^0 & & & & & & 1 \\
(A+B)^1 & & & & & 1 & & 1 \\
(A+B)^2 & & & & 1 & & 2 & & 1 \\
(A+B)^3 & & & 1 & & 3 & & 3 & & 1 \\
(A+B)^4 & & 1 & & 4 & & 6 & & 4 & & 1 \\
(A+B)^5 & 1 & & 5 & & 10 & & 10 & & 5 & & 1 \\
\end{array}
$$

Each new row in Pascal's triangle is solved by taking the top two numbers and adding them together to get the number below.

The triangle always starts with the number one and has ones on the outside. Another way to calculate the numbers in Pascal's triangle is to calculate the binomial coefficients, written C(r;c). A formula for the binomial coefficients is r! divided by c! × (r-c)!. The **variable** r represents the row and c, the column, of Pascal's triangle. The exclamation point represents the **factorial**. The factorial of a number is that number times every integer number less than it until the number one is reached. So 4! would be equal to $4 \times 3 \times 2 \times 1$ or 24.

Binomial coefficients are written C(r;c) and represent "the number of combinations of r things taken c at a time." The numbers in Pascal's triangle are simply the binomial coefficients. The importance of binomial coefficients comes from a question that arises in every day life. An example is a how to take three books from a shelf two at a time. The first two books alone would be one way to take two books from a set of three. The other ways would be to take books two and three or books one and three. This gives three ways to take two books from a set of three. For larger arrangements, listing the number of combinations can be nearly impossible. So instead, the binomial coefficient can be found instead. For the above three books taken two at a time, all that needs to be found in the binomial coefficient C(3,2), which is the third row and second column of Pascal's triangle, or three.

Blaise Pascal (1623-1662), a founder of the theory of probability, developed the earliest known calculating machine that could perform the carrying process in **addition**. The machine, finished in 1642, could add numbers mechanically using interlocking dials. Machines like these eventually led to the first punch card machines and computers. Pascal had a great influence on people like Leibniz and Newton. His father was also a mathematician, and made sure Blaise had the best education possible by introducing him to the **Marin Mersenne**'s "Academy" at the age of fourteen. The academy was one of the best places to study mathematics at the time, and his father was one of the founders. When Pascal was young, he was introduced to the work done in combinatorics and the binomial numbers. His paper compiling the work of the Chinese, Hindus and Greeks would later cause his name to be permanently attached to the combinatorial triangle forever.

A number of unsolved problems in Pascal's days encouraged the formation of **probability theory**. The Gambler's Ruin and the Problem of Point are two examples of such problems.

The Gambler's Ruin was a problem Pascal challenged the great mathematician, **Pierre de Fermat**, to solve. The problem, according to one explanation, was determining what the chances of winning were for each of two men playing a game with two dice. When an eleven was thrown on the dice by the one man, a point would be scored. When the second man threw a fourteen on the dice, he would score a point. The points only counted if the opponent's score was **zero**. Otherwise, the point scored by one of the men would be subtracted from his opponent's score. So one of the men would always have a score of zero throughout the game. The game was won when one man gained twelve points. Pascal asked, what was the probability of each man winning? Binomial coefficients can be used to answer the question.

The Problem of Points was also a game about probabilities. The question was determining how a game's winnings should be divided if the game was ended prematurely. Questions about games like these stirred the development of probability theory, and the need to understand binomial numbers completely.

PASCH, MORITZ (1843-1930)

German geometer

Moritz Pasch's mathematical work provided one of the foundations for modern mathematics, especially **geometry**. In fact, Pasch was the first mathematician since **Euclid** who presented geometric elements in relationships defined by abstract, formal axioms. By doing this, he discovered that there were a certain amount of assumptions in Euclid's work that had been previously overlooked. In conjunction with this work, Pasch also developed Pasch's **axiom**, which deals with lines and triangles. He played a central role in the development of the axiomatic method that is a central feature of 20th-century mathematics.

Pasch was born in Breslau, Germany (now Worclaw, Poland) on November 8, 1843. He attended the Elisabeth Gymnasium in Breslau, and completed his education in Berlin and Giessen. Pasch began studying chemistry then switched to mathematics at the behest of a mentor. He published his dissertation in 1865.

Pasch spent his teaching career at one institution, the University of Giessen. He began there in 1870 as a lecturer, then moved rapidly up the ranks to a full professorship in 1875. In 1888, Pasch became chair of the department. Pasch also served the university in administrative capacities, serving as dean in 1883, and, in the 1893–94 academic year, as rector (a position equivalent to that of president). Leading a relatively simple life,

Pasch also married at one point and with his wife, had two daughters, though his wife and one daughter died young.

As a mathematician, Pasch spent the first 17 years of his career studying **algebraic geometry**, then moved onto foundations. He published his most important work, *Vorlesungen über Neuere Géométrie*, in 1882. In this treatise, Pasch addressed descriptive geometry from an axiomatic point of view. His work is distinguished because he eschewed physical interpretation of mathematical terms in favor of purely formal axioms. He also defined what is now known as Pasch's axiom, which states that in a **triangle** ABC, if a line enters through side AB and does not pass through the C, then it must leave the triangle either between B and C, or between C and A.

Pasch retired from the University of Giessen in 1911 to devote himself to mathematical research full time. In 1923 his work was recognized by his peers when Pasch received two honorary doctoral degrees from the University of Freiburg and the University of Frankfurt. He died seven years later in Bad Hamburg while on vacation on September 20, 1930.

PEANO AXIOMS

In 1889, the Italian mathematician **Giuseppe Peano** (1858-1932) published the first set of axioms, or assumptions, upon which to build a rigorous system of **arithmetic**, **number theory**, and **algebra**. Peano was a professor and developer of advanced mathematics, but he had become troubled that the grand edifice of mathematics built up over the course of 2000 years rested on the shakiest of foundations, namely, ordinary arithmetic. Sure, every school child knew the rules of counting, **addition**, **subtraction**, **multiplication**, and **division**; but whence came these "rules"? What reason do we have to believe that these rules are valid? **Euclid** (c. 300 BC) had placed a firm foundation under plane **geometry** with his set of five axioms, from which the whole of classical geometry could be derived deductively. This was Peano's goal for arithmetic. To reach this goal, he began, as Euclid did, with five axioms:

Axiom 1. 0 is a number.

Axiom 2. The successor of any number is a number.

Axiom 3. If a and b are numbers and if their successors are equal, then a and b are equal.

Axiom 4. 0 is not the successor of any number.

Axiom 5. If S is a set of numbers containing 0 and if the successor of any number in S is also in S, then S contains all the numbers.

It should be noted that the word "number" as used in the Peano axioms means "non-negative integer." The fifth axiom deserves special comment. It is the first formal statement of what we now call the "induction axiom" or "the principle of mathematical induction." Since first stated by Peano, it has become the accepted principle for proving statements about the non-negative **integers** (or "whole numbers"). If there is some property P which we believe to be true of all the **whole numbers**, mathematicians agree that we cannot simply show a few examples where P is true and then conclude that it is, therefore, true for all numbers. Since the set {0, 1, 2, 3,...} is infinite, we cannot conclude from a finite number of examples, no matter how large, that property P holds for all the numbers in {0, 1, 2, 3,...}. However, Peano's fifth axiom, the principle of mathematical **induction**, gives us the means for determining whether property P is true for all numbers in {0, 1, 2, 3...}. The induction axiom says that we need carry out only two steps to determine the **truth** of P for the infinite set of whole numbers. First, we must show that 0 has property P. Second, we must show that, for any number n, if n has property P, then this implies that the succesor of n, namely n+1, also has property P. Imagine a line of an infinite number of dominoes standing upright. Two things need to happen for all the dominoes to fall with a single push: (1) the first domino must fall, and (2) each domino in the line must be close enough to the next domino so that when it falls it causes the next domino to fall. Note than both conditions must hold in order for all the dominoes to fall. If the first domino does not fall, then the chain reaction never begins. On the other hand, if the first domino does fall but somewhere along the line the separation between a domino and its successor is so great that they never make contact, then the chain reaction stops there. This is the essence of Peano's fifth axiom.

One of the great beauties of the Peano axioms is that they make possible the generation of an infinite set of numbers from a finite number of symbols. Essentially, Peano hands us two items, the number 0 and the concept "successor," and an "instruction manual," i.e, his five axioms, and promises us that with just these we can build an entire system of arithmetic.

As a first step in building this arithmetic system, we may generate all the natural numbers (positive integers) as follows. Let us agree that if n is any number, rather than writing out "the succesor of n," we will write "s(n)." Now Peano has given us 0. Axiom 1 says that 0 is a number and Axiom 2 says that the successor of any number is a number. Therefore, s(0) is a number, s(s(0)) is a number, s(s(s(0))) is a number, and so on. Clearly this process, if continued forever, is sufficient to generate all the natural numbers. As each new natural number is "born" by our process, we may want to give it a name. For instance we may wish to call s(0) "Sam" or "Samantha" or, preferably, "1." To our newly generated s(s(0)), which clearly is s(1), we will give the name "2." We shall call s(s(s(0)))=s(s(1))=s(2) by the name "3" and so forth. Thus the set {1, 2, 3,...} is generated.

Having generated the natural numbers, we can next begin to define operations on those numbers. For example, Peano defines addition as follows: For any natural numbers n and k: i. n+0=n and ii. n+s(k)=s(n+k), where s(k) still means "the successor of k." So the addition "2 + 1" is interpreted as 2 + 1 = 2 + s(0) = s(2+0) = s(2) = 3. Multiplication can be defined in a similar way and then subtraction is defined in terms of the addition of inverse elements and division is defined in terms of the multiplication of inverse elements. In order to do this, the negative integers are defined as the additive inverses (or opposites) of the natural numbers and the **rational numbers** of the form 1/n, where n is not 0, are defined as multiplicative inverses of the natural numbers and their opposites. In this way, we expand the number system to include all the integers and all the rational numbers of the form

1/k, where k is not 0. Then we must account for division of the form p/q where p/q turns out to be neither an integer nor the multiplicative inverse of an integer. Thus we bring in the rest of the rational numbers to allow for this. In this way, we can continue to build up a richer number system. We will need **irrational numbers** to account for the solutions of **equations** such as $x^2 = 2$ and **complex numbers** to account for solutions of equations such as $x^2 + 1 = 0$. The important thing is that all these expansions of the number system can be accomplished by definitions, without adding any more axioms or primitive terms to Peano's original system. That is not to say that this expansion is trivial; for example, it took some true inventiveness on the part of mathematicians such as Dedekind (1831-1916), and **Cantor** (1845-1918) to come up with an acceptable **definition** of a real number. However, once this was accomplished, mathematicians could feel more comfortable about the foundations of the real number system, which has been the setting for the development of arithmetic, algebra, geometry, and modern mathematical **analysis** including the **calculus**. Thus, starting with 0, the idea of a successor for each number, and his five axioms, Peano provided a simple but solid foundation upon which to construct the edifice of modern mathematics.

See also Giuseppe Peano; Mathematical Induction

PEANO CURVE

The Italian mathematician **Giuseppe Peano** (1858-1932) is best known for creating an **axiom** system for **arithmetic** that today remains the starting point for most rigorous developments of modern mathematics, but he is also famous for his construction of a curve that fills an entire planar region. This seems counterintuitive since our usual notion of a curve is that it is one-dimensional. So how could a one-dimensional object fill a two-dimensional region? Peano's curve was something of a curiosity when it appeared in an 1890 article, but a year later the German mathematician, **David Hilbert** (1862-1943), produced another "plane-filling" curve. At this point some mathematicians began to wring their hands and proclaim such curves to be "non-intuitive," "monstrous," and "pathological"; and they feared that these things threatened to undermine some of the most cherished concepts in mathematics. Another problem for the mathematicians, in addition to the quandary about the meaning of "dimension," was that these curves possessed the bizarre property of being everywhere continuous but nowhere differentiable. **Karl Weierstrass** (1815-1897), sometimes called the father of mathematical **analysis**, had produced such a curve, although his was not plane-filling, in 1861, to the dismay of mathematical analysts who relied on their intuition to guide their mathematical research. **Continuity** and differentiability are concepts at the heart of **calculus**. In non-technical terms, a curve is continuous if there are no "breaks" or "gaps" in it. A **function** is differentiable if it is "smooth" with no "sharp" corners or cusps. One of the most important concepts in calculus is that differentiability of a curve at a point implies continuity of the curve at that point, but the converse is not true. A curve may be continuous at a point without being differentiable at that point. Mathematicians were well aware of this—the absolute value function is continuous at the origin, but it has a sharp corner there so that it is not differentiable there. The famous cycloid curve has infinitely many cusps, and so infinitely many points at which it is not differentiable; but the cycloid curve is continuous at all of those points. Mathematicians were used to seeing curves that were continuous but not differentiable at isolated points, but the Peano and Hilbert curves, like the Weierstrass curve, were continuous at all points but not differentiable at any point. Largely as a result of the non-intuitive nature of these curves, they were regarded as exceptions and oddities. They were ignored and kept safely out of sight for about 70 years until **Benoit Mandelbrot** (1924-) showed them to be members of a class of curves now called "**fractals**," a term coined by Mandelbrot in his 1975 work *Les Objets Fractals*. In this work, and in his 1982 book, *The Fractal Geometry of Nature*, Mandelbrot defined the "fractal dimension" of an object and showed that by this definition the Peano curve had fractal **dimension** 2.

It is now customary to call all plane-filling curves Peano curves, although Peano's original plane-filler was merely the first of many such curves to be discovered in the 19th and 20th centuries. In *The Fractal Geometry of Nature,* Mandelbrot shows how to construct fractal curves using an infinitely repeated iterative process. He starts with an initial figure, called the initiator, and then shows a second construction called the generator that produces the next stage of the curve. After that the process is repeated creating the curve stage by stage. At each new stage the figure will look more and more like the fractal being constructed, but the true fractal curve is complete only after an infinite number of iterations. For example, Peano's original curve has a line segment as its initiator. The generator is formed by shrinking the initiator (a) by 1/3 (b) and placing 9 copies of the shrunken piece in the configuration shown below.

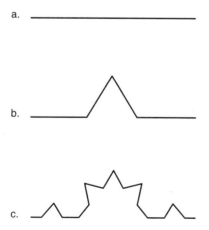

Now the process is repeated, shrinking each of the generator segments by 1/3 and placing 9 copies on each segment of the generator. Repeating this process to **infinity** produces the Peano curve.

PEANO, GIUSEPPE (1858-1932)
Italian logician

Giuseppe Peano served most of his adult life as professor of mathematics at the University of Turin. His name is probably best known today for the contributions he made to the development of **symbolic logic**. Indeed, many of the symbols that he introduced in his research on logic are still used in the science today. In Peano's own judgment, his most important work was in **infinitesimal calculus**, which he modestly described as "not... entirely useless." Some of Peano's most intriguing work involved the development of cases that ran counter to existing theorems, axioms, and concepts in mathematics.

Peano was born in Spinetta, near the city of Cuneo, Italy, on August 27, 1858. He was the second of five children born to Bartolomeo Peano and the former Rosa Cavallo. At the time of Peano's birth, his family lived on a farm about three miles from Cuneo, a **distance** that he and his brother Michele walked each day to and from school. Sometime later, the family moved to Cuneo to reduce the boys' travel time.

At the age of 12 or 13, Peano moved to Turin, some 50 miles (80 km) south of Cuneo, to study with his uncle, Michele Cavallo. Three years later he passed the entrance examination to the Cavour School in Turin, graduating in 1876. He then enrolled at the University of Turin and began an intensive study of mathematics. On July 16, 1880, he passed his final examinations with high honors and was offered a job as assistant to Enrico D'Ovidio, professor of mathematics at Turin. A year later he began an eight-year apprenticeship with another mathematics professor, Angelo Genocchi.

Peano's relationship with Genocchi involved one somewhat unusual feature. In 1883 the publishing firm of Bocca Brothers expressed an interest in having a new calculus text written by the famous Genocchi. They expressed this wish to Peano, who passed it on to his master in a letter of June 7, 1883. Peano noted to Genocchi that he would understand if the great man were not interested in writing the book himself and, should that be the case, Peano would complete the work for him using Genocchi's own lecture notes and listing Genocchi as author.

In fact, that was just Genocchi's wish. A little more than a year later, the book was published, written by Peano but carrying Genocchi's name as author. Until the full story was known, however, many of Genocchi's colleagues were convinced that Peano had used his master's name to advance his own reputation. As others became aware of Peano's contribution to the book, his own fame began to rise.

Peano's first original publications in 1881 and 1882 included an important work on the integrability of **functions**. He showed that any first-order differential equation of the form $y' = f(x, y)$ can be solved provided only that f is continuous. Some of these early works also included examples of a type of problem of which Peano was to become particularly fond, examples that contradicted widely accepted and fundamental mathematical statements. The most famous of these, published in 1890, was his work on the space-filling curve.

At the time, it was commonly believed that a curve defined by a parametric function would always be limited to an arbitrarily small region. Peano showed, however, that the two continuous parametric functions $x = x(t)$ and $y = y(t)$ could be written in such a way that as t varies through a given **interval**, the graph of the curve covers every point within a given **area**. Peano's biographer Hubert Kennedy points out that Peano "was so proud of this discovery that he had one of the curves in the sequence put on the terrace of his home, in black tiles on white."

Peano's first paper on symbolic logic was an article published in 1888 in which he continued and extended the work of **George Boole**, the founder of the subject, and other pioneers such as Ernst Schröder, H. McColl, and **Charles S. Peirce**. His magnum opus on logic, *Arithmetices principia, nova method-oexposita* (*The Principles of Arithmetic, Presented by a New Method*), was written about a year later. In it, Peano suggested a number of new notations including the familiar symbol $\in$ to represent the members of a set. He wrote in the preface to this work that progress in mathematics was hampered by the "ambiguity of ordinary language." It would be his goal, he said, to indicate "by signs all the ideas which occur in the fundamentals of **arithmetic**, so that every proposition is stated with just these signs." In succeeding pages, then, we find the introduction of now familiar symbols such as $\cap$ for "and," $\cup$ for "or," $\supset$ for "one deduces that," $\ni$ for "such that," and $\prod$ for "is prime with." Also included in this book were Peano's postulates for the natural numbers, an accomplishment that Kennedy calls "perhaps the best known of all his creations."

In 1891 Peano founded the journal *Rivista di matematica* (*Review of Mathematics*) as an outlet for his own work and that of others; he edited the journal until its demise in 1906. He also announced in 1892 the publication of a journal called *Formulario* with the ambitious goal of bringing together all known theorems in all fields of mathematics. Five editions of *Formulario* listing a total of 4,200 theorems were published between 1895 and 1908.

By 1900 Peano had become interested in quite another topic, the development of an international language. He saw the need for the creation of an "interlingua" through which people of all nations—especially scientists—would be able to communicate. He conceived of the new language as being the successor of the classical Latin in which pre-Renaissance scholars had corresponded, a *latino sine flexione,* or "Latin without grammar." He wrote a number of books on the subject, including *Vocabulario commune ad latino-italiano-français-english-deutsch* in 1915, and served as president of the Akademi Internasional de Lingu Universal from 1908 until 1932.

While still working as Genocchi's assistant, Peano was appointed professor of mathematics at the Turin Military Academy in 1886. Four years later he was also chosen to be extraordinary professor of infinitesimal calculus at the University of Turin. In 1895 he was promoted to ordinary professor. In 1901 he resigned his post at the Military Academy, but continued to hold his chair at the university until his death of a heart attack on April 20, 1932.

Peano had been married to Carla Crosio on July 21, 1887. She was the daughter of the painter Luigi Crosio and was particularly fond of the opera. Kennedy points out that the Peanos were regular visitors to the Royal Theater of Turin where they saw the premier performances of Puccini's *Manon Lescaut* and *La Bohème.* The couple had no children. Included among Peano's honors were election to a number of scientific societies and selection as knight of the Crown of Italy and of the Orders of Saints Maurizio and Lazzaro.

PEARSON, KARL (1857-1936)
English statistician

Karl Pearson is considered the founder of the science of **statistics**. He believed that a true understanding of human evolution and heredity required mathematical methods for **analysis** of the data. In developing ways to analyze and represent scientific observations, he laid the groundwork for the development of the field of statistics in the 20th century and its use in medicine, engineering, anthropology, and psychology.

Pearson was born in London, England, on March 27, 1857, to William Pearson, a lawyer, and Fanny Smith. At the age of nine, Karl attended the University College School, but was forced to withdraw at sixteen because of poor health. After a year of private tutoring, he went to Cambridge, where the distinguished King's College mathematician E. J. Routh met with him each day at 7 A.M. to study papers on advanced topics in **applied mathematics**. In 1875, he was awarded a scholarship to King's College, where he studied mathematics, philosophy, religion, and literature. At that time, students at King's College were required to attend divinity lectures. Pearson announced that he would not attend the lectures and threatened to leave the college; the requirement was dropped. Attendance at chapel services was also required, but Pearson sought and was granted an exception to the requirement. He later attended chapel services, explaining that it was not the services themselves, but the compulsory attendance to which he objected. He graduated with honors in mathematics in 1879.

After graduation, Pearson traveled in Germany and became interested in German history, religion, and folklore. A fellowship from King's College gave him financial independence for several years. He studied law in London, but returned to Germany several times during the 1880s. He lectured and published articles on Martin Luther, Baruch Spinoza, and the Reformation in Germany, and wrote essays and poetry on philosophy, art, science, and religion. Becoming interested in socialism, he lectured on Karl Marx on Sundays in the Soho district clubs of London, and wrote hymns for the Socialist Song Book. Pearson was given the name Carl at birth, but he began spelling it with a "K," possibly out of respect for Karl Marx.

During this period, Pearson maintained his interest in mathematics. He edited a book on elasticity as it applies to physical theories and taught mathematics, filling in for professors at Cambridge. In 1884, at age twenty-seven, Pearson became the Goldsmid Professor of Applied Mathematics and **Mechanics** at

University College in London. In addition to his lectures in mathematics, he taught engineering students, and showed them how to solve mathematical problems using graphs.

In 1885, Pearson became interested in the role of women in society. He gave lectures on what was then called "the woman question," advocating the scientific study of questions such as whether males and females inherit equal intellectual capacity, and whether, in the future, the "best" women would choose not to bear children, leaving it to "coarser and less intellectual" women. He joined a small club which met to discuss questions of morality and sex. There he met Maria Sharpe, whom he married in 1890. They had three children, Egon, Sigrid, and Helga. Maria died in 1928, and Pearson married Margaret V. Child, a colleague at University College, the following year.

Pearson was greatly influenced by **Francis Galton** and his 1889 work on heredity, *Natural Inheritance.* Pearson saw that there often may be a connection, or **correlation**, between two events or situations, but in only some of these cases is the correlation due not to chance but to some significant factor. By making use of the broader concept of correlation, Pearson believed that mathematicians could discover new knowledge in biology and heredity, and also in psychology, anthropology, medicine, and sociology.

An enthusiastic young professor of zoology, W. F. R. Weldon, came to University College in 1891, further influencing Pearson's direction. Weldon was interested in Darwin's theory of natural selection and, seeing the need for more sophisticated statistical methods in his research, asked Pearson for help. The two became lunch partners. From their association came many years of productive research devoted to the development and application of statistical methods for the study of problems of heredity and evolution. Pearson's goal during this period was not the development of statistical theory for its own sake. The result of his efforts, however, was the development of the new science of statistics.

Remaining at the University College, Pearson became the Gresham College Professor of **Geometry** in 1891. His lectures for two courses there became the basis for a book, *The Grammar of Science,* in which he presented his view of the nature, function, and methods of science. He dealt with the investigation and representation of statistical problems by means of graphs and diagrams, and illustrated the concepts with examples from nature and the social sciences. In later lectures, he discussed probability and chance, using games such as coin tossing, roulette, and lotteries as examples. He described frequency distributions such as the normal distribution (sometimes called the bell curve because its graph resembles the shape of a bell), skewed distributions (for which the graphed design is not symmetrical), and compound distributions (which might result from a mixture of the two). Such distributions represent the occurrence of variables such as traits, events, behaviors, or other incidents in a given population, or in a sample (subgroup) of a population. They can be graphed to illustrate where each subject falls within the continuum of the **variable** in question.

Pearson introduced the concept of the "standard deviation" as a measure of the variance within a population or sam-

ple. The **standard deviation** statistic refers to the average **distance** from the **mean** score for any score within the data set, and therefore suggests the average amount of variance to be found within the group for that variable. Pearson also formulated a method, known as the chi-square statistic, of measuring the likelihood that an observed relation is in fact due to chance, and used this method to determine the significance of the statistical difference between groups. He also developed the theory of correlation and the concept of regression analysis, used to predict the research results. His correlation coefficient, also known as the Pearson *r*, is a measure of the strength of the relationship between variables and is his best-known contribution to the field of statistics.

Between 1893 and 1901 Pearson published 35 papers in the *Proceedings* and the *Philosophical Transactions* of the Royal Society, developing new statistical methods to deal with data from a wide range of sources. This work formed the basis for much of the later development of the field of statistics. He was elected to the Royal Society in 1896, was awarded the Darwin Medal in 1898, and, in 1903, was elected an Honorary Fellow of King's College and received the Huxley Medal of the Royal Anthropological Institute.

In 1901, Pearson helped found the journal *Biometrika* for the publication of papers in statistical theory and practice. He edited the journal until his death. His research often required extensive mathematical calculation, which was carried out under his direction by students and staff mathematicians in his biometric laboratory. Since high-speed electronic computers had not yet been invented, performing the calculations by hand was tedious and time-consuming. The laboratory staff produced tables of calculations which Pearson made available to other statisticians through *Biometrika,* and later as separate volumes. Access to these tables made it possible for others to carry out statistical research without the support of a large staff, and, again, proved to be a valuable contribution to the early development of the field of statistics.

Pearson became the Galton Professor of Eugenics in 1911, and headed a new department of applied statistics as well as the biometric laboratory and a eugenics laboratory, established to study the genetic factors affecting the physical and mental improvement or impairment of future generations. During World War I, Pearson's staff served Britain's interest by preparing charts showing employment and shipping statistics, investigating stresses in airplane propellers, and calculating gun trajectories. From 1911 to 1930, Pearson produced a four-volume biography of Francis Galton. In 1925, he founded the journal *Annals of Eugenics,* which he edited until 1933. In 1932, Pearson was the first foreigner to be awarded the Rudolf Virchow Medal by the Anthropological Society of Berlin. He retired in 1933 at age 77, and received an honorary degree from the University of London in 1934. Pearson died on April 27, 1936, in Coldharbour, Surrey.

Pearson produced more than three hundred published works in his lifetime. His research focused on statistical methods in the study of heredity and evolution but dealt with a range of topics, including albinism in people and animals, alcoholism, mental deficiency, tuberculosis, mental illness, and anatomical comparisons in humans and other primates, as well as astronomy, meteorology, stresses in dam construction, inherited traits in poppies, and variance in sparrows' eggs. Pearson was described by G. U. Yule as a poet, essayist, historian, philosopher, and statistician, whose interests seemed limited only by the chance encounters of life. Colleagues remarked on his boundless energy and enthusiasm. Although some saw him as domineering and slow to admit errors, others praised him as an inspiring lecturer and noted his care in acknowledging the contributions of the members of his lab group. For Pearson, scientists were heroes. The walls of his laboratory contained quotations from **Plato**, **Blaise Pascal**, Huxley and others, including these words from **Roger Bacon**: "He who knows not Mathematics cannot know any other Science, and what is more cannot discover his own Ignorance or find its proper Remedies."

PEIRCE, CHARLES SANDERS (1839-1914)

American logician

Charles Sanders Peirce remains one of the enigmatic figures in the history of American science. He made substantial contributions to a number of fields, especially logic, but his use of unusual terminology makes it difficult to appraise much of his work. As the project of publishing his collected writings continues, it may become possible to do justice to this many-sided thinker.

Charles Sanders Peirce was born on September 10, 1839 in Cambridge, Massachusetts. His father, Benjamin Peirce, was not only a professor of mathematics at Harvard University but also perhaps the most accomplished American mathematician of his generation. Peirce's early education outside the home was at various private schools in Boston and Cambridge, and he showed an interest in puzzles and chess problems. By the age of 13, he had read Archbishop Whately's *Elements of Logic*, perhaps a hint of the interests to come. Peirce entered Harvard in 1855, and the results were not impressive. Although he succeeded in graduating four years later, it was with a class rank of 71 out of 91. Upon graduation Peirce obtained a temporary position with the United States Coast Survey, which was to be his employer for most of his working life. His contributions to geodesy were many, and his service to the Coast Survey have been recognized with a memorial.

In the early 1860s Peirce studied under Louis Agassiz at Harvard, but his work with the Coast Survey had its benefits. He had become a regular aide to the Survey in 1861, which resulted in his exemption from military service. He was an assistant to the Coast Survey from 1867 to 1891, but that did not prevent his continuing researches in other areas. In particular, not only did he observe a solar eclipse in the United States in 1869, a year later he led an expedition to Sicily to observe a solar eclipse from a position that he had selected.

Peirce had developed a technical competence in mathematics that came in handy when he turned to logic. As an example of a result in mathematics itself, he succeeded in showing that of linear associative algebras (a subject to which his father had devoted a book) the only three that had a

uniquely defined operation of **division** were **real numbers**, **complex numbers**, and the **quaternions** of Sir **William Rowan Hamilton**. Perhaps the most significant innovation he made in logic was the extension of Boolean **algebra** to include the operation of inclusion. The most widely influential treatise on the algebra of logic was produced by the German mathematician Ernst Schröder beginning in 1890, and he displayed a detailed familiarity with Peirce's work. In fact, had Peirce made the effort to produce a systematic account of the subject before Schröder, it would be easier to measure the importance of Peirce's contributions.

One of the factors that played a role in Peirce's interest in logic and its algebraic expression was his having taken a position in 1879 at Johns Hopkins University in Baltimore. During the five years that he worked at the university, he stayed on at the Coast Survey. As Nathan Houser remarked in an article about Peirce, "during those years Peirce was a frequent commuter on the B & O Railroad between Baltimore and Washington." Peirce's first paper on the algebra of logic was published in the *American Journal of Mathematics* in 1880. The period 1880 to 1885 saw Peirce's introduction of two ideas to mathematical logic: truth-functional **analysis** and quantification theory. Truth-functional analysis is the ancestor of the technique used by the Austrian philosopher Ludwig Wittgenstein to serve as the basis for logic in his *Tractatus Logico-philosophicus*. Quantification theory was at the heart of the logical apparatus introduced by **Gottlob Frege** for the reduction of mathematics to logic. It is difficult to imagine two more crucial contributions at the time, although Peirce's share of the recognition suffers by virtue of the scattered nature of his contributions.

Peirce was never one to limit his scientific investigations to a single discipline. In 1879, he determined the length of the meter based on a wavelength of **light**. This provided a natural alternative to the standard meter bar on deposit in Paris. Three years later, he worked on a mathematical study of the relationship between variations in gravity at different points on the Earth's surface and the shape of the Earth. Better known is his role in serving as an advocate of a philosophy of science called pragmatism. Peirce's pragmatism was heavily dependent on the idea of **inference** to the best explanation. In other words, what existed was determined by what was needed for successful scientific practice at the time. While neither realists nor their opponents were happy with Peirce's position, it has continued to offer an alternative. In particular, philosophers of science with an inclination to take the history of science seriously find Peirce's approach one of the few that take change in one's scientific models to heart.

In light of Peirce's contributions in so many areas of science and mathematics, the puzzle remains of why he was unable to secure an academic position commensurate to his abilities. One factor may have been domestic; he married Harriet Melusina Bay in 1862 and was divorced from her in 1883, the year of his second marriage (to Juliette Froissy of Nancy, France). Peirce and his first wife had been separated since 1876, and public sentiment was on her side. More generally, however, Peirce's personality tended to go between extremes, and it was difficult for others to adjust to his mood swings. He was quick to enter into disputes (and frequently with the wrong party) and was easily influenced by others.

Peirce spent his later years in Milford, Pennsylvania, removed from the centers of intellectual life. He had been asked to resign from the Coast Survey in 1891 and for the rest of his life his income was uncertain, despite prodigious periods of writing. Even his philosophy, to which he continued to devote his best efforts, was neglected, if only as a result of his remoteness from university settings. He died in Milford on April 19, 1914, having made contributions across the intellectual map, but more to the benefit of the discipline than his own.

PELL, JOHN (1611-1685)

English professor, diplomat, and vicar

John Pell wrote or contributed to several useful books on **algebra** and the study of mathematics itself. He was particularly known for his work on the algebraic equation, first studied by **Brahmagupta**, $y^2=ax^2+1$, where "a" is a non-square integer.

Pell was born on March 1, 1611 (some sources say 1610) in Southwick, Sussex, England, the son of a vicar. He was only a small boy when both of his parents died, but it is not clear who raised him or where. He received his early education at the Steyning School in Sussex, leaving when he was 13 to begin studying at Trinity College in Cambridge, England. Pell received a bachelor's degree from the school in 1629, having published his "Description and Use of the Quadrant" in 1628, and a master's degree the following year.

In 1630, Pell worked as the assistant master at Collyer's School in Horsham and then for another year at the Chichester Academy. He married in 1632 and in 1638 moved to London with funding arranged by the academy's headmaster. In this more cosmopolitan environment, Pell's talents in mathematics and languages soon made him a popular figure in the academic world. Also in 1638, he published his major work, *Idea of Mathematics,* which secured his reputation both in England and abroad as a mathematician of note. In it, Pell emphasized the importance of mathematics and proposed "the establishment of a public library of all mathematical books."

Due to political upheaval in London, Pell was soon compelled to move away from England in order to find an academic post in mathematics. Finally he settled in Amsterdam in 1643, becoming mathematics professor at the university there. While in Amsterdam, Pell wrote an inflammatory treatise called *Controversiae de vera circuli mensura (The Controversy over the True Way to Measure a Circle)* in 1647. His work, written to document his stance on the value of **pi,** met with reproofs from such stellar figures in the mathematics world as **René Descartes** and **Gilles Personne de Roberval.**

In 1646 Pell moved to a new university in the Dutch city of Breda and then moved back to England in 1652 to teach mathematics in London. From 1654 to 1658 he worked as a diplomat in Zurich, Switzerland, also tutoring students in mathematics. Although some sources disagree, others say that Pell should be credited for several accomplishments and

advances in the years following this period. For instance, there is some dispute as to whether *An Introduction to Algebra* (1659) is actually Pell's work or that of one of his students in Zurich. In addition, Pell may have invented the symbol for **division** still in use today (a horizontal line with a dot above and below) and the method of writing out **equations** in three columns—one for explanation and two for identification.

In 1658 Pell left Zurich for England, where he became a vicar (a deputy cleric). Apparently, he spent the next 20 years in similar religious posts and suffered from extreme poverty toward the end of his life. However, he maintained his interest in mathematics, writing down a table of factors of all **integers** up to 100,000 in 1668. One source describes him as living with a former student for a period before dying in London on December 12, 1685.

See also Algebra

PENROSE, ROGER (1931-)

English theoretical physicist and mathematics professor

A modern bridge between the worlds of physics and mathematics, Sir Roger Penrose was an important contributor to theories on black holes in the 1960s and today is a fixture in the arcane world of recreational mathematics. His particular area of interest is a field of **geometry** known as tesselation, which involves evenly covering a surface with tiles of predetermined shape. Penrose is famous for his invention, with his father, of the Penrose impossible staircase and the Penrose **triangle** (also called the "tribar"). He also created what are now known as **Penrose tilings.**

Penrose was born in Colchester, England on August 8, 1931, the son of a geneticist who was an expert on inherited mental defects. His mother was a medical doctor. Because of his upbringing, Penrose was originally more interested in medicine and biology than mathematics. However, school scheduling conflicts forced him to decide early on that mathematics was his greater love. He received his higher education at University College in London, obtaining his doctorate in **algebraic geometry** from Cambridge University in 1957.

From 1957 to 1966, Penrose held numerous temporary academic posts at schools in London, Cambridge, Princeton, Syracuse, and Texas, working as both a lecturer and a researcher. In 1966, London's Birkbeck College gave him his first appointment as full professor. During the mid-1960s, Penrose began experimenting with the development of a new cosmology based on a complex geometry that he invented as he went along. His first efforts involved what he called "twistors," objects with no mass and with both angular and linear momentum in their special kind of **space.** His main goal in using these objects was to reconstruct modern physics.

Also in the mid-1960s, Penrose and physicist Stephen Hawking proved part of **Albert Einstein's** theory of general relativity to show that at the center of black holes there exist a "space-time singularity" of infinite density and **zero volume** where our present laws of physics do not apply. On his own,

Penrose came up with his theory of "cosmic censorship," which states that such singularities must have an "event horizon" that would effectively conceal any information about the black hole. In 1969, Penrose suggested theoretical circumstances in which energy, which normally would be trapped by the black hole's crushing gravity, might be extracted from a Kerr black hole (an uncharged rotating body). In 1970, he and Hawking argued that our universe probably started as a singularity that initiated the famous "Big Bang."

In 1973, Penrose left Birkbeck College for Oxford University, which had appointed him a distinguished professor of mathematics. The following year, he introduced Penrose tiling, which is unique in that it uses quasi-symmetrical (non-repeating) tile shapes to cover a surface. Penrose found that computers were useless for this sort of application, so he reportedly did his research with only a notebook and pencil. His other creations, the staircase and the tribar, figured prominently in the revolutionary art of **Maurits Cornelius Escher**.

Penrose and Hawking shared the prestigious 1988 Wolf Prize for physics, and in 1994 Penrose received a knighthood for his work in **mathematics and physics**. He continues to work as the Rouse-Ball mathematics professor at Oxford University. On his homepage at Oxford, Penrose sums up his feelings about mathematics, saying, "Mathematics is beautiful and elegant as well as being fun and yielding insights about ourselves and our world. Mathematics should be taught in a way that communicates this fullness."

PENROSE TILINGS

Penrose tilings constitute a class of non-periodic tilings of the plane. A tiling of the plane, as the name suggests, is a covering of the entire plane by shapes (tiles), no two of which overlap. A tiling can have almost any imaginable form, but the most interesting and most carefully studied tilings are those in which all of the tiles are identical copies of just a few different tiles. A tiling of this type is said to be periodic if there is a pattern that repeats itself—more precisely, if there is some small block of the tiling that, when it is shifted about by translations, will cover the entire tiling (a **translation** is an operation on the plane that shifts the position of every point by some fixed amount in some fixed direction). Many of the designs of the celebrated artist M. C. Escher are periodic tilings; for example, in his "Fish and birds" tiling, a single block consisting of one fish and one bird will cover the entire tiling if it is shifted about.

There are many ways to construct non-periodic tilings, but for a long time it was believed that any set of tiles that could make a non-periodic tiling could also be arranged into a periodic tiling. In 1961 Hao Wang began to study possible arrangements of colored **square** tiles called Wang dominoes, and in 1964 Robert Berger built on Wang's work to prove that there is a set of Wang dominoes that can only tile the plane non-periodically. Berger was able to construct such a set, but it was extremely complicated, with over 20,000 dominoes. He was later able to reduce the number to 104, and then Raphael

Robinson reduced it to 24. By examining a different type of tile, Robinson was later able to find a set of just 6 tiles that can only tile non-periodically.

The mathematician and physicist **Roger Penrose** set out to improve on these results, and in 1974 he found a pair of tiles that will only form non-periodic tilings. The two tiles are very simple, formed by cutting a parallelogram into two pieces (in a carefully prescribed manner); the two pieces are called a "kite" and a "dart" (Penrose has also found a non-periodic tiling that uses two rhombic tiles). Copies of these tiles fit together in all sorts of interesting ways, but if certain restrictions are placed on the way the tiles can be put next to each other, or if small modifications are made to their shapes, then only non-periodic tilings can occur. Although the darts are smaller than the kites, and so one might guess that more darts than kites are needed, there are more kites than darts, in a very precise way: the ratio of kites to darts is the **golden mean**, 1.61803398.... The appearance of the golden **mean** is just one of the many ways in which Penrose tilings, despite their aperiodicity, show a remarkable degree of order. Even though they have no repeating patterns, Penrose tilings are highly pleasing to the eye.

Most of the time, if you try to lay down kites and darts so that they never overlap, you will end up with a small **space** that fits neither a kite nor a dart. In spite of that, there are in fact infinitely many different Penrose tilings, and Penrose and mathematician John Conway have shown that there are in fact uncountably many different Penrose tilings (that is, there are more Penrose tilings than there are **whole numbers**). Penrose has discovered another startling fact about his tilings: if you limit yourself to a finite piece of one of the tilings, you can always find that identical piece in every other tiling, no matter how large the piece is or what its design is. In other words, there is no way to tell which Penrose tiling you have just by looking at a finite piece.

Another unusual attribute of Penrose tilings is that some of the tilings have "five-fold symmetry." We say that an object has n-fold **symmetry** if there is some center of **rotation** for which the object remains the same if you rotate it by a 1/n-turn. For example, the square has 4-fold symmetry, since a 1/4-turn about the center will leave the square unchanged; a regular hexagon has 6-fold symmetry, since a 1/6-turn about the center leaves it unchanged. Geometers realized long ago that any periodic tiling of the plane can only have 2-fold, 3-fold, 4-fold or 6-fold symmetries; 5-fold symmetry, and 7-fold symmetry or higher, are strictly forbidden. This is not so easy to prove, but is motivated by the fact that the only regular **polygons** that tile the entire plane are the equilateral **triangle**, the square, and the regular hexagon; pentagons, septagons, and other regular polygons cannot fill the entire plane without overlaps or gaps. Thus, mathematicians were fascinated when Penrose showed that in the realm of non-periodic tilings, 5-fold symmetry can occur.

Penrose tilings at first appear to belong purely to the domain of recreational mathematics; however, they have recently proven very important in the field of crystallography. When atoms form crystals, they are arranged into periodic tilings of space. For a long time, crystallographers believed that the only two possible types of solids were crystals (periodic tilings) and glasses (completely random arrangements of atoms). When Penrose discovered his tilings, scientists began to ask whether there was anything in nature corresponding to this new geometric structure. In 1982, Dan Shechtman discovered an alloy that displayed the forbidden five-fold symmetry; his alloy and subsequently discovered alloys have been given the name "quasicrystals." The atomic structure of these quasicrystals is not yet known, but scientists have found some evidence supporting the idea that they may be 3-dimensional analogues of Penrose's tilings.

Although Penrose tilings don't seem like material for legal disputes, they found their way into the courts when Penrose, who patented his tilings, sued the Kimberley-Clark Corporation for embossing his design on rolls of Kleenex Quilted bathroom tissue. "So often we read of very large companies riding rough-shod over small businesses or individuals," said David Bradley, the director of Pentaplex, which has the exclusive rights to license the tilings. "But when it comes to population of Britain being invited by a multi-national to wipe their bottoms on the work of a Knight of the Realm, then a last stand must be made." Beyond politics of globalization and the farcical nature of this case lies an important question: should mathematicians "own" their discoveries, or do mathematical discoveries belong to everyone?

See also Maurits Cornelius Escher; Golden mean; Roger Penrose

PERFECT NUMBERS

A perfect number is a number that is the sum of its proper divisors (a proper divisor is a divisor smaller than the number itself). For example, 6 is perfect, since its proper divisors are 1, 2, and 3, and $1 + 2 + 3 = 6$. Twenty-eight is a perfect number, since its proper divisors are 1, 2, 4, 7, and 14, and $1 + 2 + 4 + 7 + 14 = 28$. Perfect numbers are very rare; the first four, 6, 28, 496, and 8128 appear to have been known since ancient times, and since then, 33 more perfect numbers have been found. The largest known perfect number, discovered in 1998, has 1819050 digits.

Perfect numbers have been studied since the time of the Greeks (perhaps even earlier). To **Pythagoras**, the perfect numbers had mystical significance. The first recorded mathematical **theorem** concerning perfect numbers is from 300 B.C., in Euclid's *Elements*. Euclid proves the following statement: If the sums of the **powers** of 2 are added until their sum is a prime number, then that sum multiplied by the largest power of 2 in the sum is a perfect number. Thus, for example, $1 + 2 + 4 = 7$, which is prime; thus by Euclid's theorem, $7 \times 4 = 28$ is a perfect number. And $1 + 2 + 4 + 8 + 16 = 31$ which is prime, so $31 \times 16 = 496$ is a perfect number. An equivalent way to formulate Euclid's theorem is that whenever $2^n - 1$ is a prime number, $2^{n-1} \times (2^n - 1)$ is a perfect number.

Although Euclid treated perfect numbers from a purely logical perspective, many of the mathematicians and philosophers that followed him associated a moral significance to the

idea. The mathematician **Nicomachus of Gerasa** published a treatise around AD 100 in which he discussed superabundant numbers, whose divisors add up to more that the number itself, and deficient numbers, whose divisors add up to less than the number. Nicomachus likened superabundant numbers to excessive behavior, and deficient numbers to insufficiencies; perfect numbers were like perfect virtue in humans. Early Christian thinkers attached a theological importance to perfect numbers, observing that God created the world in 6 days, and made the moon go around the earth in 28 days. Saint Augustine wrote, "Six is a number perfect in itself, and not because God created all things in six days; rather, the converse is true. God created all things in six days because the number is perfect."

Euclid's theorem gives a way to generate even perfect numbers. For many years after the time of **Euclid**, mathematicians asked whether the converse of Euclid's theorem is true: if an even number is perfect, is it necessarily of the form $2^{n-1} \times (2^n - 1)$, where $2^n - 1$ is prime? Many mathematicians assumed that the answer to that question is "yes." but a formal **proof** was not given until the 18th century, by **Leonhard Euler**. The question of whether there are any odd perfect numbers is still open; it is one of the oldest open questions in mathematics.

PERIMETER

Perimeter is the total length (or the arc length) along the border or outer boundary of a closed two-dimensional plane or curve. For example, the perimeter of a **circle** is the total length around its boundary, while the perimeter of a polygon is the total length (or sum) of its sides. The word perimeter is derived from the Greek words "peri" (around) and "metron" (to measure).

The perimeter of a circle is commonly called the circle's **circumference**. With regards to the radius, r, or **diameter**, d, of a circle, the equation to solve for perimeter, p, is $p = 2\pi r = \pi d$, where **pi**, π, is a constant number that is defined as the ratio of a circle's perimeter (or circumference) to its diameter, with its value approximately equal to 3.14159. As an example, if the radius of a circle is measured to be 20 centimeters and the approximate value of 3.14 is used for pi, then the perimeter is calculated to be "2 x 3.14 x 20 cm", or $p = 125.6$ cm.

Polygons are defined as closed planar figures with straight sides. Vertices (or edges) are the means by which polygons are classified. For instance, rectangles (with four vertices) and triangles (with three vertices) are particular types of polygons. The perimeter of any polygon is the total length of its sides.

The general equation for determining the perimeter of a rectangle is "$p = 2W + 2L$," where "W" is the rectangle's width and "L" is the rectangle's length. Knowing that a rectangle has four sides, with opposite sides of the rectangle having the same length, the perimeter of a rectangle having side-lengths of 3.4 cm and 8.2 cm can be solved by realizing that the rectangle has two sides of length 3.4 cm, and two remaining sides of length of 8.2 cm. Therefore, the sum of the lengths of all four sides of the rectangle is "3.4 + 3.4 + 8.2 + 8.2 = 23.2 cm." The perimeter of the rectangle is $p = 23.2$ cm.

Triangles are classified in terms of their sides and angles. Equilateral triangles have three equal sides, isosceles triangles have two equal sides, and scalene triangles have no equal sides. In acute triangles, angles are less than ninety degrees; in right triangles, one **angle** is equal to ninety degrees; and in obtuse triangles, one angle is greater than ninety degrees. But, with regards to all triangles, the perimeter of a **triangle** is calculated exactly the same way: by measuring its three sides. For example, if a triangle has side "a" with length 3 cm, side "b" with length 4 cm, and side "c" with length 5 cm, then the length of the triangle's perimeter, p, is "3 + 4 + 5." Therefore p = 12 cm.

Determining the perimeter of a bounded curve can be accomplished by referring to **geometry** that the circumference of a circle is defined as the limit of the perimeters of regular polygons inscribed in the circle. A similar method is used for a bounded curve. For a portion of the curve from the point $(x_1, \Phi(x_1))$ to the point $(x_2, \Phi(x_2))$ the formula for the **distance** between these two points is given by

$$\sqrt{\left\{(x_1 - x_2)^2 + (\Phi(x_1) - \Phi(x_2))^2\right\}}$$

This distance formula between those two points can then be expanded over the entire curve to solve (with the use of calculus) the perimeter of the bounded curve. The tools of calculus necessary to determine this integration are deferred to the calculus-related articles.

See also Circle; Circumference; Diameter; Polygon; Radius; Rectilinear figures; Triangle

PERSPECTIVE

Mathematical perspective is the realistic representation of a three-dimensional object on a two-dimensional surface. In a perspective drawing, vertical lines in an image are drawn as vertical lines on the paper or canvas. Parallel horizontal lines, however, are drawn as oblique lines that converge to a single point, known as the vanishing point. A **projection** is a **mapping** of a geometrical figure onto a plane according to certain rules. The history of perspective and projection is an interesting one whose origins lie not in the field of mathematics, but in art.

Medieval paintings are noted for their lack of visual depth. Artists from those times focused on themes, using symbolism within a painting to convey a message. They did not attempt to create visual realism. The third **dimension**—depth—was not recreated. As a result, figures in medieval paintings looked flat. As the Middle Ages gave way to the Renaissance, artists became concerned with portraying the natural world as the human eye sees it. Toward that end, artists worked to develop techniques for accurately portraying the three-dimensional world on a two-dimensional canvas. The attempts of early Renaissance artists to master the third dimension led directly to the development of mathematical rules of perspective.

One of the pioneers of mathematical perspective was Leone Battista Alberti. His *Della Pittura*, published in 1435, is the first known written account of mathematical perspective. Alberti described the following method for achieving the proper perspective in a painting. He placed an upright glass panel between the scene to be painted and his eye. Holding his eye in a fixed position, he imagined rays of **light** running from his eye to each point in the scene itself. Alberti referred to this set of rays as a projection. The intersection of this projection with the glass panel produced a set of points, which Alberti called a section. By painting that section on the glass panel, or on a canvas, Alberti was able to create a painting of a scene that gave the same impression on the human eye as the scene itself. This system worked well for creating realistic, three-dimensional paintings of actual scenes. It did not work for recreating imagined scenes, however, as no glass panel could be set up between the eye and the imagination. Alberti recognized this problem, and thus recognized the need for a mathematical system of perspective that could accurately and consistently convey realism in painting.

Alberti made great strides in developing such a system. He applied his findings to the specific task of recreating checkerboard tiling patterns that existed in the ground plane. He was able to create a painting of a checkerboard floor that gave the proper appearance of depth. This was an important milestone in the development of mathematical perspective. However, Alberti's technique had one major limitation—it only worked for figures that existed in the ground plane. Though he could create a painting of proper depth of a tiled floor, he could not do the same for non-horizontal figures, such as a person or a building. Despite the limitations of Alberti's system, it introduced two concepts crucial to the development of an accurate system of mathematical perspective—vanishing points and horizon lines.

The Italian artist Piero della Francesca expanded Alberti's technique and perfected the science of perspective. In his work *De prospectiva pingendi*, published in 1478, Francesca outlined his system for creating three-dimensional images on a two-dimensional surface. The basic tenets of his system are as follows:

- A straight line in perspective remains straight.
- Parallel lines either remain parallel or converge to a single point. This point is known as the vanishing point.
- Vanishing points exist on a line known as the horizon line.

These rules, developed and refined by artists over five hundred years ago, are the same rules used by artists today to create proper perspective in paintings.

It may seem strange that what, in hindsight, appears to be a math problem was tackled by so many artists. However, many people in Renaissance times pursued more than one discipline. Artists were not just artists. Artists of the day were also the architects, engineers, and scientists of the day. As such, it was important for them to be able to create drawings and paintings with the proper appearance of depth. Scientists during the Renaissance times were also artists. To those scientists, art was a means of understanding the world around them. Being able to accurately portray the natural world on canvas was not just an artistic problem, but a scientific one as well. Thus, the pursuit of an accurate system of perspective was a mathematical one. None other than **Leonardo da Vinci**, the Renaissance genius whose painting *The Last Supper* is considered one of the finest examples of perspective painting, regarded painting as a science, because it revealed the reality in nature.

Girard Desargue, a seventeenth-century French mathematician, extended the idea of mathematical perspective to an entirely new branch of mathematics. Desargues recognized that different perspective drawings can be created from the same scene by viewing the scene from different angles. Desargues asked a fundamental question: What do different sections of the same scene have in common? This question formed the basis for a new field of mathematics known as projective **geometry**. Projective geometry is the study of those properties of plane figures that are unchanged when a given set of points is projected onto a second plane. Though Desargues pioneered this field, his work went largely unnoticed for several centuries. French mathematician Jean Victor Poncelet helped revive interest in the field of projective geometry when he published the first known textbook on the topic in 1822.

One of the key differences between Euclidean geometry and projective geometry is their treatment of parallel lines. In Euclidean geometry, parallel lines never meet. In projective geometry, parallel lines meet at **infinity**. This difference is reconciled by the fact that Euclidean geometry is considered a special case of projective geometry.

Mathematical developments often arise out of attempts to solve particular problems. This is certainly the case with perspective and projection. A problem faced by Renaissance artists was ultimately solved through mathematical means. Not only did this result in the development of a mathematical system of perspective, it led directly to the creation of an entirely new field of mathematics—projective geometry. This, in turn, has been extended to many other fields, including cartography and aerial photography.

See also Albrecht Durer

Pi

Pi (or π) is the ratio of the **circumference** of a **circle** to its **diameter**. This ratio is the same for every circle—for instance, if you double the diameter of a circle, you double its circumference as well. The observation that this ratio is constant has been known so long that historians cannot say when it was first discovered. It was certainly known by the year 2000 B.C., when the Babylonians estimated its value at 3 1/8, and the Egyptians at 3.1605. Another early estimate of the value of pi, although a less accurate one, is found in the Old Testament, which puts the value at 3 in the following passage about the great temple of Solomon: "And he made a molten sea, ten cubits from the one brim to the other: it was round all about, and his height was five cubits: and a line of thirty cubits did compass it about." (I Kings 7:23)

Although many early civilizations realized that the ratio of the circumference to the diameter is a constant, finding the exact value of that constant turned out to be difficult. Historians believe that the estimates of the Babylonians and Egyptians were calculated simply by measuring circles. The first theoretical estimate for the value of pi was made by the Greek mathematician **Archimedes** (287-212 B.C.), who estimated the circumference of a circle by comparing it to the perimeters of simpler shapes, such as hexagons, that were superscribed or inscribed in the circle. He came up with the estimate 223/71 < pi < 22/7. The average of those two bounds is 3.1418, which is within about 0.0002 of the actual value of pi.

To form this estimate, Archimedes considered shapes with as many as 96 sides. His calculation was a remarkable one, especially considering that in his day, much of the algebraic and trigonometric notation that we now rely on had not yet been developed. Over the centuries that followed, first Arab and then European scholars carried Archimedes's calculation further, using shapes with larger and larger numbers of sides. By 1600 the value of pi was known up to 35 decimal places.

Around this time, mathematicians began to explore the mathematics of infinite sums and products of numbers. They discovered that pi, although a number that arises in a purely geometric setting, can be expressed in strikingly elegant **arithmetic** ways, as an infinite sum or product. For instance, **John Wallis** (1616-1703) proved that pi=2x(2x2x4x4x6x6...)/(3x3x5x5x7x7...).

The mathematician **James Gregory** (1638-1675) proved the following formula, which is sometimes attributed to Leibniz: pi/4=1 - 1/3 + 1/5 - 1/7 +...

And the great mathematician **Leonhard Euler** (1707-1783) discovered the famous formula pi^2/6 = 1/1^2 + 1/2^2 + 1/3^2 + 1/4^2 +...

The discovery that pi could be expressed in terms of **infinite series** gave a new way to estimate its value: calculate larger and larger portions of an infinite series. With this new technique, by the start of the 18th century mathematicians were able to calculate pi to 100 decimal places. But they also realized that the calculation of the digits of pi is a task without an end: in 1766 Lambert proved that pi is not a rational number, so that the numbers that appear in its decimal form never fall into a repeating pattern. A century later, in 1882, Lindemann proved something even stronger: pi is transcendental—that is, it is not the solution of any polynomial equation with integer coefficients. Lindemann used this result to answer the classical question of "squaring the circle" in the negative: he showed that it is impossible to construct a **square** equal in **area** to a given circle using only a straightedge and compass.

In spite of the careful analysis to which pi had been subjected by this time, among non-mathematicians its nature still appeared to present some mysteries. In 1897, the House of Representatives of Indiana unanimously passed the following legislation: "Be it enacted by the General Assembly of the State of Indiana: It has been found that a circular area is to the square on a line equal to the quadrant of the circumference, as the area of an equilateral rectangle is to the square of one side." In other words, pi is equal to 4. Fortunately, the Senate of Indiana declined to adopt this "bill for an act introducing a new mathematical truth."

With the advent of supercomputing power, it has become possible to calculate pi to levels of precision that previously were inconceivable. The current record is 50 billion decimal places.

PICARD, CHARLES EMILE (1856-1941)

French professor

Along with **Simeon-Denis Poisson**, Charles Emile Picard was the most important and distinguished French mathematician of his day. Not only was he an avid supporter of other mathematicians and a gifted teacher, he made many contributions of his own, including advances in **algebraic geometry** and linear **differential equations**.

Picard was born in Paris, France on July 24, 1856, the son of a silk factory director. His father died during the siege of Paris in 1870, prompting his mother to work to support her two sons and keep them in school. Picard attended school at Lycée Henri IV, where he was a superb student and especially enjoyed history, Latin, Greek, and literature. Picard was an active child and young man, frequently practicing gymnastics and mountain climbing. Near the end of his secondary school years, he read a book on **algebra** that convinced him to make mathematics his profession.

In 1874, Picard was accepted by two prestigious schools: the École Normale Supérieure and the École Polytechnique. He chose to study at the former, since it would let him concentrate entirely on research. Picard earned his doctorate in the sciences by 1877, when the École Normale Suérieure hired him as an assistant. Between 1878 and 1879, Picard's energy and talent were such that he produced more than 100 articles on his research. He moved to the University of Toulouse in 1879 to begin an appointment as professor at the age of 23.

Also in 1879, Picard proved the first of the algebraic theorems that would later bear his name. It states that an integral function of the complex **variable** takes every finite value, with only one possible exception. The **theorem** later became a launching point for many key studies in complex function theory. Picard's second algebraic theorem, which established a classification of regular analytic **functions**, was published in 1880.

In 1881 Picard returned to Paris to work as a lecturer in experimental and physical **mechanics** at the Sorbonne and in astronomy and mechanics at the École Normale Supérieure. The Sorbonne rewarded him for his research in 1885 with a unanimous election to its chair of differential and integral **calculus**. From 1883 to 1888, Picard concentrated on expanding Jules-Henri Poincare's work on automorphic functions, leading to his discovery of "hyperfuchsian" and "hyperabelian" functions. This work yielded Picard's insights into algebraic surfaces beginning in 1901, after which he moved on to study of the **Galois theory**, finding a group of transformations (Picard group) for a linear differential equation.

The Sorbonne granted Picard's request in 1897 to let him teach **analysis** and higher algebra. Later that year, he cowrote *The Theory of Algebraic Functions of Two Independent Variables*. This work in particular led to advances in algebraic **geometry**. Also, beginning in 1894, he taught young engineers at France's Central School of Arts and Manufacturing; by 1937 he had trained more than 10,000 of them. In 1917 Picard was elected permanent secretary for the mathematical sciences at the prestigious Academy of Sciences.

During his lifetime Picard enjoyed a great deal of professional success, but this was balanced by much tragedy in his personal life. His daughter and two sons all died in World War I, while his grandsons were wounded and captured in World War II. His last two years were shadowed by the Nazi invasion and occupation of France. Picard died on December 11, 1941 in the Palais de l'Institute, the headquarters of the Academy of Sciences.

PICARD'S METHOD

Picard's method, sometimes called the method of successive approximations, gives a means of proving the existence of solutions to **differential equations**. Emile Picard, a French mathematician, developed the method in the early 20th century. It has proven to be so powerful that it has replaced the Cauchy-Lipschitz method that was previously employed for such endeavors.

Picard developed his method while he was a professor at the University of Paris. It arose out of a study involving the Picard-Lindelof existence **theorem** that had been formulated at the end of the 19th century. Picard's method is utilized in similar situations as those that employ the Taylor series method. It is a method that converts the differential equation into an equation involving integrals.

Some differential **equations** are difficult to solve, but Picard's method provides a numerical process by which a solution can be approximated. The method consists of constructing a sequence of **functions** that will approach the desired solution upon successive **iteration**. It is similar to the Taylor series method in that successive iterations also approach the desired solution to a differential equation. Picard's method allows us to find a series solution about some fixed point. The number of terms or iterations that is required to reach the desired solution depends on how far from the chosen point the solution must apply. The closer the chosen point to the unknown point, the fewer terms that are needed. It can be shown that the series is convergent and provides a solution to the differential equation of interest although the number of terms will depend upon how rapidly the series converges as well.

The details of Picard's method involve starting with an initial value problem and expressing it as an integral equation. This is done by integrating both sides with respect to one **variable** from a defined starting point to a defined termination point, x_0 to x_1. The initial value given is substituted into the resulting integral equation. This yields the function evaluated at the initial value summed with the remaining integral. After a simple substitution and appropriate arrangements of the **limits** on the remaining integral, the result can be used to generate successive approximations of a solution to the initial equation. The number of iteration steps is determined by two factors:

how quickly the series converges and how far away from the point of interest is the value given in the initial problem.

PIGEONHOLE PRINCIPLE

The pigeonhole principle states that if you have eleven letters to put in ten pigeonholes, then at least one pigeonhole will contain at least two letters. Although this seems like an obvious enough statement, this counting principle can nevertheless be used to prove some rather non-obvious facts. It is a basic and important tool in the field of **combinatorics**.

The formal phrasing of the pigeonhole principle is as follows. Suppose N and R are finite **sets** containing n and r elements respectively, where n > r. Suppose f is a function from N into R (thus f "puts n elements into at most r pigeonholes"). Then there is some element a of R whose preimage contains at least [n/r] elements (where [n/r] denotes the ceiling function). Recall that the preimage of a is the collection of all elements in N that are sent to a by the function f (that is, all the elements of N that are put in a's pigeonhole), and that the ceiling function assigns to any number the nearest integer that is greater than or equal to that number. Thus, to use a concrete example, if we have a function that assigns to every element of a thirty-element set (N) one element of a nine-element set (R). Since 30 divided by 9 is 3 and 1/3 (which rounds up to 4), one of the nine elements of R must be associated to at least 4 elements of N. This is again obvious, since if each element of R were associated to at most three elements of N, then N could have at most 9*3=27 elements.

The pigeonhole principle can also be generalized to infinite sets, in various ways. One generalization states that if infinitely many elements are to be divided among finitely many pigeonholes, then one pigeonhole must contain infinitely many elements. A second generalization states that if uncountably many elements are to be divided among countably many pigeonholes, then one pigeonhole must contain uncountably many elements. Further generalizations can be made, using higher and higher cardinalities of sets.

Now for an example of how this principle can be used. This example is taken from *Proofs from the Book* by Martin Aigner and Gunter M. Ziegler, which contains several examples of proofs using the pigeonhole principle. Start with any positive integer n. Choose from the first 2n positive **integers** any subset S containing at least n + 1 elements. (For instance, if n is 10, we are choosing any 11-element subset of the set {1,2,...,20}.) Then there must be two numbers in the subset S such that one is a **factor** of the other. To prove this, we factor the highest possible power of 2 out of each number s in S and write it as $s = 2^m t$. Note that t must be odd (if it were even, we could factor another 2 out), and obviously t < 2n (since t is less than or equal to s). Thus each element in S is being assigned a number t from the set U = {1,3,5,...,2n-1}. Since S has n+1 elements and U has n elements, there must be one number t in U that is assigned to two (or more) elements of S. In other words,

there are two elements of S whose factorizations are $2^x t$ and $2^y t$. Thus the smaller of these two numbers is a factor of the larger.

See also Combinatorics

PITISCUS, BARTHOLOMEO (1561-1613)
German trigonometricist and theologist

A theologist by trade and a strong influence in the Calvinist government of his time, Bartholomeo Pitiscus also essentially coined the term "trigonometry." The term comes from the title of his book *Trigonometria*, which consists of three parts, including five chapters devoted to plane and spherical **geometry**, now known as plane and spherical **trigonometry**. In addition to its contribution to mathematical nomenclature, the text is highly regarded and is especially noteworthy because in it Pitiscus used all six of the trigonometric **functions**.

Pitiscus was born August 24, 1561, in Grünberg, Silesia (now Zielona Góra, Poland), but the details of his upbringing and mathematical education are not known. He studied Calvinist theology at Zerbst and Heidelberg, and, in 1584, was appointed to tutor 10-year-old Friedrich der Aufichtige, known as Frederick IV, elector Palatine of the Rhine. Pitiscus was subsequently appointed court chaplain at Breslau and court preacher to Frederick. Frederick took control of the government of the Palatinate after his uncle, John Casimir, died in 1592 and Pitiscus remained influential in the government's defense of Calvinism under his former pupil.

Trigonometria: sive de solutione triangulorum tractatus brevis et perspicuus was published in 1595 as the final section of A. Scultetus' *Sphaericorum libri tres methodicé conscripti et utilibus scholiis expositi*. The work was published in revised form in 1600 under the title *Trigonometriae sive de dimensione triangulorum libri quinque*.

Trigonometriae is divided into three sections, the first of which contains five books on plane and spherical trigonometry. The second section provides tables for all six of the **trigonometric functions**, carrying them out to five or six decimal places. The third section of the text, "Problem varia" contains 10 books of problems in geodesy, measuring of heights, geography, geometry, and astronomy. An enlarged edition containing the original first and third sections of *Trigonometriae* was published in 1609 with expanded tables at the end of the text. A third edition was published in 1612 with an expanded "Problem varia" section including problems related to architecture. The original *Trigonometria* was translated into English by R. Handson in 1614.

Further details of Pitiscus' life are not known. He died on July 2, 1613, in Heidelberg, Germany.

PLANE GEOMETRY

Plane **geometry** is that subsection of geometry that deals with figures in a two-dimensional plane. Euclidean geometry, sometimes called parabolic geometry, is divided into two subsections: plane geometry, geometry dealing with figures in a plane, and **solid geometry**, geometry dealing with solid in three-dimensional **space**. Plane geometry is sometimes called two-dimensional Euclidean geometry and deals with figures such as circles, lines, **polygons** and the like. Plane geometry is concerned with the study of figures in two-dimensional Euclidean space, which is usually denoted R², also known as the Euclidean plane. This branch of Euclidean geometry is also focused on studying the properties of flat surfaces.

Plane geometry is a branch of Euclidean geometry, developed by Greek mathematician Euclid in the 4th century B.C., that is governed by Euclid's five postulates as laid out in his work *The Elements*. In the early 20th century mathematicians recognized that Euclid's postulates were incomplete in that concepts such as between, inside, and outside were not made precise. In 1902 **David Hilbert** developed a modern **axiom** system removing the incompleteness of Euclid's system. Later, American mathematician **George Birkhoff** laid out an axiom system for plane geometry in his work, "A set of postulates for plane geometry, based on scale and protractor," published in 1923. Saunders MacLane formulated a modified version of Birkoff's axiom system that was published in 1959. Birkoff's system consists of five axioms that are similar in function to Euclid's five postulates:

- 1) Existence axiom
- 2) Ruler axiom (introduces the ruler function)
- 3) Protractor axiom (introduces the protractor function)
- 4) Betweenness axiom
- 5) **Similarity** axiom

The difference between Birkoff's axiom system and Euclid's postulates is that Birkoff's system uses **real numbers** and their properties. Euclid's *The Elements* never mentioned **distance** and angular measure but instead used concepts such as **congruence** of segments and angles. In the early 1960s Birkoff's axioms were again modified by the School Mathematics Study Group providing a new standard for teaching high school geometry.

There are six basic assumptions used in plane geometry:
- 1) Only one line can be drawn through two distinct points.
- 2) Two straight lines can intersect at only one point.
- 3) The length of the line segment joining two points is the shortest distance between those two points.
- 4) A geometric figure can be moves without altering its size or shape.
- 5) A point divides a line into two infinite subsets of points of the line.
- 6) A straight line divides a plane into three subsets of points: two half planes and the line itself.

There are also four types of **symmetry** used in plane geometry: **rotation**, **translation**, reflection, and glide reflection. Rotation in plane geometry means to turn a figure around. A rotation has a center and an **angle** associated with it. Translation in plane geometry means to move a figure without rotating or reflecting the figure. A translation has a direction and a distance associated with it. Reflection means to produce a mirror image of a figure and every reflection has a mirror line, a line between the figure before and after reflection. Glide reflection is a reflection of a figure combines with a translation along the direction of the mirror line. The six assumptions along with the four types of symmetry gives mathematicians a precise way of approaching plane geometry.

PLATO (CA. 428 B.C.–CA. 347 B.C.)
Greek philosopher

One of the best known and most influential philosophers of all time, Plato has been admired for thousands of years as a teacher, writer, and student. His interests and knowledge were wide and varied, including science, mathematics, and poetry.

Plato was born in Athens around 428 B.C. His family, both on his father's and mother's sides, was aristocratic and politically influential. It's thought that, in addition to his parents, his immediate family included two older brothers and a younger sister. Plato's original name was Aristocles, but he became known during his school days as Platon, which means broad, because of his broad forehead (and, presumably, his broad range of knowledge).

It should be noted that nearly everything believed to be true about Plato's life is based on hypotheses by generations of scholars. These scholars have pieced together historical information to form what we accept as a reasonably accurate account of Plato's life. Some of the information comes from Plato's *VIIth Letter*, which contains some autobiographical information. Most of Plato's extensive writings, however, contained no, or very little, mention of himself.

It's clear that Plato—who as a young man studied and wrote poetry, participated in military service, and contemplated politics—was tremendously influenced by his relationship with Socrates, the Athenian philosopher and teacher. Plato became a disciple of Socrates when he was about 18 years old, turning his energies and attention philosophy and the question of virtue. Plato was profoundly affected by his teacher's trial and execution by political leaders about 10 years later, in 399 B.C.

After Socrates was killed, an embittered Plato left Athens, pronouncing that nothing would be right in the world until "kings were philosophers or philosophers were kings." He traveled to Greek cities in Africa and Italy, not returning to Athens until 387 B.C. It's thought that he was captured by pirates and held for ransom on his way back to Athens. While traveling, he met with priests and prophets from many places, studying not only philosophy, but astronomy, **geometry**, geology, and religion.

After returning to Athens, Plato devoted himself to studying and teaching philosophy. At the center of his philosophy is the concept of "forms." Plato believed that the changing world he experienced, and the things within that world, are merely reflections of a separate world that contains eternal, unchanging entities called forms. Forms, Plato said, are the true objects of knowledge and understanding, and only people who seek to understand them can have lives that are truly happy and healthy. These views are best expressed in the *Republic,* one of Plato's works from the middle part of his life.

In addition to philosophy, Plato had an abiding interest in mathematics, which he said idealized abstractions and was a pure form of thought. He considered mathematical thought to be lofty and separate from "common" thought, a category in which he included science. Plato believed that mathematics, in its purest form, could be applied to the heavens. In a dialogue called *Timaeus,* Plato asserted that heavenly bodies exhibit perfect geometric form, and that they move in exact circles. His thoughts concerning mathematics and its relation to astronomy were very influential on astronomers at the time.

Around 387 B.C. Plato founded a school outside of the walls of Athens that might be considered the first university. Located west of the city on land that once belonged to a Greek named Academus, the school was called the Academy. Plato's interest in mathematics was reflected in the Academy's curriculum, in which the study was emphasized; it is said that above the school's doorway was written, "Let no one ignorant of mathematics enter here." The Academy was a functioning school for nearly one thousand years; it was closed down by the Eastern Roman Emperor Justinian in 529 A.D.

Plato left Athens twice more in his life. He went to Italy in 367 B.C. to try to tutor Dionysus II, who had recently been named supreme ruler after his father died. He agreed to tutor Dionysus II, perhaps recalling his philosopher-king statements following the death of Socrates. Plato tried, but he found the situation in Italy to be highly unsatisfactory; he returned to Athens two years later. In 361 B.C., he was persuaded to return to Italy to try tutoring the young king again; but he did so only briefly before returning to Athens for good. It was around the time that Plato was travelling to and from Italy that his most famous student, **Aristotle**, entered the Academy.

There is little known about the later years of Plato's life. It's thought that he continued on at the Academy, teaching, conversing, and writing. He died in 347 B.C. at the age of about 80, leaving the Academy to his sister's son.

Plato's extensive and influential writings are generally divided into three groups. The first group is called the "Socratic" dialogues and are thought to have been written during the 12 years or so after the death of Socrates. These works—including the *Apology,* the *Crito, Charmides, Euthyphro, Georgias, Ion, Hippias Minor* and *Major,* and *Protagoras*—adhere closely to the teachings of Socrates. It is not known whether Plato did any writing before Socrates' death, but it's clear that he began writing extensively after it.

The second group of writings come from what is known as the "middle" or "transitional" period of Plato's life. This was the time after he returned from his travels and founded the Academy. These works probably include the *Meno, Cratylus, Euthydemus, Menexenus, Phaedo, Phaerus, Republic,* and the *Symposium.* The works from Plato's middle period focus on the metaphysical speculation for which he is widely recognized.

The third group of writings, which are known as the "later" dialogues, were written after Plato returned to the Academy from his tutoring trips to Italy. These works—including the *Critias, Parmenides, Philebus, Sophist,* *Statesman, Timaeus, Theatetus,* and *Laws*—are reflective and examine the metaphysical speculation of the "middle" works.

The works, thoughts, and theories of Plato have remained influential for more than 2000 years. His philosophy was a strong influence on the Christian Church in the early Middle Ages, and his life and writings still fascinate scholars and students around the world.

PLATONIC SOLIDS

A Platonic solid is a 3-dimensional shape built from identical flat figures called regular **polygons**, with the additional requirement that the same number of polygons should meet at each corner of the solid. A polygon is a closed shape in the plane made up of line segments. It is said to be regular if all the line segments have the same length and all the corner angles are equal. Thus, a 3-sided regular polygon is an equilateral **triangle**; a 4-sided regular polygon is a **square**; a 5-sided regular polygons is a pentagon; and so forth.

There are only 5 ways to assemble regular polygons into Platonic solids. When we build a Platonic solid, we must have at least 3 polygons meeting at each corner—two polygons are not sufficient to build a solid **angle**. Let's start by considering what Platonic solids can be built out of equilateral triangles. If we put 3 triangles together at a corner, they form a base that is another equilateral triangle; we can cap that off with one more triangle to get a 4-sided solid called a tetrahedron. If we put 4 triangles together at a corner, they form a pyramid with a square base. We can put two of these pyramids together along their bases to form an 8-sided solid called an octahedron. If we put 5 triangles together at a corner, it is harder to visualize the final result, but the triangles continue to fit together as we add more, and they finally close up into a 20-sided solid called an icosahedron.

Those are the only Platonic solids that can be formed from triangles. If we try to fit 6 equilateral triangles together, they form a flat surface, since equilateral triangles have 60-degree angles, and 6 of them together will form a full 360-degree angle. And it is impossible to fit more than 6 triangles together, since the angle measurements would add up to more than 360 degrees.

If we try to make Platonic solids out of squares, we should start by trying to put together 3 squares at each corner; this produces a cube. Four squares at a corner will form a flat surface, and more than 4 squares cannot fit at a corner. So the cube is the only Platonic solid made of squares. If we fit 3 pentagons together at a corner, once again it is hard to see the shape that forms. The pentagons do fit together, though, into a 12-sided solid called a dodecahedron. More than 3 pentagons cannot fit together at a corner. Three hexagons (6-sided polygons) fit together to form a flat surface, and more than 3 will not fit together at all. And for any polygon with more than 6 sides, it is impossible to fit together even 3. Thus the 5 solids we have found are the only ones.

Platonic solids were named after **Plato**, who was one of the first philosophers to be struck by their beauty and rarity.

But Plato did more than admire them: he made them the center of his theory of the universe. Plato believed that the world was composed entirely of four elements: fire, air, water, and earth. He was one of the originators of atomic theory, believing that each of the elements was made up of tiny fundamental particles. The shapes that he chose for the elements were the Platonic solids. In Plato's system, the tetrahedron was the shape of fire, perhaps because of its sharp edges. The octahedron was air. Water was made up of icosahedra, which are the most smooth and round of the Platonic solids. And the earth consisted of cubes, which are solid and sturdy. This analysis left one solid unmatched: the dodecahedron. Plato decided that the it was the symbol of the "quintessence," writing, "God used this solid for the whole universe, embroidering figures on it." Plato's description of the universe made a deep impression on his disciples, but it failed to satisfy his most illustrious student, **Aristotle**. Aristotle reasoned that if the elements came in the forms of the Platonic solids, then each of the Platonic solids should stack together, leaving no holes, since for example water is smooth and continuous, with no gaps. But, Aristotle pointed out, the only Platonic solids that can fill **space** without gaps are the cube and the tetrahedron, hence the other solids cannot possibly be the foundation for the elements. His argument struck his followers as so cogent that the atomic theory was discarded, to be ignored for centuries.

Although Aristotle's dismissal of Plato's structure was correctly reasoned, his analysis contained a famous error: the tetrahedron does not fill space without gaps. Incredibly, Aristotle's mistake was not discovered for more than 17 centuries. Aristotle was so highly esteemed by his followers that they confined themselves to trying to calculate how many tetrahedra would fit around one corner in space, rather than considering the possibility that the great man was mistaken. In the process, they came up with many conflicting formulas, some absurd. No one seems to have taken the simple step of building some tetrahedra and observing that they do not even fit nicely around a single point, let alone fill all of space without gaps. The error was not set right until the 15th century by Regiomontanus.

Although Plato's and Aristotle's theories were both flawed, they have provided the inspiration for much successful later work. Plato's theory, in which the elements are able to decompose into "subatomic" particle and reassemble in the form of other elements, can be considered a precursor to the modern atomic theory. And Aristotle's question, what kinds of shapes fill space, has proven to be a crucial problem in the study of crystals, in which the atoms are locked into repeating geometric patterns in 3-dimensional space.

PLÜCKER, JULIUS (1801-1868)

German geometer and physicist

One of the outstanding figures of 19th–century science, Julius Plücker is remembered for his seminal work in analytical **geometry**, and for his discoveries in physics. He is regarded as the co–discoverer of cathode rays (with his student, J. W.

Hittorf), and he is credited with laying the foundation of spectroscopy.

Plücker studied in Bonn, Heidelberg, and Paris, eventually securing a teaching position at the University of Bonn. A professor of mathematics from 1836 to 1847, he taught physics from 1847 until 1868. The early decades of the 19th century were an auspicious time for mathematicians, like Plücker, who directed their creative energy to geometry. Firstly, German mathematicians greatly benefitted from the foundation of *Journal fur die reine und angewandte Mathematik* (*Journal for Pure and Applied Mathematics*) in 1826. Founded by **August Leopold Crelle**, an engineer, the journal, which strongly supported research in pure mathematics, provided Plücker with the means to make his ideas known to the European scientific community. In 1827, **Karl Gauss** established the field of differential geometry, which studies curved surfaces, or surfaces in three-dimensional **space**. Crelle's journal and Gauss's research in geometry provided a favorable intellectual atmosphere for Plücker, who focused on **analytic geometry**.

By choosing to work in analytic geometry Plücker entered a field of age-old controversy, for mathematicians, particularly in Germany, had strong opinions about geometry: the proponents of synthetic geometry believed that geometry should be practiced with straightedge and compass only, without recourse to **algebra**. It is not surprising, therefore, that these "pure" geometers questioned the algebraic methodology of analytic geometry. Unlike synthetic geometry, analytic geometry fully utilizes algebra. Thus, in analytical geometry, the plane is defined by a coordinate system, and a point has two coordinates (an x coordinate and a y coordinate). Lines are defined by equations (a straight line is represented by the linear equation $ax + by = c$), while quadratic **equations** are used for circles (e.g., $x^2 + y^2 + ax + by + c = 0$, where a and b are the coordinates of the circle's center, and c determines the radius), ellipses, parabolas, and hyperbolas. The general two-variable function used to represent a curve is $f(x, y) = 0$.

Plücker's first book, the first volume of *Analytisch–geometrische Entwicklungen* (*Developments in Analytical Geometry*) was published in 1828; the second volume appeared in 1831. In his book, Plücker essentially demonstrated that analytical geometry was capable of replicating the results of pure geometry. Furthermore, in the second volume, Plücker introduced a new system of coordinates that had been discovered, independently, by **August Möbius**, **Karl Wilhelm Feuerbach**, and E. Bobillier. What made this new system, also called homogenous coordinates, revolutionary was the fact that it added a third coordinate to the Cartesian two. Unlike his predecessors, Plücker defined the three coordinates, x, y, and t as the distances of a point within a **triangle** to the sides of that triangle. However, by introducing a third coordinate, Plücker effectively transcended the traditional concept of space: one could not count on a point having a single set of coordinates. If, for example, one imagines the third coordinate as $t = time$, it is clear that a point p has more than one set of homogenous coordinates, such as x, y, t, as well as fx, fy, ft, where f represents any **factor**.

Furthermore, the introduction of the third coordinate t permits the **transformation** of Cartesian equations into trilinear equations. For example, if a curve is represented as $f(x, y) = 0$ in the Cartesian system, in Plücker's system the equation for that curve becomes $f(x/t, y/t) = 0$. Because **zero** is not permissible as a divisor, one would assume, in the new equation, that $t \neq 0$. As long as this is the case, it is safe to assume that, in a plane, as in the Cartesian system, an **infinity** of infinite lines can be drawn through a point p. These lines radiate in all possible directions, forming a solar image. Consequently, points on an infinite line, or "points at infinity," can lie on any line going through p. Plücker, however, knowing that the coordinates x, y, 0 cannot be translated into any pair of Cartesian coordinates, nevertheless introduced the seemingly impossible idea of $t = 0$. T thus becomes a line, sort of like x and y, but with one essential distinction: t, in Plücker's system, is the *only* line containing points at infinity. The sun-like image of lines projecting from a center is replaced by the image of a single line; instead of being everywhere, points of infinity are on line t. However, the many lines from the Cartesian solar image have not disappeared: in Plücker's system, they are all represented by a single line.

In 1829, Plücker introduced another revolutionary idea into geometry: he replaced the point by the straight line as the fundamental element in geometry. As Howard Eves explains, "A point now instead of having coordinates, possesses a linear equation—namely, the equation satisfied by the coordinates of all the lines passing through that point. The double interpretation of a pair of coordinates as either point coordinates or line coordinates and of a linear equation as either the equation of a line or the equation of a point furnishes the basis of Plücker's analytical **proof** of the principle of duality of plane projective geometry." If we remember that t can be interpreted as $t = time$, the idea of interchanging a straight line and a point appears less outlandish. Indeed, if x is a point, xt can easily be imagined as a line. Plücker's geometry, more so than pure geometry, provided a solid base for work in projective geometry. However, in Germany, analytic geometry, despite Plücker's brilliant work, failed to receive the recognition it deserved, probably owing to the tremendous authority of **Jakob Steiner**, the proponent of synthetic geometry, whose allies included, surprisingly, **Karl Jacobi**, who was not in the synthetic geometry camp.

Unwilling to work in Steiner's shadow, Plücker ceased his geometrical research without fully developing his geometry into a system of more than three dimensions. Leaving his work unfinished, he took over a professorship in physics. Plücker's work in physics initially focused on magnetism, particularly the effect of magnetic fields on objects. This work was a prelude to one of the great discoveries of 19th-century physics—cathode rays. In 1854, the well-known maker of scientific instruments Johann Heinrich Wilhelm Geissler perfected his vacuum tube. Placing electrodes in the vacuum tube, one at each end, Plücker drove electric current between the two electrodes. As a result of this experiment, which Plücker performed in 1858, a green glow appeared in the tube.

Although lacking empirical proof, Plücker correctly attributed the glow to cathode rays.

In addition, Plücker conducted pioneering research in spectroscopy, finding that chemical substances had individual spectral lines, thus opening the field of spectroscopic chemical analysis. In 1862, he discovered that a chemical element's spectral pattern may be affected by temperature changes. Plücker's work in physics, particularly his discovery that cathode rays are affected by an electromagnetic field, opened new vistas in this scientific discipline, inspiring research that resulted in J. J. Thomson's discovery of the electron.

Plücker returned to mathematical research in 1865. He completed his line-based geometry of four-dimensional space and also developed an appropriate notation system. Plücker's notation marks six homogenous coordinates as l, m, n, λ, μ, ν, where $l\lambda + m\mu + n\nu = 0$. According to René Taton, the "originality and elegance of the new notation drew the attention of many mathematicians to the importance of ruled geometry, the properties of congruences and complexes, and their possible applications to the solution of certain **differential equations**, and to problems in geometrical optics, statics, and vector analysis."

Underestimated by his German colleagues, Plücker's brilliant contributions to science earned him the admiration of his English colleagues, who awarded him the Copley Medal in 1868.

POINCARÉ CONJECTURE

The Poincaré conjecture started out as a question asked by the founder of the subject of **topology**, **Henri Poincaré**, in 1904: Is a simply connected, compact, three-dimensional **manifold** necessarily a hypersphere? Poincaré's intuition that this was one of the most important questions of topology has stood the test of time. In nearly a century since then, topologists have solved a host of related problems. They have invented deep and powerful theories to scale Poincaré's fortress, and these theories have enabled them to gain potentially revolutionary insights into the shape of our universe. Nevertheless, the original question asked by Poincaré remains unanswered.

For non-specialists, the significance of the Poincaré conjecture is obscured by the thicket of terms one has to get through to understand it:

- A hypersphere is the surface of a four-dimensional ball, or the solution set to the equation $x^2 + y^2 + z^2 + w^2 = 1$ in Euclidean four-dimensional **space**. It can also be thought of as an ordinary three-dimensional universe in which all the "edges of space" have been brought around into a single point.

- A manifold is a surface, a space, or a higher-dimensional "spacetime," which on a small scale resembles an ordinary Euclidean space, but on a large scale can curve around in any conceivable fashion.

- The **dimension** of a manifold is the number of independent directions available for travel in that space. Thus, a piece of string is one-dimensional, while the surface of a table or an inner tube is two-dimensional. The space we live in in is three-dimensional, if you ignore the dimension of time; it is four-dimensional if you include time. Five- and even higher-dimensional manifolds can also be defined mathematically, even if they cannot be so easily visualized.

- A compact manifold is one that has no boundaries; it "closes up on itself."
- A simply-connected manifold is one that has no holes. A loop of string and an inner tube are not simply-connected, but a **sphere** and an ordinary tabletop are simply-connected.

Thus Poincaré's question can be rephrased as follows. Take the most featureless three-dimensional space you can think of: one with no holes and no edges. Is it necessarily deformable to the simplest three-dimensional topological space we know about—a hypersphere? Or might it have some features that we don't know about?

According to the Poincaré conjecture, if it looks like a hypersphere, then it must be a hypersphere. This conjecture is a reality check for topologists: It is analogous to understanding the number **zero** in **algebra**, or understanding the vacuum in physics. Most experts believe that the conjecture is true; however, some have worked on the problem for years and finally come to believe the opposite (although they can't prove it, either!).

In spite of their lack of success on the original conjecture, topologists have proved many related results. To begin with, a version of the Poincaré conjecture for two-dimensional manifolds, or surfaces, is known to be true. Any surface can be completely described by its number of holes (a sphere has none, a **torus** has one) and whether it is orientable (like the sphere) or unorientable (like a **Möbius strip**). The sphere is the only surface that has no holes or edges and is orientable.

In 1961, Stephen Smale proved a version of the Poincaré conjecture for manifolds of dimension $n > 4$. In spite of the fact that such high-dimensional objects are impossible to visualize, five dimensions turned out to be an advantage. They gave Smale room for a powerful technique called "surgery": removing "handles" from the manifold, one pair at a time, until there was nothing left but an n-dimensional sphere.

Unfortunately, Smale's technique would not work for 4-dimensional manifolds. The reason can be understood by a lower-dimensional analogy. If you draw a curve from one side of a **square** to the other, and another curve from the top to the bottom, the two curves have to intersect one another. They can't be pulled apart. But if you do the same thing in a cube, the curves don't have to intersect. Even if you start with two curves that intersect, there's enough "room" in the cube to pull them apart. The explanation is as simple as $1 + 1 = 2$. Since the dimensions of the two curves $(1 + 1)$ add up to the dimension of the square (2), there's no extra room to maneuver them away from each other. But the dimensions of the curves $(1 + 1)$ add up to less than the dimension of the cube (3), which provides an extra dimension to maneuver in.

In Smale's **proof**, the handles attach to the manifold along (2-dimensional) disks. In a 5-dimensional manifold, all the disks can be kept separate. But in a 4-dimensional manifold, the disks cannot be pulled apart. Because $2 + 2 = 4$, there are no extra dimensions to maneuver in. Thus the handles can get inextricably tangled, and the surgery cannot be performed successfully.

It took another generation before Michael Freedman succeeded in getting past this hurdle, in 1982. The proof took another huge leap in topological technology. Since intersections were inevitable, Freedman began by classifying all the ways that 2-dimensional surfaces can intersect inside a 4-dimensional manifold. This information he distilled into a **matrix** called the "intersection form," and he proved that this matrix was essentially all that you need to tell two 4-dimensional manifolds apart (provided they have no holes). In particular, the only 4-dimensional manifold with no holes, no edges, and the simplest possible intersection form is a sphere.

After Freedman, the only dimension for which Poincaré's conjecture remained unsolved was three, the same number of dimensions for which Poincaré originally posed the question. Three-dimensional manifolds are too roomy to be classified by their number of holes alone, but too cramped to allow surgery or to possess a well-defined intersection form. If the Poincaré conjecture is true for them—and that's a big if—the proof will almost certainly require another new generation of topological methods.

See also Simon Donaldson; Topology

POINCARÉ, JULES HENRI (1854-1912)
French number theorist, topologist, and mathematical physicist

Jules Henri Poincaré has been described as the last great universalist—"the last man," E. T. Bell wrote in *Men of Mathematics,* "to take practically all mathematics, pure and applied, as his province." He made contributions to **number theory**, theory of **functions**, **differential equations**, **topology**, and the foundations of mathematics. In addition, Poincaré was very much interested in astronomy, and some of his best-known research is his work on the three-body problem, which concerns the way planets act on each other in **space**. He worked in the area of mathematical physics and anticipated some fundamental ideas in the theory of relativity. He also participated in the debate about the nature of mathematical thought, and he wrote popular books on the general principles of his field.

Poincaré was born in Nancy, France, on April 29, 1854. The Poincaré family had made Nancy their home for many generations, and they included a number of illustrious scholars. His father was Léon Poincaré, a physician and professor of medicine at the University of Nancy. Poincaré's cousin Raymond Poincaré was later to serve as prime minister of France and as president of the republic during World War I. Poincaré was a frail child with poor coordination; his larynx was temporarily paralyzed when he was five, as a result of a bout of diphtheria. He was also very bright as a child, not necessarily an advantage in dealing with one's peers. All in all he was, according to James Newman in the *World of Mathematics,* "a suitable victim of the brutalities of children his own age."

Poincaré received his early education at home from his mother and then entered the *lycée* in Nancy. There he began to demonstrate his remarkable mathematical talent and earned a first prize in a national student competition. In 1873, he was admitted to the École Polytechnique, although he scored a **zero** on the drawing section of the entrance examination. His work was so clearly superior in every other respect that examiners were willing to forgive his perennial inability to produce legible diagrams. He continued to impress his teachers at the École, and he is reputed to have passed all his math courses without reading the textbooks or taking notes in his classes.

After completing his work at the École, Poincaré went on to the École des Mines with the intention of becoming an engineer. He continued his theoretical work in mathematics, however, and three years later he submitted a doctoral thesis. He was awarded his doctorate in 1879 and was then appointed to the faculty at the University of Caen. Two years later, he was offered a post as lecturer in mathematical **analysis** at the University of Paris, and in 1886 he was promoted to full professor, a post he would hold until his death in 1912.

One of Poincaré's earliest works dealt with a set of functions to which he gave the name Fuchsian functions, in honor of the German mathematician Lazarus Fuchs. The functions are more commonly known today as automorphic functions, or functions involving **sets** that correspond to themselves. In this work Poincaré demonstrated that the phenomenon of periodicity, or recurrence, is only a special case of a more general property; in this property, a particular function is restored when its **variable** is replaced by a number of transformations of itself. As a result of this work in automorphic functions, Poincaré was elected to the French Académie des Sciences in 1887 at the age of 32.

For all his natural brilliance and formal training, Poincaré was apparently ignorant of much of the literature on mathematics. One consequence of this fact was that each new subject Poincaré heard about drove his interests in yet another new direction. When he learned about the work of **Georg Bernhard Riemann** and **Karl Weierstrass** on Abelian functions, for example, he threw himself into that work and stayed with the subject until his death.

Two other fields of mathematics to which Poincaré contributed were topology and probability. With topology, or the **geometry** of functions, he was working with a subject which had only been treated in bare outlines, and from this he constructed the foundations of modern algebraic topology. In the case of probability, Poincaré not only contributed to the mathematical development of the subject, but he also wrote popular essays about probability which were widely read by the general public. Indeed, he was elected to membership in the literary section of the French Institute in 1908 for the literary quality of these essays.

In the field of celestial **mechanics**, Poincaré was especially concerned with two problems, the shape of rotating bodies (such as stars) and the three-body problem. In the first

of these, Poincaré was able to show that a rotating fluid goes through a series of stages, first taking a spheroidal and then an ellipsoidal shape, before assuming a pear-like form that eventually develops a bulge in it and finally breaks apart into two pieces. The three-body problem involves an analysis of the way in which three bodies, such as three planets, act on each other. The problem is very difficult, but Poincaré made some useful inroads into its solution and also developed methods for the later resolution of the problem. In 1889, his work on this problem won a competition sponsored by King Oscar II of Sweden.

The work Poincaré did in celestial mechanics was part of his interest in the application of mathematics to physical phenomena; his title at the University of Paris was actually professor of mathematical physics. Of the roughly 500 papers Poincaré wrote, about 70 dealt with topics such as **light**, electricity, capillarity, **thermodynamics**, heat, elasticity, and telegraphy. He also made contributions to the development of relativity theory. As early as 1899 he suggested that absolute **motion** did not exist. A year later he also proposed the concept that nothing could travel faster than the speed of light. These two **propositions** are, of course, important parts of **Albert Einstein**'s theory of special relativity, first announced in 1905.

As he grew older, Poincaré devoted more attention to fundamental questions about the nature of mathematics. He wrote a number of papers criticizing the logical and rational philosophies of **Bertrand Russell**, **David Hilbert**, and **Giuseppe Peano**, and to some extent his work presaged some of the intuitionist arguments of **L. E. J. Brouwer**. As E. T. Bell wrote: "Poincaré was a vigorous opponent of the theory that all mathematics can be rewritten in terms of the most elementary notions of classical logic; something more than logic, he believed, makes mathematics what it is."

Poincaré died on July 17, 1912, of complications arising from prostate surgery. He was 58 years old. During his lifetime he had received many of the honors then available to a scientist, including election to the Royal Society as a foreign member in 1894. Poincaré was married to Jeanne Louise Marie Poulain D'Andecy, with whom he had four children, one son and three daughters.

POISSON DISTRIBUTION

The Poisson distribution is a mathematical rule that assigns probabilities to the number of occurrences of a certain event. It is most commonly used to **model** the number of random occurrences of some phenomenon in a specified unit of **space** or time. The Poisson distribution is one of the most important in probability. It is usually written as: $P(x) = (\mu^x e^{-\mu})/x!$, where $P(x)$ is the probability that the outcome of the function will be x, and μ is the average number of occurrences in a specified **interval**, either of space or time. In a Poisson distribution the **mean** and variance are equal and can be described by: $E(x) = Var(x) = \mu$.

There are four assumptions made in order to apply the Poisson distribution to a problem. First, it is assumed that the probability of observing a single event over a small interval, either of space or time, is approximately proportional to the size of that interval. Also it is assumed that the probability of two events occurring in the same narrow interval is negligible. Third, the probability of an event occurring within a certain interval does not change over different intervals but remains the same. Lastly, the probability of an event in one interval is independent of the probability of an event in any other interval that is not overlapping with the first. So probabilities of intervals are not linked to each other. Violation of either of the last two assumptions can lead to overdispersion or extra variation. Aside from these assumptions that are made in order to effectively apply the Poisson distribution, there are some empirical tests that can be preformed to determine if a distribution is a Poisson distribution. To apply these tests the time/area intervals for all of the data should be the same. The simplest test is to determine if the variance is roughly equal to the mean for the data. If this is the case then the data may fit a Poisson distribution. Also a histogram graphical representation of the Poisson data should be skewed to the right, though this skewness may become less pronounced if the mean is large.

The Poisson distribution is very similar to the binomial distribution if the probability of an event occurring is very small. In some ways it is superior to the binomial distribution. For a binomial distribution one must know both the number of successful events as well as the number of unsuccessful events, whereas in the Poisson distribution one needs to know only the mean number of successful occurrences of an event. But as mentioned before, the binomial distribution can be applied in situations where the probability of an event occurring can be within a wide range but the Poisson distribution can be applied only when the probability of an event occurring is relatively large. Both methods are important in the theory of sampling.

The Poisson distribution was formulated by French mathematician **Simeon-Denis Poisson** in 1837. He published the distribution in a work entitled *Research on the Probability of Opinions*. This method was developed to calculate the probability of the success of trials in situations where the probability of success on any one trial is extremely low but where the number of trials is very large. The first application was by Ladislaus von Bortkiewicz, a German professor born in Russia in 1898, to the description of the number of deaths by horse kicking in the Prussian army.

POISSON, SIMEON-DENIS (1781-1840)

French mathematician

Remembered for his theoretical contributions in magnetism and electricity, Simeon-Denis Poisson was also credited with furthering the work in celestial **mechanics** of **Pierre-Simon Laplace** and **Joseph-Louis Lagrange**. Poisson made significant inroads in the field of probability as well, developing the **Poisson distribution** to describe the likelihood of a particular event.

Poisson was born on June 21, 1781 in Pithiviers, Loiret, France, the son of a low-ranking civil servant. He was a sickly child whose mother often put him in the care of a nurse. When

he was old enough, Poisson was made an apprentice to an uncle who was a surgeon, but the young man showed neither talent for nor interest in the profession. When he was 18, Poisson enrolled at the École Central in Fontainebleau, where his aptitude for mathematics and learning in general finally came to light after an early education of limited usefulness. In fact, he did so well that in 1798 he was admitted to the famous École Polytechnique in Paris after graduating first in his class.

With such teachers as Laplace and Lagrange, who were immediately impressed by Poisson's mathematical prowess, Poisson progressed rapidly. Apart from his relative inability to draw acceptable diagrams, which precluded him from a career in descriptive **geometry**, Poisson was so gifted in mathematics that the school made him a teacher as soon as he graduated in 1800. He became an assistant professor in 1802 and in 1806 replaced the school's illustrious **Jean Baptiste Joseph Fourier** as full professor.

As word of his mathematical skill traveled, Poisson received many offers of employment. In 1808 he left the École Polytechnique to accept an appointment as an astronomer at the Bureau des Longitudes and then as mechanics professor at the Faculty of Sciences in 1809. Two years later he published his *Treatise on Mechanics,* which became the standard reference for many years.

One of Poisson's most important works was a memoir in 1812 in which he expounded on the two-fluid theory of electricity. According to his modified version, similar fluids repel and dissimilar fluids attract based on what he called the inverse **square** law. The electrical charge of a body (i.e., positive or negative), then, depended on how the normally uniform distribution of both fluids became disrupted, causing a charge. Poisson adapted Lagrange's work on the subject, showing mathematically that this function would remain constant over the surface of an insulated conductor and gave a **proof** of the formula for the **force** at a charged conductor's surface. Some scientific historians consider that his work in this area launched a new branch of mathematical physics.

In 1815 Poisson suggested some changes to Fourier's theory of heat that served only to embitter the men's relationship, with Fourier accusing Poisson of wasting his talent on merely modifying the work of others. However, Poisson also built upon Lagrange's and Laplace's celestial mechanics—particularly concerning the stability of planets' orbits and calculations of the gravitational attraction exerted by ellipsoidal and spherical bodies—with good results. His expression for the gravitational force in terms of mass distribution inside a planet is still in use to calculate details of the Earth's shape based on paths of satellites in **orbit**.

Poisson accepted a nomination in 1820 to the Royal Council of the University, giving him the leverage and prestige he needed to defend science in France as a worthwhile and necessary discipline. (A new conservative government under the restored Louis XVIII was working to eliminate the scientific programs and policies implemented during the country's Revolutionary and Napoleonic periods.) After Lagrange's death in 1827, Poisson bore even more of this burden. Thus, the fact that he published almost all of his books in the last decade of his life is even more impressive.

In 1824, Poisson published a paper in which he discussed a theory of magnetism. This was based on the two-fluid **model**, but yielded a general expression for the magnetic potential as the sum of two integrals. In addition, he contributed Poisson's ratio to the theory of elasticity, which concerns the ratio of longitudinal extension to lateral contraction. He also wrote a memoir in 1833 on the movement of the Moon. In 1835, he published *A Mathematical Theory of Heat.*

The Poisson distribution rule (also known as the Poisson law of large numbers), which he introduced in the 1837 *Research on the Probability of Opinions,* represented one of his most significant efforts. The principle involves events that normally would be highly unlikely but that happen nevertheless due to the many chances for them to occur, such as plane crashes, for example. In modern times, the Poisson rule is helpful in the analysis of traffic, radioactivity, and random events in **space** or time. In the realm of pure mathematics, Poisson's most important contributions were papers on definite **integers** and **Fourier series**, which helped to launch the research of **Bernhard Riemann** and **Peter Gustav Lejeune Dirichlet.**

Poisson died on April 25, 1840 in Paris. He had married in 1817 and had been made a baron in 1837.

See also Poisson distribution

POLAR COORDINATES

One of the several systems for addressing points in the plane is the polar-coordinate system. In this system a point P is identified with an ordered pair (r,θ) where r is a **distance** and θ an **angle**. The angle is measured counter-clockwise from a fixed ray OA called the "polar axis."

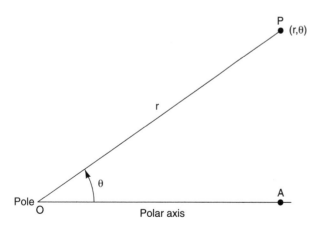

Figure 1.

The distance to P is measured from the end point O of the ray. This point is called the "pole." Thus each pair determines the location of a point precisely.

When a point P is given coordinates by this scheme, both r and θ will be positive. In working with polar coordinates, however, it occasionally happens that r, θ, or both take on negative val-

ues. To handle this one can either convert the negative values to positive ones by appropriate rules, or one can broaden the system to allow such possibilities. To do the latter, instead of a ray through O and P one can imagine a number line with θ the angle formed by OA and the positive end of the number line, as shown here.

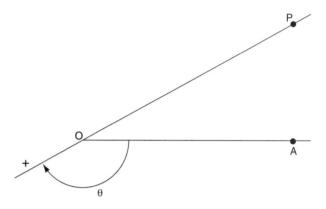

Figure 2.

One can also say that an angle measured in a clockwise direction is negative. For example, the point (5, 30°) could also be represented by (-5, -150°).

To convert r and θ to positive values, one can use these rules:

1. (-r, θ) = (r, θ ± π) or (r, θ ± 180°)
2. (r, θ) = (r, θ ± 2π) or (r, θ ± 360°)

(Notice that θ can be measured in radians, degrees, or any other measure as long as one does it consistently.) Thus one can convert (-5, -150°) to (5, 30°) by rule I alone. To convert (-7, -200°) would require two steps. Rule I would take it to (7, -20°). Rule II would convert it to (7, 340°).

Rule II can also be used to reduce or increase θ by any multiple of 2π or 360°. The point (6.5, 600°) is the same as (6.5, 240°), (6.5, 960°), (6.5, -120°), or countless others.

It often happens that one wants to convert polar coordinates to rectangular coordinates, or vice versa. Here one assumes that the polar axis coincides with the positive x-axis and the same scale is used for both.

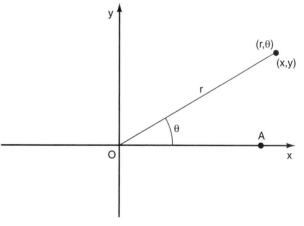

Figure 3.

The equations for doing this are:

- $r = \sqrt{(x^2 + y^2)}$
- $\theta = \arctan y/x$
- $x = r \cos \theta$
- $y = r \sin \theta$

For example, the point (3, 3) in rectangular coordinates becomes ($\sqrt{18}$, 45°) in polar coordinates. The polar point (7, 30°) becomes (6.0622, 3.5). Some scientific calculators have built-in **functions** for making these conversions.

These formulas can also be used in converting **equations** from one form to the other. The equation $r = 10$ is the polar equation of a circle with it center at the origin and a radius of 10. Substituting for r and simplifying the result gives $x^2 + y^2 = 100$. Similarly, $3x - 2y = 7$ is the equation of a line in rectangular coordinates. Substituting and simplifying gives $r = 7/(3 \cos \theta - 2 \sin \theta)$ as its polar equation.

As these examples show, the two systems differ in the ease with which they describe various curves. The Archimedean **spiral** $r = k\theta$ is simply described in polar coordinates. In rectangular coordinates, it is a mess. The **parabola** $y = x^2$ is simple. In polar form it is $r = \sin \theta/(1 - \sin^2 \theta)$. (This comparison is a little unfair. The polar forms of the **conic sections** are more simple if one puts the focus at the pole.) One particularly interesting way in which polar coordinates are used is in the design of radar systems. In such systems, a rotating antenna sends out a pulsed radio beam. If that beam strikes a reflective object the antenna will pick up the reflection. By measuring the time it takes for the reflection to return, the system can compute how far away the reflective object is. The system, therefore, has the two pieces of information it needs in order determine the position of the object. It has the angular position, θ, of the antenna, and the distance r, which it has measured. It has the object's position (r, θ) in polar coordinates.

For coordinating points in **space** a system known as cylindrical coordinates can be used. In this system, the first two coordinates are polar and the third is rectangular, representing the point's distance above or below the polar plane. Another system, called a spherical coordinate system, uses a radius and two angles, analogous to the latitude and longitude of points on earth.

Polar coordinates were first used by **Isaac Newton** and Jacob (Jacques) Bernoulli in the seventeenth century, and have been used ever since. Although they are not as widely used as rectangular coordinates, they are important enough that nearly every book on **calculus** or **analytic geometry** will include sections on them and their use; and makers of professional quality graph paper will supply paper printed with polar-coordinate grids.

Pólya, George (1887-1985)

Hungarian-born American number theorist

The career of George Pólya was distinguished by the discovery of mathematical solutions to a number of problems originating in the physical sciences. He made contributions to **probability theory**, **number theory**, the theory of **functions**, and the **calculus** of variations. He also cared about the art of teaching mathematics; he worked with educators, advocating the importance of problem solving, for which the United States gave him a distinguished service award. Pólya continued to do innovative research well into his nineties, but he is probably best known for his book on methods of problem solving, called *How to Solve It,* which has been translated into many languages and has sold more than one million copies.

Pólya was born in Budapest, Austria-Hungary (now Hungary), on December 13, 1887, the son of Jakob and Anna Deutsch Pólya. As a boy, he preferred geography, Latin, and Hungarian to mathematics. He liked the verse of German poet Heinrich Heine and translated some into Hungarian. His mother urged him to become a lawyer like his father, and he began to study law at the University of Budapest, but soon turned to languages and literature. He earned teaching certificates in Latin, Hungarian, mathematics, physics, and philosophy, and for a year he was a practice teacher in a high school. Though physics and philosophy interested Pólya, he decided to study mathematics on the advice of a philosophy professor. In an interview published in *Mathematical People: Profiles and Interviews,* Pólya explained how he chose a career in mathematics: "I came to mathematics indirectly.... It is a little shortened but not quite wrong to say: I thought I am not good enough for physics and I am too good for philosophy. Mathematics is in between."

Pólya received his doctorate in mathematics from the University of Budapest in 1912 at the age of 24 with a dissertation on the calculus of probability. Traveling to Germany and France, he was influenced by the work of eminent mathematicians at the University of Göttingen and the University of Paris. In 1914, he took his first teaching position, at the Eidgenössische Technische Hochschule in Zurich, Switzerland; he taught there for twenty-six years, becoming a full professor in 1928.

During World War I, Pólya was initially rejected by the Hungarian Army because of an old soccer injury. When the need for soldiers increased later in the war, however, he was asked to report for military service, but by this time he had been influenced by the pacifist views of British mathematician and philosopher **Bertrand Russell** and refused to serve. As a result, Pólya was unable to return to Hungary for several years, and he became a Swiss citizen. He married Stella Vera Weber, the daughter of a physics professor, in 1918.

Pólya proved an important **theorem** in probability theory in a paper published in 1921, using the term "random walk" for the first time. Many years later, a display demonstrating the concept of a random walk was featured in the IBM pavilion at the 1964 World's Fair in New York; it recognized the work of Pólya and other distinguished scholars. Pólya's work with Gabor Szegö, also a Hungarian, resulted in *Problems and Theorems in Analysis,* published in 1925. The problems in the book were grouped not according to the topic but according to the methods that could be used to solve them. Pólya and Szegö continued to work together and they published another book, *Isoperimetric Inequalities in Mathematical Physics,* in 1951.

Pólya was awarded the first international Rockefeller Grant in 1924 and spent a year in England at Oxford and Cambridge, where he worked with English mathematician **Godfrey Harold Hardy** on *Inequalities,* which was published in 1934. During the 1930s, Pólya frequently visited Paris to collaborate on papers with Gaston Julia. He received another Rockefeller Grant in 1933, this time to visit the United States, where he worked at both Princeton and Stanford universities. In 1940, after World War II had begun in Europe, Pólya left Switzerland with his wife and emigrated to the United States. He taught for two years at Brown University and for a short time at Smith College before joining Szegö at Stanford University in 1942, where he would remain until his retirement in 1953 at age 66. Pólya became a U.S. citizen in 1947.

Before leaving Europe, Pólya had begun writing a book on problem solving. Observing that Americans liked "how-to" books, Hardy suggested the title *How to Solve It.* The book, Pólya's most popular, was published in 1945; it examined discovery and invention and discussed the processes of creation and **analysis**. Although he officially retired in 1953, Pólya continued to write and teach. Another book on heuristic principles for problem solving, at a more advanced level of mathematics, was published in 1954, entitled *Mathematics and Plausible Reasoning.* A third book on problem solving, *Mathematical Discovery,* was published in 1962. In 1963, Pólya was the recipient of the distinguished service award from the Mathematical Association of America. The citation, as quoted in the *Los Angeles Times,* read: "He has given a new **dimension** to problem-solving by emphasizing the organic building up of elementary steps into a complex **proof**, and conversely, the decomposition of mathematical invention into smaller steps. Problem solving *a la Pólya* serves not only to develop mathematical skill but also teaches constructive reasoning in general."

Pólya also became interested in the teaching of mathematics teachers, and he taught in a series of teacher institutes at Stanford University supported by the National Science Foundation, General Electric, and Shell. His film, "Let Us Teach Guessing," won the Blue Ribbon from the Educational Film Library Association in 1968. In 1978, the National Council of Teachers of Mathematics held problem-solving competitions in *The Mathematics Student*; they named the awards the Pólya Prizes.

Pólya made significant contributions in many areas, including probability, **geometry**, real and **complex analysis**, **combinatorics**, number theory, and mathematical physics. Perhaps one indication of the breadth of his accomplishments is the range of fields that now contain concepts bearing his name. For example, the "Pólya criterion" and the "Pólya distribution" in probability theory; and "Pólya peaks," the "Pólya representation," and the "Pólya gap theorem" in complex function theory. Pólya's writings have been praised for their clarity and elegance; his papers were called a joy to read. The Mathematical Association of America established the Pólya Prize for Expository Writing in the *College Mathematics Journal.* Pólya's papers were collected and published by MIT Press in 1984.

Pólya received honorary degrees from the University of Wisconsin at Milwaukee, the University of Alberta, the University of Waterloo, and the Swiss Federal Institute of Technology. He was a member of the American Academy of Arts and Sciences, the National Academy of Sciences of the United States, the Hungarian Academy of Sciences, the Academie Internationale de Philosophie des Sciences in Brussels, and a corresponding member of the Académie Royale des Sciences in Paris. The Society for Industrial and **Applied Mathematics** named an honorary award after him, the Pólya Prize in Combinatorial Theory and Its Applications.

Among his colleagues, Pólya was known as a kind and gentle man, full of curiosity and enthusiasm. In honoring him, Frank Harary praised his depth, his versatility, and his speed and power as a mathematician. The mathematician N. G. de Bruijn wrote of Pólya in *The Pólya Picture Album: Encounters of a Mathematician*: "All his work radiates the cheerfulness of his personality. Wonderful taste, crystal clear methodology, simple means, powerful results. If I would be asked whether I could name just one mathematician who I would have liked to be myself, I have my answer ready at once: Pólya." Pólya suffered a stroke at age ninety-seven and died in Palo Alto, California, on September 7, 1985.

POLYGONS

Polygons are closed plane figures bounded by three or more line segments. In the world of **geometry**, polygons abound. The term refers to a multisided geometric form in the plane. The number of angles in a polygon always equals the number of sides.

Polygons are named to indicate the number of their sides or number of noncollinear points present in the polygon. For example, a triangle has three sides and three vertices, a rectangle has four sides and four vertices; others: pentagon, five and five; hexagon, six and six; heptagon, seven and seven; octagon, eight and eight; nonagon, nine and nine; decagon, ten and ten; and an *n*-gon has *n* number of sides and *n* number of vertices.

A **square** is a special type of polygon, as are triangles, parallelograms, and octagons. The prefix of the term, *poly* comes from the Greek word for many, and the root word *Gon* comes from the Greek word for **angle**.

A regular polygon is one whose whose sides and interior angles are congruent. Regular polygons can be inscribed by a **circle** such that the circle is tangent to the sides at the centers, and circumscribed by a circle such that the sides form chords of the circle. Regular polygons are named to indicate the number of their sides or number of vertices present in the figure. Thus, a hexagon has six sides, while a decagon has ten sides. Examples of regular polygons also include the familiar square and octagon.

Not all polygons are regular or symmetric. Polygons for which all interior angles are less than 180° are called convex. Polygons with one or more interior angles greater than 180° are called concave.

The most common image of a polygon is of a multisided **perimeter** enclosing a single, uninterrupted **area**. In reality, the sides of a polygon can intersect to form multiple, distinct areas. Such a polygon is classified as reflex.

In a polygon, the line running between non-adjacent points is known as a diagonal. The diagonals drawn from a single vertex to the remaining vertices in an n-sided polygon will divide the figure into n-2 triangles. The sum of the interior angles of a convex polygon is then just (n-2)* 180.

If the side of a polygon is extended past the intersecting adjacent side, it defines the *exterior* angle of the vertex. Each vertex of a convex polygon has two possible exterior angles, defined by the continuation of each of the sides. These two angles are congruent, however, so the exterior angle of a polygon is defined as one of the two angles. The sum of the exterior angles of any convex polygon is equal to 360 degrees.

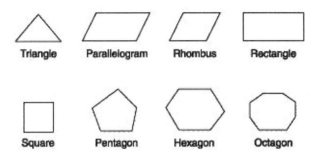

Triangle   Parallelogram   Rhombus   Rectangle

Square   Pentagon   Hexagon   Octagon

Figure 1.

POLYHEDRON

A polyhedron is a three-dimensional closed surface or solid, bounded by plane figures called **polygons**.

The word polyhedron comes from the Greek prefix *poly-*, which means "many," and the root word *hedron* which refers to "surface." A polyhedron is a solid whose boundaries consist of planes. Many common objects in the world around us are in the shape of polyhedrons. The cube is seen in everything from dice to clock-radios; CD cases, sticks of butter, or the World Trade Center towers are in the shape of polyhedrons called parallelepipeds. The pyramids are a type of polyhedron, as are geodesic domes. Most shapes formed in nature are irregular. In an interesting exception, however, crystals grow in mathematically perfect, and frequently complex, polyhedrons.

The bounding polygons of a polyhedron are called the faces. The line segments along which the faces meet are called the edges. The points at which the ends of edges intersect (think of the corner of a cereal box) are the vertices. Vertices are connected through the body of the polyhedron by an imaginary line called a diagonal.

A polyhedron is classified as convex if a diagonal contains only points inside of the polyhedron. Convex polyhedrons are also known as Euler polyhedrons, and can be defined by the equation $E=v+f-e=2$, where v is the number of vertices, f is the number of faces, and e is the number of edges. The intersection of a plane and a polyhedron is called the cross sec-

tion of the polyhedron. The cross sections of a convex polyhedron are all convex polygons.

Polyhedrons are classified and named according to the number and type of faces. A polyhedron with four sides is a tetrahedron, but is also called a pyramid. The six-sided cube is also called a hexahedron. A polyhedron with six rectangles as sides also has many names—a rectangular parallelepided, rectangular prism, or box.

A polyhedron whose faces are all regular polygons congruent to each other, whose polyhedral angels are all equal, and which has the same number of faces meet at each vertex is called a regular polyhedron. Only five regular polyhedrons exist: the tetrahedron (four triangular faces), the cube (six **square** faces), the octahedron (eight triangular faces—think of two pyramids placed bottom to bottom), the dodecahedron (twelve pentagonal faces), and the icosahedron (twenty triangular faces).

Other common polyhedrons are best described as the same as one of previously named that has part of it cut off, or truncated, by a plane. Imagine cutting off the corners of a cube to obtain a polyhedron formed of triangles and squares, for example.

POLYNOMIALS

Polynomials in one **variable** x are algebraic expressions of the form

$P(x) = c_0 + c_1 x + c_2 x^2 + c_3 x^3 + ... + c_N x^N.$

Here c_0, c_1, c_2,... c_N are called the coefficients of the polynomial. If $c_N \neq 0$ then the degree of this polynomial is the nonnegative integer N. A polynomial in one variable can also be conveniently expressed using the summation notation as

$$P(x) = \sum_{n=0}^{N} c_n x^n.$$

Then a polynomial in two variables x and y is an algebraic expression of the form

$$P(x, y) = \sum_{m=0}^{M} \sum_{n=0}^{N} c_{m,n} x^m y^n.$$

Polynomials in three or more variables are defined in a similar manner. In general the coefficients of a polynomial can belong to an arbitrary commutative ring. But in this article we will restrict our attention to polynomials in one variable and with coefficients in the **field** of **complex numbers C**, or in a familiar subring of this field. For more information on polynomials in two or more variables, see the article on **algebraic geometry**.

Let C[x] denote the set of polynomials in one variable x and having coefficients in the field C of complex numbers. We define **addition** of polynomials in this set by adding the corresponding coefficients, so that

$$\sum_{l=0}^{L} a_l x^l \Big\} + \Big\{ \sum_{m=0}^{M} b_m x^m \Big\} = \sum_{n=0}^{N} (a_n + b_n) x^n.$$

Here $N = \max\{L, M\}$, and it must be understood that $a_l = 0$ if $l > L$ and $b_m = 0$ if $m > M$. The **definition** of **multiplication** is slightly more complicated. We set

$$\Big\{ \sum_{l=0}^{L} a_l x^l \Big\} \Big\{ \sum_{m=0}^{M} b_m x^m \Big\} = \sum_{k=0}^{N} c_k x^k,$$

where $K = L + M$, and for each integer k with $0 \le k \le K$ we define

$c_k = a_0 b_k + a_1 b_{k-1} + a_2 b_{k-2} + ... + a_{k-1} b_1 + a_k b_0.$

Again it must be understood that $a_l = 0$ if $l > L$ and $b_m = 0$ if $m > M$. With these definitions of addition and multiplication the set C[x] forms a ring. And then we can identify several important subrings by restricting the coefficients in various ways. For example, Z[x] is the subring of polynomials with coefficients in the ring Z of **integers**, Q[x] is the subring of polynomials with coefficients in the field Q of **rational numbers**, and $\Re[x]$ is the subring of polynomials with coefficients in the field $\Re$ of **real numbers**. Each of these collections of polynomials has interesting mathematical properties.

If $P(x)$ is a nonzero polynomial in Z[x] it may happen that $P(x)$ can be written as a product of polynomials in Z[x]. Some examples of this are

$x^2 - 2x - 15 = (x + 3)(x - 5)$, and $x^8 - 1 = (x - 1)(x + 1)(x^2 + 1)(x^4 + 1)$.

We say that a nonzero polynomial $P(x)$ in Z[x] is *irreducible* if in any factorization of the polynomial as $P(x) = Q(x)R(x)$ one of the factors is equal to 1 or -1. Irreducible polynomials in Z[x] are similar to **prime numbers** in the ring Z. This is because every polynomial in Z[x] can be factored in essentially one way into a product of irreducible polynomials. In general it is not easy to decide if a given polynomial in Z[x] is irreducible.

Each polynomial in C[x] determines a *polynomial function* from the field C to the field C. If $P(x)$ is a polynomial in C[x] it determines the polynomial function that sends the complex number α to the complex number $P(\alpha)$. Usually a polynomial function is simply called a polynomial, and it is clear from the context if the polynomial is to be regarded as a function or as an element of a ring of polynomials. Because each polynomial $P(x)$ determines a function in this way, we may ask about the set of complex numbers α such that $P(\alpha) = 0$. This set is called the set of *roots* or the set of *zeros* of the polynomial. One of the most important results about polynomials is the *fundamental theorem of algebra*. This **theorem** asserts that every polynomial in C[x] with positive degree has a root in C. The first rigorous **proof** of the fundamental theorem of **algebra** was given by Gauss. Notice that the same

result does not hold if the field C is replaced by $\Re$: the polynomial $x^2 + 1$ has positive degree and coefficients in $\Re$, but it does not have a root in $\Re$. Also, if $P(x)$ is a polynomial in C[x] with positive degree and α_1 in C is a root, then $P(x)$ factors in the ring C[x] as $P(x) = (x - \alpha_1)Q(x)$. If $Q(x)$ has positive degree we can apply the **fundamental theorem of algebra** again to Q. If α_2 is a root of Q then we get the factorization $P(x) = (x - \alpha_1)(x - \alpha_2)R(x)$. Continuing in this manner, we find that a polynomial of degree N in **C**[x] can be written as

$P(x) = c_0 + c_1 x + c_2 x^2 + c_3 x^3 + ... + c_N x^N$
$= c_N (x - \alpha_1)(x - \alpha_2)... (x - \alpha_N),$

where the complex numbers $\alpha_1, \alpha_2, ..., \alpha_N$ are the **roots** of $P(x)$. Of course we *cannot* conclude that the roots are distinct. For example, we have the factorization

$x^6 + 2x^5 + 2x^4 - 2x^2 - 2x - 1 = (x - 1)(x + 1)(x - \omega)(x - \omega)(x - \omega^-)(x - \omega^-),$

where $\omega = -\frac{1}{2} + i\frac{1}{2}\sqrt{3}$ and $\omega^- = -\frac{1}{2} - i\frac{1}{2}\sqrt{3}$. In this case we say that ω and ω^- are roots of *multiplicity* two. More generally, if a complex number β appears exactly M times among the roots of the polynomial $P(x)$, then we say that P has a root of multiplicity M at β. Thus a polynomial of degree $N \ge 1$ in C[x] has exactly N roots in C provided the roots are *counted with multiplicity*.

Polynomials in the ring $\Re[x]$ are often used as polynomial **functions** to approximate more general continuous functions. Let $[u, v] \subseteq \Re$ be a closed **interval** in $\Re$ and let $f:[u, v] \to \Re$ be a continuous function. Then the *Weierstrass approximation theorem* states that for every $\varepsilon > 0$ there exists a polynomial $P(x)$ in $\Re[x]$ such that

$|f(x) - P(x)| < \varepsilon$

for all real numbers x with $u \le x \le v$. Thus every continuous function defined on a finite, closed interval in $\Re$ can be uniformly approximated by a polynomial. In the special case $[u, v] = [0, 1]$ there is a particularly simple type of polynomial that can be used to uniformly approximate an arbitrary continuous function. Let $f: [0, 1] \to \Re$ be continuous and then define the Nth *Bernstein polynomial* associated to f by

$$B_N(x) = \sum_{n=0}^{N} \binom{N}{n} f\left(\frac{n}{N}\right) x^n (1 - x)^{N-n}.$$

It can be shown that for every $\varepsilon > 0$ there exists an integer M such that

$|f(x) - B_N(x)| < \varepsilon$

for all real numbers x with $0 \le x \le 1$ and all integers $N \ge M$. Alternatively, the sequence $B_N(x)$ associated to f converges uniformly to f on [0, 1] as $N \to \infty$.

PONCELET, JEAN–VICTOR (1788-1867)
French geometer

Regarded as one of the fathers of modern projective **geometry**, Jean-Victor Poncelet's development of the concept of poles and polar lines, in the context of conics, was the forerunner of the principle of duality. His development of distinctions between projective and metric properties served as a foundation for modern structural concepts.

Poncelet was born in July 1, 1788, in Metz, Lorraine, France, to Claude Poncelet, a wealthy landowner and advocate at the Parliament of Metz, and Anne-Marie Perrein. While he was Claude Poncelet's natural son, he was not legitimized as such until later in life. Poncelet's earliest studies took place in the small town of Saint-Avold, where he lived with a family to which he had been entrusted. Poncelet returned to Metz in 1804 and continued his studies at the city's *lycée* (the equivalent of elementary school). A highly successful student, he entered Metz's École Polytechnique in the fall of 1807, where he spent three years, having fallen behind a year due to poor health. His instructors at the Polytechnique included **Gaspard Monge**, S. F. Lacroix, and A. M. Ampère.

In 1810 Poncelet entered the French corps of military engineers, and graduated from the École d'Application of Metz in early 1812. He then conducted fortification work on Walcheren, a Dutch island, and served in the Russian campaign as lieutenant, assigned to the engineering general staff. In November 1812 Poncelet was taken prisoner by the Russians and held hostage at a camp on the Volga River at Saratov until June 1814. During his captivity, Poncelet studied mathematics. Because he had no access to mathematics books, he reconstructed elements of pure and **analytic geometry**, and then began research on the projective properties of conics and systems of conics. This preliminary research, published in 1862 in his *Applications d'analyse et degéométrie*, served as the basis for Poncelet's most important work.

Poncelet returned to France upon his release and was appointed captain of the engineering corps at Metz. For the next 10 years, he contributed to several topography and fortification projects, while concentrating on building and organizing an engineering arsenal. During this time, Poncelet developed a new **model** for a **variable** counterweight drawbridge. He also devoted ample time to the continuation of his research on conics. In addition, he began to publish articles in the *Annales de mathématiques pures et appliquées*. His *Traité des propriétés projectives des figures*, written in consultation with Servois and published in 1822, set forth several fundamental concepts of projective geometry, including the cross-ratio, perspective, involution, and circular points at **infinity**.

During this time Poncelet also presented his theories on conics to the Académie Royale des Sciences, hoping to advance the idea that geometric concepts could be generalized using elements at infinity and imaginary elements. His theories were based on the concept of an "ideal chord" and a "principle of continuity." The referee of the Académie rebuffed Poncelet's theories, primarily the "principle of continuity," and Poncelet set out to prove his critic wrong. He was sidetracked, however, when he was offered a position as professor of **mechanics** applied to machines at the École d'Application de l'Attilerie et du Génie at Metz, which commenced in January 1825. He continued to publish in journals, but by and large abandoned his work in projective geometry.

As a professor and practitioner of mechanics, Poncelet focused on improving the efficiency of machines through a combination of mathematical theory and direct application. His work led to significant improvements in the operation of turbines and waterwheels, with his innovations more than doubling the efficiency of the waterwheel. Poncelet also became more involved in the civic life of Mertz, becoming a member of that city's municipal council and secretary of the Conseil Général of the Moselle in 1830. He soon moved from his native home, however, when in 1834 he was named scientific rapporteur for the Committee of Fortifications, editor of the *Mémorial de officier du génie,* and a member of the mechanics section of the Académie Royale des Sciences (an elected position), necessitating a move to Paris. He accepted a teaching post at the Faculty of Sciences in Paris as well in 1837, and he began to publish his courses and papers in book form at this time.

In 1842, Poncelet married Louise Palmyre Gaudin, and he planned to retire from the military soon after, in 1848. His plans were disrupted by the French Revolution of 1848, however, during which he was appointed brigadier general and assigned to a post of commandant at the École Polytechnique. There, he focused mainly on curriculum matters and management of riot situations which erupted among students in May and June of 1848. He remained at the institution until the fall of 1850. He also continued his work with the Conseil Général, although he resigned his position with the Académie. Poncelet finally retired in October 1850 and then focused on publishing his works. He died in Paris on December 22, 1867.

POSTULATE

A postulate is an assumption, that is, a proposition, or statement, that is assumed to be true without any **proof**. Postulates are the fundamental **propositions** used to prove other statements known as theorems. Once a **theorem** has been proven it may be used in the proof of other theorems. In this way, an entire branch of mathematics can be built up from a few postulates. Postulate is synonymous with **axiom**, though sometimes axiom is taken to mean an assumption that applies to all branches of mathematics, in which case a postulate is taken to be an assumption specific to a given theory or branch of mathematics. Euclidean **geometry** provides a classic example. **Euclid** based his geometry on five postulates and five "common notions," of which the postulates are assumptions specific to geometry, and the "common notions" are completely general axioms.

The five postulates of Euclid that pertain to geometry are specific assumptions about lines, angles, and other geometric concepts. They are:

1) Any two points describe a line.

2) A line is infinitely long.

3) A circle is uniquely defined by its center and a point on its **circumference**.

4) Right angles are all equal.

5) Given a point and a line not containing the point, there is one and only one parallel to the line through the point.

The five "common notions" of Euclid have application in every branch of mathematics, they are:

1) Two things that are equal to a third are equal to each other.

2) Equal things having equal things added to them remain equal.

3) Equal things having equal things subtracted from them have equal remainders.

4) Any two things that can be shown to coincide with each other are equal.

5) The whole is greater than any part.

On the basis of these ten assumptions, Euclid produced the *Elements*, a 13-volume treatise on geometry (published c. 300 B.C.) containing some 400 theorems, now referred to collectively as Euclidean geometry.

When developing a mathematical system through logical deductive reasoning any number of postulates may be assumed. Sometimes in the course of proving theorems based on these postulates a theorem turns out to be the equivalent of one of the postulates. Thus, mathematicians usually seek the minimum number of postulates on which to base their reasoning. It is interesting to note that, for centuries following publication of the *Elements*, mathematicians believed that Euclid's fifth postulate, sometimes called the **parallel postulate**, could logically be deduced from the first four. Not until the nineteenth century did mathematicians recognize that the five postulates did indeed result in a logically consistent geometry, and that replacement of the fifth postulate with different assumptions led to other consistent geometries.

Postulates figure prominently in the work of the Italian mathematician **Guiseppe Peano** (1858-1932), who formalized the language of **arithmetic** by choosing three basic concepts: **zero**; number (meaning the non-negative **integers**); and the relationship "is the successor of." In addition, Peano assumed that the three concepts obeyed the five following axioms or postulates: 1) Zero is a number.

2) If b is a number, the successor of b is a number.

3) Zero is not the successor of a number.

4) Two numbers of which the successors are equal are themselves equal.

5) If a set S of numbers contains zero and also the successor of every number in S, then every number is in S.

Based on these five postulates, Peano was able to derive the fundamental laws of arithmetic. Known as the **Peano axioms**, these five postulates provided not only a formal foundation for arithmetic but for many of the constructions upon which **algebra** depends.

Indeed, during the nineteenth century, virtually every branch of mathematics was reduced to a set of postulates and resynthesized in logical deductive fashion. The result was to change the way mathematics is viewed. Prior to the nineteenth century mathematics had been seen solely as a means of describing the physical universe. By the end of the century, however, mathematics came to be viewed more as a means of deriving the logical consequences of a collections of axioms.

In the twentieth century, a number of important discoveries in the fields of mathematics and logic showed the limita-

tion of proof from postulates, thereby invalidating Peano's axioms. The best known of these is Gödel's theorem, formulated in the 1930s by the Austrian mathematician **Kurt Gödel** (1906-1978). Gödel demonstrated that if a system contained Peano's postulates, or an equivalent, the system was either inconsistent (a statement and its opposite could be proved) or incomplete (there are true statements that cannot be derived from the postulates).

See also Arithmetic; Logic, symbolic; Proof; Theorem.

POWER FUNCTIONS

Power **functions** have the form $f(x) = ax^b$, where a and b are real constants. The **variable** x can take on all real number values. Power functions should not be confused with **exponential functions**, which have the similar looking form $f(x) = ab^x$. The variable x appears in the exponent in an exponential function, whereas x is the base of a power function. Some familiar power functions are $y = x^2$, $y = x^3$, $y = 2x^{3/2}$, and $y = x^{-1.2}$. The exponent b determines basic shape of the power function graph. If $b = 0$ or $b = 1$, then $y = ax^b$ is not really a power function; these are so-called degenerate cases because the function "degenerates" into the constant function $y = a$ or the **linear function** $y = ax$. If b is greater than 1, then the graph of $y = ax^b$ is increasing and concave up as x grows without bound. If b is between 0 and 1, then the graph of $y = ax^b$ is increasing and concave down as x grows without bound. For negative values of x, a power function can become undefined when b is rational with even denominator because even-numbered **roots** of negative **real numbers** are not real numbers. In those cases, we restrict the **domain** of the power function to nonnegative real numbers.

Power functions are frequently used as mathematical models for data collected in certain real world situations. A very famous set of data which has a power function **model** is the data collected originally by the Dutch-Swedish astronomer **Tycho Brahe** (1546-1601) and analyzed by the German astronomer **Johannes Kepler** (1571-1630). This data described the motions of the planets of the solar system in terms of their **mean distance** from the sun and the time required for them to make one complete revolution around the sun. After about twenty years of studying this data, Kepler came up with his famous three laws of planetary **motion**. The first law stated that the planets move about the sun in elliptical orbits with the sun at one focus. The second law said that a vector from the sun to a planet will sweep out **area** at a constant rate. The third law is where the power function model comes into play. It said the following: "If T is the time required for a planet to make one complete revolution about the sun and d is the planet's mean distance from the sun, then $T^2 = kd^3$ where k is a constant." By taking the positive **square root** of both sides of this equation, we can see that it is equivalent to $T = k^{1/2}d^{3/2}$, which has the form of a power function. Kepler obtained this equation without the benefit of modern technology, but today's computers and calculators come with built-in power regression features which allow one to input the data and output the power function model which best fits the data. When this is done with Kepler's data, one gets an equation of the form $y = ax^b$, where b is very close to 1.5 or 3/2, giving very strong evidence for the **truth** of Kepler's third law.

Other examples of power function models include: Newton's law for a body that is free-falling with negligible air resistance near the surface of the earth, $d = -16t^2$ where d is distance in feet and t is time in seconds; $I = kd^{-2}$ which says that the intensity of **light** is inversely proportional to the square of the distance form the observer to the light source; $V = (4/3)\pi r^3$, the formula for the **volume** of a **sphere** of radius r; $v = kp^{-1}$, Boyle's Law for gasses, where v is volume and p is pressure; and $N = s^d$, the formula for calculating the self-similarity **dimension** d of a **fractal**. This is only a small sampling of the many areas in which power function models are used by mathematicians and scientists.

Prior to the use of calculators or computers, scientists who suspected that a data set could be best modeled by a power function would use so-called "log-log" graph paper to plot the **logarithms** of both coordinates of the original data set. If the original data is power function data, the log-log graph will be very close to linear with a **slope** of b and a y-intercept of log(a). The reason for this is that if we take logarithms of both sides of the equation $y = ax^b$, we get $\log(y) = \log(ax^b)$, which, by the product property of logarithms, is equivalent to $\log(y) = \log(a) + \log(x^b)$, which, by the power property of logarithms, is equivalent to $\log(y) = \log(a) + b\log(x)$. This last equation is linear in log(x) and log(y) with a slope of b and a y-intercept of log(a) as claimed. Thus when a graphing **calculator** does power regression, it applies a logarithmic **transformation** to both the first and second coordinates of the original data, carries out linear regression on the transformed data, then transforms the linear function into the power function which best fits the original data.

See also Exponential functions; Linear functions; Logarithms

POWER SERIES

A power series expresses a function $f(x)$ as an infinite sum (or "series") of terms involving **powers** of the **variable** x. Power series can be considered a natural extension of **polynomials**, which are finite sums of such terms. All of the important **functions** of **calculus** can be represented by power series; for example:

$$e^x = 1 + x + x^2/2! + x^3/3! + ...$$
$$\sin(x) = x - x^3/3! + x^5/5! - ...$$

In addition, unknown functions, such as solutions to **differential equations**, frequently can be represented in this way as well.

When truncated after a finite number of terms, power series usually provide excellent approximations to the functions they represent, and they can easily be computed by hand. For many applications in physics and other sciences, it is easier to work with a "first-order approximation" or "second-

order approximation" to a function (in other words, to stop after the x or x^2 term in the power series) than to use the function itself. The errors inherent in this approximation may be smaller than other sources of experimental error, and in any case they can be well controlled.

The major practical limitation of power series is that they do not always converge to any real number. Without some guarantee of **convergence**, it is impossible to predict whether a higher-order approximation is really better than a lower-order one.

The usefulness of power series for theoretical mathematics depends on the fact that they can be differentiated and integrated in the same way that ordinary polynomials can. In the early years of calculus, mathematicians tended to accept this on faith, but in the nineteenth century they gradually realized that such statements are valid only under certain conditions about convergence. The drive to understand these conditions was a major force behind the evolution of calculus into the modern subjects of real and complex **analysis**.

See also Convergence; Taylor and Maclaurin series

POWERS

The term powers is most often used in mathematics to mean the product obtained by multiplying a number by itself one or more times. For instance, four (2×2), eight ($2 \times 2 \times 2$), and sixteen ($2 \times 2 \times 2 \times 2$) are all powers of 2. A compact notation has been developed to represent powers, and is of the form a^n, where "a" is the basis and "n" the exponent. This notation means that the number a is to be multiplied by itself "n" (whole number) times; and a^n is said to be the "nth power of a." If $a = 3$ and $n = 4$, then "$3^4 = 3 \times 3 \times 3 \times 3 = 81$," so 81 is a power of 3. Power is also used to mean the exponent of a basis. In the expression 3^4, three is said to be raised to the fourth power. So for the general expression a^n, the term power can be used to mean either the number represented by a^n, or the exponent n itself.

Just as repeated **addition** of the same number led to the **arithmetic** operation of **multiplication**, so the repeated multiplication of a given number led to the algebraic operation of powers (or exponentials). The ancient Egyptians, Babylonians, and Greeks encountered powers of numbers when investigating formulas describing particular geometric forms. The areas of **polygons** and the volumes of solid figures are expressed as powers of a particular length of the figure. For example, the **area** (A) of a **square** of side s is calculated as side s times side s, or $A = s^2$; that is, the second power of s, or "s squared." The volumes of many solid geometrical figures are related to the third power of a length. As examples, the **volume** (V) of a cube with sides of length x is calculated as $V = x \cdot x \cdot x$, or $V = x^3$, where x^3 is the third power of x, or "x cubed;" and the volume of a **sphere** with radius r is $(4/3)\pi r^3$.

Of particular importance are the powers of ten: 10, 100, 1000, etc., represented in exponential form as 10^1, 10^2, 10^3, etc. The predominant number system used today, the so-called Hindu-Arabic system, is a positional numbering system based on the powers of ten. In this system any real number can be expressed as the sum of products involving powers of ten. For instance, 54,321 denotes the summation: $(5)(10^4) + (4)(10^3) + (3)(10^2) + (2)(10^1) + (1)(10^0)$, where $10^0 = 1$. Some powers of ten have been given unique names. For instance, in the United States 10^3 denotes thousand, 10^6 denotes million, and 10^9 denotes billion.

As noted previously, the concept of powers is an outgrowth of repeated multiplication by the same **factor**, just as multiplication sprang from repeated additions of the same term. Because addition is the most basic arithmetic operation, it is said to be an operation of the first order. Multiplication (an extension of addition) is called an operation of the second order, and power (built upon multiplication) is an operation of the third order. Because mathematicians recognized the need for powers through rather simple formulas, the exponent was originally limited only to positive **integers**. Several important definitions for the powers of numbers (with the integer value of the exponent $n \geq 0$) are:

- For any basis "a," $a^1 = a$.
- For any non-zero basis "a," $a^0 = 1$.
- Powers of the form a^{2n} (for non-zero n) are called even powers.
- Powers of the form a^{2n+1} are called odd powers.

Several important rules pertaining to the multiplication, **division**, and exponentiation of powers are:

- Rule 1: $(a \times b)^n = a^n \times b^n$ (the power of a product is the product of powers),
- Rule 2: $(a / b)^n = a^n / b^n$ (the power of a ratio is the ratio of two powers),
- Rule 3: $a^n \times a^m = a^{n+m}$ (the product of two powers "with the same basis" is another power of that basis),
- Rule 4: $a^n / a^m = a^{n-m}$, where $n \geq m$ (similar to rule 3 but for ratios),
- Rule 5: $(a^n)^m = a^{n \times m}$ (the power of a power is itself a power of the same basis),
- Rule 6: Powers with a negative basis are positive for even **exponents** and negative for odd exponents. For instance, the even powers of "-1" are $(-1)^{2n} = 1$, while its odd powers are $(-1)^{2n+1} = -1$.

Many times in history mathematical definitions have been expanded to encompass new elements or concepts. For example, **subtraction** of "5 - 7" was undefined in some early arithmetic systems because **negative numbers** were not recognized as real quantities; so in general, the expression "a - b" was only valid for $a \geq b$, while for $b > a$ it was undefined. Eventually the number system was changed to include negative quantities so that "a - b" was valid for all values of a and b.

Rule 4 places a similar restriction on powers (i.e., negative exponents are unrecognized for powers). However, definitions may be changed in order to obtain a more generalized system. The following definition is added to those previously given:

- Rule 7: $a^{-n} = 1 / a^n$, for all non-zero a.

Rule 4 is then modified into:

- $a^n / a^m = a^{n-m} = a^n \times a^{-m}$, for all integers of n and m (and non-zero a).

To demonstrate the utility of powers with negative exponents consider the division of 7,120 by 1,000,000. These numbers may be expressed as the product of a decimal number and a power of ten (called **scientific notation**): $7,120 = 7.12 \times 10^3$ and $1,000,000 = 1 \times 10^6 = 10^6$. The ratio may then be represented as: $7,120 / 1,000,000 = (7.12 \times 10^3) / (10^6) = (7.12 \times 10^3)(10^{-6}) = (7.12)(10^3)(10^{-6}) = (7.12)(10^{3-6}) = 7.12 \times 10^{-3}$.

Finally it should be noted that just as addition and multiplication have inverse operations (subtraction and division, respectively) the operation of power has two corresponding inverse operations, that of **roots** and **logarithms**.

See also Addition; Area; Arithmetic; Decimal position system; Exponents; Integers; Logarithms; Multiplication; Polygons; Products and quotients; Roots; Sums and differences; Volume

PREDICATE

In ordinary English, the sentence "Fish swim" has a subject, "Fish," and a verb "swim." The verb, as defined by English grammar, is also called a "predicate." This is the simplest type of predicate, but, in general, a predicate can be understood as one or more words that say something about a subject. Thus, a predicate is attributive. In the sentence "John is tall," the predicate is "is tall." This predicate attributes tallness to John. In mathematical logic, the word "predicate" retains its English language sense of attributing some feature to a subject, but as we shall see, what is regarded as a predicate in the grammatical sense may not always be considered a predicate in the logical sense. Modern **symbolic logic** is generally broken into two parts: propositional logic (also called the propositional **calculus**) and predicate logic (also called the predicate calculus). Propositional logic is concerned only with the logical structure of sentences, whereas, predicate logic deals with the content of sentences as well as their structure. Thus, propositional logic deals with the correct ways to structure sentences using the so-called "sentential" connectives: "and," "or," "If... then," "If and only if," and "not." Predicate logic overlays propositional logic with the universal and existential quantifiers: "For all" and "There exists," respectively. These quantifiers make it possible to attribute qualities to all, some, or none of and entire class of subjects. For example, the statement "All men are mortal" may be rendered as "For all x, x is mortal," where x is restricted to the set of all men. Likewise, "Some men are mortal" may be rendered as "There exists x, such that x is mortal." In each of these cases, the predicate is "is mortal"; mortality is being attributed to either all men or some men.

From the point of view of mathematical logic, a predicate can be seen as a function which has the effect of attributing some quality to some **variable** or to some specific instance of the variable. So let M be the function which attributes mortality to a variable. Then M(x) is the function notation for "x is mortal." If a is a specific instance of x, then M(a) says that a is mortal.

Disputes about what qualities can and cannot be considered as legitimate predicates have been the source of controversy throughout the history of philosophy and logic. One of the most notable examples of this is the dispute over whether existence can be considered a predicate. **Aristotle** (384-322 B.C.), the founder of logic as a discipline, seemed to consider it so. The Italian scholastic philosopher St. Anselm (1033-1109), who was appointed archbishop of Canterbury in 1093, implicitly used existence as a predicate in his famous "ontological proof" for the existence of God. Anselm defined God to be "That being of which none greater can be conceived." But, he said, if such a being could be conceived without having the property of existence, then it would be possible to conceive of a being with all the same properties, but also having the property of existence. Such a being would be greater than the one without existence. Therefore, according to Anselm, "that being of which none greater can be conceived" must exist, and that being is God. The French philosopher-mathematician, **René Descartes** (1596-1650), essentially reiterated Anselm's argument, but the German philosopher Immanuel Kant (1724-1804) said that Anselm's **proof** was flawed because he had assumed that existence is a predicate. In other words, Anselm had assumed that existence was just one of the many attributes that something could possess; but Kant insisted that existence is very different from other kinds of properties that might be attributed to a being, such as tallness or strength or beauty. Existence cannot be considered as a predicate in the sense that these other qualities can. The introduction into modern logic of the quantifiers "For all" and "There exists" came, in part, to clarify the role of existence in logical discourse. Thus, the statement "Fish swim" is rendered, for logical purposes, "For all x, if x is a fish, then x swims." On the other hand, "God exists" is rendered "There exists an x such that x is God." The **truth** value of this latter statement, according to Kant, must be established by an argument which does not regard existence as a predicate. The point here is that while the sentences "Fish swim" and "God exists" have the same grammatical structure in the English language, they do not have the same logical structure. In the grammatical structure, "swim" and "exists" play the same role, whereas in the logical renderings "There exists an x" plays the same role as "For all x." They are both quantifiers.

See also Quantifier

PRIME NUMBER THEOREM

The prime number **theorem** addresses the question of what proportion of **integers** are **prime numbers**. It was proved at the end of the 19th century by the French mathematician **Jacques Hadamard** and the Belgian mathematician **Charles de la Vallée-Poussin**, working independently.

The theorem states that there are approximately N/ln N prime numbers between 2 and N. Here, ln N denotes the natural logarithm of N and "approximately" means that the ratio of the two quantities gets close to 1 as N gets large or, in other words, the limit of the ratio is 1 as N goes to **infinity**. Another, more informal, way of viewing the theorem is as saying that probability of a number of size N being prime is about 1/ln N. For example, the number of primes below $10^9 = 1,000,000,000$ (one billion) is 50,847,534, whereas $10^9/\ln(10^9)$ is approximately 48,254,942 and the ratio is about 1.05.

The theorem was first conjectured at the beginning of the 19th century by **C. F. Gauss** and **A. M. Legendre**. The question was studied in the 19th century by **P. L. Chebyshev** and **B. Riemann** among others. Riemann introduced new methods of complex **analysis** and, in particular showed how the prime number theorem would follow from properties of the **Zeta function**. Riemann also showed that an even better approximation to the prime counting function would follow from a certain property of the Zeta function which became known as the **Riemann hypothesis**. This hypothesis formed the content of Hilbert's eighth problem and it is unproved to this date and is considered to be the most important unsolved problem in mathematics.

The first proofs of the prime number theorem followed Riemann's approach and, for a while, there was a debate as to whether **complex analysis** was really necessary. The argument was settled when, in 1949, **P. Erdös** and A. Selberg produced a **proof** of the prime number theorem which did not use complex analysis, a so-called elementary proof, although it was far from simple.

The recent use of prime numbers in cryptography (the science of secret codes) has added new interest to the study of prime numbers and their distribution. For instance, it became important to know the relative frequency of prime numbers among very large numbers, say 100 to 200 digits. Since it is currently impossible to count the prime numbers in this **range**, an estimate such as given by the prime number theorem is important and guarantees that there plenty such primes. For instance, a 100-digit number is prime with probability $1/\ln(10^{100})$ which is approximately 0.004. Thus, to obtain a 100-digit prime number, one uses a fast primality test to test random 100-digit numbers until a prime is found and this should succeed after a few hundred tries.

See also Prime numbers; Zeta function

PRIME NUMBERS

A prime number is a number that has just two factors (divisors): itself and the number 1. For example, 7 is prime, since its only divisors are 7 and 1. The number 6 is not prime, since it is divisible by 2 and 3. Numbers that have more than two divisors are called composite numbers.

Prime numbers have fascinated mathematicians since the age of **Pythagoras**. By Euclid's time, mathematicians knew the fundamental **theorem** of **arithmetic**: every whole number can be written as the product of prime numbers, in a unique way (up to reordering of the factors). For example, 60 can be written as 2 x 2 x 3 x 5. Thus, prime numbers can be looked upon as the building blocks for the **whole numbers**.

Among small numbers, many are prime: 2, 3, 5, 7, 11, 13 and so forth. Among larger numbers, however, the number of primes thins out greatly, raising an important question for the ancients: is there a finite or infinite number of primes? The answer to this question was known to **Euclid**; in his *Elements,* he proves that there is an infinite collection of primes. His **proof**, which is remarkably simple and elegant, is one of the first recorded proofs to use the method called reductio ad absurdum, which works by assuming the opposite of the desired statement, and then following a logical train of reasoning until an absurdity is reached.

Thus, Euclid starts out by assuming the opposite of what he wants to prove: he assumes that there are only finitely many primes. If that is the case, he reasons, we can make a list of all the primes; let's suppose they are denoted p_1, p_2, p_3,..., p_n. Let's consider a new number: $N = p_1 p_2 p_3 ... p_n + 1$. What kind of number is N? It is not a prime number, because it is larger than all of the prime numbers on the list. But neither is N a composite number: by the Fundamental Theorem of **Algebra**, any composite number can be broken down into a product of primes, but N is not divisible by any primes (p_1 divides into N with a remainder of 1; p_2 divides into N with a remainder of 1; and so forth). There is only one number that is neither prime nor composite: the number 1. But N is certainly larger than 1. So N is a number that is neither 1, nor prime, nor composite. This is an absurdity. Hence our original assumption, that there is a finite collection of primes, must be wrong; we have proven that the number of primes is infinite.

In around 200 B.C., the Greek mathematician Eratosthenes came up with an **algorithm** for finding prime numbers, now known as the **Sieve of Eratosthenes**. It works in the following way. Suppose you want to find all the prime numbers less than 100; start with a list of all the numbers from 1 to 100. The number 2 is prime, so circle it. Now anything that is divisible by 2 is not prime, so cross out every number divisible by 2. The first uncrossed number on the list is 3. Since 3 has not been crossed out, it is not divisible by any smaller numbers, hence it is prime; circle it, and cross out all the numbers divisible by 3. Four is crossed out, so proceed to 5, circle it, and continue in this way. The circled numbers will be a complete list of the primes less than 100.

Beginning in the 17th century, prime numbers played an important role in the growing field of **number theory**, the branch of mathematics that seeks patterns in the number system. The great mathematician **Pierre de Fermat**, now most famous for his recently proven last theorem (see **Fermat's last theorem**), proved that every prime number of the form $4n + 1$ can be written as the sum of 2 squares, in a unique way. He also conjectured that when n is a power of 2, the number $2^n + 1$ is prime. This is true when $n = 1, 2, 4, 8$, and 16. But a century after Fermat, **Leonhard Euler** showed that it is false when $n = 32$: $2^{32} + 1$ is not prime.

The field of number theory abounds with open questions about prime numbers. One of the most famous is **Goldbach's conjecture**, formulated in 1742 in an exchange of letters between **Christian Goldbach** and Euler. It states that every even number greater than 2 can be written as the sum of two primes. Goldbach's conjecture has been verified for all numbers smaller than 4×10^{14}, but so far no one has come up with a proof, and it remains possible that there is some tremendously large number for which the conjecture is false. Another important open problem is the twin primes conjecture, which hypothesizes that there are infinitely many "twin primes," pairs of primes only 2 apart.

Although the prime numbers seem to be distributed in a very uneven way, in the large scale their thinning out follows a definite pattern. In the 19th century, mathematicians proved the **prime number theorem**, which states that when n is a large number, the density of primes near n is $1/\log(n)$. They also showed that when n is large, the number of primes less than n is roughly $n/\log(n)$.

The problem of factoring large numbers is extremely difficult, so even though it is known that there are infinitely many primes, finding them is not easy. The largest known prime to this day is the number $2^{6972593} - 1$, which has 209860 digits.

PROBABILITY DENSITY FUNCTION

The probability density function, often denoted as pdf, is a function that describes the probability of finding a random **variable** within a defined **range**. The significance of the pdf, $f(x)$, is that $f(x)dx$ is the probability that the random variable x' is in the **interval** $(x, x + dx)$. This is often written as $P(x \leq x' \leq x + dx) = f(x)dx$. Since $f(x)dx$ is a probability it is unitless and therefore $f(x)$ has the units of inverse random variable units. It is also possible to define the probability of finding the random variable somewhere in a finite interval. $P(a \leq x \leq b) = \int_a^b f(x)dx$, where the finite interval is $[a, b]$. This probability is equal to the **area** under the curve defined by $f(x)$ from a to b.

As with other probability distributions the pdf has some restrictions. Since $f(x)$ is a probability density it must be positive for all values of the random variable x: $f(x) \geq 0$, $-\infty < x < \infty$. The probability of finding the random variable somewhere on the real axis must be unity, that is it must be possible to find the random variable on the axis somewhere: $\int_{-\infty}^{\infty} f(x')dx' = 1$. These are the only two restrictions that must be satisfied for a function to be a probability density function. The expectation of the probability density function is defined as $E(x') = \int_{-\infty}^{\infty} xf(x)dx =$. This integral must be convergent for this to be true.

There are three common, important probability density **functions**: the normal probability density function, the exponential probability density function, and the Cauchy probability density function. The normal probability density function is important because the normal random variable is a frequently used **model** in statistical theory. This distribution is often called the bell curve or Gaussian distribution and its probability density function has the form $f(x) = (\exp(-1/2x^2))/(\sqrt{2\pi})$. This probability density function has a **mean** of 0 and a **standard deviation** of 1. The exponential probability density function arises in the study of the **Poisson distribution**. It has the form $f(x) = a \exp(-ax)$ $(0 \leq x < \infty)$ where a is positive for $x \geq 0$. The Cauchy probability density function, or as it is sometimes called the Lorentzian distribution, describes a resonance behavior and is usually written as $f(x) = 1/[(\pi(1 + x^2)](-\infty < x < \infty)$. It does not have a mean value or a variance since the integral does not converge. As a result of this it has some unique properties such that if independent random variables have a Cauchy distribution then the average of those variables also has a Cauchy distribution. The variability of the average is identical to that of a single observation.

Probability density functions play an important role in statistical analysis of events. These functions are useful in determining the likelihood of finding a random variable in a specified interval. The probability density function is related to another important idea, the cumulative distribution function, as $f(x) = F'(x)$, where $F(x)$ is the cumulative distribution function. This relationship is important in many engineering and mathematics applications.

PROBABILITY THEORY

Probability theory is a branch of mathematics concerned with determining the long-run frequency or chance that a given event will occur. This chance is determined by dividing the number of selected events by the number of total events possible. For example, each of the six faces of a die has one in six probability on a single toss. Inspired by problems encountered by 17th-century gamblers, probability theory has developed into one of the most respected and useful branches of mathematics with applications in many different industries. Perhaps what makes probability theory most valuable is that it can be used to determine the expected outcome in any situation from the chances that a plane will crash to the probability that a person will win the lottery.

The branch of mathematics known as probability theory was inspired by gambling problems. The earliest work was performed by **Girolamo Cardano** (1501-1576) an Italian mathematician, physician, and gambler. In his manual *Liber de Ludo Aleae*, Cardano discusses many of the basic concepts of probability complete with a systematic analysis of gambling problems. Unfortunately, Cardano's work had little effect on the development of probability because his manual, which did not appeared in print until 1663, received little attention.

In 1654, another gambler named Chevalier de Méré created a dice proposition which he believed would make money. He would bet even money that he could roll at least one twelve in 24 rolls of two dice. However, when the Chevalier began losing money, he asked his mathematician friend **Blaise Pascal** (1623-1662) to analyze the proposition. Pascal determined that this proposition will lose about 51% of the time. Inspired by this proposition, Pascal began studying more of these types of problems. He discussed them with another famous mathematician, **Pierre de Fermat** (1601-1665) and together they laid the foundation of probability theory.

Probability theory is concerned with determining the relationship between the number of times a certain event occurs and the number of times any event occurs. For example, the number of times a head will appear when a coin is flipped 100 times. Determining probabilities can be done in two ways; theoretically and empirically. The example of a coin toss helps illustrate the difference between the two approaches. Using a theoretical approach, we reason that in every flip there are two possibilities, a head or a tail. By assuming each event is equally likely, the probability that the coin will end up heads is 1/2 or 0.5. The empirical approach does not use assumption of equal likeliness. Instead, an actual coin flipping experiment is performed and the number of heads is counted. The probability is then equal to the number of heads divided by the total number of flips.

A theoretical approach to determine probabilities requires the ability to count the number of ways certain events can occur. In some cases, counting is simple because there is only one way for an event to occur. For example, there is only one way in which a 4 will show up on a single roll of a die. In most cases, however, counting is not always an easy matter. Imagine trying to count the number of ways of being dealt a pair in 5-card poker.

The fundamental principle of counting is often used when many selections are made from the same set of objects. Suppose we want to know the number of different ways four people can line up in a carnival line. The first spot in line can be occupied by any of the four people. The second can be occupied any of the three people who are left. The third spot can be filled by either of the two remaining people, and the fourth spot is filled by the last person. So, the total number of ways four people can create a line is equal to the product $4 \times 3 \times 2 \times 1 = 24$. This product can be abbreviated as 4! (read "4 factorial"). In general, the product of the positive **integers** from 1 to n can be denoted by n! which equals $n \times (n-1) \times (n-2) \times ...2 \times 1$. It should be noted that 0! is by definition equal to 1.

The example of the carnival line given above illustrates a situation involving permutations. A permutation is any arrangement of n objects in a definite order. Generally, the number of permutations of n objects is n! Now, suppose we want to make a line using only two of the four people. In this case, any of the four people can occupy the first **space** and any of the three remaining people can occupy the second space. Therefore, the number of possible arrangements, or permutations, of two people from a group of four, denoted as $P_{4,2}$ is equal to $4 \times 3 = 12$. In general, the number of permutations of n objects taken r at a time is

$$P_{n,r} = n \times (n-1) \times (n-2) \times ... \times (n-r+1)$$

This can be written more compactly as $P_{n,r} = n!/(n-r)!$

Many times the order in which objects are selected from a group does not matter. For instance, we may want to know how many different 3 person clubs can be formed from a student body of 125. By using permutations, some of the clubs will have the same people, just arranged in a different order. We only want to count then number of clubs that have different people. In these cases, when order is not important, we use what is known as a combination. In general, the number of combinations denoted as $C_{n,r}$ or

$$\binom{n}{r}$$

is equal to $P_{n,r}/r!$ or $C_{n,r} = n!/r! \times (n-r)!$ For our club example, the number of different three-person clubs that can be formed from a student body of 125 is $C_{125,3}$ or $125!/3! \times 122! = 317,750$.

Probability theory is concerned with determining the likelihood that a certain event will occur during a given random experiment. In this sense, an experiment is any situation which involves observation or measurement. Random experiments are those which can have different outcomes regardless of the initial conditions and will be heretofore referred to simply as experiments.

The results obtained from an experiment are known as the outcomes. When a die is rolled, the outcome is the number found on the topside. For any experiment, the set of all outcomes is called the sample space. The sample space, S, of the die example is denoted by S={1,2,3,4,5,6} which represents all of the possible numbers that can result from the roll of a die. We usually consider sample spaces in which all outcomes are equally likely.

The sample space of an experiment is classified as finite or infinite. When there is a limit to the number of outcomes in an experiment, such as choosing a single card from a deck of cards, the sample space is finite. On the other hand, an infinite sample space occurs when there is no limit to the number of outcomes, such as when a dart is thrown at a target with a continuum of points.

While a sample space describes the set of every possible outcome for an experiment, an event is any subset of the sample space. When two dice are rolled, the set of outcomes for an event such as a sum of 4 on two dice is represented by E = {(3,1),(2,2),(1,3)}.

In some experiments, multiple events are evaluated and **set theory** is needed to describe the relationship between them. Events can be compounded forming unions, intersections, and complements. The union of two events A and B is an event which contains all of the outcomes contained in event A and B. It is mathematically represented as A ∪ B. The intersection of the same two events is an event which contains only outcomes present in both A and B, and is denoted A ∩ B. The complement of event A, represented by A', is an event which contains all of the outcomes of the sample space not found in A.

Looking back at the table we can see how set theory is used to mathematically describe the outcome of real-world experiments. Suppose A represents the event in which a 4 is obtained on the first roll and B represents an event in which the total number on the dice is 5.

A = {(4,1),(4,2),(4,3),(4,4),(4,5),(4,6)} and
B = {(3,2),(2,3),(1,4)}

The compound set A∪B includes all of the outcomes from both **sets**,

{(4,1),(4,2),(4,3),(4,4),(4,5),(4,6),(3,2),(2,3),(1,4)}

The compound set A∩B includes only events common to both sets, {(4,1)}. Finally, the complement of event A

would include all of the events in which a 4 was not rolled first.

By assuming that every outcome in a sample space is equally likely, the probability of event A is then equal to the number of ways the event can occur, m, divided by the total number of outcomes that can occur, n. Symbolically, we denote the probability of event A as $P(A) = m/n$. An example of this is illustrated by drawing from a deck of cards. To find the probability of an event such as getting an ace when drawing a single card from a deck of cards, we must know the number of aces and the total number of cards. Each of the 4 aces represent an occupance of an event while all of the 52 cards represent the sample space. The probability of this event is then 4/52 or .08.

Using the characteristics of the sets of the sample space and an event, basic rules for probability can be created. First, since m is always equal to or less than n, the probability of any event will always be a number from 0 to 1. Second, if an event is certain to happen, its probability is 1. If it is certain not to occur, its probability is 0. Third, if two events are mutually exclusive, that is they can not occur at the same time, then the probability that either will occur is equal to the sum of their probabilities. For instance, if event A represents rolling a 6 on a die and event B represents rolling a 4, the probability that either will occur is 1/6 + 1/6 = 2/6 or .33. Finally, the sum of the probability that an event will occur and that it will not occur is 1.

The third rule above represents a special case of adding probabilities. In many cases, two events are not mutually exclusive. Suppose we wanted to know the probability of either picking a red card or a king. These events are not mutually exclusive because we could pick a red card that is also a king. The probability of either of these events in this case is equal to the sum of the individual probabilities minus the sum of the combined probabilities. In this example, the probability of getting a king is 4/52, the probability of getting a red card is 26/52, and the probability of getting a red king is 2/52. Therefore, the chances of drawing a red card or a king is 4/52 + 26/52 - 2/52 = .54.

Often the probability of one event is dependent on the occupance of another event. If we choose a person at random, the probability that they own a yacht is low. However, if we find out this person is rich, the probability would certainly be higher. Events such as these in which the probability of one event is dependent on another are known as conditional probabilities. Mathematically, if event A is dependent on another event B, then the conditional probability is denoted as $P(A|B)$ and equal to $P(A\cap B)/P(B)$ when $P(B) \neq 0$. Conditional probabilities are useful whenever we want to restrict our probability calculation to only those cases in which both event A and event B occur.

Events are not always dependent on each other. These independent events have the same probability regardless of whether the other event occurs. For example, probability of passing a math test is not dependent on the probability that it will rain.

Using the ideas of dependent and independent events, a rule for determining probabilities of multiple events can be developed. In general, given dependent events A and B, the probability that both events occur is P(A∩B) = P(B) × P(A|B). If events A and B are independent, P(A∩B) = P(A) × P(B). Suppose we ran an experiment in which we rolled a die and flipped a coin. These events are independent so the probability of getting a 6 and a tail would be (1/6) × 1/2 =.08.

The theoretical approach to determining probabilities has certain advantages; probabilities can be calculated exactly, and experiments with numerous trials are not needed. However, it depends on the classical notion that all the events in a situation are equally possible, and there are many instances in which this is not true. Predicting the weather is an example of such a situation. On any given day, it will be sunny or cloudy. By assuming every possibility is equally likely, the probability of a sunny day would then be 1/2 and clearly, this is nonsense.

The empirical approach to determining probabilities relies on data from actual experiments to determine approximate probabilities instead of the assumption of equal likeliness. Probabilities in these experiments are defined as the ratio of the frequency of the occupance of an event, f(E), to the number of trials in the experiment, n, written symbolically as P(E) = f(E)/n. If our experiment involves flipping a coin, the empirical probability of heads is the number of heads divided by the total number of flips.

The relationship between these empirical probabilities and the theoretical probabilities is suggested by the Law of Large Numbers. It states that as the number of trials of an experiment increases, the empirical probability approaches the theoretical probability. This makes sense as we would expect that if we roll a die numerous times, each number would come up approximately 1/6 of the time. The study of empirical probabilities is known as **statistics**.

Probability theory was originally developed to help gamblers determine the best bet to make in a given situation. Suppose a gambler had a choice between two bets; she could either wager $4 on a coin toss in which she would make $8 if it came up heads or she could bet $4 on the roll of a die and make $8 if it lands on a 6. By using the idea of mathematical expectation she could determine which is the better bet. Mathematical expectation is defined as the average outcome anticipated when an experiment, or bet, is repeated a large number of times. In its simplest form, it is equal to the product of the amount a player stands to win and the probability of the event. In our example, the gambler will expect to win $8 ×.5 = $4 on the coin flip and $8 ×.17 = $1.33 on the roll of the die. Since the expectation is higher for the coin toss, this bet is better.

When more than one winning combination is possible, the expectation is equal to the sum of the individual expectations. Consider the situation in which a person can purchase one of 500 lottery tickets where first prize is $1000 and second prize is $500. In this case, his or her expectation is $1000 × (1/500) + $500 × (1/500) = $3. This means that if the same lottery was repeated many times, one would expect to win an average of $3 on every ticket purchased.

PRODUCTS AND QUOTIENTS

Products and quotients are the results obtained by applying the operations of **multiplication** and **division**, respectively, to particular mathematical objects. Both will be examined; products first, followed by quotients.

Product is defined in mathematics as the quantity obtained by multiplying one quantity or expression by another quantity or expression. For instance, the expression "5 × 6" is the product of the **whole numbers** "5" and "6." Products are encountered in elementary **arithmetic**: children typically learn by rote the products of various whole numbers. Beyond whole numbers, there are also products of **integers**, **rational numbers**, **real numbers**, and **complex numbers**, as well as products of vectors, **functions**, matrices, tensors, and other "higher" mathematical objects.

The word "product" first appeared in use during the first half of the 15th century. Product is from the Middle English and the Medieval Latin word *productum* (something produced), which was derived from the neuter of the Latin *productus* and the past participle of the Latin *producere* (to bring forth). In its simplest form the product (the result of multiplying one quantity by another) is found by multiplying a number "a" by a number "b." The number "a" is called the multiplicator and "b" is called the multiplicand. Both the multiplicator and multiplicand are called factors because the product is identical if the multiplicator and multiplicand are switched.

Various symbols are used to indicate the product of two numbers. The symbols "×" (lying cross), "·" (raised dot), and "*" (asterisk) are all used; so for numbers "a" and "b," "a × b," "a · b," and "a * b" are all called the product of "a" and "b." The expressions "(a)(b)" and "ab" (called "bracket" notation and "juxtaposition" notation, respectively) also denote the product of numbers "a" and "b."

As previously pointed out, besides the products of whole numbers, integers, and real numbers, there are products of complex numbers, vectors, functions, matrices, tensors, etc. The symbols described previously to denote multiplication of integers and real numbers (i.e., the cross, dot, asterisk, brackets, and juxtaposition) are also used to indicate products of functions, complex numbers, etc. Different symbols, however, may denote different types of products. Consider the products of vectors. The lying cross "×" indicates the cross product of two vectors, which is itself another vector. In contrast, the dot symbol "·" indicates the dot product of two vectors, which is a scalar quantity (i.e., a real number).

The **powers** and factorials of integers are compact notations for the expression of products. Powers are used to denote the product of equal terms (normally having integer values). The nth power of a, denoted as "a^n," is the product of n factors of a. That is to say: "$a^2 = a \cdot a$," "$a^5 = a \cdot a \cdot a \cdot a \cdot a$," and so forth. In the expression "a^n," a is called the basis and n the exponent. As indicated above, both "a" and "n" customarily have integer values only.

The **factorial** of a positive integer n also denotes a product. Factorial, represented as "n!," is the product of all positive integers from 1 to n. It is written out as "$n! = 1 \cdot 2 \cdot 3 \cdot \ldots \cdot (n-$

1) · n," where by convention, 0! = 1. For example "5! = 1 · 2 · 3 · 4 · 5 = 120."

Quotient is defined in mathematics as the quantity obtained by dividing one quantity or expression by another quantity or expression. The process of determining a quotient is called division. For instance, the expression "30 / 5" is the quotient of "30" divided by "5." One may form quotients of whole numbers, integers, rational numbers, real numbers, complex numbers and functions.

The word quotient first appeared in use during the first half of the 15th century. It was derived from the Middle English *quocient*, an alteration of the Latin *quotiens* (how many times), which was derived from the Latin *quot* (how many). In its simplest form the quotient is found by dividing a number "a" by a number "b." The number "a" is called the dividend and "b" is called the divisor. The divisor must have a non-zero value. We can see the reason for this with an example: assume "6 / 0 = n," where "n" is some number. Multiplying both sides of the equation by "0" results in the equation "6 = n × 0," which violates the multiplicative rule that the product of **zero** and any number equals zero. Quotients are most often represented by the symbols "/" (called diagonal or solidus) and "÷" (obelus) so that for numbers "a" and "b" (with a non-zero value of "b") the quotient may be expressed "a / b" or "a ÷ b."

For real numbers (and for whole numbers, integers, and rational numbers, all of which are contained within the set of real numbers) the product of two quotients is the quotient of the product of the dividends (or numerators) and the product of the divisors (or denominators). Symbolically this is written "(a / b) · (c / d) = (a · c) / (b · d)," where b and d are non-zero. The quotient of two quotients may be rewritten as the product of the "top" quotient and the inverted "bottom" quotient, and the result reduced to lowest terms. Symbolically this process is "(a / b) / (c / d) = (a / b) · (d / c) = (a · d) / (b · c)," where b, c, and d are non-zero. For example "(2 / 3) / (5 / 6) = (2 / 3) · (6 / 5) = 12 / 15 = 4 / 5."

Finally, it should be noted that for the rational, real, and complex numbers, division may be defined as the inverse operation of multiplication, where "a / b = a · (1 / b)." Therefore, the quotient "6 / 2" represents the same number as "6" (the dividend of the quotient) multiplied by the reciprocal of the divisor (i.e., 1 / 2), or written as an equation: "6 / 2 = 6 · (1 / 2)." Multiplication by the inverse of a quantity is meaningful in many applications where the operation of division is not well defined (for example, matrices).

See also Complex numbers; Division; Factorials; Functions; Integers; Matrix; Multiplication; Multiplicative notation; Numbers and numerals; Powers; Real numbers; Rational numbers; Vector analysis; Whole numbers

PROJECTION

The term projection has several different, but similar, meanings. A few examples should help clarify the term. First, sup-

pose we are given an object O that is specified as a subset of Euclidean three-dimensional **space** (E^3), and a plane P in E^3. The map from O to P that send each point of O to the point in P closest to it, is called the projection map from O to P. For example, suppose that O is a knot—that is a loop of string with its ends glued together. If we spray paint directly at P but in front of the knot O, then the space that is not painted is the projection of O on P. The next example is called polar projection. It is a map f from the unit **sphere** minus the north pole to the plane z = -1 (this plane is one unit below the x-y plane) in Euclidean three-dimensional space. For any point x other than the north pole, draw the line that passes through x and the north pole. This line intersects the plane z = -1 in the point f(x). Polar projection is the reason why the sphere is sometimes referred to as the plane plus one point at **infinity**. This map is conformal, i.e. if two great circles on the sphere intersect at an **angle** of d degrees then their images under f are lines in the plane that intersect at an angle of d degrees.

In the next example, X and Y are **sets**. The Cartesian product of X and Y, denoted by XxY, is the set of all ordered pairs of the form (x, y) in which x is an element of X and y is an element of Y. There are two natural projection maps in this context: the map from XxY to X that maps the element (x, y) to x, and the map from XxY to Y that maps (x, y) to y. If X and Y are topological spaces, then the product **topology** on XxY is the smallest one for which the two projection maps are continuous. If X and Y are groups then XxY is a group under the operation (x, y)(m, n) = (xm, yn) and the two projection maps are group homomorphisms.

In the next example, V and W are Hilbert spaces and V is contained in W. It is a fact that any orthonormal basis for V can be completed to an orthonormal basis for W. This means that any element w of W has a unique decomposition w = v + t in which v is an element of V and t is an element of the orthogonal complement of V. In other words, if x is any element of V then x·t = 0. The map that sends w to v is called the projection map from W to V.

In general, a projection is a map p from a set A to a set B such that for each point b in B, $p^{-1}(b)$ has some special property. For example, the map from E^n to E^{n-1} defined by $p((x_1,..,x_n)) = (x_2,...,x_n)$ has the property that $p^{-1}(b)$ is a line for all b. Sometimes, a projection map is as above but restricted to some subset of X. This is the case of the example with the knot. If A = XxB and p is defined by p(x,b) = b then $p^{-1}(b)$ is isomorphic to B if A and B are groups, for example.

See also Functions; Hilbert space

PROOF

A proof is a logical argument demonstrating that a specific statement, proposition, or mathematical formula is true. It consists of a set of assumptions, or premises, which are combined according to logical rules to establish a valid conclusion. This validation can be achieved by direct proof that verifies the conclusion is true, or by indirect proof that establishes that it cannot be false.

The term *proof* is derived from the Latin *probare*, meaning *to test*. The Greek philosopher and mathematician **Thales** is said to have introduced the first proofs into mathematics about 600 B.C. A more complete mathematical system of testing, or proving, the **truth** of statements was set forth by the Greek mathematician **Euclid** in his **geometry** text, *Elements*, published around 300 B.C. As proposed by Euclid, a proof is a valid argument from true premises to arrive at a conclusion. It consists of a set of assumptions (called axioms) linked by statements of deductive reasoning (known as an argument) to derive the proposition that is being proved (the conclusion). If the initial statement is agreed to be true, the final statement in the proof sequence establishes the truth of the **theorem**.

Each proof begins with one or more axioms, which are statements that are accepted as facts. Also known as postulates, these facts may be well known mathematical formulae for which proofs have already been established. They are followed by a sequence of true statements known as an argument. The argument is said to be valid if the conclusion is a logical consequence of the **conjunction** of its statements. If the argument does not support the conclusion, it is said to be a fallacy. These arguments may take several forms. One frequently used form can be generally stated as follows: If a statement of the form "if p then q" is assumed to be true, and if p is known to be true, then q must be true. This form follows the rule of detachment; in logic, it is called *affirming the antecedent*; and the Latin term *modus ponens* can also be used. However, just because the conclusion is known to be true does not necessarily mean the argument is valid. For example, a math student may attempt a problem, make mistakes or leave out steps, and still get the right answer. Even though the conclusion is true, the argument may not be valid.

The two fundamental types of proofs are direct and indirect. Direct proofs begin with a basic **axiom** and reach their conclusion through a sequence of statements (arguments) such that each statement is a logical consequence of the preceding statements. In other words, the conclusion is proved through a step-by-step process based on a key set of initial statements that are known or assumed to be true. For example, given the true statement that "either John eats a pizza or John gets hungry" and that "John did not get hungry," it may be proved that John ate a pizza. In this example, let p and q denote the **propositions**:

p: John eats a pizza.

q: John gets hungry.

Using the symbols / for "intersection" and ~ for "not," the premise can be written as follows: p/q: Either John eats a pizza or John gets hungry. and ~q: John did not get hungry. (Where ~q denotes the opposite of q).

One of the fundamental laws of traditional logic, the law of contradiction, tells us that a statement must be true if its opposite is false. In this case, we are given ~q: John did not get hungry. Therefore, its opposite (q: John did get hungry) must be false. But the first axiom tells us that either p or q is true; therefore, if q is false, p must be true: John did eat a pizza.

In contrast, a statement may also be proven indirectly by invalidating its negation. This method is known as indirect proof, or proof by contradiction. This type of proof does not aim to directly validate a statement; instead, the premise is proven by showing that it cannot be false. Thus, by proving that the statement ~p is false, we indirectly prove that p is true. For example, by invalidating the statement "cats do not meow," we indirectly prove the statement "cats meow." Proof by contradiction is also known as *reductio ad absurdum*. A famous example of *reductio ad absurdum* is the proof, attributed to **Pythagoras**, that the **square root** of 2 is an irrational number.

Other methods of formal proof include proof by exhaustion (in which the conclusion is established by testing all possible cases). For example, if experience tells us that cats meow, we will conclude that all cats meow. This is an example of inductive **inference**, whereby a conclusion exceeds the information presented in the premises (we have no way of studying every individual cat). Inductive reasoning is widely used in science. Deductive reasoning, which is prominent in mathematical logic, is concerned with the formal relation between individual statements, and not with their content. In other words, the actual content of a statement is irrelevant. If the statement "if p then q" is true, q would be true if p is true, even if p and q stood for, respectively, "The Moon is a philosopher" and "Triangles never snore."

PROOF BY DEDUCTION

Proof by **deduction** is the primary method of proof used in classical mathematics. Deductive proof is the process of deriving conclusions from logical premises without resort to empirical evidence. A deductive mathematical system typically consists of some definitions, some assumptions, called axioms or postulates, some rules of **inference**, and theorems. The proving of theorems is almost the definition of what a pure mathematician does for a living. Although the results of the mathematician's labor may be applicable to problems in the physical world, the mathematician's proofs of these results must come from within the structure of the deductive system in which the mathematics is being done. In such a system a deductive proof is a step-by-step procedure following the rules of inference and using only the definitions, axioms, and previously proved theorems. A simple example of a deductive proof schema goes back to the Greek philosopher, **Aristotle** (384-322 B.C.), and is called the syllogism. A famous syllogism is "All men are mortal. Socrates is a man. Therefore, Socrates is mortal." The first two sentences are called premises and the last is called the conclusion. In **Aristotelian logic**, this is an instance of the general form "All A is B. C is A. Therefore, C is B." This is an example of a rule of inference. The general rule is always a valid way of reasoning, but,in any instance of applying the rule, the validity of the proof depends upon the **truth** of the premises. For example, one could argue that the truth of "All men are mortal" cannot be known with certainty as long as any man is left alive on the earth. In that case, "Socrates is mortal" would still be true, since we know that Socrates is dead, but we would not have proved this truth deductively. The knowledge that Socrates is mortal was gained from historical evidence and historical evidence is not allowed in a deductive proof.

The importance of deductive proof in mathematics was established early in the subject's development by Aristotle and **Euclid** (c. 300 B.C.), whose *Elements* presented the development of plane **geometry** with perhaps the most famous deductive system of all time. Modern versions of Euclidean geometry are still taught in many American high schools as a way of teaching students about deductive proof. The *Elements* is arguably the most influential work of mathematics ever written because it established deductive proof as the standard of mathematical argument. In reality, mathematicians do not really develop mathematics deductively. They use intuition, guesswork, and analogies to events in the real world; in short, whatever their creative powers suggest to them. However, once the mathematics is developed, their top priority is to show that all new results can be established deductively within the framework of existing mathematical theory. The stamp of approval by the mathematics community will only come when all results have been verified deductively by experts in the field. This insistence on deductive proof is not merely an idiosyncrasy of mathematicians. Historically, the failure to put a firm theoretical basis under newly-developed mathematics has resulted in skepticism and even scorn from people outside of mathematics. Perhaps the best example is the **calculus** developed by **Newton** and **Leibniz**. Although the calculus is commonly regarded as one of the great intellectual achievements in the history of thought, parts of it were ridiculed by philosophers, such as Bishop **George Berkeley** (1685-1753), because some of its concepts were on rather shaky ground in the theoretical sense. Although Newton is regarded as one of the greatest mathematicians who ever lived, his purpose in inventing the calculus was more utilitarian than theoretical. He needed the mathematics in developing his theories of gravitational attraction and planetary **motion**. These theories would bring about the greatest revolution in science since Copernicus, but, from the point of view of Berkeley and others, the very foundation of the calculus was flawed. The so-called "infinitesimals" used in the discussion of derivatives was mocked by Berkeley—he asked sarcastically if they were the ghosts of departed quantities. To answer this and other questions about the theoretical underpinnings of the calculus, a host of nineteenth-century mathematicians, led by Cauchy and **Weierstrass**, gave rigorous definitions of previously fuzzy notions and put the calculus on firm theoretical ground by showing how the results of Newton and Leibniz could be proved deductively.

Modern mathematics has its deductive basis in the axioms of **Giuseppe Peano** (1858-1932), an Italian mathematician who put forward five axioms upon which to build all of **arithmetic**, **algebra**, and **analysis**. Consider the following **theorem** of arithmetic: The product of two odd numbers is always an odd number. We shall give a rather short deductive proof of this theorem. First we note that from the **Peano axioms**, and theorems proved deductively from those axioms, we can establish that every even number has the form 2n,

where n is an integer. Thus every odd number has the form $2n+1$. Now consider two odd numbers s and t, which may be expressed as $2m+1$ and $2k+1$. Then the product is $st=(2m+1)(2k+1)=4mk+2m+2k+1=2(2mk+m+k)+1$ which must be odd since it is one more than $2(2mk+m+k)$ which is even. This simple example illustrates a proof by deduction. Unfortunately, most deductive proofs require a few more steps than this. As an example on the other extreme, Andrew Wiles's 1994 deductive proof of **Fermat's Last Theorem** is more than 200 pages in length!

PROOF BY INDUCTION

Inductive reasoning is sometimes compared to thinking that something will always be a certain way just because it always has been that way: When you predict that the sun will rise in the east tomorrow morning, you are confident in your forecast because the sun has never failed to so in all of written history. Sometimes, this kind of argument can carry great weight, but generally, the fact that something has been going on for a very long time does not guarantee that it will do so indefinitely. After all, the Roman Empire lasted quite a while in its day and was thought permanent and invincible, until it was overrun by Germanic tribes.

Mathematical **induction** is more precisely defined, and consequently, a more sound form of reasoning. Its most common use is in **number theory**, although it is also essential in other branches of mathematics and logic.

In number theory one can make statements about the natural numbers:

"2 is even."

"365 is divisible by 7."

"2^5 is greater than 5^2."

Mathematicians determine which of these statements are true and which are false. What if one wanted to know if a certain general proposition were true *for all* natural numbers? Consider for example the statement (P):

"The sum of all the numbers from 1 to n is equal to $n(n + 1)/2$, for all natural numbers n," i.e.,

"$1 + 2 + 3 + 4 +... + n = n(n + 1)/2$."

Since there are infinitely many natural numbers, no team of mathematicians could ever verify this proposition by individually checking each number, no matter how many mathematicians were employed, and how much time and research money they were granted.

We use mathematical induction to determine if general statements like P above are true regardless of which number n statement P is made about. First, instead of thinking about the general statement P, we look at each individual statement about each natural number. P_1 is the P statement made about the number 1, P_2 about 2, $P3$ about 3, etc., P_n about the number n.

P_1: "$1 = 1(1 + 1)/2$."

P_2: "$1 + 2 = 2(2 + 1)/2$."

P_3: "$1 + 2 + 3 = 3(3 + 1)/2$."

...

P_n: "$1 + 2 + 3 + 4 +... + n = n(n + 1)/2$."

The best image for mathematical induction is an infinite row of dominos each representing a statement P_n. With dominos, we know that if any one block falls over, the next one in line will fall over, too. So if we knock over the first block, we know with certainty that *all* subsequent blocks will fall.

Mathematical induction works similarly. We distinguish between two steps: the *base step* and the *inductive step*. In the base step we verify that the first statement, P_1, is true. This is usually very easy. In the inductive step we assume that the statement is true for some number k greater than 1, but less than n. This assumption is known as the *inductive hypothesis*. Then we prove the following conditional involving P_k and P_{k+1}, the P statement made about the number k and its successor, $k+1$:

"If P_k is true, then P_{k+1} is true."

By the inductive hypothesis, P_k is true. To prove the conditional, we must show that P_{k+1} is a logical consequence of P_k. If we succeed we have completed the **proof**. We have shown that if any statement P_k is true, then its successor statement P_{k+1} is true, too. Since we know that P_1 is true (by the basic step), P_1s successor statement (P_2) must be true, and its successor in turn, and its successor, and so on through all the natural numbers.

To illustrate induction we will complete the proof of the statement P above. It is easy to verify P_1. Now we show that if P_k is true, then P_{k+1} is true. First note that P_{k+1} is the following statement:

"$1 + 2 + 3 +... + k + (k + 1) = (k + 1)[((k+1) + 1)/2]$"

This is the equation we will be working toward. With P_{k+1} in mind, we will have a fair idea of how to manipulate P_k to get to it:

$1 + 2 + 3 +... + k = k(k + 1)/2$

(P_k)

$1 + 2 + 3 +... + k + (k + 1) = (k(k + 1)/2) + (k + 1)$

(Add $k + 1$ to both sides of the equation. Notice that the left-hand side of the new equation is already identical to the left-hand side of the $P(k+1)$ expression. The following steps involve only the right-hand side.)

$1 + 2 + 3 +... + k + (k + 1) = (k + 1)[(k/2) + 1]$

(**Factor** out $k + 1$.)

$1 + 2 + 3 +... + k + (k + 1) = (k + 1)[(k/2) + 2/2]$

(Expand 1 into 2/2.)

$1 + 2 + 3 +... + k + (k + 1) = (k + 1)[(k + 2)/2]$

(Factor out 1/2 inside the square brackets.)

$1 + 2 + 3 +... + k + (k + 1) = (k + 1)[((k+1) + 1)/2]$

(Note that $k + 2 = (k +1) + 1$.)

Which is precisely P_{k+1}. We have shown that P_{k+1} follows from P_k by simple **arithmetic**.

Induction works not only for numbers, but for many other **sets** as well. The sets on which induction works are called *inductive sets*. The precise **definition** of inductive sets is fairly complex. Roughly, inductive sets are sets whose elements are built up from basic elements through the finite repetition of some specified operations on these basic elements. The natural numbers are constructed on the basic element *0* by the repetition of the successor operation, *s*. The set of *well-formed formulas* in logic is another induc-

tive set. It is constructed from the set of basic sentences by the operations of **conjunction**, disjunction, negation, and **implication**.

See also Number theory; Proof by deduction; Proof by reductio ad absurdum

PROOF BY REDUCTIO AD ABSURDUM

Proof by *reductio ad absurdum* is also known as **proof by contradiction**. It is one of the most powerful tools of reasoning at the disposal of logicians and mathematicians. *Reductio* proofs are used for **conditionals**, **propositions** of the form "if p then q." Most mathematical theorems take this form, for instance the **Pythagorean theorem**: "If one of the angles in a **triangle** is a right **angle**, then the sum of the squares of the adjacent sides is equal to the **square** of the opposite side."

There are two key ideas in the *reductio* line of reasoning. First, a proposition must either be true or false, but not both. Therefore, if it cannot possibly be false, then it must be true. This is as simple as it is essential. Second, if a formally correct derivation from a set of assumptions leads to a contradiction, then at least one assumption must be false.

Reductio ad absurdum is Latin, and means reduction to an absurdity. That describes the proof technique quite well: Start with a particular assumption, and show that your assumption leads to something absurd—a contradiction. Let p be the proposition to be proved. The key step is to assume that p is false, i.e., that *not-p* is true. Then, using rules of **inference**, derive some consequences of *not-p*. If *not-p* leads to a contradiction, then the original assumption "*not-p* is true" must be false. So p must be true, which is just what you set out to prove.

An example from **number theory** illustrates this proof strategy nicely. Consider the conditional statement (p):

"If a^2 is an even number, then a must also be an even number."

Suppose this statement were false. Then it would be possible that for some odd integer a, a^2 is even. In formal language, we assume the negation of p which amounts to:

"a^2 is even and a is odd."

Now, $a^2 = a \times a$. This means that a^2 is the product of two odd numbers (a, and again, a). But it is a simple fact of number theory that the product of two odd numbers must also be odd. This contradicts the hypothesis that a^2 is even. Our assumption (*not-p*) must be false, so p must be true, finishing the proof.

When proving a conditional statement in mathematics, the easiest thing to do is usually to start off on a *reductio* proof. Quite often, this is the quickest way to demonstrate the desired conclusion.

See also Proof by induction; Proof by deduction

PROPORTIONS

The concept of proportion is closely related to the concept of ratio; it is nearly impossible to discuss one without defining the other. Specifically, a proportion is an equation of two equal **ratios**; if two (or more) ratios are equal, they are considered proportional. The ratios 3/5, 6/10, and 12/20, for example, are proportional; each is a constant multiple of another. If each element of the ratio is a constant multiple of the associated element in the second ratio, then the ratios are considered proportional. The ratio 1:3:8, for example, is proportional to the ratio 4:12:32, since each element in the second ratio is exactly the associated element in the first ratio multiplied by 4. However, the ratio 1:3:8 is *not* proportional to the ratio 4:12:26. In shorthand notation, proportions may be written as $x{:}y{::}ab$, which reads "x is to y as a is to b."

Deriving the basic proportion equation—$a/b = c/d$—involves basic **algebra**. It is based on the basic relationship of setting two products equal to each other, such as $a*d = b*c$. This equality implies that $a = (b*c)/d$ and finally that $a/b = c/d$. To return to the original equation, the mathematical shortcut of cross-multiplication is used in which the numerator of each ratio (e.g., a) is multiplied by the denominator of the other ratio (e.g., d). Both products are then set equal.

Proportions are applied often to everyday events and situations; it is one of the most widely applied concepts from mathematics. A percentage, which represents the number of events out of 100, is determined by setting a known ratio a/b equal to the ratio $x/100$. Then, a is said to be "$x\%$" of b. Suppose, for example, that 26 of 40 participants in a life-saving certification class at college A were women, and 64 of 160 participants in a life-saving certification class at college B were women. Based on the concept of proportion and percentage, one could equivalently say that "65% of the 40 participants at college A and 40% of the 160 participants at college B were women." Since the *total* number of participants is known at each college, it is also a matter of simple algebra to determine the *actual* number of women participants at each class.

Since percentages are ratios that are normalized to a common and well-understood standard (that is, normalized to 100), they are a good tool for comparing values when used properly. Relative to the *total* number of participants at each college, a greater percentage of the class were women at college A than college B (65% is greater than 40%). Notice, however, that percentages can also be somewhat misleading. Although 65% of the participants at college A were women and only 40% of the participants at college B were women, a greater number of women actually attended the college B session (64 at college B vs. 26 at college A). Comparison of percentages must always be considered in their proper context.

The concept of **similarity** is also based on proportion. Similar triangles, for example, are triangles in which each set of associated angles are equal—that is, if **triangle** ABC is similar to triangle DEF, then the magnitude of **angle** A equals the magnitude of angle D, the magnitude of angle B equals the magnitude of angle E, and the magnitude of angle C equals the magnitude of angle F. As a result of the similarity, the ratios of the associated line segments is are equal—AB:DE::BC:EF::CA:FD. This idea can be extended to any set of similar **polygons** and is fundamentally important **axiom** in **geometry**.

While proportion is primarily a mathematical concept—an equality of ratios—it also has aesthetic value in both the natural and manmade environments and expressions of these environments. Artistic caricatures, for example, tend to exaggerate some element of the subject to excessive proportion in order to draw attention to it, such as an overwhelmingly large hat or unfeasibly small car. The viewer notices the exaggeration precisely because the artist has inserted an unexpected proportionality into the artwork, relative to the other objects in the piece. The hat appears too large for the subject—or the car appears too small—because other objects in the caricature are sized in a normal range relative to each other. Art that captures proportion accurately is often considered quite striking; art that exaggerates proportion may even be called grotesque. Beyond its mathematical implications, then, the human experience with proportion affects its aesthetic interpretation.

See also Algebra; Equation; Golden mean; Ratios; Similarity

PROPOSITIONS

In mathematics, a proposition is a statement which must be proved by a deductive argument in order to be considered true. Propositions are also called theorems. In general, mathematical propositions have the logical form "If p, then q" where p and q may be definitions, axioms, or other propositions. The "p" part of the proposition is called the hypothesis and the "q" part is called the conclusion. For example, in the proposition "If two **integers** are both odd, then their product is odd," the hypothesis is "two integers are odd" while the conclusion is "their product is odd." When written in everyday English, propositions may not literally have the "If p, then q" form. For instance, the proposition of the preceding sentence would usually be stated as "The product of any two odd integers is odd." The point is that if a statement in English is truly a proposition in the mathematical sense, then it can be translated into "if... then" form. In a somewhat simplified view of the **proof** process, mathematicians start with the hypothesis and attempt to reach the conclusion through a step-by-step process using only truths that are already known because they are definitions, axioms, or previously proved propositions. This procedure is called a "direct" proof, since it proceeds directly from hypothesis to conclusion; but there is a second way to prove a proposition, called "indirect" proof. In an indirect proof the mathematician begins with the negation of the conclusion of the proposition and attempts to arrive at a contradiction of the hypothesis. This is a logically valid method due to an **inference** pattern known as the law of contraposition. The contrapositive of a proposition of the form "If p, then q" is the form "If not q, then not p." The contrapositive of a proposition is logically equivalent to the proposition itself. Therefore, if one proves the contrapositive of a proposition, one has also proved the proposition. Indirect proof is usually deployed when the contrapositive form is easier to prove than the original proposition. For our previously mentioned proposition, "If two integers are odd, then their product is odd," the contrapositive is "If the product of two integers is not odd, then the integers are not both odd." Of course this could also be expressed as "If the product of two integers is even, then at least one of the integers is even." There is a group of mathematicians, called "intuitionists" or "constructivists," who do not accept indirect proof as logically valid. They hold the position that all proofs must be "constructive," i.e., must construct the entity or entities of which the conclusion of the proposition speaks. Thus a direct proof of "If two integers are odd, then their product is odd" leads to an actual construction that shows that the product must have the form 2 times an integer plus 1. This is acceptable to the intuitionsts. The indirect proof, on the other hand, simply gives a contradiction of the original hypothesis without constructing the form of the product of two odd integers. Intuitionists do not accept this as a valid proof of the original proposition. The reason is that intuitionists use a non-standard logic which denies the so-called "Law of the Excluded Middle" which says that any proposition must be either true or false—there is no middle ground. Intuitionists say that there is a middle ground; some propositions are neither true nor false, but undecided or undecidable. It turns out that the law of the excluded middle and the law of contraposition are equivalent. So if exclude middle is out, then so is contraposition. Hence, indirect proof is outlawed under **intuitionism**.

The part of logic which studies the validity of propositional forms is called propositional logic or the propositional **calculus**. This is essentially the backbone of **symbolic logic**, because it determines which propositional forms are always logically valid and which are not. A propositional form that is true for all instances of the form is called a tautology. In standard logic, the law of contraposition is a tautology. Using "p→q" as symbolic for "If p, then q," the law of contraposition may be written as (p→q)→(not q→not p). This is a tautology because no matter whether p and q are true or false, the proposition is true. Another famous tautology is the law of syllogism: [(p→q) and (q→r)]→(p→r). This pattern is valid regardless of the **truth** values of p, q, and r. This pattern, in non-symbolic form, was introduced into logic by the Greek philosopher **Aristotle** (384-322 B.C.), who invented the discipline of logic. Aristotle's version was "All A is B. All C is A. Therefore, all C is B." This can be compared to the symbolic version by writing it as "If x is A, then x is B. If x is C, then x is A. Therefore, if x is C, then x is B." Aristotle viewed the syllogism as the only inference pattern needed for logical discourse, and, due to the enormous power of his reputation as a philosopher, this view held sway for more than 2000 years after his death. In modern symbolic logic the syllogism is only one of many inference patterns used by logicians in constructing the foundations of deductive mathematics.

PTOLEMY, CLAUDIUS (CA. 70-CA. 130)
Greek geometer, astronomer, and geographer

Claudius Ptolemy was a famed Greek scholar whose work in astronomy and geometry helped form the basis of trigonometry. Ptolemy's earliest and most famous work was the *Almagest*, which focused on astronomy. In it, he developed geometrical proofs, which included the first notable value for π since **Archimedes'** time, to explain the **motion** of the sun, the moon, and the planets. Although his geocentric theory of astronomy mistakenly identified the Earth instead of the sun as the center of the solar system, this work served as the authoritative text on astronomy for over 1,400 years. Ptolemy also wrote works on music, optics, and astrology. His treatise, *Geography*, survived for centuries as a primary reference source and included the first practical use of parallels and longitudes.

Ptolemy and his life remain a mystery outside of his scientific written works. Estimates of the era in which he lived are based on his recorded astronomical observations. The name Ptolemy was common in Egypt, leading to the belief that he was a native of that country. His first name, Claudius, suggests that he was recognized as a Roman citizen, probably due to a legacy handed down through his Greek ancestors. The exact date and cause of Ptolemy's dead are unknown.

Ptolemy probably lived in Alexandria, Egypt, where he made most of his astronomical observations. The city's famous library provided him with access to the largest collection of accumulated knowledge during his time. From his introduction to the *Almagest*, Ptolemy appears to adhere to the Aristotelian school of philosophy, but he was also influenced by other philosophies, such as stoicism. To his credit, Ptolemy acknowledged that he based much of his work on that of previous scholars, especially the astronomical theories of **Hipparchus**.

Like **Hipparchus** before him, Ptolemy's interest in astronomy and measurements involving the "heavenly" bodies led to his important contributions in **geometry** and **trigonometry**. Although Hipparchus is credited as the founder of trigonometry, Ptolemy's earliest known work, the *Almagest*, comprehensively systemized the sum of Greek knowledge in both trigonometry and astronomy.

Ptolemy's Greek title for the *Almagest* was *Syntaxis*, translated as "mathematical compilation." It was quickly recognized as a groundbreaking work and came to be called "the great compilation" by the Greeks, probably to distinguish it from earlier and more elementary works on astronomy. It eventually became known as the *Almagest* through early Arabic translations and subsequent Latin translations as the *almagesti* or *almagestum*.

Consisting of 13 books, the *Almagest* includes original theories and proofs, including a method of calculating the chords of a **circle**. These chords were used by the Greeks in spherical geometry and required trigonometry to make various calculations for solar, lunar, and planetary positions and eclipses of the sun and moon. Ptolemy's **model** of astronomy used combinations of circles, known as epicycles, or small circular orbits centered on a larger circle's **circumference**.

Although his geocentric theory that the heavenly bodies rotated around a central, stationary Earth was eventually discredited, the theorems and system that he advanced in this work provided an admirable and ready means to construct tables representing the movements of the sun, moon, and planets. Its geometrical representations of these movements represent a testimonial to both Ptolemy and the great mathematical thinkers he borrowed from. The *Almagest* also included a catalogue of stars based largely on the catalogue developed by Hipparchus.

Ptolemy also wrote several other works on astronomy. In his *Handy Tables*, he improved upon and systematically organized the Table of Chords from the *Almagest* into one volume for practical use. Ptolemy also wrote two treatises that demonstrated the Greeks had mastered more than basic "classical" geometry. In the *Analema* and *Planisphaerium*, Ptolemy uses both trigonometry and other mathematical techniques to discuss the **projection** of points on the celestial **sphere** and stereographic projection. These works include methods for constructing sundials and the basis for the astrolabe, an astronomical instrument used during the Middle Ages. In the *Planetary Hypotheses*, he carries on his astronomical work from the *Almagest* and includes new theories of planetary latitudes and for determining the size of planets and their distances from the Earth. Ptolemy also wrote other astronomical works, which are either lost or only partially survive.

Ptolemy's second great scientific work was *Geography*, an early attempt to map the world as it was known in his time. Although much of it is inaccurate, the book placed the accumulated Greek and Roman knowledge of the Earth's geography on a solid scientific foundation. Ptolemy's inaccuracies were inevitable, considering that modern surveying techniques were unknown and Ptolemy himself knew little of the world outside of the Roman empire. As a result, Ptolemy had to rely on information from travelers, including soldiers in the Roman army.

Much of *Geography* is based on the work of Marinus of Tyre, as well as Hipparchus and Strabo. Ptolemy's important contribution in the work was his introduction of the first systematic use of latitudes and longitudes. Not only did the book contain a number of maps, Ptolemy included directions followed by geographers until the Renaissance for creating maps. Unfortunately, Ptolemy miscalculated the Earth's circumference and many of his estimates of latitude were incorrect. Nevertheless, *Geography* became a profoundly influential work whose popularity led to the creation of opulent manuscripts for the wealthy.

Like many of the learned men of his day, Ptolemy displayed a wide-ranging interest in the sciences. His treatise, *Optics*, uses mathematical **equations** to develop theories of **light** and refraction (especially as it pertains to the planets and stars) and also discusses vision and mirrors. His work *Harmonica* is only partially intact and represents a treatise on musical theory, including the mathematical intervals of notes.

In the time of Ptolemy, astrology was also considered a science, and Ptolemy wrote a four-book work called *Tetrabibilios*. Always the educator, Ptolemy viewed this work

as the logical follow-up to the *Almagest* and created it as a textbook for casting horoscopes. To provide this work with a scientific basis, he attempted to relate his efforts in astronomy to astrology.

While modern-day science has proven many of Ptolemy's works to be inaccurate, he deserves his place in history as a great intellect. On the basis of the longtime worldwide interest in his works, Ptolemy had a profound impact on the thoughts and philosophies of many generations. A compiler of knowledge and teacher, Ptolemy exhibited a clarity of style and ingenuity, which made both the *Almagest* and *Geography* standard textbooks well into the Renaissance period and beyond. As a result, Ptolemy, who was the last great astronomer of the Alexandrian school, was long recognized as the authority on "all things in heaven and on Earth."

PYTHAGORAS OF SAMOS (CA. 580 B.C.-CA. 500 B.C.)

Greek geometer and philosopher

Pythagoras' wide-ranging interests in mathematics, music, and astronomy mark him as a seminal figure in early Western Civilization's intellectual development. In the realm of mathematics, he developed the **Pythagorean theorem** and discovered **irrational numbers**. Pythagoras also founded a philosophical and religious school steeped in mysticism and secrecy which left its mark on poets, artists, scientists, and philosophers down to the 20th century.

Pythagoras was born about 580 B.C., probably in Samos, Greece, where he grew up. His father, Mnesarchus, may have been an engraver of seals or a merchant. Living in a prosperous seaport that was the center of learning and art during the Golden Age of Greece stimulated Pythagoras' thirst for knowledge. However, conflicting reports and the lack of written records leave much of his early life a matter of conjecture. Some reports indicate that he had at least two elder brothers, was a champion athlete, and a child prodigy.

Pythagoras is said to have studied in Greece under Creophilus and Pherecydes of Syros, and **Thales** of Miletus. In his early twenties, he traveled to Egypt, where he probably learned **geometry** and was initiated in Egyptian philosophy and science. Pythagoras then may have traveled to Babylon, or have been taken their as a prisoner of the Persians following their invasion of Egypt around 525 B.C. If Pythagoras did live in Babylon, he probably gained a more in-depth understanding of mathematics and geometry. For example, Babylonians had developed reciprocals and **square roots** for solving **equations** involved in mathematical astronomy.

While it is uncertain whether Pythagoras traveled further east, many see his philosophical foundations as being closely aligned to Eastern mysticism and philosophy. Pythagoras eventually returned to his home, only to find it ruled by the tyrant Polycrates. As a result, Pythagoras eventually moved to Croton in southern Italy, where he founded his famous school.

Pythagoras established his school in Croton around 529 B.C. Focusing primarily on religion, philosophy, and mathematics, the school was known as a *homakoeion*, meaning a gathering place for people to learn. The school's success can be attributed to the charismatic personality of Pythagoras. In a relatively short period of time, he established a large following, broken up into two groups—the *akousmatikoi*, who primarily studied the philosophical teachings of Pythagoras, and the *mathematikoi*, who focused on theoretical mathematics.

Students of Pythagoras' school, known as Pythagoreans, followed a number of strict rules. For instance, they were conscientious vegetarians, took a vow of silence for the first five years of their membership, and kept no written records. Liberal in nature, Pythagoras' school was open to everyone, including women, who were allowed to share in the instructing. Following Pythagoras' own belief in simplicity and disdain for worldly honors, Pythagoreans took no credit for their own work or discoveries, attributing all findings either to their master or the group.

Numbers were the very foundation of Pythagorean philosophy, which maintained that numbers were both mystical in nature and had a reality of their own outside of the human mind. Impressed by the **ratios** that existed in musical harmonies, astronomy, and geometrical shapes, the Pythagoreans developed a theory that all things, in essence, were numbers and related through numbers. Out of this belief, they developed certain representations for numbers: 1 was a point, 2 a line, 3 a surface, and 4 a solid. Moral qualities were also numbers, with 4 representing justice and 10 (known as the tetractys) representing the sum of all nature due to its being the sum of 1+2+3+4 (the point, line, plane, and solid).

While the Pythagoreans' mystical belief in numbers is largely ignored today, their belief that one could penetrate the secrets of the universe through numbers led them to conduct a careful study of mathematical theory, focusing on the principals of geometry.

The nonexistence of written records attributable to Pythagoras or his followers has made it difficult for historians to determine what contributions were made directly by Pythagoras himself. However, most historians agree that Pythagoras was likely responsible for several basic tenets in math and science today.

Although the Babylonians knew about the relationship between the legs of a right **triangle** and the **hypotenuse** centuries before the birth of Pythagoras, he was the first to provide a general geometric **proof** of this relationship. The **theorem** states that the square of the measure of the hypotenuse of a right triangle is equal to the sum of the squares of the two legs. Considering that deductive geometry was in its infancy, this theorem was remarkable for its time. **Euclid**, who Pythagoras strongly influenced, stated the theorem, which is called the Pythagorean theorem, in book one of his *Elements*.

The Pythagorean theorem also led to the discovery of irrational numbers, which is considered to be one of the greatest discoveries of mathematical antiquity. Pythagoras discovered that the hypotenuse of the isosceles right triangle with legs of unit length is equal to the **square root** of two, which

cannot be accurately represented by a rational number (a number that can be expressed either as a whole number, integer, or fraction). The irrational number cannot be represented by any ratio of **integers**, and in its decimal form does not terminate and does not repeat a certain pattern. Interestingly, the discovery of irrational numbers led to the contradiction of many existing mathematical proofs and theorems. According to legend, the Pythagoreans viewed irrational numbers as symbolic of the unspeakable.

Pythagoras was also the first to discover the mathematical basis of music. For Pythagoras and his followers, music was more than entertainment; it was an integral part of their philosophy and religion. Music was also seen as having an almost mystical affect on humans in terms of soothing the psyche and the soul. Pythagoras found that strings of the lyre produced harmonious tones when the ratios of the lengths of the strings are **whole numbers**. As a result, Pythagoras discovered the numerical ratios which determine the concordant scale in music. This discovery strengthened the Pythagoreans' belief that all things are numbers and ordered by numbers. Perhaps the most famous outgrowth of this belief is Pythagoras' theory that the heavenly bodies in the cosmos are separated in regular intervals, like the law of harmony. As a result, Pythagoras believed in a vast universal or cosmic harmony (ratio), which led to his doctrine of the "Music of the Spheres."

In the realm of astronomy, Pythagoras was the first to declare the Earth spherical in nature. It is reported that he made this discovery when he noticed the round shadow thrown by the Earth on the moon during a lunar eclipse. Pythagoras also deduced that the sun, moon, and planets have movements of their own, that is, that they rotate on their own axes and **orbit** a central point. However, Pythagoras mistakenly identified this central point as Earth. Eventually, the Pythagoreans deduced that Earth orbited around another central point, but never identified this point as the sun. The Pythagoreans also correctly determined that the Earth's **rotation** around this "central point" took 24 hours.

In his own time and the first few centuries following his death, Pythagoras was revered by many, with some accounts of his deeds reaching mythic **proportions** (such as being able to walk on water). His influence can be found in such famous Greek scientists and philosophers as Euclid and **Plato**. However, Pythagoras was also viewed with disdain by some, primarily because of his philosophies (such as transmigration of the soul, or reincarnation), which resembled the philosophy of the Orient rather than the Greek philosophy of his day.

Although Pythagoras and his school prospered, their controversial philosophies and strong political influence created enemies. With the rise of the democratic party in southern Italy, Pythagoras and his students became the objects of persecution. The democrats considered the Pythagoreans as elitists who placed themselves above others and were also suspicious of their secret rites. This mistrust came to a head sometime between 500 B.C. and 510 B.C., resulting in the destruction of the Pythagorean school and campus. The exact fate of Pythagoras himself remains uncertain. According to some accounts, Pythagoras was killed during this attack. Other accounts indicate that he escaped and lived to be nearly 100 years old. In keeping with Pythagoras' mythic stature, he is also said to have ascended bodily into heaven.

Regardless of the fate of Pythagoras, his followers continued his teachings in other lands, eventually returning to Italy to reestablish the Pythagorean school in Tarentum. In terms of influence, the Pythagoreans outlasted all other philosophies of ancient Greece. They established a strong presence and/or influence both in Egypt and ancient Rome, where the Senate erected a statue in honor of Pythagoras as "the wisest and bravest of Greeks."

PYTHAGOREAN NUMBER THEOREM

The Pythagorean number **theorem** is the basis for modern **number theory**. Modern number theory is that branch of mathematics concerned with properties and relationships of numbers. This includes much of mathematics and more specifically mathematical **analysis** and is generally limited to the study of **integers** or other **sets** that possess properties of all integers. The Pythagoreans carried out extensive mathematical investigations concerned with numbers, mainly of odd and even numbers and of prime and **square** numbers, that established a scientific foundation for mathematics. Their results and beliefs are summarized into the Pythagorean number theorem.

Pythagoras was a Greek mathematician thought to have been born in about 500 B.C.. Although originally born on the island of Samos, Pythagoras was thought to have moved to Crotona in about 530 B.C. where he founded a movement known as Pythagoreanism. The disciples of this movement are believed to have formed the initial ideas for the Pythagorean number theorem. The Pythagoreans formulated a view from an arithmetical standpoint that believed the concept of the number was the key to the qualities of mankind and matter and that it was the ultimate principle of all proportion of the universe. They believed that everything was a composition of a number and that an object's existence could only be understood in that number. This was in stark contrast to the accepted thinking at the time that numbers were only of utilitarian use for solving problems in **calendar** construction, architecture, and commerce. The Pythagoreans believed that numbers had an importance unto themselves and that each number possessed it own special attributes. They drew distinctions between logistic, which involved the art of computation, and **arithmetic**, which involved number theory.

Originally the Pythagoreans treated numbers concretely as patterns but this eventually evolved into a refined concept of the number as an abstract entity. This concept of the number is what still exists today. To the Pythagoreans each number possessed its own special attributes, some of which are listed below:

- 1, "monad"—represents unity and is the generator of numbers.
- 2, "dyad"—represents diversity and opinion and is the first true female number (the Pythagoreans believed even numbers were female and odd numbers were male).
- 3, "triad"—represents harmony that is equal to unity plus diversity. This is the first true male number.
- 4—represents justice and retribution. It is the squaring of accounts.
- 5—represents marriage that is equal to the first female plus the first male.
- 6—represents creation that is equal to the first female plus the first male plus 1.
- 10, "tetractys"—represents the universe.

In the development of the Pythagorean number theorem the Pythagoreans valued rigor and **proof** which led them to search for essential properties and definitions of numbers. The Pythagoreans believed a number is a collection of units. They studied **prime numbers**, composite or rectilinear numbers, odd and even numbers and devoted a great deal of time to the study of the "tetractys." To them the holiest of numbers was the number 10 because it had a special significance. The first four numbers also held a special significance for the Pythagoreans. They accounted for all of the possible dimensions and their sum equals 10, the holiest number, the universe. The first four numbers were the only numbers needed to represent all known objects geometrically. It is the Pythagorean veneration of the tetractys that is responsible for the present use of the base ten.

The Pythagorean number theory led to many important theorems and models. The **model** of the universe is based on the initial attempt by the Pythagoreans to understand cosmology in terms of mathematical principles. Since the number 10 was the number of the universe they believed that there had to be 10 heavenly bodies and that the planets orbited a central fire, the Sun. In **geometry** the greatest discovery of the Pythagoreans was the **hypotenuse** theorem, which is usually called the **Pythagorean theorem**. This theorem relates the hypotenuse of a right **triangle** to the other two sides in a purely mathematical way. The Pythagorean theorem is usually written as: $a^2 + b^2 = r^2$, where a and b are the lengths of the sides of a right triangle and r is the length of the hypotenuse. Pythagorean triplets are integer solutions to the Pythagorean theorem. There also came another important theorem from the Pythagorean number theory. Basically this theory says that the diagonal of a square with sides of integral length cannot be rational. From this Pythagoras' constant, $\sqrt{2}$, was discovered. This was in conjunction with the study focused on **irrational numbers** and the golden ratio. The Pythagorean number theorem had an important influence on the formation of modern mathematics.

See also Geometry; Number theory; Pythagoras

PYTHAGOREAN THEOREM

The Pythagorean **theorem** is one of the most ancient theorems of mathematics. The Pythagorean theorem states that in a right **triangle** (a triangle with a 90-degree **angle**), the sum of the squares of the lengths of the two legs is equal to the **square** of the length of the **hypotenuse**; the legs of a right triangle are the two sides adjoining the 90-degree angle, and the hypotenuse is the opposite side. In terms of symbols, if the lengths of the legs are labeled a and b, and the hypotenuse is labeled c, the Pythagorean theorem states the famous relationship that $a^2+b^2=c^2$, the form in which the theorem is most commonly remembered. The Pythagorean theorem is one of the most powerful of the fundamental theorems of **geometry**. It is the basis for the **definition** of the **distance** between two points in the rectangular coordinate system, the key definition on which all of coordinate (or Cartesian) geometry rests.

Although the theorem is named after the Greek mathematician **Pythagoras** (circa 560-480 B.C.), it was in fact known much earlier: a statement of the theorem was discovered on a Babylonian tablet that may date as far back as 1900 B.C.. It is not believed, however, that the Babylonians gave a formal mathematical **proof** of the theorem. The Babylonians generally discovered mathematical relationships by experiment, and historians suspect that the Babylonians discovered the theorem simply by measuring the sides of a large number of right triangles, and noticing that for all of them, the relationship $a^2+b^2=c^2$ was true. From their observations they were willing to conclude that the relationship was true for every right triangle.

It was not until the time of the Greeks, more than a thousand years later, that the idea had evolved that mathematical truths could be proven by the rigors of logic. Historians believe that the first careful proof of the Pythagorean theorem, using deductive reasoning, was given either by Pythagoras himself or by one of his disciples at his school in Cortona, a Greek seaport in Southern Italy. Pythagoras was quick to recognize the significance of the theorem, and according to legend, he was so overjoyed that he offered a sacrifice of oxen to the gods in thanks for the discovery.

A natural question asked both by the Pythagoreans and by the mathematicians that came after them was, When will the lengths of all three sides of a right triangle be **whole numbers**? In other words, which whole numbers satisfy the relationship $a^2+b^2=c^2$? The numbers 3, 4 and 5 do, for example, since $3^2+4^2=5^2$. Likewise, the numbers 5, 12 and 13 satisfy the relationship, as do 7, 24 and 25. Triples of whole numbers that satisfy the relationship are known as Pythagorean triples. Over the years, mathematicians have developed many ways to generate Pythagorean triples, and in fact have proven that there is an infinite collection of Pythagorean triples. A related, more complicated question, is, For what **exponents** n will the equation $a^n+b^n=c^n$ have non-zero whole number solutions? Long before this question was finally set to rest, it was widely believed that n=2 is the only exponent for which this equation will have solutions. This statement is the famous **Fermat's last theorem**, which stumped mathematicians for more than two hundred years before finally giving way to proof in 1995, by the mathematician **Andrew Wiles**.

Ever since the school of Pythagoras produced the first proof of the Pythagorean theorem, mathematicians and amateurs have found amusement in trying to produce alternate proofs, and through the centuries dozens more proofs have been found. **Leonardo da Vinci** came up with one, as did James Garfield in 1876, a few years before becoming President of the United States. The great Hungarian mathematician **Paul Erdös**, when a seventeen-year-old mathematical prodigy, boasted that he knew 37 different proofs. Many of these proofs are very simple, but perhaps the simplest and most elegant of all is the following proof, the one that historians ascribe to Pythagoras himself.

Start with a right triangle whose legs have lengths a and b, and whose hypotenuse has length c, and build a large square. There are two different ways to calculate the **area** of the large square. Each side of the square has length a+b, so the area of the square is $(a+b)^2$. On the other hand, the large square is made up of one smaller square and four copies of the right triangle, so its area is also equal to the area of the smaller square plus four times the area of the right triangle. The area of the smaller square is c^2, and the area of the right triangle is ab/2. So we have the equation

- $(a+b)^2=c^2+4(ab/2)$

Now $(a+b)^2=a^2+b^2+2ab$, and $4(ab/2)=2ab$, so we can rewrite our equation as

- $a^2+b^2+2ab=c^2+2ab$

If we cancel the 2ab that appears on both sides of the equation then we are left with the relationship $a^2+b^2=c^2$, which is exactly what we wanted to prove.

Q

QUADRATIC EQUATIONS

A quadratic equation is a second order, univariate polynomial with constant coefficients and can usually be written in the form: $ax^2 + bx + c = 0$, where $a \neq 0$. In about 400 B.C. the Babylonians developed an algorithmic approach to solving problems that give rise to a quadratic equation. This method is based on the method of **completing the square**. Quadratic **equations**, or **polynomials** of second-degree, have two **roots** that are given by the quadratic formula: $x = (-b \pm \sqrt{(b^2 - 4ac)})/2a$. There is another form of this equation yielding the roots for a quadratic equation that is obtained by first dividing the original quadratic equation through by x: $x = (2c)/(-b \pm \sqrt{(b^2 - 4ac)})$. This equation, which provides the roots to the quadratic equation, is often useful when $b^2 > 4ac$. In these cases the usual form providing roots to the quadratic equation can yield erroneous numerical results.

The earliest solutions to quadratic equations involving an unknown are found in Babylonian mathematical texts that date back to about 2000 B.C.. At this time the Babylonians did not recognize negative or complex roots because all quadratic equations were employed in problems that had positive answers such as length. The theory involving quadratic equations, and all polynomial equations, was flawed prior to the 17th century because of this idea. In 400 B.C. the Babylonians developed the quadratic formula used to find the roots of quadratic equations. About 100 years later **Euclid** formulated a geometrical approach to solving quadratic equations. His approach involved determining a length that would be the root of a quadratic equation. There were many other methods used in ancient times to determine the roots to quadratic equations. The Egyptians employed the false position method which involved approximating x to make part of the quadratic equation easy to calculate. Then a scaling **factor** was incorporated to find the root of the original equation. Greek mathematicians employed the **iteration** method in which a positive root of a quadratic equation is approximated and substituted for the

unknown. Then this is used to form another approximation hat is substituted and calculated. This process is repeated until the real root is determined. Between 598 and 665 A.D. **Brahmagupta**, an Indian mathematician, advanced the Babylonian methods to almost modern methods. Indian and Chinese mathematicians recognized negative roots to quadratic equations. **Al-Khwarizmi**, an Arab mathematician, developed a classification of quadratic equations in the 9th century. They were classified into one of six different types depending on which coefficients were negative. He wrote six chapters with each chapter devoted to a different type of equation. The equations were composed to three types of quantities: roots, squares of roots and numbers, and numbers. In each chapter al-Khwarizmi described the rule used for solving each type of quadratic equation and then presented a **proof** for each example. Later, in 1145, Abraham bar Hiyya Ha-Nasi, also known by the Latin name Savasorda, published a book that was the first to give the complete solution of the quadratic equation. Over the next few hundred years several mathematicians advanced the study of quadratic and **cubic equations**. Near the end of the 18th century **Carl Friedrich Gauss**, a German mathematician, gave a proof that showed every polynomial equation has at least one root. The root may not be able to be expressed as an algebraic formula involving the coefficients of the equation but a root did exist. Eventually a team of three international mathematicians combined and showed that only polynomials of degree five or less could be solved via a general algebraic formula. It is this set of polynomials that the theory of equations focuses on.

QUADRATIC FUNCTIONS

Quadratic **functions** are among the most familiar in mathematics. The general form of a quadratic function is $y = ax^2 + bx + c$, where a, b, and c are real valued constants, except that a cannot be 0. The graph of a quadratic function is called a **parabola**, which looks like a "U" with the sides bent outwards. When *a* is positive, the parabola opens upward, looking somewhat like a smile; when *a* is negative, the parabola opens downward like a frown. The absolute value of *a* determines how quickly the parabola rises or falls; the larger the absolute value of *a*, the more rapidly the parabola rises or falls. The values of *b* and *c* in the quadratic equation determine the location of the parabola in the xy-plane. In particular, if b = 0, then the parabola will be symmetric to the y-axis; its highest(or lowest) point, called the vertex, will be located at the point (0,c). If *c* is also 0, then the vertex will be at the origin, i.e., (0,0). If *b* is not 0, then the axis of **symmetry** will no longer be the y-axis; it will be a vertical line with equation x = -b/(2a), so that the first coordinate of the vertex will be -b/(2a). The second coordinate of the vertex also depends on the value of *c* and may be obtained by substituting -b/(2a) into the function for *x*.

The simplest of all quadratic functions is $y = x^2$, whose axis of symmetry is the y-axis and whose vertex is (0,0). This is sometimes called the "parent" of all quadratic functions because all other quadratic functions can be generated from it through the use of transformations that scale and translate the graph of $y = x^2$. For example, if we apply the **translation** which moves the vertex 3 units to the right and 4 units up, we obtain a new parabola with equation $y - 4 = (x-3)^2$, or, with some **algebra**, $y = x^2 - 6x + 13$ in standard form. Or we could start with $y = x^2$, and scale the second coordinates by a **factor** of 2, which causes the graph to rise more rapidly for each horizontal increment and which gives the new equation $y = 2x^2$. Or we could do the scaling combined with the translation to obtain the equation $y - 4 = 2(x-3)^2$, which is equivalent to $y = 2x^2 - 12x + 22$ in standard form. In general, if second coordinates of the graph of $y = x^2$ are scaled by a factor of *a* and the vertex is translated *h* units horizontally and *k* units vertically, then we obtain the so-called vertex form for this new parabola, $y - k = a(x-h)^2$. Left in this form, the vertex of the parabola (h,k) is immediately evident. So in some cases, it is convenient to use the vertex form, while in other cases the standard form is preferred. There is a third form of the equation derivable from the geometric **definition** of a parabola. This definition says that a parabola is the set of all points which are equally distant from a fixed point, called the focus, and a fixed line, called the directrix. If the parabola has vertex at the origin, focus at the point (0,p), and directrix with equation y = -p, then it can be shown that an equation for the parabola is $y = (1/(4p))x^2$. The focus gives the parabola an interesting and useful property. Lines that are directed away from the focus are reflected off the parabola and directed away in lines parallel to the axis of symmetry. Some useful applications of this focal property of parabolas are discussed in the next paragraph.

Quadratic functions have numerous applications in physics and engineering, as well as in other fields. In physics, the quadratic function $f(t) = -16t^2$ is a mathematical **model** for the freefall of an object due to the earth's gravity when air resistance is negligible. In fact, the flight of a projectile near the earth's surface takes the path of a parabola, the graph of a quadratic function. Engineers use the focal property of parabolas to design searchlights, automobile headlights, tele-

scopes, satellite dishes, and many other devices in which it is necessary to concentrate a beam of **light** or other electromagnetic signals at a single point. In the case of a searchlight or automobile headlight, the bulb is placed at the focus of a parabolic mirror so that, when it is turned on, the light rays coming from the bulb bounce off the mirror and are all directed as parallel rays away from the mirror, providing a very concentrated beam of light. In the case of a reflecting telescope, light rays coming in from distant objects bounce off the parabolic mirror to the focus where the concentrated light is collected for the astronomer's viewing.

It is often useful to be able to find the point or points at which a quadratic function crosses the x-axis. The first coordinates of these points are called the zeros of the function. This problem is equivalent to solving the equation $ax^2 + bx + c = 0$, whose solution is given by one of the best known formulas in all of mathematics, the quadratic formula. This famous formula gives the zeros in terms of the coefficients *a*, *b*, and *c*. They are $x = (-b + \sqrt{(b^2 - 4ac)}) / (2a)$ and $x = (-b - \sqrt{(b^2 - 4ac)}) / (2a)$. The expression under the radical, b^2-4ac, is called the **discriminant** of the quadratic formula because it discriminates among three possibilities for the zeros of the quadratic function. These possiblities are: (1) if the discriminant is positive, there are two real zeros; (2) if the discriminant is 0, there is only one real **zero**; and (3) if the discriminant is negative, there are no real zeros, since the **square** root of a negative quantity is not a real number.

One final interesting fact about parabolas is that they are all similar in a geometric sense, which technically means that any parabola may be mapped onto any other parabola by a **similarity** transformation. A similarity **transformation** is some combination of dilations, translations, reflections, and rotations. This means that all parabolas "look alike" in the same sense that all circles look alike or all squares look alike. This may seem counterintuitive when you see the graphs of two different quadratic functions on the same axis system. You might say something like, "one is staying closer to the y-axis than the other one." This gives the illusion that the two parabolas have different shapes. In **truth**, however, they have the same shape in the same sense that two circles with different radii have the same shape or two squares with sides of different length have the same shape. The **proof** of this fact is not difficult, but it does require a background in transformations.

See also Parabola; Quadratic equations

QUANTIFIER

In **symbolic logic**, quantifiers are words or symbols which indicate quantity in the sense of "all," "some," "none," or "one." The two main logical quantifiers are "For all," usually symbolized by ∀, and "There exists," symbolized by ∃. Thus, ∀x is read "For all x" and ∃x is read "There exists an x." The quantifier ∀ is called the universal quantifier because, when it is used, it indicates that every member of the "universe of discourse" has some property which will follow the symbol.

The universe of discourse is just the set of all the entities of which we are making statements. So let us suppose that our universe of discourse is the set of **integers**; then "$\forall x[(x$ is even$)\rightarrow(x+1$ is odd$)]$" states that every even integer has the property that if 1 is added to it, the result is an odd integer. The quantifier $\exists$ is called the existential quantifier because it indicates that there is at least one member of the universe that has the property which follows it. For example, "$\exists x[x^2=4]$" states that there exists at least one integer whose **square** is 4. Now we know that there are, in fact, two such integers, 2 and -2, but the existential quantifier always indicates "at least one."

The **truth** or falsehood of a quantified statement depends upon the universe of discourse in which the statement occurs. For example, "$\forall x[x > -x]$" is true if the universe of discourse is the set of positive integers, but it is false if the universe of discourse is the set of all integers. The statement says that x is always greater than its opposite, but if x is negative, then its opposite, -x, will be positive, so that $x > -x$ will be false. Here is another example using the existential quantifier: $\exists x[x+5=2]$. This statement is true if the universe of discourse is all integers, but false if it is the positive integers. Thus it is very important to specify the universe of discourse for each statement that is made under quantification.

Suppose P represents a property and Px says that x has the property P. Since to say that something is true for all x is the same as saying that there is no x for which it is false, we can say that the statements "$\forall x[Px]$" and "not$\exists x[$not $Px]$" are equivalent. When two statements are equivalent, we say that the first statement is true if and only if the second statement is true. The logical symbol for "if and only if" is the double ended arrow, $\leftrightarrow$. Thus, "$\forall x[Px]\leftrightarrow$not$\exists x[$not $Px]$" expresses the **equivalence** of "$\forall x[Px]$" and "not$\exists x[$not $Px]$." With similar reasoning, we can write "$\forall x[$not $Px]\leftrightarrow$not$\exists x[Px]$," which essentially states that to say "Px is false for all x" is equivalent to saying "There is no x for which Px is true." Likewise, "not$\forall x[Px]\leftrightarrow\exists x[$not $Px]$" and "not$\forall x[$not $Px]\leftrightarrow\exists x[Px]$" are equivalences. Knowing such equivalences gives the logician more flexibility in proving **propositions**. The equivalences in this paragraph also lead to expressions for the negations of quantified statements: the negation of "$\forall x[Px]$" is "$\exists x[$not $Px]$"; the negation of "$\exists x[$not $Px]$" is "$\forall x[Px]$"; the negation of "$\forall x[$not $Px]$" is "$\exists x[Px]$"; and the negation of "$\exists x[Px]$" is "$\forall x[$not $Px]$." It can be seen from these statements that a general rule for negating a quantified statement is: "To negate a statement covered by one quantifier change the quantifier from universal to existential or vice versa and negate the statement which it quantifies." This rule can also be quite useful in proving theorems.

Many quantified statements simply represent what we might call common sense. For instance, let Ax represent "x is an animal." Then the common sense statement "If everything is an animal, then something is an animal" can be expressed as "$\forall x[Ax]\rightarrow\exists x[Ax]$." Note that this statement does not claim that everything is, in fact, an animal; it only claims that if it were the case that everything is an animal, then surely it would follow that something is an animal. We may also have state-

ments in more than one **variable** with more than one quantifier. Here is an example: "There exists an x such that for all y and z, x+y+z=y+z" may be rendered in **logical symbols** as "$\exists x\forall y\forall z[x+y+z=y+z]$." Another example from **geometry**: "For each pair of points x and y there is a point z such that z is between x and y." Here let Bxzy mean "z is between x and y"; then our statement may be written as "$\forall x\forall y\exists z[Bxzy]$."

Quantification was not always an explicit part of logic. Propositional logic uses no quantifiers and deals only with the structure of propositions. When logicians wish to speak about the content as well as the structure of propositions, they overlay propositional logic with quantification theory. The resulting system is called **predicate** logic or the predicate **calculus** because one now predicates or attributes certain qualities to all, some, or no elements in the universe of discourse. The use of quantifiers helped to clarify a number issues in logic, perhaps the most bothersome of which was the dispute over whether existence is a predicate. From **Aristotle** (384-322 BC), the inventor of logic, through the 19th century, existence was treated as a predicate, meaning, for example, that the statements "x works" and "x exists" were treated as having the same logical structure. There is no question that they have the same grammatical structure—both have the same subject-verb structure in the English language. The problem was that in English, "verb" and "predicate" are regarded as synonymous. Hence, it appears that "works" and "exists" are both predicates in the same sense. The great German philosopher, Immanuel Kant (1724-1804), said that while both "works" and "exists" may be considered predicates in the grammatical sense, "exists" may not be considered a predicate in the logical sense. He was essentially saying that logical predicates attribute qualities to things that are already in existence. "John works" says that the attribute of working is being claimed for John assuming that John exists. If John does not exist, then it is nonsense to attribute anything to him. So existence is a pre-condition for the attribution of qualities. It is very different from attributes that one predicates of existing objects. "John works," "John flies," "John is tall" attribute qualities to an already existing John. "John exists" may have the same grammatical structure as these, but it does not have the same logical structure when analyzed correctly. Here is where quantification can help. "John works" means "$\exists x(x$ is John$)\rightarrow($John works$)$." "John exists" means "$\exists x[x$ is John$]$." The first proposition is a conditional; the second is a declarative statement. The first says "If John exists, then John works." The second declares "John exists." Now this might seem irrelevant to any conversation about John that the ordinary person might have, but substitute the word "God" for x in "x exists" and you have one of the classic disputes of the ages. The so-called "ontological" **proof** for the existence of God is valid only if existence is a logical predicate. Aristotle, St. Anselm (1033-1109), and **Descartes** (1596-1650) assumed that it is. Kant, **Russell** (1872-1970), and most modern logicians say that it is not.

See also Predicate

QUANTUM MECHANICS

Quantum mechanics is the theory used to provide an understanding of the behavior of microscopic particles such as electrons and atoms. However, it is more than just a collection of formulae used by physicists and chemists to calculate, for example, where an electron might be. The quantum theory also introduced an entirely new way of thinking about very small objects that is strangely different from the way we think about macroscopic objects.

An example of a macroscopic object is a baseball. Whenever we throw a ball into the air it is a good idea to know where it will fall and how fast it will be traveling when it hits something. The most exact way to describe the ball's **motion** is by using classical **mechanics**, which predicts the position and velocity of the ball at every instant during its flight. This approach fits our everyday experience since we are accustomed to seeing a ball move in a very well-defined path.

The problem comes when we try to apply the classical approach to microscopic objects. If an electron were just an exceptionally small ball, its motion would follow a path predicted by classical mechanics. However, experiments have shown that this is not the case. The best illustration of this is the "double-slit experiment," in which electrons are sent one at a time towards a wall with two small slits or holes. On the other side of the wall is a screen of some sort which detects where electrons hit, perhaps by making a spot. If one slit is covered, electrons can pass through the uncovered hole and strike the screen. The hits on the screen are directly behind the open slit, exactly what we would expect. We get the same result by opening the first slit and covering the second, only now the pattern of spots is behind the first hole.

What if both slits are open? Taking a single baseball and throwing it at the wall, we know it will either pass through one slit or the other. Of course, it might miss the holes and never make it to the screen, but that case is not particularly interesting. Our experience with macroscopic objects makes us think that the electrons will behave the same way so the screen should show electron hits only directly behind the holes. However, this is not what happens. Instead the spots are spread out over the entire screen, even in places that should be blocked by the wall as shown here.

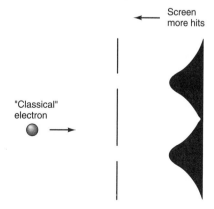

Screen
more hits

"Classical"
electron

Figure 1.

We'll see shortly that the only way for such a pattern to occur is if each electron somehow travels through both open slits! You might think that someone just made a terrible mistake when they conducted the experiment but many scientists have now verified this result (since they didn't want to believe it either after spending years learning classical mechanics). Apparently an entirely new way of thinking must be used for microscopic objects.

The pattern of electron hits that occurs when both slits are open appears very strange for particles. However, similar patterns are commonly produced by *waves* which are disturbances in a medium that carry energy. Suppose you filled a bathtub with water and then tapped the surface with your finger after the water had become still. By disturbing the water you create a wave that will move on the surface away from your finger. The water acts as the medium through which the wave moves. If you place a piece of paper in the water, you will notice that when the wave passes through that location the paper will be disturbed. Then the paper becomes still as the wave proceeds on. This is an example of a *traveling wave* which carries the energy of your tap (in the disturbance) from place to place in the medium.

Now put a barrier in the bathtub with two openings in it. What happens if you start a wave moving toward the barrier? You would see that parts of the wave pass through the openings, splitting the original wave into two separate waves. As those new waves continue on the other side of the barrier, they recombine (or *interfere*). At some locations they reinforce and produce an even larger wave (*constructive interference*) and in others they cancel (*destructive interference*). You can experience the same effect by listening to the sound waves coming from two speakers producing the same musical note. If you move around (some **distance** away from the speakers), you will find places where the sound is louder. The two speakers are producing separate but identical waves which are interfering. By marking the loudest locations on a map, you would draw an *interference pattern* similar to that made by the water wave after passing through the two-opening barrier. This is the same pattern that electrons make when both slits are open. The conclusion is that electrons act more like waves than they do like solid objects.

Let's try another wave experiment before talking more about electrons. Suppose instead of one wave, we produce a sequence of waves by tapping the water in a rhythm, or *frequency*. As the waves move through the water, the piece of paper would bob up and down at the same frequency. However, for certain rhythms the traveling waves and their reflections from the walls of the bathtub reinforce in an important way. The paper continues to bob up and down as before, but the disturbance in the water appears to be stationary (the wave does not travel). This is called a *standing wave*. The certain allowed frequencies that produce standing waves are determined by the dimensions of the bathtub whose walls confine the traveling waves. It is even easier to see a standing wave in the case of a guitar string. Striking the string sends many different traveling waves of different frequencies towards the ends which are held still. Standing waves occur for certain frequencies and those correspond to particular musical notes. Holding the string down at some point confines the waves more, and different standing waves appear on the string producing different notes. The energy of the wave can have any amount, depending on how hard the string was struck, and more energy translates into a louder sound (of an allowed frequency). We encounter these kinds of waves in our macroscopic world and because their motion can be understood using classical mechanics, they might be called *classical waves*.

Waves produce interference patterns like that of electrons in the double-slit experiment and we are accustomed to waves moving in less well-defined paths than solid objects. The new idea of quantum mechanics is to use waves to represent microscopic objects. These waves obey many of the properties of classical waves with an important conceptual difference. The "disturbance" of the new wave is not a physical motion of a medium, like the raised surface of water, it is an increased probability that the object is at a particular location. For an electron in the double-slit experiment, its traveling wave passes through both openings and interferes to produce a pattern of probability on the screen. The most probable locations (where constructive interference occurs) are the places where more electrons will hit and this is how the pattern of probabilities becomes a measurable pattern on the screen. This approach makes sense but we must give up the idea of predicting the path of objects such as electrons. Instead, in quantum mechanics we concentrate on determining the probability of obtaining a certain result when a measurement is made, for example of the position of an electron hit. This makes measurable quantities, called *observables*, particularly important.

Quantum mechanics requires advanced mathematics to give numerical predictions for the outcome of measurements. However, we can understand many significant results of the theory from the basic properties of the probability waves. An important example is the behavior of electrons within atoms. Since such electrons are confined in some manner, we expect that they must be represented by standing waves that correspond to a set of allowed frequencies. Quantum mechanics states that for this new type of wave, its frequency is proportional to the energy associated with the microscopic particle. Thus, we reach the conclusion that electrons within atoms can only exist in certain *states*, each of which corresponds to only one possible amount of energy. The energy of an electron in an atom is an example of an observable which is *quantized*, that is it comes in certain allowed amounts, called *quanta* (like quantities).

When an atom contains more than one electron, quantum mechanics predicts that two of the electrons both exist in the state with the lowest energy, called the *ground state*. The next eight electrons are in the state of the next highest energy, and so on following a specific relationship. This is the origin of the idea of electron "shells" or "orbits," although these are just convenient ways of talking about the states. The first shell is "filled" by two electrons, the second shell is filled by another eight, etc. This explains why some atoms try to combine with other atoms in chemical reactions. If an atom does not contain enough electrons to fill all of its lowest-energy shells completely, and it comes near another atom which has all its shells filled plus extra electrons, the atoms can become held together. The "bond" between the atoms comes from sharing the extra electrons. Thus quantum mechanics provides an understanding of chemistry by explaining why atoms combine, such as sodium and chlorine in salt.

This idea of electron states also explains why different atoms emit different colors of **light** when they are heated. Heating an object gives extra energy to the atoms inside it and this can transform an electron within an atom from one state to another of higher energy. The atom eventually loses the energy when the electron transforms back to the lower-energy state. Usually the extra energy is carried away in the form of light which we say was produced by the electron making a *transition*, or a change of its state. The difference in energy between the two states of the electron (before and after the transition) is the same for all atoms of the same kind. Thus, those atoms will always give off light of that energy, which corresponds to a specific color. Another kind of atom would have electron states with different energies (since the electron is confined differently) and so the same basic process would produce another color. Using this principle, we can determine which hot atoms are present in stars by measuring the exact colors in the emitted light.

Although quantum mechanics in its present form is less than 70 years old, the theory has been extremely successful in explaining a wide range of phenomena. Another example would include a description of how electrons move in materials, like those that travel through the chips in a personal computer. Quantum mechanics is also used to understand superconductivity, the decay of nuclei, and how lasers work. The list could go on and on. A great number of scientists now use quantum mechanics daily in their efforts to better understand the behavior of microscopic parts of the universe. However, the basic ideas of the theory still conflict with our everyday experience and the argument about their meaning continues among the same physicists and chemists who use the quantum theory.

If thinking about electrons and other particles as waves seems unsettling, you're in good company. That idea has been hotly debated since the beginning of quantum mechanics by some of the greatest physicists. The search for a new theory actually began in the late 1800s when several phenomena involving light were discovered which could not be understood using classical mechanics. One of those phenomena is called "blackbody radiation," the behavior of light within a box held at a certain temperature. It had been known for many years that light often behaves like a wave so physicists attempted to explain blackbody radiation as a case of standing waves within the box. Those standing waves were treated classically, with only certain allowed frequencies but any amount of energy. The results of that approach seemed promising, but did not quite agree with the experiments.

In 1900, Max Planck published a paper that explained blackbody radiation, but to do so he had to assume that the energy of the standing waves was quantized. Since the frequencies already had only certain allowed values, he made the simplest assumption possible, stating that the energy was equal to the frequency multiplied by a constant. That constant was written as $\hbar$ and it was later named after him. Planck did not know why the energy had to be quantized; he simply had to introduce the idea of energy quanta to make the theory agree with the experiments. The quantum idea was used by other physicists including **Albert Einstein** to explain more phenomena involving light which had eluded the classical approach. Einstein thought of light as being a wave "packet" which became known as a *photon*.

The concepts of energy quanta and photons were not accepted immediately, since most physicists believed that light was a classical wave. However, at least the new idea was not totally foreign since **Isaac Newton** had attempted to treat light as particles about 200 years before. Eventually it was accepted that a light wave could behave like a particle in some circumstances. However, in 1924 **Louis de Broglie** suggested that the opposite might also occur; that a solid object could correspond to what he called a *matter wave*. This was a revolutionary and unsettling idea, but many physicists set about to determine where it might lead. In 1926, Erwin Schroedinger developed an equation for those waves so that the theory could be used to make numerical predictions for the behavior of many physical systems. The Schroedinger equation proved to be extremely useful in describing microscopic objects and in fact it remains the cornerstone of quantum mechanics even now. However, the real breakthrough was that energy quanta came naturally from the mathematics instead of having to be assumed as Planck had done. This was because the equation for matter waves was similar, but different in a subtle way from the equation for classical waves. Schroedinger called his new approach *wave mechanics* and the name quantum mechanics came later.

The usefulness of quantum mechanics can hardly be argued. However, the basic concepts were a source of debate from the very beginning. That tradition is best exemplified by the historical debates that started in the late 1920s between two giants of physics, Einstein and **Niels Bohr**. In 1927 **Werner Heisenberg** had recognized an important result of treating microscopic particles as waves that described their probability of being in a certain location. Heisenberg derived the *uncertainty relation* that said the uncertainty in a measurement of a particle's position multiplied by the uncertainty in its velocity must always be larger than a specific constant, which unsurprisingly includes Planck's constant. For example, if we exactly measure where a particle is at a particular time (uncertainty in position is almost **zero**), then we do not have any information about the velocity of the particle (uncertainty in velocity is extremely large). That means we have no idea were the particle is going. The uncertainty relation restates that we cannot think of well-defined paths for microscopic particles.

Many different interpretations arose to make sense of this and other facets of quantum mechanics, and Bohr was the leading proponent of the *Copenhagen interpretation* (since that is where he was from). Today this is the accepted viewpoint among most, but not all physicists. The Copenhagen interpretation considers the uncertainty relation to express a fundamental limit to how accurately we can measure properties of microscopic particles. A way to understand this is to think how we detect a macroscopic object such as a car. We actually detect something that has bounced off the car, such as light. Suppose instead of light, we use a baseball (not really recommended) and decide whether or not a car was as at a location at a particular instant based on whether the ball bounces back to us after being thrown. The car is much larger and heavier than the baseball so we can detect where the car is without significantly disturbing the path of the automobile. However, what would happen if we tried to use a baseball to detect another baseball? We could find where the ball to be measured was located, but in the process it would be knocked off its original path into a new direction. If on the other hand we bounce a ping pong ball off a baseball, we have less exact information where the baseball is at a certain instant, but at least we disturb its path less. In every instance we have to interact (disturb) with an object to measure its position and the uncertainty relation simply reflects this. Why don't we notice this everyday? The answer is in the size of Planck's constant. It is so small that for macroscopic objects like a baseball the uncertainties in position and velocity are unmeasurable so the ball moves in a well-defined path. Quantum mechanics is actually around us all the time, but we just don't notice it.

Einstein had used quantum ideas, but he remained dissatisfied with quantum mechanics and particularly with Bohr's interpretation. Beginning in 1927 they began a public debate over the meaning of the new theory that raged for many years in scientific publications and often in person at conferences. Einstein felt that there must be some experiment which permitted exact measurements of position and velocity for microscopic objects. He continually challenged Bohr with suggestions of new experiments. However, in every instance Bohr was able to refute Einstein's experiments with arguments of his own that supported the Copenhagen interpretation. Einstein eventually gave up that approach but still maintained that quantum mechanics was somehow incomplete and that once the missing ideas were found, uncertainty would disappear. This search for a missing piece of the puzzle, if there is one, continues as physicists attempt to devise new experiments to more clearly understand the meaning of quantum mechanics.

QUARTIC EQUATIONS

Quartic **equations** are polynomial equations with one unknown **variable** (usually denoted by x), which is never raised to a power greater than 4. Symbolically, they can always be written as follows: $ax^4 + bx^3 + cx^2 + dx + e = 0$, where a, b, c, d, and e are numbers. The problem of solving polynomial equations in general, and quartic equations in particular, has been a central theme in **algebra** from antiquity to the present.

For much of this period, the interest in such problems was purely theoretical. Nowadays, polynomial equations have practical significance as well. To position the end of a robot arm correctly to screw in a bolt, for instance, the robot (or its programmer) **sets** up a system of polynomial equations. The solution to this system tells the robot what **angle** at each joint will cause the end of the arm to end up in the desired location. In essence, the robot uses mathematics to replace the senses of sight and touch.

Quadratic equations were solved by the ancient Babylonians, but mathematicians did not master **cubic equations** until 1545, when the Italian mathematician **Girolamo Cardano** published the first method for solving them. One might have expected quartic equations to be even harder, but in fact they were conquered at the same time. In remarkably offhand fashion, Cardano attributed the method to his student Luigi Ferrari, "who invented it at my request."

Though the computations can be backbreaking to do by hand, the solution method can be broken down into a simple sequence of steps that are easily programmed into a computer. Even so, the solution to a quartic equation typically requires many lines of type, and involves **square roots** inside of cube roots inside of square roots. Sadly, Ferrari's ingenious method is almost useless for practical purposes. Nevertheless, its discovery demonstrated the power of algebra, and prepared the way for a shocking discovery over 250 years later: the unsolvability of **quintic equations**.

See also Fundamental theorem of algebra; Galois theory; Polynomials; Quintic equations; Roots

QUATERNIONS

The quaternions are the second "number system" other than the **real numbers** ever discovered, and the first to be found by a deliberate search. Their immediate predecessor was the field of **complex numbers**, formed by adjoining an imaginary unit, the **square root** of -1 (usually denoted i), to the real number system. The question of whether the number i "really exists" vexed mathematicians for over 200 years, but by the early nineteenth century it was clear that complex numbers were too useful, for both **mathematics and physics**, to be ignored.

William Rowan Hamilton, a brilliant Irish astronomer and mathematician, was the first mathematician to wonder if other useful number systems existed "beyond" the complex numbers. His search was reminiscent of astronomers' search for planets beyond Uranus, and was successfully completed at almost the same time. On October 16, 1843, he suddenly realized that such a system could be constructed by adjoining not one but two more imaginary units, which he called j and k.

Thus, a quaternion is a "number" of the form $a + bi + cj + dk$, where i, j, and k are the three imaginary units. (The variables a, b, c, d represent ordinary real numbers. For example, 2 - $3i + j + 6k$ is a quaternion.) Each of the units i, j, and k is a square root of -1. Quaternions are added in the most obvious way, by collecting like terms. The rule for **multiplication** is far less obvious: The product of each two units, if taken in cyclic order, is the third unit. That is, $ij = k$, $jk = i$, and $ki = j$. But—

this was Hamilton's revolutionary insight—if the units are multiplied in the opposite order, the sign of the product must reverse. That is, $ji = -k$, $kj = -i$, and $ik = -j$. This means that quaternions do *not* obey the commutative law of multiplication. They do, however, obey every other law of **arithmetic**, and hence they constitute what is now called a division **algebra**.

Hamilton felt the quaternions were his greatest discovery, and for about 50 years they were considered a major branch of mathematics, at least in the English-speaking world. But they gradually fell from favor, because many of their properties (though not all—see below) could be rephrased in the notation of vectors. Moreover, vectors were far more flexible, because they are not restricted, as quaternions are, to four variables.

Nevertheless, quaternions should not be considered a dated nineteenth-century relic. They liberated the subject of algebra, in the same way that hyperbolic (Lobachevskian) **geometry** opened up the subject of geometry. They showed that algebraic structures could violate the laws of commutativity, associativity, and distributivity; these properties were not absolute truths, but simply features that some algebraic structures possess and others do not. This opened the floodgates to the discovery of numerous other algebraic structures, such as groups, rings, modules, and algebras.

Quaternions also have an ongoing importance to geometry. They provide the most natural way to represent rotations in three-dimensional **space**; by contrast, the vector and **matrix** approach (though more common in textbooks) is very cumbersome. A quaternion of unit length, $q = a + bi + cj + dk$, represents a **rotation** in the following way: The real part, a, is the **cosine** of *half* the **angle** of rotation, while the imaginary part, $bi + cj + dk$, is a vector that points along the axis of rotation. Curiously, quaternions have an extra sense of directionality that rotations lack. In three-dimensional space, the rotation about an axis v through an angle of x degrees is indistinguishable from a rotation about - v through an angle of -x degrees. But the two quaternions representing those motions, q and -q, are different. In the late 1920s, quantum physicists discovered that the same difference holds for subatomic particles. Electrons, protons, and neutrons (all the particles that make up ordinary matter) have not only an axis of spin but also a direction of spin along this axis, described as "spin up" or "spin down." The states of these particles are not vectors but *spinors*—quantities with a directed axis of rotation. Thus, Hamilton's discovery of quaternions—the first spinors—came 85 years ahead of its time.

See also Algebra; Complex numbers; William Rowan Hamilton; Real numbers; Rotations; Vector algebra

QUENEAU, RAYMOND (1903-1976)

French author

Raymond Queneau was a poet and writer whose intense interest in mathematics flavored many of his works. He cofounded an artistic group called Ouvroir de litterature potentielle (Oulipo—Workshop of Potential Literature) that still influences artists by imposing mathematical rules as game-like constraints on the generation of text. Some of Queneau's best-known works are *Exercises de style (Exercises in Style)* and the popular novel *Zazie dans le Metro* (1959), which director Louis Malle made into a movie in 1960. Many critics believe Queneau was one of the most important authors of the mid-20th century.

The son of hat-makers, Queneau was born in La Havre, France on February 21, 1903. He was reportedly fascinated by mathematics from an early age. Queneau attended the city's elementary and secondary schools, graduating in 1920, and in 1921 began studying at the University of Paris. This was roughly the start of his surrealist period, when he revealed a black sense of humor, became obsessed with death, and began ridiculing authority. He remained at the university until 1925, when (despite increasingly bad asthma attacks) he entered the military.

After a two-year tour of duty in Algeria and Morocco, Queneau returned to Paris in 1927 to study philosophy, although he also spent a lot of time studying mathematics with books by **Emile Borel** and **Georg Ferdinand Ludwig Philipp Cantor**. He found a job in Paris and in his spare time began spending time with artists in the famous rue du Chateau, who were surrealists and had come to be known as dissidents. However, despite his initial affinity with the sur-realists, he would soon dissociate himself from the movement because of their systematic rejection of scientific knowledge and reason. In 1930, Queneau began researching at the National Library for a project that he called *"les fous lit-teraires"* ("literary madmen"). According to Queneau, these were writers who had advanced what the mainstream consid-ered eccentric theories about various aspects of reality. By 1934 his work had resulted in the 700-page *Encyclopedia of the Inexact Sciences,* which was promptly rejected by two publishing houses.

Meanwhile, Queneau had been busy with other projects as well. From 1931 to 1934 he wrote reviews for the left-wing *Le critique social,* and in 1933 he published his first novel, *Le Chiendent (The Bark Tree)* after an inspirational trip to Greece. Also at this time, Queneau began a period of psycho-analysis that would last at least six years. During the remain-der of the decade he wrote *Guele de Pierre* (1934); *Les Derniers Jours (The Last Days)* (1936); his popular novel *Odile* (1937); a "novel-in-verse" and what would become his most famous poem, *Chêne et chien (Oak and Dog)* (1937); the novel *Les Enfants du Limon* (1938); and *Un Rude Hiver* (1939). From 1936 to 1938 Queneau worked as a reporter for a publication called *The Intransigent.* He then landed a job as a reader at the prestigious *New French Review.*

During World War II, Queneau wrote more than ever, and afterward found that he had acquired a reputation in the literary world. This was mainly due to the success of *Exercises in Style* (1947), but also to the active role he had taken in many artistic media, including film and radio. In 1950, he joined the College de Pataphysique, a group of writers and intellectuals dedicated to the science of imaginary solutions, particularly concentrating on exceptions to natural law. His 1950 poem

"Petite cosmogenie portative" ("A Small, Portable Cosmogeny") shows the influence of the school, as it describes the creation of the universe in a kind of scientific-alchemical language. Soon after joining the school, Queneau took on another project as well—editing the gigantic, multi-volume *Encyclopedie de la Pleiade,* a scholarly collection of past and present classical authors. Despite the difficulty of this uncharacteristically "serious" job, he was eager to do it; even since childhood, when he had written a catalog for his local library, Queneau had been preoccupied with ways to organize information. By 1955 he was the publication's director, and finished the task in 1956. Later that year, Queneau agreed to edit the results of a literary survey ("For an Ideal Library") asking respondents to vote for the most 100 important books of all time. His own additions included many mathematics classics as well as books by **Albert Einstein** and **John von Neumann**'s *Theory of Games.*

Queneau founded Oulipo with a literary friend as a sub-group within the College du Pataphysique in 1960. During their monthly meetings, Oulipo members would pose them-selves writing challenges. These stemmed mainly from Queneau's love of mathematics, which by then had become focused on **game theory**, **number theory**, **combinatorics**, and **set theory**. Another Oulipo-inspired work was his 1961 *Cent Mille Millards de Poemes,* a book of 10 sonnets each of whose 14 lines is printed on a single strip. Detaching and mixing the strips allows a theoretical potential of 10^{14} different poems. Queneau's 1963 book *Borders* is a collection of many of his math-based writings. His other major publications in the 1960s were the novel *The Blue Flowers* and a trilogy of poems.

Queneau died in Paris on October 25, 1976. He and his wife, a relative of pioneering surrealist André Breton, were married in 1928 and had one son.

QUETELET, ADOLPHE (1796-1874)
Belgian statistician

Lambert–Adolphe–Jacques Quetelet was one of the individu-als most responsible in the 19th century for the quantification of arguments in the physical and social sciences. He was born in Ghent on February 22, 1796, and was educated at the Lycee of Ghent. At age 19 he was appointed an instructor in mathe-matics at the Royal College in Ghent. In 1819 he was the first person to receive a doctorate from the University of Ghent for a dissertation on **conic sections**, and in the same year he moved to Brussels to take the chair of mathematics at the Athenaeum. Quetelet was elected to the Belgian Royal Academy in 1820. He married a Mlle. Curtet and they had two children.

In the early stages of his career, however, there was nothing to suggest that Quetelet's fame would spread through-out Europe. In 1824, he spent three months in Paris and learned two things: probability and how to run an observatory. As soon as he returned to Belgium, he began to campaign for the construction of a Royal Observatory. Successful in his efforts, Quetelet was appointed director of the Observatory in

1828. At the observatory, he could pursue celestial mechanicsin the same manner as **Pierre Simon Laplace**, whose work he greatly admired.

In 1835, Quetelet published *Sur l'homme et le developement de ses facultés* ("On Man and the Development of his Faculties"). He explores two topics: the importance of gathering quantitative data in order to answer questions about human problems, and the usefulness of the idea of "l'homme moyen" (the average man). Quetelet reveals a general understanding that the reliability of an average increases with the size of the population. What appears to have given the book some of its brilliance was Quetelet's vision of the average as something at which nature is aiming. Deviations from the average were seen as errors, although the notion of deviation as an object of study was familiar to Quetelet. This point also rendered Quetelet's perspective an easy subject for attack by the next generation of statisticians, who recognized the importance of variability in connection with the theory of evolution and other areas. The English statistician **Francis Galton** often criticized Quetelet's focus on the average rather than the deviations.

To some extent, Quetelet's portrayal of social phenomena as expressed in his book reflected the sources of his approach. On the one hand, he had a law of large numbers that had come down to him from more mathematically–minded statisticians, which indicated that when an experiment was repeated, the more reliable is the outcome. On the other hand, he also had a Laplacean view of some deterministic **mechanics** underlying the phenomena. As a result, in every area Quetelet found the accumulation of data essential for the purpose of recognizing the normal distribution underneath. The law of large numbersbecame the central principle of all science for Quetelet, and at the very least that encouraged him to promote data–gathering in every field.

In 1846, Quetelet published a volume of letters on the theory of probability. He was active in the reform of scientific teaching in Belgium and had become permanent secretary of the Brussels Academy in 1834. He was instrumental in the founding of the Statistical Society of London and was the first foreign member of the American Statistical Association. At a time when statisticsbegan to play a role in settings outside the calculation of annuities and games of chance, Quetelet spoke for the statistical community. In addition to his scientific work, he wrote an opera and published poetry and popular essays.

Quetelet died on February 17, 1874 in Belgium. His funeral was distinguished by the presence of many scientists from abroad and his memory was honored by the erection of a monument in Brussels. For someone who seldom used mathematics, Quetelet had acquired quite a reputation as a mathematician. The "average man" as Quetelet envisioned may have been a figment of the imagination, but the recognition of the importance of data–gathering was a timely lesson for a scientific culture about to undergo a probabilistic revolution.

QUINTIC EQUATIONS

Quintic **equations** are polynomial equations with one **variable**, customarily denoted by x, which is never raised to a power greater than the fifth. Symbolically, such an equation can be written as follows: $ax^5 + bx^4 + cx^3 + dx\mathrm{d}x^2 + ex + f = 0$. The problem of solving polynomial equations in general, and quintic equations in particular, has been a central theme in from antiquity to the present.

Babylonian mathematicians already knew how to solve **quadratic equations** (with no power of x greater than 2). In the early 1500s, Italian mathematicians discovered how to solve **cubic equations** (with no power of x greater than 3) and **quartic equations** (with no power greater than 4). In each case, the solution could be found with nothing more than the elementary operations of **addition**, **subtraction**, **multiplication** and **division**, plus the special operations of taking **square roots** or cube roots. This discovery suggested that a general method for solving all **polynomials** might be just around the corner.

However, more than 250 years went by with very little progress. Finally, in 1826, the Norwegian mathematician **Niels Henrik Abel** cleared up the mystery, in a novel and unexpected way: He showed that no formula had been found because *no such formula could ever exist*. That is, there is no "black box" that could accept the coefficients a, b, c, d, e, f as input, churn through a calculation that involves only elementary operations or n-th roots (also called "radicals"), and be *guaranteed* to produce, as output, a valid solution x to the above equation.

This was one of the first great impossibility theorems of mathematics, and its **proof** required a revolutionary approach. To prove that there is no "black box" that can solve a polynomial, one must study black boxes. That is, it is necessary to study the *operations themselves*: addition, subtraction, multiplication, division, and n-th roots.

The first four of these operations define a mathematical structure known as a **field**; and the operation of finding an n-th root amounts to constructing what is called a "cyclic extension" of a field. Quadratic, cubic, and quartic equations are solvable by radicals because their solutions always lie in cyclic extensions of cyclic extensions of cyclic extensions of the field of **rational numbers**. But for quintic equations, that is no longer true.

In 1830, **Evariste Galois** developed the method that mathematicians still use to study extension fields. Any extension field, Galois proved, possesses symmetries. For example, in a quadratic extension the two square roots can be interchanged without altering the underlying field. (This is the source of the ambiguous ± sign in the quadratic formula.) These symmetries are described by a mathematical structure called a Galois group, and the structure of the extension field is exactly mirrored in the structure of its Galois group. If the field is constructed by cyclic extensions, then the Galois group can be "deconstructed" by cyclic **factor** groups. But Galois realized that some Galois groups cannot be decomposed at all; they spring into existence fully formed, as it were. These are called simple groups, and the smallest one, called A_5, has 60 elements. It is too large to be the Galois group for a quartic polynomial (which can have 24 elements at most), but it can be the Galois group for a quintic. In fact, later mathematicians have shown that any quintic polynomial $p(x)$ with rational coefficients, which cannot be split into smaller polynomials, and which has exactly three real-number solutions, must have A_5 as its Galois group. It follows that the equation $p(x) = 0$ cannot be solved by radicals.

But quintic equations *can* be solved in other ways. One way is simply to add more tools to the "black box." In 1858, **Charles Hermite** showed how to solve all quintic polynomials by using theta **functions**, and **Felix Klein** in 1888 did the same with hypergeometric functions. But for working mathematicians and engineers, who may not want to learn about specialized and unfamiliar classes of functions, a more practical approach is to use approximate numerical methods. Equation-solving routines are now a standard part of computer **algebra** packages. Research is continuing on robust and efficient computer-based methods for solving not only quintic polynomials, but polynomials involving much higher **powers** of x.

Why, when adequate methods exist for solving even more complicated polynomials, do mathematicians still attach so much importance to theorems about inadequate methods? One might as well ask why baseball pitchers continue to throw the ball by hand, when a howitzer would work much better. Mathematics, like the game of baseball, has a certain integrity to it that can be spoiled if the "rules of the game" are changed too drastically. Moreover, the algebraic approach teaches us about the structure of *all* polynomials, while the brute-force computer approach only teaches us about one polynomial at a time.

See also Galois theory; Group theory

R

RADIAN

A radian is a unit of angular measure equal to the **angle** between two radii that enclose a section of a circle's **circumference** equal in length to the length of a radius. The entire angle of a **circle** is 2π radians and so 2π radians is equal to $360°$. The word radian was first used in the late 19th century and is derived from the English word radius. Radians are usually denoted without a symbol or with the symbol *rad*.

The radian measure of an angle is the ratio of the length of the arc subtended by that angle in a circle to the radius of the circle. The radian measure of an angle and the degree measure of the same angle are related using the circumference of a circle. The circumference of a circle is given as: $C = 2\pi r$, where r is the radius of the circle. Because the circumference of a circle contains exactly 2π of its radii, and because an arc with length equal to the length of the radius subtends one radian it goes that 2π radians = $360°$. Another concept closely related to radians is the steradian. Just as the measure of an angle in a circle is in terms of radians the fraction of a **sphere** subtended by an object can be measured in steradians. The steradian is a unit of solid-angle measure defined as the solid angle of a sphere subtended by a portion of the surface whose **area** is equal to the **square** of the sphere's radius. The entire **space** subtended by a sphere is 4π steradians.

The radian and degree are both units of angular measure but of different size but may be used interchangeably. Engineers and technicians use degrees more frequently, whereas radian measure is used almost exclusively in theoretical studies. Studies such as **calculus** use radians because of the simplicity of results such as derivatives and series expansions of **trigonometric functions**.

RAMANUJAN, S. I. (1887-1920)
Indian number theorist

S. I. Ramanujan was a self-taught prodigy from India. His introduction to the world of formal mathematics and subsequent fame arose from his correspondence and collaboration with the renowned British mathematician **Godfrey Harold Hardy**. In his short but prolific career Ramanujan made several important contributions to the field of **number theory**, an area of pure mathematics that deals with the properties of and patterns among ordinary numbers. Three quarters of a century after his death, mathematicians still work on his papers, attempting to provide logical proofs for results he apparently arrived at intuitively. Many of his theorems are now finding practical applications in areas as diverse as polymer chemistry and computer science, subjects virtually unknown during his own times.

Srinivasa Iyengar Ramanujan, born on December 22, 1887, was the eldest son of K. Srinivasa Iyengar and Komalatammal. He was born in his mother's parental home of Erode and raised in the city of Kumbakonam in southern India, where his father worked as a clerk in a clothing store. They were a poor family, and his mother often sang devotional songs with a group at a local temple to supplement the family income. Ramanujan received all his early education in Kumbakonam, where he studied English while still in primary school and then attended the town's English-language school. His mathematical talents became evident early on; at eleven he was already challenging his mathematics teachers with questions they could not always answer. Seeing his interest in the subject, some college students lent him books from their library. By the time he was thirteen, Ramanujan had mastered S. L. Loney's *Trigonometry,* a popular textbook used by students much older than he who were studying in Indian colleges and British preparatory schools. In 1904, at the age of 17, Ramanujan graduated from high school, winning a special prize in mathematics and a scholarship to attend college.

Shortly before he completed high school, Ramanujan came across a book called *A Synopsis of Elementary Results in Pure and Applied Mathematics.* This book, written by British tutor G. S. Carr in the 1880s, was a compilation of approximately five thousand mathematical results, formulae, and **equations**. The *Synopsis* did not explain these equations or provide proofs for all the results; it merely laid down various mathematical generalizations as fact. In Ramanujan, the book unleashed a passion for mathematics so great that he studied it to the exclusion of all other subjects. Because of this, although Ramanujan enrolled in the Fine Arts (F.A.) course at the local Government College, he never completed the course. He began to spend all his time on mathematics, manipulating the formulae and equations in Carr's book, and neglected all the other subjects that were part of his course work at the college. His scholarship was revoked when he failed his English composition examination. In all, he attempted the F.A. examinations four times from 1904 to 1907. Each time he failed, doing poorly in all subjects except mathematics.

During these four years, and for several more, Ramanujan pursued his passion with single-minded devotion. He continued to work independently of his teachers, filling up sheets of paper with his ideas and results which were later compiled in his famous *Notebooks*. Carr's book had merely been a springboard to launch Ramanujan's journey into mathematics. While it gave him a direction, the book did not provide him with the methods and tools to pursue his course. These he fashioned for himself, and using them he quickly meandered from established theorems into the realms of originality. He experimented with numbers to see how they behaved, and he drew generalizations and theorems based on these observations. Some of these results and conclusions had already been proved and published in the Western world, though Ramanujan, sequestered in India, could not know that. But most of his work was original.

Meanwhile, his circumstances changed. Without a degree, it was very difficult to find a job, and for many of these years Ramanujan was desperately poor, often relying on the good graces of friends and family for support. Occasionally he would tutor students in mathematics, but most of these attempts were unsuccessful because he did not stick to the rules or syllabus. He habitually compressed multiple steps of a solution, leaving his students baffled by his leaps of logic. In July 1909 he married Janaki, a girl some ten years his junior. Keeping with local customs and traditions, the marriage had been arranged by Ramanujan and Janaki's parents. Soon afterwards, he traveled to Madras, the largest city in South India, in search of a job. Because he did not have a degree, Ramanujan presented his notebooks as evidence of his work and the research he had been conducting in past years. Most people were bewildered after reading a few pages of his books, and the few who recognized them as the work of a genius did not know what to do with them. Finally, Ramachandra Rao, a professor of mathematics at the prestigious Presidency College in Madras, reviewed the books and supported him for a while. In 1912 Ramanujan secured a position as an accounts clerk at the Madras Port Trust, giving him a meager though independent salary.

During this time, Ramanujan's work caught the attention of other scholars who recognized his abilities and encouraged him to continue his research. His first contribution to mathematical literature was a paper titled "Some Properties of Bernoulli's Numbers," and it was published in the *Journal of the Indian Mathematical Society* around 1910. However, Ramanujan realized that the caliber of his work was far beyond any research being conducted in India at the time, and he began writing to leading mathematicians in England asking for their help.

The first two mathematicians he approached were eminent professors at Cambridge University, and they turned him down. On January 16, 1913, Ramanujan wrote to Godfrey Harold Hardy, who agreed to help him. Hardy was a fellow of Trinity College, Cambridge, and he specialized in number theory and **analysis**. Although he was initially inclined to dismiss Ramanujan's letter, which seemed full of wild claims and strange theorems with no supporting proofs, the very bizarreness of the theorems nagged at Hardy, and he decided to take a closer look. Along with J. E. Littlewood, he examined the theorems more thoroughly, and three hours after they began reading, they both decided the work was that of a genius. "They must be true because, if they were not true, no one would have had the imagination to invent them," Hardy is quoted as saying in Robert Kanigel's book *The Man Who Knew Infinity*.

Hardy now set about the task of bringing Ramanujan to England. In the beginning, Ramanujan resisted the idea due to religious restrictions on traveling abroad, but he was eventually persuaded to go. Ramanujan spent five years in England, from 1914 to 1919, during which time he enjoyed a productive collaboration with Hardy, who personally trained him in modern analysis. Hardy described this as the most singular experience of his life, says Kanigel in *The Man Who Knew Infinity:* "What did modern mathematics look like to one who had the deepest insight, but who had literally never heard of most of it?" Ramanujan was to receive several laurels during this period, including a B.A. degree from Cambridge, and appointments as Fellow of the Royal Society (at 30 he was one of the youngest ever to be honored thus) and Trinity College. But the English weather affected Ramanujan's health, and he contracted tuberculosis. In 1919 he returned to India, where he succumbed to the disease, dying on April 26, 1920.

Until the very end, Ramanujan remained passionately involved in mathematics, and he produced some original work even after his return to India. His great love for the subject and his genius are perhaps best exemplified in an incident described by Hardy in his book *A Mathematician's Apology*. He related that while visiting Ramanujan at a hospital outside London, where he lay ill with tuberculosis, Hardy mentioned the number of his taxicab, 1729. Hardy thought it a rather dull number. "No, Hardy! No, Hardy!" Ramanujan replied. "It is a very interesting number. It is the smallest number expressible as the sum of two cubes in two different ways." Kanigel reported another comment Hardy made later. He said that had Ramanujan been better educated, "he would have been less of a Ramanujan and more of a European professor and the loss might have been greater." Ramanujan himself attributed his mathematical gifts to his family deity, the goddess Namagiri. A deeply religious man, he combined his passion with his faith, and he once told a friend that "an equation for me has no meaning unless it expresses a thought of God."

RAMSEY, FRANK PLUMPTON (1903-1930)
English lecturer

Despite the shortness of his life, Frank Plumpton Ramsey had a strong influence on the world of mathematical logic. He launched a new specialty within his field now known as Ramsey theory, which deals with **combinatorics**. Ramsey was especially enamored of philosophy toward the end of his life and wrote several highly original works on that subject.

Ramsey was born in Cambridge, England on February 22, 1903. His father was president of the town's Magdelene College and his brother would one day become Archbishop of Canterbury. Ramsey received his secondary school education at Winchester College from 1915 to 1920, when he won an academic scholarship to Cambridge's Trinity College. There he concentrated on mathematics, becoming a senior scholar in 1921 and graduating with honors in 1923.

After a short trip to Vienna, Ramsey returned to Cambridge as a fellow at King's College in 1924. (This was testimony to his talent, since the school virtually never offered fellowships to people who had not studied there.) King's appointed him a lecturer in mathematics in 1926; he quickly earned a reputation as a brilliant and fascinating speaker. The school would soon name him director of mathematical studies.

Ramsey published his first major work in 1925. "The Foundations of Mathematics" contained his formal agreement that mathematics is a component of logic, as **Bertrand Russell** and **Alfred North Whitehead** had declared in their *Principia Mathematica*. However, Ramsey suggested improvements in the work: namely that its authors eliminate the **axiom** of reducibility (Ramsey called it "certainly not self-evident") and that Russell's theory of types be simplified by treating certain semantic paradoxes as merely linguistic (e.g., the statement "I am lying.").

In 1926, Ramsey's paper "Mathematical Logic" appeared in the *Mathematical Gazette*. It featured an attack on **Egbertus Jan Luitzen Brouwer**'s insistence that **propositions** are either true of false, and criticized **David Hilbert** for characterizing mathematics as "a meaningless game with marks on paper."

Ramsey presented perhaps his most influential paper, "On a problem of formal logic," to the London Mathematical Society in 1928. In it, he suggested methods for analyzing the consistency of a logical formula and introduced some combinatorics theorems that soon proved to be the seeds of a new branch of mathematics (Ramsey theory).

At about this time Ramsey also wrote two papers on economics: "A contribution to the theory of taxation" and "A mathematical theory of saving." Both appeared in the British *Economic Journal*. However, these were not deemed as important as his writings on philosophy—a subject on which Ramsey spent increasing amounts of time toward the end of his life. He wrote such papers as "Universals" (1925), "Facts and propositions" (1927), "Universals of law and fact" (1928), "Knowledge" (1929), "General propositions and causality" (1929), and "Theories" (1929). Some historians have suggested that Ramsey, had he lived longer, would have been one of the world's foremost philosophers.

Ramsey died at age 27 on January 19, 1930 in Cambridge after undergoing surgery for jaundice in London. He and his wife had married in 1925 and had two daughters. At six feet three inches, Ramsey was a physically imposing man, although he was known for his quiet, modest, and easy-going nature. He was a staunch atheist, enjoyed hiking, loved classical music, had a strong sense of humor, and limited his work hours to four each morning.

RANDOM

The word "random "is used in mathematics much as it is in ordinary speech. A random number is one whose choice from a set of numbers is purely a matter of chance; a random walk is a sequence of steps whose direction after each step is a matter of chance; a random **variable** (in **statistics**) is one whose size depends on events which take place as a matter of chance.

Random numbers and other random entities play an important role in everyday life. People who frequent gambling casinos are relieved of their money by slot machines, dice games, roulette, blackjack games, and other forms of gambling in which the winner is determined by the fall of a card, by a ball landing in a wheel 's numbered slot, and so on. Part of what makes gambling attractive is the randomness of the outcomes, outcomes which are usually beyond the control of the house or the player.

Children playing tag determine who is "it" by guessing which fist conceals the rock. Who does the dishes is determined by the toss of a coin.

Medical researchers use random numbers to decide which subjects are to receive an experimental treatment and which are to receive a placebo. Quality control engineers test products at random as they come off the line. Demographers base conclusions about a whole population on the basis of a randomly chosen sample. Mathematicians use **Monte Carlo methods**, based on random samples, to solve problems which are too difficult to solve by ordinary means.

For absolutely unbreakable ciphers, cryptographers use pages of random numbers called one-time pads.

Because numbers are easy to handle, many randomizations are effected by means of random numbers. Video poker machines "deal" the cards by using randomly selected numbers from the set 1, 2,..., 52, where each number stands for a particular card in the deck. Computer simulations of traffic patterns use random numbers to mark the arrival or non-arrival of an automobile at an intersection.

A familiar use of random numbers is to be seen in the lotteries which many states run. In Delaware's "Play 3" lottery, for example, the winning three-digit number is determined by three randomly selected numbered balls. The machine that selects them is designed so that the operator cannot favor one ball over another, and the balls themselves, being nearly identical in size and weight, are equally likely to be near the release mechanism when it is activated.

Random numbers can be obtained in a variety of ways. They can be generated by physical means such as tossing coins, rolling dice, spinning roulette wheels, or releasing balls from a lottery machine. Such devices must be designed, manufactured, and used with great care however. An unbalanced coin can favor one side; dice which are rolled rather than tumbled can favor the faces on which they roll; and so on. Furthermore, mathematicians have shown that many sequences that appear random are not.

One notorious case of faulty randomization occurred during the draft lottery of 1969. The numbers which were to indicate the order in which men would be drafted were written on slips and enclosed in capsules. These capsules were then mixed and drawn in sequence. They were not well mixed, however, and, as a consequence, the order in which men were drafted was scandalously lacking in randomness.

An interesting source of random numbers is the last three digits of the "handle" at a particular track on a particular day. The handle, which is the total amount bet that day, is likely to be a very large number, perhaps close to a million dollars. It is made up of thousands of individual bets in varying amounts. The first three digits of the handle are anything but random, but the last three digits, vary from 000 to 999 by almost pure chance. They therefore make a well-publicized, unbiased source of winning numbers for both those running and those playing illegal "numbers" games.

Cards are very poor generators of random numbers. They can be bent, trimmed, and marked. They can be dealt out of sequence. They can be poorly shuffled. Even when well shuffled, their arrangement is far from random. In fact, if a 52-card deck is given eight perfect shuffles, it will be returned to its original order.

Even a good physical means of generating random numbers has severe limitations, possibly in terms of cost, and certainly in terms of speed. A researcher who needs thousands of randomly generated numbers would find it impractical to depend on a mechanical means of generating them.

One alternative is to turn to a table of random digits which can be found in books on statistics and elsewhere. To use such tables, one starts from some randomly chosen point in the table and reads the digits as they come. If, for example, one wanted random numbers in the **range** 1 to 52, and found 22693 35089... in the table, the numbers would be 22, 69, 33, 50, 89,... The numbers 69 and 89 are out of the desired range and would be discarded.

Another alternative is to use a **calculator** or a computer. Even an inexpensive calculator will sometimes have a key for calling up random numbers. Computer languages such as Pascal and BASIC include random number generators among the available **functions**.

The danger in using computer generated random numbers is that such numbers are not genuinely random. They are based on an **algorithm** that generates numbers in a very erratic sequence, but by computation, not chance.

For most purposes this does not matter. Slot machines, for example, succeed in making money for their owners in spite of any subtle bias or regularity they may show. There are times, however where computer-generated "random" numbers are really not random enough.

Mathematicians have devised many tests for randomness. One is to count the frequency with which the individual digits occur, then the frequency with which pairs, triples, and other combinations occur. If the list is long enough the "law of large numbers," says that each digit should occur with roughly the same frequency. So should each pair, each triple, each quadruple, and so on. Often, lists of numbers expected to be random fail such tests.

One interesting list of numbers tested for randomness is the digits in the decimal approximation for **pi**, which has been computed to more than two and a quarter billion places. The digits are not random in the sense that they occur by chance, but they are in the sense that they pass the tests of randomness. In fact, the decimal approximation for pi has been described as the "most nearly perfect random sequence of digits ever discovered." A failure to appreciate the true meaning of "random" can have significant consequences. This is particularly true for people who gamble. The gambler who plays hunches, who believes that past outcomes can influence forthcoming ones, who thinks that inanimate machines can distinguish a "lucky" person from an "unlucky" one is in danger of being quickly parted from his money. Gambling casinos win billions of dollars every year from people who have faith that the next number in a random sequence can somehow be predicted. If the sequence is truly random, it cannot.

RANGE

Range is a term used in different branches of mathematics including **algebra**, **set theory** and **probability theory**. It is usually first encountered in algebra in the study of **functions**, and is typically defined as the set of all values attained by a given function throughout its **domain**. One can think of a function, denoted by the letter f, as a rule that takes a given value and yields another, unique value. Within algebra this rule takes the form of an algebraic expression. For instance, the expression "x^2" transforms the number values of "x" into another set of number values. This rule can be denoted by the equation $y = x^2$. In general, a function f can be thought of as consisting of all the values that x possesses, as well as all the values that y possesses.

A function has a domain and a range. The domain is the set of all values of "x", and the range is the set of all values of "y". Because the values of y are calculated from the values of x, x is called the **independent variable** and y the **dependent variable**. The following example uses the equation $y = x^2$ in order to illustrate the relationship between a function and its domain and range: for x = -2, then $x^2 = 4$; for x = 0, then $x^2 = 0$; and for x = 2, then $x^2 = 4$. In this illustration, the domain of x consists of the set {-2, 0, 2}, while the range of y is the set {0, 4}. Obviously the domain could have many more elements than just three; indeed, in some cases the domain and range of a function form infinite **sets**. In this example, if the domain of f is allowed to be the set of **integers**, then by **definition** the domain is an infinite set, and the range of f (calculated by $f(x)$

$= y = x^2$) is also an infinite set, the values of which are all positive (except for **zero**).

The description of function, range, and domain given above is essentially the same as the one provided by German mathematician **Peter Gustav Lejeune Dirichlet** (1805-1859) in the mid-nineteenth century. However, a more general meaning for range was developed with the advent of set theory in the latter part of the nineteenth century. In order to motivate this 'set-theoretic' concept of range, a closer look at the concepts of 'relation' and 'function' as found in set theory is presented.

Consider the set A consisting of the first three lowercase letters of the alphabet; i.e., A = {a, b, c}. Generally speaking, the order in which the elements of a set are written is inconsequential, so that A = {a, b, c} = {b, c, a} = {c, a, b}, etc. There are, however, sets whose elements are arranged with respect to one another; such sets are called "ordered". Ordered sets having two elements are called "ordered pairs". An ordered pair is denoted by parentheses, as in (a, b). Here "a" is the first element of the ordered pair and "b" the second element. A relation is a type of set whose elements are all ordered pairs. A relation R on a set S is a set of ordered pairs of S. For instance, for set A = {a, b, c}, the set {(a, b), (a, c)} is a set of ordered pairs of A and therefore constitutes a relation on A. The set {(c, b)} is a different relation on set A. Indeed, any collection of ordered pairs of set A constitutes a relation on A. Using these ideas and definitions from set theory, range is then defined as: the range of a relation R is the set of all the "second elements" of the ordered pairs in R. The domain of R is the set of all the "first elements" of the ordered pairs in R.

From this concept of relation a function is defined as a "right unique" relation. A right unique relation is one that assigns to each element of the domain one, and only one, element in the range. Referring to the function f defined by the example above ($f(x) = x^2$), f is the set of ordered pairs given by $f = \{(-2, 4), (0, 0), (2, 4)\}$. Note that each element of the function's domain corresponds to only one element of the range. For example, the domain element "-2" is paired only with "4", and with no other element of the range. The same is true for the domain elements "0" and "2". As an everyday example of a relation and its range, consider a set P of people, each of whom has a spouse within P. The relationship of marriage is then a relation on set P and consists of married couples. One could define this relation R as: R = {(x, y) | where x is a husband with wife y}. The range of this relation R is the set of all wives in P while the domain of R is the set of all husbands in P. For a set P of monogamous people, R is a function because each husband in the domain has only one wife. However, in some societies polygamy is practiced. In those cases R is not necessarily a function because to each husband x in the domain there could be multiple wives.

This treatment of range within the context of set theory has shown a broader, more general view of function, namely that a function is simply a particular type of relation (specifically, a right unique relation), meaning that a function is a set of ordered pairs whose range and domain consist of the sets of the "second" and "first" elements of these pairs, respectively.

Although the term range is most often used in mathematics with respect to relations and functions, it is also used in statistical distributions as the name given to **limits** between which the probability takes on non-zero values. Thus the range of a sample is the difference between the highest and lowest observations. Range is an elementary measure of **dispersion** and, in terms of the **mean** range in repeated sampling, provides a reasonable estimate of the population **standard deviation**.

See also Algebra; Dependent variable; Peter Gustav Lejeune Dirichlet; Dispersion; Domain; Functions; Independent variable; Limits; Probability theory; Set theory; Standard deviation; Statistics

Rational numbers

A rational number is one which can be expressed as the ratio of two **integers** such as 3/4 (the ratio of 3 to 4) or -5 : 10 (the ratio of -5 to 10). Among the infinitely many rational numbers are 1.345, 1 7/8, 0, -75, $\sqrt{25}$, $\sqrt{.125}$, and 1. These numbers are rational because they can be expressed as 1345:1000, 15:8, 0:1, -75:1, 5:1, 1:2, and 1:1 respectively. The numbers π, $\sqrt{2}$, i, and $\sqrt[3]{4}$ are not rational because none of them can be written as the ratio of two integers. Thus any integer, any common fraction, any mixed number, any finite decimal, or any repeating decimal is rational. A rational number that is the ratio of a to b is usually written as the fraction a/b.

Rational numbers are needed because there are many quantities or measures which natural numbers or integers alone will not adequately describe. Measurement of quantities, whether length, mass, or time, is the most common situation. Rational numbers are needed, for example, if a farmer produces and wants to sell part of a bushel of wheat or a workman needs part of a pound of copper.

The reason that rational numbers have this flexibility is that they are two-part numbers with one part available for designating the size of the increments and the other for counting them. When a rational number is written as a fraction, these two parts are clearly apparent, and are given the names "denominator" and "numerator "which specify these roles. In rational numbers such as 7 or 1.02, the second part is missing or obscure, but it is readily supplied or brought to light. As an integer, 7 needs no second part; as a rational number it does, and the second part is supplied by the obvious relationship 7 $\leftrightarrow$ 7/1. In the case of 1.02, it is the decimal point which designates the second part, in this case 100. Because the only information the decimal point has to offer is its position, the numbers it can designate are limited to **powers** of ten: 1, 10, 100, etc. For that reason, there are many rational numbers which **decimal fractions** cannot represent, 1/3 for example.

Rational numbers have two kinds of **arithmetic**, the arithmetic of decimals and the arithmetic of common **fractions**. The arithmetic of decimals is built with the arithmetic of integers and the rules for locating the decimal point. In multiplying 1.92 by.57, integral arithmetic yields 10944, and the decimal point rules convert it to 1.0944.

Common fraction arithmetic is considerably more complex and is governed by the familiar rules

ac/bc = a/b

a/b + c/d = (ad + bc)/bd

a/b - c/d = (ad - bc)/bd

(a/b)(c/d) = ac/bd

(a/b) ÷ (c/d) = (a/b)(d/c)

a/b =c/d if and only if ad = bc

If one looks closely at these rules, one sees that each rule converts rational-number arithmetic into integer arithmetic. None of the rules, however, ties the value of a rational number to the value of the integers that make it up. For this the rule (a/b)b = a, b ≠ 0 is needed. It says, for example, that two 1/2s make 1, or twenty 3/20s make 3.

The rule would also say that **zero** 5/0s make 5, if zero were not excluded as a denominator. It is to avoid such absurdities that zero denominators are ruled out.

Between any two rational numbers there is another rational number. For instance, between 1/3 and 1/2 is the number 5/12. Between 5/12 and 1/2 is the number 11/24, and so on. If one plots the rational numbers on a number line, there are no gaps; they appear to fill it up.

But they do not. In the fifth century B.C. followers of the Greek mathematician **Pythagoras** discovered that the diagonal of a **square** one unit on a side was irrational, that no segment, no matter how small, which measured the side would also measure the diagonal. So, no matter how many rational points are plotted on a number line, none of them will ever land on √2, or on any of the countless other **irrational numbers**.

Irrational numbers show up in a variety of formulas. The **circumference** of a **circle** is π times its **diameter**. The longer leg of a 30°-60°-90° **triangle** is √3 times its shorter leg. If one needs to compute the exact length of either of these, the task is hopeless. If one uses a number which is close to π or close to √3, one can obtain a length which is also close. Such a number would have to be rational, however, because it is with rational numbers only that we have computational procedures. For π one can use 22/7, 3.14, 3.14159, or an even closer approximation.

More than 4,000 years ago the Babylonians coped with the need for numbers that would measure fractional or continuously **variable** quantities. They did this by extending their system for representing natural numbers, which was already in place. Theirs was a base-60 system, and the extension they made was similar to the one we currently use with our decimal system. Numbers to the left of what would be a "sexagesimal point" had place value and represented successive units, 60s, 3600s, and so on. Numbers smaller than 1 were placed to the right of the imaginary sexagesimal point and represented 60ths, 3600ths, and so on. Their system had two deficiencies which make it hard for contemporary archaeologists to interpret what they wrote (and probably made it hard for the Babylonians themselves). They had no zero to act as a place holder and they had no symbol to act as a sexagesimal point. All this had to be figured out from the context in which the number was used. Nevertheless, they had an approximation

for √2 which was correct to four decimal places, and approximations for other irrational numbers as well. In fact, their system was so good that vestiges of it are to be seen today. We still break hours down sexagesimally, and the degree measure of angles as well.

The Egyptians, who lived in a later period, also found a way to represent fractional values. Theirs was not a place-value system, so the Babylonian method did not suggest itself. Instead they created unit fractions. They did not do it with a ratio, such as 1/4, however. Their symbolism was analogous to writing the unit fraction as 4^{-1} or 7^{-1}. For that reason, what we would write as 2/5 had to be written as a sum of unit fractions, typically 3^{-1} + 15^{-1}. Clearly their system was much more awkward that of the Babylonians.

The study of rational numbers really flowered under the Greeks. Pythagoras, **Eudoxus**, **Euclid**, and many others worked extensively with **ratios**. Their work was limited, however, by the fact that it was almost entirely geometric. Numbers were represented by line segments, ratios by pairs of segments. The Greek astronomer **Ptolemy**, who lived in the second century, found it better to turn to the sexagesimal system of the Babylonians (but not their clumsy cuneiform characters) in making his extensive astronomical calculations.

RATIOS

Put simply, a ratio is a comparison of two numbers. Ratios are generally used to relate the magnitude of two like parameters of an object (or two or more similar objects) relative to one another. For example, the parameters used in a ratio may describe the same object, such as the length and width of a cardboard box, or differing objects, such as the weight, in pounds, of two bags of fruit. Alternatively, a ratio may provide a means to convert **units of measurement** that describe the same object, such as 12 inches of string equaling 1 foot of string or 100 centimeters of pavement equaling 1 meter of pavement. The magnitude may represent size, **distance**, weight, amount, or any other quantifiable parameter of the object(s); the parameters being compared within the ratio are generally quantities (such as the number of red marbles compared to the number of blue marbles in a jar), measurements (such the length of a rowboat compared to the length of a cruise ship), or conversion factors (such as 1000 grams equaling one kilogram). In each case, the ratio provides a means of *comparison* for two or more values.

Ratios are most often written in one of three forms. If *a* and *b* are the values to be compared, the ratio can be written as a fraction (that is, *a* / *b*). In fact, all **fractions** can be considered ratios, and in fact, ratios extend the concept of fractions to a more general form. Alternatively, the ratio may be written in the shorthand notation of *a:b*, or the language construct of "*a* to *b*." Expressing ratios as fractions, however, is feasible only when two values are being compared; writing *a* / *b* / *c* makes no sense. When three or more values are compared, the shorthand notation *a:b:c* is normally used.

Like fractions, ratios are often reduced to the lowest form possible, based on the greatest common **factor**. If a piggybank contains 10 quarters, six dimes, and eight pennies, for example, a reasonable ratio is 10:6:8. Since two is a common factor to all three numbers, an equal ratio is 5:3:4—for every five quarters in the piggyback, there are three dimes and four pennies. Reducing the ratio, however, results in a loss of information—one still knows how many quarters are in the bank relative to the number of dimes, but one does not know exactly how many quarters are in the bank. The statement that "the ratio of adults to children attending the soccer match was 5:3," for example, means that five adults attended the match for every three children present; it does *not* mean that only five adults and three children attended the match. For ease of use, ratios are commonly reduced to some number x relative to 1, that is x:1. Intuitively, an event described by the ratio 1:100 seems more likely to occur than an event described by the ratio 1:10,000; these two ratios are easily compared. Although exactly similar to these two ratios, comparisons between the ratios 20:2,000 and 0.1:1,000 are not intuitive.

Two of the most useful ratios are scale and rate. As a two-dimensional representation of a particular area, a map will not be accurate unless it uses a standard scale to represent distances—such as one inch to 10 miles—in both latitude and longitude. Using this ratio, any two objects that are 10 miles apart on the earth will appear one inch apart on the map, ensuring easy comprehension by the reader and application to his environment. Similarly, construction blueprints must have an accurate scale applied throughout in order to be of practical use to a builder. Using scale also applies to three-dimensional **modeling**. Aeronautical engineers, for example, often build scale models of design aircraft—that is, models made in the exact shape of the proposed aircraft that have been equally reduced in height, width, and length—for use in wind tunnels.

Rate is another special ratio; it relates distance to time. It is an especially useful tool that enables cone to compare speeds objects as well as estimate how much time will be required to travel a particular distance at a particular speed. If a truck travels 225 miles in 5 hours, for example, the rate is 225 miles per 5 hours, or more commonly, an average speed of 45 miles per hour. Rates can be either instantaneous—that is, the speed of an object at an event moment in time—or it can be a mean, as in the truck example above. It is likely that the truck did not travel at exactly 45 miles per hour during the entire five hours, perhaps stopping for fuel or slowing during periods of heavy traffic.

See also Fractions; Golden mean; Proportions; Units of measurement

REAL NUMBERS

The set of real numbers is one of the one of the most fundamental objects in mathematics. In order to describe the real numbers we will introduce notation for some related objects. Let Z denote the set of **integers**. That is,

$$Z = \{\ldots,-3,-2,-1,0,1,2,3,\ldots\}.$$

Then let Q denote the set of **rational numbers**. Thus Q consists of all **fractions**

$$\frac{p}{q}$$

where p and q are integers and $q \neq 0$. There is also an **equivalence** relation in the **field** Q which we use automatically when doing **arithmetic**. Two fractions

$$\frac{p}{q}$$

and

$$\frac{r}{s}$$

are equal in Q, that is they determine the same rational number, whenever the identity $ps = rq$ holds. As usual we regard Z as a subset of Q by recognizing that each integer n can be written as $n/1$, and so the integer n may be viewed as a rational number. We assume that the reader is familiar with the binary operations of **addition**, denoted by +, and **multiplication**, denoted by ·, in Z and Q. In the language of abstract **algebra**, (Z, +, ·) is a commutative ring with an **identity element** and without **zero** divisors. Such a ring is often called an *integral domain*. Then (Q, +, ·) is the *field of fractions* associated to Z, (see the article on fields). The set $\Re$ of real numbers also forms a field and it contains Q as a proper subfield. Thus the binary operations + and · in Q extend to binary operations in $\Re$, and then $(\Re, +, ·)$ forms a field. As is usual, we will refer more simply to the fields Q and $\Re$ without indicating the well-known field operations + and ·. And if x and y are real numbers we often write xy for their product rather than $x \cdot y$. In most respects $\Re$ is much larger than Q. For example, Q forms a countably infinite set but $\Re$ is an uncountable set.

Another important aspect of the field $\Re$ is that it has a distinguished subset P, called the *positive* real numbers, that induces an order relation in $\Re$. We can define the subset P in $\Re$ to be the set of nonzero squares. That is, x belongs to P if $x \neq 0$ and if there exists y in $\Re$ such that $y^2 = x$. Then the subset P satisfies the following conditions:

- (1) if x belongs to P then $-x$ is not an element of P,
- (2) if x belongs to $\Re$ then either $x = 0$ or x belongs to P or $-x$ belongs to P,
- (3) if x and y belong to P then $x + y$ belongs to P,
- (4) if x and y belong to P then xy belongs to P.

In general, a field containing a subset P that satisfies these axioms is called an *ordered field*. For example, the set of positive rational numbers verifies these conditions and so Q is an ordered field. The set P in an ordered field induces a binary relation $<$ into the ordered field. We say that $x < y$ holds exactly when $y-x$ belongs to P. And then we write $x \leq y$ if either $x=y$ or if $x < y$. The familiar properties of the relation $<$ can then be deduced from the four axioms for P. Alternatively, we could state axioms for the order relation $<$ and then define P to be the set of points x such that $0 < x$.

There is one further **axiom** that is required in order to distinguish $\mathfrak{R}$ from all other ordered fields. Let S be a nonempty subset of $\mathfrak{R}$. We say that S is *bounded from above* if there exists an element t in $\mathfrak{R}$ such that $s \leq t$ holds for all elements s in S. Similarly, we say that S is *bounded from below* if there exists an element r in $\mathfrak{R}$ such that $r \leq s$ holds for all elements s in S. We say that t is a *least upper bound* for S if t is an upper bound and every upper bound u of S satisfies $t \leq u$. The concept of a *greatest lower bound* is defined in a similar manner. Thus r is a greatest lower bound for S if r is a lower bound and every lower bound q of S satisfies $q \leq r$. We are now in position to state the third axiom that is satisfied by the real numbers: if S is a nonempty subset of $\mathfrak{R}$ and S is bounded from above then S has a least upper bound in $\mathfrak{R}$. An equivalent form of this axiom would be: if S is a nonempty subset of $\mathfrak{R}$ and S is bounded from below then S has a greatest lower bound in $\mathfrak{R}$. An ordered field that satisfies either (and therefore both) of these axioms is called a *complete* ordered field. It can be shown that there is essentially only one complete ordered field called the field of real numbers. If $(F, +, \cdot)$ is a complete ordered field then there exists a bijection $\boldsymbol{\varphi}\colon F \to \mathfrak{R}$ that is an **isomorphism** of fields and that preserves the order relation in F and $\mathfrak{R}$.

There are several more informal observations that we can make about the real numbers that clarify the way mathematicians think about them. Each real number can be associated with a unique point on a line. This allows us to use a geometrical language when working with real numbers. Thus we may speak of the **distance** between two real numbers. Alternatively, by using the set P of positive real numbers, we can define the absolute value function $\|\colon \mathfrak{R} \to \{0\} \cup P$. Then the function $(x,y) \to |x - y|$ defines a metric in $\mathfrak{R}$ and so a metric **topology**. In particular, we find that $|x - y|$ is the distance between x and y. Another important property of real numbers is that each real number has a decimal expansion (see the article on **decimal fractions**), and also a continued fraction expansion. These expansions can provide useful information about the number. For example, a real number is rational if and only if its decimal expansion is eventually periodic.

RECORDE, ROBERT (1510-1558)

Welsh-English mathematician, physician, and author

Recorde is considered the founder of the English school of mathematical writers, and was the first English writer on **arithmetic**, geometry and astronomy. Recorde was the first mathematician to introduce **algebra** in England. Most of his works were prevalent until the end of the 16th century, and some until the end of the 17th century. Many of his writings were in poetic form as an aid to students in remembering the rules of operation.

Recorde was the second son of Thomas Recorde, a second–generation Welshman, and Rose Johns of Montgomeryshire. He received a B.A. from Oxford University in 1531 and in the same year became a member of All Souls College—a chantry and graduate foundation for theology, civil and canon law, and medicine. He later moved to Cambridge, where he received his M.D. degree in 1545, and became physician to Edward VI and Queen Mary. Records indicate that Recorde had been licensed in medicine twelve years previously at Oxford. His university career and any other degrees are lost to history.

Three physicians during this period contributed greatly to mathematics: Nicolas Chuquet, **Girolamo Cardano** and Recorde. Recorde was the most influential within his own country, as he established the English mathematical school. He wrote in the vernacular, which may have limited his effect on the Continent. Recorde's first existent mathematical work was the *Grounde of Artes* (1541), which was made popular for its use of the abacusand algorithmfor commercial use. Recorde's first edition dealt with only **whole numbers** with fundamental operations, progression, and counter reckoning. In 1552, the text was expanded to include **fractions** and alligation. He wrote *The Pathway to Knowledge* (1551), a translation of **Euclid**'s first four volumes of *Elements*. The text separated construction from theorems, on the premise students need to understand at the onset, both what is being taught and why. His *The Gate of Knowledge* was a text on measurement and use of the quadrant, but has been lost, and perhaps never published. *Castle of Knowledge* (1556) dealt with construction and use of the **sphere**. The text was a study of standard authorities, and offered corrections of errors he felt were caused by others' lack of knowledge of Greek.

Recorde is best known for his *The Whetstone of Witte* (1557), which contained the "second part of arithmetic" promised in *The Ground of Artes*, and contained elementary algebra through **quadratic equations**. The text contains his introduction of the sign "=" for equals, which he selected because nothing could be more equal than two parallel straight lines. In this text he also explained how the **square root** in algebra could be extracted.

Aside from his mathematical textbooks, he wrote *The Urinal of Physick* (1547), which was a short but systematic work which included figures and descriptions. The text was a complete assessment of urines, and contained nursing practices intended for the medical profession.

Recorde was a skillful teacher, as well as an outstanding scholar of mid–16th century English. He was as familiar with Greek and medieval texts as he was with contemporary developments. Recorde emphasized learning Greek in order to understand other sources, and developed English equivalents for Latin and Greek terms to simplify his teachings. Recorde was not internationally known, mainly because his works were in English and on an elementary level, but his texts were the standard in Elizabethan England. While the majority of his time was spent in the mathematical sciences, Recorde was also highly skilled in rhetoric, philosophy, literature, history, cosmography (the science that studies the whole order of nature), and all areas of natural history.

Earlier in his career, Recorde served his government in other capacities. In 1549, he was appointed comptroller of the Bristol Mint. Trouble ensued when he refused to divert monies intended for King Edward to the armies under Lord John

Russell and Sir William Herbert (later earl of Pembroke). Herbert accused him of treason and he was confined for 60 days, and the mints were forced to stop production. The incident lead to later career problems.

From 1551 to 1553, Recorde was surveyor of the mines in Ireland, in charge of the Wexford silver mines, and technical supervisor of the Dublin Mint. The mines were unprofitable and Recorde was recalled in 1553. In 1556, he tried to regain reinstatement at Court, but earlier he had charged malfeasance as commissioner of mints against Pembroke. Unfortunately, for him, Queen Mary and King Philip, sided with Pembroke, and Pembroke sued for libel with damages against Recorde. Recorde apparently was unable to pay the sum and was imprisoned at King's Bench prison, where he penned his will, leaving a little money to his relatives, and died in prison. Not until 1570 was his estate compensated for monies due him when he was recalled as surveyor of mines in 1553.

Recorde had five daughters and four sons. He was also an active supporter of the Protestant Reformation.

RECURSIVE FUNCTION

A recursive function is a function whose **domain** is generated by previous values of the function. A recursive function is typically defined by giving some initial value and then stating a recursion rule which defines how a given function value depends on a previous value, or values, of the function. For example, here is a recursive function: $f(0) = 2$ and $f(n) = f(n - 1) + 3$ where n is a natural number (positive integer). This function starts with an initial value of 2, after which each new function value is generated by adding 3 to the function value preceding it. So this function will generate the set $\{2,5,8,11,...\}$. Another example is: $f(0) = 1$ and $f(n) = nf(n - 1)$, where n is a natural number. This recursive function generates the set $\{1,1,2,6,24,120,...\}$ and is called the **factorial** function. It is usually defined using the factorial symbol (!) as follows: $0! = 1$ and $n! = n(n - 1)!$, where n is a natural number. This is just one of many important mathematical **functions** which are defined recursively.

Recursive functions are often used in **modeling** real world processes such as the accumulation of money in a bank account, creating tables for paying off a loan over time, population growth, radioactive decay, and many others. Consider a $1000 certificate of deposit that is to earn 6% interest compounded annually. This means that $1000 is deposited initially and at the end of one year 6% of $1000 or $60 interest is added to the original $1000. At the end of the second year 6% of $1060 or $63.60 is added to the $1060 for a total of $1123.60, and this process repeats itself over and over for the life of the certificate. Processes that repeat themselves using previous values of the process are good candidates for modeling with recursive functions. In this case, let us define $A(n)$ to be the amount in the account after n years and $A(0)$ will be the initial amount deposited. Then a recursive function for this process may be defined as: $A(0) = 1000$ and $A(n) = A(n - 1) + .06A(n - 1)$ for $n = 1, 2, 3,...$. Note that an equivalent and slightly simpler form for $A(n)$ is $1.06A(n - 1)$, but the version originally given more clearly demonstrates the process which is being repeated again and again. Let us say that the term of this certificate is 5 years. Then with the recursive function defined we can calculate how much money is in the account at the end of each year: $A(1) = \$1060$, $A(2) = \$1123.60$, $A(3) = \$1191.02$, $A(4) = \$1262.48$, and $A(5) = \$1338.23$. Most modern graphing calculators have a "sequence" **mode** that allows the user to input an initial value and a recursion formula to generate tables of values for recursive functions. Most computer languages either have built-in recursion capability or allow the programmer to create "loops" in which a function calls itself repeatedly using previous values of the function. Real world systems that change in discrete time units are called discrete dynamical systems. Such systems are frequently modeled by recursive functions, which, in this context, are called finite difference **equations** to emphasize that they are the discrete counterparts of the continuous **differential equations** of the **calculus**. One can think of a differential equation as resulting from taking the limit of a difference equation as the time intervals between each step approach 0.

So far we have looked only at recursion functions in which each new value of the function depends upon only one previous value, namely, the one immediately preceding it. Such functions are the simplest type and are sometimes called "first order" recursive functions; but it is also possible to define higher order recursive functions. In fact, a very famous sequence can be generated by a second order recursive function as follows: $f(0) = 1$, $f(1) = 1$, and $f(n) = f(n - 1) + f(n - 2)$, where n is a natural number. The famous sequence so generated is $\{1,1,2,3,5,8,13,21,34,...\}$, the so-called **Fibonacci sequence** named for Leonardo of Pisa (c.1170-1240) who had the nickname **Fibonacci**. Notice that for a second order recursive function we need two "initial" values because the function has to "call" two previous values of itself on each **iteration**. This is true in general: for an nth order recursive function, we need n "initial" values.

Recursion is not merely an effective technique for modeling certain natural processes; it is also indispensable in the logical development of the natural number system. The Italian mathematician **Giuseppe Peano** (1858-1932) showed how to generate all natural numbers from a set of simple axioms and recursion. He started with axioms asserting that 0 is a number; that a recursive function s, called "the successor of" exists; and that the successor of any number is a number. Let s(k)=the successor of the number k. Then starting with 0, we recursively obtain s(0), the successor of 0; s(s(0)), the successor of the successor of 0; s(s(s(0))), the successor of the successor of the successor of 0; and so on. Now all of these successors are numbers because the successor of a number is a number. Furthermore, we can give these numbers names: s(0) we will call 1; s(s(0)) we will call 2; s(s(s(0))) we will call 3; and so on. Thus we may generate the natural numbers with this recursive process. Here we could define the recursive function as $s(0) = 1$ and $s(n) = s(s(n - 1))$.

See also Peano axioms

REGIOMONTANUS (1436-1476)

German trigonometrist and astronomer

Regiomontanus was the pen name of Johann Müller, responsible for what some call the "rebirth of trigonometry" during the century following his death. Although he was a practicing astrologer, he also was one of the first Europeans to closely observe certain celestial objects and events previously tied to superstitions. Regiomontanus published his own and others' translations of classical and Arabic texts with his then novel printing press. He predicted that sailors would one day be able to navigate according to the position of the moon, a process that eventually did arise during the Age of Exploration. (Some speculate that Regiomontanus' early death prevented him from anticipating Copernicus.) Once dismissed as a comparatively undisciplined technician, Regiomontanus is now recognized as the West's answer to Nasir al–Din or Nasir Eddin, having brought plane and spherical **trigonometry** within the purview of pure mathematics and out of its apprenticeship to astronomy.

Regiomontanus was born in an area once known variously as Franconia or East Prussia, now part of Russia, on June 6, 1436. His father was a miller, but nothing else is known of the family. Regiomontanus studied dialectics for two or three years at the University of Leipzig before entering the University of Vienna at age 14 under the name Johannes Molitoris de Königsberg, but quickly came to be known by a Latin variant of his hometown. Königsberg meant King's Mountain, so the translation *regio monte* yielded Regiomontanus. He was awarded a bachelor's degree in one year. After two years at Vienna Regiomontanus elected to study the mathematics and protocol of astronomy, then still indistinguishable from astrology.

Müller's teacher, Georg von Peurbach, was a reformist who decried existing Latin translations of Greek texts from which he was forced to teach. Peurbach was particularly unsatisfied with a reference called *The Alfonsine Tables*. Regiomontanus became part of the faculty during 1457 and joined forces with his former teacher to improve these source materials. A visiting Church official, Cardinal Bessarion, came to their aid in 1460 by contracting Peurbach to produce a new Latin translation of **Claudius Ptolemy**'s *Amalgest* from the Greek instead of from its intermediate Arabic translation.

By the spring of 1461 Peurbach died at the age of 38, before the work was completed, and his last wish was that his student carry on his work. Bessarion became Regiomontanus' benefactor, teaching him Greek and taking him to Rome in search of ancient manuscripts. The cardinal's deeper motive was to forge common ground for the Greek and Latin factions of the Catholic Church. At various intellectual centers of Italy, they consulted with leading astronomers, particularly members of the Averroists. Regiomontanus was appointed professor of astronomy to take Peurbach's place, and finished the *Epitome* as promised. He even added commentary and revised calculations in some places, restoring omitted mathematical passages in others. Ironically, it would not see publication for decades after Regiomontanus' own death.

Regiomontanus' 1464 effort, *De triangulis omnimodis libri quinque*, presented what was likely the first trigonometrical formula for the area of a **triangle**, spearheading the application of algebraic and algorithmic solutions to geometric problems. This publication sported the still innovative notations for **sine** and **cosine**. Simultaneously, Regiomontanus was drawing up charts for **trigonometric functions**, particularly sine tables for use in the days prior to the invention of **logarithms**. His longitudinal studies bore the *Tabulae Directionum*, or "tables of directions" for fixed celestial objects in relation to the apparent movement of the heavens. This work introduced the tangent function, with which Regiomontanus would later compute a table of tangents. After moving back to Nuremberg to assume the post of Royal Astronomer to King Matthias Corvinus of Hungary, he found yet another sponsor. Bernard Walther was an amateur astronomer who funded an observatory and workshop, and commissioned various scientific tools of Regiomontanus' design for their use. Upon securing a printing press, Regiomontanus gave his old teacher's work precedence, making Peurbach's *Theoriae planetarum novae* one of the earliest mass–produced science texts. In 1474 Regiomontanus printed his own *Ephemerides*, an almanac of daily planetary positions and eclipses for the years 1475 to 1506. This was the reference Christopher Columbus used as leverage against the Jamaican aborigines, who found his prediction of a lunar eclipse awe–inspiring. More notably, one of the objects Regiomontanus had recorded during his research was the comet of 1472, renamed Halley's Comet upon its third return 210 years later.

As recorded in the *Nuremberg Chronicle*, Regiomontanus was summoned to Rome with hopes of reforming the Julian **calendar**. This was an ongoing problem many scientists, including Regiomontanus, had already pondered. Pope Sixtus IV awarded him the bishopric of Ratisbon, but he fell victim to a plague in the wake of the Tiber river flood of 1476. Regiomontanus' papers were inherited by Walther, who unfortunately did not disseminate them. Upon his death they were acquired by the Nuremberg Senate, where they languished for nearly 30 more years. Regiomontanus' projects in mathematics and science were extended by intellectual successors like Germany's Johann Werner and Belgium's Philip van Lansberge.

Regiomontanus was also a university professor, lecturing on Virgil and Cicero. Major classical works translated by Regiomontanus included the *Conics* of **Apollonius** and some writings of **Archimedes**. He also salvaged the only known extant portion of an uncopied manuscript by **Diophantus**. Regiomontanus' mathematical writing is generally classified as rhetorical, even careless, because of his habit of referring to proofs without actually presenting them. Of one in *De triangulis* he signs off, "...I leave it to you for homework."

Regiomontanus rejected the possibility that the Earth rotated around the sun; however, being an early Renaissance man, not all of his beliefs were so fixed. It is suggested that Novara, a teacher of **Copernicus**, came into possession of a letter from Regiomontanus that implied a primitive Copernicanism, but this has not been proven. Regardless, Copernicus was inspired by Regiomontanus' critical notes to

Ptolemy's *Amalgest* to revise the millennium–old view of the planet Earth and the solar systemit shares.

RELATIVITY, GENERAL

Einstein's theory of relativity consists of two major portions: The special theory of relativity (see the next entry, **Relativity, special**) and the general theory of relativity. Special relativity deals with phenomena that become noticeable when traveling near the speed of **light**, and with reference frames that are moving at a constant velocity (inertial reference frames). General relativity deals with reference frames that are accelerating (noninertial reference frames), and with phenomena that occur in strong gravitational fields. General relativity also uses the curvature of **space** to explain gravity.

History

In the seventeenth century, **Isaac Newton** (1642-1727) completed a grand synthesis of physics that used three laws of **motion** and the law of gravity to explain motions we observe both on the Earth and in the heavens. These laws worked very well, but by the end of the nineteenth century, physicists began to notice experiments that did not work quite the way they should according to Newton's understanding. These anomalies led to the development of both relativity and **quantum mechanics** in the early part of the twentieth century.

One such experiment was the Michelson-Morley experiment, which disproved the hypothesis that propagation of light waves requires a special medium (which had been known as *ether*). Einstein took the result of this experiment as the basic assumption (namely, that the speed of light is constant) that led to the special theory of relativity.

The **orbit** of the planet Mercury around the sun has some peculiarities that cannot by explained by Newton's classical laws of physics. As Mercury orbits the Sun, the position where Mercury is closest to the Sun is called the perihelion. The perihelion migrates a small amount each orbit. This very small but measurable effect was reported by the astronomers Urbain Leverrier (1811-1877) and Simon Newcomb (1835-1909) in 1859 and 1895, respectively. The migration rate is 43 seconds of arc per century (one second of arc is 1/3,600 of a degree), so that it takes nearly 8,400 years for this migration to add up to one degree. This precession of Mercury's perihelion can not be easily explained by **Newton's laws**, but it is a natural consequence of Einstein's general theory of relativity. The effect is noticeable for Mercury but not the other planets, because Mercury is closest to the Sun, where the gravitational **field** is strongest. General relativity differs most from Newton's law of gravity in strong gravitational fields. The rate of this migration is what general relativity predicts and provides an important experimental confirmation to general relativity.

Preliminary concepts

To understand many concepts in relativity one needs to first understand the concept of a reference frame. A reference frame is a system for locating an object's (or event's) position in both space and time. It consists of both a set of coordinate axes and a clock. An object's position and motion will vary in different reference frames. If for example you are riding in a car, you are at rest in the reference frame of the car. You are, however, moving in the reference frame of the road, which is fixed to the reference frame of the Earth. The reference frames are moving relative to each other, but there is no absolute reference frame. Either reference frame is as valid as the other. A reference frame that is moving at a constant velocity is an inertial reference frame. A noninertial reference frame is accelerating or rotating. The theory of general relativity expands on the theory of special relativity by including the case of noninertial reference frames.

Special relativity combined our concepts of space and time into the unified concept of spacetime. In essence time is a fourth **dimension** and must be included with the three space dimensions when we talk about the location of an object or event. General relativity allows for the possibility that **spacetime** is curved. Gravity is a manifestation of the **geometry** of curved spacetime.

General relativity

PRINCIPLE OF EQUIVALENCE Einstein's General Theory of Relativity, published in 1916, uses the principle of **equivalence** to explain the **force** of gravity. There are two logically equivalent statements of this principle. For the first statement, consider an enclosed room on the Earth. One feels a downward gravitational force. This force causes what we feel as weight, and causes falling objects to accelerate downward at a rate of 32 ft/s (9.8m/s). Now imagine the same enclosed room but in space far from any masses. There will be no gravitational forces, but if the room is accelerating at $9.8 m/s^2$, then one will feel an apparent force. This apparent force will cause objects to fall at a rate of $9.8 m/s^2$ and will cause one to feel normal Earth weight. We feel a similar phenomenon when we are pushed back into the seat of a rapidly accelerating car. This type of apparent force is an inertial force and is a result of an accelerating (noninertial) reference frame. The inertial force is in the opposite direction of the acceleration producing it. Is it possible to distinguish between the above two situations from within the room? No. According to the first statement of the Principle of Equivalence it is not possible without looking outside the room. Gravitational forces are indistinguishable from inertial forces caused by an accelerating reference frame.

What if the room in space is not accelerating? There will be no gravitational forces, so objects in the room will not fall, and the occupants will be weightless. The same room is now magically transported back to Earth, but by a slight error it ends up 100 feet above the ground rather than on the surface. The Earth's gravity will accelerate the room downward at $9.8 m/s^2$. Just as when the room is accelerating in space, this acceleration will produce an inertial force that is indistinguishable from the gravitational force. But in this case the inertial force is upward, and the gravitational force is downward. Because there is no way to distinguish between inertial and gravitational forces, and they are in the opposite direction, they cancel out exactly. Hence the occupants of the room are weightless. In general, objects that are in free fall will be weightless. This prediction allows us to experimentally test the Principle of Equivalence. Simply let an object fall freely and see if it is weightless. Astronauts in the space shuttle are weightless, not because there is no gravity, but because they are in free fall. You can show yourself that freely falling objects will be weightless. Put a small hole in the bottom of an empty plastic milk jug and fill the jug with water. Drop the jug. While it is falling, no water will leak out the bottom, because as a consequence of the principle of equivalence, freely falling objects will be weightless.

The second statement of the principle of equivalence involves the concept of mass. Mass appears in two distinct ways in Newton's Laws. In Newton's second law, the amount of force required to accelerate an object increases as its mass increases. It takes more force to accelerate a refrigerator than the can of soda that is in the refrigerator. The mass in Newton's second law is the inertial mass. In Newton's law of gravity, the gravitational force between two objects increases as the mass of the objects increases. That is why you will weigh more on a massive planet, such as Jupiter, than on the Earth. The mass in the law of gravity is the gravitational mass. Newton did not seriously consider the possibility that these two masses might be different. Einstein did. Are the inertial mass and the gravitational mass identically the same thing? Yes. According to the second statement of the principle of equivalence, the inertial mass and the gravitational mass are equal.

These two statements of the principle of equivalence are logically equivalent. That means that it is possible to use either statement to prove the other. This principle is the basic assumption behind the general theory of relativity.

Geometrical nature of gravity

From this principle, Einstein was able to derive his general theory of relativity, which explains the force of gravity as a result of the geometry of spacetime. To see how Einstein did this, consider the example above of the enclosed room being accelerated in space far from any masses. The person in the room throws a ball perpendicular to the direction of acceleration. Because the ball is not being pushed by whatever is accelerating the room, it follows a curved path as seen by the person in the room. You would see the same curved path if you threw a ball sideways in a moving car, but be careful not to hit the driver. Now replace the ball by a light beam shining sideways in the enclosed room. The person in the room sees the light beam follow a curved path, just as the ball does and for the same reason. Be careful, though—the deflection of the light beam is very much smaller than the ball, because the light is moving so fast it gets to the wall of the room before the room can move very far.

Now consider the same enclosed room at rest on the surface of the Earth. The ball thrown sideways will follow a downward curved path because of the Earth's gravitational field. What will the light beam do? The principle of equivalence states that it is not possible to distinguish between grav-

itational forces and inertial forces. Hence, any experiment will have the same result in the room at rest on the Earth as in the room accelerated in space. The light beam will therefore be deflected downward in the room on the Earth just as it would in the accelerated room in space. So, in the room on Earth, gravity deflects the light beam.

Light is deflected by a gravitational force! How? Light has no mass. According to Newton's law of gravity only objects having mass are affected by a gravitational force. What if spacetime is curved? Then we would see light and other objects follow an apparently curved path. Einstein therefore concluded that the presence of a mass curves spacetime and that gravity is a manifestation of this curvature.

Prior to Einstein, people thought of spacetime as being flat and having a Euclidean geometry. This geometry is the geometry that applies to flat surfaces and that is studied in most high school geometry classes. In general relativity however, spacetime is not always Euclidean. The presence of a mass curves or warps spacetime near the mass. The warping is similar to the curvature in a sheet of rubber that is stretched out with a weight in the center. The curvature of spacetime is harder to visualize, because it is four-dimensional spacetime rather than a two-dimensional surface. This curvature of spacetime produces the effects we see as gravity. When we travel long distances on the surface of the Earth, we must follow a curved path because the Earth is not flat. Similarly an object traveling in curved spacetime near a mass follows what we see as a curved path. For example, the Earth orbits the Sun because the spacetime near the Sun is curved. The Earth travels in a nearly circular path around the Sun as a small marble would in a circular path around a curved funnel. An object falling near the surface of the Earth is then like the marble rolling straight down the funnel.

Experimental verification

BENDING OF LIGHT The first experimental confirmation of general relativity occurred in 1919, shortly after the theory was published. Because light has no mass, Newton's law of gravity predicts that a strong gravitational field will not bend light rays. However as discussed above, general relativity predicts that a strong gravitational field will bend light rays. The curved spacetime will cause even massless light to travel in a curved path. The most convenient mass large enough to have a noticeable effect is the Sun. The apparent position of a star almost directly behind the Sun should be shifted a very small amount as the light rays are bent by the Sun. But, we normally can not see these stars. It is daytime. We must wait until a total solar eclipse to be able to see the stars that are almost directly behind the Sun. Shortly after Einstein published his general theory Arthur Eddington (1882-1944) mounted an expedition to observe the total eclipse of May 29, 1919. Einstein was right. The apparent positions of the stars shifted a small amount. Subsequent eclipse expeditions have further confirmed Einstein's prediction.

More recently, we see this effect with gravitational lenses. If a very distant quasar is almost directly behind a not-as-distant galaxy, the mass of the galaxy can bend the light coming from the more distant quasar. When this occurs we see a double image of the quasar with one image on each side of the nearer galaxy. A number of these gravitational lenses have been observed.

Binary pulsar

The 1993 Nobel Prize in physics was awarded to Joseph Taylor and Russell Hulse for their 1974 discovery of a binary pulsar. A pulsar, or rapidly rotating neutron star, is the final corpse for some stars that occurs when it collapses to about the size of a small city. A binary pulsar is simply two pulsars orbiting each other. Because pulsars are so collapsed they have strong enough gravitational fields that general relativity must apply. Binary pulsars can therefore provide an excellent experimental test of general relativity. About 40 binary pulsars have been discovered since Hulse and Taylor's original discovery.

Mercury's orbit shows a migration in its perihelion as it orbits the Sun. The binary pulsar displays a similar effect as the pulsars orbit each other. The effect is much greater as expected from general relativity because the pulsars have a much stronger gravitational field.

General relativity also predicts that gravity waves should exist in a way that is analogous to electromagnetic waves. Light, radio waves, x rays, infrared light, and ultraviolet light are all examples of electromagnetic waves that are oscillations in electric and magnetic fields. These oscillations can be caused by an oscillating electron. As the electron oscillates, the electron's electric field oscillates causing electromagnetic waves. Similarly as the pulsars oscillate by orbiting each other, general relativity predicts that they should cause the gravitational field to oscillate and produce gravity waves. Gravity waves have so far not been detected directly, even though several groups have been trying for over 20 years. But the binary pulsar is slowing down at a rate that suggests it is losing energy by emitting gravity waves. So, it could be said that the gravity waves predicted by general relativity have been detected indirectly.

Consequences of General Relativity

Karl Schwarzschild first used general relativity to predict the existence of black holes, which are stars that are so highly collapsed that not even light can escape. Because the gravitational field around a black hole is so strong, we must use general relativity to understand the properties of black holes. Most of what we know about black holes comes from theoretical studies based on general relativity. Ordinarily we think of black holes as having been formed from the collapse of a massive star, but Stephen Hawking has combined general relativity with **quantum mechanics** to predict the existence of primordial quantum black holes. These primordial black holes were formed by the extreme turbulence of the big bang during the formation of the universe. Hawking predicts that over sufficiently long times these quantum black holes can evaporate.

General relativity also has important implications for cosmology, the study of the origin of the universe. The **equations** of general relativity predict that the universe is expand-

ing. Einstein noticed this result of his theory, but did not believe it. He therefore added a "cosmological constant" to his equations. This cosmological constant was basically a fudge factor that Einstein was able adjust so that his equations predicted that the universe was not expanding. Later Edwin Hubble (1889-1953), after whom the Hubble Space Telescope was named, discovered that the universe is expanding. Einstein visited Hubble, examined Hubble's data, and admitted that Hubble was right. Einstein later called his cosmological constant the biggest blunder of his life. Modern cosmology uses general relativity as the theoretical foundation to understand the expansion of the universe and the properties of the universe during its early history.

Albert Einstein's general theory of relativity fundamentally changed the way we understand gravity and the universe in general. So far, it has passed all experimental tests. This, however, does not mean that Newton's law of gravity is wrong. Newton's law is an approximation of general relativity. In the approximation of small gravitational fields, general relativity reduces to Newton's law of gravity. If at some time in the future someone does an experiment that does not agree with the theory of relativity, we will have to modify the theory just as relativity modified Newton's classical physics.

RELATIVITY, SPECIAL

Einstein's theory of relativity consists of two major portions: The special theory of relativity and the general theory of relativity (see the previous entry, **Relativity, general**). Special relativity deals with phenomena that become noticeable when traveling near the speed of **light** and reference frames that are moving at a constant velocity, inertial reference frames. General relativity deals with reference frames that are accelerating, noninertial reference frames, and with phenomena that occur in strong gravitational fields. General relativity also uses the curvature of **space** to explain gravity.

History

In the 17th century, **Isaac Newton** completed a grand synthesis of physics that used three laws of **motion** and the law of gravity to explain motions we observe both on the Earth and in the heavens. These laws worked very well, but by the end of the 19th century, physicists began to notice experiments that did not work quite the way they should according to Newton's classical understanding. These anomalies led to the development of both relativity and **quantum mechanics** in the early part of the twentieth century.

One such experiment was the Michelson-Morley experiment. To understand this experiment, imagine a bored brother and sister on a long train ride. (Einstein liked thought experiments using trains.) To pass the time, they get up and start throwing a baseball up and down the aisle of the train. The boy is in the front and the girl in the back. The train is traveling at 60 mph, and they can each throw the ball at 30 mph. As seen by an observer standing on the bank outside the train, the ball appears to be traveling 30 mph (60-30) when the boy throws the ball to the girl and 90 mph (60+30) when the girl throws it back. The Michelson-Morley experiment was designed to look for similar behavior in light. The Earth orbiting the Sun takes the place of the train, and the measured speed of light (like the baseball's speed) should vary by the Earth's orbital speed depending on the direction the light is traveling. The experiment did not work as expected; the speed of light did not vary. Because Einstein took this result as the basic assumption that led to the special theory of relativity, the Michelson-Morley experiment is sometimes referred to as the most significant negative experiment in the history of science.

The **orbit** of the planet Mercury around the sun has some peculiarities that can not by explained by Newton's classical laws of physics. The general theory of relativity can explain these peculiarities, so they are described in the article on general relativity.

Special Relativity

To understand many concepts in relativity one first needs to understand the concept of a reference frame. A reference frame is a system for locating an object's (or event's) position in both space and time. It consists of both a set of coordinate axes and a clock. An object's position and motion will vary in different reference frames. Go back to the example above of the boy and girl tossing the ball back and forth in a train. The boy and girl are in the reference frame of the train; the observer on the bank is in the reference frame of the Earth. The reference frames are moving relative to each other, but there is no absolute reference frame. Either reference frame is as valid as the other.

For his special theory of relativity, published in 1905, Einstein assumed the result of the Michelson-Morley experiment. The speed of light will be the same for any observer in any inertial reference frame, regardless of how fast the observer's reference frame is moving. Einstein also assumed that the laws of physics are the same in all reference frames. In the special theory, Einstein limited himself to the case of non-accelerating, nonrotating reference frames (moving at a constant velocity), which are called inertial reference frames.

From these assumptions, Einstein was able to find several interesting consequences that are noticeable at speeds close to the speed of light (usually taken as greater than one tenth the speed of light). These consequences may violate our everyday common sense, which is based on the sum total of our experiences. Because we have never traveled close to the speed of light we have never experienced these effects. We can, however, accelerate atomic particles to speeds near the speed of light, and they behave as special relativity predicts.

SPACETIME Special relativity unified our concepts of space and time into the unified concept of **spacetime**. In essence time is a fourth **dimension** and must be included with the three space dimensions when we talk about the location of an object or event. As a consequence of this unification of space and time the concept of simultaneous events has no absolute meaning. Whether or not two events occur simultaneously and the order

in which different events occurs depends on the reference frame of the observer.

If, for example, you want to meet a friend for lunch, you have to decide both which restaurant to eat at and when to eat lunch. If you get either the time or the restaurant wrong you are not able to have lunch with your friend. You are in essence specifying the spacetime coordinates of an event, a shared lunch. Note that both the space and time coordinates are needed, so space and time are unified into the single concept of spacetime.

UNUSUAL EFFECTS OF MOTION Imagine a rocket ship traveling close to the speed of light. A number of unusual effects occur: Lorentz contraction, time dilation, and mass increase. These effects are as seen by an outside observer at rest. To the pilot in the reference frame of the rocket ship all appears normal. These effects will occur for objects other than rocket ships and do not depend on there being someone inside the moving object. Additionally, they are not the result of faulty measuring devices (clocks or rulers); they result from the fundamental properties of spacetime.

A rocket moving close to the speed of light will appear shorter as seen by an outside observer at rest. All will appear normal to an observer such as the pilot moving close to the speed of light inside the rocket. As the speed gets closer to the speed of light, this effect increases. If the speed of light were attainable the object would appear to have a length of **zero** to an observer at rest. The length of the rocket (or other moving object) measured by an observer at rest in the reference frame of the rocket, such as the pilot riding in the rocket, is called the proper length. This apparent contraction of a moving object as seen by an outside observer is called the Lorentz contraction.

A similar effect, time dilation, occurs for time. As seen by an outside observer at rest, a clock inside a rocket moving close to the speed of light will move more slowly. The same clock appears normal to the pilot moving along with the rocket. The clock is not defective; the rate at which time flows changes. Observers in different reference frames will measure different time intervals between events. The time **interval** between events measured both at rest in the reference frame of the events and with the events happening at the same place is called the proper time. This time dilation effect increases as the rocket gets closer to the speed of light. Traveling at the speed of light or faster is not possible according to special relativity, but if it were, time would appear to the outside observer to stop for an object moving at the speed of light and to flow backward for an object moving faster than light. The idea of time dilation is amusingly summarized in a famous limerick: "There was a young lady named Bright, Whose speed was much faster than light. She set out one day, In a relative way, And returned on the previous night." As seen by an outside observer, the mass of the rocket moving close to the speed of light increases. This effect increases as the speed increases so that if the rocket could reach the speed of light it would have an infinite mass. As for the previous two effects to an observer in the rocket, all is normal. The mass of an object measured by an observer in the reference frame in which the object is at rest is called the rest mass of the object.

These three effects are usually thought of in terms of an object, such as a rocket, moving near the speed of light with an outside observer who is at rest. But it is important to remember that according to relativity there is no preferred or absolute reference frame. Therefore the viewpoint of the pilot in the reference frame of the rocket is equally valid. To the pilot, the rocket is at rest and the outside observer is moving near the speed of light in the opposite direction. The pilot therefore sees these effects for the outside observer. Who is right? Both are.

SPEED OF LIGHT LIMIT Think about accelerating the rocket in the above example. To accelerate the rocket (or anything else) an outside **force** must push on it. As the speed increases, the mass appears to increase as seen by outside observers including the one supplying the force (the one doing the pushing). As the mass increases, the force required to accelerate the rocket also increases. (It takes more force to accelerate a refrigerator than a feather.) As the speed approaches the speed of light the mass and hence the force required to accelerate that mass approaches **infinity**. It would take an infinite force to accelerate the object to the speed of light. Because there are no infinite forces no object can travel at the speed of light. An object can be accelerated arbitrarily close to the speed of light, but the speed of light can not be reached. Light can travel at the speed of light only because it has no mass. The speed of light is the ultimate speed limit in the universe.

$$E = mc^2$$

This famous equation means that matter and energy are interchangeable. Matter can be directly converted to energy, and energy can be converted to matter. The equation, $E=mc^2$, is then a formula for the amount of energy corresponding to a certain amount of matter. E represents the amount of energy, m the mass, and c the speed of light. Because the speed of light is very large a small amount of matter can be converted to a large amount of energy. This change from matter into energy takes place in nuclear reactions such as those occurring in the sun, nuclear reactors, and nuclear weapons. Nuclear reactions release so much energy and nuclear weapons are so devastating because only a small amount of mass produces a large amount of energy.

A pair of paradoxes

A paradox is an apparent contradiction that upon closer examination has a noncontradictory explanation. Several paradoxes arise from the special theory of relativity. The paradoxes are interesting puzzles, but more importantly, help illustrate some of the concepts of special relativity.

Perhaps the most famous is the twin paradox. Two twins are initially the same age, as is customary for twins. One of the twins becomes an astronaut and joins the first interstellar expedition, while the other twin stays home. The astronaut travels at nearly the speed of light to another star, stops for a visit, and returns home at nearly the speed of light. From the point of view of the twin who stayed home, the astronaut was

traveling at nearly the speed of light. Because of time dilation the homebound twin sees time as moving more slowly for the astronaut, and is therefore much older than the astronaut when they meet after the trip. The exact age difference depends on the **distance** to the star and the exact speed the astronaut travels. Now think about the astronaut's reference frame. The astronaut is at rest in this frame. The Earth moved away and the star approached at nearly the speed of light. Then the Earth and star returned to their original position at nearly the speed of light. So, the astronaut expects to be old and reunite with a much younger twin after the trip. The resolution to this paradox lies in the fact that for the twins to reunite, one of them must accelerate by slowing down, turning around and speeding up. This acceleration violates the limitation of special relativity to inertial (non-accelerating) reference frames. The astronaut, who is in the noninertial frame, is therefore the younger twin when they reunite after the trip. Unlike much science fiction in which star ships go into a fictional warp drive, real interstellar travel will have to deal with the realities of the twin paradox and the speed of light limit.

The garage paradox involves a very fast car and a garage with both a front and back door. When they are both at rest, the car is slightly longer than the garage, so it is not possible to park the car in the garage with both doors closed. Now imagine a reckless driver and a doorman who can open and close both garage doors as fast as he wants but wants only one door open at a time. The driver drives up the driveway at nearly the speed of light. The doorman sees the car as shorter than the garage, opens the front door, allows the car to drive in, closes the front door, opens the back door, allows the car to drive out without crashing, and closes the back door. The driver on the other hand, sees the garage as moving and the car as at rest. Hence, to the driver the garage is shorter than the car. How was it possible, in the driver's reference frame, to drive through the garage without a crash? The driver sees the same events but in a different order. The front door opens, the car drives in, the back door opens, the car drives through, the front door closes, and finally the back door closes. The key lies in the fact that the order in which events appear to occur depends on the reference frame of the observer. (See the section on spacetime.)

Experimental verification

Like any scientific theory, the theory of relativity must be confirmed by experiment. So far, relativity has passed all its experimental tests. The special theory predicts unusual behavior for objects traveling near the speed of light. So far no human has traveled near the speed of light. Physicists do, however, regularly accelerate subatomic particles with large particle accelerators like the recently canceled Superconducting Super Collider (SSC). Physicists also observe cosmic rays which are particles traveling near the speed of light coming from space. When these physicists try to predict the behavior of rapidly moving particles using classical Newtonian physics, the predictions are wrong. When they use the corrections for Lorentz contraction, time dilation, and mass increase required by special relativity, it works. For example, muons are very short lived subatomic particles with an average lifetime of about 2 millionths of a second. However when they are traveling near the speed of light physicists observe much longer apparent lifetimes for muons. Time dilation is occurring for the muons. As seen by the observer in the lab time moves more slowly for the muons traveling near the speed of light.

Time dilation and other relativistic effects are normally too small to measure at ordinary velocities. But what if we had sufficiently accurate clocks? In 1971 two physicists, J. C. Hafele and R. E. Keating used atomic clocks accurate to about one billionth of a second (1 nanosecond) to measure the small time dilation that occurs while flying in a jet plane. They flew atomic clocks in a jet for 45 hours then compared the clock readings to a clock at rest in the laboratory. To within the accuracy of the clocks they used time dilation occurred for the clocks in the jet as predicted by relativity. Relativistic effects occur at ordinary velocities, but they are too small to measure without very precise instruments.

The formula $E=mc^2$ predicts that matter can be converted directly to energy. Nuclear reactions that occur in the sun, in nuclear reactors, and in nuclear weapons confirm this prediction experimentally.

Albert Einstein's Special theory of relativity fundamentally changed the way we understand time and space. So far it has passed all experimental tests. It does not however mean that Newton's law of physics is wrong. **Newton's laws** are an approximation of relativity. In the approximation of small velocities, special relativity reduces to Newton's laws. If at some time in the future someone does an experiment that does not agree with the theory of relativity, we will have to modify the theory just as relativity modified Newton's classical physics.

RELIABILITY OF DIGITS AND CALCULATIONS

Determining the reliability of digits and calculations is important, because physical measurements have mathematical limitations. For example, suppose a ruler marked with tenths of inches is used to measure the length of an object. If the measurement is reported as 1.54 inches, the first two digits are certain, and the last digit (4) is an estimate. This represents the most reliable length that can reasonably be obtained for that object using that measuring device. Estimating any additional digits following the 4 would misrepresent the accuracy of the measurement and reduce its reliability.

If a calibrated gravimetric scale is used to measure a weight, and provides a digital answer of 2.328 grams, the measurement is assumed to be precise to one thousandth of a gram and can be reported with reliability. In this case, the reliability of the answer is not limited by the estimating ability of the measurement taker, but by the precision (and correct calibration) of the measuring device.

When measured values are added or subtracted, the result can be no more precise than the precision of the least precise measurement. This means that the answer should contain no more digits to the right of the decimal point than does the value with the least number of digits to the right of the decimal point.

For example, assume that a second digital scale, which is calibrated and precise only to a tenth of a gram, is used to weigh a second object and indicates a weight of 1.2 grams. The sum of the two weights, 2.328 grams + 1.2 grams, should be reported as 3.5 grams, not 3.528 grams.

When measured values are multiplied or divided, the number of significant figures is a determining factor regarding the reliability of the answer. Significant figures are numbers that are known with some degree of reliability. They include:

- Nonzero digits (e.g., 3.758 contains four significant figures)
- Zeroes between nonzeroes (e.g., 10.004 contains five significant figures)
- Zeroes to the right of nonzeroes that are right of the decimal point (e.g., 12.50 contains four significant figures)

For values less than 1, zeroes to the left of a decimal point and zeroes to the left of the first nonzero are NOT significant figures (e.g., 0.8 and 0.0000008 each contain only one significant figure). These zeroes indicate the location of the decimal point. They do not provide information about the reliability of the value.

For measurements reported in **whole numbers**, zeroes to the right of nonzeroes may or may not be significant (e.g., a weight of 65,000 pounds could have several different numbers of significant figures, depending on the precision of the scale). This uncertainty is eliminated by using exponential or **scientific notation** to specify the number of significant figures. Thus, 65,000 pounds reported as 6.5×10^4 pounds contains two significant figures and indicates that the scale is reliable to the nearest 1000 pounds. Likewise, 65,000 written as 6.50×10^4 pounds contains three significant figures, and indicates reliability to the nearest 100 pounds, and so forth.

The result of **multiplication** or **division** can contain no more significant figures than does the value with the least number of significant figures. For example, the product of 1.2 meters and 1.867 meters should be reported as 2.2 **square** meters, not 2.2404 square meters. The answer must contain only two significant figures. The product of 13.0 feet and 0.487 feet should be reported as 6.33 square feet, rather than 6.331 square feet, because 13.0 contains three significant figures.

One exception to these calculation rules applies when known exact numbers are used, e.g., 12 inches to a foot or 35 people in a room. Exact numbers, which are usually **integers**, are assumed to have an infinite number of significant figures. Therefore, they place no limiting factors on the calculation results, (e.g., 12 inches/foot multiplied by 3.456 feet should be reported as 41.47 inches, rather than 41 inches).

Extreme care must be taken when using a **calculator** or computer to calculate an answer based on measured observations. These devices can provide answers that include many more decimal places or significant figures than are actually reliable. The final answer must be adjusted according to the rules given above.

A slightly more sophisticated technique for indicating the reliability of measurements is called error analysis, and is based on statistical analysis of the random errors in the measurements. Random errors are deviations from the true value introduced by the observer's operation and reading of the instrument. For example, assume that 20 length measurements of the same object are taken and calculated to have a **mean** of 11.7 inches and a **standard deviation** of 2.683 inches. The standard error of the mean is calculated from $2.683/\sqrt{20} = 0.6$ inches. Therefore, the final answer would be reported as 11.7 ± 0.6 inches. If this point were shown on a graph, error bars would be placed on both sides of the point to indicate the extent of its standard error. If several values ! containing standard errors are used in numerical calculations, (e.g., 11.7 ± 0.6 inches $* 25.3 \pm 1.1$ inches), error propagation formulas are used to calculate the standard error in the final answer.

RENARD, CHARLES (1847-1905)

French military engineer

Charles Renard was one of the pioneers of air travel, using his mathematical and engineering expertise to create functional designs for a glider and a dirigible.

Very little is known about Renard's personal life. He was born in France in 1847 and studied mathematics and engineering at a French university. His interest and talent in the field of aerodynamics were widely known, as he succeeded in designing and flying a pilotless, 10-winged, heavier-than-air machine in 1871.

In 1884, Renard was a captain in the French Army stationed at the Aerostation Militaire in Chalais-Meudon. He and another captain, Arthur Krebs, came up with engineering plans for a dirigible they dubbed *La France*. By definition, a dirigible is an aircraft that can be steered in any direction regardless of wind and can return to its point of departure under its own power. Working with this definition, Renard and Krebs prepared to see if they had created the first true dirigible.

On August 9, 1884, *La France* took its first successful flight. With Renard aboard, the dirigible flew in a circle of about four or five miles, returning to its launch area with no mishaps. Although other people had designed dirigibles, Renard's was the first to meet this prime requirement.

The engineer died in 1905 in France at age 58.

REUTERSVARD, OSCAR (1915-   )

Swedish artist

Known for his artistic renderings of shapes that are not physically possible, Oscar Reutersvard is widely acknowledged as "the father of impossible figures." His work has inspired such other great artist-mathematicians as **Maurits Cornelius Escher** and **Roger Penrose**.

Despite his fame, very little information about Reutersvard's personal life exists. He was born in Sweden in 1915.

Reutersvard's contributions to the art world began early, but did not show the originality for which he would become known until he started deliberately drawing three-dimensional objects that could not exist in real life. He is believed to be the first artist to do this; previous impossible figures were apparently merely the result of mistakes in artistic perspective.

In 1934, Reutersvard created an impossible figure (also known as an "undecidable figure") that he named the impossible **triangle**. Formed from a series of cubes, the triangle at first seems like the simple geometrical shape with which all schoolchildren are familiar. However, as the eye tries to follow its outlines, the triangle abruptly becomes a dizzying experience as its bottom link plays havoc with the brain's intuitive knowledge of physical laws.

The impossible triangle and Reutersvard's other works are so popular in his homeland that the government commissioned him to make a series of stamps using his art in 1982. Over his lifetime, he has created thousands of impossible figures, and numerous books have been published on his works in Russian, Polish, and Swedish.

Reutersvard currently works in the Art History Department of Sweden's Lund University.

See also Maurits Cornelius Escher; Roger Penrose

RHETICUS, GEORG JOACHIM (1514-1574)

German mathematician and astronomer

Born in Feldkirch, Austria, to the town physician, Rheticus was exposed to science throughout his childhood. However, in 1528, his father was tried and convicted on a charge of sorcery, and subsequently beheaded. This family tragedy did not discourage Rheticus, for he continued his studies and ultimately received his degree from the University of Wittenberg on 27 April 1536. Shortly thereafter, through the influence of Philip Melanchthon, the great German educational reformer and close associate of Martin Luther, Rheticus was appointed to a chair of mathematics and astronomy at the University. It was here that he became familiar and highly intrigued with **Nicolaus Copernicus'** theory that the Earth revolves around the Sun. So impressed was he over this new concept, that in 1539 Rheticus called upon Melanchthon once again to make arrangements to travel to Frauenberg in Ermland (now Frombork, Poland) to study with Copernicus for two years.

Deeply impressed by the Copernican system, Rheticus committed himself to getting the astronomer's theory published. However, in Copernicus' eyes, a complete explanation of the theory still needed to be conceived. Undaunted, Rheticus proposed to publish an introduction of the theory under his own name. With permission from Copernicus and financial assistance from the mayor of Danzig, Rheticus published under his own name the first account of the new views in his *Narratio Prima* (1540; "The First Account of the Book on the Revolutions by Nicolaus Copernicus").

With Rheticus' work well received, he persuaded Copernicus to complete his work under the title *De revolu-*

tionibus orbium coelestium ("On the revolutions of the heavenly spheres"). Shortly thereafter, Rheticus received permission from Duke Albert of Prussia to publish *De Revolutionibus*. The Duke also requested that Rheticus return to his chair at the University of Wittenberg, and in 1541, Rheticus returned to the University where he was elected dean of the Faculty of Arts.

From his stay at Wittenberg until his death, Rheticus also worked on his great treatise, which was completed and published in two volumes after his death by his pupil Valentin Otto as *Opus Palatinum de triangulis* (1596; "The Palatine Work on Triangles"). Many historians consider this work to be where **trigonometry** came of age.

This classic work challenged the geocentric cosmology that has been accepted since the time of **Aristotle**. In an age of religious intolerance, coupled with the fate of his father, it is a testament to his courage and commitment to knowledge that Rheticus enthusiastically supported such a volatile theory. Copernicus' heliocentric theory of planetary **motion** was not only in sharp scientific contrast to the prevailing geocentric Ptolemaic system, but was also a volatile theological issue — as Galileo was to discover when he faced the Inquisition on charges of heresy in 1615.

See also Nicolaus Copernicus; Trigonometry

RHIND PAPYRUS

The oldest known mathematical text is a document usually referred to as the Rhind Papyrus, written sometime around 1650 BC. It is named for the Egyptologist Alexander Rhind who purchased it in Luxor in 1858. It was not deciphered until 1877, when its formal title was discovered to be, "Accurate Reckoning of Entering into Things, Knowledge of Existing Things, Knowledge of All Obscure Secrets." The author, a scribe called **Ahmes** the Moonborn, stated that the mathematical problems set forth were not of his own devising, that he was in fact transcribing another text that was hundreds of years old even in his time. The Rhind Papyrus is a collection of 85 mathematical word problems written in a cursive form of hieroglyphics called hieratic, on a parchment measuring 18 feet long by 13 inches wide.

The Rhind Papyrus appears to be a practical instructional tool, its purpose might have been to train government officials in agrarian administration. Many of its problems deal directly with the harvesting and storing of grain or with the **division** of food and other resources. And though it does not directly address the processes of calculation, instead merely showing problems and solutions, the Papyrus clearly shows a knowledge of additive **arithmetic**, **multiplication** and division, plane **geometry** and volumes of cylinders and prisms. There is at least one problem that suggests geometric progression, which reads in part, "In one woman's house there are seven store rooms, each has 7 cats; each catches 7 mice, each of which has eaten 7 ears of barley, which each grew 7 measures of grain." Another problem shows how much barley is needed

to distribute among a number of people, when each person's share is increased by 1/8 more than the last.

The ancient Egyptians did not recognize mathematics as an academic discipline in of itself; indeed, their language had no word for "mathematician." Instead, mathematics was one of the many domains of the priestly caste, which served as a tool to allow them to pursue their numerous "mysterious" interests, such as astronomy. Since the priests were expected to train the aristocracy in the arts and sciences of administration, texts such as the Rhind Papyrus became integral instructional devices in demonstrating how limited resources might best be distributed among the populus. Their success in this shows in the longevity of the many Egyptian dynasties, and their advances in mathematically based enterprises like astronomy and architecture. More important to us, though, more than three millennia later, is this vital evidence of the long tradition of practical **applied mathematics**.

See also History of mathematics; Mathematics and culture

RIEMANN, GEORG FRIEDRICH BERNHARD (1826-1866)
German topologist and mathematical physicist

Intuitive and original, Georg Friedrich Bernhard Riemann had a gift for understanding connections among apparently heterogeneous phenomena. Almost all his works proved to be the beginning of new, productive research. He anticipated the work of **James Clerk Maxwell** in electrodynamics, and introduced the novel concept of curvature in a two–dimensional direction, providing the **non–Euclidean geometry** of **space** necessary for **Albert Einstein**'s theory of general **relativity**.

Riemann worked for many years in failing health in a university system where abject poverty was the lot of those without independent means. When he died of tuberculosis at age 40, however, he had known, worked with, and influenced the great mathematicians and physicists of his day. Riemann was shy and self–effacing and recognition for his work came slowly during his lifetime; awareness of his truly striking achievements was to come later as his work was validated and as it stimulated the work of others. Today, Riemann's work, which often incorporates methods of physics, is accepted as some of the most remarkable in the history of all science, not only in mathematics. A number of theorems, methods, concepts, and conjectures carry his name.

Riemann was born September 17, 1826, in Breselenz in the kingdom of Hanover, now Germany. His parents were Friedrich Bernhard Riemann, a Protestant minister who had served in the Napoleonic wars, and Charlotte Ebell, the daughter of a court councillor. He was the second child of six, two boys and four girls. Riemann's family suffered from tuberculosis. His mother and three sisters, and eventually Riemann himself, died of the disease.

Riemann was taught primarily by his father and began making mathematical calculations as early as age six. By age ten he had surpassed his father and also his schoolmaster

Shulz, who was employed to tutor him in mathematics. In 1840, at age 14, Riemann went to live with his grandmother in Hanover, where he studied a classical curriculum at the gymnasium, or secondary school. In 1842, his grandmother died and he transferred to the gymnasium at Löneburg, closer to his father's parish at Quickborn. Schmalfuss, the director at Löneburg, noted Riemann's interest in mathematics and lent him several classic books, including **Leonhard Euler**'s works and **Adrien–Marie Legendre**'s *Theory of Numbers*, which Riemann studied, mastered and returned in a few days. He was to use some of this information in his later work.

In 1846, Riemann entered Göttingen University as student of theology and philology, but his talent and superior knowledge in mathematics was obvious, and he soon persuaded his father to let him devote himself to mathematics. The faculty at Göttingen included the "king of mathematicians" **Karl Friedrich Gauss**, who lectured on the method of least squares. It also included Moritz A. Stern, who lectured on the numerical solution of **equations** and definite integrals, and the astronomer Carl Wolfgang Benjamin Goldschmidt, who lectured on terrestrial magnetism. Stern was later to recall that Riemann "already sang like a canary." The quality of teaching at Göttingen was low and elementary, however, and students were not included in the creative thinking of the teachers.

After only a year at Göttingen, Riemann spent two years in Berlin, where democratic ideas were afoot and the faculty at the University of Berlin shared ideas in open discussions with students. The mathematics faculty included **Peter Dirichlet, Karl Jacobi, Jakob Steiner**, and Ferdinand Eisenstein. Riemann attended Dirichlet's lectures on the theory of numbers, definite integrals, and partial **differential equations**, and Jacobi's on analytical **mechanics** and higher **algebra**. With Eisenstein, Riemann discussed the possibility of introducing complex variables into investigations of elliptic **functions**. Riemann was attracted to Dirichlet's methods of logical **analysis** of fundamental questions and avoidance of long computations. Riemann's stay in Berlin was coincident with the March 1848 insurrection in Prussia, and at the height of the revolution Riemann joined other students in guarding the palace of King Friedrich IV for 24 hours. When the king was deemed safe, he returned to the university.

In 1849, Riemann returned to Göttingen to work on his doctoral dissertation under the supervision of Gauss. He also attended the lectures of physicist Wilhelm Weber and assisted in his laboratory, and studied with the natural philosopher Johann Friedrich Herbart, absorbing concepts that would appear in his later mathematical works. The university system in which he found himself had no provision for talented scientists without independent means, and Riemann spent the next ten years in a period of "honored starvation" and declining health, obtaining very little money for his lectures. During this time, he developed his ideas about the nature of complex functions and Abelian integrals, manifolds, and the foundations of geometry. He also worked on a group of papers in mathematical physics.

In 1851, Riemann completed his doctoral dissertation on a general theory of functions of a complex **variable**, intro-

ducing topological methods into the theory. In this paper, which introduced what are known as Riemann surfaces, he gave a magnificently simple and panoramic view of a subject to which Gauss and **Augustin–Louis Cauchy** had contributed, but which had been previously tortuous and difficult to access. Gauss, who was usually aloof and critical, was very pleased with his student's work, describing Riemann has having a "gloriously fertile originality." In his theory, Riemann employed a modification of a procedure attributed to Dirichlet. A criticism of this procedure by **Karl Weierstrass** delayed full acceptance of Riemann's approach. Riemann's explanations had been clear to physicists but because of the nature of the proofs had presented difficulties for mathematicians. Only in 1900 was Riemann's work fully validated by mathematician **David Hilbert**.

In 1853, to qualify as a "Privatdozent" or university lecturer, Riemann was required to present a postdoctoral thesis to the Council of Göttingen University. He chose to present a thesis on **Fourier series**, a problem Dirichlet had addressed, and in it introduced the Riemann integral. Masterfully written, this work established Riemann as Dirichlet's equal. Riemann was also required to give a lecture, for which he had to submit three topics to the university faculty. Two were on electricity, something Riemann had thought about for a long time. To his surprise and dismay, at Gauss' recommendation the faculty chose his third topic, on the foundations of geometry. Riemann's lecture, given in 1854, was brilliant, however. In it he introduced the main ideas of modern differential geometry.

As a new lecturer, Riemann was delighted when a relatively large class of eight students appeared to hear him speak about partial differential equations and their application to physical problems. Riemann was attracted by many problems at once. In the ensuing few years he developed material which did for **algebraic geometry** what he had done for differential geometry in his postdoctoral lecture. He also wrote on electrodynamics and the propagation of waves.

In 1855, Gauss died and Dirichlet was brought in to take his place. Dirichlet, Riemann's friend and collaborator managed to obtain a small government stipend for Riemann with some difficulty. At 200 thalers, it amounted to about one–tenth of a full professor's salary. Two years later, Riemann, whose health was rapidly declining, was made extraordinary (or assistant) professor and his salary was raised to 300 thalers. Shortly thereafter, his father died and the family home at Quickborn was broken up.

When Dirichlet died in 1859, Riemann succeeded him as full professor. For the first time he had a salary sufficient to support himself, and he began to receive well–deserved honors. He was elected a corresponding member of the Berlin Academy of Sciences in "physical–mathematics" based on his doctoral dissertation and his theory of Abelian functions. In gratitude he submitted to the Academy an article "On the Number of Primes Not Exceeding a Given Bound." He also became a corresponding member of the Bavarian Academy of Sciences and a full member of the Göttingen Academy of Sciences. In 1860 he visited Paris where he met with the eminent French mathematicians **Charles Hermite**, Joseph

Bertrand, Victor Puiseux, Charles Briot, and Jean Claude Bouquet.

In 1862, Riemann married Elise Koch, a friend of his sister's. A month thereafter he had an attack of pleurisy and his tuberculosis achieved the upper hand. His friends secured a leave and money for travel, and Riemann continued his studies in the less–harsh climate of Italy, with intermittent returns to Göttingen. In Pisa, he was welcomed by the mathematicians **Enrico Betti** and Eugenio Beltrami, who were inspired by his ideas in their own work. In 1863, Riemann's daughter, Ida, was born in Pisa. In November of 1865 he returned again to Göttingen and was able to work a few hours each day through the cold, wet winter. In March of 1866 he was elected a foreign member of the Paris Académie Royale des Sciences, and in June he was elected a member of the Royal Society of London. It was clear that Riemann was dying, however, and to conserve his remaining days, he returned to Italy.

On July 20, 1866, Riemann died in Selasca, Italy, on Lake Maggione, with his wife beside him. He was buried in nearby Biganzola. At the time of his death he was attempting a unified explanation of gravity and **light** (a pursuit that is still elusive) and a theory of mechanics of the human ear, an investigation inspired by Hermann von Helmholtz.

In 1876, Riemann's works were collected and published by Heinrich Weber and Riemann's friend, **Richard Dedekind**, who was one of three who attended Riemann's lectures of 1855–1856. Although his published work consists of only one volume, his ideas continue to inspire mathematical inquiry.

Riemann hypothesis

The Riemann hypothesis is a question in the field of **number theory** that is perhaps the most famous unsolved problem in mathematics. First posed by the great mathematician **Bernhard Riemann** in 1859, it has captured the imagination of generations of mathematicians. **David Hilbert**, one of the most important mathematicians of the early twentieth century, once said that if he could look 500 years into the future, his first question would be, "Has someone solved the Riemann hypothesis?" In 1900, at a meeting of the International Congress of Mathematicians in Paris, Hilbert placed the Riemann hypothesis on his famous list of the most important mathematics problems for the next century, a list that has set the course of twentieth-century mathematics. And in April 2000, the Clay Mathematics Institute (in Paris, as a tribute to Hilbert) announced that the Riemann hypothesis is to be one of its seven Millennium Prize Problems, whose solutions will each garner a $1 million prize.

The Riemann hypothesis concerns the Riemann **zeta function**, a function that transforms each number s into the new number $1/1^s + 1/2^s + 1/3^s + 1/4^s + \ldots$. So for example, when s is equal to 2, the Riemann zeta function gives out $1/1^2 + 1/2^2 + 1/3^2 + \ldots$, which, surprisingly, **Leonhard Euler** proved in 1734 is equal to $\pi^2/6$. More than a century after Euler, Riemann examined the zeta function from the standpoint of the emerging field of complex **analysis**: he studied the behavior of the zeta function when the input s is allowed to be a complex number. When s is complex, something unexpected can happen: even though the numbers 1, 2, 3,... are all positive, the sum $1/1^s + 1/2^s + 1/3^s + \ldots$ may come out to **zero**. Riemann calculated the first few of these zero-producing **exponents**, called zeros for short, and they all lay on the same line in the complex plane: a vertical line whose x-coordinate (the real coordinate) is equal to 1/2. Riemann proceeded to make his famous hypothesis: except for certain well-understood exceptions, all the zeros of the zeta function lie on that 1/2-line.

One of the reasons that the Riemann hypothesis has fascinated mathematicians is that the location of the Riemann zeros is exactly what controls the distribution of **prime numbers**, those numbers that are only divisible by 1 and themselves. As early as the 4th century B.C., **Euclid** proved that the number of primes is infinite. But even though there are infinitely many of them, they are they exception, not the rule, and the task of finding them is complicated by the fact that they crop up randomly, seemingly unpredictably. For that reason, searching for large prime numbers is a difficult task, and the discovery of a new prime number is a cause for celebration. But although on the local scale the distribution of prime numbers is haphazard, on the large scale the primes display a surprising order. In 1896, **Jacques Hadamard** and **Charles-Jean-Gustave de la Vallée Poussin** proved a remarkable **theorem**: the probability that a given large number x is prime is roughly $1/\log(x)$. Thus, the primes maintain a delicate balance between order and randomness.

The **prime number theorem** of Hadamard and de la Vallée Poussin gives the probability that a number is prime, but it does not say how far the actual distribution of primes strays from this rough probability law. It was Riemann who realized that the location of the Riemann zeros holds the key to just how closely the primes follow the $1/\log(x)$ distribution. A century earlier, Euler had discovered a connection between the zeta function and the prime numbers, in the famous formula $1/1^s + 1/2^s + 1/3^s + \ldots = (2^s 3^s 5^s 7^s 11^s \ldots)/(2^s-1)(3^s-1)(5^s-1)(7^s-1)(11^s-1)\ldots$. Using this formula, Riemann was able to show that if the Riemann zeros are located on the 1/2-line, as he hypothesized, then the prime numbers stay close to the $1/\log(x)$ distribution, with about as much digression from the distribution as you would expect if you were flipping a coin that was weighted to have a $1/\log(x)$ probability. If the zeros do not all lie on the 1/2-line, then the primes are much more unpredictable than has hitherto been imagined.

The first 1,500,000,000 zeros have been calculated by computer, and they do all lie on the 1/2-line, evidence that substantially bolsters Riemann's claim. But since the Riemann hypothesis is a statement about an infinite collection of numbers, it cannot be proved simply by calculating more and more zeros; there will always be more zeros to check.

Although the Riemann hypothesis appears to deal with objects that lie firmly within the realm of pure mathematics, recently researchers have discovered a surprising connection with an emerging branch of physics: quantum chaos. A huge amount of numerical evidence points to the likelihood that the Riemann zeros correspond to the energy levels of a quantum mechanical system whose classical counterpart is chaotic. This new connection is important for both physicists and mathematicians. Energy levels of quantum chaotic systems are generally difficult to calculate, while Riemann zeros are comparatively easy to calculate; thus the link between the two gives physicists a way to generate a wealth of useful date. And for mathematicians, the connection to physics gives a potential way finally to prove the Riemann hypothesis: if mathematicians could find a quantum system whose energy levels correspond to the Riemann zeros, then the symmetries of the system would prove the Riemann hypothesis instantly.

See also Prime numbers

RIEMANN MAPPING THEOREM

One of the central results in the subject of **complex analysis**, the Riemann mapping theorem unifies the areas of **algebra**, **geometry**, and **topology**. Today it is recognized as one of the most important theorems of nineteenth-century mathematics, even though its original **proof** by **Georg Freidrich Bernhard Riemann** was flawed. The search for generalizations and a better understanding of this theorem has continued throughout the twentieth century.

Early mapmakers, from **Hipparchus** in ancient Greece to Gerhardus Mercator in 16th-century Flanders, discovered that certain types of maps of the globe preserve shape information on a small scale. This means that the map accurately represents small circles as circles (rather than as ellipses or ovals), and that it accurately represents angles. Such maps are useful for navigation, even though they may grossly distort large-scale features. (For example, Greenland is disproportionately large in the Mercator projection.) A locally shape-preserving map is called *conformal*, and there are many conformal projections besides Mercator's.

A truly systematic theory of conformal mappings became possible in the early 1800s, when mathematicians discovered these maps can be represented by complex analytic **functions**. Any point in the region to be mapped can be labeled by coordinates (x, y), and these coordinates can be transformed into a complex number $x + iy$. Similarly, any point in the image of this region can be described by coordinates (x', y'), which can be thought of a complex number $x' + iy'$. The function that assigns each point $x + iy$ to its image $x' + iy'$ turns out to be differentiable, which means that all the tools of **calculus** can be applied to it. However, it is not merely differentiable as a function of **real numbers**; it is also differentiable as a function of **complex numbers**. The extra structure of complex numbers (specifically, the fact that they constitute a **field** rather than merely a **vector space**) has a profound effect in the theory of such functions, which are therefore called "analytic" rather than "differentiable."

For many years, complex analysts perfected laborious formulas for **mapping** one particular type of planar region, such as the interior of a **triangle**, to another by means of analytic (conformal) functions. These formulas were used by engineers

and physicists, for example to draw the equipotential curves of a distribution of electric charge. However, in his 1851 doctoral thesis, Riemann turned what had formerly been done on a case-by-case basis into a general theory. He proved that all bounded planar region were conformally equivalent, provided they had no holes in them. That is, any region with no holes (such regions are called "simply connected") could be mapped conformally to the the interior of a **circle** of radius 1, a "unit disk." Later mathematicians removed even the hypothesis of boundedness. The only simply connected planar region that cannot be conformally mapped to a unit disk is the plane itself.

Riemann's proof was controversial because of his assumption of what became known as the "Dirichlet principle." This principle could be summed up by the aphorism, "You can't fool Mother Nature." For example, no matter what distribution of electric charge you place on the boundary of a region, "Mother Nature" can find an electrostatic potential in the interior that matches the charge distribution. This potential turns out to be the real part of a complex-analytic function. To physicists this fact was self-evident; however, mathematicians considered it entirely too empirical to be used in a mathematical proof. Over the next 80 years, they found other ways to prove Riemann's celebrated theorem. Finally, however, **David Hilbert** vindicated Riemann by proving that the Dirichlet principle is true for the types of regions that Riemann was considering. It is no exaggeration to say that the entire field of elliptic partial **differential equations** has grown out of the quest to prove the Dirichlet principle.

The concept of conformal **equivalence** can be applied to curved surfaces, as well as planar regions. The "uniformization theorem" says that any closed surface with no boundary has one of only three types of conformal geometry. For example, the equation $z^2 + w^2 = 1$ (where z and w are *complex*, not real, variables) defines a surface that is conformally equivalent to a **sphere**. The equation $z^2 = w^3 - w$ defines a **torus**, which has the same conformal geometry as a plane. Most polynomial **equations** involving **exponents** greater than three, when defined over the complex numbers, define toruses with two or more holes. These have the conformal geometry of a unit disk. The absence of other possibilities stems directly from the Riemann mapping theorem, because the geometry of the surface is first "lifted" to a planar covering space and then mapped by Riemann's theorem to the unit disk. In this way the theorem provides a great deal of information on the oldest problem of **algebraic geometry**, to link the geometry of a surface to the algebraic equation that defines it.

Other directions of research inspired by the Riemann mapping theorem include:

- three-dimensional topology, where William Thurston has suggested a "Geometrization Conjecture" that classifies three-dimensional manifolds into eight distinct geometric types;
- the theory of several complex variables, which classifies the conformal types of regions in two-dimensional complex space (four-dimensional Euclidean space) and higher;

- and the theory of circle-packings, a computer-friendly method of computing the mappings that are guaranteed to exist by the Riemann mapping theorem.

See also Differential calculus; Peter Gustav Lejeune Dirichlet

RIEMANNIAN GEOMETRY

Riemannian **geometry**, also called differential geometry, is the study of curved **space**. It is also the language of general **relativity** theory (which posits that our universe is a curved space). Finally, it is the most active branch of geometry in contemporary mathematics. Few geometers today study Euclidean circles and triangles and spheres; most of them study manifolds and bundles, which are the basic concepts of Riemannian geometry.

In the 1820s, **Carl Friedrich Gauss**, **Nikolai Lobachevsky**, and **Janos Bolyai** discovered the first **non-Euclidean geometry**, called the hyperbolic plane. In this geometry, the Euclidean **parallel postulate** and many of the theorems of Euclidean geometry do not hold; for example, the sum of the angles of a **triangle** is always less than 180 degrees.

Few mathematicians understood this discovery at first, but the Italian mathematician Eugenio Beltrami, in the 1860s, proved that the hyperbolic plane is no more exotic than a **sphere**. For example, imagine a triangle drawn on the Earth's surface with one vertex at the North Pole, another on the equator at **zero** degrees longitude (just off the coast of Africa), and a third vertex on the equator at 90 degrees west longitude (near the Galapagos). With the help of a globe, it is easy to see that this spherical triangle has three right angles, and thus an **angle** sum of 270 degrees. The excess comes from the positive curvature of the Earth's surface, which makes the lines bulge outward.

Similarly, in the **hyperbolic geometry** of Gauss, Lobachevsky and Bolyai, the lines of a triangle tend to bulge inward, because the hyperbolic plane is negatively curved. Beltrami even succeeded in drawing a piece of the hyperbolic plane, called a "pseudosphere." Unfortunately, it is impossible to imbed the entire hyperbolic plane in three-dimensional space (a fact proved by **David Hilbert**). Thus we cannot see it as we can see the surface of a sphere; but we can still imagine it.

In 1854, **Bernhard Riemann** revolutionized geometry by focusing attention for the first time on *space itself*, rather than objects in space. The sphere or the hyperbolic plane are very special spaces, because the curvature of space at every point is the same. However, there are many spaces in which the curvature varies from point to point. For example, a **torus** is positively curved along its outer surface but negatively curved, or saddle-shaped, along its inner surface.

One of Riemann's most important insights was to define the curvature as an *intrinsic* feature of the space. When we look at a sphere or a torus, we can see its curvature *extrinsically,* from the way these surfaces bend in the three-dimensional space around them. But Riemann took the point of view of a bug crawling along the surface of the sphere or the torus,

which cannot perceive its space from outside. Nevertheless, according to Riemann, it can still observe all the essential geometric features of its space.

The shift from an extrinsic to an intrinsic viewpoint required a whole new conception of what a space is; accordingly, Riemann defined the concept of a **manifold**, which is still used by mathematicians today. It is a set that can be mapped to a Euclidean space on a small scale, just as cartographers map the globe. Riemann further required the **mapping functions** to be differentiable (or "smooth"); this allows one to use the tools of **calculus**, such as derivatives and integrals. Finally, Riemann required the manifold to have an **infinitesimal** distance scale or *metric* at every point.

Using these few ingredients, Riemann showed how, from the bug's-eye view, to compute every important geometric feature of a space. He showed how to find geodesics (the equivalent of straight lines), and how to compute distances and volumes inside the manifold. Most importantly, he derived a formula for computing the curvature of a space from its metric alone. For a surface (a two-dimensional manifold), the Riemann curvature at any given point turns out to be a single number. For a three-dimensional manifold the *Riemann curvature tensor* has 6 components, and for a four-dimensional manifold it has 20. These represent the number of distinct measurements an inhabitant of these spaces would need to make in order to determine what kind of space he or she lives in. Ironically, a curve (a one-dimensional manifold) has no curvature at all. More precisely, the bending of a curve is an extrinsic phenomenon governed by the way it is imbedded in space; it is not detectable by a bug living on the curve.

A great triumph of nineteenth-century differential geometry was the Gauss-Bonnet **Theorem**, which states that the integral of the Riemann curvature over an entire closed surface (or the "total curvature") is always an integer times 2 **pi**. The total curvature does not change if the surface is deformed, and is therefore a topological **invariant** of the surface. It is 4 pi if the surface is a sphere, 0 if it is a torus, and so on. In principle, therefore, an inhabitant of a two-dimensional universe can determine the global structure of his or her universe by integrating local measurements.

Riemann's ideas unexpectedly acquired much greater importance when **Albert Einstein**, in his theory of general relativity, argued that the universe we live in is curved. Geodesics in this curved space are the paths of **light** rays; the curvature of space gives rise to gravity. Suddenly, to appreciate the importance of differential geometry, we no longer need to imagine bugs living on surfaces; we are the bugs, and the universe is the four-dimensional manifold we would like to know the shape of. In particular, one solution to Einstein's **equations** for the shape of space leads to the theory of black holes. Another solution leads to the Big Bang **model** of cosmology.

A short survey cannot possibly do justice to the many developments in Riemannian geometry in the twentieth century. It is worth mentioning that the century has seen a gradual shift of viewpoint, strongly influenced by physics, toward considering the fundamental object in geometry to be not a manifold but a bundle. Bundles are (roughly speaking) manifolds with a vector space attached at every point. The vector space is a sort of repository for electromagnetic, gravitational, and any other fields.

See also Michael Francis Atiyah; Enrico Betti; Elie-Joseph Cartan; Shiing-shen Chern; Simon Kirwan Donaldson; Tulio Levi-Civita; Lie group; Hermann Minkowski; Minkowski space-time; Girolamo Saccheri; Space; Space-time; Hermann Weyl; Edward Witten

RINGEL, GERHARD (1919-)
Austrian-American mathematician

Gerhard Ringel is one of the world's foremost experts on **combinatorics** and graph theory.

Ringel was born in Kollnbrunn, Austria, on October 28, 1919. He received a doctorate from Friedrich-Wilhelms University in Bonn, Germany, in 1951. After earning teaching credentials there in 1953, Ringel began lecturing in mathematics at the university. He was promoted to assistant professor in 1956, remaining in that post until 1960. Meanwhile, in 1957, Ringel had taken a simultaneous job as a mathematics lecturer at Johann-Wolfgang-Goethe University in Frankfurt. He also left that post in 1960.

Starting a new position as associate professor at Berlin's Free University soon thereafter, Ringel rose rapidly through the academic ranks. He was appointed a full professor of mathematics in 1966 and director of the school's Mathematics Institute in 1970. In 1967, he spent a year as visiting professor at the University of California, Santa Cruz.

Ringel returned to UC Santa Cruz in 1970, accepting an appointment as professor of mathematics. He spent the next 20 years actively teaching at the school, achieving emeritus status in 1990. Ringel continues his research there, concentrating chiefly on methods of current graphs and their applicability to embedding graphs into surfaces, to some problems in elementary **geometry**, and to **group theory**.

See also Combinatorics

ROBERVAL, GILLES PERSONNE DE (1602-1675)
French mathematician

Gilles Personne de Roberval is generally considered to be the founder of kinematic **geometry** (a branch of **mechanics** that deals with pure **motion**) because of his discoveries about plane curves and the method he developed for drawing the tangent to a curve. He was also important historically because of his close contact and correspondence with many of the other important mathematicians of his day. Roberval was a leading expert in the geometry of infinitesimals.

Roberval was born near Senlis, France on August 10, 1602, the son of a poor Catholic farmer. According to historical records, he had no formal schooling until he became intensely interested in mathematics at age 14. At that point, or perhaps before, he left his family and began traveling around the country making a living as a tutor. Roberval took classes at the universities he found himself near, in this manner gaining a more complete education in mathematics, physics, and mechanics.

Arriving in Paris in 1628, Roberval started making the acquaintance of other brilliant mathematicians, including **Marin Mersenne** and **Blaise Pascal**. However, Roberval would become the only professional mathematician among his new group of friends. In 1632 the Collége de Maitre Gervaise in Paris appointed him professor of philosophy, but he would also serve as a private mathematics tutor there for the rest of his life. Roberval won a prestigious competition for a mathematics teaching position at the Royal College in 1634. He kept this post for the remainder of his life as well. In fact, many scientific historians believe that Roberval's tendency to keep his research secret was due to his wish to retain the position, which was reopened to competition every three years. In 1655 Roberval received an appointment to the school's mathematics chair.

In 1636, Roberval wrote *The Features of Mechanics*. He had a highly publicized debate with **René Descartes**, who was reportedly a longtime adversary, about the center of oscillation of the compound pendulum in 1647. In 1669, Roberval contributed a paper on infinitesimals ("Features of Indivisibles") to a collection of works by members of the French Royal Academy of Sciences (of which he was a founding member in 1666). This paper became a powerful **addition** to methods in the early study of integers—especially since it made discussion of them more logical and precise.

Roberval invented what is now known as the Roberval balance, used for weighing scales of the balancing sort, in 1669. At this late stage in his life, the mathematician also worked in the field of cartography, writing several works on the enormous project of **mapping** France. He also studied the vacuum and made equipment that Pascal used for his experiments on putting a "vacuum within a vacuum."

Roberval died in Paris on October 27, 1675 at the age of 73.

ROLLE, MICHEL (1652-1719)
French algebraist

Michel Rolle was a largely self–educated mathematician who taught himself both **algebra** and Diophantine **analysis**, a method for solving **equations** with no unique solution. He won early repute when he solved a problem set by the mathematician Jacques Ozanam, but his real passion lay in the field of the algebra of equations. His most famous work was the *Traite d'algebre* of 1690, in which he not only invented the modern notation for the *n*th root of x, but also expounded on his "cascade" method to separate the **roots** of an algebraic equation. He is best remembered for the **theorem** which bears his name, **Rolle's Theorem**, which determines the position of roots in an equation.

Little is known about Rolle's youth. He was born in Ambert, Bass–Auvergne in France on April 21, 1652. His father was a shopkeeper and the young Rolle had only a basic elementary education before he began to earn a living as a scribe for a notary, and then for various attorneys near Ambert. The allure of the big city brought Rolle to Paris by age 24, and an early marriage and family kept him in secretarial and accountancy work. But his curious mind led him to a study of algebra and of Diophantine analysis. The work of the third–century Alexandrian mathematician, **Diophantus**, had been translated into Latin by Bachet de Meziriac, and Rolle made himself familiar with the solutions to both determinate equations, for which there exists a unique solution, and indeterminate equations. Diophantus had provided a method of analysis for indeterminate equations, and Rolle became an expert in this field.

Rolle was also familiar with the work of another Frenchman, a contemporary of his and also a self–taught mathematician, Jacques Ozanam whose interests were in recreational mathematics—puzzles and tricks—and analysis. In 1682, Rolle published "an elegant solution" to a problem set by Ozanam, according to Jean Itard in *Dictionary of Scientific Biography*. The problem was as complex as the solution. Ozanam proposed finding four numbers which corresponded in the following ways: the difference of any two of these numbers not only make a perfect **square**, but also equal the sum of the first three. Rolle's solution as published in *Journal des scavans*, brought him notoriety and some patronage. The controller general of finance, Jean–Baptiste Colbert, took notice of him and arranged an honorary pension for Rolle so that he could continue with his mathematical work. Rolle also became the tutor to the son of a powerful minister, Louvois, and for a short time held an administrative post at the ministry of war.

Rolle became a member of the Academie Royal des Sciences in 1685. He published in 1690 his most famous work, *Traite d'algebre,* a book that has remained well known to this day. Algebra was a departure for Rolle, away from Diophantine analysis in which he had won early recognition, and into the algebra of equations. Rolle broke new ground in many aspects with this work. His new notation for the *n*th root of a number was first published in *Traite d'algebre*, and thereafter became the standard notation. Rolle also made important advances in systems of affine equations, which assign finite values to finite quantities, building on the work of Bachet de Meziriac, another contemporary whose work on mathematical tricks and puzzles led to the development of the field of recreational mathematics. But primarily the *Traite d'algebre* has retained its fame for the exposition of Rolle's method of "cascades." Utilizing a technique developed by Dutch mathematician Johann van Waveren Hudde to find multiple roots of an equation, or the highest common **factor** of a polynomial, Rolle was able to separate the roots of an algebraic equation. His "cascades" correspond to the derivatives of the highest common factor of a polynomial. Though never clearly defined, it is implied that such "cascades" are the result of the following: if in an equation $f(x) = 0$, $f(x)$ is multiplied by a progression, then the simplified result equated to **zero** would be such a "cascade." However, mathematicians of the day complained that Rolle's theory of cascades was given with too little **proof**.

To rectify that, Rolle published *Demonstration d'une Methode pour resoudre les Egalitez de tous les degrez* in 1691,

essentially a little known work which further demonstrated the method of "cascades." However, in the process, Rolle developed the **calculus** theorem which bears his name. According to this theorem, if the function $y = f(x)$ is differentiable on the open **interval** from a to b and continuous on the closed interval from a to b, and if $f(a) = f(b) = 0$, then $f'(x) = 0$ has at least one real root between a and b. It was not for another century and a half that Rolle's name was given to the theorem he developed, a theorem that is a special case of the **mean–value theorem**, one of the most important theorems in calculus.

Rolle's work encompassed also a certain flair for Cartesian **geometry**, though he did break with Cartesian techniques in other respects. In 1691, he was one of a vanguard to go against the Cartesian grain in the order relation for the set of **real numbers**, noting, for example, that –2a was a larger quantity than –5a. He also described the calculus as a collection of ingenious fallacies. Rolle published another important work on solutions of indeterminate equations in 1699, *Methode pour resoudre les equations indeterminees de l'algebre,* the year that he became a *pensionnaire geometre* at the Academie. This was a post that carried with it a regular salary which further enabled him to devote more time to mathematics. At this time, members of the Academie were drawing up sides on the value of **infinitesimal** analysis, in which a **variable** has zero as a limit. Rolle became one of its most outspoken critics. Though by 1706 he finally and formally recognized the value of the new techniques, his very opposition and criticism had served to help develop the new discipline of infinitesimal analysis.

Two years later, in 1708, Rolle suffered a stroke, which diminished his mental powers. Though he recovered, a second stroke in 1719 killed him. He is remembered not only for Rolle's Theorem and the method of "cascades" for separating and determining the position of roots in an algebraic equation, but also for his break with Cartesian techniques.

ROLLE'S THEOREM

Rolle's **theorem** implies that if we have a function *f* that is continuous on an **interval** [*a, b*] and *f(a) = f(b)*, then there always exists at least one critical point of the function in (*a, b*). The theorem is usually written symbolically as: Assume *f* is a function that is continuous on [*a, b*] and differentiable on (*a, b*). If *f(a) = f(b)* then there is a number *c* in (*a, b*) such that the derivative of *f* at *c*, *f'(c) = 0*. From Rolle's theorem it follows that *c* is at least one critical point of *f* in (*a, b*). This means that at some point (*c, f(c)*) on the graph of *f* that the **slope** of the tangent line is 0. There may be more than one such point *c* but the theorem provides that there exist at least one. The line tangent to point (*c, f(c)*) is parallel to the line joining (*a, f(a)*) to (*b, f(b)*).

Rolle's theorem is named after the French mathematician **Michel Rolle** who developed the theorem during the 17th century. Rolle's theorem is very closely related to the **mean value theorem**. In the mean value theorem *f(a)* does not have to equal *f(b)* but there is still at lest one point in between these two on *f* where the slope of the tangent line is parallel to that of the line connecting (*a, f(a)*) and (*b, f(b)*). The mean value theorem is a generalized form of Rolle's theorem that is applicable to a wider variety of situations but has a similar meaning.

ROMAN NOTATION

Roman notation is an additive (and subtractive) system of numerical notation, originally used within the ancient Roman empire that extended far past the Italian peninsula, in which letters are used to denote certain "base" numbers. Arbitrary numbers are then denoted using combinations of these base symbols. Roman numerals were developed about 500 B.C. and were at least partially derived from old Greek alphabet symbols that were not used in the new Latin alphabet. Other symbols were more creatively derived. Some Roman numerals probably came about from counting on fingers. The "I" for one came from holding up one finger, as were the symbols "II" and "III" likewise created from holding up two and three fingers, respectively. The number five was derived from holding up five fingers with the "V" shape made by the thumb and the forefinger. Although not known for sure, the symbol "X" for 10 is speculated to have been created by combining a regular "V" on top of an upside down "V", thereby making it two times five, or ten. The seven letters (or symbols) of the Roman numeral system are I, V, X, L, C, D, and M, that stand, respectively, for 1, 5, 10, 50, 100, 500, and 1,000 in the Arabic numeral system. The Roman numerals representing one to nine are in order: I, II, III, IV, V, VI, VII, VIII, and IX. Roman numerals may be written in lowercase letters, though they more commonly appear in capitals. A bar over a letter multiplies it by 1,000; thus,

$$\overline{\text{X}}$$

equals 10,000 (10 x 1,000). For the Romans, multiple bars were possible but, in practice, only one bar was usually used; with two bars rarely used and more than two bars almost never used. The ancient Roman numerals at one time went no higher than 100,000. However, large debts accumulated under the rule of Vespian made the use of the number 1,000,000 (million) necessary. At that time the Romans wrote it as "10 x 100,000", or an X with a frame around it, that is |X|. So with just the aforementioned seven number symbols, and modifiers like the "bar", the Romans expressed numbers from 1 to 1,000,000. Examples of large numbers (such as **calendar** years) expressed with Roman numerals are MDCCCLXXXII for 1882, MCMIL for 1949, and MCMLXXIV for 1974.

Today Roman numerals are read according to five basic rules that in some cases have only recently been incorporated into use. In rule one, a letter repeated once or twice repeats its value that many times. For example, XXX = 30 and CC = 200. In rule two, if one or more letters are placed immediately *after* another one of equal or lesser value, the two values are added. For example, II = 2, XV = 15, LX = 60, and DM = 1,500. In rule three, if a letter is placed immediately *before* another one of greater value, the first is sub-

tracted from the second. For example, IV = 4, XL = 40, CD = 400, and CM = 900. (As a note, the practice of placing smaller digits before large ones to indicate **subtraction** was rarely used by Romans but came into popularity in Europe after the invention of the printing press.) In rule four, instead of using four symbols to represent, for instance, 4 (IIII) and 40 (XXXX), such numbers are instead denoted by preceding the symbol for 5, 50, 10, 100, etc., with a symbol indicating *subtraction*. During the ancient Roman days VIIII for 9, for instance, was commonly found. Today IX would be used to denote 9. In rule five, a bar placed on top of a letter or string of letters increases the numeral's value by 1,000 times. For example, XV = 15, while

$$\overline{\text{XV}}$$

= 15,000 (15 x 1,000).

The history of Roman numerals is not well documented and many written articles are contradictory. It is known that the Roman numerals were part of the customs and culture of the Roman people. The Roman numeral system was abandoned by AD 1500 for the newer Arabic numbers. But, up until the eighteenth century, Roman numerals were still used in Europe for bookkeeping even though the Arabic numerals were known and widely used from around AD 1000. Roman numerals do not lend themselves easily to complex **multiplication** or subtraction. Thus, the switchover to the new Arabic "figure" numbers was a necessary development as money became more commonly used than barter, and as trade merchants desired more efficient ways to keep their books. Roman numerals are still in used today, more than 2000 years after their introduction, but only for certain limited purposes. Examples of today's uses are in the release year and copyright date of movies, television programs and videos; pagination of preliminary pages of books before the main page numbering begins and numbering of paragraphs in complex documents; numerals on the faces of many time pieces; inscription dates on public buildings, monuments, and gravestones; names of monarchs (e.g., King Edward VII of England) and Popes (e.g., John Paul II); and wars (e.g., World War I and WWII). These uses are based mostly on formality or variety.

See also History of mathematics

ROOT, EXPONENTIAL, AND LOGARITHMIC EQUATIONS

A root is a value (X) that when multiplied by itself (n) number of times provides another value (Y), i.e., $X^n = Y$. In this exponential equation, X is called a root of Y, n is called the exponent, and Y is called a power of X.

For example, if X=3 and n=2, the equation becomes $3^2 = 9$. In this case, 9 is the second power of 3 and 3 is the **square** root of 9. Written another way, this becomes $\sqrt{9} = 3$ or $9^{1/2} = 3$. Of course, -3 is also a root of 9 because $-3^2 = 9$. In practice, an equation written as $\sqrt{9}$ refers only to the positive root 3 (which is called the principal root), whereas an equation written as $\pm\sqrt{9}$ refers to positive and negative **roots**, i.e., 3 and -3.

Logarithms are used as an aid in simplifying expressions for extracting roots and expanding **powers**. For the case above, the logarithmic equation is $\log_3 9 = 2$, where 3 is the root (or base) and 2 is the exponent. Logarithms are always defined in terms of a base, which must be greater than 0 and cannot equal 1. The logarithm of a number is the exponent required to raise the base to the desired power (i.e., the logarithm is the number of times the base must be multiplied by itself to provide the number).

Although any positive base except 1 can be used, logarithms are often given in base 10. A base 10 logarithm is called a common or Briggs logarithm. Therefore, $\log_{10} 100 = 2$. The inverse of a logarithmic equation is the corresponding exponential equation, which is called the antilog: $10^2 = 100$.

Because, base 10 logs can be easily calculated on most hand-held calculators, it is often desirable to convert a log equation in another base into base 10 using the following equation: $\log_A B = \log_{10} B / \log_{10} A$. To convert our earlier problem to a base 10 problem, $\log_3 9$ is equivalent to $\log_{10} 9 / \log_{10} 3$. From a logarithm table or **calculator**, $\log_{10} 9$ is 0.95424 and $\log_{10} 3$ is 0.47712 and 0.95424/0.47712=2. Therefore, $\log_3 9 = 2$, which is correct.

Logarithms are useful for finding the positive roots of very large numbers. For example to find the fifth root of 32,768 ($\sqrt[5]{32,768}=X$ or $X^5 = 32,768$), the logarithmic equation may be written $\log_X 32,768 = 5$. Converting to log base 10 provides $5 = \log_{10} 32,768 / \log_{10} X$. From a calculator, $\log_{10} 32,768 = 4.51545$. Therefore, $5 = 4.51545 / \log_{10} X$ or $\log_{10} X = 0.90309$. Remember that a logarithmic equation is the inverse of an exponential equation in the same base. Therefore, $\log_{10} X = 0.9039$ is equivalent to $10^{0.90309} = X$. Now it is a simple matter to find that X=8. Therefore, $8^5 = 32,768$ and 8 is the fifth root of 32,768.

Usually, the base 10 is not written out, thus, log 100 is assumed to mean $\log_{10} 100$. Note that logarithmic **equations** provide only the positive roots of positive numbers. The roots of **negative numbers** are determined using **imaginary numbers**.

ROOTS

A root of a function f is a number x such that $f(x) = 0$. The **fundamental theorem of algebra** states that if f is a polynomial of degree n with complex number coefficients then it has at most n roots. The problem of finding the roots of f has a four thousand year history. The ancient Egyptians (2000 BC) knew how to find the roots of a degree 2 polynomial equation (or quadratic equation) using the quadratic formula. In the 1500s, **Girolamo Cardano** discovered how to find the roots of any degree 3 polynomial (or cubic equation). Degree 4 **polynomials (quartic equations)** were soon easy to solve as well. However, **Abel** proved in 1827 that there is no general formula

that can solve a polynomial of degree 5 (**quintic equations**) or higher. His discovery, in part, led to the development of modern **algebra** and **group theory**. Because of Abel's discovery, mathematicians gave up the problem of finding the exact roots of a polynomial and, instead, focused their energy on finding approximate solutions to polynomials.

Isaac Newton used the following method that works for all differentiable **functions** (see **differential calculus**). Let x_0 be a number that is guessed to be close to a root of f. Let $x_1 = F(x_0)$ where $F(x) = x - f(x)/f'(x)$. Here, f' stands for the derivative of f. Let $x_2 = F(x_1)$. In general, let $x_n = F(x_{n-1})$. Taylor's theorem implies that in a neighborhood of x_0, $f(x)$ is approximately equal to $f'(x_0)*(x - x_0) + f(x_0)$ which is **zero** when $x = F(x_0)$. If x_0 is chosen close enough to a root, then the sequence of numbers $\{x_n\}$ converges rapidly (quadratic to be precise) to a root of f. However, if x_0 is badly chosen, the sequence will not converge at all. Consequently, it is often necessary to use some other method for approximating the roots before applying Newton's method. Also, Newton's method requires computing f' which may be difficult for non-polynomial functions. For these functions, there are ways to approximate f'. The secant method uses the fact that $f'(x_n)$ is approximately $(f(x_n) - f(x_{n-1}))/(x_n - x_{n-1})$.

In 20th century, many new methods were discovered for approximating the roots of polynomial equations. Most use the methods of **numerical analysis**. Some of them find one root at a time. These include the Jenkins-Traub method, Larkin's method and Muller's method. Methods that find all the roots simultaneously include Weyl's method, the Durand-Kerner method, and the divide-and-conquer algorithms. These last ones are well suited for parallel processing which gives them a speed advantage over the iterative methods.

Weyl's method goes like this: Find a **square** in the complex plane that contains all the zeros of f. Next divide the square into four smaller squares and determine which of these four squares contains zeros of f. If a square contains zeros of f, then divide it into four smaller squares and determine which of the smaller squares contains zeros of f. Continue this process until the desired degree of accuracy is achieved. This **algorithm** can easily be modified to find only the zeros of f within a specified **area**.

See also Cubic equations; Fundamental theorem of algebra; Group theory; Polynomial; Quadratic equations; Quartic equations; Quintic equations; Numerical analysis

ROTATION

A rotation is one of three rigid motions that move a figure in a plane without changing its size or shape. As its name implies, a rotation moves a figure by rotating it around a center somewhere on a plane. This center can be somewhere inside or on the figure, or outside the figure completely. The two other rigid motions are reflections and **translations**.

Figure 1 illustrates a rotation of 30° around a point C.

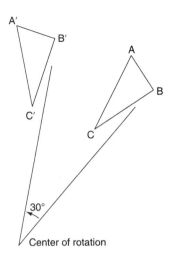

Figure 1.

This rotation is counterclockwise, which is considered positive. Clockwise rotations are negative. The "product" of two rotations, that is, following one rotation with another, is also a rotation. This assumes that the center of rotation is the same for both. When one moves a heavy box across the room by rotating it first on one corner then on the other, that "product" is not a rotation.

Rotations are so commonplace that it is easy to forget how important they are. A person orients a map by rotating it. A clock shows time by the rotation of its hands. A person fits a key in a lock by rotating the key until its grooves match the pattern on the keyhole. Rotating an M 180° changes it into a W; 6s and 9s are alike except for a rotation.

Rotary motions are one of the two basic motions of parts in a machine. An automobile wheel converts rotary **motion** into translational motion, and propels the car. A drill bores a hole by cutting away material as it turns. The earth rotates on its axis. The earth and the moon rotate around their centers of gravity, and so on.

Astronomy prior to **Copernicus** was greatly complicated by trying to use the earth as the center of the rotation of the planets. When **Kepler** and Copernicus made the sun the gravitational center, the motions of the planets became far easier to predict and explain (but even with the sun as the center, planetary motion is not strictly rotational).

When points are represented by coordinates, a rotation can be effected algebraically. How hard this is to do depends upon the location of the center of rotation and on the kind of coordinate system which is used. In the two most commonly employed systems, the rectangular **Cartesian coordinate system** and the polar coordinate system, the center of choice is the origin or pole.

In either of these systems a rotation can be thought of as moving the points and leaving the axes fixed, or vice versa. The mathematical connection between these alternatives is a simple one: rotating a set of points clockwise is equivalent to rotating the axes, particularly with reflections, it is usually preferable to leave the axes in place and move the points.

When a point or a set of points is represented with polar coordinates, the **equations** that connect a point (r, θ) with the rotated image (r', θ') are particularly simple. If θ_1 is the **angle** of rotation:

$$r' = r$$
$$\theta' = \theta + \theta$$

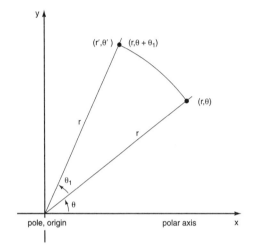

Figure 2.

Thus, if the points are rotated 30° counterclockwise, $(7,80°)$ is the image of $7,50°$. If the set of points described by the equation $r = \theta/2$ is rotated π units clockwise, its image is described by $r = (\theta - \pi)/2$. Rectangular coordinates are related to polar coordinates by the equations $x = r \cos \theta$ and $y = r \sin \theta$.

Therefore the equations which connect a point (x, y) with its rotated image (x', y') are $x' = r \cos (\theta + \theta_1)$ and $y' = r \sin (\theta + \theta_1)$.

Using the **trigonometric identities** for $\cos (\theta + \theta_1)$ and $\sin (\theta + \theta_1)$, these can be written $x' = x \cos \theta_1 - y \sin \theta_1$ and $y' = x \sin \theta_1 + y \cos \theta_1$ or, after solving for x and y: $x = x' \cos \theta_1 + y \sin \theta_1$ and $y = -x' \sin \theta_1 + y \cos \theta_1$.

To use these equations one must resort to a table of sines and cosines, or use a **calculator** with SIN and COS keys.

One can use the equations for a rotation many ways. One use is to simplify an equation such as $x^2 - xy + y^2 = 5$. For any second-degree polynomialequation in x and y there is a rotation which will eliminate the xy term. In this case the rotation is 45°, and the resulting equation, after dropping the primes, is $3x^2 + y^2 = 10$.

Another area in which rotations play an important part is in rotational **symmetry**. A figure has rotational symmetry if there is a rotation such that the original figure and its image coincide. A **square**, for example, has rotational symmetry because any rotation about the square's center which is a multiple of 90° will result in a square that coincides with the original. An ordinary gear has rotational symmetry. So do the numerous objects such as vases and bowls which are decorated repetitively around the edges. Actual objects can be checked for rotational symmetry by looking at them. Geometric figures described analytically can be tested using the equations for rotations. For example, the **spiral** $r = 2\theta$ has two-fold rotational symmetry. When the spiral is rotated 180°, the image coincides with the original spiral.

RUFFINI, PAOLO (1765-1822)
Italian algebraist and educator

Paolo Ruffini made significant contributions in the areas of medicine and philosophy, as well as mathematics, where he developed the theory that a quintic equation cannot be solved by radicals. This theory later came to be known as the Abel–Ruffini **theorem**, named after Ruffini and the Norwegian mathematician **Niels Abel**, who published the first accepted version of this theorem. Ruffini also played a crucial role in the development of **group theory**.

Ruffini was born September 22, 1765, in Valentano, Papal States (also known as Valentano, Viterbo and now part of Italy), to Basilio Ruffini, a physician, and his wife, Maria Francesca Ippoliti. He moved with his parents to Modena, Duchy of Modena (now also part of Italy) when he was a teenager and attended the University of Modena, where he studied medicine, philosophy, literature, and mathematics. He practiced mathematics with many well–known instructors and mathematicians, including Luigi Fantini, who taught **geometry**, and Paolo Cassiani, who taught infinitesimal calculus (see **infinitesimal** and **calculus**). The well–rounded Ruffini excelled in all his classes, so much so that when Cassiani took a leave from his foundations of **analysis** course in the fall of 1787 to accept a position as councillor of the Este domains, Ruffini, at that time still a student, instructed the class himself.

Ruffini graduated from the university with a degree in philosophy and medicine the following summer, and soon after that obtained a degree in mathematics as well. That fall he was appointed professor of the foundations of analysis. He remained in that position until 1791, when he replaced Fantini as professor of the elements of mathematics. That same year Ruffini obtained his license to practice medicine by the Collegiate Medical Court of Modena. In addition to work in the academic field, he also began to practice medicine.

Ruffini's professional situation changed drastically in 1796, however, when Napoleon's troops occupied Modena and he was appointed representative from the department of Panaro to the Junior Council of the Cisalpine Republic, against his wishes. Ruffini was relieved of his post in 1798 and returned to scientific research, but was banned from teaching or holding a public office after he refused to swear an oath of allegiance to the republic, citing religious reasons. Ruffini continued to practice medicine and conduct mathematical research, and during this time published what later became known as the Abel–Ruffini theorem, which stated that a general algebraic equation of greater than the fourth degree could not be solved using only radicals, such as **square roots**, cube roots, etc.

The theorem was first published in *Teoria generale delle equazioni* in 1799, and later in revised form in *Riflessioni intorno alla soluzione delle equazioni algebriche generalie* in 1813. Upon publication, the theorem was met with skepticism among other mathematicians, with the exception of **Augustin–Louis Cauchy**, who told Ruffini in an 1821 letter that his work deserved the attention of the mathematics community and that he wholeheartedly agreed with Ruffini's findings. The theorem gained general acceptance when it was proven by Abel in 1824.

Ruffini also developed the theory of substitutions, considered a significant contribution to the theory of **equations**. This theory laid the foundation for **Èvariste Galois'** general theory of the solubility of algebraic equations and served as an important forerunner of group theory.

Ruffini later developed the basic rule for determining the quotient and the remainder resulting from the **division** of a polynomial in the **variable** x by a binomial of the form x–a, using approximation by means of infinite algorithms. Despite a growing trend toward acknowledging the uncertainty of the foundations of infinitesimal analysis, Ruffini, with the support of Cauchy and Abel, demanded rigor of himself and others. During his tenure as president of the Societa Italiana dei Quaranta, he refused to accept to papers by Giuliano Frullani because they relied on series for which the **convergence** had not been demonstrated.

Ruffini moved on to apply his manner of thinking to philosophical and biological matters, determining that the faculties of the soul could not be measured because they do not correspond to magnitudes. Ruffini returned to academia with the fall of Napoleon in Modena in 1814. At this time, Ruffini was appointed rector of the university, as well as chair of the mathematics and practical medicine departments.

Ruffini continued as a practicing physician as well, treating impoverished and royal citizens of Modena alike. He contracted typhus while aiding victims of the 1817–18 typhus epidemic, but recovered enough to write about the experience from a medical and personal perspective. In 1822, however, he contracted chronic pericarditis, accompanied by a serious fever, and he died on May 10 of that year.

RUSSELL, BERTRAND (1872-1970)
English logician and philosopher

A seminal figure in 20th-century mathematical logic and philosophy, Bertrand Russell produced more than seventy books and pamphlets on topics ranging from mathematics and logic to philosophy, religion, and politics. He valued reason, clarity, fearlessness, and independence of judgment, and held the conviction that it was the duty of the educated and privileged classes to lead. Having protested against Britain's participation in World War I and against the development of nuclear weapons, Russell was imprisoned twice for his convictions. In writing his own obituary, he described himself as a man of unusual principles, who was always prepared to live up to them.

Bertrand Arthur William Russell was born on May 18, 1872, in Ravenscroft, Trelleck, Monmouthshire, England. His family tree can be traced back to John Russell, a favorite courtier of Henry VIII. His grandfather, Lord John Russell, served as Prime Minister under Queen Victoria. Bertrand

Russell's father, Lord Amberley, served in Parliament briefly, but was defeated because of his rejection of Christianity and his advocacy of voting rights for women and birth control. His mother was Kate Stanley, the daughter of a Liberal politician. Orphaned before the age of five, Russell was raised by his paternal grandmother, a Scotch Presbyterian with strong moral standards who gave duty and virtue greater priority than love and affection. His grandmother did not have confidence in the moral and religious environment at boarding schools, so, after briefly attending a local kindergarten, he was taught by a series of governesses and tutors. Later he wrote in his *Autobiography* that he had been influenced by his grandmother's fearlessness, her public spirit, her contempt for convention, and her indifference to the opinion of the majority. On his twelfth birthday she gave him a Bible in which she had written her favorite texts, including "Thou shalt not follow a multitude to do evil." It was due to her influence, he felt, that in later life he was not afraid to champion unpopular causes.

When Russell was eleven, his brother Frank began to teach him **Euclidean geometry**. As he recounts in his *Autobiography,* "This was one of the great events of my life, as dazzling as first love. I had not imagined that there was anything so delicious in the world." Russell was greatly disappointed, however, to find out that **Euclid** started with axioms which were not proved but were simply accepted. Russell refused to accept the axioms unless his brother could give a good reason for them. When his brother told him that they couldn't continue unless he accepted the axioms, Russell relented, but with reluctance and doubt about the foundations of mathematics. That doubt remained with him and determined the course of much of his later work in mathematics.

At age 15, Bertrand left home to take a training course to prepare for the scholarship examination at Trinity College, Cambridge. Lonely and miserable, he considered suicide but rejected the idea because, as he later said, he wanted to learn more mathematics. He kept a secret diary, written using letters from the Greek alphabet, in which he questioned the religious ideas he had been taught. After a year and a half, he took the Trinity College examination and won a scholarship. One of the scholarship examiners was **Alfred North Whitehead**. Apparently impressed with Russell's ability, Whitehead arranged for him to meet several students who soon became his close friends. He was invited to join a group, "The Apostles," which met weekly for intense discussions of philosophy and history. His chief interest and chief source of happiness, he said, was mathematics; he received a first class degree in mathematics in 1893. He found, however, that his work in preparing for the exams had led him to think of mathematics as a set of tricks and ingenious devices, too much like a crossword puzzle; this disillusioned him. Hoping to find some reason for supposing mathematics to be true, he turned to philosophy. He studied the philosophy of idealism, which was popular in Cambridge at the time, concluding that time, **space** and matter are all illusions and the world resides in the mind of the beholder. He took the Moral Science examination in 1894 and received an honors degree.

At 17, Russell had fallen in love with Alys Pearsall Smith, an American from a wealthy Philadelphia Quaker family, and after graduation they became engaged. His grandmother did not approve. Hoping he would become interested in politics and lose interest in Alys, she arranged for him to become an attaché at the British Embassy in Paris. He found his work at the Embassy boring, and upon his return three months later, he and Alys were married.

In 1895, Russell wrote a dissertation on the foundations of **geometry**, which won him a fellowship at Trinity College and enabled him to travel and study for six years. He and Alys went to Berlin to study the German Socialist movement and the writings of Karl Marx. During his travels, he formulated a plan to write a series of books on the philosophy of the sciences, starting with mathematics and ending with biology, growing gradually more concrete, and a second series of books on social and political questions, working to more abstract issues. During his long life, he managed to fulfill much of this plan. As a result, according to biographer Ronald Clark, no other Englishman of the twentieth century was to gain such high regard in both academic and non-academic worlds.

Russell returned to London in 1896 and lectured on his experiences to students of the London School of Economics and the Fabian Society, publishing his studies as *German Social Democracy,* the first of his numerous books and pamphlets. In this publication he demonstrated his ability to investigate a subject quickly and then present it in clear and convincing language. He and Alys traveled to the United States, where they visited Alys's friend, the poet Walt Whitman. Russell lectured on **non-Euclidean geometry** at Bryn Mawr College, where Alys had been a student, and at Johns Hopkins University. In 1900, he was asked to lecture on German mathematician **Gottfried Wilhelm von Leibniz** at Cambridge. He wanted to show that mathematical truths did not depend on the mathematician's point of view. He re-examined the philosophy of idealism and abandoned it, concluding that matter, space and time really did exist. He published his views in *A Critical Exposition of the Philosophy of Leibniz.*

Russell then started work on an ambitious task—devising a structure which would allow both the simplest laws of logic and the most complex mathematical theorems to be developed from a small number of basic ideas. If this could be done, then the axioms which mathematicians accepted would no longer be needed, and both logic and mathematics would be part of a single system. At a conference in Paris in 1900, he met **Giuseppe Peano**, whose book on **symbolic logic** held that mathematics was merely a highly developed form of logic. Russell became interested in the possibility of analyzing the fundamental notions of mathematics, such as order and cardinal numbers, using Peano's approach. He published in 1903; its fundamental thesis was that mathematics and logic are identical.

Applying logic to basic mathematical concepts and working with **Georg Cantor**'s **proof** that a class of all classes cannot exist, Russell formulated his famous paradox concerning classes that are members of themselves—such as the class of classes. According to Russell's paradox, it is impossible to answer the question of whether the class containing all classes that are not members of themselves is a member of itself—for if it is a member of itself, then it does not meet the terms of the class; and if it is not, then it does. (This is the same kind of paradox as is found in the statement "Everything I say is a lie.") In developing this Theory of Types, however, Russell realized the absurdity of asking whether a class can be a member of itself. He concluded that if classes belong to a particular type, and if they consist of homogenous members, then a class cannot be a member of itself. In planning *The Principles of Mathematics,* an exposition of his ideas on mathematics and logic, Russell decided on two volumes, the first containing explanations of his claims and the second containing mathematical proofs. His former teacher, Alfred Whitehead, had been working on similar problems. Consequently, the two scholars decided to collaborate on the second part of the task. For nearly a decade they worked together, often sharing the same house, sending each other drafts and revising each other's work. The result, *Principia Mathematica,* was a separate work published in three volumes, the last in 1913. Before it was completed, Russell was elected to the Royal Society, of which Whitehead was already a member.

Russell was persuaded to run for Parliament in 1907 as a supporter of voting rights for women; he lost the election. Appointed a lecturer in logic and the principles of mathematics at Trinity College in 1910, he published a short introduction to philosophy, *The Problems of Philosophy,* in 1912, and *Our Knowledge of the External World* in 1914. Invited to Cambridge, Massachusetts, to give the Lowell Lectures and a course at Harvard University in 1914, he also lectured in New York, Chicago, and Ann Arbor, Michigan.

Russell was an outspoken critic of England's participation in World War I. He worked with the No-Conscription Fellowship to protest the drafting of young men into the army. In 1916 he gave a series of lectures in London which were published as *Principles of Social Reconstruction.* They were also published in the United States as *Why Men Fight: A Method of Abolishing the International Duel.* He was invited to give a lecture series at Harvard based on the book, but he was denied a passport because he had been convicted of writing a leaflet criticizing the imprisonment of a young conscientious objector. As a result of his conviction, he was dismissed from his lectureship at Trinity College. He wrote an open letter to Woodrow Wilson, the President of the United States, urging him to seek a negotiated peace rather than go to war. Since Russell's mail was being intercepted by the British government, he sent the letter with a young American woman. The story made the headlines in the *New York Times.*

Russell wrote articles for *The Tribunal,* published by the No-Conscription Fellowship. In one article, he predicted that the consequences of the war would include widespread famine and the presence of American soldiers capable of intimidating striking workers. He was charged with making statements likely to prejudice Britain's relations with the United States and was sentenced to six months in prison. Before his imprisonment, he wrote *Roads to Freedom: Socialism, Anarchism and Syndicalism.* While in prison he wrote *Introduction to Mathematical Philosophy,* which

explained the ideas of *The Principles of Mathematics* and *Principia Mathematica* in relatively simple terms.

In 1920, Russell visited Russia, where he was disappointed with the results of the 1917 revolution, and China, where he lectured at the National University of Peking. The following year, Russell divorced his wife Alys and married Dora Black. His son John Conrad was born, and in 1922 his daughter Katharine Jane was born. In 1927 he and Dora established Beacon Hill School for their children and others. He traveled to the United States for lecture tours in 1924, 1927, 1929 and 1931, speaking on political and social issues. He divorced Dora in 1935 and married Patricia Spence the following year. A son, Conrad Sebastian Robert, was born in 1937.

Russell lived in the United States during World War II, from 1938 to 1944, lecturing at the University of Chicago, the University of California at Los Angeles, Bryn Mawr, Princeton, and the Barnes Foundation at Merion, Pennsylvania. He was invited to teach at New York City College, but the invitation was revoked due to objections to his atheism and unconventional personal morality. He continued to publish works on philosophy, logic, politics, economics, religion, morality and education. In 1944, he returned to Trinity College, where he had been offered a fellowship. He published *A History of Western Philosophy,* participated in radio broadcasts in England, and lectured in Norway and Australia. He was awarded the Order of Merit by the King in 1949 and the Nobel Prize for Literature in 1950. At the age of 80, he divorced Patricia Spence and married Edith Finch.

Acutely aware of the dangers of nuclear war, Russell served as the first president of the Campaign for Nuclear Disarmament in 1958, and as president of the Committee of 100 in 1960. As a member of the Committee of 100, he encouraged demonstrations against the British government's nuclear arms policies; for this he was sentenced to two months in prison, which was reduced to one week for health reasons. In his ninth decade, Russell established the Bertrand Russell Peace Foundation, published his *Autobiography,* appealed on behalf of political prisoners in several countries, protested nuclear weapons testing, criticized the war in Vietnam, and established the War Crimes Tribunal. He died on February 2, 1970, at the age of 98.

RUTHERFORD, ERNEST (1871-1937)
New Zealand physicist

Ernest Rutherford was one of the giants in the field of atomic physics as science was just beginning to understand this world of **infinitesimal** yet powerful reactions. He did important research on radiation and coined such words as "alpha," "beta," and "gamma" waves, "proton," "half-life," "neutron," and "daughter atom." In addition, Rutherford showed conclusively that, despite the dominant scientific belief of his day that atoms were immutable and indivisible, radioactive elements can shed atoms to become different elements.

The son of successful farmers and one of 12 children, Rutherford was born in Bridgewater, a small town near Nelson,

New Zealand. He enjoyed life on the farm and considered making it his livelihood as well, but his intellectual talents were apparent from an early age as he made his way through the local government-run schools. When he won an academic scholarship to Canterbury College in Christchurch, New Zealand in 1889 as he was finishing up at Nelson Collegiate School, Rutherford decided to take that route instead.

Rutherford received a degree from Canterbury in 1893, having majored in **mathematics and physics**. He stayed on at the college to do research, some of which resulted in a radio wave detector that worked by using iron's magnetic properties. In 1895 he won another scholarship—this time to England's Trinity College at Cambridge University. There he worked as a research student under the famous physicist J.J. Thomson in the Cavendish Laboratory. Rutherford remained at Cambridge for three years, doing research on the conductivity of gas ionized by radiation and discovering the existence of alpha and beta rays in uranium radiation. He left in 1898 to accept the physics chair at Canada's prestigious McGill University. He was only 28 years old.

Once he settled in at McGill, Rutherford continued his research on radioactivity. During this period he identified radiation of the alpha, beta, and gamma varieties. Then, in 1902, introduced and proved the "spontaneous transformation" theory of radioactive decay, showing that an atom of one radioactive substance can become a different atom by emitting radiation. In 1904 Rutherford published his first book on radiation, *Radioactivity,* which brought his work to the attention of the wider scientific world.

In 1907, Rutherford accepted a new position as professor of physics at the University of Manchester in England, where he soon established a center dedicated to the study of radiation. For his work in that field, he received the 1908 Nobel Prize in Chemistry. It was at the center in 1909 that Rutherford discovered the atomic nucleus and invented a **model** of the atom that looked something like our solar system, with electrons revolving around the sun-like nucleus. His discovery became the basis of the new nuclear science, and is the basic model that physicists still use today. Rutherford later predicted the presence of an uncharged particle (the neutron) as well.

During World War I (1914-1918), Rutherford assisted the British Navy with developing ways to detect enemy submarines, but afterward he returned to his research at the Manchester radiation lab. His next breakthrough was to cause a nonradioactive atom to fall apart by dislodging a single particle, thus creating the first man-made nuclear reaction. He named the particle "proton," since it had a positive charge.

Rutherford left Manchester to became director of the Cavendish Laboratory and the Cavendish professor of physics in 1919. That year he published *Radiation from Radioactive Substances.* Rutherford remained active in research until the end of his life, and continued publishing at an energetic pace. In 1926 he released *The Electrical Structure of Matter;* he published *The Artificial Transmutation of the Elements* in 1933 and *The Newer Alchemy* in 1937. Meanwhile, he took an active stance against the Nazis during World War II, serving as president of the Academic Assistance Council, which helped innumerable German refugees.

Married in 1900, Rutherford and his wife had one daughter. His favorite hobbies were golf and driving cars. Rutherford died on October 19, 1937 and is buried at London's Westminster Abbey between **Isaac Newton** and Lord Kelvin.

S

SACCHERI, GIOVANNI GIROLAMO (1667-1733)

Italian priest and teacher

Giovanni Girolamo Saccheri was a Jesuit priest who did pioneering work in the areas of mathematical logic and **non-Euclidian geometry**.

The son of a lawyer, Saccheri was born in San Remo, Genoa (now Italy) on September 5, 1667. He began academic training with the Jesuits in Genoa in 1685 and five years later enrolled at the Jesuit College of Brera to study philosophy and theology. In the meantime, he made a modest living by tutoring at the school. One of his teachers, a poet and mathematician named Tommaso Ceva, convinced Saccheri to direct his energies toward mathematics and became the young man's academic mentor. With Ceva's guidance, Saccheri published his first book, *Quaesita geometrica,* in 1693. In it, Saccheri solved numerous problems in elementary and coordinate geometry. He was ordained a priest in 1694 at Como, Italy.

That same year, Saccheri began teaching philosophy at the University of Turin, where he remained until 1697. While there, he published *Logica demonstrativa,* one of his most important works. *Logica* treats logic in the Euclidian style, with definitions, demonstrations, and postulates. In 1697 Saccheri changed jobs again, this time moving to the Jesuit College of Pavia (also known as the Universita Ticinese), where he would teach for the rest of his life. Two years later, he became the mathematics chair at the school, appointed by the Senate of Milan.

In 1708, Saccheri published *Neo-statica,* based on a branch of **mechanics** (statics) that deals with objects at rest or forces in equilibrium. His most famous work, however, did not come until 1733. *Euclides ab Omni Naevo Vindicatus (Euclid Cleared of Every Flaw)* attempted to prove **Euclid**'s **parallel postulate**. In this work of synthesis, Saccheri gave a complete analysis of the parallels problems in terms of **Omar Khayyam**'s quadrilateral. He used the Euclidian assumption that straight lines cannot enclose an **area**, and thus felt justified in excluding geometries that contain no parallels. Yet he also had to show that a unique parallel through a point to a given line exists. To do this Saccheri assumed the existence of multiple parallels and tried to find a contradiction in that. He reportedly managed to convince himself of a contradiction, but no one else. Nevertheless, Saccheri's logical and mathematical reasoning are now integral facets of modern mathematical logic and non-Euclidian geometry.

Saccheri died in Milan, Italy on October 25, 1733.

SCHWARTZ, LAURENT (1915-   )

French mathematician

Laurent Schwartz has won many prestigious awards for his contributions to mathematics, but his most influential work has been in the areas of functional **analysis**, **integral calculus**, and **differential calculus**. In particular, it was his work on the theory of distributions in the late 1940s for which Schwartz will always be remembered.

Schwartz was born on March 5, 1915 in Paris, France. His father was a well-known surgeon who helped Schwartz recover from the polio he contracted when he was 11. As a boy, Schwartz was an excellent student, doing especially well in Greek, Latin, and mathematics. He attended the École Normale Supérieure (ENS) for his early education. It was there that Schwartz became politically active—a characteristic that would stay with him for life. The basis of his political beliefs became nonviolence, working against oppression, service to humanity, and intellectual honesty. He remained a devoted Trotskyist until about 1947.

Schwartz graduated from the ENS in 1937, after which he performed his obligatory service in the military. Serving as an officer until 1939, he remained for a third year of active service when World War II started. The war presented a real danger to Schwartz because of his Jewish heritage; he changed his name to Laurent-Marie Selimartin, a traditionally upper-class Protestant name, to try to fool the Nazis. Some of his academic friends were not so resourceful or lucky.

In 1940, Schwartz traveled to Toulouse, France to be with his parents and there found a position with the National Science Academy. This ended in 1942, but a research stipend from Michelin sustained Schwartz from 1943 until the end of the war in 1945. Again finding himself without a way to make a living at that point, he decided to move to the University of Clermont-Ferand. He met **André Weil** at the school, which helped inspire Schwartz to write his doctoral thesis ("Real Exponential Sums," 1943) in the following two years.

Despite the mental and physical exhaustion produced by the war, Schwartz nevertheless came up with his most brilliant mathematical work just as the conflict was ending. Taking a position as professor in the Faculty of Science at France's University of Nancy in 1945, he soon finished formalizing his theory of distributions, which replaced the "delta function" of **Paul Adrien Maurice Dirac**. Prior to Schwartz's breakthrough, physicists used the Dirac delta in their work with mass distributions. However, while it was a useful tool, it was limited. Schwartz's theory of distributions for functional analysis introduced a more classically mathematical function, which also came to be used heavily in **differential equations**, spectral theory, and potential theory. He published his findings in the 1948 paper "Theory of Distributions," which earned him the coveted Fields Medal. He published a book on the topic in 1950-1951.

Schwartz left the University of Nancy in 1952, joining the faculty of the Sorbonne in Paris the following year. In 1959 he also began working at the Polytechnical School in Palaiseau as professor of analysis. He remained at both schools until his retirement from active teaching in 1983; Schwartz was especially helpful in initiating fundamental reforms in the training programs of the Polytechnical School and making it a major center of mathematical research. His efficacy in this respect was reflected in the French Government's appointment of Schwartz as president of the National Committee for the Evaluation of French Universities in 1985. He held that post until 1989.

Still intensely concerned with world politics, Schwartz remains busy and involved with the latest mathematical developments. He wrote an autobiography in 1997. Schwartz and his wife, herself a professional mathematician, were married in 1938 and have had two children, one of whom died in 1971.

SCIENTIFIC METHOD

Scientific thought aims to make correct predictions about events in nature. Although the predictive nature of scientific thought may not at first always be apparent, a little reflection usually reveals the predictive nature of any scientific activity. Just as the engineer who designs abridge ensures that it will withstand the forces of nature, so the scientist considers the ability of any new scientific **model** to hold up under scientific scrutiny as new scientific data become available.

It is often said that the scientist attempts to understand nature. But ultimately, understanding something means being able to predict its behavior. Scientists therefore usually agree that events are not understandable unless they are predictable. Although the word science describes many activities, the notion of prediction or predictability is always implied when the word science is used.

Until the seventeenth century, scientific prediction simply amounted to observing the changing events of the world, noting any irregularities, and making predictions based upon those regularities. The Irish philosopher and bishop **George Berkeley** (1685-1753) was the first to rethink this notion of predictability.

Berkeley noted that each person experiences directly only the signals of his or her five senses. An individual can infer that a natural world exists as the source of his sensations, but he or she can never know the natural world directly. One can only know it through one's senses. In everyday life people tend to forget that their knowledge of the external world comes to them through their five senses.

The physicists of the nineteenth century described the atom as though they could see it directly. Their descriptions changed constantly as new data arrived, and these physicists had to remind themselves that they were only working with a mental picture built with fragmentary information.

Scientific models

In 1913, **Niels Bohr** used the term *model* for his published description of the hydrogen atom. This term is now used to characterize theories developed long before Bohr's time. Essentially, a model implies some **correspondence** between the model itself and its object. A single correspondence is often enough to provide a very useful model, but it should never be forgotten that the intent of creating the model is to make predictions.

There are many types of models. A conceptual model refers to a mental picture of a model that is introspectively present when one thinks about it. A geometrical model refers to diagrams or drawings that are used to describe a model. A mathematical model refers to **equations** or other relationships that provide quantitative predictions.

It is an interesting fact that if a mathematical model predicts the future accurately, there may be no need for interpretation or visualization of the process described by the mathematical equations. Many mathematical models have more than one interpretation. But the interpretations and visualization of the mathematical model should facilitate the creation of new models.

New models are not constructed from observations of facts and previous models; they are postulated. That is to say that the statements that describe a model are assumed and predictions are made from them. The predictions are checked against the measurements or observations of actual events in nature. If the predictions prove accurate, the model is said to be validated. If the predictions fail, the model is discarded or adjusted until it can make accurate predictions.

The formulation of the scientific model is subject to no limitations in technique; the scientist is at liberty to use any method he can come up with, conscious or unconscious, to develop a model. Validation of the model, however, follows a single, recurrent pattern. Note that this pattern does not constitute a method for making new discoveries in science; rather it provides a way of validating new models after they have been postulated. This method is called the scientific method.

The scientific method 1) postulates a model consistent with existing experimental observations; 2) checks the predictions of this model against further observations or measurements; 3) adjusts or discards the model to agree with new observations or measurements.

The third step leads back to the second, so, in principle, the process continues without end. (Such a process is said to be recursive.) No assumptions are made about the reality of the model. The model that ultimately prevails may be the simplest, most convenient, or most satisfying model; but it will certainly be the one that best explains those problems that scientists have come to regard as most acute.

Paradigms are models that are sufficiently unprecedented to attract an enduring group of adherents away from competing scientific models. A paradigm must be sufficiently open-ended to leave many problems for its adherents to solve. The paradigm is thus a theory from which springs a coherent tradition of scientific research. Examples of such traditions include Ptolemaic astronomy, Copernican astronomy, Aristotelian dynamics, Newtonian dynamics, etc.

To be accepted as a paradigm, a model must be better than its competitors, but it need not and cannot explain all the facts with which it is confronted. Paradigms acquire status because they are more successful than their competitors in solving a few problems that scientists have come to regard as acute. Normal science consists of extending the knowledge of those facts that are key to understanding the paradigm, and in further articulating the paradigm itself.

Scientific thought should in principle be cumulative; a new model should be capable of explaining everything the old model did. In some sense the old model may appear to be a special case of the new model. In fact, whether this is so seems to be open to debate.

The descriptive phase of normal science involves the acquisition of experimental data. Much of science involves classification of these facts. Classification systems constitute abstract models, and it is often the case that examples are found that do not precisely fit in classification schemes. Whether these anomalies warrant reconstruction of the classification system depends on the consensus of the scientists involved.

Predictions that do not include numbers are called qualitative predictions. Only qualitative predictions can be made from qualitative observations. Predictions that include numbers are called quantitative predictions. Quantitative predictions are often expressed in terms of probabilities, and may contain estimates of the accuracy of the prediction.

Historical evolution of the scientific method

The Greeks constructed a model in which the stars were lights fastened to the inside of a large, hollow **sphere** (the sky), and the sphere rotated about the Earth as a center. This model predicts that all of the stars will remain fixed in position relative to each other. But certain bright stars were found to wander about the sky. These stars were called planets (from the Greek word for wanderer). The model had to be modified to account for **motion** of the planets. In **Ptolemy**'s (90-168 A.D.) model of the solar system, each planet moves in a small circular **orbit**, and the center of the small **circle** moves in a large circle around the Earth as center.

Copernicus (1473-1543) assumed the Sun was near the center of a system of circular orbits in which the Earth and planets moved with fair regularity. Like many new scientific ideas, Copernicus's idea was initially greeted as nonsense, but over time it eventually took hold. One of the factors that led astronomers to accept Copernicus's model was that Ptolemaic astronomy could not explain a number of astronomical discoveries.

In the case of Copernicus, the problems of **calendar** design and astrology evoked questions among contemporary scientists. In fact, Copernicus's theory did not lead directly to any improvement in the calendar. Copernicus's theory suggested that the planets should be like the earth, that Venus should show phases, and that the universe should be vastly larger than previously supposed. Sixty years after Copernicus's death, when the telescope suddenly displayed mountains on the moon, the phases of Venus, and an immense number of previously unsuspected stars, the new theory received a great many converts, particularly from non-astronomers.

The change from the Ptolemaic model to Copernicus's model is a particularly famous case of a paradigm change. As the Ptolemaic system evolved between 200 B.C. and 200 A.D., it eventually became highly successful in predicting changing positions of the stars and planets. No other ancient system had performed as well. In fact the Ptolemaic astronomy is still used today as an engineering approximation. Ptolemy's predictions for the planets were as good as Copernicus's. But with respect to planetary position and precession of the equinoxes, the predictions made with Ptolemy's model were not quite consistent with the best available observations. Given a particular inconsistency, astronomers were for many centuries satisfied to make minor adjustments in the Ptolemaic model to account for it. But eventually, it became apparent that the web of complexity resulting from the minor adjustments was increasing more rapidly than the accuracy, and a discrepancy corrected in one place was likely to show up in another place.

Tycho Brahe (1546-1601) made a lifelong study of the planets. In the course of doing so he acquired the data needed to demonstrate certain shortcomings in Copernicus's model. But it was left to **Johannes Kepler** (1571-1630), using Brahe's data after the latter's death, to come up with a set of laws consistent with the data. It is worth noting that the quantitative superiority of Kepler's astronomical tables to those computed from the Ptolemaic theory was a major factor in the conversion of many astronomers to Copernicanism.

In fact, simple quantitative telescopic observations indicate that the planets do not quite obey **Kepler's Laws**, and **Isaac Newton** (1642-1727) proposed a theory that shows why they should not. To redefine Kepler's Laws, Newton had to neglect all gravitational attraction except that between individual planets and the sun. Since planets also attract each other, only approximate agreement between Kepler's Laws and telescopic observation could be expected.

Newton thus generalized Kepler's laws in the sense that they could now describe the motion of any object moving in any sort of path. It is now known that objects moving almost as fast as the speed of **light** require a modification of **Newton's laws**, but such objects were unknown in Newton's day.

Newton's first law says that a body at rest remains at rest unless acted upon by an external **force**. His second law states quantitatively what happens when a force is applied to an object. The third law states that if a body A exerts a force F on body B, then body B exerts on body A a force that is equal in magnitude but opposite in direction to force F. Newton's fourth law is his law of gravitational attraction.

Newton's success in predicting quantitative astronomical observations was probably the single most important factor leading to acceptance of his theory over more reasonable but uniformly qualitative competitors.

It is often pointed out that Newton's model includes Kepler's Laws as a special case. This permits scientists to say they understand Kepler's model as a special case of Newton's model. But when one considers the case of Newton's Laws and relativistic theory, the special case argument does not hold up. Newton's Laws can only be derived from **Albert Einstein**'s (1876-1955) relativistic theory if the laws are reinterpreted in a way that would have only been possible after Einstein's work (see **Relativity, general** and **Relativity, special**).

The variables and parameters that in Einstein's theory represent spatial position, time, mass, etc. appear in Newton's theory, and there still represent **space**, time, and mass. But the physical natures of the Einsteinian concepts differ from those of the Newtonian model. In Newtonian theory, mass is conserved; in Einstein's theory, mass is convertible with energy. The two ideas converge only at low velocities, but even then they are not exactly the same.

Scientific theories are often felt to be better than their predecessors because they are better instruments for solving puzzles and problems, but also for their superior abilities to represent what nature is really like. In this sense, it is often felt that successive theories come ever closer to representing **truth**, or what is "really there." Thomas Kuhn, the historian of science whose writings include the seminal book *The Structure of Scientific Revolution* (1962), found this idea implausible. He pointed out that although Newton's **mechanics** improve on Ptolemy's mechanics, and Einstein's mechanics improve on Newton's as instruments for puzzle-solving, there does not appear to be any coherent direction of development. In some important respects, Professor Kuhn has argued, Einstein's general theory of relativity is closer to early Greek ideas than relativistic or ancient Greek ideas are to Newton's.

SCIENTIFIC NOTATION

When it is necessary to write very large or very small numbers, scientists and mathematicians use a method of notation that combines the significant digits of the number multiplied by ten with appropriate **exponents** that are **integers**. This is called scientific notation. When using this method, the numbers are often said to be multiplied by a power of ten. Positive **powers** of ten each add a **zero** to a number to the left of the decimal point, negative powers add a zero to the right of a decimal point. To illustrate, although these would not normally be used because they can be written and read more easily the usual way, in scientific notation, 1 would be written as 1×10^0, because $10^0=1$. The number 10 would be written as 1×10^1, and 100 as 1×10^2. Working with numbers smaller than 1, 0.1 would be 1×10^{-1} in scientific notation, $0.01=1 \times 10^{-2}$ and so on.

This method is a kind of shorthand that not only requires less writing, but also makes numbers more clear to the reader. With standard numeric notation, for example, one might have to count zeros to interpret a number like 0.00000000452, which, in scientific notation, takes fewer symbols to write and is easier, at a glance, to interpret: 4.52×10^{-9}. This makes scientific notation very useful in working with the kinds of numbers often found especially in sciences where very large or small sizes and time frames are encountered.

SECOND-ORDER ORDINARY DIFFERENTIAL EQUATIONS

An ordinary differential equation (ODE) is an equality involving a function containing an unknown and the derivative(s) of that function. The order of an ordinary differential equation is determined by the order of the highest-order derivative of the function appearing in the equality. A second-order ordinary differential equation contains the second derivative of the function $f(x, y)$ and is usually written as: $d^2y/dx^2 + Bdy/dx = f(x, y)$, where d^2y/dx^2 is the second derivative of the function f with respect to x, and dy/dx is the first derivative of the function f with respect to x. A solution to a second-order ordinary differential equation is any function y that satisfies that differential equation. Second-order ordinary **differential equations** have two linearly independent solutions and any linear combinations of those linearly independent solutions are also solutions. Second-order ordinary differential **equations** have a wide variety of uses including calculating the changing size of a population, determining the flow of current in an electric circuit, describing the **motion** of a pendulum, and describing the motion of a weight attached to the end of a spring. These uses are of interest in chemistry, physics, economics, engineering, and electronics.

There are different classifications of second-order ordinary differential equations that help in determining which methods are most effective for solving them. For second-order differential equations there are two types of solutions: the particular solutions and the general or complementary solution. Second-order ordinary differential equations can also be expressed as a system of two first-order differential equations. This property is used to solve them. If a second-order ordinary differential equation has the form: $d^2y/dx^2 + P(x)dy/dx + Q(x)y = G(x)$ where $P(x)$, $Q(x)$ and $G(x)$ are continuous then it is said to be a linear second-order ordinary differential equation. If the equation is similar to the form above but has the form: $d^2y/dx^2 + bdy/dx + cy = G(x)$ where b and c are constants and $G(x) = 0$ then it is called homogeneous as well as linear. This type of equation can be solved by employing its characteristic or auxiliary equation and determining the **roots** by the quadratic formula. The resulting solution will be the complementary solution. If $G(x) \neq 0$ the linear second-order ordinary differential equation is considered nonhomogeneous and another method must be employed to find the solutions. The general solution of a nonhomogeneous linear second-order differential equation is the sum of a particular solution and the complementary solution. Since the complementary solution can be found as previously described it remains to find only the particular solution. The method of variation of parameters is employed to find the particular solution.

Some linear second-order ordinary differential equations have **variable** coefficients, that is $P(x)$ and $Q(x)$ from above are **functions** rather than constants. For certain equations the variable coefficients can be transformed into constant coefficients by substituting another function z that is equivalent to the functions acting as coefficients. These particular differential equations for which this can be done have: $(Q'(x) + 2P(x)Q(x)) / (2(Q(x)^{3/2})) =$ constant, where $Q'(x)$ is the first derivative of the function $Q(x)$ with respect to x.

There are other special classes of second-order ordinary differential equations that include x-missing, which is just an equation that does not involve a function of x and has the form: $d^2y/dx^2 = f(y, dy/dx)$. Similarly there are y-missing equations that have the form: $d^2y/dx^2 = f(x, dy/dx)$. Some linear second-order differential equations can be simplified by eliminating the first-order term via substitution so that it is transformed into standard form: $d^2z/dx^2 + q(x)z = 0$.

The solutions of second-order ordinary differential equations can be classified as to their singularity at the origin. A solution is said to be singular if at that point, the singular point, the function fails to be analytic. Linearly independent solutions to second-order ordinary differential equations that are singular at the origin are classified as "of the second kind". If the solution is nonsingular at the origin it is said to be "of the first kind".

Although there are many methods for solving different kinds of second-order ordinary differential equations the only practical solution method for very complex equations is to use numerical methods. All solutions to second-order ordinary differential equations satisfy existence and uniqueness properties. One last note concerning the solutions to second-order ordinary differential equations that makes determination of the general solution easier is that if one solution of a particular equation is known then the other solution can be found using the reduction of order method. This is because of the relationship between the particular solutions and the general solution.

SEKI, KOWA (1642-1708)
Japanese public servant and teacher

The fact that Seki Kowa was known during his own lifetime as the "Arithmetical Sage," a tribute that later graced his tombstone, indicates the magnitude of his genius. He is generally considered to be the founder of the Japanese mathematical tradition, known as Wasan. Among his many accomplishments, he is best known for being the first person to study **determinants** and for discovering what came to be known as Bernoulli numbers even before **Jacob Bernoulli** did. Because there was a tradition of strict secrecy in the Japanese schools of the period (due to competition among them and the constraints of modesty imposed by his noble upbringing), some of Seki's contributions may still be unknown. However, there is no doubt that he was a major factor in the popularization and development of modern mathematics. He is also credited with transforming the discipline from an esoteric art form practiced by aristocrats into a science worthy of respect.

Seki's natural parents were hereditary samurai warriors who lived in Fujioka, Kozuke, Japan, although some sources say Seki was born in Edo, now Tokyo. He was born the second son of Nagaakira Utiyama. When he was a young child, a nobleman and accountant named Seki Gorozayemon adopted him, giving him the name by which he is known today. In fact, some sources give Seki's first name as Takakazu. The circumstances surrounding the adoption are not clear today. Seki's prodigious talent for mathematics was apparent from the time he was an infant. An educated household servant began the boy's formal introduction to the discipline when he realized the young nobleman's interest and astounding grasp of mathematical concepts.

As Seki's understanding of and passion for mathematics increased with the years, he began studying under his first academic mentor, a man named Yositane Takahara, who himself was a disciple of the great Sigeyosi Mori (author of *A Book on Division* in 1622). One of Seki's most famous mathematical feats came in his thirties, when he took up a challenge posed by Chu Shih-chieh. Chu had written *An Introduction to Mathematical Studies* in 1299, which contained numerous problems that he had solved using the "method of the celestial element." This technique was based on changing the problem into an algebraic equation with one **variable**. At the end of his book, Chu presented 150 problems that he believed could not be solved with this technique. Another mathematician of the period, Kazuyuki Sawaguchi, solved 135 of those, claiming in 1670 that the last 15 were truly unsolvable. However, Seki stunned the academic world in 1674 when he published his solutions to these stubborn problems using entirely new means.

The solutions appeared in his book *Hatubi sanpo*, which revealed Seki's design of a completely new system of notation. This system was so innovative that most mathematicians did not even understand it until a Seki disciple published a guide to the notation in 1685. Seki's system, which he called *endan*, was based on Chinese ideographs and allowed him to eliminate the cumbersome traditional use of Chinese calculating rods. The rods would not allow calculation of algebraic expressions with more than one variable, whereas ideographs permitted the representation of known and unknown quantities and the use of several variables.

Seki's method was limited only by his algebraic alphabet, which allowed him to work with **equations** only from the second through fifth degrees. Seki discovered a procedure, very similar to that used much later by **William George Homer**, that let him solve second-degree algebraic equations with numerical coefficients; he also introduced the concept of a **discriminant**. In addition, in about 1683 Seki discovered determinants, the mathematical notations that are still used today in virtually every branch of mathematics and in many of the natural sciences. Also in 1683, Seki wrote about **magic squares**, becoming the first Japanese to treat the topic. Among the mathematician's other important contributions were his discovery of Bernoulli numbers and his exact **definition** ("rectification") of a **circle**'s **circumference**, for which he obtained a value for **pi** (π) that was correct to eighteen decimal places.

Late in life, Seki made significant contributions to the area of **calculus**. Mainly, he is credited with developing a way to find the approximate value of a numerical equation's **root**. He passed on many of his methods to disciples at his school of mathematics, which had long been renowned as the best in Japan. Like the other Japanese schools, however, Seki's had a code of strict secrecy and permitted only his very best students to learn about his discoveries and ideas.

For the last four years of his life, Seki served as master of ceremonies in the household of a shogun, having become a shogunate samurai himself some time earlier. As a descendant of the samurai class, he had also spent years serving the public as examiner of accounts to the Lord of Koshu. Seki died in Edo on October 24, 1708.

See also Jacob Bernoulli

SEQUENCES

A sequence is an ordered list of numbers. It can be thought of as a function, f(n), where the argument, n, takes on the natural-number values 1, 2, 3, 4,... (or occasionally 0, 1, 2, 3, 4,...). A sequence can follow a regular pattern or an arbitrary one. It may be possible to compute the value of f(n) with a formula, or it may not.

The terms of a sequence are often represented by letters with subscripts, a_n, for example. In such a representation, the subscript n is the argument and tells where in the sequence the term a_n falls. When the individual terms are represented in this fashion, the entire sequence can be thought of as the set, or the set where n is a natural number. This set can have a finite number of elements, or an infinite number of elements, depending on the wishes of the person who is using it.

One particularly interesting and widely studied sequence is the **Fibonacci sequence**: 1, 1, 2, 3, 5, 8,.... It is usually defined recursively: $a_n = a_{n-2} + a_{n-1}$. In a recursive **definition**, each term in the sequence is defined in terms of one or more of its predecessors (recursive definitions can also be

called "iterative"). For example, a_6 in this sequence is the sum of 3 and 5, which are the values of a_4 and a_5, respectively.

Another very common sequence is 1, 4, 9, 16, 25,..., the sequence of **square** numbers. This sequence can be defined with the simple formula $a_n = n^2$, or it can be defined recursively: $a_n = a_{n-1} + 2n - 1$.

Another sequence is the sequence of **prime numbers**: 2, 3, 5, 7, 11, 13,.... Mathematicians have searched for centuries for a formula which would generate this sequence, but no such formula has ever been found.

One mistake that is made frequently in working with sequences is to assume that a pattern that is apparent in the first few terms must continue in subsequent terms. For example, one might think from seeing the five terms 1, 3, 5, 7, 9 that the next term must be 11. It can, in fact, be any number whatsoever. The sequence can have been generated by some random process such as reading from a table of random digits, or it can have been generated by some obscure or complicated formula. For this reason a sequence is not really pinned down unless the generating principle is stated explicitly. (Psychologists who measure a subject's intelligence by asking him or her to figure out the next term in a sequence are really testing the subject's ability to read the psychologist's mind.) Sequences are used in a variety of ways. One example is to be seen in the divide-and-average method for computing square **roots**. In this method one finds the **square root** of N by computing a sequence of approximations with the formula $a_n = (a_{n-1} + N/a_{n-1})/2$. One can start the sequence using any value for a_1 except **zero** (a negative value will find the negative root). For example, when N = 4 and $a_1 = 1$

$a_1 = 1.0$

$a_2 = 2.5$

$a_3 = 2.05$

$a_4 = 2.0006$

$a_5 = 2.0000$

This example illustrates several features that are often encountered in using sequences. For one, it often only the last term in the sequence that matters. Second, the terms can converge to a single number. Third, the iterative process is one that is particularly suitable for a computer program. In fact, if one were programming a computer in BASIC, the recursive formula above would translate into a statement such as R = (R + N/R)/2.

Not all sequences converge in this way. In fact, this one does not when a negative value of N is used. Whether a convergent sequence is needed or not depends on the use to which it is put. If one is using a sequence defined recursively to compute a value of a particular number only a convergent sequence will do. For other uses a divergent sequence may be suitable.

Mortgage companies often provide their customers with a computer print-out showing the balance due after each regular payment. These balances are computed recursively with a formula such as $A_n = (A_{n-1})(1.0075) - P$, where A_n stands for the balance due after the n-th payment. In the formula $(A_{n-1})(1.0075)$ computes the amount on a 9% mortgage after one month's interest has been added, and $(A_{n-1})(1.0075) - P$ the amount after the payment P has been credited. The sequence would start with A_0, which would be the initial amount of the loan. On a 30-year mortgage the size of P would be chosen to bring A_{360} down to zero. As anyone who has bought a house knows, this sequence converges, but *very* slowly for the first few years.

Tables, such as tables of **logarithms**, square roots, **trigonometric functions**, and the like are essentially paired sequences. In a table of square roots, for instance

1.0	1.00000
1.1	1.04881
1.2	1.09545

the column on the left is a sequence and the column on the right the sequence where each b_n equals the square root of a_n. By juxtaposing these two sequences, one creates a handy way of finding square roots.

Sequences are closely allied with (and sometimes confused with) series. A sequence is a list of numbers; a series is a sum. For instance 1/1, 1/2, 1/3, 1/4,... is a harmonic sequence; while 1/1 + 1/2 + 1/3 + 1/4 +... is a harmonic series.

SERRE, JEAN-PIERRE (1926-)
French topologist

Jean-Pierre Serre received a Fields Medal for his work in **topology**, the study of geometric figures whose properties are unaffected by physical manipulation. He has received international acclaim for both his theoretical contributions to mathematics and the clarity of his writings. Serre has authored a dozen books and numerous technical articles in the areas of topology, **analytic geometry**, **algebraic geometry**, **group theory**, and **number theory**.

Serre was born in Bages, France, on September 15, 1926, to Jean and Adèle (Diet) Serre. Both pharmacists, his parents instilled in him an early interest in chemistry. That interest eventually gave way to mathematics, however, when Serre began reading his mother's **calculus** books. By the age of 15, he was teaching himself the fundamentals of such topics as derivatives, integrals, and series. During high school at the Lycée de Nîmes, Serre found a practical use for his mathematical talents. He told C. T. Chong and Y. K. Leong (for an article published in *The Mathematical Intelligencer*) that some of the older students at the boarding house where he lived had a tendency to bully him. Even though they were taking more advanced classes than he was, he pacified them by doing their math homework for them. "It was as good a training as any," he recalled. In 1944, Serre won first prize in the Concours Général, a national mathematics competition.

In 1945, having passed the competitive entrance examination, Serre entered the prestigious École Normale Supérieure in Paris. Soon after enrolling at the institution, he decided to abandon his plans to become a high-school mathematics teacher and to concentrate instead on research. It was not until then, he later told Chong and Leong, that he realized he could earn a living as a research mathematician. Serre mar-

ried Josiane Heulot, a chemist, in 1948, and they have one daughter. Between 1948 and 1954, he held various positions at the Centre National de la Recherche Scientifique in Paris. His earliest research was in the field of topology; he was awarded his doctorate from the Sorbonne in this subject in 1951 after writing a dissertation on homotopy groups. After two years on the faculty of the University of Nancy, he became professor and chair of the department of **algebra** and **geometry** at the Collège de France, which remained his home throughout his career. Serre retired in 1994, assuming the position of honorary professor.

The Fields Medal, given once every four years by the International Congress of Mathematics, is intended to recognize outstanding contributions to mathematics by a researcher under age 40. It was awarded to Serre in 1954, when he was 28 years old. The citation read that Serre "achieved major results on the homotopy groups of spheres, especially in his use of the method of spectral sequences. Reformulated and extended some of the main results of complex **variable** theory in terms of sheaves" (a sheaf is a group of planes that pass through a common point). In fact, Serre's work on homotopy groups (classes of **functions** that are equivalent under a continuous deformation) was the origin of the loop **space** method in algebraic topology, which led quickly to numerous results. In 1952, he lectured on homotopy groups at Princeton University, discussing the topic's extension, called C-theory.

Since receiving the Fields Medal, Serre has explored other topics in mathematics, including complex variables, cohomology, algebraic geometry, and number fields. He explained to Chong and Leong that he finds it easy and natural to move gradually from one topic to another as he perceives their relationships to each other. Serre hesitates to categorize some of his work as being in one field or another, saying "such questions are not group theory, nor topology, nor number theory: they are just mathematics."

Serre is a member of the French, Dutch, Swedish, and U. S. academies of science and has been made an honorary fellow of the Royal Society of London and the London Mathematical Society. In addition to the Fields Medal, he has been awarded the Balzan Prize in 1985 and the Medaille d'Or of the Centre National de la Recherche Scientifique.

In 1995, Serre was awarded the Leroy P. Steele Prize for Mathematical Exposition by the American Mathematical Society. Specifically, the prize was given for his book *A Course in Arithmetic,* which was originally published in 1970. However, the citation noted: "Any one of Serre's numerous other books might have served as the basis of this award. Each of his books is beautifully written, with a great deal of original material by the author, and everything smoothly polished." It went on to mention that many of Serre's books have become standard references in their respective topic areas. In general, the citation continued, "they are alive with the breath of real mathematics, and are an example to all of how to write for effect, clarity, and impact." *A Course in Arithmetic* was praised for both its presentation of basic topics in algebra and its explanation of the link between function theory and the combinatorial aspects of elementary number theory, which formed a basis for modern developments in number theory and the geometry of numbers.

In his acceptance of the Steele Prize, Serre explained that the book had been a compilation of notes for various lectures he had given during the early 1960s. "Strangely enough," he said, "the different pieces fitted well together; I was especially pleased with the way algebraic and analytic arguments complemented each other." With characteristic humor, he recalled some problems, typographical as well as technical, that had plagued both the original French and English versions of *A Course in Arithmetic,* leading him to refer to the earlier editions as "A Curse in Arithmetic."

Another of Serre's books, *Abelian L-Adic Representations and Elliptic Curves,* was reissued in 1989 (it was originally published in 1968). In reviewing the book for the *Bulletin of the American Mathematical Society,* Kenneth Ribet noted that, although numerous advances had arisen in the field since it was first published, the original text was certainly not outmoded. "For one thing,... it's the only book on the subject. More importantly, it can be viewed as a toolbox which contains clear and concise explanations of fundamental facts about a series of related topics.... The tools introduced in this book have been, and will continue to be, extremely useful in other contexts."

In a more recent book, published in 1992, Serre presents a historically based explanation of the inverse Galois problem and explores its applications in algebraic number theory, **arithmetic** geometry, and coding theory. According to Michael Fried, who reviewed *Topics in Galois Theory* for the *Bulletin of the American Mathematical Society,* Serre was heard to remark at a friend's birthday celebration in 1990, "The inverse Galois problem gives us excuses for learning a lot of new mathematics."

Not only have Serre's books proven to be popular. Serre is often invited to lecture at universities around the world, including Bonn, Göttingen, Mexico, Moscow, and Singapore. He has made numerous visits to Harvard and the Institute for Advanced Study at Princeton.

When Chong and Leong asked Serre about the style of writing he prefers, he replied, "precision combined with informality!" Although noting that such writing is rare, he mentioned **Michael Atiyah** and John Milnor as examples of mathematicians whose work is accessible to nonmathematicians. Elaborating on his views about good exposition, he suggested that mathematical papers should contain more side remarks and open questions, which are often more interesting than the theorems themselves. "Alas," he said, "most people are afraid to admit that they don't know the answer to some question, and as a consequence they refrain from mentioning the question, even if it is a very natural one. What a pity! As for myself, I enjoy saying 'I do not know.'" Indeed, it is that admission that leads to investigation and discovery.

SET THEORY

Set theory is concerned with understanding those properties of **sets** that are independent of the particular elements that make up the sets. Thus the axioms and theorems of set theory apply to all sets in general, whether they are composed of numbers or physical objects. The foundations of set theory were largely developed by the German mathematician **George Cantor** in the latter part of the nineteenth century. The generality of set theory leads to few direct practical applications. Instead, precisely because of its generality, portions of the theory are used in developing the **algebra** of groups, rings, and fields, as well as, in developing a logical basis for **calculus**, **geometry**, and **topology**. These branches of mathematics are all applied extensively in the fields of physics, chemistry, biology, and electrical and computer engineering.

Definitions

A set is a collection. As with any collection, a set is composed of objects, called members or elements. The elements of a set may be physical objects or mathematical objects. A set may be composed of baseball cards, salt shakers, tropical fish, numbers, geometric shapes, or abstract mathematical constructs such as **functions**. Even ideas may be elements of a set. In fact, the elements of a set are not required to have anything in common except that they belong to the same set. The collection of all the junk at a rummage sale is a perfectly good set, but one in which few of the elements have anything in common, except that someone has gathered them up and put them in a rummage sale.

In order to specify a set and its elements as completely and unambiguously as possible, standard forms of notation (sometimes called *set-builder notation*) have been adopted by mathematicians. For brevity a set is usually named using an uppercase Roman letter, such a S. When defining the set S, curly brackets { } are used to enclose the contents, and the elements are specified, inside the brackets. When convenient, the elements are listed individually. For instance, suppose there are 5 items at a rummage sale. Then the set of items at the rummage sale might be specified by R={basketball, horseshoe, scooter, bow tie, hockey puck}. If the list of elements is long, the set may be specified by defining the condition that an object must satisfy in order to be considered an element of the set. For example, if the rummage sale has hundreds of items, then the set R may be specified by R = {I: I is an item in the rummage sale}. In this notation, I corresponds to an element of the set. The **definition** is read "R equals the set of all I such that I is an item in the rummage sale." If the set has an infinite number of elements it is specified similarly, such as S = {x: x is a real number and $0 < x < 1$}. This is the set of all x such that x is a real number, and 0 is less than x, and x is less than 1. The special symbol ø is given to the set with no elements, called the empty set or null set. Finally, $x \in A$ means that x is an element of the set A, and $x \notin A$ means that x is not an element of the set A.

Properties

Two sets S and T are equal, if every element of the set S is also an element of the set T, and if every element of the set T is also an element of the set S. This means that two sets are equal only if they both have exactly the same elements. A set T is called a proper subset of S if every element of T is contained in S, but not every element of S is in T. That is, the set T is a partial collection of the elements in S.

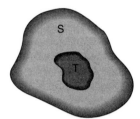

T is a subset of S

Figure 1a.

In set notation this is written T ⊂ S and read "T is contained in S." S is sometimes referred to as the parent or universal set. Also, S is a subset of itself, called an improper subset. The complement of a subset T is that part of S that is not contained in T, and is written T'. Note that if T' is the empty set, then S and T are equal.

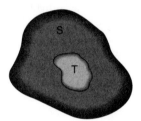

The complement of T

Figure 1b.

Sets are classified by size, according to the number of elements they contain. A set may be finite or infinite. A finite set has a whole number of elements, called the *cardinal number* of the set. Two sets with the same number of elements have the same **cardinal number**. To determine whether two sets, S and T, have the same number of elements, a one-to-one **correspondence** must exist between the elements of S and the elements of T. In order to associate a cardinal number with an infinite set, the **transfinite numbers** were developed. The first transfinite number $\aleph_0$, is the cardinal number of the set of **integers**, and of any set that can be placed in one-to-one correspondence with the integers. For example, it can be shown that a one-to-one correspondence exists between the set of **rational numbers** and the set of integers. Any set with cardinal number $\aleph_0$ is said to be a countable set. The second transfinite number $\aleph_1$ is the cardinal number of the **real numbers**. Any set in one-to-one correspondence with the real numbers has a cardinal number of $\aleph_1$, and is referred to as uncountable. The **irrational numbers** have cardinal number $\aleph_1$. Some interesting differences exist between subsets of finite sets and subsets of infinite sets. In particular, every proper subset of a finite set has a smaller cardinal number than its parent set. For example, the set S = {1,2,3,4,5,6,7,8,9,10} has a cardinal number of 10, but

every proper subset of S (such as {1,2,3,4}) has fewer elements than S and so has a smaller **cardinality**. In the case of infinite sets, however, this is not true. For instance, the set of all odd integers is a proper subset of the set of all integers, but it can be shown that a one-to-one correspondence exists between these two sets, so that they each have the same cardinality.

A set is said to be ordered if a relation (symbolized by <) between its elements can be defined, such that for any two elements of the set:

1) either b < c or c < b for any two elements

2) b < b has no meaning

3) if b < c and c < d then b < d.

In other words, an ordering relation is a rule by which the members of a set can be sorted. Examples of ordered sets are: the set of positive integers, where the symbol (<) is taken to mean less than; or the set of entries in an encyclopedia, where the symbol (<) means alphabetical ordering; or the set of U.S. World Cup soccer players, where the symbol (<) is taken to mean shorter than. In this last example the symbol (<) could also mean faster than, or scored more goals than, so that for some sets more than one ordering relation can be defined.

Operations

In addition to the general properties of sets, there are three important set operations, they are union, intersection, and difference. The union of two sets S and T, written S∪T, is defined as the collection of those elements that belong to either S or T or both. The union of two sets corresponds to their sum.

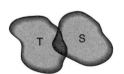

The union of T and S

Figure 2a.

The intersection of the sets S and T is defined as the collection of elements that belong to both S and T, and is written S∩T. The intersection of two sets corresponds to the set of elements they have in common, or in some sense to their product.

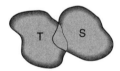

The intersection of T and S

Figure 2b.

The difference between two sets, written S-T, is the set of elements that are contained in S but not contained in T.

The difference S - T

Figure 2c.

If S is a subset of T, then S-T = ø, and if the intersection of S and T (S∩T) is the null set, then S-T = S.

Applications of set theory

Because of its very general or abstract nature, set theory has many applications in other branches of mathematics. In the branch called **analysis**, of which differential and **integral calculus** are important parts, an understanding of limit points and what is meant by the **continuity** of a function are based on set theory. The algebraic treatment of set operations leads to **Boolean Algebra**, in which the operations of intersection, union, and difference are interpreted as corresponding to the logical operations "and," "or," and "not," respectively. Boolean Algebra in turn is used extensively in the design of digital electronic circuitry, such as that found in calculators and personal computers. Set theory provides the basis of topology, the study of sets together with the properties of various collections of subsets.

SETS

In mathematics, a set can be thought of as a well defined collection of objects. The "objects" with which we will be concerned are abstract entities such as numbers or other sets. Until late in the nineteenth century, there were few references to sets in the mathematical literature, but this was to change dramatically due to the work of one brilliant German mathematician, **Georg Cantor** (1845-1918). The official birth of **set theory** can be traced to an 1874 article by Cantor in *Crelle's Journal,* one of the most influential mathematics journals of its time. In this article, Cantor suggested that there were different kinds of infinite sets and different orders of **infinity**. This idea stirred up tremendous controversy among some editors of the journal and among mathematicians in general. Prior to this publication, all infinite sets had been considered alike. No one had ever suggested that there were different orders of infinity. During the period from 1874 through 1897, Cantor continued to develop his set theory, stirring up the mathematical community as it had seldom been stirred before.

If Cantor had restricted his writings to finite sets, nobody would have taken much notice. After all, finite sets, even very large ones, are not mysterious. If you want to know how many elements are in a finite set, you just count them. It might be tedious for very large sets, but, in principle, you can finish the job and report the results of your counting. Determining which of two finite sets is larger is no problem; just count the number of elements in each and declare that the one with the most elements is larger. If they have the same

number of elements then they are the same size. Just to make this a little more precise, Cantor introduced the idea of a one-to-one **correspondence** between the elements of two finite sets: If you can pair each element of set A with exactly one element in set B and vice-versa, then, of course, the sets must have the same number of elements and this pairing of elements is called a one-to-one correspondence. No one objected to this. If every element of a finite set A is also a member of finite set B, then, says Cantor, A is a subset of B. There was no controversy about that. Under this **definition**, every set is a subset of itself. That was okay with everyone. If a set has no members, then call it the empty set or the "null" set, if you prefer. That was fine. Cantor went on to define the intersection of two sets: The intersection of sets A and B is the set of all elements which are in both set A and set B. He defined union: The union of sets A and B is the set of all elements that are in A or in B or in both. No one had any objections to those definitions. He defined the notion of **cardinality** for finite sets: The cardinality of a finite set was equal to the number of elements in the set. This ruffled no feathers. None of Cantor's newly minted theory gave anyone pause so long as it applied only to finite sets. Unfortunately, mathematics deals with both the finite and the infinite; and Cantor knew that any set theory that failed to account for infinite sets was virtually useless as a foundation for mathematics. Thus he extended his theory to include the infinite, and the revolution began.

Cantor wanted his theory of finite sets to be logically consistent with whatever he had to say about infinite sets. So, for example, he brought his one-to-one correspondence notion into the world of infinite sets and strange things began to happen. In the finite world, if a set A is a subset of a set B, and if there are elements in B that are not in A; then set B has more elements than set A, and a one-to-one correspondence cannot be set up between the elements of A and the elements of B. In such cases, set A is said to be a "proper" subset of B. Another way of saying this, in Cantor's new language, is to say that the cardinality of set B is greater than the cardinality of set A. Translate this to the infinite world. The set of natural numbers, {1,2,3,...} is infinite. So is the set of all even natural numbers, {2,4,6,...}, which is clearly a subset of the natural numbers by Cantor's definition. There are obviously elements in {1,2,3...} that are not in {2,4,6,...}; for example 1,3,5, and so on. Therefore, {2,4,6,...} is a proper subset of {1,2,3,...}. So should we not say that the set of natural numbers has a cardinality that is larger than the cardinality of the set of even natural numbers? But wait! It is quite simple to establish a one-to-one correspondence between the natural numbers and the even natural numbers,.e.g., 1 in the natural numbers corresponds to 2 in the even natural numbers, 2 in the naturals corresponds to 4 in the evens, 3 natural to 6 even, and so on. In general, each natural number n corresponds to the even number 2n. Therefore, Cantor says that, since a one-to-one correspondence can be set up between {1,2,3,...} and {2,4,6,...}, these two sets must have the same cardinality. Cantor decided to call this cardinality $\aleph_0$, or "aleph null". $\aleph$ ("aleph") is the first letter in the Hebrew alphabet. Thus any set that can be placed in one-to-one correspondence with the set of natural numbers has cardinality $\aleph_0$. Cantor called such sets "denumerably infinite" or "countably infinite." But Cantor was not finished. Next he produced a set, quite familiar to every mathematician, which was infinite but could not be placed in one-to-one correspondence with the set of natural numbers. That set was the **interval** of **real numbers** [0,1]. In fact, any interval of real numbers will do, including the entire real number line or "continuum" as Cantor called it. Cantor produced an ingenious **proof** of this by assuming that a one-to-one correspondence between the natural numbers and the interval [0,1] exists and then constructing a number in that interval which had been left out of the alleged one-to-one correspondence. This proof can be seen in the article (**Non-denumerably Infinite Sets**). This was the shot heard round the mathematical world. Cantor had exhibited an infinite set that had a greater cardinality than the natural numbers. All infinite sets were not alike. All "infinities" are not the same. Some infinities are greater than other infinities. Cantor went on to show that there are infinities larger than the infinity of real numbers. In fact he showed that there is an infinity of infinities having cardinalities which he called $\aleph_0$, $\aleph_1$, $\aleph_2$,.... One very influential mathematician, **Leopold Kronecker** (1823-1891), believed that any such study of infinite sets was absurd because nothing meaningful, in the mathematical sense, could be said about infinity. Kronecker thought that the study of infinity was outside the discipline of mathematics. Because of Kronecker's standing in the mathematical community, Cantor was stung by his criticism. For a time, Cantor even began to doubt himself and the importance of his work. Kronecker died in 1891 and the furor created by his criticism of Cantor's program gradually faded. In 1897, Cantor published the second of a two-volume treatise bringing together all of his ideas about sets and infinity. This treatise looks very much like a present-day textbook on set theory. In fact, it became a cornerstone in the foundation of twentieth century mathematics. Also in 1897, Cantor was honored for his work on the theory of sets at the first International Conference of Mathematicians held in Zurich, Switzerland. With the creation of set theory, Cantor, almost single-handedly, brought forth a revolution in the foundations of mathematics, the impact of which still shapes the structure of the subject at the beginning of the twenty-first century. Today, Cantor's set theory is universally accepted as one of the greatest achievements in the **history of mathematics**.

See also Set theory

SEXAGESIMAL NUMERATION

Sexagesimal numeration is a numeral system in which all derived units are based on the number 60 and the **powers** of 60. The word sexagesimal is derived from the Latin word *sexagesimus* (sixty). Between 4000-3000 B.C. the Sumerians developed a sexagesimal number system that had an additive decimal (base-10) sequence of "1 to 10, 10 to 60, 60 to 600, 600 to 3600, and so forth". For example, the Sumerian number of 100 was expressed as "1, 40"; that is, one unit of 60 plus

40. The year 2000 would be expressed in this system as three units of 600, plus three units of 60, plus 10 plus 10. Ptolemy and other Alexandrian astronomers also used sexagesimal **fractions** that are believed to have been one way that led to the persistent use of the base 60 for measures of time and angles. This is in spite of the nearly worldwide adoption of base-10 numeration for most other measurement needs (currency, the **metric system**, etc.). For instance, 60 seconds equals one minute and 60 arc-minutes equals one arc-degree. Between 3000-2000 B.C. the Babylonians developed a place-value system (from the older Sumerian system) with a base of 60 that contained an incomplete sexagesimal positional notation. The reason why such a system was developed is unclear, but mathematical common sense indicates that one possibility may have been that it contained many divisors (2, 3, 4, and 5, and some multiples) in order to facilitate **division**. With such a large base it was awkward to have unrelated names for the digits 0, 1,... 59, so a simple grouping system from base 10 was used for these numbers. It used only two symbols ("a vertical wedge derived from a small cone" for 1 and "a corner wedge derived from a small circle" for 10) instead of the 60 distinct symbols that a base-60 system would normally use. Therefore, for the number 33 the symbol grouping "three corner wedges and three vertical wedges" was used. For larger numbers the vertical wedge was also used as powers of 60 (such as 1 (60^0, 60 (60^1), 3600 (60^2), and so forth). Thus, for the number 90 the grouping of symbols "one vertical wedge and three corner wedges" was used to denote (60 + 10 + 10 + 10). As a result, the system suffered from ambiguities in representing values that could be resolved only by analyzing the context. In addition, the Babylonians had long used empty spaces to separate one sexagesimal order from the next in written numbers, but special symbols were sometimes also employed for this purpose. Similar symbols were eventually used to indicate the absence of certain orders. Astronomers may have also used the symbols to indicate fractions. The complete development of the system was, therefore, delayed by the lack of a consistent use of these symbols and for a symbol for empty places, i.e. the **zero**. The Babylonians appear to have developed a placeholder symbol that functioned as a zero by the third century B.C., but its precise meaning and use is still uncertain. The world of mathematics and astronomy owes much to the Babylonians for their development of the sexagesimal system in order to calculate time (the 12 double-hours in a day, minutes, and seconds) and angles (degrees, arc-minutes, and arc-seconds) that is still in practical use today. In fact, the Babylonian year that was composed of 360 days (of 12 months with 30 days each) most likely created the 360 degrees of a **circle** where each degree of the circle is further divided into 60 arc-minutes, and each arc-minute into 60 arc-seconds.

The sexagesimal number system was familiar to both the Indian and Mesopotamian cultures (where ancient Babylonia was located within Mesopotamia). In these societies it is likely that the sexagesimal system was based on the observation of the planets, specifically Jupiter and Saturn. Every 60 years, Jupiter and Saturn return to the same relative place in the sky. Jupiter takes 12 years to cross all the constellations in the night sky, from which the zodiac signs have been derived. Saturn takes an average 30 years to cross the zodiac. So Jupiter makes 5 complete crossings of the zodiac, while Saturn makes 2 transits in the same sixty-year period. During the sixty-year cycle, the four numbers involved in the revolutions of the planets (2 and 5) and in the number of years for each revolution (12 and 30) are divisible into 60, allowing for a sensible use of the sexagesimal number system.

Today, the sexagesimal number system is an archaic number system. It has been replaced by numbering systems that are much easier to use. The prime example of which is the dominant decimal number system (base 10); and other systems of use in special applications, such as the binary (base 2) and hexadecimal (base 16) number systems that are especially needed for the operation of digital computers.

See also Babylonian mathematics; Binary number system; Binary, octal, and hexadecimal numeration; Decimal position system; Egyptian mathematics

SHELL METHOD

The Shell method is a method of computing the **volume** of a surface of revolution. It is also commonly referred to as the method of cylindrical shells. This method is most convenient when, for instance, a region in the x,y-plane below the graph of a function of x is being revolved about the y-axis, or a region below the graph of a function of x is being revolved about the y-axis. In such cases, it is usually preferable to the most familiar method, the disk method.

To derive the shell-method formula, first suppose that we have a nonnegative continuous function f(x) defined on an **interval** [a,b]. (The derivation is analogous for a non-positive function, or for a function of y instead of x.) Suppose that we are revolving the **area** below the graph of f (and above the x-axis) around the y-axis, and wish to compute the volume of the resulting region.

As always when using the techniques of **calculus** to measure some geometric quantity (area, arc-length, volume, etc.), we begin with an approximation. Divide the interval [a,b] into subintervals: $a = x_0 < x_1 < x_2 < ... < x_{n-1} < x_n = b$. We restrict our attention to one subinterval $[x_{k-1}, x_k]$, and try to approximate the volume that results from revolving around the y-axis only the region below the graph of f and between x_{k-1} and x_k.

Suppose for the moment that the function f has the constant value $f(x_k)$ on this interval. Then the surface of revolution is in fact a cylindrical shell (that is, the volume between two concentric cylinders, one with radius x_{k-1} and one with radius x_k). The height of this shell is $f(x_k)$, and so its volume is $p(x_k^2 - x_{k-1}^2)f(x_k)$. If we follow the same procedure over each interval, we obtain the approximation

$\text{Sum}_{k=1}^n \pi (x_k^2 - x_{k-1}^2)f(x_k)$,

or, after factoring the difference of two squares and rearranging terms,

$\text{Sum}_{k=1}^n \pi f(x_k) (x_k + x_{k-1}) (x_k - x_{k-1})$

This is, of course, not accurate for most **functions**, because in general the function f will not be constant over the interval $[x_{k-1}, x_k]$. However, if we allow the length of this interval to shrink, the approximation will become more and more accurate, because the function f will vary less over a small interval than over a larger interval. Therefore as we let the number of subintervals approach **infinity**, the sum above approaches the actual volume. But that sum is a Riemann sum for the integral

$(\text{Int})_a^b \pi f(x)(2x)dx$.

Thus this integral gives the volume of the surface of revolution.

It is plain to see that this formula is more manageable than the disk formula: first and most importantly, to use the disk method we would have to compute the **inverse function** of f (which can be difficult or, occasionally, even impossible). We must also consider the region being rotated as the area between two curves, thus requiring us to compute two integrals instead of one (one simple, but one possibly difficult).

As with the disk method, the shell method can be generalized to the case in which the region being rotated is the area between the graphs of two functions f and g. The generalized formula is (assuming the graph of f lies above the graph of g)

$(\text{Int})_a^b 2 \pi x (f(x) - g(x))dx$.

See also Surfaces and solids of revolution

SHIFTING ORTHOGONAL COORDINATES

Length, **area**, and **volume** are all calculated differently in different orthogonal coordinate systems. The gradient of a function and the divergence and curl of vector fields, likewise, are calculated according to the orthogonal coordinates which the function or vector **field** is given in.

For example, in **polar coordinates** the point (r, a) is the point (rcos(a), rsin(a)) in **Cartesian coordinates**. Hence the **distance** from the point (r, a) to the point (s, b) given in polar coordinates is equal to $(rcos(a) - scos(b))^2 + (rsin(a) - ssin(b))^2)^{1/2}$. But the distance from (w,x) to (y,z) in Cartesian coordinates is $((w-y)^2 + (x-z)^2)^{1/2}$.

Let u_1, u_2, and u_3 be **functions** defining a three-dimensional **orthogonal coordinate system**. Then at every point (x, y, z) in **space**, the surfaces defined by the **equations** $u_1 = x$, $u_2 = y$, and $u_3 = z$ are orthogonal. The intersection of any two of these surfaces is called a coordinate curve. Let $c_1(t) = (t, y, z)$. So, the image of c is a coordinate curve. Let $s_1(t)$ be the length along the curve c_1 from $c_1(0)$ to $c_1(t)$. Similarly define functions s_2 and s_3 to be the arc length functions of the other coordinate curves. Then define h_1 to be the derivative of s_1 with respect to u_1. Similarly, let h_2 and h_3 be the derivatives of s_2 and s_3 with respect to u_2 and u_3 respectively. These functions are called the scale factors. Then the arc length of a curve S is given by the line integral $\int ((h_1 du_1)^2 + (h_2 du_2)^2 + (h_3 du_3)^2)^{1/2}$ where the integral is taken over S.

To make sense of this, suppose that S is a small curve which begins at some point P = (0, 0, 0) and ends at a point Q = (Q_1, Q_2, Q_3) where both P and Q are given in u_i coordinates. Then the length of S is approximately given by the distance (in Cartesian coordinates) from (0, 0, 0) to $(Q_1 h_1(P), Q_2 h_2(P), Q_3 h_3(P))$.

Also the volume of an object O is given by the volume integral $\int (h_1 h_2 h_3)\, du_1\, du_2\, du_3$ where the integral is taken over O.

To make sense of this, suppose that O is a small parallelepiped with vertices (in u_i coordinates) given by (0,0,0), (A,0,0), (0,B,0), (0,0,C), (A,B,0), (0,B,C), (A,0,C). Then its volume is approximately equal to the volume of the Euclidean parallelepiped with dimensions $h_1(P)$xA, $h_2(P)$xB, and $h_3(P)$xC.

The gradient of a function f given in u_i coordinates is
Grad f = $(1/h_1\ (\partial f/\partial u_1),\ 1/h_2\ (\partial f/\partial u_2),\ 1/h_3\ (\partial f/\partial u_3)\ )$
(with respect to u_i coordinates). The divergence and curl of a vector field $\mathbf{F} = (\mathbf{F}_1, \mathbf{F}_2, \mathbf{F}_3)$ (given in u_i coordinates) are
Div $\mathbf{F} = 1/h_1 h_2 h_3\ [\partial/\partial u_1 (\mathbf{F}_1 h_2 h_3) + \partial/\partial u_2 (\mathbf{F}_2 h_1 h_3) + \partial/\partial u_3 (\mathbf{F}_3 h_1 h_2)\ ]$.
Curl $\mathbf{F} = (1/h_2 h_3\ [\partial/\partial u_2 (\mathbf{F}_3 h_3) - \partial/\partial u_3 (\mathbf{F}_2 h_2)],\ 1/h_1 h_3\ [\partial/\partial u_3 (\mathbf{F}_1 h_1) - \partial/\partial u_1 (\mathbf{F}_3 h_3)],\ 1/h_1 h_2\ [\partial/\partial u_1 (\mathbf{F}_2 h_2) - \partial/\partial u_2 (\mathbf{F}_1 h_1)])$.
(with respect to u_i coordinates).

See also Cartesian coordinates system; Orthogonal coordinate system; Polar coordinates system; Vector analysis

SIERPIŃSKI, WACLAW (1882-1969)
Polish mathematician

Waclaw Sierpinski and his colleagues are credited with revolutionizing Polish mathematics during the first half of the 20th century. They took a couple of relatively new fields of mathematics and devoted whole journals to them. Although detractors had opined that such an experiment could not succeed, the mathematical heritage of the Polish community between the world wars has left a legacy of results, problems, and personalities, chief among them being Sierpinski.

Sierpinski was born in Warsaw on March 14, 1882. His father was Constantine Sierpinski, a successful physician, and his mother was Louise Lapinska. Sierpinski received his secondary education at the Fifth Grammar School in Warsaw, where he studied under an influential teacher named Wlodarski. From there, he entered the University of Warsaw and began studying **number theory** under the guidance of G. Voronoi. In 1903, Sierpinski's work in mathematics was recognized by a gold medal from the university, from which he graduated the next year. After graduation, he taught in secondary schools, a standard career path due to the shortage of positions available to Poles under Russian rule. In that capacity, he was involved in the school strike that occurred during the revolution of 1905. Even though the strike was not wholly unsuccessful, Sierpinski resigned his teaching position and moved to Krakow.

In 1906 Sierpinski received his doctorate from the Jagiellonian University in Krakow. Two years later he passed the qualifying examination to earn the right to teach at Jan Kazimierz University in Lwow, to which he had gone at the invitation of one of the faculty. There, Sierpinski offered perhaps the first systematic course in **set theory**, the subject of his investigations for the next 50 years. In 1912 he gathered his lecture notes and published them as *Zarys Teorii Mnogosci* ("Outline of Set Theory"). Sierpinski's texts were recognized by prizes from the Academy of Learning in Krakow.

With the outbreak of World War I in 1914, Sierpinski was interned by the Russians, first at Vyatka, then in Moscow. This internment was not particularly severe, for while he was in Moscow, Sierpinski was accorded a cordial reception by the leading Russian mathematicians of the era. In fact, he was able to conduct some joint research with **Nicolai Lusin** during this period in the field of set-theoretic **topology**.

The area of set-theoretic topology in which Sierpinski worked depended on a few basic notions. One of these is that of a closed set, or a set that includes its boundary. A simple example is the **interval** of **real numbers** between 0 and 1, including both endpoints. Related to the notion of a closed set is that of an open set, one which does not include its boundary. An example of an open set is the interval between 0 and 1, not including either 0 or 1. If one takes that interval and includes 0 but not 1, then the set is neither closed nor open. Much time was spent investigating the results of combining open and closed **sets** in various infinite combinations. The entrance of **infinity** is what required the use of methods and ideas from set theory.

When the war ended Sierpinski returned to Lwow, but in the fall of that year he was appointed to a position at Warsaw. He devoted a number of papers to the set-theoretic topics of the continuum hypothesis and the **axiom** of choice. The continuum hypothesis, which had been known to **Georg Cantor**, claimed that there were no infinite numbers between the number of **integers** and the number of real numbers. If the continuum hypothesis is present, it reduces the complexity of the hierarchy of infinite numbers. The axiom of choice had been a matter for much discussion at the turn of the 19th century, allowing for the possibility of making an infinite number of choices simultaneously. Some distinguished French mathematicians like **Emile Borel** questioned the meaningfulness of such a choice, and one of the early consequences of investigation into the axioms of set theory was the discovery that the axiom of choice was equivalent to a number of other **propositions** of set theory. Sierpinski took an agnostic position with respect to the axiom, using it in proofs and also trying to eliminate it wherever possible.

It was during the period between the world wars that Sierpinski, in conjunction with several Polish colleagues, created what has since become known as the Polish school of mathematics. The subjects that dominated the Polish school were logic and set theory, topology, and the application of these subjects to questions in **analysis**. To make certain that there would be an audience for the work in these areas, the journal *Fundamenta Mathematicae* was founded in 1919. Although by the end of the 20th century a profusion of specialized journals within mathematics had sprung up,

Fundamenta was the first of its kind and was greeted with some suspicion about its likelihood for survival. The quality of its papers was high, the contributors were international, and the problems proposed and solved were substantial. As a permanent record of the Polish mathematical school, *Fundamenta Mathematicae* supplements the reminiscences of those who took part in the work.

Sierpinski often led the Polish delegations to international congresses and conferences of mathematicians. One of the most ambitious projects in which he was involved was a Congress of Mathematicians of Slavic Countries, whose very existence attests to a political consciousness side-by-side with the mathematical one. The event took place in Warsaw in 1929 and was chaired by Sierpinski.

During World War II Sierpinski was in Warsaw, holding classes in whatever secret settings were available. A good deal of the discussion went on in Sierpinski's home, where his wife did her best to make guests feel as comfortable as possible in such troubled times. In 1944, the Nazis took control of Warsaw and Sierpinski was taken by the Germans to a site near Krakow. After the latter city was liberated by the Allies, he held lectures at the Jagiellonian University there before returning to Warsaw.

The period after the war was marked by further honors to Sierpinski as his students dominated the mathematical landscape. He served as vice president of the Polish Academy of Sciences from its inception and was awarded the Scientific Prize (First Class) in 1949, as well as the Grand Cross of the Order of Polonia Restituta in 1958. There were not many mathematicians in any country who approached his publication records of more than 600 papers in set theory and approximately 100 articles about number theory.

Sierpinski died in Warsaw on October 21, 1969. His mathematical textbooks educated an entire generation, and he helped to lay the foundations of the discipline of set-theoretic topology. Sierpinski's legacy was in establishing a Polish mathematical community, a contribution at the same time to mathematics and to national identity.

SIERPIŃSKI'S TRIANGLE, CARPET, AND SPONGE

The Sierpinski **triangle** (also know as the Sierpinski gasket or sieve) is a **fractal** first described by Sierpinski in 1915. To construct it, draw the outline of an equilateral triangle on white paper. In its middle draw a black upside down triangle with side length 1/3 the side of the original. Now there are 3 small white right side up triangles inside the large one. In the middle of each of these triangles draw a black upside down triangle. Repeat this process for the 9 small white triangle that are left. Continue this process forever to get the Sierpinski triangle.

Here is a way to obtain something that is a lot like Sierpinski's triangle. Start with any object whatsoever. Make three copies of it and place them one of them symmetrically above the other two. Now shrink all three object by half and make three copies of their union. Place the three copies so that

one of them is symmetrically above the other two. Continue the process of shrinking, making three copies and placing indefinitely. If we had started with a triangle then we would end up with Sierpinski's triangle but regardless of what object we choose the result "looks like" Sierpinski's triangle. Probably for this reason, the Sierpinski triangle is commonly used to illustrate fractal properties. One such property is that the white part of the triangle has no **area** yet it has infinite **perimeter**. Another property is self-similarity, that is if we look through a microscope at any triangle inside Sierpinski's triangle then what we see looks a lot the whole triangle. One other property defines a fractal: a geometric object whose topological **dimension** is not equal to its **Hausdorff dimension** (i.e., fractal dimension; **Felix Hausdorff** was a German mathematician who helped develop the field of **topology**). The topological dimension of an object is defined as follows. Let the object be covered by balls (a "ball" is the set of all point y that are at a **distance** at most r from some central point c for some fixed number r and point c). A refinement of the cover is a collection balls so that each ball fits inside one of the balls of the cover and the union of all balls of the collection covers the object. The order of the refinement is the largest number n such that there is a point in the object that is contained in n balls of the refinement. The topological dimension is equal to minus one plus the smallest number n such that for any cover of the object there is a refinement of order n. It can be shown that the topological dimension of n-dimensional real **space** is n as expected. The Hausdorff dimension is defined as the minus the limit as the radius r goes to **zero** of the quantity $\log(N)/\log(r)$ where N is the smallest number of balls of radius r required to cover the object. The Hausdorff dimension of the Sierpinski triangle is $\log(3)/\log(2) = 1.58...$

To obtain the Sierpinski carpet, draw an outline of a **square** on white paper. In its middle draw a black square whose side length is a third as long as that of the first square. The white space left over is made up of eight squares each equal in size to the middle square. Treat each of these white squares like the first square: draw a black square in the middle of each whose side length is a third that of the white square. Now the remaining space is made up of $8^2 = 64$ small white squares. Treat each of these like the first white square. Continue this process forever. Although it is very similar to the Sierpinski triangle, the Sierpinski carpet has one property that the triangle does not have: every compact one-dimensional curve that can be drawn in the plane has a topological equivalent inside the Sierpinski carpet. This means the following: suppose we draw a curve (or overlapping curves) on a piece of paper. Now suppose that the curves drawn are made of rubber and so can be stretched, compressed or bent at will. Then they can be deformed so that they fit inside the white part of the Sierpinski carpet.

What is called the Sierpinski sponge should really be called the Menger sponge (or Menger curve) after **Karl Menger** who first described it in 1926. It a three dimensional analogue of the Sierpinski carpet. To construct it, start with a wooden cube. Draw a three by three grid on each its faces. Drill a square hole that starts with the middle-square of some face and continues to the other side. Do this for each face. The object is now like twenty small cubes that have been glued together. For each of these smaller cubes do the same thing that you did with the original cube. Continue this process forever. The result is the Menger sponge. Unlike the Sierpinski carpet, every compact one-dimensional curve has a topological equivalent inside the Menger sponge, not just those that can be drawn on paper.

SIEVE OF ERATOSTHENES

The sieve of **Eratosthenes** is a useful method for making a list of **prime numbers**. Let d and n be positive **integers**. We say that d *divides* n, or that d is a *divisor* of n, if the fraction n/d is an integer. For example, 4 divides 12 and 1 divides 17, but 3 does not divide 14. The set of divisors of 12 is {1, 2, 3, 4, 6, 12}, and the set of divisors of 17 is {1, 17}. Sometimes it is important to include negative integers in a discussion of divisors. But in this article we will consider only positive integers and positive divisors. A positive integer $n > 1$ is called a *prime* number if the only divisors of n are 1 and n. A positive integer $n > 1$ is said to be a *composite* number if it is not prime. The sequence of prime numbers begins with 2, 3, 5, 7, 11, 13, 17, 19, 23, 29, 31, 37, 41,.... Prime numbers are important in **number theory** because every positive integer $n > 1$ can be written as a product of prime numbers. For example,

$$(2)(2)(5)(5) = 100 \text{ and } (7)(7)(11)(181619) = 97892641.$$

In fact each positive integer $n > 1$ can be written as a product of prime numbers in essentially one way. Here is a more precise statement of this fact: suppose that $n > 1$ is a positive integer, $p_1, p_2, p_3,...,p_M$ is a collection of prime numbers arranged so that

$$p_1 \leq p_2 \leq p_3 \leq ... \leq p_M,$$

and $q_1, q_2, q_3,...,q_N$ is a collection of prime numbers arranged so that

$$q_1 \leq q_2 \leq q_3 \leq ... \leq q_N.$$

Assume that n factors into prime numbers as both

$$(p_1)(p_2)(p_3) ... (p_M) = n \text{ and } (q_1)(q_2)(q_3) ... (q_N) = n.$$

Then we have $M = N$ and $p_m = q_m$ for each $m=1, 2, 3, ...,M$. This fact is often called the *Fundamental Theorem of Arithmetic*. The distinct prime numbers which appear in the factorization of n are called the *prime divisors* of n. For example, the prime divisors of 100 are 2 and 5, and the prime divisors of 97892641 are 7, 11 and 181619. Clearly the number n is itself a prime number if its only prime divisor is n. And a number $n > 1$ is composite if it has a prime divisor p with $1 < p < n$.

Now suppose that we want to decide if a large positive integer n is a prime. By our previous remarks it suffices to test each prime number p with $1 < p < n$ to see if p is a prime divisor of n. If no prime divisor p with $1 < p < n$ is found, then we can conclude that n itself is prime. The sieve of Eratosthenes is based on the simple observation that if n is composite then its smallest prime divisor p_1 must satisfy $1 < p_1 \leq \sqrt{n}$. To see why this is so, assume as before that $p_1, p_2, p_3,...,p_M$ is a collection of prime numbers arranged so that

$$p_1 \leq p_2 \leq p_3 \leq ... \leq p_M,$$

and let

$$(p_1)(p_2)(p_3) ... (p_M) = n$$

be the factorization of n as a product of prime numbers. Here p_1 is the smallest prime divisor of n and therefore

$$(p_1)^M \leq (p_1)(p_2)(p_3) ... (p_M) = n.$$

Now if n is composite then we must have $2 \leq M$. But then it follows that

$$p_1 \leq n^{1/M} \leq n^{1/2} = \sqrt{n}.$$

This shows that the smallest prime divisor p_1 of n satisfies $1 < p_1 \leq \sqrt{n}$. Therefore, in order to determine if n is prime we only need to check for prime divisors p in the **range** $1 < p \leq \sqrt{n}$.

We can turn these observations into a simple **algorithm** for making a list of prime numbers. Suppose that we have already determined that the prime numbers less than or equal to $10 = \sqrt{100}$ are 2, 3, 5, and 7. Write down the integers from 2 to 100. Then cross out each integer on the list that is divisible by 2. Next cross out each integer on the list that is divisible by 3. Continue in this manner to cross out each integer on the list that is divisible by 5 and then each integer that is divisible by 7. The integers that have not been crossed out have the property that they do not have a prime divisor less than or equal to their **square root**. Hence the numbers that have not been crossed out are exactly the prime numbers greater than 10 and less than or equal to 100.

SIMILARITY

Geometric figures that are considered to be similar have the same exact shape, but differ in size. Similarity in geometric figures means the ratio of lengths of any two corresponding sides in the figures is the same, and all corresponding angles are equal. For triangles, there are several theorems that allow triangles to be proven similar. The first is the "angle-angle similarity" **theorem**. This theorem states that if two angles in one **triangle** are congruent to two corresponding angles in a second triangle, the triangles are similar. It might also be that the triangles are congruent, but based on this particular theorem, there is not enough information to prove that. The second similarity theorem is the "side-included-angle-side similarity" theorem. This theorem states that if two **sets** of sides in two triangles are in proportion and their included angles are congruent, then the triangles are similar. Finally, the third similarity theorem is the "side-side-side similarity" theorem. This theorem says that if all three sets of sides in two triangles are in proportion, then the triangles are similar. If the proportion is one, the triangles are also congruent.

Similarity plays a large role in the study of **fractals**. This branch of mathematics studies the irregular patterns made of parts of a figure that are in some way similar to the whole figure. The leaves on a fern are an example of self-similarity. If a leaf is examined, it is possible to see that the pattern present in the larger leaf is present as the area examined is further and further reduced. It is possible to see the continued replication of the pattern in smaller and smaller form.

See also Congruence; Fractals

SIMPSON, THOMAS (1710-1761)

English mathematician

An almost entirely self-educated man, Thomas Simpson is most famous for the advances he made in the areas of **interpolation** and numerical **methods of integration**, although he also did work on **probability theory**. His enthusiasm and energy led him to produce many books that were popular in his day.

Simpson, the son of a weaver, was born in Market Bosworth, Leicestershire, England on August 20, 1710. Although his father pressured Simpson to follow in his professional footsteps, Simpson wanted to do something different. At a young age he moved to the town of Nuneaton, where he began supplementing his meager formal education by studying on his own and weaving to make a living. A peddler wandering through the town loaned Simpson books on mathematics and astrology, which Simpson studied intently. He was also motivated in his studies by wanting to understand an eclipse that he witnessed in 1724.

Anecdotal records suggest that Simpson soon became so knowledgeable of mathematics and astronomy that he acquired a reputation as a fortuneteller or astrologer. He turned to that profession to make money and also married his widowed landlady, although her son was older than he was. In 1733, according to one source, an "unfortunate accident" caused Simpson to leave his family behind and move to another town, Derby, where he tutored at an evening school and resumed weaving. By 1736, he had moved to London, where there was an audience for his mathematical talents and where he joined a group of wandering lecturers who used the city's coffeehouses as their classrooms. He began publishing his first papers, on **fluxions**, in the city's well-known *Ladies' Diary*. The following year, he published *A New Treatise of Fluxions*. The success of his first book and the academic community's rising awareness of the new mathematician's promise allowed Simpson to bring his family to London.

Simpson seems to have been an energetic scholar, since in addition to his tutorial and teaching responsibilities, he also found time to write *The Laws of Chance* (1740), *Annuities and Reversions* (1742), and *Mathematical Dissertations* (1743). Later in 1743, Simpson accepted a job as a mathematics teacher at the Royal Military Academy in Woolwich, which he would keep for the rest of his life. Also that year, he formulated what would become known as Simpson's rule, which allows estimation of the **area** under a curve using parabolic arcs. Simpson maintained a strenuous writing schedule as well, publishing the popular textbooks *Algebra* in 1745, *Geometry* in 1747, *Trigonometry* in 1748, and *Doctrine and Applications of Fluxions* in 1750.

In 1754 Simpson took on the job of editor of the *Ladies' Diary* in addition to his teaching position at the academy. The intense demands of the job reportedly weakened Simpson's health, however, and when he left the *Ladies' Diary* in 1760 to serve as a consultant for the firm hired to build a bridge across the Thames River, he only hastened his demise. He died in the town of his birth on May 14, 1761.

SINE

Sine is a trigonometric function derived from the Latin *sinus* meaning curve. In a right **triangle** the sine of an **angle** ($\sin\theta$) is defined as the length of the side of the triangle opposite the angle divided by the length of the **hypotenuse**. $\sin\theta = \text{opp/hyp}$ The values of the sine of an angle are always between **zero** and one for a right triangle. This function can be extended beyond $90°$ by using Cartesian coordinates. Values of the sine function can then be either positive or negative but still **range** between -1 and 1; these values repeat every $360°$ and therefore this function is considered a periodic function.

The sine function is related to the other six basic trigonometric **functions** in special relationships. The sine function is related to the **cosine** function as: $\cos\theta = \sin(90° - \theta)$. The sine function is also related to the tangent function and can be written as: $\tan\theta = \sin\theta/\cos\theta$. Using these relationships in conjunction with the following the sine function can be related to the other three **trigonometric functions**, cosecant (csc), secant (sec), and cotangent (cot). $\csc\theta = 1/\sin\theta$ The **Pythagorean theorem** for right triangles can be translated into a Pythagorean identity for sines and cosines and often appears as: $\sin^2\theta + \cos^2\theta = 1$.

The periodic nature of the sine function makes it important in applications involving the study of periodic phenomena such as **light** and electricity. A general triangle, not necessarily containing a right angle, can also be analyzed using trigonometric functions. In spherical **trigonometry**, the sine of an angle of a triangle on the surface of a **sphere** is important in surveying, navigation and astronomy.

SLIDE RULE

With the publication of logarithmic tables by **John Napier** (1550-1617) in 1614, astronomers and mathematicians were freed from much of the drudgery long associated with their professions. Some scholars realized that **logarithms** could be used in mechanical devices to take automatic calculation one step further. In 1620 **Edmund Gunter** (1581-1626), a professor at Gresham College, London, created a forerunner of the slide rule, which he described as his "logarithmic line of numbers."

Around that same time, English mathematician William Oughtred (1574-1660), began developing his own device employing Napier's logarithms. Although he was an ordained minister, Oughtred was most interested in mathematics and he spent every spare minute studying and planning his slide rule. In 1621 Oughtred introduced the device he had been perfecting—the first linear (or straight) slide rule. His device consisted of two sliding rulers marked with graduated scales representing logarithms, allowing calculations to be performed mechanically by sliding one ruler against the other. The invention caused quite a quarrel between Oughtred and his former pupil, Richard Delamain, because in 1630 Delamain described plans for his own design, a circular slide rule. Delays in publishing caused accounts of Delamain's device to appear in print before write-ups of Oughtred's slide rule appeared. While

Oughtred accused Delamain of stealing his ideas, it has been generally concluded that the devices were both developed independently. In addition to his slide rule, Oughtred was also responsible for the introduction of the **multiplication** sign "x" and the trigonometric abbreviations still commonly used today—sin, cos and tan, for **sine, cosine,** and tangent.

Few innovations were made to slide rule design until 1902 when Norwegian Carl George Lange Barth copyrighted a design for a circular slide rule. An advanced **model**, it was intended to speed up mathematical calculations and offered several advantages, perhaps the most important of which was it did not "run off the scale," as the more traditional straight line device was known to do.

While early slide rules were commonly used by tax collectors, masons and carpenters, slide rules today are most often used in engineering. Simple, yet versatile instruments, they perform calculations quickly and with reasonable accuracy.

SLOPE

The slope of a line corresponds to the idea of the steepness of an inclined plane or the steepness of a stairway. If all the steps of a stairway are uniform, the steepness of the stairway can be defined in terms of the "rise" and "run" of one of the steps. The steepness of a uniform stairway is the number obtained by dividing the rise by the run of one of the steps. If the rise (vertical component) is 6 inches and the run (horizontal component) is 8 inches, then the steepness of the stairway is 6/8 (or 3/4). Likewise, in formal **geometry**, slope, in rectangular coordinates, is defined as the ratio of the change of the ordinate (vertical component) to the corresponding change of the abscissa (horizontal component) of a point moving along a nonvertical line.

The origin of the term slope as used in mathematics most likely came from the Swedish word for slope "riktningskoefficient" and the Dutch word for slope "richtingscoëfficiënt" that both mean "direction coefficient". According to most mathematical historians, the origin for the symbol "m" that symbolizes slope is unknown. The French word "monter" means "to climb", but there is no evidence to substantiate that this word can be associated with the background of the slope symbol "m". In fact, it is known that the French mathematician **René Descartes** did not use "m" to reference slope. The earliest known use of "m" for slope is from an 1844 British text by Matthew O'Brien entitled *A Treatise on Plane Co-Ordinate Geometry*. Later in 1848 George Salmon (1819-1904) referred to O'Brien's 1844 article within his *A Treatise on Conic Sections* and used the slope-intercept formula "y = mx + b", where "b" is the ordinate (vertical component) of the point where the line intersects the y-axis. It is also known that the four authors Isaac Todhunter in 1855 (*Treatise on Plane Co-Ordinate Geometry*), George A. Osborne in 1891 (*Differential and Integral Calculus*), and Arthur M. Harding and George W. Mullins in 1924 (*Analytic*

Geometry) each used "m" to refer to slope in their mathematical writings.

If a straight line, or line segment, is determined by the distinct points $A = (x_1, y_1)$ and $B = (x_2, y_2)$ in an x-y plane with the origin at (0, 0) and x_1 not equaling x_2, then the slope, m, of the line segment AB is given by $(y_2 - y_1) / (x_2 - x_1)$. This equation can also be stated $m = \Delta y / \Delta x$, where $\Delta x = x_2 - x_1$ and $\Delta y = y_2 - y_1$.

As an example for the calculation of slope, if on a line segment within the x-y plane the two endpoints are $A = (x_1, y_1) = (1, 1)$ and $B = (x_2, y_2) = (3, 2)$, then the slope, m, of the line AB is (2 - 1) / (3 - 1), or m = 1/2. This value means that the slope of the line increases one-unit step on the y-axis for each two-unit step on the x-axis. As mentioned earlier, a straight line is defined by the slope-intercept equation "y = mx + b". The value of the constant "b" indicates where the line segment intercepts the y-axis. In the previous example one may insert "y = 1", "x = 1", and "m = 1/2" to find that "b = 1/2". This means that at the y-intercept (i.e., where "x = 0") the value of y is 1/2.

The sign of the slope indicates whether the line segment slopes "upward" or "downward". For the x-y coordinate system previously described, a positive slope is considered to slope "upward" from left to right, and a negative slope is considered to slope "downward" as viewed from left to right. If $y_1 = y_2$ then the line is horizontal with its slope valued at **zero**. If $x_1 = x_2$ then the line is vertical with an undefined slope.

In functional notion the **dependent variable** y is a function of the **variable** x, most often expressed as y = $f(x)$. If the line is not straight (but is a curve) the **definition** of slope must be modified. At a point $(x_1, f(x_1))$ on the curve a straight-line tangent to the curve at that point is constructed. The slope of the curve at point $(x_1, f(x_1))$ is then just the slope m of this tangential line. In differential **calculus** the slope m of the line tangent to a curve at point $(x_1, f(x_1))$ is defined formally by:

$$m(x_1, f(x_1)) = \lim_{\Delta x \to 0} (f(x_1 + \Delta x) - f(x_1)) / \Delta x,$$

where the function f is continuous in the neighborhood around the point $(x_1, f(x_1))$, and Δx does not equal zero. The equation given above for the slope m at the point $(x_1, f(x_1))$ may also be called the derivative of y with respect to x, and is symbolized in **differential calculus** by the notation "dy / dx".

As an example of using differentiation to determine slope, consider the curve given by y = $f(x)$ = x^2 - 4x + 5. For the ordered pair $(x_1, f(x_1))$ the "limit equation" previously defined will yield the slope m at that point; i.e., $m(x_1, f(x_1))$. Substituting "x^2 - 4x + 5" into the limit equation for $f(x)$ gives:

$$m(x_1, f(x_1)) = \lim_{\Delta x \to 0} (((x_1 + \Delta x)^2 - 4(x_1 + \Delta x) + 5) - (x_1^2 - 4x 4x_1 + 5)) / \Delta x.$$

The tools of calculus necessary to determine this differentiation are deferred to the calculus-related articles. Nevertheless, by performing the appropriate calculations, the result is m(x_1, $f(x_1)$) = 2x_1 - 4. For instance, at the point (x, $f(x)$) = (5, 10), the slope m(5, 10) = (2 x 5) - 4 = 6. In this case, the value "6" means that the tangential line at the point (x = 5, y = 10) on the curve changes 6 units on the y-axis for a 1 unit change on the x-axis. Since the slope is positive, the tangent line slopes upward from left to right in the x-y plane.

See also Analytic geometry; Derivatives and differentials; Differential calculus; Limits; Line; Polygon

SNELL, WILLEBRORD (1580-1626)

Dutch mathematical physicist

Willebrord Snell is best known for his discovery regarding the refraction of **light** rays. This discovery, known as Snell's law, demonstrates that when a ray of light passes from a thinner element such as air, into a denser element, such as water or glass, the **angle** of the ray bends to the vertical. Snell's law—a key revelation in the science of optics—was formulated after much experimentation in 1621. This is expressed as sin i =μ sin r (i=angle of incidence, r=angle of refraction and μ = a constant). However, he did not publish his findings, and the law did not appear in print until **René Descartes** discussed it (without giving Snell credit) in his *Dioptrique* in 1637. Snell also determined a formula to measure distances using trigonometric triangulation. His method, developed in 1615, used his home and the spires of Leiden churches as reference points. (In 1960, a plaque recognizing his work was placed on his home.) In an age of world exploration, this was very important work, because it contributed to improved accuracy in the art of mapmaking. Using the triangulation method, Snell measured the Earth's meridian for the first time, and also attempted to measure the size of the Earth using this method. He set down the principles of spherical **trigonometry** that determine the length of a meridian arc when measuring any base line. Snell's writings on his triangulation method were presented in *Eratosthenes batavus* (1617). His observations of comets sighted in 1585 and 1618 are described in his *Cyclometricus de circuli dimensione* (1621). His last works, *Canon triangulorum* (1626) and *Doctrina triangulorum*, published after his death in 1627, also addressed the measuring of distances through plane and spherical trigonometry.

Snell was born in Leiden, Holland, in 1591. He was the son of Rudolph Snellius (Snel van Royen), a professor of mathematics at the University of Leiden, and Machteld Cornelisdochter. He became interested in mathematics at an early age, but initially entered the University of Leiden to study law. Snell married Maria De Lange, daughter of a Schoonhoven burgher master, in 1608. They had eighteen children; only three survived. In 1613, after his father's death, he became mathematics professor at the University. He also taught astronomy and optics.

Snell occupied himself in his early career with travel throughout Europe and the translation of numerous mathematical treatises. In his travels, he consulted with such eminent scientists as **Tycho Brahe**, **Johannes Kepler**, and Michael

Maestlin, Kepler's astronomy teacher. Snell's Latin translation of the German *Wisconstighe Ghedachtenissen* was published as *Hypomnemata mathematica* in 1608; books of **Apollonius** based on plane loci were also translated during this time. Other translations of books on mathematics published by Snell were Ramus' *Arithmetica* (1613), and *Geometria* (1622).

The book translations were done at a time when Snell was very involved in applying mathematics to determine the shape and size of the Earth and the exact position of points on its surface. This work was to occupy him from 1615 onward. The mathematics he used is called plane trigonometry, which deals with calculations in two dimensions for measuring distances that cannot be measured directly, such as across lakes or vast expanses of land. Plane trigonometry began to be used in the mid–15th century—not surprising that a method so useful in geography, surveying, and mapmaking would be needed at a time when Europeans were undertaking their systematic maritime journeys of discovery. Snell's method, triangulation, had been used by Tycho Brahe and also by Gemma Frisius in 1533, but Snell advanced it to such a higher level he became known as the father of triangulation. Using his house and the spires of churches as starting points, he determined angles with a quadrant more than 7 feet long (213 cm). Building on a network of triangles, he calculated the **distance** between two towns—Alkmaar and Bergen-op-Zoom—located on the same meridian approximately 80 miles (130 km) from each other. Taking this calculation further, he was able to determine the radius of the Earth. The calculations for this difficult work were tedious, given that he computed his measurements without the help of **logarithms**. In spite of that, they remain surprisingly accurate. In 1624, he published *Tiphys batavus*, a work on navigation which parallels his work on determining land measurements. In it, his geometric theories hint of the differential **triangle** developed by later mathematicians, especially **Blaise Pascal**.

Snell's work in astronomy resulted in two books: *Cyclometricus de circuli dimensione*, and *Concerning the Comet*, the latter published in 1618. His observations demonstrate his shorter method for determining comet distance and movements. Coinciding with this work was his study of light refraction. In his experimentation, he referred to Kepler's writings, as well as those of Risner. The phenomenon of refraction was known as far back as **Claudius Ptolemy**, but Snell's explanation—which states that the ratio of the sines of the angles of the incident and refracted rays to the normal is a constant—became known as Snell's law of optical refraction. The constant determines how much a ray of light will be bent as it travels through one medium to another. Some controversy surrounds the fact that, although Snell never published his findings, they appear uncredited in **René Descartes** *Dioptrique* in 1637, more than ten years after Snell's death. Some have accused Descartes of plagiarism, but it is unknown whether Descartes knew of Snell's discovery and passed it off as his own, or if his knowledge was gained independent of Snell's work.

Snell died in Leiden on October 30, 1626, and he was buried in the Pieterskerk. A monument was erected there in his honor.

SNOWFLAKE CURVE

The snowflake curve is a fractal. It is also known as the Koch snowflake or Koch's island after Helge von Koch who first described it in 1904. To construct it, first draw a line segment AD. Next draw point B and C on AD so that AB, BC and CD all have the same length. Now draw a point E so that the **triangle** BCE is equilateral. Now erase segment BC. The shape that is left is called the motif of the snowflake curve.

On a separate piece of paper, draw an equilateral triangle with side length equal to the length of AD. Now replace each edge of the triangle with a copy of the motif so that the motif points away from the triangle. Now replace each of edge of what you have with a reduced (by one third) copy of the motif so that the motifs point outwards. Again, replace each edge of what you now have with a reduced (by one ninth) copy of the motif so that the motifs point outward. Continue in this way forever. The result is the Koch curve.

At each new step of the construction, the **perimeter** of the curve is four thirds times the perimeter of the curve at the previous step. Thus, the perimeter of the Koch curve is infinite.

If s is the number of sides of the island at the nth step then the **area** of the island at the (n+1)st step is equal to its area at the nth step + s times $(1/9)^{n+1}$ times the area of the original equilateral triangle. This is because all the new triangles are lengthwise 1/3 the size of the previous triangles. Thus the area of the new triangles is 1/9 that of the previous. Also there are s new triangles at the n+1st step. The number s is equal to $3*4^n$ since at each step the number of sides increases by a **factor** of four and the number of sides of the original triangle is three. These facts imply that the area of the Koch snowflake is equal to the area of the original triangle times one plus one third of the sum as n goes from 0 to **infinity** of $(4/9)^n$. The last sum is the sum of a **geometric series**. Its value is nine fifths. Hence the area of the Koch snowflake is eight fifths of the area of the original triangle.

The Koch curve is nondifferentiable everywhere (see **Derivatives and differentials**). This means that for any point x on the curve, and any sequence of points y_n on the curve, such that the **distance** between x and y_n decreases to **zero** as n increases towards infinity, then the **slope** of the line segments xy_n does not approach any finite limit. The set of points on the original triangle that are contained in the Koch curve form a **Cantor set**. Because this set is totally disconnected, the Koch curve contains no line segments.

The snowflake curve can be partitioned into three congruent curves. Each curve is formed from the motif by repeatedly substituting small copies of the motif for each line segment at each step. Each of these curves can then be partitioned into four smaller curves, each a copy of the big curve but one third its size. Then each of these smaller curves can be partitioned into four smaller curves. Of course, this process never ends. This process is similar to another process. A line segment can be divided into two line segments. Then each of the two halves can be divided into two and so on. Also a **square** can be divided into four squares, each a copy of the original square reduced by one half. A cube can be divided into eight cubes, each a copy of the original cube reduced by one half. Thus, we have that a d- dimensional "cube" can be divided into 2^d "cubes", each a copy of the original reduced by one half. If the reduction factor is s, the number of copies is s^{-d}. Following this train of thought, the self-similarity **dimension** of the Koch curve is a number d such that $3^d = 4$. This implies that d = log(4)/log(3) which is approximately 1.262. Incidentally, the self-similarity dimension of the Koch curve is equal to its **Hausdorff dimension**.

See also Cantor set; Fractals; Hausdorff dimension; Sierpinski triangle, carpet, and sponge

SOLID GEOMETRY

Solid **geometry** is that subsection of geometry that deals with figures in three-dimensional **space**. **Euclidean geometry**, sometimes called parabolic geometry, is divided into two subsections: **plane geometry**, geometry dealing with figures in a plane, and solid geometry, geometry dealing with solids in three-dimensional space. Solid geometry is sometimes called three-dimensional Euclidean geometry and deals with solids, such as polyhedra and spheres, and lines and planes in three-dimensional space. Solid geometry is concerned with the study of figures in three-dimensional Euclidean space, which is usually denoted R^3.

Solid geometry is a branch of Euclidean geometry, developed by Greek mathematician **Euclid** in the 4th century B.C., that is governed by Euclid's five postulates as laid out in his work *The Elements*. More specifically the basics of solid geometry are described in Book XI of *The Elements*. Book XI consists of 28 definitions, in which Euclid departed from traditional definitions and instead used motion-based definitions, and 39 **propositions**. These motion-based definitions were formulated as revolutions of figures about a **diameter**. The definitions cover everything from the **definition** of a solid, inclination of figures in three-dimensional space, **similarity** of those figures, definitions of pyramids, spheres, cones, cylinders, cubes, octahedrons, icosahedrons, and dodecahedrons. There are also several definitions that deal with angles and angular relations of figures relative to each other. The first 18 propositions deal with lines and planes and their relations to figures in Euclidean space. The elementary theorems of three-dimensional geometry, solid geometry, are laid out in this book. The rest of the propositions deal with planes and solids in Euclidean space. Book XII of Euclid's *The Elements* is concerned with the measurement of figures and contains 18 propositions dealing with this endeavor. Euclid used **Eudoxus'** method of exhaustion in this book and so many believe that most of this work is due to Eudoxus. Book XIII, the final book in Euclid's original version of *The Elements*, contains 18 propositions describing the construction of regular solids. It is thought that this book is based on *Comparison of the Five Figures*, a work by Aristaeus. The first six propositions involve the golden ratio of cut lines. The remainder of the propositions describes the construction of the five regular solids, pyramid, octahedron, cube, icosahedron, and dodecahedron, and proves that no other regular polyhedra are possible. After Euclid wrote *The Elements* a fourteenth and fifteenth books were added by Hypsicles in about 170 B.C. and Isadorus of Miletus in about 530 A.D.. In the Book XIV Hypsicles, working from treatises by Aristaeus and Apollonius, compares the five regular solids with respect to their faces, surface areas, and volumes. In Book XV, inferior to the previous one because of its impreciseness and inaccurateness, Isadorus deals with several topics including inscribing certain regular solids within others, determination of the number of edges and vertices on each solid, and determining the **angle** of inclination between adjacent faces in each solid.

Solid geometry is an extension and reinforcement of the propositions describing plane geometry. It forms the basis for the foundations of other branches of mathematics including geometry and **trigonometry**. The development of geometry in general was driven by three classical problems in Greek mathematics: plane, solid, and linear problems. Those problems whose solutions employ the use of one or more sections of a cone are solid problems. **René Descartes** advanced solid geometry in the 17th century by inventing **Cartesian coordinates** to express geometric relations in algebraic form. These advances led to the development of **analytic geometry**, descriptive geometry, differential geometry and projective geometry which are important to engineering and the natural sciences as well as mathematics.

SOLIDS WITH KNOWN CROSS SECTION AREAS

A cross section of a solid is mathematically defined as the plane surface that results from the intersection of a plane with the solid. Therefore, a cross section is an infinitesimally thin slice of the solid. Solids commonly encountered in solid **geometry** include the **sphere**, ellipsoid, **cylinder**, cube, cone, and pyramid. Because solids are 3-dimensional figures, they have volumes that can be calculated. A cross section of a solid is a 2-dimensional shape, for which an **area** (called the cross-sectional area of the solid) can be calculated.

Although a cross section of a solid is mathematically a "slice" of the solid, it is easier to visualize a cross section as one of the cut ends produced when the solid is cut completely in two. For example, imagine that the earth were cut in two

along the equator. Each cut surface (cross section) of the earth would be flat and in the shape of a **circle**. The area of a circle is determined from πr^2, where r is the radius of the circle. Therefore, the cross-sectional area of the earth could be calculated using this formula.

Imagine that the earth is perfectly round, like a sphere. Then, a cross section produced by cutting from the North Pole to the South Pole would be identical in area to the cross section obtained from the cut across the equator. So would any complete cut that goes through the center of the earth. This is because of the **symmetry** of a sphere at all points around its center. Unless otherwise specified, a cross section is generally considered to be a section including the center of the solid (i.e., the intersection of its axis).

An ellipsoid can be visualized as a sphere stretched along one of its axis. Imagine an earth stretched way out of proportion and shaped like a bullet, with the North Pole at one end of the bullet, and the South Pole at the other end. If this Earth is cut along the equator, the resulting cross section is a circle. However, if this Earth is cut from the North Pole to the South Pole, the resulting cross section is an **ellipse**. The formula for the area of an ellipse is $\pi r_1 r_2$, where r_1 and r_2 are the radii of the ellipse along its short and long axis, respectively.

A cylinder also has two different cross sections, depending on which axis is cut along. A cut through the curved surfaces (usually the short axis) provides a circular cross section. A cut from one flat end to the other flat end (usually the long axis) provides a rectangular cross section. The area of a rectangle is calculated from its width times its length.

A cube cut along the x, y or z axis has a **square** cross section. No matter which of these axis the cube is cut along, the resulting cross sections will be squares with the same area. The area of a square is calculated from its width times its length (which are equal to each other).

All of the solids described so far have been cut through their centers to provide a cross section. For the cone and the pyramid, cross sections are sometimes better defined as being parallel or perpendicular to their bases. For example, if a cone is cut across its curved surfaces along any plane parallel to its base, a circular cross section is obtained. This is known as one of the **conic sections**. A cross section taken near the base will have a larger area than one taken near the point of the cone. If the cone is cut from its point to its base (through the center of the base), a triangular cross section is obtained. The area of a **triangle** is calculated from ½ times its base times its height.

Consider a pyramid with a square for its base and four triangular sides meeting at a point at the top (the apex). If this pyramid is cut across a plane parallel to its base, a square cross section will be obtained. Just as for the cone, a cross section taken near the base will have a larger area than one taken near the apex. If this pyramid is cut from its apex to its base (through the center of the base), the resulting cross section is triangular.

Remember that a cross section is defined by the intersection of a plane with a solid. The orientation of that plane to the solid must be defined in order to obtain the correct cross section and calculate the cross-sectional area.

SPACE

Space is the three-dimensional extension in which all things exist and move. We intuitively feel that we live in an unchanging space. In this space, the height of a tree or the length of a table is exactly the same for everybody. Einstein's *special theory of relativity* (see **Relativity, special**) tells us that this intuitive feeling is really an illusion. Neither space nor time is the same for two people moving relative to each other. Only a combination of space and time, called *space-time*, is unchanged for everyone. Einstein's *general theory of relativity* (see **Relativity, general**) tells us that the **force** of gravity is a result of a warping of this **space-time** by heavy objects, such as planets. According to the *big bang theory* of the origin of the universe, the expansion of the universe began from infinitely curved space-time. We still do not know whether this expansion will continue indefinitely or whether the universe will collapse again in a big crunch. Meanwhile, astronomers are learning more and more about outer space from terrestrial and orbiting telescopes, *space probes* sent to other planets in the solar system, and other scientific observations. This is just the beginning of the exploration of the unimaginably vast void, beyond the Earth's outer atmosphere, in which a journey to the nearest star would take 3,000 years at a million miles an hour.

The difference in the perception of space and time, predicted by the special theory of relativity, can be observed only at very high velocities close to that of **light**. A man driving past at 50 mph (80 kph) will appear only a hundred million millionth of an inch thinner as you stand watching on the sidewalk. By themselves, three-dimensional space and one-dimensional time are different for different people. Taken together, however, they form a four-dimensional space-time in which distances are same for all observers. We can understand this idea by using a two-dimensional analogy. Let us suppose your **definition** of south and east is not the same as mine. I travel from city A to city B by going 10 miles along my south and then 5 miles along my east. You travel from A to B by going 2 miles along your south and 11 miles along your east. Both of us, however, move exactly the same **distance** of 11.2 miles south-east from city A to B. In the same way, if we think of space as south and time as east, space-time is something like south-east.

The general theory of relativity tells us that gravity is the result of the curving of this four-dimensional space-time by objects with large mass. A flat stretched rubber membrane will sag if a heavy iron ball is placed on it. If you now place another ball on the membrane, the second ball will roll towards the first. This can be interpreted in two ways; as a consequence of the curvature of the membrane, or as the result of an attractive force exerted by the first ball on the second one. Similarly, the curvature of space-time is another way of interpreting the attraction of gravity. An extremely massive object can curve space-time around so much that not even light can escape from its attractive force. Such objects, called *black holes*, could very well exist in the universe. Astronomers believe that the disk found in 1994 by the Hubble telescope, at the center of the elliptical galaxy M87 near the center of the Virgo cluster, is

material falling into a supermassive black hole estimated to have a mass three billion times the mass of the Sun.

The relativity of space and time and the curvature of space-time do not affect our daily lives. The high velocities and huge concentrations of matter, needed to manifest the effects of relativity, are found only in outer space on the scale of planets, stars, and galaxies. Our own Milky Way galaxy is a mere speck, 100,000 light years across, in a universe that spans ten billion light years. Though astronomers have studied this outer space with telescopes for hundreds of years, the modern space age began only in 1957 when the Soviet Union put the first artificial satellite, *Sputnik 1*, into **orbit** around the Earth. At present, there are hundreds of satellites in orbit gathering information from distant stars, free of the distorting effect of the Earth's atmosphere. Even though no manned spacecraft has landed on other worlds since the Apollo moon landings, several space probes, such as the *Voyager 2* and the *Magellan*, have sent back photographs and information from the moon and from other planets in the solar system. There are many questions to be answered and much to be achieved in the exploration of space. The Hubble telescope, repaired in space in 1993, has sent back data that has raised new questions about the age, origin, and nature of the universe. The launch of an United States astronaut to the Russian *Mir* space station in March 1995, the planned docking of the United States space shuttle *Atlantis* with *Mir*, and the proposed international space station, have opened up exciting possibilities for space exploration.

SPACE-TIME

Space-time is the coordinate system used to describe the universe. The simplest physical view of space-time, and the one most commonly used, is four-dimensional, assuming three dimensions of **space** and one of time. Other modern models **postulate** a larger number of dimensions unperceived by humans in their daily lives. Ten, eleven, and twenty-six are the numbers of dimensions used in some popular but as yet unprovable theories, such as superstring theory.

Until **Albert Einstein** formulated his theory of **relativity** in the early part of the twentieth century, the term space-time was almost never used. Spatial dimensions and time were thought to be independent and wholly separable. While the **equations** for a particle's spatial attributes often depended upon time, time itself was thought to be a great constant, passing at apparently equal rates for all observers. Since the codification and testing of relativity, however, it is known that the apparent passage of time is not constant and varies with the speed of the observer. Time and the three spatial dimensions are inextricably related.

Relativity also allows for the possibility—indeed, the certainty—that space-time is non-Euclidean (see **non-Euclidean geometries**). One of its major postulates is that mass is equivalent to energy, and that this composite mass-energy warps space-time. In some regions, then, space-time will be warped like a saddle; in other regions, like a **sphere**, depending upon the distribution of mass. The most common image of warped space-time used for visualization purposes is a series of masses on a sheet of rubber, spandex, or some other stretchy surface. The total **geometry** of space-time is as yet unknown, although it is very close to "flat" on the average. The exact geometry of space-time should be of continuing interest in years and decades to come, as it will determine the ultimate end of the universe.

See also Minkowski space-time; Relativity, general; Relativity, special

SPHERE

A sphere is a three dimensional figure that is the set of all points equidistant from a fixed point, called the center. The **diameter** of a sphere is a line segment which passes through the center and whose endpoints lie on the sphere. The radius of a sphere is a line segment whose one endpoint lies on the sphere and whose other endpoint is the center.

A great **circle** of a sphere is the intersection of a plane that contains the center of the sphere with the sphere. Its diameter is called an axis and the endpoint of the axes are called poles. (Think of the north and south poles on a globe of the earth.) A meridian of a sphere is any part of a great circle.

A sphere of radius r has a surface **area** of $4\pi r^2$ and a **volume** of $4/3 \pi r^3$.

A sphere is determined by any four points in **space** that do not lie in the same plane. Thus there is a unique sphere that can be circumscribed around a tetrahedron. The equation in Cartesian coordinates x, y, and z of a sphere with center at (a, b, c) and radius r is $(x-a)^2 + (y-b)^2 + (z-c)^2 = r^2$.

SPIRAL

A spiral is a curve formed by a point revolving around a fixed axis at an ever-increasing **distance**. It can be defined by a mathematical function which relates the distance of a point from its origin to the **angle** at which it is rotated. Some common spirals include the spiral of Archimedes and the hyperbolic spiral. Another type of spiral, called a logarithmic spiral, is found in many instances in nature.

Characteristics of a spiral

A spiral is a function which relates the distance of a point from the origin to its angle with the positive x axis. The equation for a spiral is typically given in terms of its polar coordinates. The polar coordinate system is another way in which points on a graph can be located. In the rectangular coordinate system, each point is defined by its x and y distance from the origin. For example, the point (4,3) would be located 4 units over on the x axis, and 3 units up on the y axis. Unlike the rectangular coordinate system, the polar coordinate system uses the distance and angle from the origin of a point to define its location. The common notation for this system is

(r,θ) where r represents the length of a ray drawn from the origin to the point and θ represents the angle which this ray makes with the x axis. This ray is often known as a vector.

Like all other geometric shapes, a spiral has certain characteristics which help define it. The center, starting point, of a spiral is known as its origin or nucleus. The line winding away from the nucleus is called the tail. Most spirals are also infinite, that is they do not have a finite ending point.

Types of spirals

Spirals are classified by the mathematical relationship between the length r of the radius vector, and the vector angle q, which is made with the positive x axis. Some of the most common include the spiral of Archimedes, the logarithmic spiral, parabolic spiral, and the hyperbolic spiral.

The simplest of all spirals was discovered by the ancient Greek mathematician **Archimedes of Syracuse** (287-212 BC). The spiral of Archimedes conforms to the equation $r = a\theta$, where r and θ represent the polar coordinates of the point plotted as the length of the radius a, uniformly changes. In this case, r is proportional to θ.

The logarithmic, or equiangular spiral was first suggested by **René Descartes** (1596-1650) in 1638. Another mathematician, **Jakob Bernoulli** (1654-1705), who made important contributions to the subject of probability, is also credited with describing significant aspects of this spiral. A logarithmic spiral is defined by the equation $r = e^{a\theta}$, where e is the natural logarithmic constant, r and θ represent the polar coordinates, and a is the length of the changing radius. These spirals are similar to a **circle** because they cross their radii at a constant angle. However, unlike a circle, the angle at which its points cross its radii is not a right angle. Also, these spirals are different from a circle in that the length of the radii increases, while in a circle, the length of the radius is constant. Examples of the logarithmic spiral are found throughout nature. The shell of a *Nautilus* and the seed patterns of sunflower seeds are both in the shape of a logarithmic spiral.

A parabolic spiral can be represented by the mathematical equation $r^2 = a^2\theta$. This spiral discovered by **Bonaventura Cavalieri** (1598-1647) creates a curve commonly known as a **parabola**. Another spiral, the hyperbolic spiral, conforms to the equation $r = a/\theta$.

Another type of curve similar to a spiral is a helix. A helix is like a spiral in that it is a curve made by rotating around a point at an ever-increasing distance. However, unlike the two dimensional plane curves of a spiral, a helix is a three dimensional **space** curve which lies on the surface of a **cylinder**. Its points are such that it makes a constant angle with the cross sections of the cylinder. An example of this curve is the threads of a bolt.

SQUARE

A square is a rectangle with all sides equal. A square with side *a* has **perimeter** 4 *a* and **area** *a2*.

The square is used as the unit of area; that is, a figure's area is expressed as the number of equal squares of some standard, such as square inches or square meters, that the figure can contain. In Greek **geometry**, the area of a figure was determined by converting the figure to a square of the same area. This is easily accomplished for triangles, rectangles, and other **polygons**, but is often impossible for other figures, such as circles.

SQUARE ROOT

The number k is a **square** root of the number n if $k2 = n$. For example, 4 and -4 are the square **roots** of 16 since $4 \times 4 = 16$ and $(-4) \times (-4) = 16$. The square root symbol is $\sqrt{}$ and for n greater than **zero** the symbol is understood to be a positive number. Thus $\sqrt{16} = 4$ and $-\sqrt{16} = -4$.

When n is a negative number, the square root $\sqrt{n}$ is called *imaginary*. Customarily, $\sqrt{-1}$ is designated by i so that the square root of any negative number can be expressed as ai where a is a real number. Thus $\sqrt{-5} = 5i$.

SQUARING THE CIRCLE

Squaring the **circle** is one of the most ancient problems of mathematics: given a circle, construct a **square** of equal **area**. This problem is closely related to the question of finding the value of **pi**, since for a circle of radius 1 has area equal to pi. Attempts to square the circle and find the value of pi go back as far as the oldest mathematical documents that have been discovered.

The earliest reference to the problem of squaring the circle is in the **Rhind papyrus**, an Egyptian document written about 1650 B.C., describing work that was performed 200 years earlier; some historians believe that it is based on work dating as far back as 3400 B.C.. The papyrus gives a technique for squaring the circle based on the assumption that pi is equal to 3.16. This is an excellent approximation, but it is not the exact value. Thus the Egyptians' method, although it probably served them well in practical applications, cannot be considered a true solution.

It was not until the time of the Greeks, when the idea of mathematical **proof** had evolved, that the problem of squaring the circle was stated in precise mathematical language. There are many different questions associated to the phrase 'squaring the circle'; the problem that became most intensely studied was to construct the required square using 'plane' techniques, that is, using only a compass and a straightedge.

The first recorded formal attempt to square the circle was made by **Anaxagoras of Clazomenae** (500-428 B.C.), who worked on the problem while imprisoned. Centuries later, Plutarch wrote in his work *On Exile*, "There is no place that can take away the happiness of a man, nor yet his virtue or wisdom. Anaxagoras, indeed, wrote on the squaring of the circle while in prison." The problem soon became quite popular, and in fact there is a reference to it in Aristophanes' play *The Birds*, written in 414 B.C.. The Greeks, who recognized the dif-

ficulty of the problem, came to refer to people who attempted the impossible as 'circle-squarers', even inventing a new word to describe such futile attempts.

The Greek mathematicians were apparently convinced that it was impossible to square the circle using plane techniques, although a proof of that fact was beyond their reach. On the other hand, the Greeks found many ways to square the circle using other techniques, and the problem stimulated many advances in understanding. One of the most important ideas was that of Antiphon the Sophist, a contemporary of Socrates, who proposed the following technique: start with a polygon (say, a **triangle**) that is inscribed in the circle, and then construct more and more inscribed **polygons** by doubling the number of sides, first to 6, then to 12, and so on. He claimed that eventually the polygons would be so close to the circle that they would be indistinguishable from it. The Greek mathematicians had methods for squaring polygons, so by this method they could produce squares whose area got closer and closer to the desired area, until the difference was not noticeable. This is of course not an acceptable plane method, since the exact square is never actually reached, only approximated. However, it introduced an important new concept: the idea of approximating a measurement 'by exhaustion', that is, by measuring larger and larger objects inside the object of interest. Another elegant approach to the problem was that of Archimedes, who used the **geometry** of the **spiral** to square the circle.

As the center of mathematical thought shifted from the Greeks to the Arabs in the Middle Ages, and to the Europeans in the Renaissance, many new methods were created to square the circle, including several by **Leonardo da Vinci**. However, these later mathematicians, unlike their Greek counterparts, did not realize that squaring the circle using plane techniques was impossible, and many fallacious 'proofs' were circulated that the circle could be squared.

James Gregory (1638-1675) was one of the first mathematicians to catch a glimpse of the correct reasoning behind the impossibility of squaring the circle. Mathematicians realized that the question of squaring the circle depends on the algebraic properties of the number pi, since it had been proven that all constructible lengths were what are called 'algebraic numbers', numbers that are the solutions of polynomial **equations** with integer coefficients. Gregory developed many of the ideas of infinite sequences, and one of the uses he found for them was to approximate pi. He became convinced that pi was not an algebraic number, which is correct. However, his proof had errors, and many mathematicians remained certain that pi was an algebraic number.

In 1761, **Johann Lambert** proved that pi is an irrational number, that is, it is not the ratio of two **whole numbers**. While this strengthened the case that pi was not an algebraic number, it fell short of a proof. Instead of ending the debate, it had the opposite effect: it stimulated many amateur mathematicians to try to square the circle. These attempts were so numerous that in 1775 the Paris Academy passed a resolution refusing to examine any more 'proofs' of squaring the circle, and a few years later the Royal Society in London followed suit.

In 1882, **Carl Lindemann** finally put the problem to rest by proving that pi is not an algebraic number. In spite of his proof, many amateur mathematicians have remained convinced that they can find methods to square the circle, and their fallacious proofs continue to plague mathematicians to this day.

STANDARD DEVIATION

The standard deviation is a statistical measure of the **dispersion** or uncertainty in a random **variable**. The standard deviation is the **square root** of the variance, a measure of how spread out a distribution is, and is written for a random variable x as: $\Sigma = \sqrt{[\Sigma^N_{i=1}(x_i - x)^2/(N - 1)]}$, Σ is the standard deviation, x_i is each individual value of x, x is the **mean** of all of the data points, and N is the number of data points. In this form the standard deviation is an unbiased estimate of the variance and is sometimes said to be corrected for a loss of a degree of freedom. On occasion the standard deviation is written as a biased estimate of the variance when the denominator is just N rather than N - 1. The value of x_i - x is called the residual for each x value. The standard deviation of a set of data is the most commonly used measure of the spread of that data. If the standard deviation is small then the data points are tightly clustered around the mean value. If it is large then they are widely scattered relative to the mean value.

The variance is a somewhat abstract measure of the variability in a set of data. Unlike the variability the standard deviation can be easily conceptualized by plotting it along with the individual points in the set. It is easy to visualize the standard deviation in this way along with the data set.

An important characteristic of the standard deviation is as a measure of the percentile rank. Standard deviation is a measure of the spread of a set of data. If that set is a normal distribution, a set of data with the concentration of points in the middle of the distribution, and the mean is known, then it is possible to compute the percentile rank associated with any particular data point value of that set. The percentile rank is the proportion of values in a distribution that one specific value is greater than or equal to.

Because of its mathematical tractability the standard deviation is a useful measure of the spread in a set of data and is often employed in inferential **statistics**. Inferential statistics are a type of statistics used to draw inferences about populations from only a sample of that population. Where there is a possibility of extreme values being present in a set of data the standard deviation should be supplemented by the semi-interquartile **range**. This is because the standard deviation is more sensitive than the semi-interquartile range to extreme values. The standard deviation can be thrown off by an extreme value in a set of data.

There are special ways to calculate the standard deviation for particular cases. If f(x) is a linear **transformation** of x such that f(x) = bx + a, then the standard deviation of f(x) is $b\Sigma x$ where $\Sigma^2 x$ is the variance of x. In this case the variance of f(x) is $b^2\Sigma^2 x$.

Standard deviation has many uses in statistical analysis of data **sets**. Along with its usefulness in the mathematical world it is also widely used in the financial world. If a financial variable is highly volatile then it has a high standard deviation. Often times the standard deviation is used as a measure of the volatility of a random financial variable. Standard deviation is also widely used by the United States census bureau in calculating and interpreting the census results from year to year.

STATISTICS

Statistics is that branch of mathematics devoted to the collection, compilation, display, and interpretation of numerical data. In general, the field can be divided into two major subgroups, *descriptive statistics* and *inferential statistics*. The former subject deals primarily with the accumulation and presentation of numerical data, while the latter focuses on predictions that can be made based on those data.

Some fundamental concepts

Two fundamental concepts used in statistical analysis are population and sample. The term *population* refers to a complete set of individuals, objects, or events that belong to some category. For example, all of the players who are employed by Major League Baseball teams make up the population of professional major league baseball players. The term *sample* refers to some subset of a population that is representative of the total population. For example, one might go down the complete list of all major league baseball players and select every tenth name. That subset of every tenth name would then make up a sample of all professional major league baseball players.

Another concept of importance in statistics is the distinction between discrete and continuous data. Discrete variables are numbers that can have only certain specific numerical value that can be clearly separated from each other. For example, the number of professional major league baseball players is a discrete **variable**. There may be 400 or 410 or 475 or 615 professional baseball players, but never 400.5, 410.75, or 615.895.

Continuous variables may take any value whatsoever. The readings on a thermometer are an example of a continuous variable. The temperature can range from 10°C to 10.1°C to 10.2°C to 10.3°C (about 50°F) and so on upward or downward. Also, if a thermometer accurate enough is available, even finer divisions, such as 10.11°C, 10.12°C, and 10.13°C, can be made. Methods for dealing with discrete and continuous variables are somewhat different from each other in statistics.

In some cases, it is useful to treat continuous variable as discrete variables, and vice versa. For example, it might be helpful in some kind of statistical analysis to assume that temperatures can assume only discrete values, such as 5°C, 10°C, 15°C (41°F, 50°F, 59°F) and so on. It is important in making use of that statistical analysis, then, to recognize that this kind of assumption has been made.

Collecting data

The first step in doing a statistical study is usually to obtain raw data. As an example, suppose that a researcher wants to know the number of female African-Americans in each of six age groups (1-19; 20-29; 30-39; 40-49; 50-59; and 60+) there are in the United States. One way to answer that question would be to do a population survey, that is, to interview every single female African-American in the United States and ask what their age is. Quite obviously, such a study would be very difficult and very expensive to complete. In fact, it would probably be impossible to do.

A more reasonable approach is to select a sample of female African-Americans that is smaller than the total population and to interview this sample. Then, if the sample is drawn so as to be truly representative of the total population, the researcher can draw some conclusions about the total population based on the findings obtained from the smaller sample.

Descriptive statistics

Perhaps the simplest way to report the results of the study described above is to make a table. The advantage of constructing a table of data is that a reader can get a general idea about the findings of the study in a brief glance.

TABLE 1. STATISTICS

Number of Female African Americans in Various Age Groups

Age	Number
0-19	5,382,025
20-29	2,982,305
30-39	2,587,550
40-49	1,567,735
50-59	1,335,235
60+	1,606,335

Graphical representation

The table shown above is one way of representing the frequency distribution of a sample or population. A *frequency distribution* is any method for summarizing data that shows the number of individuals or individual cases present in each given **interval** of measurement. In the table above, there are 5,382,025 female African-Americans in the age group 0-19; 2,982,305 in the age group 20-29; 2,587,550 in the age group 30-39; and so on.

A common method for expressing frequency distributions in an easy-to-read form is a graph. Among the kinds of graph used for the display of data are histograms, bar graphs, and line graphs. A histogram is a graphs that consists of solid bars without any space between them. The width of the bars corresponds to one of the variables being presented and the height of the bars, to a second variable. If we constructed a histogram based on the table shown above, the graph would have six bars, one for each of the six age groups included in the study. The height of the six bars would correspond to the frequency found for each group. The first bar (ages 0-19) would be nearly twice as high as the second (20-29) and third (30-39)

bars since there are nearly twice as many individuals in the first group as in the second or third. The fourth, fifth, and six bars would be nearly the same height since there are about the same numbers of individuals in each of these three groups.

Another kind of graph that can be constructed from a histogram is a frequency polygon. A frequency polygon can be made by joining the midpoints of the top lines of each bar in a histogram to each other.

Distribution curves

Finally, think of a histogram in which the vertical bars are very narrow... and then very, very narrow. As one connects the midpoints of these bars, the frequency polygon begins to look like a smooth curve, perhaps like a high, smoothly shaped hill. A curve of this kind is known as a *distribution curve*.

Probably the most familiar kind of distribution curve is one with a peak in the middle of the graph that falls off equally on both sides of the peak. This kind of distribution curve is known as a "normal" curve. Normal curves result from a number of random events that occur in the world. For example, suppose you were to flip a penny a thousand times and count how many times heads and how many times tails came up. What you would find would be a normal distribution curve, with a peak at equal heads and tails. That means that, if you were to flip a penny many times, you would most commonly expect equal numbers of heads and tails. But the likelihood of some other distribution of heads and tails-such as 10% heads and 90% tails-would occur much less often.

Frequency distributions that are not normal are said to be skewed. In a skewed distribution curve, the number of cases on one side of the maximum is much smaller than the number of cases on the other side of the maximum. The graph might start out at **zero** and rise very sharply to its maximum point and then drop down again on a very gradual **slope** to zero on the other side. Depending on where the gradual slope is, the graph is said to be skewed to the left or to the right.

Other kinds of frequency distributions

Bar graphs look very much like histograms except that gaps are left between adjacent bars. This difference is based on the fact that bar graphs are usually used to represent discrete data and the space between bars is a reminder of the discrete character of the data represented.

Line graphs can also be used to represent continuous data. If one were to record the temperature once an hour all day long, a line graph could be constructed with the hours of day along the horizontal axis of the graph and the various temperatures along the vertical axis. The temperature found for each hour could then be plotted on the graph as a point and the points then connected with each other. The assumption of such a graph is that the temperature varied continuously between the observed readings and that those temperatures would fall along the continuous line drawn on the graph.

A **circle** graph, or "pie chart," can also be used to graph data. A circle graph shows how the total number of individuals, cases or events is divided up into various categories. For example, a circle graph showing the population of female

African-Americans in the United States would be divided into pie-shaped segments, one (0-19) twice as large as the next two (20-20 and 30-39), and three about equal in size and smaller than the other three.

Measures of central tendency

Both statisticians and non-statisticians talk about "averages" all the time. But the term average can have a number of different meanings. In the field of statistics, therefore, workers prefer to use the term measure of central tendency for the concept of an "average." One way to understand how various measures of central tendency (different kinds of "average") differ from each other is to consider a classroom consisting of only six students. A study of the six students shows that their family incomes are as follows: $20,000; $25,000; $20,000; $30,000; $27,500; $150,000. What is the "average" income for the students in this classroom?

The measure of central tendency that most students learn in school is the *mean*. The **mean** for any set of numbers is found by adding all the numbers and dividing by the number of numbers. In this example, the mean would be equal to $20,000 + $25,000 + $20,000 + $30,000 + $27,500 + $150,000 ÷ 6 = $45,417. But how much useful information does this answer give about the six students in the classroom? The mean that has been calculated ($45,417) is greater than the household income of five of the six students.

Another way of calculating central tendency is known as the *median*. The **median** value of a set of measurements is the middle value when the measurements are arranged in order from least to greatest. When there are an even number of measurements, the median is half way between the middle two measurements. In the above example, the measurements can be rearranged from least to greatest: $20,000; $20,000; $25,000; $27,500; $30,000; $150,000. In this case, the middle two measurements are $25,000 and $27,500, and half way between them is $26,250, the median in this case. You can see that the median in this example gives a better view of the household incomes for the classroom than does the mean.

A third measure of central tendency is the *mode*. The **mode** is the value most frequently observed in a study. In the household income study, the mode is $20,000 since it is the value found most often in the study. Each measure of central tendency has certain advantages and disadvantages and is used, therefore, under certain special circumstances.

Measures of variability

Suppose that a teacher gave the same test to two different classes and obtained the following results: Class 1: 80%, 80%, 80%, 80%, 80% Class 2: 60%, 70%, 80%, 90%, 100% If you calculate the mean for both **sets** of scores, you get the same answer: 80%. But the collection of scores from which this mean was obtained was very different in the two cases. The way that statisticians have of distinguishing cases such as this is known as measuring the *variability* of the sample. As with measures of central tendency, there are a number of ways of measuring the variability of a sample.

Probably the simplest method is to find the range of the sample, that is, the difference between the largest and smallest observation. The range of measurements in Class 1 is 0, and the range in class 2 is 40%. Simply knowing that fact gives a much better understanding of the data obtained from the two classes. In class 1, the mean was 80%, and the range was 0, but in class 2, the mean was 80%, and the range was 40%.

Other measures of variability are based on the difference between any one measurement and the mean of the set of scores. This measure is known as the *deviation*. As you can imagine, the greater the difference among measurements, the greater the variability. In the case of Class 2 above, the deviation for the first measurement is 20% (80%-60%), and the deviation for the second measurement is 10% (80%-70%).

Probably the most common measures of variability used by statisticians are the variance and **standard deviation**. Variance is defined as the mean of the squared deviations of a set of measurements. Calculating the variance is a somewhat complicated task. One has to find each of the deviations in the set of measurements, **square** each one, add all the squares and divide by the number of measurements. In the example above, the variance would be equal to $[(20)^2 + (10)^2 + (0)^2 + (10)^2 + (20)^2]$ 4 5 = 200.

For a number of reasons, the variance is used less often in statistics than is the standard deviation. The standard deviation is the **square root** of the variance, in this case, $\sqrt{+200} = 14.1$. The standard deviation is useful because in any normal distribution, a large fraction of the measurements (about 68%) are located within one standard deviation of the mean. Another 27 percent (for a total of 95 percent of all measurements) lie within two standard deviations of the mean.

Inferential statistics

Expressing a collection of data in some useful form, as described above, is often only the first step in a statistician's work. The next step will be to decide what conclusions, predictions, and other statements, if any, can be made based on those data. A number of sophisticated mathematical techniques have now been developed to make these judgments.

An important fundamental concept used in inferential statistics is that of the null hypothesis. A null hypothesis is a statement made by a researcher at the beginning of an experiment that says, essentially, that nothing is happening in the experiment. That is, nothing other than natural events are going on during the experiment. At the conclusion of the experiment, the researcher submits his or her data to some kind of statistical analysis to see if the null hypothesis is true, that is, if nothing other than normal statistical variability has taken place in the experiment. If the null hypothesis is shown to be true, than the experiment truly did not have any effect on the subjects. If the null hypothesis is shown to be false, then the researcher is justified in putting forth some alternative hypothesis that will explain the effects that were observed. The role of statistics in this process is to provide mathematical tests to find out whether or not the null hypothesis is true or false.

A simple example of this process is deciding about the effectiveness of a new medication. In testing such medications, researchers usually select two groups, one the control group and one the experimental group. The control group does not receive the new medication; it receives a neutral substance instead. The experimental group receives the medication. The null hypothesis in an experiment of this kind is that the medication will have no effect and that both groups will respond in exactly the same way, whether they have been given the medication or not.

Suppose that the results of one experiment of this kind was as follows, with the numbers shown being the number of individuals who improved or did not improve after taking part in the experiment.

	Improved	Not Improved	Total
Experimental Group	62	38	100
Control Group	45	55	100
Total	107	93	200

At first glance, it would appear that the new medication was at least partially successful since the number of those who took it and improved (62) was greater than the number who took it and did not improve (38). But a statistical test is available that will give a more precise answer, one that will express the probability (90%, 75%, 50%, etc.) that the null hypothesis is true. This test, called the *chi square test*, involves comparing the observed frequencies in the table above with a set of expected frequencies that can be calculated from the number of individuals taking the tests. The value of chi square calculated can then be compared to values in a table to see how likely the results were due to chance and how likely to some real affect of the medication.

Another example of a statistical test is called the *Pearson correlation coefficient*. The Pearson **correlation** coefficient is a way of determining the extent to which two variables are somehow associated, or correlated, with each other. For example, many medical studies have attempted to determine the connection between smoking and lung cancer. One way to do such studies is to measure the amount of smoking a person has done in her or his lifetime and compare the rate of lung cancer among those individuals. A mathematical formula allows the researcher to calculate the Pearson correlation coefficient between these two sets of data-rate of smoking and risk for lung cancer. That coefficient can range between 1.0, meaning the two are perfectly correlated, and -1.0, meaning the two have an inverse relationship (when one is high, the other is low).

The correlation test is a good example of the limitations of statistical analysis. Suppose that the Pearson correlation coefficient in the example above turned out to be 1.0. That number would mean that people who smoke the most are always the most likely to develop lung cancer. But what the correlation coefficient does not say is what the cause and effect relationship, if any, might be. It does not say that smoking causes cancer.

Chi square and correlation coefficient are only two of dozens of statistical tests now available for use by researchers. The specific kinds of data collected and the kinds of information a researcher wants to obtain from these data determine the specific test to be used.

STEINER, JAKOB (1796-1863)
Swiss geometer

Jakob Steiner, unschooled until age 18, was one of the founders of and greatest contributors to the field of projective or modern **geometry**. Building on the work of such greats as **Gerard Desargues** and **Gaspard Monge**, the Swiss geometer discovered both the Steiner surface and the Steiner **theorem**, and extrapolated the work of the French geometer, **Jean Poncelet**, to develop the Poncelet–Steiner theorem, all of which helped to build the developing field of synthetic or projective geometry. A firm believer in the power of geometry, he pronounced that the calculations involved in **algebra** and **analysis** simply replaced real thinking, while geometry stimulated it. Made wealthy by his lectures on geometry, at his death Steiner bequeathed a third of his fortune to the Berlin Academy to establish the Steiner Prize.

Steiner was born on March 18, 1796, in the village of Utzensdorf, near Bern, Switzerland. The last of eight children of Niklaus Steiner and Anna Barbara Weber, Steiner grew up without formal schooling, working instead on the family farm and business where his natural skill with numbers was greatly needed. He did not learn to write until age 14, and against his parents' will, he finally left home four years later to attend a school in Yverdon run by Johann Heinrich Pestalozzi, the Swiss educational reformer whose theories laid the foundation for modern elementary education. The Pestalozzi method reacts to the highly individual needs of each learner and scorns rote memorization in favor of concrete experience and critical thinking. It was a method Steiner took to eagerly, so eagerly that within a year–and–a–half of studying with Pestalozzi, Steiner himself was teaching mathematics. His studies and teaching experience at Yverdon left a lasting legacy with Steiner: the desire, as a researcher, to discover a scientific unity in his chosen field of mathematics, and the ability, as a teacher, to encourage independent thinking on the part of his students.

Steiner left Yverdon, Switzerland, in 1818 for Heidelberg, Germany, where he studied at the university. To pay his way as a student, Steiner gave private lessons in mathematics. At the university, he attended lectures by Ferdinand Schweins in combinatorial analysis, and also studied algebra as well as differential and integral **calculus**. Early papers from 1821, 1824, and 1825 demonstrate the influence of his studies in Heidelberg. At the suggestion of a friend, however, Steiner decided to leave Heidelberg for the Prussian capital of Berlin in 1821.

His lack of education and stubborn will played against Steiner initially in Berlin. To receive a teaching license, he was forced to take examinations in various subjects. Steiner was not totally successful, receiving a partial license in mathematics and employment for a short time at a gymnasium, or high school. He gave private instruction to make a full living, and attended the University of Berlin from 1822 to 1824. Thereafter, he taught at a technical school in Berlin until 1834, when he was appointed a professor at the University of Berlin. He held this post until his death 29 years later.

A blunt, somewhat corrosive personality, Steiner was known for both the startling nature of his lectures and the originality of his research. The important phase of the latter began with an 1826 publication of "Einige geometrische Betrachtungen" in the new *Journal fur die reine and angewandte Mathematik,* founded by his friend, August Crelle. In total, Steiner contributed 62 articles to that journal. Those articles, in addition to his longer works, *Systematische Entwicklung der abhangigkeit geometrischer Gestalten voneinander* of 1832, and the posthumously published *Vorlesungen uber synthetische Geometrie* and *Allgemeine Theorie uber das Beruhren und Schneiden der Kreise und der Kugeln,* together dealt with Steiner's passion for reforming geometry and also made public the basics of what has come to be called projective geometry. Steiner's articles and books have been gathered in his *Gesammelte Werke.*

It was Steiner's great dream to discover the commonality between seemingly unrelated theorems, providing a simple and logical way of deducing such theorems. As he noted in his *Systematische Entwicklung,* "Here the main thing is neither the synthetic nor the analytic method, but the discovery of the mutual dependence of the figures and of the way in which their properties are carried over from the simplest to the more complex ones." All of Steiner's work draws from one basic principle, or so contended F. Gonseth in the foreword to Steiner's posthumously published *Allgemeine Theorie.* That principle is the stereographic **projection** of the plane onto the **sphere**.

Steiner noted in *Systematische Entwicklung* the principle of duality in projective geometry is a principle normally considered a property of algebra. But Steiner showed that in projective geometry that if two operations are interchangeable or dual, then whatever results are true for one are also true for the other. He contributed a plethora of original concepts that created the scaffolding of projective geometry. Among others, he discovered the Steiner surface, containing an **infinity** of **conic sections**; the Steiner theorem, proving that the intersection points of corresponding lines of two projective pencils (**sets** of geometric objects) of lines form a conic section. He also built on the work of Poncelet, in his Poncelet–Steiner theorem which shows that just one given **circle** with its center and straight edge are needed for any Euclidian construction. His investigation of conic sections and surfaces is at the very heart of modern geometry.

Steiner never married. Over the course of a long career, many honors came his way. He received an honorary doctorate from the University of Königsberg in 1833 and was elected a member of the Prussian Academy of Science in 1834. He also became a corresponding member of the French Académie Royale des Sciences after a winter of lecturing in Paris in 1844 and 1845, and held membership in the Accademia dei Lincei as well. In the final decade of his life, a kidney ailment limited his lecturing to the winter months only, and he died in 1863, revered as both a mathematician and educator.

Over the years, Steiner was able to accumulate a relatively large estate. His lecturing proved popular and most of his savings came from this source. In addition to the bequest he left to the Berlin Academy for the establishment of a mathematics prize in his name, and to the large chunk of money left to his relatives, Steiner also left money to the school in his native village of Utzensdorf to establish prizes for students adept at mathematics. To his death, he never forgot the value of education, nor the bitter memory of the lack of it in his own youth.

STEVIN, SIMON (1548-1620)
Dutch applied mathematician, physicist, and engineer

An encyclopedic mind and a prolific writer on many topics, Stevin, known as "the Dutch **Archimedes**," contributed to many areas of knowledge, particularly **arithmetic**, **algebra**, **geometry**, physics, optics, **mechanics**, astronomy, hydrostatics, musical theory, and military engineering. A dedicated empiricist, he experimentally refuted, years before **Galileo** (who received credit), the Aristotelian doctrine that heavy objects fall faster than light–weight objects. He also improved mathematical notation and championed the decimal system, suggesting that decimal mensuration could be used in all spheres of life. In addition, Stevin was an exceptional prose stylist. He pioneered the use of his vernacular, Dutch, at a time when Latin was dominant as the language of science. Stevin significantly enriched the Dutch scientific vocabulary, coining words and phrases which have been become part of the Dutch lexicon.

Born in Bruges (now Belgium), in the southern Netherlands, Stevin was employed by the city administration as a bookkeeper. He made several trips abroad in the 1570s, settling in Leiden, in the north, in 1581. Already in his thirties, he proceeded to obtain a formal education at the University of Leiden, adding to independently acquired knowledge of science and engineering. Having left the Spanish–occupied southern provinces, Stevin prospered in the northern Netherlands, participating in the national struggle for liberation from Spanish rule. His engineering skills were recognized by the leaders of the Dutch liberation war and he became quartermaster–general of the army in the early 1600s. He was also engaged by Prince Maurice of Nassau, the stadholder (head of state), as his personal science and mathematics teacher. Always willing to put his scientific expertise in the service of the war effort, Stevin was very highly regarded by his compatriots. He married Catharine Cray in 1610; the couple had four children, one of whom, Hendrick, also became a scientist.

Although **decimal fractions** had been known centuries before Stevin, he is credited with explaining the concept of decimal **fractions**, which, although known to proficient mathematicians, seemed unfamiliar to most mathematicians. He did not view decimal fractions as fractions, and proceeded to write their numerators without the denominators. For example, Stevin represented the value of π (in modern notation: 3.1416...) as 3 0 1 1 4 2 1 3 6 4. The second digit—around which he drew a circle—of each pair was the power of ten of the implied divisor. Thus, 1 1 is 1/10; 4 2 is 4/100, etc. Stevin's notation may have been cumbersome, but his idea made sense, for, to cite Carl Boyer's example, it seems more natural to imagine 3 seconds as an integer (3) than as 3/60 of one hour.

Unlike his predecessors, who used words to represent operations, Stevin insisted whenever this was possible, on notational symbols. For example, he used circled numbers to represent **exponents**, circled 2 replacing Q (for **square**), and circled 3 instead of C (for cube). While still imperfect, Stevin's notation, as scholars suggest, points to the modern algebraic notation which was firmly established not long after his death.

In 1586, Stevin described his experiment with two leaden spheres, one ten times heavier than the other. Having dropped the spheres simultaneously from a height of 30 ft (914 cm) onto a wooden surface, Stevin observed that they both hit the surface at the same time. Historians, who agree that Stevin and not Galileo, should be credited for this revolutionary experiment, nevertheless accept Galileo as the founder of modern mechanics, because it was the Italian scientist who formulated the final theoretical consequences of Stevin's discovery. However, according to Stephen F. Mason, Stevin, who described his falling spheres, as the "experiment against Aristotle," effectively demolished Aristotelian mechanics, thereby creating the foundations of modern mechanics.

Stevin's book, *De Beginselen der Weegconst* (*Principles of the Art of Weighing*), continues the work begun by Archimedes (for example, refining the Greek scientist's explanation of buoyancy) and tackles such Archimedean subjects as levers and inclined planes, including his famous wreath of spheres. Stevin placed a wreath of spheres on two inclined planes AB and BC, AB being twice the length of BC. If we imagine ABC as a **triangle**, we will notice that, the spheres being spaced evenly and weighing the same, and if there are two spheres on BC, there will naturally be four on AB. Despite the different numbers of spheres on each side, the wreath does not move, because there is an inverse proportion between the effect of gravity and the length of an inclined plane. A different triangle, with AB three times the length of BC, would, according to Stevin's principle, would balance two spheres on one side BC, and six on the other. "Stevin," Mason has written, "... obtained an intuitive understanding of the parallelogram of forces, a method of finding the resultant action of a combination of two forces that are not in the same straight line nor parallel. The method... consists in the representation of the two forces, in magnitude and direction, by two straight lines originating from a common point: the resultant is then given by the diagonal of the parallelogram formed by drawing two other lines parallel to the first two." Once again, Stevin made a great discovery, leaving the precise theoretical formulation to later scientists. The parallelogram of forces was clearly explained by **Isaac Newton** and by the French mathematician **Pierre Varignon** in 1687.

In Stevin's time, the musical scale was not evenly tempered, which posed a serious problem for players of stringed instruments who had to contend with the paradox exemplified by the fact that a sequence of four pure fifths and a sequence of two octaves should, but do not result in, a perfect third: the **interval** created when the final tones of the two sequences are

played together is not a perfect minor third. The interval of a pure octave is represented by the ratio 2 : 1, which means that if a frequency f is doubled, the new frequency is an octave higher than f. The pure fifth corresponds to the ratio 3 : 2, and the pure third to the **ratios** 5 : 4 (major) and 6 : 5 (minor). To resolve this paradoxical situation, which obliged musicians to constantly change tuning strategies, mathematicians and musicians proposed a variety of solutions, including the **division** of the octave into 12 mathematically, but perhaps not acoustically, equal semitones. Scholars believe that the modern, equally–tempered, scale originated with Stevin's idea that all semitones should be absolutely equal—their frequencies determined by the mathematical formula $2^{n/12}$, even if the consequences included less–than–perfect intervals.

Stevin may have been familiar with *Dialogo della musica antica e della moderna* (*Dialogue Concerning Ancient and Modern Music* [1581]) by Vincenzo Galilei, father of Galileo, which, among many other subjects, discussed the equally–tempered scale. However, scholars have suggested that the scale proposed by Stevin is the scale that was gradually accepted in the 17th and 18th centuries, particularly by players of keyboard instruments, and by composers such as Johann Sebastian Bach. His *Well–Tempered Clavier*, which contains preludes and fugues in all the 24 keys, constitutes the foundation of the classical, mathematical, conception of tonality, the embodiment of which is the modern piano. In science, Stevin's idea was further refined by the 19th–century English mathematician and phonetician A.J. Ellis, who divided the octave (2 : 1) into 1,200 cents, 100 cents representing an equally–tempered semitone.

STIRLING, JAMES (1692-1770)
Scottish number theorist

James Stirling was a mathematician best known for the formula on series expansion which bears his name, Stirling's formula. Stirling's work contributed greatly to the development of the theory of **infinite series** and to the branch of **calculus** known as **infinitesimal** calculus. Although his formal education was interrupted over political quarrels, Stirling worked both in Venice and London, doing early work on the extension of **Isaac Newton**'s theory of plane curves. His most influential work, also inspired by theories of Newton, *Methodus differentialis*, was published in 1730. In 1735 Stirling returned to his native Scotland, where he took a position at the Leadhills mining company, married, and concerned himself almost totally with mining affairs for the rest of his life.

Stirling was born in Garden, Stirlingshire, Scotland, in 1692. The third son of Archibald Stirling and Anna Hamilton, Stirling came from a staunchly Jacobite family, supporters of the restoration of the Stuarts after their overthrow in 1688. His father had once been imprisoned on charges of treason for his beliefs, but was later released. Such a background did not serve the young Stirling well when he left college in Glasgow to be admitted to Balliol College at Oxford University in 1711, the recipient of a scholarship donated by John Snell of

Ayrshire. With the onset of the First Jacobite Rebellion of 1715, and Stirling's refusal to take an oath of allegiance to the king, he was deprived of his scholarship, and left Oxford, without graduating, in 1716.

Stirling's time at Oxford had not been wasted, however. Connections with such notables as the mathematician and surgeon, John Arbuthnot, helped to get his first book published in 1717, with Newton himself among the names on the list of subscribers. *Lineae tertii ordinis Neutonianae*, to give the book its short title, not only recounted but also extended Newton's theory of plane curves of the third order. In this work, Stirling proved the 72 types of cubic curves that Newton had enumerated, and added four more of his own. He also solved a problem of **Gottfried Wilhelm von Leibniz** on curves. This book was dedicated to Nicholas Tron, another connection made during Stirling's Oxford years. Tron, the Venetian ambassador to London and a member of the Royal Society of London, secured a teaching position for Stirling in the Venetian Republic. Stirling left for Venice in 1718, but the position fell through. It is unclear just how long Stirling stayed in Venice or what his activities were while there. He did submit to the Royal Society an important early paper, "Methodus differentialis Newtoniana illustrata," from Venice in 1718, a paper which helped inspire his later, major work. There is also evidence that he attended the University of Padua in 1721. Additionally, it is believed that Stirling's departure from Venice may have been caused by him learning some of the trade secrets of the famous glass blowers. Such secrets were as closely guarded as computer innovations today, and Stirling's life may well have been in danger if he had access to such knowledge.

Whatever the reason, it seems Stirling was back in Glasgow by 1722, but there is a gap in the biographical data until he reappears in London in 1724 or 1725. Part of the difficulties of accurately tracing the events of Stirling's life is that the name James Stirling is a rather common one in Scotland, and at least two other James Stirlings have been confused with the mathematician in some historical records. It is clear that Stirling remained in London for ten years, first teaching at and later becoming a partner in Watt's Academy in Covent Garden, also known as the Little Tower Street Academy. The academy was a respected school and affiliation with it helped to buoy Stirling's usual poor financial situation. Stirling also won, with Arbuthnot's support, a fellowship in the Royal Society in 1726.

While in London, Stirling continued his mathematical researches, moving from the study of cubic curves to the field of differences, also inspired by earlier work by Newton, particularly his *Analysis* of 1711. This work saw earlier publication in Stirling's 1718 paper, but by 1730 he had a full book on the subject, *Methodus differentialis: sive tractatus de summatione et interpolatione serierum infinitarum*. Published in Latin initially, the book was translated into English as well.

Stirling's *Methodus differentialis* is divided in two parts, which follow a short introductory section. One part deals with summation, or the method of associating a sum with a divergent series, and the other with **interpolation**, or the estimation of an intermediate term in a sequence. The book also deals with infinite series and quadrature, the process of finding a **square** equal in **area** to any given surface, especially that bounded by a curve. It was in Proposition 28, Example 2, of *Methodus differentialis* that Stirling proposed an asymptotic, or increasingly exact formula, for $n!$ which bears his name: $n! = n^n e^{-n} \sqrt{(2\pi n)}$. This series expansion formula was also derived by **Abraham de Moivre**, although credit usually goes to Stirling who also supplied tables of coefficients for recurring series, creating what are known as Stirling Numbers of the first or second kind.

Stirling found a new interest in the early 1730s with the study of the Earth and its gravitational forces, culminating with the 1735 publication of *On the Figure of the Earth and On the Variation of the Force of Gravity on Its Surface*. He also spent the summers of 1734–1736 in Scotland, in the employ of the Scottish Mining Company in their lead mines at Leadhills, Lanarkshire. In 1737, Stirling accepted a permanent appointment with the mines, at a salary of 210 pounds per year, a position he held for the rest of his life. Though involved primarily in administration, at which he was by all accounts notably successful, Stirling also developed a system of shaft ventilation and of air blasts for smelters, which he published in 1745 in a paper entitled "Description of a Machine to Blow Fire by the Fall of Water." This paper may have been influenced by his earlier work with the Venetian glass blowers.

Stirling married Barbara Watson of Thirty Acres, a village near Stirlingshire. The exact date of the marriage is not known, though it is assumed it could not have taken place before 1745. The couple had one child, a daughter, who later married her cousin, Archibald Stirling, the succeeding manager at Leadhills. Stirling continued to communicate with other mathematicians, though his Jacobite principles prevented him from returning to the world of professional mathematics when a chair opened in Edinburgh in 1746 upon the death of his friend **Colin Maclaurin**. That same year, Stirling was honored by election to the Berlin Academy of Sciences. His last work of a scientific nature took place in 1752 when he conducted the first survey of the River Clyde for the Corporation of Glasgow, one of a series of surveys which were intended to improve the navigability of the river. Stirling was presented with a silver teapot by the Corporation in recognition of his services. The following year, Stirling's wife died, and in 1754 he resigned from the Royal Society, an event which appears to have been prompted by financial concerns. Stirling was heavily in arrears to the Society for his annual subscriptions.

Later in life, Stirling fell into ill health. Visiting Edinburgh for medical treatment in 1770, he died and was buried at Greyfriars Churchyard. He is remembered by a small plaque in the cemetery wall, and by his work on infinite series theory. By the time of Stirling's death, his most significant work was far behind him, relegated to the 1720s and 1730s, before taking his position at the lead mines.

STOCHASTIC

Stochastic means involving chance or probability. A stochastic process is one that involves a random **variable** and depends on probability. It is a process that involves random behavior or is subject to probabilistic behavior. A dynamical system is deterministic if the future of that system is completely predictable from knowledge of its present state. If there is some intrinsic randomness in the system it makes the perfect prediction of the future of that system impossible and so it is considered a stochastic system. There may exist strong trends or correlations in such systems but there is always some element of uncertainty.

The word stochastic is derived from the Greek *stokhastikos* which is derived from *stokhos* meaning aim or goal. This word was originally derived from an Indo-European word, *stegh*, that is also the ancestor of the English word.

Stochastic **calculus** involves **functions** that contain a random variable or variables. Ordinary **differential equations** are normally interpreted as determining dynamical systems since they often describe evolution in time. These functions are deterministic, completely predictable, but there are circumstances where there is some intrinsic randomness and understanding these systems would involve stochastic calculus. Some stochastic processes can be called Markov processes if certain constraints are upheld.

SUBTRACTION

As **addition** is based upon the idea of combining groups of things to yield a larger group, subtraction, the opposite of addition, is based on the idea of removing objects from a group, thereby reducing its size. These intuitive ideas of combining and reducing were undoubtedly engrained in daily life for even the earliest humans. When numbers are used to represent the quantity of objects composing groups, subtraction becomes an operation upon those numbers. Indeed, subtraction is one of four fundamental **arithmetic** operations (the others being addition, **multiplication**, and **division**). The symbol that denotes subtraction is "-," read "minus." For instance, "A minus B" means that quantity B is to be subtracted from quantity A, where "A" is the minuend and "B" is the subtrahend. The word minus comes from the Latin word "minus", which means "less".

By calling subtraction an operation, it is meant that any number subtracted from another number yields a third unique number. For instance, "7 - 2" represents "5," unique among the **whole numbers** $\{1, 2, 3, 4, 5,...\}$. However, many of the arithmetic laws that govern the operation of addition do not hold for subtraction. For example, the commutative law of addition "a + b = b + a" is true for whole numbers, **rational numbers**, **real numbers**, etc., but does not hold for subtraction (e.g., "5 - 7 ≠ 7 - 5"). Within arithmetic the only fundamental addition laws that are valid for subtraction are the distributive and monotonic laws: the distributive law for subtraction of products states that for numbers a, b, and c, "a · (b - c) = (a · b) - (a · c)," while the monotonic law for subtraction states that "if

a < b then (a - c) < (b - c)." So as an arithmetic operation, subtraction does not generally possess the **symmetry** that addition possesses. However, as shall be shown, subtraction can be modified to conform to the addition laws.

In the historical development of number systems subtraction played an important part. For people using number systems that lacked terms for **zero** and/or **negative numbers**, subtraction leads to confounding mathematical situations. For example, subtraction of whole numbers leads to expressions like "5 - 5" and "7 - 9." Without zero or negative numbers, expressions such as these are impossible. For some ancient peoples the answer was to expand their number system to include zero and negative numbers. Thus the set of **integers** $\{..., -3, -2, -1, 0, 1, 2, 3,...\}$ became an extension of the whole numbers. By including negatives and zero in the number system, the operation of subtraction can be recast as the addition of positive and negative numbers. As was pointed out earlier, "5 - 7" and "7 - 5" represent two distinct numbers (i.e., subtraction is not commutative). However, if one views subtraction as the addition of positives and negatives, then the subtraction of "7 from 5" becomes equivalent to the addition of "5 and (-7)." As a result, the commutative law holds (i.e., 5 + (-7) = (-7) + 5). The pertinent algebraic law is the law of additive inverse, which states that for any number "a" there exists another unique number "-a" such that: "a + (-a) = 0." This law holds for the set of integers, rationals, reals, etc. It does not hold for the whole or natural numbers, since neither system contains negatives.

By treating subtraction of one number from another (a - b) as simply the addition of the number "a" with the additive inverse of "b" (that is, a + (-b)), the additive laws of arithmetic may be applied. These laws are valid on the set of integers, rationals, reals, etc.; and subtraction of quantities beyond the real numbers is now considered.

Subtraction on vectors is performed by defining the negative of a vector as being equal to a vector of equal magnitude but opposite direction. The corresponding components of the two vectors are subtracted to form the resultant vector. Specifically, subtracting the vector "$b = b_x i + b_y j$" from the vector "$a = a_x i + a_y j$" yields a vector "r," such that "$r = a - b = a + (-b) = (a_x - b_x) i + (a_y - b_y) j$," where i and j are unit vectors in the x-direction and y-direction (respectively) of the x-y plane, a_x and a_y are scalar components of vector a, and b_x and b_y are scalar components of vector b.

The subtraction of **functions** is accomplished by subtracting common terms. For example, the function $g(x) = x_3 - 2x_2 + x + 3$ can be subtracted from the function $f(x) = x_2 + 3x - 4$ to result in $f(x) - g(x) = (x_2 + 3x - 4) - (x_3 - 2x_2 + x + 3) = (-x_3 + 3x_2 + 2x - 7)$.

Subtraction on matrices is valid only if each of the matrices to be subtracted is of the same order (contain the same number of rows and columns). Subtraction of matrices is performed by subtracting the elements of a particular **matrix** from the corresponding elements of other matrices. Therefore, the resulting matrix is of the same order as the subtracted matrices.

Subtraction can be applied to **complex numbers**, which are composed of a real part followed by an imaginary part. It is defined as the subtracting of complex quantities in which the individual real and imaginary parts are separately subtracted. Specifically, subtracting "y = c + id" from "x = a + ib" results in "x - y = (a - c) + i(b - d)". For example, subtracting "y = 1 + i(6)" from "x = 2 + i(3)" yields: "x - y = (2 + i(3)) - (1 + i(6)) = (2 - 1) + i(3 - 6) = 1 + i(-3)."

As is the case with addition, subtraction is founded upon concepts found in the branch of mathematics known as **set theory**. Under set theory, the arithmetic operation of subtraction on the natural (or whole) numbers can be recast as the subtraction (or difference) of **sets**, denoted symbolically for sets "A" and "B" as "A \ B." The various laws for subtraction between whole numbers can then be derived.

See also Numbers and numerals; Set theory; Sums and differences; Vector analysis

SUMS AND DIFFERENCES

Sums and differences are the results obtained by applying the operations of **addition** and **subtraction**, respectively, to particular mathematical objects. For the sake of clarity, the discussion is divided into two parts, with sums examined first, followed by differences.

A sum is defined as the result obtained by the addition of two or more quantities or expressions. Students first encounter sums in **arithmetic**, and early on memorize the sums of various **whole numbers**. Beyond the whole numbers of elementary school arithmetic, one can consider the sums of **integers, rational numbers, real numbers** and **complex numbers**; and (following the applicable rules) the sums of vectors, matrices, tensors, and other higher mathematical objects.

The process of determining a sum is called summation (or addition). The word sum originally came from Latin for "summa" (highest) and later became the word "summe" in Middle English. An early use of the mathematical meaning of the word sum came in the form of "some" within Nicolas Chuquet's 1484 document *Triparty en la Science des Nombres*. A sum is represented in its simplest form by a "+" sign between two variables; that is, "a + b" where "+" is read plus.

Besides the "+" sign, the operation of summation (or addition) can also be represented by the Greek capital letter sigma (Σ), being especially useful when the addition of many terms is necessary. The sigma notation, which corresponds to the English letter S, was introduced to facilitate the writing of these sums. Swiss mathematician **Leonhard Euler** (1707-1783) first used the summation symbol (Σ) in 1755. A summation using sigma notation is of the form:

$$\text{``}\sum_{i=1}^{i=n} a_i = a_1 + a_2 + a_3 + \ldots + a_{n-1} + a_n\text{'',}$$

where "i" is called the index of summation and "n" is called the upper limit, where both "i" and "n" possess only integer values, and where $n \geq i$.

A particular type of sigma summation that is of great importance in mathematics is called a series. A series is a sequence of quantities, called terms, in which the relationship between consecutive terms is the same. There are different types of series. An **arithmetic series** is a series in which the difference between successive terms is a constant, commonly called the common difference. Another type of series is called a **geometric series** and is one for which each term equals the previous term multiplied by a constant, commonly called the common ratio. Series can be used to find the value of constants like **pi** (π) and Euler's number "*e*," and to construct tables of **logarithms** and trigonometric **functions**. When the upper bound of a summation is **infinity**, the result is called an **infinite series**. Important examples of infinite series are the binomial series, Taylor's series, and Maclaurin's series. Because an infinite series means that an infinite number of terms are to be summed together, in practice an approximation is made by truncating the number of terms to be summed. In other words, in the case of 'real-world' problems that are encountered in physics and engineering, the upper limit is changed from infinity to some integer value "n."

A last example of sums is defined by integral **calculus**. The **definite integral** can be thought of as the limiting value of a sum. In practice, the definite integral is used to evaluate the length of, and the **area** under, plane curves; the area of surfaces of revolution; the **volume** of solids of revolution; and is used to find solutions to many other problems in physics, chemistry, and engineering.

Difference is defined in arithmetic and mathematics as the result obtained by subtracting one quantity or expression from another. The process of determining a difference is subtraction. The difference is represented in its most general form by two letter symbols with a "-" sign in between, that is "a - b" where "-" is read "minus." The word difference originally came from the Latin word "differentia" (meaning difference or diversity).

The term difference is often used to mean sequences of differences. Perhaps this concept of sequences of differences is best developed through use of an example. Take, for instance, the set of positive integers {1, 2, 3, 4, 5, 6,...}. Squaring each term produces the sequence {1, 4, 9, 16, 25, 36,...}; called the "original sequence." Starting from the original sequence, the "sequence of first differences" is the sequence: {3, 5, 7, 9, 11,...}. In a sequence of first differences each term is the difference between two successive terms in the original sequence; in this case, "3 = 4 - 1", "5 = 9 - 4", "7 = 16 - 9", and so on. Continuing along this line of reasoning, one may consider differences between consecutive terms of the sequence {3, 5, 7, 9, 11,...} (remembering that this sequence is called a "first differences" sequence because it was derived from differences of the original sequence {1, 4, 9, 16, 25, 36,...}). Differences of the sequence {3, 5, 7, 9, 11,...} are called "second differences", and the resulting "second differences" sequence is {2, 2, 2, 2, 2, 2,...}. Sequences of third, fourth, or nth sequences are possible depending upon the original sequence. The study of these sequences of differences is called the calculus of finite differences. In a practical sense, certain calculating machines, particularly difference engines, are based on the principle of differences.

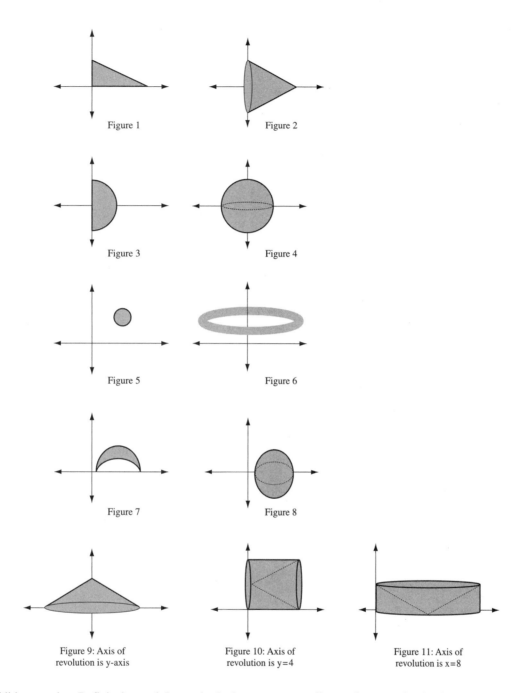

Figure 1

Figure 2

Figure 3

Figure 4

Figure 5

Figure 6

Figure 7

Figure 8

Figure 9: Axis of
revolution is y-axis

Figure 10: Axis of
revolution is y=4

Figure 11: Axis of
revolution is x=8

See also Additive notation; Definite integral; Integral calculus; Logarithms; Matrix; Trigonometric functions; Numbers and numerals; Sequences and series; Taylor's and Maclaurin's series; Vector analysis

SURFACES AND SOLIDS OF REVOLUTION

Surfaces of revolution are formed when a shape outlined in the x-y plane of the **Cartesian coordinate system** is revolved 360° around an axis of revolution to form a 3-dimensional figure. The original shape can be described mathematically in various ways. It can be defined by a single equation bordered by x and

y coordinates, for example, the shape shown in Figure 1 is the **area** bounded by a line and the x and y-axis. Revolving this 2-dimensional **triangle** around the x-axis provides the 3-dimensional cone shown in Figure 2.

A semi-circle, centered at the origin, and bounded by the y-axis is shown in Figure 3. Revolving this semi-circle around the y-axis provides the **sphere** shown in Figure 4. A **circle** centered at (x,y) coordinates of (5,5) is shown in Figure 5. Revolution of this circle around the y-axis provides the dough-nut-shaped **torus** shown in Figure 6.

The original shape can also be described by the intersection of two **equations**. Figure 7 shows the intersection between two curves, bounded by the x-axis. Revolving this shape around the x-axis provides the hollow solid shown in Figure 8. A hollow solid can also be obtained by revolving the **space** between shapes around an axis. For example, imagine that a smaller circle lies within the circle shown in Figure 5. Revolving only the space between the two circles around the y-axis would provide a hollow torus.

The axis of revolution does not have to be the x or y-axis, it can be another line in the [x,y] plane. Consider the triangle shown in Figure 1. It is bounded by the line defined by y=-0.5x+4, which intersects the x-axis at point (8,0) and intersects the y-axis at point (0,4). The triangle can be revolved around several different lines to provide different 3-dimensional solids, as shown in Figures 9-11.

Principles of **calculus** and **analytic geometry** are used to determine the **volume** of a solid of revolution. There are two general methods that are used: the disk method and the cylindrical **shell method**.

In the disk method, the original shape is divided into many small inscribed rectangles sliced perpendicular to the axis of revolution. This is similar to the method of inscribed rectangles used to find the area under a curve. Consider the triangle shown in Figure 1. If this triangle is sliced into n inscribed rectangles perpendicular to the x-axis, and each inscribed rectangle is revolved around the x-axis, a series of solid circular disks is obtained that together approximately form the cone shown in Figure 2. The volume of each disk can be obtained by multiplying its area (πr^2) by its thickness (Δx). Summing the volumes of the disks will provide an approximation of the cone volume. This approximation will improve as Δx approaches **zero** and n approaches **infinity**.

Therefore, the volume of a solid defined by a function $f(x)$ that is continuous over a closed **interval** [a,b], where $f(x) \geq 0$ for all values of x in the interval [a,b] is defined by a **definite integral** as follows: $V = \pi \int_a^b [f(x)^2]\,dx$. Because the cone in figure 2 is bounded by x=0 and x=8, its values for a and b are 0 and 8, respectively. If the original shape is a region bounded by two equations, $f(x)$ and $g(x)$ over the closed interval [a,b], and $f(x) \geq g(x) \geq 0$ for all of the x values in the interval [a,b], then the volume of the revolved solid is given by $V = \pi \int_a^b ([f(x)]^2 - [g(x)]^2)\,dx$. Similar equations are used when the axis of revolution is the y-axis or another line in the x-y plane.

In the second method, called the method of cylindrical shells, slices are considered parallel to the axis of revolution, but integration is performed along an axis perpendicular to the axis of revolution. In other words, each slice is parallel to the axis of revolution, but the slice is revolved around the axis of revolution. This method is more difficult to visualize. Consider the triangle shown in Figure 1 divided into inscribed rectangles sliced parallel to the y-axis. Mentally revolve one of these rectangles around the y-axis. The result will be a hollow **cylinder** (like a roll of toilet paper) centered around the y-axis. All of these cylindrical shells placed one inside another will approximately form the cone shown in Figure 9.

Once again, integration can be used to determine the volume of the resulting solid. The volume of each cylindrical shell can be calculated from the difference between the vol-

ume of the outer cylinder and the inner cylinder from which the shell is formed (i.e., the space between two concentric cylinders forms a cylindrical shell). Therefore, the volume of the resulting solid is given by: $V = 2\pi \int_a^b x f(x)\,dx$. The term $2\pi x$ can be thought of as a shell **circumference**, while dx represents shell thickness, and $f(x)$ represents shell length (or **altitude**). A similar equation is used when the original shape is revolved around a different axis of revolution.

SYLLOGISMS AND PROOFS

The syllogism is a form of logical argument invented by the Greek philosopher **Aristotle** (384-322 B.C.). Aristotle was the first philosopher to put forward a formal system of logic (see **Aristotelian logic**), a system which would stand virtually unchanged and unchallenged for more than 2,000 years after his death. For Aristotle, logic was the primary tool for discourse about any of the sciences. At the heart of Aristotle's logic was the syllogism. To illustrate the form that a syllogism takes, here is one that Aristotle used: "All humans are mortal. Socrates is a human. Therefore, Socrates is mortal." This has the more general form "All B is A. Some C is B. Therefore, some C is A." The first sentence in the syllogism is called the major premise. The second sentence is called the minor premise. The third sentence is called the conclusion. All syllogisms in Aristotelian logic have this "major premise/minor premise/conclusion" structure, but within this structure there are many different forms. Aristotle identified 21, but we shall not list them all here. Rather, we shall look at the four syllogistic forms from which the rest may be derived as theorems. They are: (1) All B is A. All C is B. Therefore, all C is A. (2) No B is A. All C is B. Therefore, no C is A. (3) All B is A. Some C is B. Therefore, some C is A. (4) No B is A. Some C is B. Therefore, some C is not A. Aristotle called these the syllogisms of the "first figure." Notice that if you read each of these four syllogisms from left to right, the order of the letters, or terms, is always BA CB CA. This is how Aristotle classified his syllogisms into "figures." There were four figures altogether. The order of terms in the second figure is AB CB CA; in the third figure it is BA BC CA; and in the fourth figure it is AB BC CA. The four figures were just Aristotle's way of "cataloging" his syllogisms. One does not need to know this cataloging scheme to follow Aristotle's logic.

Aristotle painstakingly developed his syllogistic logic to guide mathematicians, philosophers, scientists, and others in their quest for knowledge, **truth**, and certainty. His desire was to create a system of rules for discourse that one could follow in a mechanical way to construct an argument, or **proof**, for any given proposition. If a mathematician or philosopher followed the rules of this system, he would always construct valid proofs and be able to proclaim his result with certainty. The notion of producing valid proofs for statements, or theorems, is one of the hallmarks of mathematics. However, one cannot start from nothing. There have to be certain rules stated without proof in order to get the process started. These rules are called axioms or postulates. Since these axioms are to be taken as true without proof, they need to be statements about which reasonable people will agree; they need to be "obvious" or "self-evident." Also, the logician wants to assume as little as possible and prove as much as possible to give strong credibility to her system. Thus she desires no more axioms than are absolutely necessary to begin the process of proving theorems. In the preceding paragraph it was stated that Aristotle was able to start with the four syllogisms of the first figure and prove those of the second, third, and fourth figures by assuming just those four. In truth, he did better than that. He ultimately showed that syllogisms (1) and (2) implied syllogisms (3) and (4), thereby eliminating the need to assume (3) and (4). Thus Aristotle was able to trim the number of axioms to just two: (1) All B is A. All C is B. Therefore, all C is A. (2) No B is A. All C is B. Therefore, no C is A. As an example, let us see the "proof" that syllogism (1) implies syllogism (3). Note that "All B is A" clearly implies that "All B is A." Similarly, "All C is B implies that some C is B"; and "All C is A" implies that "some C is A". Then we just substitute for each statement in syllogism (1) its **implication** and we have syllogism (3) "All B is A. Some C is B. Therefore, some C is A." Now, that we have established syllogism (3), we may use it to prove other syllogisms; and, similarly, each new syllogism proved may be used in the proofs of others. This is the essential nature of a logical or mathematical proof. Assume some axioms, as few as necessary; use those axioms to prove theorems; and use the newly proven theorems to prove others. Although modern mathematicians and logicians do not rely solely, or even primarily, on the syllogism, they do continue to use this "axiom/theorem/proof" method to reach the "truths" of their discipline.

Aristotle's syllogistic method dominated the study of logic until the latter part of the nineteenth century. It is worth noting that the syllogisms of Aristotle may be translated into the "if-then" form with which modern readers are more familiar. For example, "All B is A" may be rendered as "If every x has property B, then every x has property A." "Some C is B" becomes "If some x has property C, then some x has property B." Thus, "All equilateral triangles are equiangular" becomes "If **triangle** x is equilateral, then triangle x is equiangular." So even though it may appear as though modern mathematics has abandoned the syllogism, it has not. The influence of Aristotle's syllogistic logic is to be seen each time a twenty-first century mathematician proves a **theorem**.

SYLVESTER, JAMES JOSEPH (1814-1897)

English algebraist, combinatoricist

James Joseph Sylvester was one of the most colorful mathematicians of the Victorian age. His role in giving an impetus to the American mathematical community in the late 19th century provided a sense of national independence from European scholarship.

Sylvester was born on September 3, 1814 in London, the son of Abraham Joseph. He was the youngest son out of a total of nine children and received his early education at Jewish schools in Highgate and Islington. Sylvester entered St. John's College at Cambridge in 1831. His undergraduate career was interrupted by illness that kept him at home for several years.

Finally, in 1837 he took the Tripos (the final mathematical examinations) and placed second. Ordinarily, a student in that situation would have proceeded to take his degree and to compete for the Smith's Prize. Since, however, Sylvester was Jewish, he could not subscribe to the ThirtyNine Articles of the Church of England, a requirement at that time for receiving a degree from Oxford or Cambridge. Sylvester later earned both bachelor and master's degrees from Trinity College in Dublin, which did not have the same requirements. Some 35 years after passing the Tripos, Sylvester received his B.A. from Cambridge when the doctrinal requirements were dropped.

Sylvester began teaching at University College, London, as the colleague of **Augustus De Morgan**, but did not find the chair of natural philosophy a congenial one. By 1839, he was a Fellow of the Royal Society and two years later made his first venture at transatlantic teaching, accepting a position at the University of Virginia. This proved to be a complete fiasco, as Sylvester encountered intense antiSemitism. In the absence of offers from other American universities, Sylvester returned to England and took a position as actuary and secretary to the Equity and Law Life Assurance Company ("assurance" is the English term for "insurance"). He took private pupils in mathematics as well, the most notable of them being Florence Nightingale. In 1846, three years after returning from the United States, Sylvester entered the Inner Temple, one of the Inns of Court designed to prepare students for the bar. He was called to the bar in 1850, but never made a success at it as did his colleague **Arthur Cayley**.

In 1855, Sylvester finally managed to secure a position teaching mathematics in England. He was appointed to the chair of mathematics at the Royal Military Academy at Woolwich. That same year he became the editor of the Quarterly Journal of Pure and Applied Mathematics, which he transformed into one of the leading periodicals about mathematical research in England. He remained its editor until 1877, during which period he also served as president of the London Mathematical Society from 1866 to 1868.

As a mathematician Sylvester occasionally indulged in hasty generalizations, but his imagination at its best could create whole new fields of study. He did not take the trouble to study the contributions of contemporary mathematicians outside his circle, with the result that some of his work amounts to an independent rediscovery. He did a good deal of work in the area of **invariant** theory, the study of algebraic expressions that remain fixed under various classes of transformations. He certainly introduced an enormous amount of terminologyincluding "discriminant"an activity of which he was especially proud.

In addition to his work on invariant theory, Sylvester deserves much credit for the early study of the subject of partitions. A partition is a way of breaking up a positive whole number as a sum of other positive **whole numbers**, so that 3 + 3, 2 + 2 + 2, and 5 + 1 are all partitions of the number 6. Counting the number of partitions of various kinds required some inventiveness and Sylvester helped to get the subject welllaunched. The subject of partitions turned out to have a good deal of importance in **analysis**, although Sylvester did not push this aspect of his studies very far.

Sylvester had resigned from Woolwich in 1870 and was busily engaged in his mathematical work, but he could not resist an invitation from the eminent American physicist Joseph Henry to come to the newlyfounded Johns Hopkins University in Baltimore. Henry assured Sylvester that he would be involved with serious students and have ample opportunities for research. He arrived in Baltimore in 1876 to find his expectations fulfilled. In addition to his work with some of the best students in the country, Sylvester also founded the American Journal of Mathematics. At this point in history there was no serious American publication devoted to mathematics and Sylvester's journal filled the void. In particular, to make certain that it would be taken seriously, he contributed 30 of his own papers to the journal. Despite his successes at Johns Hopkins, Sylvester could not pass up the opportunity to succeed H.J.S. Smith as Savilian Professor of **Geometry** at Oxford University.

Among the many honors Sylvester had received was the Royal Medal in 1861 and the Copley Medal in 1880, both from the Royal Society, and the De Morgan Medal from the London Mathematical Society in 1887. He was a corresponding member of numerous foreign societies and received four honorary degrees.

Sylvester died on March 15, 1897 and was buried at the Jewish Cemetery at Ball's Pond, London. In 1997, the London Mathematical Society held a memorial service on the centenary of his death. There was also a day's conference devoted during that centenary year to his life and the areas of mathematics to which he contributed. Without question, the mathematical communities on both sides of the Atlantic are beneficiaries of Sylvester's ideas and energy.

SYMBOLIC LOGIC

Logic is the study of the rules which underlie plausible reasoning in mathematics, science, law, and other disciplines.

Symbolic logic is a system for expressing logical rules in an abstract, easily manipulated form.

In **algebra**, a letter such as x represents a number. Although the symbol gives no clue as to the value of the number, it can be used nevertheless in the formation of sums, products, etc. Similarly P, in **geometry**, stands for a point and can be used in describing segments, intersections, and the like.

In symbolic logic, a letter such as p stands for an entire statement. It may, for example, represent the statement, "A **triangle** has three sides." In algebra, the plus sign joins two numbers to form a third number. In symbolic logic, a sign such as V connects two statements to form a third statement. For example, V replaces the word "or" and $\wedge$ replaces the word "and." The following is a list of the symbols commonly encountered:

p, q, r,...	statements
$\vee$	"or"
$\wedge$	"and"
~	"it is not the case that"

$\Rightarrow$	"implies" or "If..., then...."
$\leftrightarrow$	"implies and is implied by" or "... if and only if...."

Logic deals with statements, and statements vary extensively in the precision with which they may be made. If someone says, "That is a good book," that is a statement. It is far less precise, however, than a statement such as "Albany is the capital of New York." A good book could be good because it is well printed and bound. It could be good because it is written in good style. It could tell a good story. It could be good in the opinion of one person but mediocre in the opinion of another.

The statements which logic handles with the greatest certainty are those which obey the law of the excluded middle, i.e., which are unambiguously true or false, not somewhere in between. It doesn't offer much help in areas such as literary criticism or history where statements simple enough to be unequivocally true or false tend also to be of little significance. As an antidote to illogical thinking, however, logic can be of value in any discipline.

By a "statement" in logic one means an assertion which is true or false. One may not know whether the statement is true or false, but it must be one or the other. For example, the Goldbach conjecture, "Every even number greater than two is the sum of two primes," is either true or false, but no one knows which. It is a suitable statement for logical **analysis**.

Other words that are synonyms for "statement" are "sentence," "premise," and "proposition."

If p stands for the statement, "All right angles are equal," and q the statement, "Parallel lines never meet," one can make a single statement by joining them with "and': "All right angles are equal and parallel lines never meet." This can be symbolized p $\wedge$ q, using the inverted V-shaped symbol to stand for the **conjunction** "and." Both the combined statement and the word "and" itself are called "conjunctions." In ordinary English, there are several words in addition to "and" which can used for joining two statements conjunctively, for example, "but." "But" is the preferred conjunction when one wants to alert the reader to a relationship which otherwise might seem contradictory. For example, "He is 6 ft (1.8 m) tall, but he weighs 120 lb (54 kg)." In logic the only conjunctive term is "and."

Negation is another logical "operation." Unlike conjunction and disjunction, however, it is applied to a single statement. If one were to say, "She is friendly," the negation of that statement would be, "She is not friendly." The symbol for negation is "~." It is placed in front of the statement to be negated, as in ~(p$\wedge$q) or ~ p. If p were the statement, "She is friendly," ~ p means "She is not friendly," or more formally, "It is not the case that she is friendly." Prefacing the statement with, "It is not the case that...," avoids embedding the negation in the middle of the statement to be negated. The symbol lips is read "not p."

The statement ~ p is true when p is false, and false when p is true. For example, if p is the statement "x $<$ 4," ~ p is the statement "x $\geq$ 4." Replacing x with S makes p false but ~ p true. If a boy, snubbed by the girl in "She is friendly," were to hear the statement, he would say that it was false. He would say, "She is not friendly," and mean it.

The fact that someone says something doesn't make it true. Statements can be false as well as true. In logic, they must be one or the other, but not both and not neither. They must have a "truth value," true or false, abbreviated T or F.

p	q	~p	p∧q	p∨q
T	T	F	T	T
T	F	F	F	T
F	T	T	F	T
F	F	T	F	F

Whether a conjunction is true depends on the statements which make it up. If both of them are true, then the conjunction is true. If either one or both of them are false, the conjunction is false. For example, the familiar expression 3 < x < 7, which means "x > 3 and x < 7" is true only when both conditions are satisfied simultaneously, that is for numbers between 3 and 7.

Another word used in both ordinary English and in logic is "or." Someone who says, "Either he didn't hear me, or he is being rude," is saying that at least one of those two possibilities is true. By connecting the two possibilities about which he or she is unsure, the speaker can make a statement of which he or she is sure.

In logic, "or" means "and/or." If p and q are statements, p V q is the statement, called a "disjunction," formed by connecting p and q with "or," symbolized by "V."

For example if p is the statement, "Mary Doe may draw money from this account," and q is the statement, "John Doe may draw money from this account," then p V q is the statement, "Mary Doe may draw money from this account, or John Doe may draw money from this account."

The disjunction p V q is true when p, q, or both are true. In the example above, for instance, an account set up in the name of Mary or John Doe may be drawn on by both while they are alive and by the survivor if one of them should die. Had their account been set up in the name Mary and John Doe, both of them would have to sign the withdrawal slip, and the death of either one would freeze the account. Bankers, who tend to be careful about money, use "and" and "or" as one does in logic.

One use of **truth** tables is to test the **equivalence** of two symbolic expressions. Two expressions such as p and q are equivalent if whenever one is true the other is true, and whenever one is false the other is false. One can test the equivalence of ~(p V q) and ~p ∧ A ~ q (as with the minus sign in algebra, "~" applies only to the statement which immediately follows it. If it is to apply to more than a single statement, parentheses must be used to indicate it):

p	q	~p	~q	p∨q	~(p∨q)	~p∧~q
T	T	F	F	T	F	F
T	F	F	T	T	F	F
F	T	T	F	T	F	F
F	F	T	T	F	T	T

The expressions have the same truth values for all the possible values of p and q, and are therefore equivalent.

For instance, if p is the statement "x > 2" and q the statement "x < 2," p V q is true when x is any number except 2. Then (p V q) is true only when x = 2. The negations p and q are "x 2" and "x 2" respectively. The only number for which ~ p ∧ ~ q is true is also 2.

Equivalent **propositions** or statements can be symbolized with the two-headed arrow "↔." In the preceding section we showed the first of De Morgan's rules:

1. ~(p V q) ↔ ~p ∧ ~q
2. ~(p ∧ q) ↔ ~p V ~q

Rules such as these are useful for simplifying and clarifying complicated expressions. Other useful rules are

3. p ↔ ~(~p)
4. p ∧(q V r) ↔ (p ∧ q) V (p ∧ r) (a distributive law for "and" applied to a disjunction)
5. p ∧ q ↔ q ∧ p p V q ↔ q V p
6. (p ∧ q) V r ↔ p V (q V r) (p V q) V r ↔ p V (q V r)

Each of these rules can be verified by writing out its **truth table**.

A truth table traces each of the various possibilities. To check rule 4 with its three different statements, p, q, and r, would require a truth table with eight lines. On occasion one may want to know the truth value of an expression such as ((T V F) L A (F V T)) V ~ F where the truth values of particular statements have been entered in place of p1 q, etc. The steps in evaluating such an expression are as follows:

((T ∨ F) ∧ (F ∨ T)) ∨ ~F Given

(T A T) ∨ T Truth tables for ∨,~

T ∨ T Truth table for ∧

T Truth table for ∨

Such a compound expression might come from the run-on sentence, "Roses are red or daisies are blue, and February has 30 days or March has 31 days; or it is not the case that May is in the fall." Admittedly, one is not likely to encounter such a sentence in ordinary conversation, but it illustrates how the rules of symbolic logic can be used to determine the ultimate truth of a complex statement. It also illustrates the process of replacing statements with known truth values instead of filling out a truth table for all the possible truth values. Since this example incorporates five different statements, a truth table of 32 lines would have been needed to run down every possibility.

In any discipline one seeks to establish facts and to draw conclusions based on observations and theories. One can do so deductively or inductively. In inductive reasoning, one starts with many observations and formulates an explanation that seems to fit. In deductive reasoning, one starts with premises and, using the rules of logical **inference**, draws conclusions from them. In disciplines such as mathematics, deductive reasoning is the predominant means of drawing conclusions. In fields such as psychology, inductive reasoning predominates, but once a theory has been formulated, it is both tested and applied through the processes of deductive thinking. It is in this that logic plays a role.

Basic to deductive thinking is the word "implies," symbolized by "⇒." A statement p⇒ q means that whenever p is true, q is true also. For example, if p is the statement, "x is in Illinois," and q is the statement "x is in the United States," then p⇒ q is the statement, "If x is in Illinois, then x is in the United States."

In logic as well as in ordinary English, there are many ways of translating p⇒ q into words: "If p is true, then q is true"; "q is implied by p"; "p is true only if q is true"; "q is a **necessary condition** for p"; "p is a sufficient condition for q."

The **implication** p ⇒ q has a truth table some find a little perplexing:

p	q	p⇒q
T	T	T
T	F	F
F	T	T
F	F	T

The perplexing part occurs in the next to last line where a false value of p seems to imply a true value of q. The fact that p is false doesn't imply anything at all. The implication says only that q is true whenever p is. It doesn't say what happens when p is false. In the example given earlier, replacing x with Montreal makes both p and q false, but the implication itself is still true.

Implication has two properties which resemble the reflexive and transitive properties of equality. One, p⇒ p, is called a "tautology." **Tautologies**, although widely used, don't add much to understanding. "Why is the water salty?" asks the little boy.

"Because ocean water is salty," says his father.

The other property, "If p⇒ q and q⇒ r, then p⇒r," is also widely used. In connecting two implications to form a third, it characterizes a lot of reasoning, formal and informal. "If we take our vacation in January, there will be snow. If there is snow, we can go skiing. Let's take it in January." This property is called a "syllogism."

A third property of equality1 "If a = b, then b = a," called **symmetry** may or may not be shared by the implication p⇒q. When it is, it is symbolized by the two-headed arrow used earlier, "p ↔ q." p ↔ q means (p⇒ q) ∧ (q⇒ p). It can be read "p and q are equivalent'; "p is true if and only if q is true"; "p implies and is implied by q"; "p is a necessary and sufficient condition for q"; and "p implies q, and conversely."

In p⇒ q, p is called the "antecedent" and q the "consequent." If the antecedent and consequent are interchanged, the resulting implication, q⇒ p, is called the "converse." If one is talking about triangles, for example, there is a **theorem**, "If two sides are equal, then the angles opposite the sides are equal." The converse is, "If two angles are equal, then the sides opposite the angles are equal."

If an implication is true, it is never safe to assume that the converse is true as well. For example, "If x lives in Illinois, then x lives in the United States," is a true implication, but its converse is obviously false. In fact, assuming that the converse of an implication is true is a significant source of fallacious reasoning. If the battery is shot, then the car won't start."

True enough, but it is a good idea to check the battery itself instead of assuming the converse and buying a new one.

Implications are involved in three powerful lines of reasoning. One, known as the Rule of Detachment or by the Latin *modus ponendo ponens,* states simply "If p⇒ q and p are both true, then q is true." This rule shows up in all sorts of areas. "If x dies, then y is to receive $100,000. "When x dies and **proof** is submitted to the insurance company, y gets a check for the money. The statements p⇒ q and p are called the "premises" and q the "conclusion."

A second rule, known as *modus tollendo tollens,* says if p⇒q is true and q is false, then p is false. "If x ate the cake, then x was home." If x was not at home, then someone else ate the cake.

A third rule, *modus tollerdo ponens,* says that if p V q and not p are true, then q is true. Mary or Ann broke the pitcher.

Ann did not; so Mary did. Of course, the validity of the argument depends upon establishing that both premises are true.

It may have been the cat.

Another type of argument is known as *reductio ad absurdum*, again from the Latin. Here, if one can show that ~p⇒ (q ∧ ~q), then p must be true. That is, if assuming the negation of p leads to the absurdity of a statement which is both true and false at the same time, then p itself must be true.

SYMMETRY

Symmetry is an intrinsic characteristic of an object that means the object remains unchanged after a symmetry operation is performed on it. **Group theory** is the area of mathematics that is concerned with the systematic study and formalization of symmetry. A symmetry operation is a mathematical **transformation** that, when performed on a symmetric object, produces an object that is identical to the original object. Symmetry operations are defined relative to a given point, center of symmetry, line, axis of symmetry, or plane, plane of symmetry, and when performed on symmetric objects preserve distances, angles, sizes, and shapes. In mathematics there are several kinds of symmetries but plane symmetry, those whose operations take place in a plane on two-dimensional figures, and spatial symmetry, those whose operations take place in three-dimensional **space** on solid shapes, are the most common.

Although symmetry is an important conceptual tool of modern science the history of this concept can be dated back to the times of early astronomy when scientists believed in the Ptolemaic system, the theory that the earth was the center of the universe. Early astronomers believed that because the **circle** is the most perfect of geometrical shapes due to its high symmetry, that astronomical objects must move in circles. They found that in this **model** the motions of the planets were much to complex to be considered simple circles. In 1513 **Nicolaus Copernicus** wrote the beginnings of the Copernican theory, a theory that put the Sun, not the Earth, at rest in the center of the universe. Using this theory it was possible to construct a model of planetary motions in terms of simple circles around the Sun. This was the first application of symmetry to a scientific problem. Group theory, a method employed in the **analysis** of abstract symmetrical physical systems, was formally developed by **Evariste Galois** near the end of his life in 1832. Although he is generally considered to be the father of group theory Carl Friedrich Gauss developed the theory about 1800 but never published it. In the early 1900s German mathematician **Herman Weyl** formulated the most commonly used definition of symmetry: "An object is said to be symmetrical if one can subject it to a certain operation and it appears exactly the same after the operation as before. Any such operation is called a symmetry of the object." Later, scientists went on to formulate the symmetry of the laws of nature. The first symmetry is known as time **translation** symmetry that assumes that physical laws do not change with time. The second is that the laws of physics must be the same everywhere, on this side of the solar system as well as the other, and this is called translational symmetry. Lastly, the laws of physics to not change regardless of the direction that one faces. This is called rotational symmetry. Although these three symmetries have been accepted as valid symmetries of nature, **Albert Einstein** employed a symmetry that said the laws of physics should be the same for all observers regardless of their state of **motion**. He used this symmetry as the basis for developing his theory of relativity. This is the ultimate example of a symmetry principle leading to a radical revision of fundamental concepts.

There are four basic symmetry operations: **rotation**, inversion, reflection, and translation. Rotation means to turn an object by an **angle** around a defined point or line for a solid. Rotation about an n-fold symmetry axis is usually indicated as C_n. An inversion is a symmetry operation that creates a new set of inverted points P' such that $OP \cdot OP' = OQ^2 = k^2$ with respect to the inversion circle:

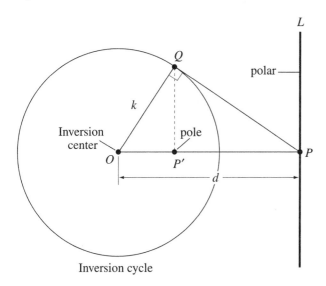

Inversion cycle

In three-dimensional space an inversion is carried out with respect to an inversion **sphere**. The center of symmetry under inversion is represented by i. Reflection is the symmetry operation by which an object is reflected in a mirror so that the signs of its coordinates are reversed to form an image.

Reflection can be horizontal or vertical and is usually represented by $n\sigma_h$ or $n\sigma_v$ depending upon whether the reflection is horizontal or vertical. Translation is the symmetry operation by which the points of an object undergo a constant offset without rotation or distortion. Translation is signified by E. An improper rotation is a combination of a rotation followed by an inversion and is sometimes called a rotoinversion. An improper rotation about an n-fold symmetry axis is signified as S_n. If one object remains symmetric after more symmetry operations than another object then we say it is more symmetrical. For example, a circle would be more symmetric than a **square**.

These symmetry operations together produce 32 crystal classes corresponding to 32 point groups in group theory. Not only can the study of symmetry be applied to arrangements of objects but also to coefficients of **equations**. **Quintic equations** can be proved to be unsolvable by applying group theory to polynomial equations. There are two main symmetry principles that are of the utmost importance in **mathematics and physics**, the Noether's symmetry **theorem** and the symmetry principle. Noether's symmetry theorem states that each symmetry of a system leads to a physically conserved quantity. This important theorem in physics shows that symmetry under translation corresponds to conservation of momentum, symmetry under rotation leads to conservation of angular momentum, and that symmetry in time corresponds to conservation of energy. The other important principle, the symmetry principle, sometimes called the Schwarz reflection principle, indicates that symmetric points are preserved under a Möbius transformation.

SYSTEMS OF EQUATIONS

Systems of **equations** are a group of relationships between various unknown variables which can be expressed in terms of algebraic expressions. The solutions for a simple system of equation can be obtained by graphing, substitution, and elimination **addition**. These methods became too cumbersome to be used for more complex systems however, and a method involving matrices is used to find solutions. Systems of equations have played an important part in the development of business and quicker methods for solutions continue to be explored.

Unknowns and linear equations

Many times, mathematical problems involve relationships between two variables. For example, the **distance** that a car moving 55 mph travels in a unit of time can be described by the equation $y = 55x$. In this case, y is the distance traveled, x is the time and the equation is known as a linear equation in two variables. Note that for every value of x, there is a value of y which makes the equation true. For instance, when x is 1, y is 55. Similarly, when x is 4, y is 220. Any pair of values, or ordered pair, which make the equation true are known as the solution of the equation. The set of all ordered pairs which make the equation true are called the solution set. **Linear equations** are more generally written as ax + by = c where a, b and c represent constants and x and y represent unknowns.

Often, two unknowns can be related to each other by more than one equation. A system of equations includes all of the linear equations which relate the unknowns. An example of a system of equations can be described by the following problem involving the ages of two people. Suppose Lynn is twice as old as Ruthie, but two years ago, Lynn was three times as old as Ruthie. Two equations can be written for this problem. If we let x = Lynn's age and y = Ruthie's age, then the two equations relating the unknown ages would be x = 2y and x - 2 = 3(y - 2). The relationships can be rewritten in the general format for linear equations to obtain,

$$\text{(Eq. 1)} \quad x - 2y = 0$$

$$\text{(Eq. 2)} \quad 2x - 3y = -4$$

The solution of this system of equations will be any ordered pair which makes both equations true. This system has only solution, the ordered pair of x = 8 and y = 4, and is thus called consistent.

Solutions of linear equations

Since the previous age problem represents a system with two equations and two unknowns, it is called a system in two variables. Typically, three methods are used for determining the solutions for a system in two variables, including graphical, substitution and elimination.

By graphing the lines formed by each of the linear equations in the system, the solution to the age problem could have been obtained. The coordinates of any point in which the graphs intersect, or meet, represent a solution to the system because they must satisfy both equations. From a graph of these equations, it is obvious that there is only one solution to the system. In general, straight lines on a coordinate system are related in only three ways. First, they can be parallel lines which never cross and thus represent an inconsistent system without a solution. Second, they can intersect at one point, as in the previous example, representing a consistent system with one solution. And third, they can coincide, or intersect at on all points indicating a dependent system which has an infinite number of solutions. Although it can provide some useful information, the graphical method is often difficult to use because it usually provides us with only approximate values for the solution of a system of equations.

The methods of substitution and elimination by addition give results with a good degree of accuracy. The substitution method involves using one of the equations in the system to solve for one **variable** in terms of the other. This value is then substituted into the first equation and a solution is obtained. Applying this method to the system of equations in the age problem, we would first rearrange the equation 1 in terms of x so it would become x = 2y. This value for x could then be substituted into equation 2 which would become 2y - 3y = -4, or simply y = 4. The value for x is then obtained by substituting y = 4 into either equation.

Probably the most important method of solution of a system of equations is the elimination method because it can be used for higher order systems. The method of elimination by addition involves replacing systems of equations with simpler equations, called equivalent systems. Consider the system with the following equations equation 1 x - y = 1 equation 2 x + y = 5 By the method of elimination, one of the variable is eliminated by adding together both equations and obtaining a simpler form. Thus equation 1 + equation 2 results in the simpler equation 2x = 6 or x = 3. This value is then put back into the first equation to get y = 2.

Often, it is necessary to multiply equations by other variables or numbers to use the method of elimination. This can be illustrated by the system represented by the following equations:

$$\text{equation 1} \quad 2x - y = 2$$

$$\text{equation 2} \quad x + 2y = 10$$

In this case, addition of the equations will not result in a single equation with a single variable. However, by multiplying both sides of equation 2 by -2, it is transformed into -2x - 4y = -20. Now, this equivalent equation can be added to the first equation to obtain the simple equation, -3y = - 18 or y = 6.

Systems in three or more variables

Systems of equations with more than two variables are possible. A linear equation in three variables could be represented by the equation ax + by + cz = k, where a, b, c, and k are constants and x, y, and z are variables. For these systems, the solution set would contain all the number triplets which make the equation true. To obtain the solution to any system of equations, the number of unknowns must be equal to the number of equations available. Thus, to solve a system in three variables, there must exist three different equations which relate the unknowns.

The methods for solving a system of equations in three variables is analogous to the methods used to solve a two variable system and include graphical, substitution, and elimination. It should be noted that the graphs of these systems are represented by geometric planes instead of lines. The solutions by substitution and elimination, though more complex, are similar to the two variable system counterparts.

For systems of equations with more than three equations and three unknowns, the methods of graphing and substitution are not practical for determining a solution. Solutions for these types of systems are determined by using a mathematical invention known as a **matrix**. A matrix is represented by a rectangle array of numbers written in brackets. Each number in a matrix is known as an element. Matrices are categorized by their number of rows and columns.

By letting the elements in a matrix represent the constants in a system of equation, values for the variables which solve the equations can be obtained.

Systems of equations have played an important part in the development of business, industry and the military since the time of World War II. In these fields, solutions for systems of equations are obtained using computers and a method of maximizing parameters of the system called **linear programming**.

SYSTEMS OF LINEAR EQUATIONS

A linear equation has the form $a_1y_1 + a_2y_2 +... + a_ny_n = b$. The numbers a_i are called the coefficients of the equation and the y_i are the variables. A solution to the equation is an ordered n-tuple of numbers $(s_1,...,s_n)$ such that if the s_i are substituted for the y_i then the equation is true. For example, $3y_1 + 5y_2 = 0$ is a linear equation. The coefficients are 3 and 5. The variables are y_1 and y_2. One solution to this equation is (5, -3) because "3 x 5 + 5 x (-3) = 0" is true. The set of all solutions to this equation is the set of all ordered pairs of the form (5a - 3b, -3a + 5b) where a and b are any numbers.

A system of linear **equations** is set of **linear equations** like so:

$$a_{11} \times y_1 + a_{12} \times y_2 +... + a_{1n} \times y_n = b_1$$
$$a_{21} \times y_1 + a_{22} \times y_2 +... + a_{2n} \times y_n = b_2$$
$$...$$
$$a_{m1} \times y_1 + a_{m2} \times y_2 +... + a_{mn} \times y_n = b_m.$$

A solution to this system of linear equations is an n-ordered n-tuple of numbers $(s_1,...,s_n)$ such that if the s_i are substituted for the y_i, then all of the equations are true. Systems of linear equations occur in economics, geology, data transmission, electrical networks, mathematical **modeling**, hydraulic systems and more.

Once researchers have a system of linear equations, they usually want to know if any solutions exist and if so, how many. Often they want to find an explicit example. For some applications, the system under analysis may have hundreds or thousands of variables in as many equations. To solve systems this large, it is necessary to use computers. Most solution finding algorithms today are based on Gaussian elimination. First, the system is represented by an augmented **matrix** like so:

$$|a_{11} a_{12}... a_{1n} b_1| |a_{21} a_{22}... a_{2n} b_2|... |a_{m1} a_{m2}... a_{mn} b_m|$$

In this matrix, the (i,j)-entry is equal to a_{ij} if $j < n+1$ and the (i,n+1)-entry is equal to bi. This matrix is said to be upper triangular if all (i,j)-entries in which $i > j$ are equal to **zero**. If this is the case, then the solutions can easily be determined. First eliminate all row in which all entries are zero. If all the bottom row entries except b_m are zero then there is no solution. Otherwise, the set of solutions is a vector **space** whose **dimension** is equal to the number of columns minus the number of rows minus one. If this dimension is zero, then there is only one solution. If this is the case, since the only nonzero entries

on the bottom row are a_{mn} and b_m, any solution $(s_1,...,s_n)$ must have $s_n = b_m/a_{mn}$. Now substitute b_m/a_{mn} for each y_n is the system and subtract b_m from both sides of each equation. The result is a system of equations in fewer variables, that is also upper triangular. If this substitution process is repeated n-1 more times, the solution will be known. Gaussian elimination works by changing one system of linear equations into another such that the new one is upper triangular and has the same set of solutions as the original. There are three elementary operations on a system that transforms it into a new system with the same solution set. These are:

- exchange two rows of the augmented matrix
- multiply a row by a nonzero number
- add a row to another row. For example, adding row 1 to row 2 changes every (2,i)-entry to the sum of itself with the (1,i)-entry. All other entries stay the same.

If a_{m1} is nonzero and all other (i,1)-entries are zero then exchange the mth row with the first row. If, on the other hand, a_{m1} is nonzero and there is a_{ni} not equal to m with ai1 nonzero the multiply the ith row by $-(a_{m1}/ai1)$ and add it the mth row. Now, the (m,1)-entry is zero. This process can be continued by moving up the rows until all the entries in the first column are zero except possibly the (1,1)-entry. Then the process can be started again on the second column until all the entries of the second column are zero except possibly the top two. The process can be continued on each successive column until the matrix is upper triangular. This is Gaussian elimination.

For an example, consider the matrix

$$\begin{vmatrix} 4 & 4 & 8 & 12 \\ 2 & 1 & 2 & 4 \\ 1 & 1 & 0 & 1 \end{vmatrix}$$

For example, multiply the top row by -1/4 and add it to the bottom row to get

$$\begin{vmatrix} -1 & -1 & -2 & -3 \\ 2 & 1 & 2 & 4 \\ 0 & 0 & -2 & -2 \end{vmatrix}$$

Now, multiply the top row by 2 and add it to the second row to get

$$\begin{vmatrix} -2 & -2 & -4 & -6 \\ 0 & -1 & -2 & -2 \\ 0 & 0 & -2 & -2 \end{vmatrix}$$

Now the second row and the third row are the same so the bottom row can be eliminated. The resulting matrix is upper triangular.

See also Linear algebra

T

TARSKI, ALFRED (1901-1983)

Polish American logician and algebraist

Alfred Tarski made considerable contributions to several areas of mathematics, including **set theory** and **algebra**, and his work as a logician led to important breakthroughs in semantics—the study of symbols and meaning in written and verbal communication. Tarski's research in this area yielded a mathematical **definition** of **truth** in language, and also made him a pioneer in studying models of linguistic communication, a subject that became known as **model** theory. Tarski's research also proved useful in the development of computer science, and he became an influential mentor to later mathematicians as a professor at the University of California at Berkeley.

Born Alfred Tajtelbaum in Warsaw, Poland (then part of Russian Poland), on January 14, 1901, Tarski was the elder of two sons born to Ignacy Tajtelbaum, a shopkeeper of modest means, and Rose (Iuussak) Tajtelbaum, who was known to have an exceptional memory. During his teens Tarski helped supplement the family income by tutoring. He attended an excellent secondary school, and although he was an outstanding student, he, surprisingly, did not get his best marks in logic. Biology was his favorite subject in high school, and he intended to major in this discipline when he first attended the University of Warsaw. However, as Steven R. Givant pointed out in *Mathematical Intelligence,* "what derailed him was success." In an early mathematics course at the university, Tarski was able to solve a challenging problem in set theory posed by the professor. The solution led to his first published paper, and Tarski, at the professor's urging, decided to switch his emphasis to mathematics.

Tarski received a Ph.D. from the University of Warsaw in 1924, the same year he met his future wife, Maria Witkowski. They got married on June 23, 1929, and later had two children, Jan and Ina. It is believed that the young mathematician was in his early-twenties when he changed his name from Tajtelbaum to Tarski. His son, Jan, told interviewer Jeanne Spriter James that this step was taken because Tarski believed that his new Polish-sounding name would be held in higher regard at the university than his original Jewish moniker. When Tarski was married, he was baptized a Catholic, his wife's religion.

Tarski served in the Polish army for short periods of time in 1918 and 1920. While working on his Ph.D. he was employed as an instructor in logic at the Polish Pedagogical Institute in Warsaw beginning in 1922. After graduating he became a docent and then adjunct professor of mathematics and logic at the University of Warsaw beginning in 1925. That same year he also took a full-time teaching position at Zeromski's Lycee, a high school in Warsaw, since his income from the university was inadequate to support his family. Tarski remained at both jobs until 1939, despite repeated attempts to secure a permanent university professorship. Some have attributed Tarski's employment difficulties to anti-semitism, but whatever the reason, his lack of academic prominence created problems for the young mathematician. Burdened by his teaching load at the high school and college, Tarski was unable to devote as much time to his research as he would have liked. He later said that his creative output was greatly reduced during these years because of his employment situation. The papers he did publish in this period, however, quickly marked Tarski as one of the premiere logicians of the century. His early work was often concentrated in the area of set theory. He also worked in conjunction with Polish mathematician **Stefan Banach** to produce the Banach-Tarski paradox, which illustrated the limitations of mathematical theories that break a **space** down into a number of pieces. Other research in the 1920s and 1930s addressed the **axiom** of choice, large cardinal numbers, the decidability of Euclidean **geometry**, and **Boolean algebra**.

Tarski's initial research on semantics took place in the early–1930s. He was concerned here with problems of language and meaning, and his work resulted in a mathematical definition of truth as it is expressed in symbolic languages. He also provided a **proof** that demonstrated that any such definition of truth in a language results in contradictions. A London *Times* obituary on Tarski noted the groundbreaking nature of his work in this area, proclaiming that the mathematician's findings "set the direction for all modern philosophical discussions of truth." Tarski expanded this early work in semantics over the ensuing years, eventually developing a new field of study—model theory—which would become a major research subject for logicians. This area of study examines the mathematic properties of grammatical sentences and compares them with various models of linguistic communication.

Additionally, Tarski pursued research in many other areas of math and logic during his career, including closure algebras, binary relations and the algebra of relations, cylindrical algebra, and undecidable theories. He also made a lasting contribution to the field of computer science. As early as 1930 he produced an **algorithm** that was capable of deciding whether any sentence in basic Euclidian geometry is either true or false. This pointed the way toward later machine calculations, and has also had relevance in determining more recent computer applications.

In 1939 Tarski left Poland for a conference and speaking tour in the United States, intending to be gone for only a short time. Shortly after his departure, however, the German Army invaded and conquered Poland, beginning World War II. Unable to return to his homeland, Tarski found himself stranded in the United States without money, without a job, and without his wife and children who had remained in Warsaw. The family would not be reunited until after the war, and in the meantime, Tarski set about finding work in America. He first served as a research associate in mathematics at Harvard University from 1939 to 1941. In 1940 he also taught as a visiting professor at the City College of New York. He had a temporary position at the Institute for Advanced Study at Princeton beginning in 1941, and in 1942 he obtained his first permanent position in the United States when he was hired as a lecturer at the University of California at Berkeley. The university would remain his professional home for the rest of his career.

Tarski became an associate professor at the university in 1945, was appointed to the position of full professor the following year, and was named professor emeritus in 1968. Tarski's contributions to mathematics and science were enhanced by his role as an educator. He established the renown Group in Logic and the Methodology of Science at Berkeley, and over his long tenure he taught some of the most-influential mathematicians and logicians to emerge after World War II, including **Julia Bowman Robinson** and Robert Montague. His stature was further enhanced through his service as a visiting professor and lecturer at numerous U.S. and international universities. In 1973 Tarski ended his formal teaching duties at Berkeley, but he continued to supervise doctoral students and conduct research during the final decade of his life. He died in 1983 from a lung condition caused by smoking.

Tarski received many awards and honors throughout his career. He was elected to the National Academy of Sciences and the Royal Netherlands Academy of Sciences and Letters, and was also made a corresponding fellow in the British Academy. In 1966 he received the Alfred Jurzykowski Foundation Award, and in 1981 he was presented with the Berkeley Citation, the university's highest faculty honor. He also was awarded numerous fellowships and honorary

degrees, and was a member in many professional organizations, including the Polish Logic Society, the American Mathematical Society, and the International Union for the History and Philosophy of Science.

TARTAGLIA (NICOLÒ FONTANA) (1499-1557)
Italian algebraist

Tartaglia was a self–taught mathematician whose work in **algebra** led to the first generalized solution of **cubic equations**. He was neither prominent nor well connected and spent most of his life earning a meager salary as a school teacher. In addition to his work in algebra, Tartaglia also made contributions to the fields of fundamental **arithmetic**, ballistics, military engineering, cartography, and the development of certain instruments used in surveying. His Italian translations of some of the works of **Euclid** and **Archimedes** represented the first time this information had been available in a modern language.

Tartaglia's given name was Nicolò (or Niccolò) Fontana. He was born in Brescia, Italy, in 1499 or possibly 1500. His father, Michele, was a postal courier who worked for the government of Brescia. When his father died in about 1506, the family was left in poverty. In 1512, the French army attacked Brescia. During the attack a French soldier set upon young Nicolò with a saber and left him with five head wounds. Although Nicolò eventually recovered, his mouth was so disfigured that he could no longer speak clearly. As a result of this speech impediment, he was given the nickname Tartaglia, from the Italian word *tartagliare*, which means "to stammer." Rather than being offended, Nicolò kept the nickname as a surname and used it for the rest of his life.

When he was about 14 years old, Tartaglia went to a tutor to learn the alphabet. By the time he reached the letter "k," he was no longer able to pay for lessons and subsequently taught himself from books. Tartaglia learned quickly and acquired enough skill at mathematics that he was able to obtain a position as a teacher of practical mathematics in Verona sometime between 1516 and 1518. He stayed in Verona until 1534 and rose to the position of headmaster of a school there. Some evidence indicates that Tartaglia had a family in Verona, although his income was barely enough to support them. In 1534, he moved to Venice to become a professor of mathematics. Except for a period of 18 months in 1548–1549, Tartaglia remained in Venice until his death.

Tartaglia's most notable contribution to mathematics came shortly after he moved to Venice. As was the custom of the times, persons in academic positions were often subject to challenges of their knowledge and skills. In 1535, Tartaglia was challenged to a contest by Antonio Fior. Fior had been an aspiring but below average student of the mathematician **Scipione dal Ferro**, before dal Ferro's death in 1526. Just before his death, dal Ferro had given Fior a method for solving the so–called "depressed cubic" equation, in which there was no second power term. In turn, Fior had kept this method secret until he decided to use it to boost his floundering career by challenging Tartaglia. For his part, Tartaglia already knew how to solve cubic **equations** that lacked the first power term, and so he readily accepted Fior's challenge. Each person was to submit 30 problems for the other to solve. Rather than give Tartaglia problems covering a wide range of mathematical topics, Fior gambled and gave Tartaglia nothing but depressed cubics. After many days, with the deadline drawing near, Tartaglia discovered the solution and quickly solved all 30 problems. Fior, on the other hand, solved only a few of the problems posed by Tartaglia and suffered a humiliating defeat.

Tartaglia now knew the solutions to two types of cubic equations, and this knowledge put him in an enviable position among his fellow mathematicians. One in particular, **Girolamo Cardano**, repeatedly beseeched Tartaglia to share his secrets. Finally, in 1539, Tartaglia gave Cardano the solutions on the condition that Cardano not reveal them until Tartaglia had published them himself. Armed with this new knowledge, Cardano set about to discover the solution to a generalized cubic equation, in which all of the terms were present. Working with his new student, Ludovico Ferrari, Cardano discovered a method for reducing any generalized cubic equation to a depressed cubic. Then, using Tartaglia's methods, they could solve the equation. Unfortunately, Cardano was still bound by his promise not to reveal Tartaglia's methods. Then, in 1543, Cardano discovered that it was dal Ferro who had originally solved the depressed cubic equation, not Tartaglia. With this information Cardano went ahead and published his solution to a generalized cubic equation in 1545, giving full credit to dal Ferro and Tartaglia for their work. Tartaglia was outraged at this perceived breech of trust and wrote hotly and offensively about Cardano. Ferrari rose to Cardano's defense and began a long series of acrimonious correspondence with Tartaglia, ending in a public debate in Ferrari's home town of Milan in 1548. Backed by a sympathetic and unruly audience, Ferrari easily defeated Tartaglia, who was forced to withdraw in fear for his life.

In addition to his work in mathematics, Tartaglia's work drew interest in other areas. His work in ballistics included the determination that the maximum theoretical range of a projectile is obtained when the firing **angle** is 45 degrees. He also proposed the concept of firing tables in which range is correlated to firing angle, powder charge, and other factors. In surveying, he showed how to apply the compass and proposed the design of two instruments for determining the heights and distances to inaccessible points.

Tartaglia published his Italian translation of Euclid's *Elements* in 1543, and produced his translation of a Latin version of some of Archimedes's works the same year. He made additional translations of Archimedes' works in 1551, adding his own commentary. Tartaglia's translations contributed to the dissemination of knowledge by making these ancient works available in a modern language.

Tartaglia died impoverished in his home in Venice on December 13, 1557.

TAUTOLOGIES

A tautology is a logical proposition with a special property: it is true under all circumstances. This property can be defined more precisely. Logical **propositions** are made up of basic propositions (usually symbolized by lower-case letters, sometimes italicized; p, q, etc.) joined together by connectives. Typically these connectives include "not" (negation), "and" (**conjunction**) "or" (disjunction), and the "if-then" relation (**implication**). Each basic proposition can be assigned a truth-value—either true (T) or false (F). Then the **truth** of a compound sentence built out of basic propositions and connectives can be calculated with the help of truth **functions** or truth tables.

A logical proposition is a tautology if it is true regardless of the truth-values of its basic propositions. The final column in a tautology's **truth table** would consist entirely of T's.

Consider an example: "p or *not-p*" (e.g. "Elvis is alive or Elvis is not alive."). The only basic proposition here is p. The second part of the disjunction, *not-p*, is the negation of a basic proposition. Suppose p is true, then *not-p* is false. If just one part of a disjunction is true, then the whole disjunction is true, so the sentence "p or *not-p*" is true. Suppose on the other hand that p is false. Then *not-p* is true, and so part of the disjunction is true, making the whole disjunction true. No matter what truth-value p assumes, "p or *not-p*" is true. A more complicated tautology is the statement: "If p, then p or q."

Sentences that are not tautologies are either **contingencies**, which are true for some, but not all combinations of basic proposition truth-values, or **contradictions**, which are always false, no matter what the truth-values of the constituent basic propositions.

In a formal deductive argument, tautologies can be inserted at any point without special justification. This trick can be quite helpful. We use it implicitly when doing a **proof by contradiction**. In that **proof** strategy we assume the negation of the sentence p we want to prove, then show that the negation leads to a contradiction. Therefore the original sentence p must be true. Technically, what we are doing here is inserting the sentence "p or p" into our set of premises. By doing this we gain access to the crucial negation which may not appear in any form in our original set of premises. We can do this unabashedly, since as a tautology it is certainly true.

See also Proof by reductio ad absurdum; Truth function

TAYLOR, BROOK (1685-1731)

English mathematician

Brook Taylor was born in Edmonton, Middlesex, England, on August 18, 1865, the oldest child of a family considered part of the minor nobility. His mother, Olivia, was the daughter of Sir Nicholas Tempest. His father, John Taylor of Bifrons House, Kent, was a strict parent who alienated Taylor from the family when Taylor married a woman without his father's approval in 1721. The estrangement ended after two years when Taylor's wife died in childbirth and he returned to the family home. Taylor remarried in 1725, this time to a woman who enjoyed the family's approval. Unfortunately, Taylor's second wife also died in childbirth in 1730, leaving him to raise their daughter, Elizabeth, alone. Following his father's death in 1729, Taylor inherited the family estate in Kent.

After being tutored at home in mathematics and the classics, Taylor was admitted to St. John's College, Cambridge, in 1709, where he completed a Bachelor of Laws degree. He was elected to Great Britain's oldest scientific community, the Royal Society, in 1712 and earned a Doctorate of Laws in 1714. He served as secretary to the Royal Society for four years then resigned, citing poor health as the impetus for his decision. It was rumored that in addition to suffering some health problems, Taylor also found the duties as a chronicler too restrictive and less than satisfying for his active mind.

While serving as secretary to the Royal Society, Taylor wrote many articles on such diverse subject as magnetism, capillarity (the study of the surface tensions in liquids which come in contact with solids), the barometer (using logarithmic function to explain atmospheric pressure at different altitudes), and the thermometer. Thirteen articles in all appeared in *Philosophical Transactions of the Royal Society* between 1712 and 1724 addressing theories on dynamics, **hydrodynamics**, and heat, among others. His correspondence to fellow mathematicians during this time contained a solution to astronomer **Johannes Kepler**'s second law of planetary **motion**. Taylor's first significant paper also appeared in *Philosophical Transactions* in 1714 and produced a solution to the problem of the center of oscillation. Taylor's claim to priority of the solution was unjustly challenged by Swiss mathematician **Johann Bernoulli**. Although the dispute was later resolved, Taylor would again face accusations by Bernoulli concerning further mathematical work.

In 1715, Taylor published *Methodus Incrementorum Directa et Inversa* ("Direct and Indirect Methods of Incrementation"), which contained the first discourse dealing with the calculusof finite differences. The dissertation also contained the **theorem** for which Taylor is most famous—the "Taylor series." The Taylor serieswas the first general expression for the expansions of **functions** of a single **variable** in infinite series. Both **Isaac Newton** and Nicolaus Mercator had devoted time to developing the series, but their work had been confined to particular cases. Again, **Johann Bernoulli** stepped in to dispute the priority of Taylor's work, claiming that he had made the discovery. Bernoulli's claim was again proven to be unfounded since his work had never reached the fruition of Taylor's. The importance of the Taylor series remained unrecognized until 1772. It was then that **Joseph–Louis La Grange** realized its importance and declared it the basic principle of differential **calculus**. The series has also been called the Maclaurin series and this series holds validity when *0* is substituted for *a* in the Taylor theorem. Still, Taylor deserves the credit since his treatise preceded **Colin Maclaurin**'s by 27 years, and

Maclaurin's theorem was another case of addressing specifics instead of developing a basis just as Newton and others had done. Taylor did have one predecessor to his discovery. **James Gregory** had worked on the theorem some 40 years before Taylor, but Gregory's work was unknown to Taylor, hence eliminating any allegations of misappropriation on the part of Taylor.

In *Methodus*, Taylor also applied the calculus for the solution of several problems which had baffled previous investigators. He obtained a formula showing that the rapidity of **vibration** of a string varies directly as the weight stretching it and inversely concerning its own weight and length. Also, in *Methodus*, Taylor supplied a determination of the differential equation of the path of a ray of **light** passing through a heterogeneous medium. No further advance was made in the theory of refraction for more than 80 years.

Also published in 1715 was Taylor's *Linear Perspective*, later followed by an updated version (1719) entitled *New Linear Perspective*, both of which set forth the basic principles of perspective, and contained the first general treatment of the principle of vanishing points. As with *Methodus*, much of this work to went unnoticed in Taylor's lifetime. Taylor's father had a keen interest in art and music so Taylor was introduced to and instructed in these arts throughout his childhood. As it turned out, he was himself a gifted artist and musician, leaving small wonder that he would eventually turn his attention to the study of artistic theory. St. John's College, Cambridge, is said to house an unpublished manuscript entitled *On Musick* among its Taylor archives.

The fact that so much of Taylor's work was recognized following his death can be directly attributed, at least in part, to Taylor himself. The brevity and obscurity of his writing made it difficult to comprehend. Concerning this too, Taylor's tireless rival Johann Bernoulli had something to say: "[The writing is] abstruse to all and unintelligible to artists for whom it was more especially written." As offensive as Bernoulli was to Taylor throughout their respective careers, it is interesting to note that Bernoulli cited some of Taylor's work dealing with string vibration in letters to his son Daniel. Still, Taylor often did leave out steps in his theorems, believing that others would recognize the end result as easily as he had. Regrettably, this was not the case, and it took the close examination of his successors to fully illuminate his contributions to the field of mathematics as well as the other sciences.

Taylor spent the last part of his writing career producing religious and philosophical texts. Several unfinished treatises were found among his papers concerning various religions. Taylor wrote nothing during the last ten years of his life, but his third and final completed book, entitled *Comtemplatio Philosophica,*was published and circulated after his death in 1731. His daughter, Elizabeth, had married Sir William Young, and it was their son who made certain that his grandfather's manuscript was published. Young authored a biography of his grandfather's life and works and attached it to *Comtemplatio* upon publication.

TAYLOR, RICHARD LAWRENCE (1962-)

English mathematician, number theorist, and algebraist

Richard Taylor is a number theorist who is perhaps best known for his collaboration with Princeton mathematician **Andrew Wiles** in constructing a **proof** of **Fermat's last theorem.**

This famous **theorem**, written by lawyer and amateur mathematician **Pierre de Fermat** in the 17th century, states that for the equation $x^n + y^n = z^n$, if n is > 2, there are no non-zero integer solutions for x, y, and z. Although Fermat stated in his writings he had a proof for the theorem, none was ever discovered in Fermat's written works. Mathematicians had puzzled over a possible proof to the the theorem for centuries, and in that time, proofs of certain cases of the theorem had been presented, However, it was not until Andrew Wiles presented his proof of Fermat's last theorem in June of 1993 that the first feasible solution appeared. When Wiles discovered a flaw in his original work several months later, he called on his former student Richard Taylor to repair the gap. Together, Taylor and Wiles synthesized a new approach to the problem area of the proof and published the results in a short paper entitled "Ring-theoretic properties of certain Hecke algebras," published in the *Annals of Mathematics* in 1995. Since that time, the proof has been widely reviewed and accepted by the mathematical community.

Taylor earned his undergraduate degree at Cambridge University in Cambridge, England and his Ph.D. at Princeton. After a one-year post-doctoral fellowship in Paris, Taylor returned to Princeton for nine months to collaborate with Wiles on the proof for Fermat's last theorem. In 1995, Taylor was appointed to the Savilian Chair of Geometry at the University of Oxford, and in 1996 he moved to the U.S. to join the faculty of the department of mathematics at Harvard University. Recently, Taylor has continued his work with modular forms, collaborating with other mathematicians in 1999 to construct proofs for both the Taniyama-Shimura conjecture and the Local Langlands conjecture.

TAYLOR'S AND MACLAURIN'S SERIES

A Taylor's series is a series expansion that acts as a representation of a **function**. A series expansion is a representation of a function as a sum of **powers** in one of its **variables** or a sum of powers of another function. A more specific form of a Taylor's series is the Maclaurin's series. The Taylor's series is an expansion about an arbitrary point, $x = a$, whereas a Maclaurin's series is an expansion about zero in particular, $x = 0$. The main advantage of using a **power series** representation of a function is that the value of the function at any point is equal to a convergent series and so can be approximated by its partial sums. These power series contributed greatly to the growth of **calculus**. They allowed mathematicians to analyze properties of functions with a single theory and to approximate values of functions easily. Taylor's series can be used to estimate values for e, ln 2, and π.

The Taylor's series of a function f is written as $\sum_{n=0}^{\infty} [f^{(n)}(0)/n!] \, x^n$. For $x = a$: $f(x) = f(a) + f'(a)(x - a) + ... [f^{(n)}(a)/n!](x - a)^n + r_n(x)$, where $r_n(x)$ is the Lagrange remainder formula and is $r_n(x) = [f^{(x+1)}(t_x)/(n + 1)!](x - a)^{(n+1)}$. The Lagrange remainder is sometimes called the error after n terms in the Taylor's series. There are power series expansions for many functions; but if a function is to have a power series expansion, then it must have **derivatives** of all orders on the defined interval of interest. Because a Taylor's series is a power series, it has a circle of convergence whose radius extends to the nearest singularity.

Brook Taylor, an English mathematician, invented the method for expanding functions in terms of **polynomials** about an arbitrary point. In 1715 he published this method in *Methodus in crementorum directa et inversa* and so this method was named in his honor, Taylor's series. The Taylor's series is just a generalized form of the Maclaurin series formulated by the Scottish mathematician **Colin Maclaurin** in about 1742. Although Taylor published his method for expanding functions in 1715, **James Gregory** (sometimes spelled **Gregorie**) and **J. Bernoulli** knew these methods long before. **Joseph-Louis Lagrange** isolated Taylor series as the idea fundamental to calculus. He considered these expansions so important to calculus that he maintained that in order to understand a continuous function one needed to know only the derivatives of a function at a given number of points. His error was not having a clear concept of **convergence** and realizing that all continuous functions have a Taylor's series. Later it became recognized that the Taylor series did not provide the sole key to understanding continuous functions. In spite of limitations, Taylor's series have had a major impact on the development of calculus.

TESSELATION

Generally, a tesselation (sometimes called a tiling or mosaic) is a **division** of **space** into pieces with finite **area** (or **volume**) called tiles. Tesselations and patterns have been in art for at least two thousand years. Ancient artifacts decorated with repeating patterns, tiled buildings like the Alhambra, and mosaics on the frames of artwork are all evidence of the widespread and age-old use of tesselations. Tesselations in nature include the bee's honeycomb, atomic structures of crystals, cell arrangements in life forms, patches of dried up mud and the design of spider's web. The manufacture of components, by stamping them out of sheets of metal, is made efficient by minimizing the unused portion of the metal. Thus tiling has applications to industry. Also communication theorists use "random tesselations" for image enhancement and coding. Crystallography uses tesselations of three-dimensional space to understand the structures of molecules.

The mathematical theory of tesselations, however, is less than a century old. In mathematics, tesselation usually means a plane-filling arrangement of **polygons** such that the intersection of any two of the polygons is empty, a point, or a line segment. This **definition** easily generalizes to higher dimensions. For example, the standard grid is a tiling by squares and the cubic lattice is a tiling by cubes. An isometry is a map, f say, from Euclidean space to itself that preserves distances, i.e. for any points x any y, |x-y| = |f(x) - f(y)|. If an isometry maps the tiling to itself (i.e. if x is a corner point of a tile then so is f(x) and vice versa), then it is called a **symmetry** of the tiling. The set of all such symmetries is called the symmetry group. For example, the symmetry group of the standard tiling by 1x1 squares is equal to the set of all compositions of the following two maps and their inverses: the map that takes any arbitrary point (x, y) to (x + 1,y) and the map that takes (x, y) to (x, y + 1). A fundamental **domain** for the tiling is a polygon with the property that its images under the symmetry group form a tiling of the plane. In the standard tiling by squares, any 1x1 **square** is a fundamental domain. Fedorov (1891) classified the symmetry groups of all two-dimensional and three-dimensional tilings. The crystallographic groups are the symmetry groups for which there is a fundamental domain with finite area (or volume). Fedorov showed that there are 17 two-dimensional crystallographic groups and 230 three-dimensional ones (up to **isomorphism**). Bieberbach(1911) solved part of the eighteenth problem on Hilbert's list when he showed that the number of crystallographic groups up to isomorphism is finite in every **dimension**. Brown and others proved in 1978 that there are 4,783 classes of four-dimensional crystallographic groups. The numbers of crystallographic groups in dimensions five and higher are unknown at this time.

An isohedral tiling is one in which every tile is a fundamental domain. Two-dimensional isohedral tilings have been classified by Grunbaum and Shepherd in their famous book **Tilings and Patterns** (1987). This book is the most comprehensive work on mathematical tilings to date. Polygons that can tile the plane isohedrally were classified by Reinhardt in 1918. If the tiles of a tiling are all congruent to each other, then the tiling is said to be monohedral. Polygons that can tile the plane monohedrally are far from classified. For example, it is not known which convex pentagons can tile the plane monohedrally although fourteen types have been discovered. It is not even known whether or not there an **algorithm** exists that can determine whether a polygon can tile the plane monohedrally. One attempt at finding such an algorithm is the following. The Heesch number of the polygon is the maximum number of times the polygon can be surrounded by copies of itself. If the polygon can tile the plane then its Heesch number is **infinity**. Polygons with Heesch number 0, 1, 2, 3, 4, and 5 have been found. It is unknown whether there are polygons with Heesch number greater than 5 but less than infinity. But if there are none, then there is an algorithm to decide whether a polygon can tile the plane—place copies of itself around itself to see if it has at least six coronas such that each completely surrounds the previous corona.

In 1966 Robert Berger proved that there is no algorithm that can determine whether any arbitrary set of polygons can tile the plane. This does not prove that there is no algorithm for determining whether a single polygon can tile the plane however because Berger used **sets** of more than one polygon.

He found a way for a set of polygons to mimic a **Turing machine**. He then used the fact that the halting problem for Turing machines is undecidable to prove that it is also undecidable whether or not his tiles can tile the whole plane. He also gave the first example of an aperiodic set of polygons. These are sets of polygons that can tile the plane but only in such a way that the symmetry group of any of their tilings is finite. Berger's aperiodic set consisted of 20,426 polygons. By 1971, Robert Robinson had found a set consisting of only 6 tiles. In 1974, **Roger Penrose** discovered an aperiodic set of 2 polygons, the so-called Penrose tiles. It's unknown at this time whether there exists an aperiodic set of only one polygon.

There are many areas of current research in tilings of which we have mentioned only a few. For a more complete introduction into tilings, see Grunbaum and Shepherd's book Tilings and Patterns.

See also Hilbert's problems; Mathematics and art; Penrose tilings; Symmetry

THALES OF MILETUS (CA. 640 B.C.-CA. 546 B.C.)
Greek geometer, astronomer, and philosopher

Thales of Miletuswas the first known Greek philosopher and mathematician who introduced the basic concepts of geometryto the Greeks. A man of many talents, Thales was a successful merchant and founded his own school of philosophy in Miletus, which included teachings in mathematics and astronomy. Notably, Thales emphasis on "natural" rather than "supernatural" explanations for phenomena marks the beginning of scientific methodology and theoretical reasoning. Honored and revered as a man of knowledge and wisdom in his own time, Thales was named one of the Seven Wise Men of Greece.

Thales was born circa 640 B.C. to 620 B.C. in Miletus, a Greek island along the coast of Asia Minor. Since Thales left behind no writings, little is known about his life. As with many ancient thinkers, fact, fiction, and myth have become intermixed. His father was Examius and his mother Cleobuline. Growing up in a strategic caravan route from the East, Thales first made his mark as a highly successful merchant. According to one account, Thales made a fortune when he foresaw a good season for the olive crop and bought all the local olive presses, monopolizing olive oil market.

With his financial success, Thales traveled to Egypt, either to pursue business opportunities or to study Egyptian science and philosophy. Certainly, Thales learned the fundamentals of **geometry** while in Egypt. His aptitude for deductive reasoning is illustrated through the story of Thales' ability to calculate the height of the pyramids. According to legend, Thales was observing the Great Pyramid one day when the Pharaoh and his entourage came by. Intrigued by the foreigner's reputation as a man of wisdom, the Pharaoh asked Thales if he could calculate the pyramid's height. Thales may have accomplished the feat by using the "magic" of the

Egyptian shadow stick, which, when placed in the ground, one could use it to calculate the height of an object by establishing a relationship between the height of the stick's shadow and the shadow of the object to be measured. A more plausible explanation is that Thales measured the pyramid's shadow at the exact time when his own shadow appeared to be the same length as himself, thus ensuring that the Great Pyramid's shadow accurately represented its true height.

Thales may also have traveled to Babylon and studied geometry and astronomy. He eventually returned to Miletus (possibly around 590 B.C.) and established a school where he taught science, astronomy, mathematics, and other subjects. Eventually, the philosophy of Thales became known as the Ionian School of Philosophy.

Thales' reputation as one of the founders of mathematics and applied geometry stems not only from his introduction of geometry to Greece, but in his establishing the need for deductive reasoning and proofs for demonstrating a statement or theorem's validity. The first written account of Thales' mathematical acumen comes from Proclus, who wrote a famous commentary on **Euclid**'s *Elements*. Prior to Thales, geometry was concerned primarily with measuring surfaces and solids. Thales developed geometry in both a practical and theoretical way by focusing on lines, circles, and triangles. Proclus credits Thales with five basic theorems of elementary geometry: (1) The **circle** is bisected by its **diameter**; (2) the base angles of an isosceles **triangle** are equal; (3) pairs of vertical angles formed by two intersecting straight lines are equal; (4) an **angle** inscribed in a semicircle is a right angle; (5) two triangles are congruent if they have two angles and one side that are equal.

Whether or not Thales actually proved geometrical theorems is unknown. Of his first three theorems, historians debate whether Thales developed them through intuition or whether he discovered and demonstrated them scientifically. However, Proclus indicates that Thales did prove **theorem** number two even though by the time Euclid wrote his *Elements* this theorem was accepted as obvious and needed no **proof**. As for theorem number four, it is not known whether Thales developed a proof or demonstrated the theorem empirically.

Many scholars believe that Thales made practical use of his fifth, or **congruence**, theorem in his renowned ability to measure the **distance** of a ship from shore. One theory of how Thales accomplished this remarkable measurement for his day says that Thales used a tower of known height on shore, a plumb line, and a carpenter's **square** (known as a gnomon). Driving a nail in the tower that corresponded to a line of sight to the ship, Thales determined the distance from the right–angled base of the gnomon to where it intersected the line of sight, thus creating an easily measured right triangle. Then, by dropping a plumb to the ground and determining the height of the tower, he was able to calculate the size of a larger right triangle that was proportional to the smaller triangle. The distance to the ship would equal the calculated length of the larger right triangle's base. To accomplish this feat, it is assumed that Thales knew that the sides of equiangular triangles are proportional.

In addition to his interest in geometry and mathematics, Thales was fascinated by astronomy. A famous story concerning Thales' love of astronomy has him falling into a well while gazing at the stars. He is soon discovered by a slave girl who wryly comments that Thales was "so interested in the heavens he could not see what was in front of his own feet."

Thales may have developed his interest in astronomy during his reported travels to Babylon and studies under the Chaldean magi. He is reported to have written works focusing on the equinoxand the solstice and is sometimes attributed with explaining solar eclipses. He is also reported to have written on nautical astronomy and to have advised navigators to depend on Ursa Minor (Little Bear) rather than Ursa Major (Great Bear) for accurately navigating their destinations.

In his own time, Thales' most noted feat of astronomical observation was his prediction of a solar eclipse in 585 B.C. as reported by Herodotus. Thales possibly learned the secret of predicting such events through the Babylonians, who had kept detailed records of astronomical events for many centuries. With these records, Thales may have been able to calculate that an eclipse might occur in that year. No matter how Thales made his famous prediction, he had a tremendous amount of luck, not the least of which was the fact that the eclipse happened to be visible in Miletus. Regardless, this prediction bolstered Thales' fame to new heights.

While he is remembered today primarily for his efforts in geometry and astronomy, Thales was also the first great Greek philosopher. In his Ionian school, philosophy, or the love of wisdom, took precedence over all other studies. Thales was renowned for his intimate knowledge of human nature and his humor. For example, the maxim "know thyself" is purported to originate with Thales. Other sage advice credited to Thales include his counsel that people could lead righteous lives by "refraining from doing what we blame in others" and his pronouncement that the strangest thing he had ever seen was "an aged tyrant."

Like most of his contemporaries, Thales mistakenly believed that the Earth was a flat disc on an infinite ocean. His belief that water was the origin of all life and matter was probably based on the knowledge that water is the essential element needed for the growth and nurturing of all organisms. Still, his belief that supposedly supernatural occurrences could be explained by natural events profoundly influenced the thinking and scientific explorations of those scholars who came after him. His most famous pupil was Anaximander, who was the first natural philosopher in the Milesian school of philosophy.

As a wealthy and learned man, Thales also played a vital role as a political counselor and statesman. Thales is reported to have convinced the Greek colonies (or Ionian city–states) to form an alliance to fight off Persian invaders. As a military advisor, he used his mathematical and engineering skills to help a Greek army cross a river by constructing a channel and diverting the river into it. Since the "Seven Wise Men of Greece" were predominantly statesmen, Thales was probably named one of the seven for these efforts. According to some accounts, Thales, who died circa 546 B.C., was the only one of the seven to be declared a Wise Man twice by the Oracle at Delphi, the second time for his prediction of the solar eclipse.

THEOREM

A theorem (the term is derived from the Greek *theoreo*, which means *I look at*) denotes either a proposition yet to be proven, or a proposition proven correct on the basis of accepted results from some area of mathematics. Since the time of the ancient Greeks, proven theorems have represented the foundation of mathematics. Perhaps the most famous of all theorems is the **Pythagorean theorem**.

Mathematicians develop new theorems by suggesting a proposition based on experience and observation which seems to be true. These original statements are only given the status of a theorem when they are proven correct by logical **deduction**. Consequently, many **propositions** exist which are believed to be correct, but are not theorems because they can not be proven using deductive reasoning alone.

Historical background

The concept of a theorem was first used by the ancient Greeks. To derive new theorems, Greek mathematicians used logical deduction from premises they believed to be self-evident truths. Since theorems were a direct result of deductive reasoning, which yields unquestionably true conclusions, they believed their theorems were undoubtedly true. The early mathematician and philosopher **Thales** suggested many early theorems, and is typically credited with beginning the tradition of a rigorous, logical **proof** before the general acceptance of a theorem. The first major collection of mathematical theorems was developed by Euclid around 300 B.C. in a book called *The Elements*.

The absolute **truth** of theorems was readily accepted up until the eighteenth century. At this time mathematicians, such as Karl Friedrich Gauss (1777-1855), began to realize that all of the theorems suggested by Euclid could be derived by using a set of different premises, and that a consistent non-Euclidean structure of theorems could be derived from Euclidean premises. It then became obvious that the starting premises used to develop theorems were not self-evident truths. They were in fact, conclusions based on experience and observation, and not necessarily true. In light of this evidence, theorems are no longer thought of as absolutely true. They are only described as correct or incorrect based on the initial assumptions.

Characteristics of a theorem

The initial premises on which all theorems are based are called axioms. An **axiom**, or **postulate**, is a basic fact which is not subject to formal proof. For example, the statement that there is an infinite number of even **integers** is a simple axiom. Another is that two points can be joined to form a line. When developing a theorem, mathematicians choose axioms which seem most reliable based on their experience. In this way, they can be certain that the theorems which are proved are as near to the truth as possible. However, absolute truth is not possible because axioms are not absolutely true.

To develop theorems, mathematicians also use definitions. Definitions state the meaning of lengthy concepts in a single word or phrase. In this way, when we talk about a figure made by the set of all points which are a certain distance from a central point, we can just use the word *circle*.

THERMODYNAMICS

Thermodynamics is the science that deals with work and heat, and the transformation of one into the other. It is a macroscopic theory, dealing with matter in bulk, disregarding the molecular nature of materials. The corresponding microscopic theory, based on the fact that materials are made up of a vast number of molecules, is called statistical **mechanics**.

Historical background

Benjamin Thompson, Count von Rumford (1753-1814) recognized from observing the boring of cannon that the work (or mechanical energy) involved in the boring process was being converted to heat by friction, causing the temperature of the cannon to rise. With the experiments of James Joule (1818-1889), it was recognized that heat is a form of energy that is transferred from one object to another, and that work can be converted to heat without limit. However, the opposite is found not to be true: that is, there are limiting factors in the conversion of heat to work. The research of Sadi Carnot (1796-1832), of Lord Kelvin (1824-1907), and of Rudolph Clausius (1822-1888), among others, has led to an understanding of these limitations.

Temperature

The idea of temperature is well known to everyone, but the need to define it so that it can be used for measurements is far more complex than the simple concepts of "hot" and "cold." If a rod of metal is placed in an ice-water bath and the length is measured, and then placed in a steam bath and the length again measured, it will be found that the rod has lengthened. This is an illustration of the fact that, in general, materials expand when heated, and contract when cooled (however, under some conditions rubber can do the opposite, while water is a very special case and is treated below). One could therefore use the length of a rod as a measure of temperature, but that would not be useful, since different materials expand different amounts for the same increase in temperature, so that everyone would need to have exactly the same type of rod to make certain that they obtained the same value of temperature under the same conditions.

However, it turns out that practically all gases, at sufficiently low pressures, expand in **volume** exactly the same amount with a given increase in temperature. This has given rise to the constant volume gas thermometer, which consists of a flask to hold the gas, attached to a system of glass and rubber tubes containing mercury. A small amount of any gas is introduced into the (otherwise empty) flask, and the top of the mercury in the glass column on the left is placed at some mark on the glass (by moving the right hand glass column up or down). The difference between the heights of the two mercury columns gives the difference between atmospheric pressure and the pressure of the gas in the flask. The gas pressure changes with a change in temperature of the flask, and can be used as a **definition** of the temperature by taking the temperature to be proportional to the pressure; the proportionality **factor** can be found in the following manner. If the temperature at the freezing point of water is assigned the value 0° and that at the boiling point is called 100°, the temperature scale is called the Celsius scale (formerly called Centigrade); if those points are taken at 32° and 212°, it is known as the Fahrenheit scale. The relationship between them can be found as follows. If the temperature in the Celsius scale is T(°C), and that in the Fahrenheit scale is T(°F), they are related by T(°F)=(9/5)T(°C)+32°. The importance of using the constant volume gas thermometer to define the temperature is that it gives the same value for the temperature no matter what gas is used (as long as the gas is used at a very low pressure), so that anyone at any laboratory would be able to find the same temperature under the same conditions. Of course, a variety of other types of thermometers are used in practice (mercury-in-glass, or the change in the electrical resistance of a wire, for example), but they all must be calibrated against a constant volume gas thermometer as the standard.

Expansion coefficients

An important characteristic of a material is how much it expands for a given increase in temperature. The amount that a rod of material lengthens is given by $L=L_0[1+\alpha(T-T_0)]$, where L_0 is the length of the rod at some temperature T_0, and L is the length at some other temperature T; α (Greek alpha) is called the coefficient of linear expansion. Some typical values for $\alpha \times 10^6$ (per °C) are: aluminum, 24.0; copper, 16.8; glass, 8.5; steel, 29.0 (this notation means that, for example, aluminum expands at a rate of 24.0/1,000,000 for each degree Celsius change in temperature). Volumes, of course, also expand with a rise in temperature, obeying a law similar to that for linear expansion; coefficients of volume expansion are approximately three times as large as that for linear expansion for the same material. It is interesting to note that, if a hole is cut in a piece of material, the hole expands just as if there were the same material filling it!

Thermostats

Since various metals expand at different rates, a thermostat can be made to measure changes in temperature by securely fastening together two strips of metal with different expansion coefficients. If they are straight at one temperature, they will be bent at any other temperature, since one will have expanded or contracted more than the other. These are used in many homes to regulate the temperature by causing an electrical contact to be made or broken as temperature changes cause the end of the strips to move.

Water

Water has the usual property of contracting when the temperature decreases, but only down to 39.2°F (4°C); below that temperature it expands until it reaches 32°F (0°C). It then

forms ice at 0°C, expanding considerably in the process; the ice then behaves "normally," contracting as the temperature decreases. Since the density of a substance varies inversely to the volume (as a given mass of a substance expands, its density decreases), this means that the density of water increases as the temperature decreases until 4°C, when it reaches its maximum density. The density of the water then decreases from 4°C to 0°C; the formation of the ice also involves a decrease in density. The ice then increases its density as its temperature falls below 0°C. Thus, as a lake gets colder, the water at the top cools off and, since its density is increasing, this colder water sinks to the bottom. However, when the temperature of the water at the top becomes lower than 4°C, it remains at the top since its density is lower than that of the water below it. The pond then ices over, with the ice remaining at the top, while the water below remains at 4°C (until, if ever, the entire lake freezes). Fish are thus able to live in lakes even when ice forms at the top, since they have the 4°C water below it to live in.

The First Law of Thermodynamics

The conservation of energy is well known from mechanics, where energy does not disappear but only changes its form. For example, the potential energy of an object at some height is converted to the kinetic energy of its **motion** as it falls. Thermodynamics is concerned with the internal energy of an object and those things that affect it; conservation ofenergy applies in this case, as well.

Heat

As noted in the introduction, doing work on an object (for example, by drilling a hole in a piece of metal, or by repeatedly bending it) causes its temperature to rise. If this object is placed in contact with a cooler object it is found that they eventually come to the same temperature, and remain that way as long as there are no outside influences (this is known as thermal equilibrium). This series of events is viewed as follows. Consistent with the concept of the conservation of energy, the energy due to the work done on the object is considered to be "stored" in the object as (what may be called) internal energy. In the particular example above, the increase in the internal energy of the object is recognized by the increase in temperature, but there are processes where the internal energy increases without a change in temperature. By then placing it in contact with an object of lower temperature, energy flows from the hotter to the colder one in the form of heat, until the temperatures become the same. Thus heat should be viewed as a type of energy which can flow from one object to another by virtue of a temperature difference. It makes no sense to talk of an object having a certain amount of heat in it; whenever it is placed in contact with a lower-temperature object, heat will flow from the hotter to the cooler one.

The First Law of Thermodynamics

These considerations may be summarized in the First law of thermodynamics: the internal energy of an object is increased by the amount of work done on it, and by the amount of heat added to it. Mathematically, if U_f is the internal energy of an object at the end of some process, and U_i is the internal energy at the beginning of the process, then $U_f - U_i = W + Q$, where W is the amount of work done on the object, and Q is the amount of heat added to the object (negative values are used if work is done by the object, or heat is transferred from the object). As is usual for an equation, all quantities must be expressed in the same units; the usual mechanical unit for energy (in the International System of Units-formerly the MKS system) is the joule, where 1 joule equals 1 kg-m²/s².

Specific heats; the calorie

An important characteristic of materials is how much energy in the form of heat it takes to raise the temperature of some material by 1g. It depends upon the type of material being heated as well as its amount. The traditional basic unit, the calorie, is defined as the amount of heat that is needed to raise 1 g of water by 1°C. In terms of mechanical energy units, one calorie equals 4.186 joules (J).

The corresponding amount of heat necessary to raise the temperature of other materials is given by the specific heat capacity of a material, usually denoted by c. It is the number of kilojoules (kJ) needed to raise 1 kg of the material by 1°C. By definition, the value for water is 4.186 kilojoules. Typical values for c in kilojoules per kg (kJ/kg), at 0°C, are: ice, 2.11; aluminum, 0.88; copper, 0.38; iron, 0.45. It should be noted that water needs more heat to bring about a given rise in temperature than most other common substances.

Change of phase

The process of water changing to ice or to steam is a familiar one, and each is an example of a change in phase. Suppose a piece of ice were placed in a container and heated at a uniform rate, that is, a constant amount of heat per second is transferred to the material in the container. The ice (the solid phase of water) first rises in temperature at a uniform rate until its temperature reaches 32°F (0°C), when it begins to melt, that is, some of the ice changes to water (in its liquid phase); this temperature is called the melting point. It is important to note that the temperature of the ice-water mixture remains at 32°F (0°C) until all the ice has turned to water. The water temperature then rises until it reaches 212°F (100°C), when it begins to vaporize, that is, turns to steam (the gaseous phase of water); this temperature is called the boiling point. Again, the water-steam mixture remains at 212°F (100°C) until all the liquid water turns into steam. Thereafter, the temperature of the steam rises as more heat is transferred to the container. It is important to recognize that during a change in phase the temperature of the mixture remains constant. (The energy being transferred to the mixture goes into breaking molecular bonds rather than in increasing the temperature.) Many substances undergo similar changes in phase as heat is applied, going from solid to liquid to gas, with the temperature remaining constant during each phase change. (Some substances, such as glass, do not have such a well-defined melting point.) The amount of heat needed to melt a gram of a material is known as the heat of fusion; that to vaporize it is the heat of

vaporization. On the other hand, if steam is cooled at a uniform rate, it would turn to liquid water at the condensation temperature (equal to the boiling point, 212°F [100°C]), and then turn to ice at the solidification temperature (equal to the melting point, 32°F [0°C]). The heat of condensation is the amount of heat needed to be taken from a gram of a gas to change it to its liquid phase; it is equal to the heat of vaporization. Similarly, there is a heat of solidification which is equal to the heat of fusion. Some typical values are shown here.

Material	Melting Point °C	Heat of Fusion cal/gm	Boiling Point °C	Heat of Vaporization cal/gm
Water	0	79.7	100	539
Ethyl Alcohol	-114	24.9	78	204
Oxygen	-219	3.3	-183	51
Nitrogen	-210	6.1	-196	48
Mercury	-39	2.8	357	65

It is interesting to note that water has much larger heats of fusion and of vaporization than many other usual substances. The melting and boiling points depend upon the pressure (the values given in the table are for atmospheric pressure). It is for this reason that water boils at a lower temperature in high-altitude Denver than at sea level.

Finally, below certain pressures it is possible for a substance to change directly from the solid phase to the gaseous one; this case of sublimation is best illustrated by the "disappearance" of dry ice when it is exposed to the atmosphere.

Equations of state; work

When an object of interest (usually called the system) is left alone for a sufficiently long time, and is subject to no outside influences from the surroundings, measurements of the properties of the object do not change with time; it is in a state of thermal equilibrium. It is found experimentally that there are certain measurable quantities which give complete information about the state of the system in thermal equilibrium (this is similar to the idea that measurements of the velocity and acceleration of an object give complete information about the mechanical state of a system). For each such state relationships can be found which hold true over a wide range of values of the quantities. These relationships are known as **equations** of state.

Equations of state

Thermodynamics applies to many different types of systems; gases, elastic solids (solids which can be stretched and which return to their original form when the stretching **force** is removed), and mixtures of chemicals are all examples of such systems. Each system has its own equation of state which depends upon the variables that need to be measured in order to describe its internal state. The relevant variables for a system can only be determined by experiment, but one of those variables will always be the temperature.

The system usually given as an example is a gas, where the relevant thermodynamic variables are the pressure of the gas (P), its volume (V), and, of course, the temperature. (These variables are the relevant ones for any simple chemical system, e.g., water, in any of its phases.) The amount of gas may be specified in grams or kilograms, but the usual way of measuring mass in thermodynamics (as well as in some other fields) is in terms of the number of moles. One kilomole (kmol) is defined as equal to M kilograms, where M is the molecular weight of the substance, with carbon-12 being taken as M=12. (One mole of any substance contains 6.02×10^{23} molecules, known as Avogadro's number.) Thus one kilomole of oxygen has a mass of 70.56 lb (32 kg); of nitrogen, 61.76 lb (28.01 kg); the molar mass of air (which is, of course, actually a mixture of gases) is commonly taken as 63.87 lb (28.97 kg). It is found, by experiment, that most gases at sufficiently low pressures have an equation of state of the form: PV=NRT, where P is in Newtons/m^2, V is in m^3, N is the number of kilomoles of the gas, T is the temperature in K, and R=8.31 kJ/kmol-K is known as the universal gas constant. The temperature is in degrees Kelvin (K), which is given in terms of the Celsius temperature as T(°K)=T(°C)+273.15°C. It should be noted that real gases obey this ideal gas equation of state to within a few percent accuracy at atmospheric pressure and below.

The equation of state of substances other than gases is more complicated than the above ideal gas law. For example, an elastic solid has an equation of state which involves the length of the stretched material, the stretching force, and the temperature, in a relationship somewhat more complex than the ideal gas law.

Work

Work is defined in mechanics in terms of force acting over a **distance**; that definition is exactly the same in thermodynamics. This is best illustrated by calculating the work done by a force F in compressing a volume of gas. If a volume of gas V is contained in a **cylinder** at pressure P, the force needed on the piston is (by the definition of pressure) equal to PA, where A is the **area** of the piston. Let the gas now be compressed in a manner which keeps the pressure constant (by letting heat flow out, so that the temperature also decreases); suppose the piston moves a distance d. Then the work done is W=Fd=PAd. But Ad is the amount that the volume has decreased, $V_i - V_f$, where V_i is the initial volume and V_f is the final volume. (Note that this volume difference gives a positive value for the distance, in keeping with the fact that work done on a gas is taken as positive.) Therefore, the work done on a gas during a compression at constant pressure is $P(V_i - V_f)$.

The first law thus gives a straightforward means to determine changes in the internal energy of an object (and it is only changes in the internal energy that can be measured), since the change in internal energy is just equal to the work done on the object in the absence of any heat flow. Heat flow to or from the object can be minimized by using insulating materials, such as fiberglass or, even better, styrofoam. The idealized process where there is **zero** heat flow is called an adiabatic process.

The Second Law of Thermodynamics

One of the most remarkable facts of nature is that certain processes take place in only one direction. For example, if a high temperature object is placed in contact with one of lower temperature, heat flows from the hotter to the cooler until the temperatures become equal. In this case (where there is no work done), the First law simply requires that the energy lost by one object should be equal to that gained by the other object (through the mechanism of heat flow), but does not prescribe the direction of the energy flow. Yet, in a situation like this, heat never flows from the cooler to the hotter object. Similarly, when a drop of ink is placed in a glass of water which is then stirred, the ink distributes itself throughout the water. Yet no amount of stirring will make the uniformly-distributed ink go back into a single drop. An open bottle of perfume placed in the corner of a room will soon fill the room with its scent, yet a room filled with perfume scent will never become scent-free with the perfume having gone back into the bottle. These are all examples of the Second Law of Thermodynamics, which is usually stated in two different ways. Although the two statements appear quite different, it can be shown that they are equivalent and that each one implies the other.

Clausius statement of the Second Law

The Clausius statement of the Second law is: No process is possible whose only result is the transfer of heat from a cooler to a hotter object. The most common example of the transfer of heat from a cooler object to a hotter one is the refrigerator (air conditioners and heat pumps work the same way). When, for example, a bottle of milk is placed in a refrigerator, the refrigerator takes the heat from the bottle of milk and transfers it to the warmer kitchen. (Similarly, a heat pump takes heat from the cool ground and transfers it to the warmer interior of a house.) An idealized view of the refrigerator is as follows. The heat transfer is accomplished by having a motor, driven by an electrical current, run a compressor. A gas is compressed to a liquid, a phase change which generates heat (heat is taken from the gas to turn it into its liquid state). This heat is dissipated to the kitchen by passing through tubes (the condenser) in the back of (or underneath) the refrigerator. The liquid passes through a valve into a low pressure region, where it expands and becomes a gas, and flows through tubes inside the refrigerator. This change in phase from a liquid to a gas is a process which absorbs heat, thus cooling whatever is in the refrigerator. The gas then returns to the compressor where it is again turned into a liquid. The Clausius statement of the Second law asserts that the process can only take place by doing work on the system; this work is provided by the motor which drives the compressor. However, the process can be quite efficient, and considerably more energy in the form of heat can be taken from the cold object than the work required to do it.

Kelvin-Planck statement of the Second Law

Another statement of the Second law is due to Lord Kelvin and Max Planck (1858-1947): No process is possible whose only result is the conversion of heat into an equivalent amount of work. Suppose that a cylinder of gas fitted with a piston had heat added, which caused the gas to expand. Such an expansion could, for example, raise a weight, resulting in work being done. However, at the end of that process the gas would be in a different state (expanded) than the one in which it started, so that this conversion of all the heat into work had the additional result of expanding the "working fluid" (in this case, the gas). If the gas were, on the other hand, then made to return to its original volume, it could do so in three possible ways: (a) the same amount of work could be used to compress the gas, and the same amount of heat as was originally added would then be released from the cylinder; (b) if the cylinder were insulated so that no heat could escape, then the end result would be that the gas is at a higher temperature than originally; (c) something in-between. In the first case, there is no net work output or heat input. In the second, all the work was used to increase the internal energy of the gas, so that there is no net work and the gas is in a different state from which it started. Finally, in the third case, the gas could be returned to its original state by allowing some heat to be transferred from the cylinder. In this case the amount of heat originally added to the gas would equal the work done by the gas plus the heat removed (the First law requires this). Thus, the only way in which heat could be (partially) turned into work and the working fluid returned to its original state is if some heat were rejected to an object having a temperature lower than the heating object (so that the change of heat into work is not the only result). This is the principle of the heat engine (an internal combustion engine or a steam engine are examples).

Heat engines

The working fluid (say, water for a steam engine) of the heat engine receives heat Q_h from the burning fuel (diesel oil, for example) which converts it to steam. The steam expands, pushing on the piston so that it does work W; as it expands, it cools and the pressure decreases. It then traverses a condenser, where it loses an amount of heat Q_c to the coolant (cooling water or the atmosphere, for example), which returns it to the liquid state. The Second law says that, if the working fluid (in this case the water) is to be returned to its original state so that the heat-work process could begin all over again, then some heat must be rejected to the coolant. Since the working fluid is returned to its original state, there is no change in its internal energy, so that the First law demands that $Q_h - Q_c = W$. The efficiency of the process is the amount of work obtained for a given cost in heat input: $E = W/Q_h$. Thus, combining the two laws, $E = (Q_h - Q_c)/Q_h$. It can be seen therefore that a heat engine can never run at 100% efficiency.

It is important to note that the laws of thermodynamics are of very great generality, and are of importance in understanding such diverse subjects as chemical reactions, very low temperature phenomena, and the changes in the internal structure of solids with changes in temperature, as well as engines of various kinds.

THOM, RENÉ FRÉDÉRIC (1923-   )
French topologist and mathematical philosopher

René Frédéric Thom is best known as the founder of **catastrophe theory**, a field with numerous applications in the exact and social sciences. Catastrophe theory provides models for the description of continuous processes that experience abrupt change. Thom is the recipient of the Fields Medal and the Grand Prix Scientifique de la Ville de Paris. He is a member of the French Académie Royale des Sciences and has been named a Knight of the Legion of Honor in France.

Thom was born on September 2, 1923, at Montbéliard, France, to Gustav Thom, a pharmacist, and Louise Ramel. He married Suzanne Heimlinger on April 9, 1949, and they have three children, Françoise, Elizabeth, and Christian. Thom's formal education took place at the Collége Cuvier in Montbeliard and later in Paris at the Lycée Saint-Louis. After earning a master's degree in mathematics and a doctorate in science from the École Normale Supérieure in Paris, Thom went to Princeton University in 1951. He taught in Grenoble from 1953 to 1954 and in Strasbourg for the next decade.

In 1958, at the age of 35, Thom received the Fields Medal, the most prestigious prize in mathematics. When awarding the prize, the International Mathematical Union (IMU) cited Thom's 1954 invention of the theory of cobordism in algebraic **topology**. Cobordism is a classification scheme for manifolds (multidimensional topological surfaces) based on homotopy (a continuous deformation of one function into another). The IMU recognized Thom's use of this technique as a "prime example of a general cohomology theory." In a 1991 address to the Symposium on the Current State and Prospects of Mathematics, Thom recalled that after he received the Fields Medal, he experienced doubts about his ability to continue developing meaningful mathematical results. He decided to turn his attention from the more algebraic types of work and concentrate on singularities of differentiable maps, a topic he considered "more flexible and more concrete."

While he was teaching at Strasbourg University, Thom collaborated with a physicist on an investigation of caustics in optical **geometry** (i.e., rays reflected or refracted by curved solid surfaces), studying both their singularities and their deformations. He told the 1991 symposium that "to my surprise, I found that in caustics, organized by some very simple optical instruments like spherical mirrors and rectilinear diopters, one may find a singularity that should not theoretically exist." He eventually connected this phenomenon to **Pierre de Fermat**'s principle of optimality.

Since 1964, Thom has been a professor at the Institut des Hautes Etudes Scientifiques (IHES) at Bures-sur-Yvette, where he concentrates on research, with no teaching or administrative duties. In this environment, he wrote a carefully reasoned article opposing the popular movement to discard geometry from general mathematics education. His argument concerning the importance of geometry centered on the potential value of studying singularities of function maps. As a result of this exercise, Thom developed the theory of singularities.

During the 1960s, Thom turned his attention from optical problems in physics to the biological topic of embryology. His objective was to apply transversality theory to natural science. Out of his mathematical **analysis** of current theories of cellular differentiation, he developed what he later called "my famous list of seven elementary catastrophes in space-time," a result that fit naturally into singularity theory. Applying a very abstract **theorem** about universal deformations of analytic **sets** proved by **Alexander Grothendieck**, a colleague at the IHES, Thom wrote a book attempting to explain the origin of natural forms. Delayed by the bankruptcy of the original publishing house, *Stabilité Structurelle et Morphogénése (Structural Stability and Morphogenesis)* finally appeared in print in 1972, although a few copies had already circulated in the mathematical community. E. Christopher Zeeman of the University of Warwick was fascinated with the contents of the book. Thom later told the 1991 symposium that "[Zeeman's] reflections on this subject led to a grandiose extension of the theory," extending it from four-dimensional **space-time** to any locally Euclidean **space**. Zeeman gave a couple of lectures on what he called "catastrophe theory,"and wrote a fascinating account about it for *Scientific American* magazine.

Generally classified as a branch of geometry because variables and results are shown as curves or surfaces, catastrophe theory attempts to explain predictable discontinuities in output for systems characterized by continuous inputs—a task that cannot be done using differential **calculus**. Contrary to the implications of the theory's name, the "catastrophes" studied are not necessarily negative in nature; the term simply refers to sudden change. For example, when a balloon is steadily filled with air, it expands and changes its shape in a relatively uniform manner until the pressure in the balloon's interior reaches a critical value. Then, when the balloon undergoes more abrupt, but predictable, change, it pops. More complex phenomena, such as the refraction of **light** through moving water, the amount of stress that can be placed on a bridge, and the synergistic effects of the ingredients in drugs can also be effectively studied using catastrophe theory. The theory provides a universal method for the study of all jump transitions, discontinuities, and sudden qualitative changes.

Thom recalled, "as a result [of Zeeman's publicity], catastrophe theory... took off like a rocket, propelled by the principal media all over the world. This glory was short-lived, however, and the brief success of [catastrophe theory] soon fell to the slings and arrows of trans-Atlantic criticism." In fact, some scientists have hailed Thom's catastrophe theory as a tool more valuable to humanity than Newtonian theory, which considers only smooth, continuous processes. However, the theory became something of a fad in the 1970s and 1980s and was used in applications that the theory does not support. As a result of such indiscriminate application, the theory has at times been unjustly criticized as a cultural phenomenon or a metaphysical view rather than a legitimate branch of mathematics. Although it has been presented in a metaphysical vein as **proof** of the deterministic nature of the universe, catastrophe theory does not purport to abolish the indeterminacy that is central to nuclear physics. Nevertheless, catastrophe theory

has been used to study abrupt systems changes in such diverse fields as **hydrodynamics**, geology, particle physics, industrial relations, embryology, economics, linguistics, civil engineering, and medicine.

By his own account, Thom has been primarily a mathematical philosopher since the early 1970s, largely due to the controversy surrounding catastrophe theory. A major criticism of the theory was that it allows only qualitative rather than quantitative predictions. That is, given a specific application, the type of abrupt change could be predicted, but it could not be quantified in numerical terms. While admitting that this is a legitimate concern, Thom still sees value in the **modeling** techniques catastrophe theory makes available to fields such as psychology. A classical example described by Zeeman, for instance, involved predicting the action of a dog experiencing a combination of the emotions of rage and fear—emotions that (if acting alone) would cause the dog to attack or flee, respectively. What catastrophe theory provides is a mathematical framework for clearly describing the situation being modeled, much as **algebra** provides a shorthand for expressing relationships between numbers and variables in a way that facilitates manipulating them to obtain results.

Speaking from the perspective of 20 years of philosophical reflection, Thom said in his 1991 symposium address: "What is important in science is not the distinction between true and false. This might seem strange to mathematicians, but I will say that if I had the choice between an error which has an organizing power of reality (this could exist) and a **truth** which is isolated and meaningless in itself, I would choose the error and not the truth. There are many examples of errors which are scientifically important, and there are many, many examples of meaningless truths in science."

TOPOLOGICAL EQUIVALENCE

Two objects are topologically equivalent if one object can be continuously deformed to the other. In one or two dimensions, this is something we can visualize: to continuously deform a surface means to stretch it, bend it, shrink it, expand it, etc.— anything that we can do without actually tearing the surface or gluing parts of it together. Intuitively, we imagine that the surface is made of infinitely flexible rubber, and any shape that we can transform this surface into without tearing or gluing the rubber is topologically equivalent to it.

In two dimensions, there are many examples of surfaces that are topologically equivalent, and also many examples of surfaces that are topologically distinct (i.e., surfaces that cannot be continuously deformed to each other). To begin with, consider an ordinary piece of paper, without the edges (that is, the edges of the paper are not considered to be part of the surface). Now crumple it up. The crumpled paper has ridges and valleys that were not originally there, but it is topologically equivalent to the flat paper. The flat paper (and hence the crumpled paper) is also topologically equivalent to an entire (i.e. infinite) plane, because we can stretch the paper—in theory—as much as we like. It is also equivalent to the interior of a **circle**, rectangle, hexagon, etc., but not to a washer (i.e. the **area** between two concentric circles), because we cannot change a circle into a washer without punching a hole in it. A cone, however, is equivalent to a plane, even though it sits in three-dimensional **space**, because we can flatten the cone down into the plane and get a circle. (To see this, imagine standing an empty ice cream cone on a table, and then pushing each point of the cone down onto the table. Remember, this ice cream cone is made of infinitely stretchable rubber!) On the other hand, if we take a flat piece of paper and roll it up into a **cylinder**, we obtain a topologically distinct surface, because parts of the paper that were not originally attached to each other (the opposite edges of the paper) are attached now. Thus a cylinder is distinct from a plane. A cylinder is, in fact, topologically equivalent to a washer.

There are other objects, such as a hollow **sphere**, that are "locally" two-dimensional, because even though the surface itself does not lie in any two-dimensional plane, each point has a neighborhood that is topologically equivalent to a plane. Among such surfaces, the sphere is topologically distinct from the plane (which is, obviously, globally two-dimensional as well as locally), and a **torus** (that is, the "shell" of a doughnut) is topologically distinct from both the sphere and the plane. We can obtain an infinite collection of mutually topologically distinct surfaces by adding more and more holes, since adding a hole automatically changes the topological "type" of a surface: a two-holed torus (imagine doughnut dough with two holes punched out, instead of one), a three-holed torus (like the shell of a pretzel), a four-holed torus, etc. In fact, one of the basic ways that **topology** distinguishes among surfaces is by counting their holes. A standard joke about topologists is that they are mathematicians who can't tell the difference between a coffee cup and a doughnut.

Topological **equivalence** can also be defined in dimensions other than two. In one **dimension**, the intuition is the same as in two—for example, a circle is topologically distinct from a line segment. In three dimensions or higher, however, we cannot visualize objects other than subsets of our own three-dimensional environment—for instance, what does a three-dimensional sphere look like, and how can we tell what it can or cannot be deformed into? (Remember that the "usual" hollow sphere is two-dimensional, even though it sits in three-dimensional space.) For three or more dimensions, we need a more rigorous notion of topological equivalence.

To come up with a **definition**, we need to see how to express mathematically the concept of deforming a surface. If surface A is being deformed to surface B, that means that we have a function f that takes each point on A to a point on B, and moreover this map is what is called a **homeomorphism**: it is continuous (intuitively, this says that points that are near each other on A will be fairly near each other on B—this is the condition that allows for stretching and shrinking, but not tearing) and one-to-one (no two points on A are sent to the same point on B—intuitively, this is the "no gluing" condition), and moreover its inverse map from B to A is also continuous. However, a deformation involves more than just this

function f. Imagine that while A is being deformed to B, we freeze the process several times. We then see a collection of surfaces that are "in between" A and B. Suppose there are nine of them. First A is deformed to a surface A_1 that is one-tenth of the way towards B, then A_1 is deformed to a surface A_2 that is two tenths of the way, and so on, until A_9 is deformed to B. So this gives us a sequence of homeomorphisms that are "in between" the identity map on A (i.e. the map that leaves A unchanged—think of this as the zero-th stage of the process) and the map f:

The identity map takes A to A $f_{1/10}$ takes A to A_1 $f_{2/10}$ takes A to A_2 (in two steps, "via" the surface A_1) $f_{3/10}$ takes A to A_3 (in three steps, via A_1 and A_2)... $f_{9/10}$ takes A to A_9 9 (in nine steps, via A_1 through A_9) f takes A to B (in ten steps, via A_1 through A_9).

Of course, there is nothing special about the number nine. We could freeze the procedure 90 times, 1000 times, 10 million times, etc.; we can freeze the procedure at any specific fraction of the way between the identity map and f and obtain an intermediate surface and an intermediate function. Thus a deformation is actually a class of homeomorphisms f_t that vary continuously with t, where t ranges between 0 and 1, f_0 is the identity map on A and f_1 is the function f from A to B.

The definition we just constructed works just as well in twelve dimensions as in two. Thus the general, n-dimensional definition of topological equivalence is: A is topologically equivalent to B if there exist a family of homeomorphisms f_t varying continuously with t in [0,1] such that f_0 is the identity map on A and f_1 takes A to B.

See also Continuity; Manifold

TOPOLOGY

Topology is the study of properties of objects that do not depend on geometric measurements, and that do not change when the object is stretched or distorted without tearing. Topology is divided into three subdisciplines, which are largely different from each other: point-set topology, algebraic topology, and differential topology.

Point-set topology is the broadest and most fundamental of the different types of topology, and also the most axiomatic. The basic notion in point-set topology is the idea of **continuity**. The most classic **definition** of continuity is the one that arises in single-variable **calculus**, concerning **functions** that map the **real numbers** to themselves: a function is continuous, loosely speaking, if its graph has no jumps. This definition can be extended to a more general setting, for any function that maps one topological **space** to another. These more general functions are called continuous when they map one space to the other space without any tearing. Thus, a function that sends the western hemisphere of the earth onto a flat disk in way that we see on maps is continuous, since this **mapping** distorts the hemisphere but does not tear it. A function that flattens the entire globe onto a map on a sheet of paper is not continuous, since it must cut the **sphere** in order to flatten it.

To study continuity in its most general sense, mathematicians had to decide what kind of spaces permit the notion of continuous maps. The mathematical definition of 'not tearing,' again speaking loosely, is that points that are close to each other in the first space should still be close to each other when they are mapped to the second space. Thus, in order for a space to be able to have continuous maps, the space must come equipped with a notion of closeness. It is hard to imagine a space that does not come with a natural definition of closeness, but when a space is defined in an abstract way, instead of by a concrete picture, there is not always one definition of closeness that is more natural than others. The study of spaces that come equipped with a notion of closeness is point-set topology. In this discipline, mathematicians use an axiomatic approach—they try to make as few assumptions as possible about the nature of the spaces they study (in particular, they try not to rely on their intuition, which is often misleading), and prove theorems that are as strong as possible. In this stripped-to-the-basics approach, even obvious-sounding statements must be proved from scratch, and their proofs can be surprisingly difficult. For example, one of the main theorems of point-set topology is the **Jordan curve theorem**, which states that a closed loop in the plane always divides the plane into an inside and an outside—this sounds straightforward, but it is not, since it is a statement about every possible loop, no matter how complicated.

Algebraic topology, which uses algebraic techniques to distinguish one topological space from another, is perhaps the oldest branch of topology. The first work that can be considered to be topology is a 1736 paper of **Leonhard Euler**, in which he solved the problem of the **Königsberg bridges**, a question about whether it is possible to walk through the city of Königsberg in such a way that each of its seven bridges is crossed once and only once. This is a question of topology, not **geometry**, since it is more important how the bridges are connected to each other than their precise geometric measurements. Euler proved that it is impossible to make such a walk, using algebraic methods: he showed that in a city where such a walk is possible, each island must have an even number of bridges coming out of it, except the islands on which the walk started and ended; this was not the case in Königsberg, so the walk was impossible.

In 1750, Euler published an even more influential paper in which he studied polyhedra from an algebraic point of view. In a convex **polyhedron** (a polyhedron for which any line connecting two points of the polyhedron is contained in the polyhedron), Euler discovered the relationship v−e+f=2, where v is the number of vertices (endpoints), e is the number of edges, and f is the number of faces (sides). In 1813, Antoine-Jean L'Huilier realized that Euler's formula was wrong for polyhedra that have holes in them, such as polyhedra that are shaped roughly like the surface of a donut. L'Huilier discovered that if the number of holes (or tunnels) in the polyhedron is denoted g, then v−e+f=2−2g. This was the first topological invariant-a number that could be attached to a topological space, that only depended on its topology, not its geometry. This number, called the **Euler characteristic** of the surface,

gave a way to distinguish between two topological surfaces: if they had different Euler characteristics, then they must be different spaces. This **invariant** was the beginning of a branch of algebraic topology called homology theory, which tries to distinguish between spaces by examining polyhedra (perhaps higher-dimensional) that can be placed on (or in) the space, and by using them to study what kind of 'holes' the space has. Another similar technique is called homotopy theory: it uses circles, spheres, and higher-dimensional spheres to measure how many holes a given space has.

The third branch of topology is differential topology, which applies the notions of calculus to topological spaces. Differential topology is studied on a more restricted class of spaces than are point-set or algebraic topology: spaces called 'smooth manifolds.' A smooth **manifold** is a space for which at every point, there is a small neighborhood of that point that can be mapped to a small disk in Euclidean space, and, when two of those neighborhoods intersect, the corresponding maps of Euclidean space are smooth (differentiable). Although this is a substantial restriction on the kinds of spaces that are studied, the power of the techniques of differential topology make this a worthwhile restriction. Many of the topological spaces that are most important to physics are smooth manifolds, and can be understood using these techniques.

TORRICELLI, EVANGELISTA (1608-1647)
Italian royal scholar

Remembered primarily for his invention of the barometer and his discovery of atmospheric pressure, Evangelista Torricelli worked extensively with **Galileo Galilei**, refined the method of indivisibles introduced by **Bonaventura Cavalieri**, and wrote the famous *Opera Geometrica*.

Torricelli was born in Faenza, Italy on October 15, 1608, the oldest of three children and the son of a modestly successful textile craftsman. From a young age, the boy demonstrated impressive mathematical talents, so his family sent him to be tutored by an uncle who was a monk of the Camaldolese order. The uncle gave Torricelli much of his early education, after which Torricelli began attending philosophy and mathematics courses in 1625 at the Jesuit school, Sapienza College (the University of Rome), back in Faenza.

His Jesuit teachers were so awed by Torricelli's aptitude that in 1626 they sent him on to more advanced studies in Rome at a school run by a former Galileo student, Castelli. Immediately recognizing the new student's abilities, Castelli began teaching Torricelli everything Galileo had taught him, including the laws of **motion**. Castelli sent Torricelli's thesis, a discussion of the paths of projectiles ("On Movement"), to Galileo, who then invited Torricelli to visit. However, Torricelli did not accept the invitation for some time; in fact, little is known about what he did for the next decade. The most probable explanation, according to one source, is that from 1632 to about 1641 Torricelli served as secretary to another Galileo friend who was a governor. Meanwhile, historical records suggest that he did much of his most important work on the **geometry** of indivisibles (a term Galileo had introduced) in 1635.

Thus, it was not until 1641 that Torricelli finally visited the great scientist at his home in Florence, shortly afterward accepting a position as his secretary. Galileo died three months later, and Torricelli took his place as leader of the city's science academy and court mathematician to Grand Duke Ferdinando II of Tuscany. One of the benefits of the latter position, which Torricelli kept for the rest of his life, is that he lived in the ducal palace in Florence. It was at the court that Torricelli learned to make telescope lenses of a very high quality. Indeed, demand for his lenses, which he machined using a secret patented process, kept him quite busy with orders.

In 1644 Torricelli published *Opera Geometrica,* which soon came to be known throughout Europe for its clear, concise explanation of many aspects of geometry. The book quickly brought him fame and led to correspondence with some of the other brilliant mathematicians of the day, including **Marin Mersenne** and **Gilles Personne de Roberval**. Even Cavalieri, whose work on indivisibles it expanded and built upon, admired the book.

Torricelli's communications with Roberval led to a famous argument in the 1640s over which of them had first discovered certain aspects of the cycloid. For instance, Torricelli claimed that it was he who had found the length of the cycloid's arc (the curve traced by a point on the **circumference** of a rotating circle.) Torricelli's work in geometry has been credited with leading to the development of integral **calculus**.

Also in 1644, Torricelli designed what would come to be known as a barometer. He had been encouraged to study the subject of the behavior of liquids in a vacuum when the duke of Tuscany complained that water in his well could not be brought higher than 30 ft (9.14 m). Building on Galileo's earlier theory, Torricelli proposed that the limiting **factor** was actually the "weight" of the atmosphere, which not only forced water 30 ft up a pipe, but then prevented it from rising any further, the external pressure having been equalized. To test his idea, Torricelli substituted mercury for water, inverting a tube of it over a dish, and noticed that it rose only one-fourteenth as high as water would have. (Mercury is 14 times heavier than water.) Soon, he also realized that the height of the mercury in the tube varied from time to time, and surmised that this must be due to changes in atmospheric pressure. Thus, the barometer was born. To this day, the vacuum above the mercury in a barometer is known as a Torricellian vacuum, while the unit of pressure involved is still called a "torr."

The scientist's other contributions include the Torricelli **theorem**: that the flow of a liquid through an opening is proportional to the **square** root of the liquid's height. He also used infinitesimals to find the point in a triangle's plane so that the sum of its distances from the vertices is a minimum (the isogonic center).

Torricelli died at the duke's palace in Florence on October 25, 1647, three days after contracting a violent illness (perhaps typhoid fever).

TORUS

A torus is a doughnut-shaped, three-dimensional figure formed when a **circle** is rotated through 360° about a line in its plane, but not passing through the circle itself. Imagine, for example, that the circle lies in **space** such that its **diameter** is parallel to a straight line. The figure that is formed is a hollow, circular tube, a torus. A torus is sometimes referred to as an anchor ring.

The surface **area** and **volume** of a torus can be calculated if one knows the radius of the circle and the radius of the torus itself, that is, the **distance** from the furthest part of the circle from the line about which it is rotated. If the former **dimension** is represented by the letter r, and the latter dimension by R, then the surface area of the torus is given by $4\pi^2 Rr$, and the volume is given by $2\pi^2 Rr_2$.

Problems involving the torus were well known to and studied by the ancient Greeks. For example, the formula for determining the surface area and volume of the torus came about as the result of the work of the Greek mathematician **Pappus of Alexandria**, who lived around the third century A.D. Today, problems involving the torus are of special interest to topologists.

TOWERS OF HANOI

The towers of Hanoi is an ancient mathematical puzzle likely to have originated in India. It consists of three poles, in which one is surrounded by a certain number of discs with a decreasing **diameter**. The object of the puzzle is to move all of the discs from one pole onto another pole. The movement of any disc is restricted by two rules. First, discs can only be moved one at a time. Second, a larger disc can not be placed on top of a smaller disc.

TRANSCENDENTAL NUMBERS

Transcendental numbers, named after the Latin expression meaning *to climb beyond,* are numbers which exist beyond the realm of **algebraic numbers**. Mathematicians have defined algebraic numbers as those which can function as a solution to polynomial **equations** consisting of x and **powers** of x. In 1744, the Swiss mathematician **Leonhard Euler** (1707-1783) established that a large variety of numbers (for example, **whole numbers, fractions, imaginary numbers, irrational numbers, negative numbers**, etc.) can function as a solution to a polynomial equation, thereby earning the attribute *algebraic*. However, Euler pointed to the existence of certain irrational numbers which cannot be defined as algebraic. Thus, $\sqrt{2}$, π, and e are all irrationals, but they are nevertheless divided into two fundamentally different classes. The first number is algebraic, which means that it can be a solution to a polynomial equation. For example, $\sqrt{2}$ is the solution of $x^2 - 2 = 0$. But π and e cannot solve a polynomial equation, and are therefore defined as transcendental. While π, which represents the ratio of the **circumference** of a **circle** to its **diameter**, had been known since antiquity, its transcendence took many centuries to prove: in 1882, Ferdinand Lindemann (1852-1939) finally solved the problem of "squaring the circle" by establishing that there was no solution. There are infinitely many transcendental numbers, as there are infinitely many algebraic numbers. However, in 1874, Georg Cantor (1845-1919) showed that the former are more numerous than the latter, suggesting that there is more than one kind of **infinity**.

TRANSFINITE NUMBERS

Transfinite numbers are infinite **ordinal numbers**. Informally, ordinal numbers may be represented by strings of *'s. For example, the ordinal number 0 is the string of no stars. Number 1 is *. Number 2 is ** and so on. The first transfinite number is then represented by ***.... Here the three dots after the *** imply that the sequence repeats infinitely. This number is traditionally called omega and is written ω. Ordinal numbers are added by juxtaposition. For example $1 + 2 = (*) + (**) = *** = 3$. So $1 + \omega = ****... = \omega$. But $\omega + 1 = ****...*$. These two numbers are different for the following reason. If a frog is on the first star of ***..., and it can leap onto successive stars, then it can reach any star in ***.... But, a frog on the first star of ***...* cannot get to the last star by hopping consecutive stars. So, **addition** of transfinite numbers is not commutative.

Suppose that x and y are two transfinite numbers. Then x times y is represented by replacing every * in y by a copy of x. For example, 2 times ω is ******... = ***... = ω. But ω times 2 is ***...***.... So **multiplication** of transfinite numbers is not commutative either. Exponentiation is harder to define so consider these examples. In order to make the notation easy, we will write $x\{y\}$ instead of x^y.

- $2\{\omega\} = [**]... = \omega$
- $\omega\{2\} = [***...]...$
- $\omega\{3\} = [[***...]...]...$
- $(\omega +1)\{2\} = [***...*]...[***...*] = [***...]...[***...*] = \omega\{2\} + 1$.

The three dots after a pair of **square** brackets means that everything in the square brackets is repeated an infinite number of times. To get $\omega\{\omega\}$, the pattern started by $\omega\{2\}$ and $\omega\{3\}$ must be repeated ω times. Notice that ω times $\omega\{\omega\} = \omega\{\omega\}$. The number $\omega\{\omega\{\omega\{\omega...\}\}\}$, that is ω raised to the power of ω raised to the power of ω and so on ω times, is commonly denoted by ε_0. It is the first ordinal number that cannot be obtained by a finite number of additions, multiplications, and exponentiations. For this reason, it is called the first inaccessible number.

Formally, ordinal numbers are defined in terms of well-ordered **sets** and order types. A set W is well ordered if there is a relation (commonly denoted by $\leq$) on it that satisfies these five properties:

- (1) for every w in W, $w \leq w$
- (2) if $w \leq v$ and $v \leq x$ then $w \leq x$

- (3) for every pair of elements w and v in W, either $w \leq v$ or $v \leq w$
- (4) there is an element (called the least element) l such that if w is any element of W then $l \leq w$.
- (5) if w is in W then there is an element v (called the successor of w) such that $w \leq v$ and if x is any element not equal to w and $w \leq x$ then $v \leq; x$.

For example, the positive **integers** are well-ordered but the positive **rational numbers** are not since no element has a successor. The words of the English language are well-ordered by "dictionary" order. In this case, the least element is 'a'. If V and W are well-ordered sets and there is a one-to-one **correspondence** f from V to W such that $v \leq; w$ if and only if $f(v) \leq; f(w)$ then V and W are said to be of the same order type. An ordinal number is an order type of a well-ordered set.

If x and y are two ordinal numbers then they are represented by two well-ordered sets X and Y whose order type is x and y respectively. Then $x + y$ is defined to be the order type of the set $Q = X \cup Y$ with the following order relation. If v and w are both in X and $v \leq; w$ as elements in X, then $v \leq; w$ as elements of Q. A similar statement holds for Y. If v is an element of X and w is an element of Y, then $x \leq; w$.

x times y is defined to be the order type of the set P, of all ordered pairs of the form (v, w) in which v is an element of X and w is an element of W with the following order relation. If (v, w) and (a, b) are elements of P then $(v, w) \leq; (a, b)$ if either $v < a$ or $v = a$ and $w \leq; b$. This is sometimes called "dictionary" order.

x to the power of y is defined to be the order type of the set R of all **functions** from X to Y with the following order relation. If f and g are in R, then $f \leq; g$ means that if v is the first element v of X such that f(v) is not equal to g(v), then $f(v) \leq; g(v)$. For example, $2\{\omega\}$ is represented by the set of all functions the set $\{0,1\}$ to the positive integers.

If x and y are ordinal numbers then $x \leq y$ means that there is function f from X to Y that maps the least element of X to the least element of Y and if w is the successor of v then f(w) is the successor of f(v). Therefore x is represented by the subset f(X) of Y. For example $0 < 1 < 2 <... < \omega < \omega + 1 <... < \omega$ times $2 <... < \omega\{\omega\} < \varepsilon_0 <....$

So the ordinal numbers can all be represented by sets having the property that $x \leq y$ means $X \subset Y$. Thus the union of all these sets has an order type (called P) that is larger than any ordinal number. But then $P + 1$ is an ordinal larger than P. This contradiction is called the Burali-Forti paradox. Many mathematicians were suspicious of transfinite numbers for this reason. Set theorists resolved this paradox by stating that the union of all those sets is not a set itself. It is a class of objects, and thus, not an ordinal number.

Cantor's student Zermelo proved that every set can be well-ordered if the **Axiom** of choice is true. In other words, the ordinal numbers can be used to "count" the elements of any set. The axiom of choice was later proved to be equivalent to **Zorn's lemma**.

See also Cantor, Georg; Cardinality; Zermelo, Ernst Friedrich Ferdinand

TRANSFORMATION

A transformation is a one-to-one function that maps a set of points onto another set of points. The "points" mentioned here may be one-dimensional points on the real number line, two-dimensional points in the plane, three-dimensional points in Euclidean **space**, or any higher dimensional points in abstract space. We shall confine ourselves to one and two dimensions in this article. The points in the **domain** of the **translation** are often called "pre-images," while points in the **range** are called "images." The mathematical formula that describes how to map pre-images to images is sometimes called a "transformation image formula." From a practical point of view, we may say that a transformation is a function that repositions points. For example, the one-dimensional transformation $T(x) = x + 3$ repositions any point on the real number line three units to the right. This particular transformation is called a translation. The transformation $T(x,y) = (x - 3, y + 2)$ is also a translation but in two-dimensions. It repositions points in the plane three units horizontally to the left and two-units vertically upward. The transformation $T(x,y) = (x,y)$ "repositions" points right back where they were. This is called the identity transformation in two-dimensions. $T(x) = x$ is the identity transformation in one-dimension. Some transformations reposition points without changing distances; these are called "isometries" or distance-preserving transformations. Other transformations have as their purpose to change distances; these are the scale changes. We will consider both types in the paragraphs below.

There are three types of isometries or distance-preserving transformations: translations, rotations, and reflections. Translations were mentioned in the preceding paragraph. They slide points in same direction. The general mathematical formula for a translation in the plane is $T(x,y) = (x + h, y + k)$. This slides points h units horizontally and k units vertically. Rotations reposition points by turning them about some central point called the center of **rotation**. Rotations about (0,0), called origin-centered rotations, give image formulas of the form $R(x,y) = (x\cos(t) - y\sin(t), x\sin(t) + y\cos(t))$, where t is the **angle** of rotation. For a rotation of 90 degrees about the origin, this formula becomes $R(x,y) = (-y,x)$. Reflections are transformations which "flip" points over some given line. The image formula in the general case is complicated, but in special cases it is less cumbersome. For instance, if r represents a reflection over the line $y = x$, the formula is $r(x,y) = (y,x)$. This says that when points are reflected over the $y = x$ line, the pre-image coordinates are reversed to obtain the image points. Sometimes a fourth transformation, the glide reflection, is included among the isometries. The glide reflection slides points in one direction and then flips them over a line. Thus it is not a simple "one-step" transformation; rather it is a "composite" of a translation followed by a reflection. It is also interesting to note that both translations and rotations can be created by the composition of two reflections. A composite of two reflections across parallel lines produces a translation, whereas a composite of two reflections across intersecting lines produces a rotation. In this sense, reflections are the only isometries really needed. Rotations and translations are introduced for convenience of notation.

Scale changes do not preserve **distance** and so are not isometries. The two basic types of scale changes are the **similarity** scale change, also called a "dilation" and the non-similarity scale change. The dilation centered at the origin has an image formula of the form $S(x,y) = (ax,ay)$ where a is not 0 or 1. The number a is called the scale **factor** and is applied to both coordinates of the pre-image to produce the image. The fact that the same scale factor is applied to both coordinates means that all figures mapped with a dilation will be "similar" to their images, hence the name similarity transformation. In **geometry**, two figures are said to be similar if the lengths of pieces in one figure are proportional to lengths of the corresponding pieces in the other. So a **triangle** with sides of length 3, 4, and 5 meters is similar to a triangle with sides of length 6, 8, and 10 meters. In this case, the scale factor is 2. Non-similarity scale changes centered at the origin have image formulas of the form $S(x,y) = (ax,by)$ where a and b are not equal in absolute value. The number a is called the horizontal scale factor and the number b is called the vertical scale factor. A scale change of this type will not produce similar figures. To illustrate this, suppose the unit **circle**, a circle with radius 1 centered at the origin, is transformed with the dilation $S(x,y) = (3x,3y)$. The image will be another circle centered at the origin but with radius 3. Since both the pre-image and image are circles, they are similar. Now let the unit circle be transformed by the two-dimensional scale change $S(x,y) = (4x,3y)$. The result will be an **ellipse** with semi-major axis of length 4 and semi-minor axis of length 3. An ellipse is not similar to a circle, illustrating that $S(x,y) = (4x,3y)$ is not a similarity transformation.

With the exception of translations, all the transformations discussed above fall into the category of "linear" transformations, so called because they all have the general form $T(x,y) = (ax + by, cx + dy)$ where a, b, c, and d are **real numbers**. The forms $ax+by$ and $cx + dy$ have the form of the left side of the general equation for a straight line, $Ax+By=C$. For example, the transformation $S(x,y)=(4x,3y)$ can be written as $S(x,y) = (4x + 0y, 0x + 3y)$. Such transformations can be associated with 2 by 2 matrices with the numbers a and b in the first row and the numbers c and d in the second row. Representing transformations as matrices is advantageous for use in computer programs. Computer programmers often create animations using software that manipulates figures by means of transformations written in **matrix** form.

TRANSLATION

A translation is one of the three transformations that move a figure in the plane without changing its size or shape. (The other two are rotations and reflections.) In a translation, the figure is moved in a single direction without turning it or flipping it over.

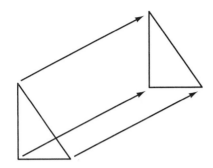

Figure 1.

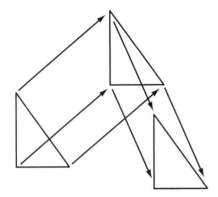

Figure 2.

A translation can, of course, be combined with the two other rigid motions (as transformations which preserve a figure's size and shape are called), and it can in particular be combined with another translation. The "product" of two translations is also a translation, as illustrated in Figure 2.

If a set of points is drawn on a coordinate plane, it is a simple matter to write **equations** which will connect a point (x,y) with its translated "image" (x¹.y¹). If a point has been moved a units to the right or left and b units up or down, a will be added to its x-coordinate and b to its y-coordinate.

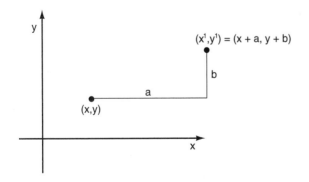

Figure 3.

(If a < 0, the **motion** will be to the left; and if b <0, down.) Therefore

$$x^1 = x + a \quad x = x^1 - a$$

$$y^1 = y + b \text{ or } y = y^1 - b$$

In these equations, the axes are fixed and the points are moved. If one wishes, the points can be kept fixed and the axes moved. This is called a **translation of axes**. If the axes are moved so that the new origin is at the former point (a,b), then the new coordinates, (x¹, y¹) of a point (x, y) will be (x - a, y - b).

If one has two translations:

$$x' = x = + 3 \quad y' = y - 6$$

$$x'' = x' - 2 \quad y'' = y' + 1$$

they can be combined into a single **transformation**. By substitution one has

$$x'' = x + 1 \quad y'' = y -5$$

which is another translation. This illustrates that the "product" of two translations is itself a translation, as claimed earlier.

The idea of a translation is a very common one in the practical world. Many machines are translational in their operation. The machinist who cranks the cutting-tool holder up and down the bed of the lathe, is "translating" it. The piston of an automobile engine is translated up and down in its **cylinder**. The chain of a bicycle is translated from one sprocket wheel to another as the cyclist pedals, and so on.

The bicycle chain is not only translated, it works because it has translational **symmetry**. After a translation of one link, it looks exactly as it did before. Because of this symmetry, it continues to fit over the teeth of the sprocket wheel (which itself has rotational symmetry) and to turn it.

One important use of translations is to simplify an equation which represents a set of points. The equation xy - 2x + 3y -13 = 0 can be written in factored form (x + 3)(y - 2) = 7. Then, letting x¹ = x + 3 and y¹ = y - 2, the equation is simply x¹y¹ = 7, which is a much simpler and more easily recognized form.

Such transformations are useful in drawing graphs where many points have to be plotted. The graph of x¹y¹ = 7 is a **hyperbola** whose branches lie entirely in the first and third quadrants with the axes as asymptotes. It is readily sketched. The graph of the original equation is also a hyperbola, but that fact may not be immediately apparent, and it will have points in all four quadrants. Many points may have to be plotted before the shape takes form.

If one has an equation of the form ax² + by² + cx + dy + e = 0 it is always possible to find a translation which will simplify it to an equation of the form ax² + by² + E = 0.

For example, the transformation x = x¹ - 2 and y = y¹ + 1 will transform x²+ 3y² + 4x - 6y - 2 = 0

into x² + 3y² - 9 = 0 which is recognizable as an **ellipse** with its center at the origin.

Transformations are particularly helpful in integrating **functions** such as integral(x + 5)⁴ dx because integralx⁴ dx is very easy to integrate, while the original is not. After the translated integral has been figured out, the result can be translated back, substituting x + 5 for x.

Translational symmetry is sometimes the result of the way in which things are made; it is sometimes the goal.

Newspapers, coming off a web press, have translational symmetry because the press prints the same page over and over again. Picket fences have translational symmetry because they are made from pickets all cut in the same shape. Ornamental borders, however, have translational symmetry because such symmetry adds to their attractiveness. The gardener could as easily **space** the plants irregularly, or use random varieties, as to make the border symmetric, but a symmetric border is often viewed as esthetically pleasing.

TRANSLATION OF AXES

In the plane, using Cartesian coordinates, a **circle** with radius 1 with center at the origin is described by the equation $x^2 + y^2 = 1$. Rather, this circle is the set of all points of the form (x, y) in the plane that satisfy the above equation. In order to describe a circle whose center is not at the origin, we can translate the axes. In other words, we move the origin to a new point say (x_0, y_0). So if (p, q) is a point with respect to the old coordinates, then it is $(p + x_0, q + y_0)$ with respect to the new coordinates. Thus the equation for the circle in the new coordinates is $(x + x_0)^2 + (y + y_0)^2 = 1$. In general, if a curve is described by the equation f(x, y) = c for some function f and some real constant c, then the curve is described in the new coordinates by the equation $f(x + x_0, y + y_0) = c$. **Translation** preserves the **distance** between points. In other words, the distance between point (p, q) are (r, s) is the same as the distance between the points $(p + x_0, q + y_0)$ and $(r + x_0, s + y_0)$. These ideas extend in a natural way to n-dimensional **space** for any positive integer n.

TRIANGLE

A triangle is a closed figure formed by connecting three non-collinear points, called vertices, by line segments to result in three sides and three interior angles (hence the word "triangle"). If the three angles all lie in the same plane, the triangle is called a plane, or Euclidean, triangle. If the three angles do not all lie in the same plane, the triangle is called a spherical, or curvilinear, triangle. The word triangle, by itself, is usually taken to mean a plane triangle, while a curvilinear triangle is stated as such. The triangle is the simplest polygon because it is the closed figure having the fewest number of angles and sides. The word triangle can be traced back to the Latin "triangulum," and to the neuter of "triangulus" (three-angled).

Some important facts and definitions pertaining to triangles are:

- Any one of the sides of a triangle may be considered the base.
- The perpendicular **distance** from the base to the opposite vertex is called the **altitude** of the triangle.
- A triangle's **area** is equal to one-half the product of the base and the corresponding altitude.

- The **perimeter** of a triangle is the sum of its three sides.
- Any triangle has three medians, which are defined to be the line segments joining the **midpoint** of a particular side to the opposite vertex.
- In Euclidean **geometry** the sum of the three angles of a triangle is equal to two right angles (180°).

Triangles may be classified by the characteristics that they possess. Some common classifications are:

- Acute triangle: each of its three interior angles is less than 90°;
- Equiangular triangle: all three interior angles are equal;
- Equilateral triangle: all three sides are equal in linear length (equiangular and equilateral triangles are equivalent);
- Isosceles triangle: two of its sides possess equal length;
- Oblique triangle: does not contain a right **angle** (90°);
- Obtuse triangle: possesses an angle between 90° and 180°;
- Right triangle: has one right angle (90°) and the side opposite it is called the **hypotenuse**;
- Scalene triangle: no two angles are equal.

Trigonometry is the mathematical study of triangles. When trigonometry is restricted to the plane it is called plane trigonometry. One of the aims of plane trigonometry is "solving the triangle." which means finding its unknown parts (sides and angles) from given data. Any triangle may be specified by six parts: three angles and the lengths of its three sides. If three of the six parts of a triangle are specified (where one side and two angles are given or two sides and one angle are given) then the rules of trigonometry can be used to determine the three unknown parts. For a right triangle, one angle (other than the right angle) and the length of a side must be known. Let A, B, and C be the angles of a right triangle and a, b, and c, respectively, be the sides opposite them. If C is the right angle then side c must be the hypotenuse; and the remaining parts of the triangle can then be found using the equation "A + B = 90°," the **Pythagorean theorem** "$a^2 + b^2 = c^2$," and the following trigonometric **functions** (which relate the two smaller angles of a right triangle to the **ratios** of particular sides):

- $\sin(A) = \cos(B) = a / c$,
- $\cos(A) = \sin(B) = b / c$,
- $\tan(A) = \cot(B) = a / b$,
- $\cot(A) = \tan(B) = b / a$,
- $\sec(A) = \csc(B) = c / b$,
- $\csc(A) = \sec(B) = c / a$.

For triangles that do not possess a right angle (called oblique), one may "solve the triangle" by first dividing it into two right triangles; then the resulting component triangles can be solved according to the procedures given above. The solution to all oblique triangles can also be found with one or more of the following three **equations**, plus the equation "180° = A + B + C," where a, b, and c are the sides opposite the angles A, B, and C (respectively):

- Law of sines: [a / sin(A)] = [b / sin(B)] = [c / sin(C)].
- Law of cosines:
  - $a^2 = b^2 + c^2 - 2bc \cos(A)$,
  - $b^2 = c^2 + a^2 - 2ca \cos(B)$,
  - $c^2 = a^2 + b^2 - 2ab \cos(C)$.
- Law of Tangents:
  - $[(a - b) / (a + b)] = \{\tan[1/2(A - B)]\} / \{\tan[1/2(A + B)]\}$,
  - $[(b - c) / (b + c)] = \{\tan[1/2(B - C)]\} / \{\tan[1/2(B + C)]\}$,
  - $[(a - b) / (a + b)] = \{\tan[1/2(A - B)]\} / \{\tan[1/2(A + B)]\}$.

These three laws can be used to solve any oblique triangle; that is, the unknown sides or angles can be found when:

- (Two angles and one side are given.) The third angle is found by "180° = A + B + C," and the two other sides are solved using the Law of Sines.
- (Two sides and one angle [opposite one of the known sides] are given, say a, c, and C.) Knowing from the Law of sines that sin(A) = (a / c) sin(C) (where (a / c) sin(C) ≤ 1), B = 180° - A - C, and b = [c sin(B)] / sin(C), several solutions are possible to solve the unknown side and two angles:
  - a = c, there is one solution (because A = C),
  - a<c, there is one solution (because A<C),
  - a>c,
    - a is so large that sin(A) ≤ 1 is not satisfied, thus no solution exists,
    - sin(A) = 1, so C is a right angle and there is one solution,
    - sin(A)<1, so angles "A_1" and "$A_2 = 180° - A_1$" can be calculated and there are two solutions.
- (Two sides and the included angle are given.) To solve for the two unknown angles, use the Law of tangents and "180° = A + B + C"; and to solve for the unknown side, use the Law of sines.
- (Three sides are given.) To solve for the three unknown angles, use the Law of cosines.

See also Angle; Euclidean geometries; Fundamental trigonometric functions; Line; Plane geometry; Plane surfaces; Polygon; Trigonometric equations; Trigonometric functions; Trigonometric tables

TRIGONOMETRIC FUNCTIONS

There are six basic trigonometric **functions** that are employed in the study of angles and angular relationships in planar and three-dimensional figures, also known as **trigonometry**. They are **sine** (abbreviated as sin x), **cosine** (abbreviated as cos x), tangent (abbreviated as tan x), cosecant (abbreviated as csc x), secant (abbreviated as sec x), and

cotangent (abbreviated as cot x). The inverses of these functions are denoted as $\sin^{-1} x$, $\cos^{-1} x$, $\tan^{-1} x$, $\csc^{-1} x$, $\sec^{-1} x$, and $\cot^{-1} x$. Note that this notation denotes inverse and not to the power of -1. In order to express a positive integer power of a trigonometric function the exponent is written directly after the name of the function such as $\cos^2 x = [\cos x]^2$. Trigonometric functions are also sometimes called circular functions because they are functions describing the horizontal and vertical positions of a point on a **circle** as a function of **angle**. These functions link **calculus** to **geometry**. In calculus all of the trigonometric functions are in terms of radians rather than any other angular measure.

Although the trigonometric functions can be defined algebraically in terms of complex exponentials they are more simply defined using a unit circle or a right **triangle**. A right triangle has three sides that are identified as **hypotenuse**, the side adjacent to a given angle, and the side opposite a given angle. The hypotenuse is the side of the triangle opposite the right angle. The trigonometric functions can be defined as **ratios** of the lengths of these three sides. The sine of an angle Θ is equal to the ratio of the length of the opposite side divided by the length of the hypotenuse. The cosine of an angle Θ is equal to the ratio of the length of the adjacent side divided by the length of the hypotenuse. The tangent of and angle Θ is defined as the ratio of the length of the opposite side divided by the length of the adjacent side. These ratios are the same for any right triangle with an angle Θ since when triangles have equal angles they are similar triangles.

The trigonometric functions are interrelated to each other. The ratio of sin Θ to cos Θ is equal to tan Θ. More complex relations are called **trigonometric identities** and are discussed in detail in another section. One such identity is the **Pythagorean theorem**, which gives us that $\cos^2 \Theta + \sin^2 \Theta = 1$. The other relations include csc Θ = 1/sin Θ, sec Θ = 1/cos Θ, and cot Θ = 1/tan Θ = cos Θ/sin Θ. There are some values for which the trigonometric functions are not defined. Tan Θ and sec Θ are not defined when the length of the adjacent side is equal to **zero**, at Θ = +/-90°, +/-450°,.... Also cos Θ and cot Θ are not defined when the length of the opposite side is equal to zero, at Θ = 0, +/-180°, +/-360°,.... The trigonometric functions sin and cos are periodic with period 2π or 360°. The maximum value attained by the cos or sin function is 1; the minimum value of -1. Both of these functions oscillate in a regular manner with amplitude of 1. Sin Θ is an odd function and so sin $-\Theta$ = -sin Θ and its graph is symmetric to the origin. Cos Θ is an even function and so cos $-\Theta$ = cos Θ and its graph is symmetric to the y-axis. The sin and cos graphs are exactly the same shape but horizontally shifted by 90°. The function tan is also periodic but it's period is 180°.

TRIGONOMETRIC IDENTITIES

Trigonometric identities are **equations** that describe relationships between the various trigonometric **functions**. The **trigonometric functions** are a set of functions that are employed in the study of angles and angular relationships in planar and three-dimensional figures. Development of the trigonometric identities was simultaneous with the development of the functions themselves. **Claudius Ptolemy**, one of the most influential Greek astronomers, was developing the identities in terms of chords of a **circle** or arcs as early as about 130. He showed that he knew one of the most important identities, $\sin(x + y) = \sin x \cos y + \cos x \sin y$ although instead of writing it in terms of sin and cos he used chords as they did in those times, very early in his career while studying astronomy.

The trigonometric identities are derived by expressing the **sine** or **cosine** of a sum of difference of angles in terms of the sines and cosines of the individual angles. The identities are true whenever they are meaningful, which distinguishes them from equations which are true only for particular values of x. Since all of the trigonometric functions can be defined in terms of sin and cos, the identities are often written involving these functions.

The most important trigonometric identity is $\sin^2 x + \cos^2 x = 1$. This identity has a name of its own and is called the Pythagorean identity. It is derived from the **Pythagorean theorem** and has been extended to the other trigonometric functions by dividing the original equation by $\cos^2 x$ and $\sin^2 x$ respectively to yield: $\tan^2 x + 1 = \sec^2 x$ and $1 + \cot^2 x = \csc^2 x$.

The next most important trigonometric identities involve the sum or difference of two different angles. The two most important of these identities are: $\sin(x + y) = \sin x \cos y + \cos x \sin y$ which was known by Ptolemy but in terms of chords and $\cos(x + y) = \cos x \cos y - \sin x \sin y$. The other two related trigonometric identities are: $\sin(x - y) = \sin x \cos y - \cos x \sin y$, and $\cos(x - y) = \cos x \cos y + \sin x \sin y$. From these **sets** of relations it is possible to derive the identities for $\sin(2x)$ and $\cos(2x)$ which are left to the reader.

The last two most important trigonometric identities that allow one to derive all of the other identities are: $\sin(-x) = -\sin x$ and $\cos(-x) = \cos x$. Using all of these identities it is possible to derive all of the other relationships between the trigonometric functions.

TRIGONOMETRIC TABLES

Trigonometric tables provide the numerical values for the fundamental trigonometric **functions** of angles, such as the **cosine** of 30° or the **sine** of 60°. The tables are usually arranged with the angles listed in degrees (or **sexagesimal numeration**) and radians. The intersection of a function name with the **angle** provides the numerical value.

Many sources provide a trigonometric table for every whole number angle from 0 to 90° (or 0 to π/2 radians). In these tables the angles from 0 to 45° are listed in the left-hand column and are used with the functions listed across the top row. The angles from 45 to 90° are listed in the right-hand column in descending order and are used with the func-

tions listed across the bottom row. Here is an excerpt from such a table:

Degrees	Radians	Sin	Cos	Tan	Cot	Sec	Csc		
0°	0.0000	0.0000	1.0000	0.0000	--------	1.0000	--------	1.7508	90°
1°	0.0175	0.0175	0.9998	0.0175	57.2900	1.0002	57.2987	1.5553	89°
2°	0.0349	0.0349	0.9994	0.0349	28.6363	1.0006	28.6537	1.5359	88°
3°	0.0524	0.0523	0.9986	0.0524	19.0811	1.0014	19.1037	1.5184	87°
etc.									etc.
30°	0.5236	0.5000	0.8660	0.5774	1.7321	1.1547	2.0000	1.0472	60°
etc.									etc.
45°	0.7854	0.7071	0.7071	1.0000	1.0000	1.4142	1.4142	0.7854	45°
		Cos	Sin	Cot	Tan	Csc	Sec	Radians	Degrees

From this table, the sine of 30° is 0.5000. Using the functions listed across the bottom row, the cosine of 60° also is 0.5000. In fact, this table arrangement is possible because of the mathematical relationship between certain functions for complementary angles (i.e., two angles that sum to 90°). Thus, $\tan 1° = \cot 89°$, $\sec 45° = \csc 45°$, etc.

Note that cot 0°, tan 90°, csc 0°, and sec 90° are not defined. This is because there are certain angles for which the **trigonometric functions** do not exist.

This same type of table can be used to find the trigonometric functions of any angle Θ with the help of a reference angle and the proper sign conventions.

Reference angles are calculated as follows:
- For $0° < \Theta < 90°$, Reference angle $= \Theta$
- For $90° < \Theta < 180°$, Reference angle $= 180° - \Theta$
- For $180° < \Theta < 270°$, Reference angle $= \Theta - 180°$
- For $270° < \Theta < 360°$, Reference angle $= 360° - \Theta$

The numerical value of a trigonometric function of angle Θ is equivalent to the absolute value of the same function of the reference angle. The sign of the numerical value is determined by the quadrant in which the terminal side of Θ falls.

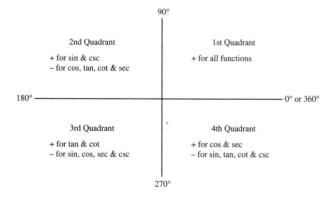

For example, to find cos 210°, the reference angle = 210° - 180° = 30°. From the table, cos 30° = 0.8660. Because the terminal side of a 210° angle falls in the 3rd quadrant, the sign is negative. Therefore, cos 210° = -.8660.

Likewise, to find sec 315°, the reference angle = 360° - 315° = 45°. From the table, sec 45° = 1.4142. Because the terminal side of a 315° angle falls in the 4th quadrant, the sign is positive. Therefore, sec 315° = 1.4142.

For negative angles (i.e., those that rotate clockwise, instead of counterclockwise), use the absolute value of the desired angle Θ to determine the reference angle. For example, to determine cos -120°, use Θ = 120°. The reference angle = 180°-120° = 60°. From the table, cos 60° = 0.5000. The terminal side of a -120° angle falls in the 3rd quadrant, therefore, cos -120° = -0.5000.

For angles greater than 360°, subtract multiples of 360° until Θ is less than 360°. For example, cos 765° is the same as cos 765°-360°-360° or cos 45°, which is 0.7071.

TRIGONOMETRY

Trigonometry is a branch of **applied mathematics** concerned with the relationship between angles and their sides and the calculations based on them. First developed as a branch of **geometry** focusing on triangles during the third century B.C., trigonometry was used extensively for astronomical measurements. The major trigonometric **functions**, including **sine**, **cosine**, and tangent, were first defined as **ratios** of sides in a right **triangle**. Since **trigonometric functions** are intrinsically related, they can be used to determine the dimensions of any triangle given limited information. In the eighteenth century, the definitions of trigonometric functions were broadened by being defined as points on a unit **circle**. This allowed the development of graphs of functions related to the angles they represent which were periodic. Today, using the periodic nature of trigonometric functions, mathematicians and scientists have developed mathematical models to predict many natural periodic phenomena.

The word trigonometry stems from the Greek words *trigonon*, which means triangle, and *metrein*, which means to measure. It began as a branch of geometry and was utilized extensively by early Greek mathematicians to determine unknown distances. The most notable examples are the use by **Aristarchus** (310-250 B.C.) to determine the **distance** to the Moon and Sun, and by **Eratosthenes** (c. 276-195 B.C.) to calculate the Earth's **circumference**. The general principles of trigonometry were formulated by the Greek astronomer, **Hipparchus** of Nicaea (active 162-127 B.C.), who is generally credited as the founder of trigonometry. His ideas were worked out by Ptolemy of Alexandria (c. 90-168 A.D.), who used them to develop the influential Ptolemaic theory of astronomy. Much of the information we know about the work of Hipparchus and Ptolemy comes from Ptolemy's compendium *The Almagest* written around 150.

Trigonometry was initially considered a field of the science of astronomy. It was later established as a separate branch of mathematics, largely through the work of the mathematicians **Johann Bernoulli** and **Leonhard Euler**.

Central to the study of trigonometry is the concept of an **angle**. An angle is defined as a geometric figure created by two lines drawn from the same point, known as the vertex. The lines are called the sides of an angle and their length is one defining characteristic of an angle. Another characteristic of an angle is its measurement or magnitude, which is determined by the amount of **rotation**, around the vertex, required to transpose one side on top of the other. If one side is rotated completely around the point, the distance travelled is known as a revolution and the path it traces is a circle.

Angle measurements are typically given in units of degrees or radians. The unit of degrees, invented by the ancient Babylonians, divides one revolution into 360° (degrees). Angles which are greater than 360° represent a magnitude greater than one revolution. **Radian** units, which relate angle size to the radius of the circle formed by one revolution, divide a revolution into 2π units. For most theoretical trigonometric work, the radian is the primary unit of angle measurement.

The principles of trigonometry were originally developed around the relationship between the sides of a triangle and its angles. The idea was that the unknown length of a side or size of an angle could be determined if the length or magnitude of some of the other sides or angles were known. Recall that a triangle is a geometric figure made up of three sides and three angles, whose sum is equal to 180°. The three points of a triangle, known as its vertices, are usually denoted by capital letters.

Triangles can be classified by the lengths of their sides or magnitude of their angles. Isosceles triangles have two equal sides and two congruent (equal) angles. Equilateral, or equiangular, triangles have three equal sides and angles. If no sides are equal, the triangle is a scalene triangle. All of the angles in an acute triangle are less than 90° and at least one of the angles in an obtuse triangle is greater than 90°. Triangles, such as these, which do not contain a 90° angle, are generally known as oblique triangles. Right triangles, the most important ones to trigonometry, are those which contain one 90°angle.

Triangles which have proportional sides and congruent angles are called similar triangles. The concept of similar triangles, one of the basic insights in trigonometry, allows us to determine the length of a side of one triangle if we know the length of certain sides of the other triangle. For example, if we wanted to know the height of a tree, we could use the idea of similar triangles to find it without actually having to measure it. Suppose a person is 6 ft (183 cm) tall and casts an 8 ft (2.44 m) long shadow. The tree, whose height is unknown, casts a shadow that is 20 ft (6.1 m) long. The triangles that could be drawn using the shadows and objects as sides are similar. Since the sides of similar triangles are proportional, the height of the tree is determined by setting up the mathematical equality: "height of tree" ÷ "length of tree's shadow" = "height of person" ÷ length of person's shadow" = x/20 = 6/8. By solving this equation, the height of the tree is found to be 15 ft (4.57 m).

The triangles used in the previous example were right triangles. During the development of trigonometry, the parts of a right triangle were given certain names. The longest side of the triangle, which is directly across from the right angle, is known as the **hypotenuse**. The sides which form the right angle are the legs of the triangle. For either acute angle in the triangle, the leg that forms the angle with the hypotenuse is known as the adjacent side. The side across from this angle is known as the opposite side. Typically, the length of each side is denoted by a lower case letter. In the diagram of triangle ABC, the length of the hypotenuse is indicated by c, the adjacent side is represented by b, and the opposite side by a. The angle of interest is usually represented by θ.

The ratios of the sides of a right triangle to each other are dependent on the magnitude of its acute angles. In mathematics, whenever one value depends on some other value, the relationship is known as a function. Therefore, the ratios in a right triangle are trigonometric functions of its acute angles. Since these relationships are of most importance in trigonometry, they are given special names. The ratio or number obtained by dividing the length of the opposite side by the hypotenuse is known as the sine of the angle θ (abbreviated sin θ). The ratio of the adjacent side to the hypotenuse is called the cosine of the angle θ (abbreviated cos θ). Finally, the ratio of the opposite side to the adjacent side is called the tangent of θ, or tan θ. In the triangle ABC, the trigonometric functions are represented by the following **equations**: sin θ = a/c, cos θ = b/c, tan θ = a/b, sin θ/cos θ.

These ratios represent the fundamental functions of trigonometry and should be committed to memory. Many mnemonic devices have been developed to help people remember the names of the functions and the ratios they represent. One of the easiest is the phrase "SOH, CAH, TOA." This means: sine is the opposite over the hypotenuse, cosine is adjacent over hypotenuse, and tangent is opposite over adjacent.

In addition to the three fundamental functions, three reciprocal functions are also defined. The inverse of sin θ, or 1/sin θ, is known as the secant of the angle or sec θ. The inverse of the cos θ is the cosecant or csc θ. Finally, the inverse of the tangent is called the cotangent of cot θ. These functions are typically used in special instances.

The values of the trigonometric functions can be found in various ways. They can often be looked up in tables which have been compiled over the years. They can also be determined by using **infinite series** formulas. Conveniently, most calculators and computers have the values of trigonometric functions preprogrammed in.

One immediate application for trigonometric functions is the simple determination of the dimensions of a right triangle, also known as the solution of a triangle, when only a few are known. For example, if the sides of a right triangle are known, then the magnitude of both acute angles can be found. Suppose we have a right triangle whose sides are 2 in (5 cm) and 4.7 in (12 cm), and whose hypotenuse is 5.1 in (13 cm). The unknown angles could be found by using any trigonometric function. Since the sine of one of the angles is equal to the length of the opposite side divided by the hypotenuse, this angle can be determined. The sine of one angle is 5/13, or 0.385. With the help of a trigonometric function table or **calculator**, it will be found that the angle which has a sine of 0.385 is 22.6°. Using the fact that the sum of the angles in a triangle is 180°, we can establish that the other angle is 180° - 90° - 22.6° = 67.4°.

In addition to solving a right triangle, trigonometric functions can also be used in the determination of the **area** when given only limited information. The standard method of finding the area of a triangle is by using the formula, area = 1/2b (base) × h(altitude). Often, the **altitude** of a triangle is not known, but the sides and an angle are known. Using the side-angle-side (SAS) **theorem**, the formula for the area of a triangle then becomes, area = 1/2 (one side) × (another side) × (sine of the included angle). For a triangle with sides of 5 cm and 3 cm respectively and an included angle of 60°, the area of the triangle would be equal to 1/2 × 5 × 3 × sin 60° = 13 cm².

The formula for the area of a triangle leads to an important concept in trigonometry known as the Law of sines which says that for any triangle, the sine of each angle is proportional to the opposite its opposite side, symbolically written: in triangle ABC, sin A ÷ a = sin B ÷ b = sin C ÷ c.

Using the Law of sines, we can solve any triangle if we know the length of one side and magnitude of two angles, or two sides and one angle. Suppose we have a triangle with angles of 45° and 70°, and an included side of 15.7 in (40 cm). The third angle is found to be 180° - 45° - 70° = 65°. The unknown sides, x and y, are found with the Law of Sines because sin 45 ÷ x = sin 70 ÷ y = sin 65 ÷ 40.

The lengths of the unknown sides are then x = 12.29 in (31.2 cm) and y = 16.35 in (41.5 cm).

The Law of sines can not be used to solve a triangle unless at least one angle is known. However, a triangle can be solved if only the sides are known by using the Law of cosines which is stated in triangle ABC, c² = a² + b² - 2ab cos C, or can be written

$$\cos C = (a^2 + b^2 - c^2) \div 2ab$$

which is more convenient when using only the sides to solve a triangle. As an example, consider a triangle with sides equal to 2 in, 3.5 in, and 3.9 in (5 cm, 9 cm, and 10 cm). The cosine of one angle would be equal to (5² + 9² - 10²)/(2 × 59) = 0.067, which corresponds to the angle 86.2°. Similarly, the other two angles are found to be 29.9° and 63.9°.

In addition to the reciprocal relationships of certain trigonometric functions, two other types of relationships exist. These relationships, known as **trigonometric identities**, include cofunctional relationships and Pythagorean relationships. Cofunctional relationships relate functions by their complementary angles. Pythagorean relationships relate functions by application of the **Pythagorean theorem**.

The sine and cosine of an angle are considered cofunctions, as are the secant and cosecant, and the tangent and cotangent.

The Pythagorean theorem states that the sum of the squares of the sides of a right triangle is equal to the **square** of the hypotenuse. For a triangle with sides of x and y and a hypotenuse of z, the equation for the Pythagorean theorem is x² + y² = z². Applying this theorem to the trigonometric functions of an angle, we find that sin² θ + cos² θ = 1. Similarly, 1 + tan² θ = sec² θ and 1 + cot² θ = csc² θ. The terms such as sin²θ or tan²θ traditionally have meant (sinθ) × (sinθ) or (tanθ) × (tanθ).

In some instances, it is desirable to know the trigonometric function of the sum or difference of two angles. If we have two unknown angles, θ and φ, then sin (θ + φ) is equal to sinθcosφ + cosθsinφ. In a similar manner, their difference, sin(θ-φ) is sinθcosφ - cosθsinφ. Equations for determining the sum or differences of the cosine and tangent also exist and can be stated as follows:

cos(θ ± φ) = cosθcosφ ± sinθsinφ tan (θ ± φ) = (tan θ ± tan φ)/(1 ± tanθtanφ)

These relationships can be used to develop formulas for double angles and half angles. Therefore, the sin 2θ = 2sinθ–cosθ and cos 2θ = 2cos²θ - 1 which could also be written cos θ2 = 1 - 2sin²θ.

For hundreds of years, trigonometry was only considered useful for determining sides and angles of a triangle. However, when mathematicians developed more general definitions for sine, cosine and tangent, trigonometry became much more important in mathematics and science alike. The general definitions for the trigonometric functions were developed by considering these values as points on a unit circle.

A unit circle is one which has a radius of one unit which means x² + y² = 1. If we consider the circle to represent the rotation of a side of an angle, then the trigonometric functions can be defined by the x and y coordinates of the point of rotation. For example, coordinates of point P(x,y) can be used to define a right triangle with a hypotenuse of length r. The trigonometric functions could then be represented by these equations: sin θ = y/r; cos θ = x/r; tan θ = y/x.

With the trigonometric functions defined as such, a graph of each can be developed by plotting its value versus the magnitude of the angle it represents.

Since the value for x and y can never be greater than one on a unit circle, the **range** for the sine and cosine graphs is between 1 and -1. The magnitude of an angle can be any real number, so the **domain** of the graphs is all **real numbers**. (Angles which are greater than 360° or 2π radians represent an angle with more than one revolution of rotation). The sine and cosine graphs are periodic because they repeat their values, or have a period, every 360° or 2π radians. They also have an amplitude of one which is defined as half the difference between the maximum (1) and minimum (-1) values.

Graphs of the other trigonometric functions are possible. Of these, the most important is the graph of the tangent function. Like the sine and cosine graphs, the tangent function is periodic, but it has a period of 180° or π radians. Since the tangent is equal to y/x, its range is -∞ to ∞ and its amplitude is ∞.

The periodicity of trigonometric functions is more important to modern trigonometry than the ratios they represent. Mathematicians and scientists are now able to describe many types of natural phenomena which reoccur periodically with trigonometric functions. For example, the times of sunsets, sunrises, and comets can all be calculated thanks to trigonometric functions. Also, they can be used to describe seasonal temperature changes, the movement of waves in the ocean, and even the quality of a musical sound.

TRISECTING AN ANGLE

Trisecting an **angle** is one of the three classical problems of Greek **geometry**, together with doubling the cube and squaring the **circle**. Although it is one of the oldest problems of mathematics, it is also one of the most misunderstood. The problem of trisecting an angle has been proven to be impossible, but mathematicians daily receive "proofs" from amateur mathematicians who believe they have found a way to trisect the angle.

The original problem considered by Greek mathematicians was to find a way to trisect any given angle using only "plane" methods, that is, using only two pieces of equipment: a compass and an unmarked straightedge. Most of the people who believe they have found a way to trisect the angle have inadvertently used the straightedge for measuring; often their error is buried so deep that it is difficult to detect.

It is important to note that there are several angles that can be trisected. For example, it is simple to trisect a right angle. A 27° angle can also be trisected. But mathematicians have proven that it is impossible to trisect a 60° angle, for example. And if it is impossible to trisect a 60° angle, then there can be no general angle trisection method.

The ancient Greeks were probably interested in the trisection problem because dividing angles is necessary for constructing regular **polygons**. It is impossible to trisect a 120° angle, and so it is impossible to construct a nine-sided regular polygon, since such a polygon has 40° exterior angles.

It is not known when the angle trisection problem was first considered. It was known to Hippocrates in the fifth century B.C., and it was later studied by Archimedes. Hippocrates was familiar with the following "method" for trisecting an angle. Let CAB be an angle to be trisected. Drop a perpendicular from C to AB, hitting AB at D. Draw a line through C perpendicular to CD, to form a line segment CE; let F be the point of intersection of the lines AE and CD. Choose the point E so that EF=2AC. Then it is a simple matter to show that the line AE trisects angle CAB. Why is this not a solution to the problem of the Greeks? Because the step in which the point E is chosen cannot be executed precisely with only a straightedge and compass. On the other hand, this method is quite useful for practical purposes.

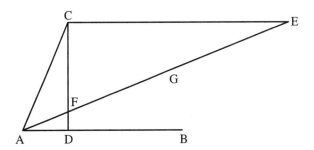

The **proof** that there is no straightedge-and-compass method to trisect the angle did not come until the 1837, when Pierre Wantzel gave a proof. The idea of the proof was based on a surprising source: **algebra**. Mathematicians had begun asking the question, if you start with a line segment that **sets** a unit of length, what other lengths can be constructed using straightedge and compass? Since these lengths are numbers, they could ask, what are the algebraic properties of all the constructible lengths? One important discovery they made was that constructible numbers cannot be the solutions of any cubic polynomial **equations**, like $8x^3-6x-1=0$, that satisfy a condition called irreducibility.

Since 60° is a constructible angle, if there were a trisection method then it would also be possible to construct a 20° angle. But 20° angles are not constructible! It is easy to show that if an angle is constructible, then so is its **cosine**. But cos(20°) is the solution of an irreducible cubic polynomial equation (in fact, it is the cubic equation in the previous paragraph). Therefore cos(20°) is not a constructible length, so 20° is not a constructible angle. Hence there is no method for trisecting an angle using straightedge and compass.

See also Squaring the circle

TRISTRAM SHANDY PARADOX

The Tristram Shandy paradox has its origins in an eighteenth century Lawrence Sterne novel in which the narrator, Tristram Shandy, is attempting to write his autobiography. However, it takes him one year to record the events of a single day of his life. Tristram laments that, at this rate, he will never finish. **Bertrand Russell**, the twentieth century philosopher, argued that if Tristram Shandy were immortal he would be able to finish his autobiography. Russell's assertion that Tristram would be able to complete this seemingly impossible task is the source of the Tristram Shandy paradox. It is also the source of debate among mathematicians, philosophers, and theologians.

There are two key components in the Tristram Shandy paradox—the number days that Tristram lives and the number of days required to write about those days. The sum of those two quantities yields the total number of days Tristram needs to complete his autobiography. If Tristram were indeed immortal, the first quantity, the number of days in his life, would be infinite. If that were the case, the second quantity, the number of days it takes him to write about his life, would also be infinite. The sum of those two quantities is infinite, since adding two infinite quantities results in an equal infinite quantity. This is represented symbolically by the following equation: $\infty + \infty = \infty$. Thus, Tristram needs an infinite number of days to finish his autobiography. If he were immortal, he would have an infinite number of days in which to write. Therefore, Russell argued, an immortal Tristram Shandy could finish his autobiography, since the number of days in his life (which is infinite) is equivalent to the number of days required to write about his life (which is also infinite).

Russell's argument is not universally accepted, however. Many of his critics contend that Tristram Shandy could not possibly finish his autobiography—even if he were immortal. Again, it takes Tristram one year to record the events of one day of his life. Thus, each day that Tristram lives adds 365 days to the time needed to complete his task. This causes him to fall

another year behind with each passing day. As a result, the amount of time needed for Tristram to write his autobiography is increasing faster than the amount of time he actually has in which to write. Granting immortality would not allow him to complete his task, according to critics of Russell's argument. It would merely cause him to fall infinitely far behind.

The two arguments outlined above are diametrically opposed. However, each argument is internally consistent. That is, each argument's conclusion can be logically drawn from its given set of assumptions. Though it may seem impossible for two opposing conclusions to both be considered logically valid, that is the case with the Tristram Shandy paradox. That each of the two arguments outlined above is internally consistent does not mean they are both correct. In fact, neither argument is universally accepted, due to the uncertain nature of **infinity** itself.

There has been debate throughout history as to whether infinity is a reality or an idea. **Aristotle**, the ancient Greek philosopher, introduced the terms *actual infinite* and *potential infinite* in an attempt to distinguish between the two. An actual infinite suggests that infinity is a quantity. As such, it can be manipulated, as can any finite mathematical quantity. A potential infinite regards infinity as a theoretical value that can be approached but never attained. Thus, it cannot be manipulated in the same manner as finite values. Russell's argument, that it is theoretically possible for Tristram Shandy to complete his autobiography, depends on the existence of an actual infinite. This view of infinity allows him to argue that the sum of two infinite values is an equal infinite value. Hence, the equation $\infty + \infty = \infty$. A potential infinite does not allow this calculation. The notion of a potential infinite holds that infinity is a process rather than a value. Therefore, it is not possible to manipulate infinite values in the same manner as finite values. This definition of infinity is used in the refutation of Russell's argument.

In the context of the Tristram Shandy paradox, discussions of infinity lead to discussions of time. Granting Tristram immortality implies that time can be extended infinitely forward. When discussing this paradox, some argue that Tristram could also have an infinite past. This implies that time could be extended infinitely backward as well. This is of natural interest in the fields of **mathematics and philosophy**. This, perhaps surprisingly, also has implications in another field—theology. An infinite past implies a universe with no beginning and, hence, no point of creation. Theists and atheists alike have cited the Tristram Shandy paradox when discussing this very issue.

The problem facing the character Tristram Shandy can be simply stated. This apparent simplicity belies the amount of debate the problem has inspired. It is an example of the counterintuitive results that can occur when discussing infinity. The Tristram Shandy paradox is also a wonderful example of the interconnectedness of academic disciplines. Though it arose from a work of literature, it has since become important in mathematics, philosophy, and theology. It has been cited in discussions pertaining to the meaning of infinity, the nature of time, and the origins of the universe.

See also Hilbert's paradox; Zeno's paradox

TRUTH

In their search for truth, many of the great philosophers have looked to mathematics as a paradigm for how such a search should be undertaken. The reason is that mathematicians have a very precise notion of what truth is within their discipline. A proposition in mathematics is true if and only if it has been rigorously proved according to the very stringent laws of logic. Not everything can be proved, of course; we must start somewhere by making a few definitions and assumptions, but logicians and mathematicians keep these to a bare minimum. Moreover, they assume only those things that any reasonable person would agree to. Beyond that, deductive mathematics is essentially about seeking the truth, or falsity, of **propositions** by **proof**. In no other discipline is the standard of truth so high. No proposition, no matter how obvious, can be accepted as true until it has been proved. A perfect example of this is **Fermat's last theorem**.

The French mathematician **Pierre de Fermat** had written in the margin of a book that he had a marvelous proof that for all **integers** greater than 2, the equation $a^n + b^n = c^n$ has no integer solutions for a, b, and c. Unfortunately, Fermat said that his proof would not fit in the margin of the book. No one ever discovered Fermat's alleged proof and for the next 350 years, some of the greatest mathematicians of all time tried and failed to prove Fermat's last **theorem**. Nevertheless, some did show that the theorem was true in specific cases. By 1839, it had been proved for n=3,4,5, and 7, so the evidence for the truth of the theorem was starting to build. By 1970, Fermat's conjecture had been established for all $n < 4,003$ by brute force with computers. In a court of law, such evidence would be considered proof beyond a reasonable doubt. If a scientist tested a hypothesis 4,002 times and had it always confirmed, she would proclaim it a law of nature. But, unlike in the courtroom, "evidence" is not good enough in mathematics. Unlike in the experimental sciences, a finite number, however large, of confirmed hypotheses is not sufficient to make a claim for truth in mathematics. There must be a proof. The 1970s and 1980s saw the advent of the supercomputer and the confirmation of the Fermat hypothesis for n up to 4,000,000, but there was still no proof of the general case. Fermat's last theorem remained only a conjecture. In the late 80s, unbeknownst to anyone else, a quiet, reserved Princeton University mathematics professor named **Andrew Wiles** began a journey to immortality. Using only pen, paper, and his mind, Wiles spent the next seven years of his life working on a problem that had vexed him since childhood—Fermat's last theorem. Using existing mathematics and mathematics that he created especially for this problem, Wiles completed the proof of Fermat's last theorem in 1994. His proof could not have been Fermat's proof, although Wiles could also claim that his proof would not fit in the margin—it was 200 pages long and had to be checked and rechecked by other mathematicians before it was proclaimed a valid proof. Fermat's last theorem is now considered to be one of the truths of mathematics. The point here is that by 1990, there were most likely no mathematicians who did not believe that Fermat's last theorem was true; yet not a single one would proclaim it to be true until a correct proof had been given. This is the standard for truth in mathematics.

Given that mathematicians know clearly what truth is for their discipline, does mathematical truth have anything to do with truth outside of mathematics? Does the fact that Fermat's last theorem is true, have any relevance to anything in the physical world? Does mathematical truth imply truth about events in the real world? Physicists would most likely answer the last question in the affirmative, because it is difficult, if not impossible, to do physics without mathematics. All of the laws of physics are stated in the language of mathematics. The spectacular success of the **calculus** as a tool to study the physical world, suggests the possibility that truth in mathematics is related to truth in nature. On the other hand, many pure mathematicians claim that their work is only about mathematics, that the theorems they prove are true as a result of careful adherence to the laws of logic which have nothing to do with objects or events in nature. One school of mathematicians, the formalists, claim that the objects of mathematical discourse are nothing more than meaningless symbols which are moved about and strung together according to accepted logical **inference** patterns. They see mathematics as a "game" to be played by the rules of **symbolic logic**. In this view, mathematics is not "about" anything outside the game itself and the great usefulness of mathematics in physics is merely a fortuitous coincidence.

Another group of mathematicians, called intuitionists, say that the objects of mathematical study are ideas or "intuitions" in the minds of individual mathematicians. Intuitionists are notable for denying the so-called "Law of Excluded Middle," a fundamental rule of standard logic, which states that any statement is either true or false—there is no middle ground. Intuitionists say that there is a middle ground, namely, the set of propositions which are either as yet undecided or perhaps undecidable. Under this view, Fermat's last theorem is neither true nor false until a correct proof is presented, and true intuitionists cannot accept Andrew Wiles' proof because it is an indirect proof. Wiles' assumed that Fermat's last theorem was false and showed that this led to a contradiction of a theorem which had already been proved. Indirect proof, a staple of traditional mathematics, is based upon the Law of Contraposition, which states that the propositions "If p, then q" and "If not q, then not p" are equivalent. Thus under standard logic, if either of these propositions is true, so is the other one. The problem for the intuitionists is that the Law of contraposition is equivalent to the Law of excluded middle, hence indirect proof is outlawed under intuitionist logic.

It is probably the case that most mathematicians are neither formalists nor intuitionists. Most tend to regard themselves as realists. Realism in mathematics is derived from Platonism, named for the Greek philosopher **Plato**, who believed that the objects of the physical world were mere "shadows" of the real objects, which he called "Forms." So, according to this view, an earthly man is only a shadow, a poor copy, of the ideal man. For the realist mathematician, the objects of mathematics are much like Plato's Forms—they have an independent existence, and physical entities that are modeled after them are mere approximations. Thus, what we usually call the real world, the world studied by the physicist, is only a shadow of the true reality, which is mathematics. This allows the realist mathematician to say that whenever the physicist's experiments do not agree with the mathematics, it is because the physical universe can only be an approximation to the real universe of mathematics. The world studied by physics is rampant with experimental error, but no such error can occur in the realm of mathematics. There are no experiments in mathematics, only pure forms following the dictates of pure logic, which for the realist is the only realm in which it is possible to find truth with complete certainty.

In this discussion of truth in mathematics, it is only fitting to give the last word to the great mathematician and logician, **Bertrand Russell**. Russell was a member of yet another school of mathematicians called logicists because they claim that mathematics is actually derived from pure logic. Like the formalists, they see mathematics as the correct manipulation of symbols according to logical principles; but, unlike the formalists, they believe that these symbols and rules of inference have an independent existence that is part of the fabric of the universe. Mathematics is more than just a game for the logicists. Nevertheless, logicists say that within the realm of mathematics, neither the symbols nor the rules of inference have an interpretation related to the natural world. Any interpretation or application of mathematics is not mathematics. Russell and **Alfred North Whitehead** wrote a massive three-volume work called *Principia Mathematica* claiming to derive all of mathematics from pure logic. They and most of the mathematical community believed that they had succeeded until 1931, when the brilliant Austrian logician **Kurt Gödel** (1906-1978) published a paper entitled "On Formally Undecidable Propositions of *Principia Mathematica* and Related Systems." In this paper, one of the most important of the twentieth Century, Gödel showed that in any axiomatic system based on **arithmetic**, there will always be propositions that cannot be proved within the system. This came to be known as "Gödel's undecidability theorem" or "Gödel's incompleteness theorem." The theorem showed that such systems can be either consistent or complete, but not both. This shattered Russell's attempts to show that *Principia Mathematica* was both consistent and complete. Furthermore, it called into question the very concept of mathematical truth, leaving Russell to say that "mathematics is that subject in which we never know what we are talking about nor whether what we are saying is true."

TRUTH FUNCTION

In the language of mathematics and logic there are essentially two types of statements. There are basic **propositions** (abbreviated here by lower-case, italicized letters, e.g., p, q, etc.), like "the **triangle** ABC is equilateral," or "the integer a is odd." Then there are complex propositions which are constructed from the basic propositions by connectives: **conjunction** ("p and q"), **disjunction** ("p or q"), **negation** (not-p), and **implica-**

tion ("if p then q). These connectives are **truth-functional**, meaning that the truth-value of the complex sentence is a function of the truth-values of the connected basic sentences. Each connective can be considered a **truth function**.

Truth functions are formally defined. Like any other function, a truth function is associated with two **sets**, a **domain** set and a **range** set. A function is a rule that associates each unique object in the first set with exactly one object in the second. In the case of truth functions, the first set is the set of finite sequences of truth-values; the second is simply the set of truth-values.

Usually, systems of logic only permit two possible truth-values: **true** (T) and **false** (F). Such systems are known as **binary logics**. In some philosophical circles, other truth-values are permitted; for instance "unknowable" is sometimes considered a truth-value in its own right.

A generic binary truth function would take a sequence of Ts and Fs (e.g., TTFTFFTTT) and assign to it a value of either T or F. For example, let v be a truth function, where: $v(\text{TTFTFFTTT}) = \text{T}$.

This **definition** of the truth function may seem obscure, but it is actually just an abstract formalization of a pattern of reasoning that we feel quite comfortable with. Consider negation. Suppose we have a set of propositions p, q, etc., that can be made about the world. For example, sentences of the sort "The sky is blue," "Jack loves Jill," or "Elvis is alive." Each of these can be assigned a truth-value, T or F, depending of course on whether or not these sentences are true or false. Suppose p is true (T). Then we quite naturally reason that its negation (not-p) must be false (F). If "the sky is blue" is a true statement, then "the sky is not blue" is a false statement. Conversely, if p is false, then not-p is true.

This reasoning is mathematically modeled by the truth function. Recall our definition. Consider the following truth function "n." It tells us what to do with a one-place sequence of truth-values (i.e., T or F). Whenever you feed either a T or an F into the n function, it spits out the opposite of what you put in. So

- $n(\text{T}) = \text{F}$, and
- $n(\text{F}) = \text{T}$.

In other words, if we put the truth-value of a sentence into the function n, we get the truth-value of the negation of the sentence! This is what we do when we reason that the negation of a true sentence must be false. Essentially, we apply the n function - even though we do not think about it in formal mathematical terms.

Like negation, the other logical connectives (**conjunction**, disjunction, **implication**, etc.) can also be understood as truth functions. The conjunction of two propositions (e.g., p **and** q) is true whenever both p and q are true; otherwise it is false. Think about the sentence: "The sky is blue and Elvis is alive." Clearly this sentence is true only if the sky really is blue and Elvis really is alive. If the sky is not blue, or if Elvis is dead, then the sentence, as a whole, is false. The corresponding conjunction truth function (call it c) takes two-place sequences of T's and F's (i.e., TT, TF, FT, FF) and converts them into either true or false according to the following plan:

- $c(TT) = T$
- $c(TF) = F$
- $c(FT) = F$
- $c(FF) = F$

When we figure out the truth-value of a conjunction from the truth-values of its constituent parts (**conjuncts**), we are effectively applying the c function. Analogously, implication disjunction are also representable by truth functions. In fact, in formal logic, it makes sense to simply define the meaning of the connectives in terms of their truth functions.

Complex compound propositions can also consist of basic propositions joined together by multiple connectives ("The sky is blue and Elvis is alive, or the sky is beige."). We can calculate the truth-value of a compound proposition from the truth-values of the basic propositions using truth functions. These complex functions would be **compositions** of many "simple" truth functions. The domain set of these functions must consist of longer truth-value sequences. For every basic proposition there is one truth-value, and so one letter in the sequence.

The truth function is closely related to the **truth table**. Truth tables are schematic representations of truth functions.

TRUTH TABLE

A **truth** table is a table used to establish the meaning of a logical connective as well as determining the validity of an argument. It is written as a two-dimensional array with $n + 1$ columns where n corresponds to the number of possible inputs. The last column, $n + 1$, is the column associated with the operation being performed. The number of rows in a truth table is dependent upon the type of statements given, either simple statements or compound statements. Truth tables are used in everything from binary logic to logic circuit tables in conjunction with Boolean operators.

During the late 1800s formal logic attempted to devise a complete, consistent formulation of mathematics. This formulation would be such that **propositions** could be formally stated and proved using a small number of symbols. Alfred North Whitehead's and Bertrand Russell's *Principia Mathematica*, published in 1925, showed that the problems with formal logic were too great to be overcome and that such a formulation was impossible. In lieu of formal logic a very simple form of logic was developed that relied on the study of truth tables and digital **logic circuits**. In the early 1900s Emil Leon Post published a paper on truth-table methods that introduced the concepts of completeness and consistency. Although Post, in this paper, attributed these methods to C. Keyser, it was previously accepted that C. Peirce and E. Schröder had formulated them. The studies of truth tables involve one or more outputs that depend on a combination of **logical symbols** and the input values. In such a formulation values at each step can take on values of only true or false. A useful principle for

the **analysis** of truth tables is de Morgan's duality law. This law states that for every proposition involving logical **addition** and **multiplication** (logical operators for "or" and "and"), there are comparable statements that include the words addition and multiplication.

To construct a truth table, the required components need to be fully understood. A statement is a declaration whose validity can be determined to be true or false. In two-valued logic, the truth-value of a statement is either T if it is true or F if it is false. So for example, the statement "10 - 8 = 2" has a truth value of T. Compound statements are simple statements whose truth value is altered by one or more logical operators. The truth-value of a compound statement is dependent in a well-defined manner upon the truth-values of its simple components. Below is an example of a truth table that contains two simple statements, P and Q, and a conjunction $\wedge$ (the logical operator for "and") of those two statements:

P	Q	$P \wedge Q$
T	T	T
F	T	F
T	F	F
F	F	F

As the truth table illustrates, for the conjunction of P and Q to be true, both simple statements have to have a truth-value of true. It is clear to see that truth tables can become very complicated when dealing with multiple statements. One of the most difficult parts in constructing a truth table is remembering the truth tables for different logical operators or connectives. Practically, truth tables can be used as **proof** tools only when there are no more than four different statements and the type of logic used is a two-valued logic, that is, that there are only two possible values: true or false.

TURING, ALAN (1912-1954)

English algebraist and logician

Alan Turing is recognized as a pioneer in computer theory. His classic 1936 paper, "On Computable Numbers, with an Application to the Entscheidungs Problem," detailed a machine that served as a **model** for the first working computers. During World War II, Turing took part in the top-secret ULTRA project and helped decipher German military codes. During this same time, Turing conducted groundbreaking work that led to the first operational digital electronic computers. Another notable paper was published in 1950 and offered what became known as the "Turing Test" to determine if a machine possessed intelligence.

Alan Mathison Turing was born on June 23, 1912, in Paddington, England, to Julius Mathison Turing and Ethel Sara Stoney. Turing's father served in the British civil service in India, and his wife generally accompanied him. Thus, for the majority of their childhoods, Alan and his older brother, John, saw very little of their parents. While in elementary school, the young Turing boys were raised by a retired military couple, the Wards. At the age of 13, Turing entered Sherbourne school, a boys' boarding school in Dorset. His record at Sherbourne was not generally outstanding; he was later remembered as untidy and disinterested in scholastic learning. He did, however, distinguish himself in mathematics and science, showing a particular facility for **calculus**. Turing also developed an interest in competitive running while at Sherbourne.

Turing twice failed to gain entry to Trinity College in Cambridge, but was accepted on scholarship at King's College (also in Cambridge). He graduated in 1934 with a master's degree in mathematics. In 1936 Turing produced his first, and perhaps greatest, work. His paper "On Computable Numbers, with an Application to the Entscheidungs Problem," answered a logical problem staged by German mathematician **David Hilbert**. The question involved the completeness of logic—whether all mathematical problems could, in principle, be solved. Turing's paper, presented in 1937 to the London Mathematical Society, proved that some could not be solved. Turing's paper also contained a footnote describing a theoretical automatic machine, which came to be known as the **Turing machine**, that could solve any mathematical problem—provided it was give the proper algorithms, or problem-solving **equations** or instructions. Although it may not have been Turing's intent at the time, his Turing machine defined the modern computer.

After graduating from Cambridge, Turing was invited to spend a year in the United States studying at Princeton University. He returned to Princeton for a second year—on a Proctor Fellowship—to finish his doctorate. While there, he worked on the subject of computability with **Alonso Church** and other mathematicians. Turing and his associates worked with binary numbers (1 and 0) and Boolean **algebra**, developed by **George Boole**, to develop a system of equations called logic gates. These logic gates were useful for producing problem-solving algorithms such as would be needed by an automatic computing machine. From the initial paper exercise, it was a simple matter to develop logic gates into electrical hardware, using relays and switches, which could—theoretically, and in huge quantities—actually perform the work of a computing machine. As a sideline, Turing put together the first three or four stages of an electric multiplier, using relays he constructed himself. After receiving his doctorate, Turing had an opportunity to remain at Princeton, but decided to accept a Cambridge fellowship instead. He returned to England in 1938.

Cryptology, the making and breaking of coded messages, was greatly advanced in England after World War I. The German high command, however, had modified a device called the Enigma machine that mechanically enciphered messages. The English found little success in defeating this method. The original Enigma machine was not new, or even secret; a basic Enigma machine had been in operation for several years, mostly used to produce commercial codes. The Germans' alterations, though, greatly increased the number of possible letter combinations in a message. The Allies were able to duplicate the modifications, but it was a continual cat-

and-mouse game; each time Allied analysts figured out a message, the Germans' changes made all of their work useless.

In the fall of 1939, Turing found himself in a top-secret installation in Bletchley, where he played a critical role in the development of a machine that deciphered the Enigma's messages by testing key codes until it found the correct combinations. This substitution method was uncomplicated, but impractical to apply because possible combinations could range into the tens of millions. Here Turing was able to put his experience at Princeton to good use; no one else had bridged the gap between abstract logic theory and electric hardware as he had with his electric multiplier. Turing helped construct relay-driven decoders (which were called Bombes, after the ticking noise of the relays) that shortened the code-breaking time from weeks to hours. The Bombes helped uncover German movements, particularly the U-boat war in the Atlantic, for almost two years. Eventually, however, the Germans changed their codes and the new level of complexity was too high to be solved practically by electrical decoders. British scientists agreed that although a Bombe of sufficient size could be made for further deciphering work, the machine would be slow and impractical.

Yet other advances would prove advantageous for the decoding machine. Vacuum tubes used as switches (the British called them thermiotic valves) used no moving parts and were a thousand times faster than electrical relays. A decoder made with tubes could do in minutes what it took a Bombe several hours to accomplish. Thus work began on a device which was later named Collosus. Based on the same theoretical principles as earlier Bombes, Collosus was the first operational digital electronic computer. It used 1800 vacuum tubes, proving the practicality of this approach. Much information concerning Collosus remained classified by the British government in the early 1990s. Some claimed that Turing supervised the construction of the first Collosus.

Many stories were circulated about Turing during the war; mostly surrounding his eccentricity. Andrew Hodges noted in his book *Alan Turing, The Enigma,* "With holes in his sports jacket, shiny grey flannel trousers held up with an ancient tie, and hair sticking out at the back, he became the cartoonist's 'boffin'—an impression accentuated by his manner of practical work, in which he would grunt and swear as solder failed to stick, scratch his head and make a strange squelching noise as he thought to himself." Unconvinced of England's chances to win the war, Turing converted all of his funds to two silver bars, buried them, and was later unable to locate them. He was horrified at the sight of blood, was an outspoken atheist, and was a homosexual. Still, for his unquestionably vital role in the British war effort, he was later awarded the Order of the British Empire, a high honor for someone not in the combat military.

In the waning months of the war, Turing turned his thoughts back to computing machines. He conceived of a device, built with vacuum tubes, that would be able to perform any function described in mathematical terms and would carry instructions in electronic symbols in its memory. This universal machine, clearly an embodiment of the Turing machine described in his 1936 paper, would not require separate hardware for different functions, only a change of instructions.

Turing was not alone in his ambition to construct a computing machine. A group at the University of Pennsylvania had built a computer called ENIAC (Electronic Numerical Integrator and Computer) that was similar to, but more complex, than Colossus. In the process, they had concluded that a better machine was possible. Turing's design was possibly more remarkable because he was working alone out of his home while they were a large university research group with the full backing of the American military. The American group published well before Turing did, but the British government subsequently took a greater interest in Turing's work.

In June of 1945 Turing joined the newly formed Mathematics Division of the National Physical Laboratory (NPL). Here he finalized plans for his Automatic Computing Engine (ACE). The rather archaic term "engine" was chosen by NPL management as a tribute to **Charles Babbage**'s Analytical Engine (and also because it made a pleasing acronym). Turing, however, was unprepared for the inertia and politics of a bureaucratic government foundation. All of his previous engineering projects had been conducted during wartime, when time was of the essence and no budget constraints existed. More than a year after the ACE project was approved, though, no engineering work had been completed and there was little cooperation between participants. A scaled-down version of the ACE was finally completed in 1950. But Turing had already left NPL in 1948, frustrated at the slow pace of the computer's development.

In 1950 Turing produced a widely read paper titled "Computing Machinery and Intelligence." This classic paper expanded on one of Turing's interests—if computers could possess intelligence. He proposed a test called the "Imitation Game," still used today under the name "Turing Test." In the test, an interrogator was connected by teletype (later, by computer keyboard) to either a human or a computer at a remote location. The interrogator is allowed to pose any questions and, based on the replies, the interrogator must decide whether a human or a computer is at the other end of the line. If the interrogator cannot distinguish between the two in a statistically significant number of cases, then artificial intelligence has been achieved. Turing predicted that within fifty years, computers could be programmed to play the game so effectively that after a five-minute question period the interrogator would have no more than a seventy-percent chance of making the proper identification.

Turing's personal life deteriorated in the early 1950s. After leaving NPL he took a position with Manchester College as deputy director of the newly formed Royal Society Computing Laboratory. But he was not involved in designing or building the computer on which they were working. By this time, Turing was no longer a world-class mathematician, having for too long been sidetracked by electronic engineering, nor was he engineer: the scientific world seemed to be passing him by.

While at Manchester, Turing had an affair with a young street person named Arnold Murray, which led to a burglary at his house by one of Murray's associates. The investigating police learned of the relationship between Turing and Murray; in fact, Turing did nothing to hide it. Homosexuality was a felony in England at the time, and Turing was tried and convicted of "gross indecency" in 1952. Because of his social class and relative prominence, he was sentenced to a year's probation and given treatments of the female hormone estrogen in lieu of serving a year in jail.

Turing committed suicide by eating a cyanide-laced apple on June 7, 1954. His death puzzled his associates; he had been free of the hormone treatments for a year, and, although a stigma remained, he had weathered the incident with his career intact. He left no note, nor had he given any hint that he had contemplated this act. His mother tried for years to have his death declared accidental, but the official cause of death was never seriously questioned.

TURING MACHINE

Turing machines were invented by **Alan Turing**, the father of computer science, in 1936. He wanted to define what an **algorithm** is in precise mathematical terms. His **definition** turned out to also be the most useful **model** of a computer to this date.

A Turing machine consists of a processor and an infinite tape with a "start" **space** on which symbols can be written or erased. The machine begins when it is fed a tape with zeros and ones already written on it (or it could be blank). The processor first reads the symbol in the "start" space. The processor remembers only one number, called its state. It moves back and forth along the tape, reading and writing symbols and changing its state. At each step in the calculation, the action of the processor depends on only two things: its current state number and the symbol currently being read. It stops processing only if its state changes to the "halt" state. The output can then be read from the tape. The number of possible states for the processor is finite. So, any Turing machine can be specified by the following finite table. The table has the processor's states listed horizontally at the top and the symbols 0, 1 and blank are written vertically on the side. Then for each state and symbol pair, the action (either move left, move right, halt, write 0, write 1, or erase the current symbol) is written in the appropriate space.

Turing also proved the existence of a Universal Turing machine. Such a machine can mimic any Turing machine. Since Turing machines can all be specified by a finite amount of information, the number of Turing machines is **denumerable**. Hence each can be assigned to a natural number so that no two Turing machines are assigned the same number. When you want a universal Turing machine to mimic a Turing Machine named Bob, say, then you find Bob's number and translate it into 1s and 0s. You feed this to the Universal Machine and then whatever input you put into the Universal Machine it will process just as Bob would. It is **Church's thesis** that a Universal Turing machine can do anything any modern computer can do.

Turing also proved that no Turing machine can determine whether any given Turing machine will halt on a given input. This fact is referred to as the undecidability of the halting problem. It is equivalent to the statement that there is no Turing machine that can determine whether any two given Turing machines always produce the same output when given the same input. This fact has profound applications to logic, **group theory**, **tesselations**, and many other areas of mathematics.

Turing also defined computable numbers. A Turing machine's input is just a string of 1s, 0s and blanks and so is its output. Therefore every Turing machine can be represented as a function f say, from a subset of natural numbers to the natural numbers. The **domain** of f is the subset on which the Turing machine halts. Since the rationals can be put into **correspondence** with the natural numbers by some function, g say, we can compose g with f to obtain a function from a subset of the natural numbers to the rationals. If the domain of f is the whole set of natural numbers and the sequence {f(n)} converges to a real number x then x is said to be computable. Turing proved that the set of computable numbers is a **field** and that **pi** is computable. The set of computable numbers contains the set of constructible numbers since those are just numbers which are computed from a special type of algorithm involving ruler and compass.

Turing machines are used to estimate computing speed. For example, suppose that there is a function f from the natural numbers to the rationals that we wish to compute. We find or make a Turing machine that computes f. We then note how many cells the processor must visit in order to compute f(n) for each n. The number of cells is called the cost of computation (for this example). We then multiply by a suitable constant that depends on how fast our computer really is. For example, the constant might be 1/100 of a second. Now we have an idea of how fast our algorithm is, even before writing it into a computer. The study of computing speed is fundamental to abstract complexity theory, an important part of computer science.

See also Decidable; Alan Turing

U

ULAM, STANISLAW MARCIN (1909-1984)

Polish-American mathematician

Stanislaw Marcin Ulam was one of the many gifted scientists involved in the effort to create a hydrogen bomb at the Los Alamos National Laboratory in New Mexico in the 1950s. As a professional mathematician, he was integral to the bomb development program because of his expertise in thermonuclear reactions and mathematical physics, which allowed him to solve the problem of how to start fusion in the hydrogen bomb. Along with **John von Neumann**, Ulam also invented the **Monte Carlo method**. In addition, his work on the new computers helped them become more flexible and useful.

Ulam, the son of a lawyer, was born in Lemberg, Poland, in the Austrian Empire (now Lvov, Ukraine) on April 13, 1909. When he was 10, he began attending the gymnasium (a classical college-preparatory school) in his hometown, and soon thereafter became interested in astronomy and physics. A brilliant boy, he became determined to understand Albert Einstein's theory of relativity, but when he realized he would need extensive knowledge of mathematics to accomplish his goal, Ulam directed his energies chiefly toward that subject.

Having taught himself much of the mathematics he knew, Ulam entered Lvov's Polytechnic Institute in 1927. He received his master's degree from the school in 1932 and his doctorate the following year. Ulam spent 1934 doing postdoctoral studies at Zurich, Vienna, and Cambridge University, after which he accepted von Neumann's invitation to work at the Institute for Advanced Studies (IAS) at Princeton University in 1935. He began lecturing there in 1936, and in 1939 decided to move to the United States. (Some of his close family members died soon afterward during the Holocaust.) Ulam also lectured at Harvard University until 1940, and then left the IAS and moved to the University of Wisconsin to serve as assistant professor of mathematics. He remained there until leaving for the Los Alamos lab in 1943 at the request of physicist Hans Bethe.

That same year, Ulam officially became an American citizen. At Los Alamos, he became intensely involved in the hydrogen (fusion) bomb program. Working with physicist Edward Teller, Ulam was credited with making huge progress toward creating a "superbomb" by suggesting that compression was the key element necessary for explosion and that the shock waves from an atomic (fission) bomb could produce that compression. He also showed how he could design the bomb so that the mechanical shock waves would be focused and promote quick ignition of the fusion fuel. Teller proposed instead that radiation, rather than mechanical, implosion be used to compress the fuel. This turned out to be the best solution, and led to the use of the "Teller-Ulam configuration" to build modern thermonuclear weapons.

Ulam left Los Alamos in 1946 to work briefly as professor of mathematics at the University of California, Los Angeles. Shortly after arriving there, he contracted encephalitis, which briefly rendered him unable to speak. His treatment succeeded, but afterward friends and colleagues found him a changed man. Apparently, Ulam's intellectual capacity had somehow been enhanced, although he seemed unable or at least reluctant to discuss the particulars of how he arrived at certain astonishing conclusions. While he rested in the hospital, von Neumann came to visit and the two mathematicians passed the time by playing solitaire. Ulam discovered the Monte Carlo method while doing so, and later developed it into an excellent calculation tool based on using approximations to solve complex problems. In fact, the method was so useful that Los Alamos scientists immediately adopted it for the national nuclear weapon program. Today, Monte Carlo is essential to applications in mathematical economy, weapons design, and operations research.

Returning to Los Alamos in 1946, Ulam served as staff member, research adviser, and group leader until 1967. He remained a consultant to the lab for the rest of his life, however. In the meantime, he published *A Collection of Mathematical Problems* in 1960 and continued to indulge his lifelong interest in **set theory**.

Two years prior to leaving Los Alamos, Ulam took a position at University of Colorado, Boulder as professor and chairperson of the Mathematics Department. He retained that post until 1975, publishing *Sets, Numbers, and Universes* in 1974 and his autobiography, *Adventures of a Mathematician,* in 1976. In the interim, he worked as a graduate research professor at the University of Florida from 1973 on and beginning in 1979 was professor of biomathematics at the University of Colorado Medical Center.

Ulam died in Santa Fe, New Mexico on May 13, 1984. He and his wife were married in 1941 and had one child.

See also Monte Carlo method

UNITS AND STANDARDS

A unit of measurement is some specific quantity that has been chosen as the standard against which other measurements of the same kind are made. For example, the meter is the unit of measurement for length in the **metric system** . When an object is said to be four meters long (4 m), that means that the object is four times as long as the unit standard (one meter [1 m]).

The term ``standard'' refers to the physical object on which the unit of measurement is based. For example, for many years the standard used in measuring length in the metric system was the **distance** between two scratches on a platinum-iridium bar kept at the Bureau of Standards in S8Fvres, France. A standard serves as a **model** against which other measuring devices of the same kind are made. The meter stick in your classroom or home is thought to be exactly one meter long because it was made from a permanent model kept at the manufacturing plant that was originally copied from the standard meter in France.

All measurements consist of two parts: a scalar (numerical) quantity and the unit designation. In the measurement 8.5 m, the scalar quantity is 8.5 and the unit designation is meters.

History

The need for units and standards developed at a point in human history when people needed to know how much of something they were buying, selling, or exchanging. A farmer might want to sell a bushel of wheat, for example, for 10 dollars, but he or she could do so only if the unit ``bushel'' was known to potential buyers. Furthermore, the unit ``bushel'' had to have the same meaning for everyone who used the term.

The measuring system that most Americans know best is the British system, with units including the foot, yard, second, pound, and gallon. The British system grew up informally and in a disorganized way over many centuries. The first units of measurement probably came into use shortly after 1215. These units were tied to easily obtained or produced standards. The yard, for example, was defined as the distance from King Henry II's nose to the thumb of his outstretched hand.

The British system of measurement consists of a complex, irrational collection of units whose only advantage is its familiarity. As an example of the problems it poses, the British system has three different units known as the quart. These are the British quart, the United States dry quart, and the United States liquid quart. The exact size of each of these quarts differs.

In addition, a number of different units are in use for specific purposes. Among the units of **volume** in use in the British system, (in addition to those mentioned above) are the bag, barrel (of which there are three types-British and United States dry, United States liquid, and United States petroleum), bushel, butt, cord, drachm, firkin, gill, hogshead, kilderkin, last, noggin, peck, perch, pint, and quarter.

The metric system

In an effort to bring some rationality to systems of measurement, the French National Assembly established a committee in 1790 to propose a new system of measurement, with new units and new standards. That system has come to be known as the metric system and is now the only system of measurement used by all scientists and in every country of the world except the United States and the Myanmar Republic. The units of measurement chosen for the metric system were the gram (abbreviated g) for mass, the liter (l) for volume, the meter (m) for length, and the second (s) for time.

A specific standard was chosen for each of these basic units. The meter was originally defined as one ten-millionth the distance from the north pole to the equator along the prime meridian. As a **definition**, this standard is perfectly acceptable, but it has one major disadvantage: a person who wants to make a meter stick would have difficulty using that standard to construct a meter stick of his or her own.

As a result, new and more suitable standards were selected over time. One improvement was to construct the platinum-iridium bar standard mentioned above. Manufacturers of measuring devices could ask for copies of the fundamental standard kept in France and then make their own copies from those. As you can imagine, the more copies of copies that had to be made, the less accurate the final measuring device would be.

The most recent standard adopted for the meter solves this problem. In 1983, the international Conference on Weights and Measures defined the meter as the distance that **light** travels in 1/299,792,458 second. The standard is useful because it depends on the most accurate physical measurement known-the second-and because anyone in the world is able, given the proper equipment, to determine the true length of a meter.

Le Système International d'Unités (the SI system)

In 1960, the metric system was modified somewhat with the adoption of new units of measurement. The modification was given the name of Le Système International d'Unités, or the International System of Units—more commonly known as the SI system.

Nine fundamental units make up the SI system. These are the meter (abbreviated m) for length, the kilogram (kg) for mass, the second (s) for time, the ampere (A) for electric current, the kelvin (K) for temperature, the candela (cd) for light intensity, the mole (mol) for quantity of a substance, the **radian** (rad) for plane angles, and the steradian (sr) for solid angles.

Derived Units

Many physical phenomena are measured in units that are derived from SI units. As an example, frequency is measured in a unit known as the hertz (Hz). The hertz is the number of vibrations made by a wave in a second. It can be expressed in terms of the basic SI unit as s^{-1}. Pressure is another derived unit. Pressure is defined as the **force** per unit **area**. In the metric system, the unit of pressure is the Pascal (Pa) and can be expressed as kilograms per meter per second squared, or $kg/m \times^2$. Even units that appear to have little or no relationship to the nine fundamental units can, nonetheless, be expressed in these terms. The absorbed dose, for example, indicates that amount of radiation received by a person or object. In the metric system, the unit for this measurement is the gray. One gray can be defined in terms of the fundamental units as meters squared per second squared, or $m^2 \times s^2$.

Many other commonly used units can also be expressed in terms of the nine fundamental units. Some of the most familiar are the units for area (**square** meter: m^2), volume (cubic meter: m^3), velocity (meters per second: m/s), concentration (moles per cubic meter: mol/m^3), density (kilogram per cubic meter: kg/m^3), luminance (candela per square meter: cd/m^2), and magnetic **field** strength (amperes per meter: A/m).

A set of prefixes is available that makes it possible to use the fundamental SI units to express larger or smaller amounts of the same quantity. Among the most commonly used prefixes are milli- (m) for one-thousandth, centi- (c) for one-hundredth, micro- (æ) for one-millionth, kilo- (k) for one thousand times, and mega- (M) for one million times. Thus, any volume can be expressed by using some combination of the fundamental unit (liter) and the appropriate prefix. One million liters, using this system, would be a megaliter (ML) and one millionth of a liter, a microliter (æL).

Natural Units

One characteristic of all of the above units is that they have been selected arbitrarily. The committee that established the metric system could, for example, have defined the meter as one one-hundredth the distance between Paris and S8Fvres. It was completely free to choose any standard it wished.

Some measurements, however, suggested ``natural'' units. In the field of electricity, for example, the charge carried by a single electron would appear to be a natural unit of measurement. That quantity is known as the elementary charge (e) and has the value of 1.6021892 B1 10^{-19}coulomb. Other natural units of measurement include the speed of light (c: 2.99792458 B1 10^8 m/s), the Planck constant (6.626176 B1 10^{-34} joule per hertz), the mass of an electron (m_e: 0.9109534 B1 10^{-30} kg), and the mass of a proton (m_p: 1.6726485 B1 10^{-27} kg). As you can see, each of these natural units can be expressed in terms of SI units, but they are often used as basic units in specialized fields of science.

Unit conversions between systems

For many years, an effort has been made to have the metric system, including SI units, adopted worldwide. As early as 1866, the United States Congress legalized the use of the metric system. More than a hundred years later, in 1976, the Congress adopted the Metric Conversion Act declaring it the policy of the nation to increase the use of the metric system in the United States.

In fact, little progress has been made in that direction. Indeed, elements of the British system of measurement continue in use for specialized purposes throughout the world. All flight navigation, for example, is expressed in terms of feet, not meters. As a consequence, it is still necessary for an educated person to be able to convert from one system of measurement to the other.

In 1959, English-speaking countries around the world met to adopt standard conversion factors between British and metric systems. To convert from the pound to the kilogram, for example, it is necessary to multiply the given quantity (in pounds) by the **factor** 0.45359237. A conversion in the reverse direction, from kilograms to pounds, involves multiplying the given quantity (in kilograms) by the factor 2.2046226. Other relevant conversion factors are 1 inch—2.54 centimeters and 1 yard—0.9144 meter.

V

VALLÉE–POUSSIN, CHARLES JEAN GUSTAVE NICOLAS DE LA (1866-1962)

Belgian number theorist

Charles Jean Gustave Nicolas de la Vallée–Poussin was responsible for proving the **prime number theorem**. A prime number is a number that can be divided by only one and itself without producing a remainder, and de la Vallée–Poussin—like many others—set out to prove the relationship between **prime numbers**. In an article for *MAA Online* dated December 23, 1996, Ivars Peterson asserts: "In effect, [the prime number theorem] states that the average gap between two consecutive primes near the number x is close to the natural logarithm of x. Thus, when x is close to 100, the natural logarithm of x is approximately 4.6, which means that in this **range**, roughly every fifth number should be a prime." De la Vallée–Poussin was additionally known for his writings about the **zeta function**, Lebesgue and Stieltjes integrals, conformal representation, algebraic and trigonometric polynomial approximation, trigonometric series, analytic and quasi–analytic **functions**, and complex variables. His writings and research, which were—and are—considered clear, stylish, and precise, were highly respected by his peers in academia and other well–placed individuals in Western society.

Despite the historical confusion posed by de la Vallée–Poussin's name (it is often rendered as Charles–Jean–Gustave–Nicolas, Charles–Jean Gustave Nicolas, Charles–Joseph, Vallée Poussin, etc.), the facts surrounding his origins are well known. De la Vallée–Poussin was born on August 14, 1866, in Louvain, Belgium. A distant relative of the French painter Nicolas Poussin, de la Vallée–Poussin's father was, like himself, an esteemed teacher at the University of Louvain. (The elder de la Vallée–Poussin, however, specialized in geology and mineralogy.) De la Vallée–Poussin's family was well–off, and as a child, he found encouragement and inspiration in fellow mathematician Louis–Philippe Gilbert (some sources identify him as Louis Claude Gilbert), with whom he would eventually work.

De la Vallée–Poussin enrolled at the Jesuit College at Mons in southwestern Belgium, where it is said he originally intended to pursue a career in the clergy. He ultimately, however, obtained a *diplôme d'ingenieur* and began to pursue a career in mathematics. In 1891, like his father and Gilbert, he became employed at the University of Louvain, where he initially worked as Gilbert's assistant. Gilbert's death the following year created an academic opening to which de la Vallée–Poussin was appointed in 1893, thereby earning him the title of professor of mathematics.

Although de la Vallée–Poussin was gaining recognition as early as 1892, when he won a prize for an essay on **differential equations**, he earned his first widespread fame four years later. In 1896, de la Vallée–Poussin capitalized on the ideas set forth by earlier mathematicians, notably **Karl Friedrich Gauss, Adrien Marie Legendre, Leonhard Euler, Peter Gustav Lejeune Dirichlet, Pafnuty Lvovich Chebyshev,** and **Georg Friedrich Bernhard Riemann,** and proved what is now known as the prime number **theorem**. (De la Vallée–Poussin shares this honor with **Jacques Hadamard**, who revealed his finding in the same year. Historians note, however, that de la Vallée–Poussin's and Hadamard's achievements were performed independently and that although both mathematicians used the Riemann zeta function in their work, they came to their conclusions in different ways.)

De la Vallée–Poussin revealed much of his groundbreaking work in a series of celebrated books and papers. His two–volume *Cours d'analyse infinitésimale* went through several printings, and the work was consistently edited between printings to offer updated information. Initially, the book was directed toward both mathematicians and students, and de la Vallée–Poussin used different fonts and sizes of types to differentiate between the audiences to whom a particular passage was directed. In the 1910s, de la Vallée–Poussin was preparing the third edition of this work, but this was destroyed by German forces, who invaded

Louvain during World War I. De la Vallée–Poussin subsequently dedicated his 1916 *Intégrales de Lebesgue fonctions d'ensemble, classes de baire* to the **Lebesgue integral** to compensate for the material destroyed by the Germans. De la Vallée–Poussin continued to publish well into his eighties, and, like *Cours d'analyse infinitésimale,* many of his writings went through various reprintings and revisions. Almost all of de la Vallée–Poussin's writing have been praised for their originality and the clarity of his writing style.

De la Vallée–Poussin, who married a Belgian woman whom he met while vacationing in Norway in the late 1890s, died on March 2, 1962, in the city of his birth. During his lifetime, he was accorded many honors. In addition to the celebrations commemorating his 35th and 50th anniversaries as chair of mathematics at the University of Louvain, he was elected to various prestigious institutions including the French Académie Royales des Sciences, the International Mathematical Union, the London Mathematical Society, the Belgian Royal Academy, and the Legion of Honor. In 1928, the king of Belgium also awarded him the title of baron in recognition of his years of academic tenure and his professional achievements.

VANDERMONDE, ALEXANDRE THÉOPHILE (1735-1796)

French academic

Although Alexandre Théophile Vandermonde's work in mathematics was limited to only two years, he is still remembered for producing such advances as the Vandermonde determinant. Most scientific historians consider Vandermonde to be the founder of the theory of **determinants**. He also made progress toward solving the classic "knight's tour problem."

Vandermonde was born in Paris, France on February 28, 1735, the son of a physician. Because he was a sickly child, Vandermonde's physician father encouraged the obviously bright boy to study music, which he believed would be less strenuous than other academic pursuits. Vandermonde went along with this plan, since he loved music, for all of his childhood and well into his adult life. However, when he met and became friends with a respected mathematician named Fontaine at age 35, Vandermonde eagerly turned to the study of mathematics.

Thus, in about 1770, Vandermonde discovered a new talent. In fact, he applied himself so intensely to his studies and made such an impression with his work that by 1771 he had been elected to the prestigious Academy of Sciences. It was to this body that he presented four papers—his total contribution to mathematics—in 1771 and 1772.

Vandermonde was reportedly the first mathematician to prepare a systematic treatment of the theory of determinants. The determinant named after him is one in which the elements of each row or column are: $1, r, r^2 ..., r^{n-1}$ of a geometric progression. Ironically, this determinant never appeared in Vandermonde's published work. Another of his papers to the Academy dealt with the knight's tour, a recreational math

problem. The object of this puzzle is to come up with a sequence of moves by a knight chess piece on a chessboard (or any other grid) so that the piece lands on each **square** of the board exactly once.

After 1772, Vandermonde dropped out of the mathematics world as precipitously as he entered it. There are records that he collaborated with **Etienne Bezout** on experiments with cold in 1776 and on the manufacture of steel with his close friend **Gaspard Monge** in 1786. However, aside from these involvements, it appears that Vandermonde's real contributions to mathematics took place during a two-year period of his life.

In 1782 Vandermonde became director of a large conservatory and in 1792 he took a top bureaucratic position with the French Army. Also in his later years, Vandermonde returned to his first love, music, writing several influential papers on harmony. He died on January 1, 1796 in Paris.

VARIABLE

The term variable is developed very early in introductory **algebra** coursework. It is defined as a letter or symbol that can be replaced by any number or expression. Examples of variables would be as follows: x, t, d, y, *, or z. The most commonly used variable in algebra is x. If a person earns $8.00 for every hour worked and s/he works 5 hours, the person would earn $8 × 5 = $40. If the person works for n hours, that person would earn $8 × n or $8n.

In the application of **functions**, variables are used to represent the relationship between two things. The **independent variable** is referred to as x and the **dependent variable** is referred to as y. It is common to write functions in terms of the independent variable.

In probability work, a random variable is a numeric quantity that can be measured in a random experiment. It is a function of the possible outcomes of an experiment. If a baseball player is up at bat ten times, the random variable can be assigned to either hits or misses. If x represented the number of hits, the outcomes would be 10, 9, 8, 7, 6, 5, 4, 3, 2, 1, or 0.

See also Probability

VARIGNON, PIERRE (1654-1722)

French mathematician

Although Pierre Varignon is principally remembered for his contributions to the area of statics, a branch of **mechanics** that concerns resting objects or forces in equilibrium, he also made advances in **calculus**.

The son of a poor mason, Varignon was born in Caen, France on an unknown date in 1654. He began studying at a Jesuit college in his hometown at a late age. There are few historical records of his life at this point, except those confirming that he entered religious life by becoming a monk in 1676.

Varignon apparently became interested in mathematics after reading the works of Euclid and **René Descartes**.

As a member of a religious order, Varignon was eligible to study at the University of Caen, where he completed a master's degree in 1682. The following year he was ordained as a priest, and in 1686 he left Caen and traveled to Paris. Paris provided Varignon with a wider audience for his talents, and in 1687 he was able to publish his *Project on the New Mechanics.* He accepted a nomination to the Academy of Sciences a year later, and was also appointed a professor of mathematics at the College Mazarin, where he would remain for the rest of his life.

Varignon was so busy with teaching that from the 1690s on he only had time to write articles and memoirs for publication. However, his frequent correspondence with **Gottfried Wilhelm von Leibniz** and **Johann Bernoulli** left important records of his work. In addition to his job at Mazarin, Varignon began teaching mathematics at the Royal College in 1704.

Varignon was one of the first French scholars to realize the value of calculus and, by adapting Leibniz's calculus to the inertial mechanics in Isaac Newton's *Principia,* helped to develop analytic dynamics. Varignon's other works include a 1699 publication on applying **differential calculus** to fluid flow and water clocks, while in 1702 he used calculus to investigate spring-driven clocks. Meanwhile, at the end of the century, he had refuted Michel Rolle's objections to the "new" calculus, helping to speed progress in that area.

In his last year at the Royal College, Varignon reportedly planned to discuss the basics of **infinitesimal** calculus, but before he could present more than a mere outline of his ideas, he died on December 22, 1723. In 1731 Varignon's College Mazarin lectures were compiled and printed in *Élémens de mathematiques.*

VEBLEN, OSWALD (1880-1960)

American mathematician

Oswald Veblen is remembered mainly for his contributions to **topology** and differential and projective **geometry**, many of which scientists later found useful in atomic physics and relativity. He taught for many years at Princeton University and helped found the university's Institute for Advanced Study (IAS).

Veblen was born in Decorah, Iowa, on June 24, 1880, the son of Norwegian immigrants. His father taught physics at the University of Iowa. Veblen attended school in Iowa City, enrolling at the University of Iowa in 1894. He earned a bachelor's degree in 1898 and then worked as a laboratory assistant at the school for a year before transferring to Harvard University. There he completed another B.A. in 1900 before going on to the University of Chicago for his doctorate, which he earned in 1903. His doctoral thesis was "A System of Axioms for Geometry."

Veblen started his long career at Princeton University in 1905. In 1910 he compiled much of his research into the two-volume *Projective Geometry.* Critics highlighted the work's expert explanation of the axiomatic method, which would immediately have a strong influence on other **algebra** and geometry researchers.

Also in 1905, **Albert Einstein** made public his special theory of relativity, which had a huge impact on Veblen. He began dedicating most of his research efforts to differential geometry, with results that would later lead to important advances in relativity theory and atomic physics.

Kept busy with his teaching duties, the next book Veblen published did not come out until 1922. *Analysis Situs* was the product of his 1916 Colloquium Lectures to the American Mathematical Society. In the lectures he had discussed this branch of geometry, which concerns the algebraic and numerical measures of geometric figures' "connectivity." *Analysis Situs* elaborated on and clarified the work of **Jules-Henri Poincaré**, which was pioneering, yet difficult. Veblen's book, the first systematic coverage of the main principles of topology, was so effective in discussing the topic that several generations of mathematicians studied it intensively and considered it the premiere work on topology. It was this book that laid the groundwork for the world-renowned Princeton school of topological research.

Based on his continuing study of Einstein's work, Veblen published a systematic treatment of **Riemannian geometry**, *The Invariants of Quadratic Differential Forms,* in 1927. In the academic year 1928-1929, Veblen taught at Oxford University. Aside from that short time, however, he remained at Princeton for his entire career. He helped create the IAS in 1932, becoming a professor there the same year, and was integral to the formation of the Institute's School of Mathematics.

In 1933, Veblen produced two important books. The first, written with student John Henry Constantine Whitehead, was *The Foundations of Differential Geometry,* which provided the first **definition** of a differentiable **manifold**. The second was *Projective Relativity Theory.* In the latter, demonstrating his mathematical influence on the field of physics, Veblen offered a new treatment of "spinors," which physicists use to depict the spin of electrons.

Becoming professor emeritus at the IAS in 1950, Veblen eventually moved to Brooklyn, Maine. He died there on August 10, 1960. Veblen and his wife had married in 1908.

VECTOR ALGEBRA

A vector is formally, an element of a vector **space**. A vector has two components: a direction (or vector) component and a scalar component. For example, gravity is vector: the direction of gravitational **force** on the Earth is towards the center of the Earth and the scalar component is approximately 9.8 meters per second squared (10.71 yd/sec^2). As a rule of thumb, any thing that has both a quantity and a direction component can be represented with a vector. For example, velocity, acceleration, force, and momentum are all vectors. Speed and mass, on the other hand, are quantities without directions. They are not vectors but they are scalars. A scalar is a quantity without a direction. A scalar and a vector can multiply to produce a vector in the same direction as the original vector. For example, if a car heads west at 60 mph (96.6 k/w) then a bus moving at half the car's velocity is moving west at 30 mph (48.3 k/hr). In this case, the scalar (1/2) was multiplied by the vector (60 mph west) and the result is the vector (30 mph west).

To illustrate properties of vectors, let us consider the vector space R^3 consisting of all ordered triplets of **real numbers**. In R^3, there is a dot product (also called scalar product) of vectors denoted with the symbol $\cdot$. If $v = (v_1, v_2, v_3)$ and $w=(w_1, w_2, w_3)$ are vectors in R^3. Then $v \cdot w = v_1 w_1 + v_2 w_2 + v_3 w_3$. The length of a vector v is defined to be the **square** root of $v \cdot v$. In R^3, the length of v is denoted by $\|v\|$ and is equal to the **distance** between $(0,0,0)$ and the point (v_1,v_2,v_3). So the **definition** of length makes sense. The **angle** between nonzero vectors v and w is defined to be the unique number a such that the **cosine** of a equals $v \cdot w$ divided by the length of v times the length of w. This angle happens to equal the angle between the line that contains $(0,0,0)$ and (v_1,v_2,v_3) with the line that contains $(0,0,0)$ and (w_1,w_2,w_3). So this definition also makes sense. A scalar in R^3 is just a real number. A vector in R^3 can be multiplied by a scalar like so: if r is real number then $rv = (rv_1, rv_2, rv_3)$. The length of rv equals the r times the length of v. Also, $(rv) \cdot w = r(v \cdot w)$. So, the angle between rv and w is the same as the angle between v and w. The vector rv is said to be a scalar multiple of v. Oftentimes, it is convenient to work with vectors that have length one. These are called unit vectors. The set of unit vectors in R^3 corresponds naturally to the **sphere** S^2, that it is the set of all points distance one away from the origin. In this way, the sphere is said to be the "set of directions" in R^3. If $v \cdot w = 0$ then v and w are said to be orthogonal. If v and w are also unit vectors, then they are orthonormal.

Vectors can be added together by adding their components. So $v + w = (v_1 + w_1, v_2 + w_2, v_3 + w_3)$. The vector $v + w$ can also be determined by the **parallelogram rule**. Draw the line segment from $(0,0,0)$ to (v_1, v_2, v_3) and the line segment from $(0,0,0)$ to (w_1, w_2, w_3) in three-dimensional space. Now draw the parallelogram that these two segments determine. The diagonal of the parallelogram, that is the segment from $(0,0,0)$ to $(v_1 + w_1, v_2 + w_2, v_3 + w_3)$ represents the vector $v + w$. As an application, consider a billiard ball that is moving north at 3 mph. Another billiard ball moving west at 4 mph strikes it and stops. If there is very little friction and air resistance, the first billiard ball will now be moving northwest at about 5 mph. In vector notation, the first billiard's original velocity is given by $(0,3)$, the second's by $(-4,0)$ and the first billiard's final velocity by $(-4, 3)$.

The cross product of two vectors in R^3 is written with an X sign. It is defined by $v \times w = (v_2 w_3 - v_3 w_2, v_3 w_1 - v_1 w_3, v_1 w_2 - v_2 w_1)$. In other words, if $v \times w = z = (z_1, z_2, z_3)$, then the determinant of the **matrix** whose rows are (a,b,c), v, w is $z_1 a + z_2 b + z_3 c$. In yet other words, if v and w are linearly independent then $v \times w$ is a vector perpendicular to plane spanned by v_1 and v_2. Its length is equal to the **area** of the parallelogram spanned by v and w, and its direction is given by the right-hand rule. This rule is the following: open your right hand so that your fingers point in the direction of v. Close your fist in the direction of w and stick your thumb out. Your thumb

points in the direction of $v \times w$. Here are the main identities concerning the cross product:

- $\|v \times w\| = \|v\|\|w\| \sin(a)$
- $v \times w = - w \times v$
- $v \times (w + y) = v \times w + v \times y$
- $r(v \times w) = (rv) \times w$
- $(v \times w) \cdot y$ = the determinant of the matrix whose rows are v, w, and y (in order).
- $(v \times w) \times y = (v \cdot y)w - (w \cdot y)v$

There are many ways in which the cross product is used. The definitions of the physical torque and angular momentum, both of which have to do with rotational **motion**, are defined with the cross product. If a straight wire carrying a current is in a magnetic **field** then the force exerted by the field on the wire is given by the vector equation $F = iL \times B$ in which I is the current, L is the length of the wire and B is the force of the magnetic field. The area of a surface in three-dimensional real space is the integral over the surface of the function $(V(x) \times W(x)) \cdot n(x)$ with respect to x. Here V and W are orthonormal vector fields and n is the unit normal vector to the surface at x. A vector field is a continuous function from the surface to the vector space R^2 and a normal vector is orthogonal to the tangent plane at x.

See also Linear algebra; Mathematics and physics; Vector analysis; Vector spaces; Velocity vectors and acceleration vectors

VECTOR ANALYSIS

Vector **analysis** is the multi-dimensional analogue of single-variable **calculus**. It is used to represent (and analyze) the motions of the solar system, fluid flows, the flow of electric charge, and other phenomena.

In single-variable calculus, the derivative of a function f from the **real numbers** to the real numbers is its rate of change. For example, if f(t) represents the **distance** traveled at time t, the derivative of f at t represents speed. Vector analysis on the other hand typically concerns **functions** from three-dimensional **space** to the real numbers. Derivatives of such functions are taken with respect to a direction. For example, suppose that the function f represents the temperature of a room. Then its derivative in the upwards direction is the rate of temperature change as we move upwards. Often this direction is given a name like positive z-direction. Then the derivative of f in the positive z-direction is written as $\partial f / \partial z$. In words, it is called the partial derivative of f in the z-direction. A point in three-dimensional space is typically represented by an ordered triplets of numbers called the coordinates of the point. The first number in the triplet is called the x-coordinate, the second is the y-coordinate, and the third is the z-coordinate. $\partial f / \partial z$ at a point (x_1,y_1,z_1) then measures the rate at which the value $f(x_1,y_1,z_1)$ changes when the third coordinate z_1 is changed. In other words, consider the single **variable** function $g(z) = f(x_1,y_1,z_1)$. Then $\partial f / \partial z$ = the derivative of g at z_1.

The gradient of f at a point p is the vector given by the coordinates ($\partial f / \partial x$ at p, $\partial f / \partial y$ at p, $\partial f / \partial z$ at p). Thus the gradient of f is a function from three-dimensional space to three-dimensional space. It is also called a vector **field**.

The pictures are to be interpreted thusly. At each point the value of the vector field is the vector whose tail is at that point, whose direction is given by the arrow and whose magnitude is the length of the arrow.

Suppose that a vector field represents the movement of a gas. That is, every vector shows the direction and speed of a particle of gas located near the vector's tail. The divergence of the vector field is a function that gives the "time rate of change of **volume** per unit volume" of the gas. Consider a small volume of gas near a point P in space. As time progresses, this gas will move in the direction that the vector field determines. If more gas is going out of this region around P than is coming in, then the divergence is negative. If less gas comes in then goes out, the divergence is positive. So the divergence measures the change in volume (per unit volume) as time progresses. A useful fact in **hydrodynamics** is that water is virtually incompressible and this means that if a vector field represents the flow of water then its divergence is practically **zero** at every point. Also, the divergence of electric field intensity is proportional to the charge density. If the vector field is represented by a function F such that $F(x,y,z) = (I(x,y,z), J(x,y,z), K(x,y,z))$ where I, J, and K are all functions of a single variable, then the divergence of F is equal to $\partial I / \partial x + \partial J / \partial y + \partial K / \partial z$. The two top vector fields shown above have positive divergence everywhere and the two bottom vector fields have zero divergence everywhere (in other words they are solenoidal).

The curl of a vector field is another vector field. It is defined as follows. Imagine a small paddle (something like 4 rectangular blocks of wood glued along their edges so that from one perspective it looks like a plus sign with a small **square** hole in the middle). Now put this small paddle in the vector field near a point P. We imagine that the vector field represents the flow of a fluid. Then, depending on how the axis of the paddle is situated, it will spin. In general, there will be exactly one way to position the paddle to maximize the amount of spinning. In this position, the axis of the paddle is parallel to the curl vector at P. The direction of the curl vector is given by the right-hand rule. Position your right hand so that your thumb I parallel to the axis of the paddle and the fingers point in the direction of spinning. Then, your thumb points in the direction of the curl vector. The magnitude of the curl vector is twice the angular speed of the paddle. Precisely, the curl of at a vector field F is equal to ($\partial K / \partial y - \partial J / \partial z$, $\partial I / \partial z - \partial K / \partial x$, $\partial J / \partial x - \partial I / \partial y$). Imagine that the vector fields in the picture above are extended to three-dimensional space in a such a way that their z-components are zero everywhere. Then the ones on the left have zero curl everywhere (in other words, they are irrotational). The curl of the vector field on the bottom right is pointing directly at the reader. The curl of the vector field on the top right is more complicated: it is zero on the two diagonal x = y and x = -y. It points directly at the reader when y is greater than x and directly away from the reader when x is greater than y.

In single-variable calculus, the integral of a function over the **interval** [a,b] on the real number line is its average value on that interval multiplied by the length (b-a) of the interval. The **fundamental theorem of calculus** states that the integral of the derivative of a function f over an interval [a,b] is equal to f(b) - f(a). For example, the distance traveled over an hour is equal to the average speed during that hour times one hour.

In vector analysis, the functions of interest might be defined on a surface or curve inside three-dimensional space. The integral of such a function over a finite curve or surface is its average value on the curve or surface multiplied by the curve's length or the surface's **area**. The fundamental **theorem** of vector analysis is called Stokes' theorem. It states that the integral of the (normal component of the) curl of a vector field over a bounded surface is equal to the integral of the (tangential component of the) vector field along the boundary of the surface.

See also Differential calculus; Integral calculus; Vector algebra; Vector spaces; Velocity vectors and acceleration vectors

VECTOR SPACES

The primary example of a vector **space** is R^n, the set of ordered n-tuples of **real numbers**. If $v = (v_1,...,v_n)$ $w = (w_1,..,w_n)$ are elements of R^n then $v + w$ is defined to be $(v_1 + w_1,...,v_n + w_n)$. If r is a real number, then rv is defined to be $(rv_1, rv_2,..., rv_n)$. The set of real numbers is the **field** of scalars for R^n. In general, a vector space V over a field K (called the field of scalars) is a set with an operation and an 'action of K.' The operation is denoted by +. It defines vector **addition**. The 'action of K' defines how an element of K can be multiplied with an element of V. The following rules must also be satisfied for any elements x, y, z in V and k in K:

- $x + y$ is an element of V.
- kx is an element of V, where kx denotes k "times" x.
- $kx = xk$.
- $x + y = y + x$.
- $x + (y + z) = (x + y) + z$.
- $k(x + y) = kx + ky$.
- There is an element denoted by 0 in V such that $0 + v = = $ for all v in V.
- $0x = 0$.
- $1x = x$.

When working with vectors, it is often useful to have a basis. This is a set of vectors $B_1 = \{v_1, v_2,...\}$ such that any vector w in V can be written uniquely as $w = k_1v_1 + k_2v_2 +...$ in which the k_i's are scalars. The uniqueness referred to means that if w also equals $r_1v_1 + r_2v_2 +...$, then $r_1 = k_1$, $r_2 = v_2$, and so on. If two bases for V are finite, then they have the same number of elements. This fact can be proved using Gaussian elimination. In fact, any two bases for V have the same **cardinality** but this fact is harder to prove if the bases are infinite. The cardinality of any base for V is called the **dimension** of V. For example, one basis for R^n is $\{e_1,..., e_n\}$ where, for any i, e_i is the vector whose coordinates are all 0 except for the i-th coordinate which is one. So the dimension of R^n is n. Infinite dimensional vector spaces are common in functional **analysis**. One such vector space, F, is the set of all integrable **functions** from [0,1] to the real numbers. Two such functions, f and g, say, can be added by adding their values at each point. So (f+g)(x) is defined to be f(x) + g(x). For any real number r, (rf)(x) is defined to be rf(x). The vector space F is also a **Hilbert space**. Banach spaces and Hilbert spaces are vector spaces with additional properties that are used frequently in functional analysis.

Vector space homomorphisms, also known as linear transformations, are maps between vector spaces (which are defined over the same field K). They are used to show relationships among vector spaces. Precisely, a homomorphism from a vector space V to a vector space W is a map H from V to W that satisfies the following property. For all vectors x and y in V and for all scalars r, H(x + y) = H(x) + H(y) and H(rx) = rH(x). For example, there is a natural homomorphism from R^2 to R^3 given by H((x,y)) = (x,y,0). If B_1={v_1, v_2,...} and B_2 = {w_1, w_2,...} are bases for a vector space V, then there is a homomorphism H from V to V such that H(vi) = wi for each i. Thus if x = $a1v_1$ + $a2v_2$ +... is any vector in V, then H(x) = $a1w_1$ + $a2w_2$ +... In this case, H can be referred to as the change-of-basis **transformation**. If V is n-dimensional, then H can be represented as an n x n **matrix** with nonzero determinant. In fact, any n x n matrix M with nonzero determinant is a change of basis matrix since the columns of M form a basis for V.

Every vector space V has a dual vector space that is usually denoted by V*. V* is the set of all homomorphisms from V to the field K. If f and g are in V*, then (f + g)(v) is defined to be f(v) + g(v). If k is in K, then (kf)(v) is defined to be kf(v). For example, in R^3, if v is any vector, then the map that sends every x in R^3 to v·x is in R^3*. In fact, every element of R^3* is of this form. The map that sends each element f of F to the integral of f over [0,1] is an element of F*.

See also Banach space; Definition; Systems of linear equations; Vector algebra; Vector analysis

VELOCITY AND ACCELERATION VECTORS

Mathematically a vector in the plane is just an ordered pair of **real numbers**. Physicists tend to embellish this rather spare **definition** so that their vectors are represented by arrows with direction and magnitude indicating the direction and strength of forces, velocities, and accelerations among other things.

Consider a point moving in the xy coordinate plane so that it traces out some curve as the path of its **motion**. As the point moves along this curve, the x and y coordinates are changing as **functions** of time. Suppose that x=f(t) and y=g(t). Now the mathematician will say that the position at any time t is (f(t),g(t)) and will say that the position vector for the point is R(t)=(f(t),g(t)). The physicist will say that the position vector R(t) is an arrow starting at the origin and ending with the tip of the arrow at the point (f(t),g(t)).

It is now possible to define the velocity and acceleration vectors for this motion in terms of ideas from **calculus**. In calculus, the derivative of a function is defined to be the instantaneous rate of change of this function with respect to some **variable**. So, for example, the derivative of f(t) with respect to t (time), denoted by f'(t) is the instantaneous rate of change of f with respect to time, also called the (instantaneous) velocity of f when f represents position. The term "instantaneous" refers to the fact that this velocity is not an average velocity over a given time **interval**, but a velocity computed at one instant of time. In fact, the instantaneous velocity at time t is the limiting value of average velocities of the form (f(t+h)-f(t))/h over time intervals of the form [t,t+h] as h approaches 0. This is an instance of the derivative of a function.

Returning to our curve with position vector R(t)=(f(t),g(t)), it is reasonable to define an instantaneous velocity vector as the derivative of the position vector R(t). Again, we regard this as the limiting value of average velocities of the form (R(t+h)-R(t))/h as h approaches 0. Since R(t)=(f(t),g(t)), it follows that (R(t+h)-R(t))/h=((f(t+h)-f(t))/h,(g(t+h)-g(t))/h) and as h approaches 0 we have R'(t)=(f'(t),g'(t)) and this is the velocity vector of the point moving along the curve at time t.

Now the physicists' arrow for the velocity vector is placed with its tail (the non-pointed end) at the point on the curve and lies along the tangent line to the curve at the point (f(t),g(t)). This physically symbolizes the fact that point is moving in the direction of its velocity vector at any time t. Since acceleration is the rate of change of velocity, the acceleration vector is the derivative of the velocity vector or the second derivative of the position vector. This is denoted by R''(t)=(f''(t),g''(t)).

The physicists' arrow for acceleration is also drawn from the point (f(t),g(t)) and points in the direction of the concave side of the curve. An interesting special case is when the point is moving on a **circle**. In this case, the position vector arrow points outward from the center of the circle along a radius with the tip of the arrow ending at the point on the circle. If the circle has radius r and is centered at the origin, then R(t)=(rcos(t),rsin(t)). Using derivative rules for the **cosine** and **sine** functions proved in calculus, the velocity vector will be R'(t)=(-rsin(t),rcos(t)). It can be shown that this vector's arrow is perpendicular to the position vector's arrow at the point (rcos(t),rsin(t)). Again using calculus, the acceleration vector R''(t)=(-rcos(t),-rsin(t))=-R(t), so that the acceleration vector's arrow points back toward the center of the circle. This is a **model** of spinning a ball attached to a string in a circular motion above one's head. The ball "wants" to fly off in a straight line in the direction of the velocity vector, but is being pulled back towards the center by it's acceleration resulting in a circular motion.

The account given above can be extended to three dimensional **space** with the following modifications. The position vector will have the form R(t)=(f(t),g(t),h(t)); the velocity vector will then be R'(t)=(f'(t),g'(t),h'(t)); and the acceleration vector will be R''(t)=(f''(t),g''(t),h''(t)).

VENN DIAGRAMS

A Venn diagram is a schematic representation used in depicting collections of **sets** and the interrelationships between those sets. Although Venn diagrams are often employed in literary studies they are more common in mathematical studies involving the validity of **deduction** and **proof**. Physically, Venn diagrams are collections of intersecting, simple, two-dimensional **polygons**. The order of a Venn diagram n is often specified and refers to the number of simple closed curves.

A simple Venn diagram is one in which no two curves intersect in more than a finite number of points and no three or more curves intersect at one common point. This says that the curves defining sets meet at points and not in segments of curves. Below is an example of an order-three Venn diagram:

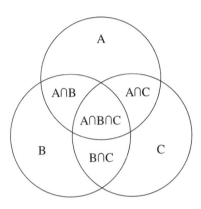

This particular diagram consists of three intersecting circles that are symmetrically placed and comprise a total of eight regions. A, B, and C are regions that consist of members of a set that are not common to any other set. The regions labeled A∩B, A∩C, and B∩C consist of members that are common to two sets but not the third. The region labeled A∩B∩C consists of members that are common to all three sets. In this particular case, since the circles are symmetrically placed such that the center of each **circle** is located at the intersection of the other two, the region A∩B∩C is a geometric shape known as a Reuleaux **triangle**. The last region is the region outside of the three sets and is indicative of the empty set, the set that contains no members. So an order-n Venn diagram is a collection of n sets represented by simple closed curves in a plane that partition the plane into 2^n connected regions called subsets (including the region outside of the curves). Each of these subsets represents a unique region formed upon the intersection of the interiors of the closed curves.

The introduction of Venn diagrams is attributed to **John Venn**, an English mathematician, although there is strong proof that they originated earlier. **Aristotle**, a Greek scientist, used a diagram called the tree of Porphyry to represent logical concepts as early as 350B.C. Swiss mathematician **Leonhard Euler** formulated a geometric system that generated class logic solutions in 1761. This system was

replaced by Venn's system in 1880 in a paper published in the *Philosophical Magazine* and *Journal of Science*. Venn's formulation of this type of system was a natural evolution of Boolean **algebra** introduced by **George Boole** in 1847 and 1854. **Boolean algebra** represents logical expressions in a mathematical form using a standard set of operators. Venn diagrams were initially based on the relationships between overlapping circles or ellipses. In 1881 Allan Marquand introduced the first logical diagrams based on squares or rectangles. Since that time the study of Venn diagrams and their uses has been expanded to literary uses.

There are several different aspects of particular Venn diagrams that have earned specific terminology. On the subject of extending Venn diagrams they can be reducible, irreducible, or extendible. A Venn diagram is called reducible if the removal of one of the curves results in a Venn diagram with n-1 curves. If this is not the case then the diagram is called irreducible. Many elliptical Venn diagrams are irreducible and it was unknown for a long time whether a reducible one existed. An article by Hamburger and Pippert appearing in a 1996 **American Scientist** showed that there was such a Venn diagram. If there is a curve such that when added to an order-n Venn diagram C produces an order-$(n + 1)$ Venn diagram then C is called extendible. The new Venn diagram is said to be an extension of C.

There are other interesting aspects of particular Venn diagrams that are associated with their **geometry**. A Venn diagram is called convex if the interiors of all of its curves are convex. A Venn diagram is called exposed if each of the curves that compose it touches the outer face, the empty set, at some point of non-intersection. Every convex Venn diagram is also exposed. A Venn diagram is said to have a hidden curve if it has a curve that does not touch the outer empty region. Every simple Venn diagram with five or less curves are exposed.

Three other interesting characteristics of particular Venn diagrams are **congruence**, **symmetry**, and monotonicity. A congruent Venn diagram is one that is composed of congruent curves. The first congruent Venn diagrams were constructed by Grünbaum in 1975. Symmetric Venn diagrams are ones that are first congruent and that also have an n-fold rotational symmetry. That is that there is a point about which the diagrams may be rotated by $2i\pi/n$, where i = 0, 1,...n - 1, and remain **invariant**. All symmetric Venn diagrams are exposed. A Venn diagram can be described as monotone if every k-region, for $0 < k < n$, is adjacent to both a $(k - 1)$-region and a $(k + 1)$-region. All Venn diagrams constructed according to Venn's original paper describing such diagrams are monotone.

The last interesting feature of the Venn diagram that will be discussed here is another graph that is associated with the Venn diagram. This graph is called the Venn graph and is another two-dimensional graph that is the planar dual of the Venn diagram. The vertices of the Venn graph are the connected open regions of the Venn diagram. A line connects two vertices if they share a common boundary. Below is shown the Venn graph for another three-order Venn diagram.

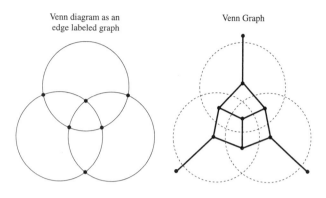

Venn diagram as an edge labeled graph

Venn Graph

VENN, JOHN (1834-1923)

English lecturer and author

John Venn is most famous for his development of diagrams, later named after him, that depict relationships between **sets**. Although Gottfried Wilhelm von Liebniz and **Leonhard Euler** had used similar diagrams, Venn's were considered more descriptive and easier to understand. He also helped to develop **George Boole**'s system of mathematical logic.

Venn was born in Hull, England on August 4, 1834, a descendant of a long line of Church of England evangelicals. He received his early education at two schools in London, at Highgate and Islington, but historical records indicate that either he was so poor a student or the schools were so incompetent that he was ill prepared for college. Nevertheless, he entered Gonville and Caius College at Cambridge in 1853 and earned a degree in mathematics there in 1857. The school made him a fellow upon graduation—he would retain that status for the rest of his life.

Becoming a priest in 1859, Venn went to work as a curate in the town of Mortlake. However, by 1862 he was back in the world of academia with a job at Cambridge University as a lecturer in moral science. His courses' main topics were **probability theory** and logic. It was at this point that Venn began developing Boole's mathematical logic, using what would become known as **Venn diagrams** to do so.

A Venn diagram is a pictorial representation of the relationships among sets. There is an outer rectangle that stands for the universal set, within which are circles or ellipses representing subsets of the universal set. For instance, Venn called three circles (R, S, and T) subsets of set U. The intersections of these circles and their complements split set U into eight nonoverlapping areas, the unions of which produced 256 distinct Boolean combinations of sets R, S, and T.

In 1866 Venn wrote *The Logic of Chance,* which had major influence on the evolution of the theory of **statistics** and developed an aspect of probability theory called frequency theory. Meanwhile, he was becoming dissatisfied with the Anglican Church, which he decided to leave in 1870. Afterward, although Venn continued to be a devout churchgoer, he dedicated himself mainly to his academic career.

Venn published *Symbolic Logic,* an attempt to correct and interpret Boole's work, in 1881. His *Principles of Empirical Logic* came out in 1889, but critics largely agreed that the first work was Venn's most original. Meanwhile, Venn had become enamored of history and had written one for his alma mater in 1897. More impressive, however, was his compilation (with his son) of a history of Cambridge University. An enormous undertaking, the first of two volumes appeared in 1922.

Aside from his academic endeavors, Venn also enjoyed building machines. His talent for such extended to a device that bowled balls for cricket; the machine was so effective that the top players of an Australian team could not even make contact with the balls during a trial run in 1909. Venn died in Cambridge, England, in 1923.

VIBRATION

Vibration is back-and-forth or up-and-down **motion** in a material or medium. The meaning of the word also includes any periodic physical process, such as the vibration of which we hear as sound, or the cyclical change of intensity of the electric field in a common alternating current circuit. A tuning fork is used for precise emission of a particular audio tone. When struck the fingers of the fork begin to vibrate at the specific frequency to which they are tuned. Such a device causes the air immediately around it to vibrate at a precise frequency. A musical cymbal, on the other hand, while it also vibrates, emits many frequencies, which are dependent upon the size, shape, and composition of the cymbal. While a tuning fork will sound like a pure tone, a cymbal emits a "crash" sound due to the various frequencies at which it vibrates. As materials around the vibrating device and the structural integrity of the device itself resist the vibrations, they slowly subside, causing the vibrations and the sound to eventually stop. A cymbal player during a concert will often help this dampening process along by grabbing the cymbal to stop the sound of it suddenly.

Even more complex vibrations take place when an elastic object, such as a ball is struck (or bounced on the ground), its **perimeter** begins to vibrate, but not in the symmetrical fashion seen in a tuning fork. Despite this, as with all other types of vibration, the amplitude of the vibration is the maximum **distance** a surface of the object is displaced and, mathematically, of vibration, while many cases are much more complex than the examples used here, is carried out in the same way no matter whether it be waves of **light** travelling through the vacuum of **space** or simple sound waves in air.

VIÈTE (ALSO KNOWN AS FRANCISCUS VIETA), FRANÇOIS [FRANCISCUS VIETA] (1540-1603)

French algebraist

Many scholars consider François Viète the seminal algebraist of the 16th century. Although he worked in mathematics in his spare time and was regarded as an amateur mathematician, he has contributed numerous improvements, solutions, and other vital developments to mathematics. For example, Viète had a hand in the development of modern **algebra** by devising algebraic notation, using letters to denote quantities, both unknown and known. He also contributed to the theory of **equations**, and introduced algebraic terms, such as "coefficient," that are still used today. Among Viète's other significant mathematical accomplishments were: using algebra to solve longstanding geometrical problems; calculating π to 10 places, an important accomplishment of his time; playing a role in the improvements brought by the development of Julian calendars; and having a public hand in the mathematical controversies of his era.

Viète was born in Fontenay–le–Comte, Poitou, France, to Étienne (a lawyer and notary) and his wife, Marguerite Dupont. After receiving his education in Fontenay—partly from monks at a local Franciscan cloister—he studied law at the university in Poitiers beginning in 1556. Upon graduation with a bachelor's degree, he worked as a lawyer in Fontenay for four years. Viète was apparently successful in this career; he counted among his clients Queen Eleanor of Austria and Mary Stuart. During this period, Viète also began conducting mathematical research on the side, as well as cosmological and astronomical studies.

In 1564, Viète became closely involved with the aristocratic Soubise family, including Antoinette d'Aubeterre, who consulted him on legal matters. Viète stopped practicing law to serve as private secretary to d'Aubeterre. He moved with the family to La Rochelle where he tutored d'Aubeterre's daughter, Catherine de Parthenay, until her marriage in 1570. During this time, Viète decided to become a Huguenot (French Protestant). These years also mark the first period where Viète had a significant amount of time to devote to his mathematical research.

After his charge's marriage, Viète's primary occupation became service in the royal courts. He was appointed by King Charles IX to serve as a councilor to Brittany's parliament at Rennes in 1573, a post he held until 1580. He then served Charles IX and his successor, Henry III, in Paris as member of the privy counsel and the maître de requêtes (Master of Requests). In 1584, soon after Henry III came to the throne, Viète was forced from the royal court because of his Huguenot sympathies. He spent the next five years in cities such as Garnache, Fontenay, and Bigotière, where he devoted his full attention to mathematics and related research. This marks Viète's second fruitful period of mathematical work.

Viète's ties to the royal court allowed his work to be printed by the royal printer, Jean Mettayer, who published most of Viète's mathematical treatises. Viète's first important book, *Canon mathematicus seu ad triangula cum appendibus* ("Mathematical Laws Applied to Triangles") was published in 1579. One of Viète's goals as a mathematician was to give more credence to trigonometryin the mathematical world and this text contributed significantly to this cause. Using all six trigonomic **functions**, he solved problems on plane and spherical triangles.

In 1589, Viète was invited to the court of Henry III (then in Tours) as a counselor to the French parliament. In the same year, he was appointed a royal Privy Councilor, and broke the code used by Philip II of Spain for internal messages when France was at war with Spain in 1589–90. Henry IV ascended to the French throne in 1589 upon Henry III's death, and, like his royal patron, Viète declared himself Catholic. He stayed in service at the court—following it to Paris in 1594—and remained there until his retirement in 1602.

These years also saw Viète publishing a series of important works. The most influential of these was his 1591 *In artem analyticam Isagoge* ("Introduction to the Analytical Arts"), an algebra text. In this work he introduced the first systematic use of symbolic algebraic notation. Viète demonstrated, for example, the value of symbols by using the plus and minus signs for operations, vowels for unknown quantities (variables), and consonants for known quantities (parameters). He also delineated new ways of solving **cubic equations**, using, among other approaches, **trigonometry**. This book helped algebra develop further, and is perhaps one of the earliest identifiable "modern" algebra textbooks.

Two years later, in 1593, *Supplementum geometriae* was published. Although published in 1593, Viète had probably written this treatise in 1591. Among the many geometric topics in this and a second volume published the same year were solutions to the problem of the trisection of an **angle** and the corresponding cubic equation, the doubling of the cube, the construction of the regular heptagon inscribed in a **circle**, and the earliest known explicit expression for π as an infinite product. The last significant work of Viète's published in his lifetime was *De numerosa potestatum puarum atque ad fectarum ad exegesin resolution e tractatcus* (1600). In it, he delineates a way of approximating **roots** of numerical equations. He also solved the fourth Apollonian problem, which concerns three circles and the construction of a fourth tangent to the other three.

Twelve years after his death, *De aquationem recognitione et emedatione libri duo* ("Concerning the Recognition and Emendation of Equations ") was published. It was Viète's primary text dealing with the theory of equations; he developed methodology for solutions of second, third, and fourth degree equations. Viète also introduced the term "coefficient" in this treatise.

Viète died December 13, 1603 (or February 23, 1603, depending on the source) in Paris, France. It is known that he was twice married, first to Barbe Cotherau and later, after becoming a widower, to Juliette Leclerc. He had at least one child, but details concerning his marriages and children are sketchy.

VIGESIMAL NUMERATION

Vigesimal numeration is a numeral system in which all derived units are based on the number 20 and the **powers** of 20. Vigesimal is derived from Latin word *vicesimus* (twentieth), based on *viginti* (twenty) that itself descended from a Sanskrit word, *vimsatih* (twenty). Other related words are *vicennial* (once every 20 years) and *vicenary* that has the same meaning as vigesimal. The use of 20 as a grouping (or base) number was used by many cultures throughout our human history (most likely) because people have twenty digits, or the number of fingers and toes.

The Aztecs and Mayans used a base 20 number system as did almost all Eskimo tribes, some native North American societies, almost all peoples native to Central and South America, and some cultures in northern Siberia and Africa. Mayan mathematics brought about the most sophisticated counting system ever developed in the Americas. Mayan numerical systems originated in concepts inherited from the Olmecs, an ancient civilization from Mexico and the adjacent Central American region. The Mayan system was a positional number system with a quasi-vigesimal base. It was "quasi-vigesimal" because successive powers of the base number (i.e., "b") had the values: $20^0 = 1$, $20^1 = 20$, $18 \times 20^1 = 360$, $18 \times 20^2 = 7200$, $18 \times 20^3 = 144000$, and so forth, rather than strict powers of 20 ($20^0 = 1$, $20^1 = 20$, $20^2 = 400$, $20^3 = 8000$, $20^4 = 160000$, and so forth). The reason for the curious use of "18 x 20" as a base in the Mayan number system was most likely due to the official Mayan year consisting of 360 days ($18 \times 20 = 360$). The Mayan Indians compiled extensive observations of planetary positions in base 20 notation. The recording of time was an extremely complicated affair. It was based on the superimposition of two **calendar** cycles called the "TZOLKIN," or Sacred Almanac, and the "HAAB." The TZOLKIN cycle consists of 20 periods of 13 days each, while the HAAB consisted of 18 months of 20 days each, plus 5 "unlucky" days. The Mayan number system was notable in its development of the **zero** as a placeholder hundreds of years before A.D. 876, its earliest known use in India.

The Maya used a system of dots and bars (most likely representing pebbles and sticks, respectively) as a way to count. The symbol of a "dot" stood for one and a symbol of a "bar" stood for five, while an ovular shell with vertical bars inside stood for zero. For example, one dot represented "1," two dots for "2,"..., one bar for "5," one dot and one bar for "6,"..., two bars for "10," one dot and two bars for "11,"..., three bars for "15," one dot and three bars for "16,"..., four dots and three bars for "19," and so forth. Mayan numbers were written from bottom to top, rather than horizontally. In writing these symbols, the bars are placed horizontally and the dots placed on top of them, and the vigesimal positions develop upward from the base. As an example of representing a number in the Mayan vigesimal number system, the number 34 would be stated as "20 + 14," or two bars and four dots [on the bottom to represent 14 as "(2 x 5) + (4 x 1)"] followed by one dot [on the top to represent 20 as "1 x 20"].

The Maya considered some numbers more sacred than others. One of these special numbers was 20, as it represented the number of fingers and toes a human being could count on. Another special number was five, as this represented the number of digits on a hand or foot. Thirteen was religiously sacred as the number of original Maya gods. Another sacred number was 52, representing the number of years in a "bundle," a unit

similar in concept to a century. Another number, 400, had sacred meaning as the number of Maya gods of the night.

The modern-day vigesimal number system is based on the powers of the base 20, that is $(..., 160000, 8000, 400, 20, 1, 1/20, 1/400,...) = (..., 20^4, 20^3, 20^2, 20^1, 20^0, 20^{-1}, 20^{-2},...)$. Any real number can be represented in a positional number system of base "b". For the base 20 (i.e., "b = 20") numbering system a real number x can be represented as $x = a_n b^n + a_{n-1} b^{n-1} +... + a_1 b^1 + a_0 b^0 + a_{-1} b^{-1} +... + a_{-(n-1)} b^{-(n-1)} + a_{-n} b^{-n}$. The equation of "x" must satisfy the requirements on a_i, where i = {0, 1, 2,..., n} and $0 \le a_i < b$.

The (modern) base 20 notational system uses the twenty digits of 0, 1, 2, 3, 4, 5, 6, 7, 8, 9, A, B, C, D, E, F, G, H, I, and J. This notation is to be understood as follows: A represents ten, B represents eleven, C represents twelve, and so forth in order up to the letter J which represents nineteen. As an example of expressing a number in vigesimal form, consider the number (in decimal form) "x = 4,265." In the vigesimal system it is represented as "AD5." Using the vigesimal numeric scheme described previously we see that AD5 = (A x 20^2) + (D x 20^1) + (5 x 20^0) = (10 x 400) + (13 x 20) + (5 x 1) = 4,265.

See also Decimal position system; Positional notation

VOLTERRA, VITO (1860-1940)
Italian mathematician

Vito Volterra had a major impact on the development of **calculus**, and he originated the concept of a theory of **functions**. Perhaps his most famous work was on integral **equations**.

Volterra was born to a poor cloth salesman on May 3, 1860 in Rome, Italy. The Volterras became virtually destitute when Mr. Volterra died when Vito was about two, so the boy and his mother moved in with her brother. Volterra grew up mainly in Florence, attending the Istituto Tecnico **Galileo Galilei** and the Scuola Tecnica Dante Aligheri. An extremely precocious child, Volterra became interested in the **geometry** of **Adrien-Marie Legendre** at age 11, and was soon devising original problems that he would then try to solve.

With his obvious interest in science and mathematics, Volterra rebelled against his family's insistence that he become a bank clerk or similar professional. However, an uncle intervened and persuaded Volterra's relatives to let the gifted boy continue his studies, even helping him to get a job as an assistant in the physics laboratory at the University of Florence. Meanwhile, Volterra was still in high school, from which he graduated in 1878. Later that year, he enrolled at the University of Florence's Department of Natural Sciences.

In 1880 Volterra won a contest to become a resident student at the prestigious Scuola Normale Superiore in Pisa, Italy. There he took classes in **mathematics and physics**, gradually narrowing his interest to **mechanics** and mathematical physics. He graduated from Pisa with a doctorate in physics in 1882, having written his thesis on **hydrodynamics**. Soon afterward, one of his professors, **Enrico Betti**, hired Volterra as an assistant. A year later, Volterra, still only 23, won a competition to become the mechanics professor at the university. When Betti died, Volterra assumed the chair of mathematical physics.

In about 1883, Volterra thought of the idea of a theory of functions that depend on a continuous set of values. This work would lead to the development of such fields of **analysis** as the solution of integral and integro-differential equations. In addition, he contributed the seeds of what would become the key concept of harmonic integrals. The following year, Volterra began studying integral equations and by 1892 he was publishing influential papers on partial **differential equations**. These concentrated particularly on the equation of cylindrical waves.

Volterra accepted a new post—as professor of mechanics at the University of Turin—in 1892. Four years later, he published a series of papers on what mathematicians would later call "integral equations of a Volterra type." In 1900 he became chair of mathematical physics at the University of Rome. When World War I began, however, Volterra interrupted his academic career briefly to enroll in the Italian Army Corps of Engineers. Although he was 55 at this point, the army gratefully accepted technical assistance from such an illustrious scholar, inducting him as an officer in the air branch. In this position, Volterra was the first to suggest using helium instead of hydrogen in dirigibles, thus helping perfect this new type of airship.

After the war, Volterra returned to his post at the University of Rome. His interests began turning to mathematical biology at this point, and he produced works on such pertinent topics as predator-prey equations and mathematical models of other biological associations. In 1922, his attention was again diverted by political matters as fascism began spreading across Italy. As a member of the Senate (appointed in 1905), he fought against what he and a few others quickly realized would be death to Italian democracy. Eventually, Volterra was forced to leave the University of Rome when he refused to take an oath of loyalty to the fascist government of Benito Mussolini in 1931. In 1932, he had to surrender his membership in numerous Italian scientific academies.

After being banished from Italian academia, Volterra lived mainly in Paris and Spain. He wrote one of his most influential books, *The Theory of Functionals and of Integral and Integro-Differential Equations,* in 1930. Starting in 1931, he accepted numerous invitations to lecture all around Europe, spending only short periods of time at his Italian country home in the Alban Hills near Rome. Afflicted with phlebitis beginning in 1938, Volterra nevertheless continued his pursuit of scientific knowledge.

Volterra, who did not dedicate himself exclusively to research but also enjoyed associating with many of the prominent artists, politicians, and writers of his time, died on October 11, 1940 in Rome. He and his wife had married in 1940.

VOLUME

Volume is the amount of **space** occupied by an object or a material. Volume is said to be a derived unit, since the volume

of an object can be known from other measurements. In order to find the volume of a rectangular box, for example, one only needs to know the length, width, and depth of the box. Then the volume can be calculated from the formula, $V = l \times w \times d$.

Volume of most physical objects is a function of two other factors, temperature and pressure. In general, the volume of an object increases with an increase in temperature and decreases with an increase in pressure. Some exceptions exist to this general rule. For example, when water is heated from a temperature of 32°F (0°C) to 39°F (4°C), it decreases in volume. Above 39°F (4°C), however, further heating of water results in an increase in volume that is more characteristic of matter.

Units of volume

The term unit volume refers to the volume of one something: one quart, one milliliter, or one cubic inch, for example. Every measuring system that exists defines a unit volume for that system. Then, when one speaks about the volume of an object in that system, what he or she means is how many times that unit volume is contained within the object. If the volume of a glass of water is said to be 35.6 in³ (90.4 cm³), for example, what is meant is that 35.6 cubic-inch unit volumes could be placed into that glass.

Mathematically, volume would seem to be a simple extension of the concept of **area**, but it is actually more complicated. The volume of simple figures with integral sides is found by determining the number of unit cubes that fit into the figure. When this idea is extended to include all possible positive **real numbers**, however, paradoxes of volume occur. It theoretically is possible to take a solid figure apart into a few pieces and reassemble it so that it has a different volume.

The units in which volume is measured depend on a variety of factors, such as the system of measurement being used and the type of material being measured. For example, volume in the British system of measurement may be measured in barrels, bushels, drams, gills, pecks, teaspoons, or other units. Each of these units may have more than one meaning, depending on the material being measured. For example, the precise size of a "barrel" ranges anywhere from 31 to 42 gal (117 to 159 l), depending on federal and state statutes. The more standard units used in the British system, however, are the cubic inch or cubic foot and the gallon.

Variability in the basic units also exists. For example, the "quart" differs in size depending on whether it is being used to measure a liquid or dry volume and whether it is a measurement made in the British or customary U.S. system. As an example, 1 customary liquid quart is equivalent to 57.75 cubic inches, while 1 customary dry quart is equivalent to 67.201 cubic inches. In contrast, 1 British quart is equivalent to 69.354 cubic inches.

The basic unit of volume in the international system (often called the **metric system**) is the liter (abbreviated as L), although the cubic centimeter (cc or cm³) and milliliter (mL) are also widely used as units for measuring volume. The fundamental relationship between units in the two systems is given by the fact that 1 U.S. liquid quart is equivalent to 0.946 L or, conversely, 1 liter is equivalent to 1.057 customary liquid quarts.

The volume of solids

The volume of solids is relatively less affected by pressure and temperature changes than is that of liquids or gases. For example, heating a liter of iron from 32°F (0°C) to 212°F (100°C) causes an increase in volume of less than 1%, and heating a liter of water through the same temperature range causes an increase in volume of less than 5%. But heating a liter of air from 32°F (0°C) to 212°F (100°C) causes an increase in volume of nearly 140%.

The volume of a solid object can be determined in one of two general ways, depending on whether or not a mathematical formula can be written for the object. For example, the volume of a cube can be determined if one knows the length of one side. In such a case, $V = s^3$, or the volume of the cube is equal to the cube of the length of any one side (all sides being equal in length). The volume of a **cylinder**, on the other hand, is equal to the product of the area of the base multiplied by the **altitude** of the cylinder. For a right circular cylinder, the volume is equal to the product of the radius of the circular base (r) squared multiplied by the height (h) of the cone and by **pi** (π), or $V = \pi r^2 h$.

Many solid objects have irregular shapes for which no mathematical formula exists. One way to find the volume of such objects is to sub-divide them into recognizable shapes for which formulas do exist (such as many small cubes) and then approximate the total volume by summing the volumes of individual sub-divisions. This method of approximation can become exact by using **calculus**. Another way is to calculate the volume by water displacement, or the displacement of some other liquid.

Suppose, for example, that one wishes to calculate the volume of an irregularly shaped piece of rock. One way to determine that volume is first to add water to some volume-measuring instrument, such as a graduated cylinder. The exact volume of water added to the cylinder is recorded. Then, the object whose volume is to be determined is also added to the cylinder. The water in the cylinder will rise by an amount equivalent to the volume of the object. Thus, the final volume read on the cylinder less the original volume is equal to the volume of the submerged object.

This method is applicable, of course, only if the object is insoluble in water. If the object is soluble in water, then another liquid, such as alcohol or cyclohexane, can be substituted for the water.

The volume of liquids and solids

Measuring the volume of a liquid is relatively straight forward. Since liquids take the shape of the container in which they are placed, a liquid whose volume is to be found can simply be poured into a graduated container, that is, a container on which some scale has been etched. Graduated cylinders of various sizes, ranging from 10 mL to 1 L are commonly available in science laboratories for measuring the volumes of liq-

uids. Other devices, such as pipettes and burettes, are available for measuring exact volumes, especially small volumes.

The volume of a liquid is only moderately affected by pressure, but it is often quite sensitive to changes in temperature. For this reason, volume measurements made at temperatures other than ambient temperature are generally so indicated when they are reported, as V = 35.89 mL (95°F; 35°C).

The volume of gases is very much influenced by temperature and pressure. Thus, any attempt to measure or report the volume of the gas must always include an indication of the pressure and temperature under which that volume was measured. Indeed, since gases expand to fill any container into which they are placed, the term volume has meaning for a gas *only* when temperature and pressure are indicated.

VON KOCH, NIELS FABIAN HELGE (1870-1924)
Swedish mathematician

Commonly known by the name Helge, Niels Fabian Helge von Koch is best remembered for devising geometrical constructs that are now called the Koch curve and the Koch snowflake (or star). He was also an expert on **number theory** and wrote extensively on the **prime number theorem.**

Von Koch, the son of a career soldier, was born in Stockholm, Sweden on January 25, 1870. After completing his early education in that city in 1887, he went on to study at the University of Stockholm. There he had the opportunity to study under **Gosta Mittag-Leffler**, who was the school's first mathematics professor.

In 1888 von Koch took some classes at Uppsala University, where he worked on linear **equations**. His first major paper, published in 1891, was on the application of infinite **determinants** to solving **differential equations** with analytic coefficients.

Von Koch earned his doctorate in mathematics from the University of Stockholm in 1892, writing a thesis that would contribute to the development of functional **analysis**. In 1893 he accepted a job as assistant professor of mathematics at an unknown school (perhaps the University of Stockholm). He had several such low-ranking appointments between then and 1905, and suffered another disappointment when he was turned down as chair of number theory and **algebra** at Uppsala University. However, in 1905 von Koch finally achieved a promotion when a colleague resigned his professorship at the Royal Technological Institute in Stockholm. The school offered von Koch the chair of pure mathematics, which he promptly accepted.

In 1901, von Koch published *On the Distribution of Prime Numbers,* which concentrated on the prime number **theorem**. In 1906, he released his work on curves and snowflakes. The von Koch curve is made by taking an equilateral **triangle** and attaching another equilateral triangle to each of the three sides. This first **iteration** produces a Star of David-like shape, but as one repeats the same process over and over, the effect becomes increasingly fractal and jagged, eventually taking on the traditional snowflake shape. The snowflake is actually a continuous curve without a tangent at any point. Von Koch curves and snowflakes are also unusual in that they have infinite perimeters, but finite areas.

After writing another book on the prime number theorem in 1910, von Koch succeeded Mittag-Leffler as mathematics professor at the University of Stockholm in 1911. He died in Stockholm on March 11, 1924, having taught for most of the remainder of his life.

See also Snowflake curve

VON NEUMANN, JOHN (1903-1957)

Hungarian American mathematician

John von Neumann, considered one of the most creative mathematicians of the 20th century, made important contributions to quantum physics, **game theory**, economics, meteorology, the development of the atomic bomb, and computer design. He was known for his problem-solving ability, his encyclopedic memory, and his ability to reduce complex problems to a mathematically tractable form. Von Neumann served as a consultant to the United States government on scientific and military matters, and was a member of the Atomic Energy Commission. According to mathematician Peter D. Lax, von Neumann combined extreme quickness, very broad interests, and a fearsome technical prowess; the popular saying was, "Most mathematicians prove what they can; von Neumann proves what he wants." The Nobel Laureate physicist Hans Albrecht Bethe said, "I have sometimes wondered whether a brain like von Neumann's does not indicate a species superior to that of man."

Max and Margaret von Neumann's son Janos was born in Budapest, Hungary, on December 28, 1903. As a child he was called Jancsi, which later became Johnny in the United States. His father was a prosperous banker. Von Neumann was tutored at home until age ten, when he was enrolled in the Lutheran Gymnasium for boys. His early interests included literature, music, science and psychology. His teachers recognized his talent in mathematics and arranged for him to be tutored by a young mathematician at the University of Budapest, Michael Fekete. Von Neumann and Fekete wrote a mathematical paper which was published in 1921.

Von Neumann entered the University of Budapest in 1921 to study mathematics; he also studied chemical engineering at the Eidgenössische Technische Hochschule in Zurich, receiving a diploma in 1925. In those same years, he spent much of his time in Berlin, where he was influenced by eminent scientists and mathematicians. In 1926 he received a Ph.D. in mathematics from the University of Budapest, with a doctoral thesis in **set theory**. He was named *Privatdozent* at the University of Berlin (a position comparable to that of assistant professor in an American university), reportedly the youngest person to hold the position in the history of the university. In 1926 he also received a Rockefeller grant for postdoctoral work under mathematician **David Hilbert** at the University of Göttingen. In 1929 he transferred to the University of Hamburg. By this time, he had become known to mathematicians through his publications in set theory, **algebra**, and quantum theory, and was regarded as a young genius.

In his early career, von Neumann focused on two research areas: first, set theory and the logical foundations of mathematics; and second, **Hilbert space** theory, operator theory, and the mathematical foundations of quantum **mechanics**. During the 1920s, von Neumann published seven papers on mathematical logic. He formulated a rigorous **definition** of **ordinal numbers** and presented a new system of axioms for set theory. With Hilbert, he worked on a formalist approach to the foundations of mathematics, attempting to prove the consistency of **arithmetic**. In about 300 B.C., **Euclid**'s *Elements of Geometry* had proved mathematical theorems using a limited number of axioms. Between 1910 and 1913, **Bertrand Russell** and **Alfred North Whitehead** had published *Principia Mathematica,* which showed that much of the newer math could similarly be derived from a few axioms. With Hilbert, von Neumann worked to carry this approach further, although in 1931 **Kurt Gödel** proved that no formal system could be both complete and consistent.

Hilbert was interested in the axiomatic foundations of modern physics, and he gave a seminar on the subject at Göttingen. The two approaches to quantum mechanics—the wave theory of Erwin Schrödinger and the particle theory of **Werner Karl Heisenberg**—had not been successfully reconciled. Working with Hilbert, von Neumann developed a finite set of axioms that satisfied both the Heisenberg and Schrödinger approaches. Von Neumann's axiomatization represented an abstract unification of the wave and particle theories.

During this period, some physicists believed that the probabilistic character of measurements in quantum theory was due to parameters that were not yet clearly understood and that further investigation could result in a deterministic quantum theory. However, von Neumann successfully argued that the indeterminism was inherent and arose from the interaction between the observer and the observed.

In 1929 von Neumann was invited to teach at Princeton University in New Jersey. He accepted the offer and taught mathematics classes from 1930 until 1933, when he joined the elite research group at the newly established Princeton Institute for Advanced Study. The atmosphere at Princeton was informal yet intense. According to mathematician Stanislaw Ulam, writing in the *Bulletin of the American Mathematical Society,* the group "quite possibly constituted one of the greatest concentrations of brains in **mathematics and physics** at any time and place." During the 1930s von Neumann developed algebraic theories derived from his research into **quantum mechanics**. These theories were later known as von Neumann algebras. He also conducted research into Hilbert **space**, ergodic theory, Haar measure, and noncommutative algebras. In 1932 he published a book on quantum physics, *The Mathematical Foundations of Quantum Mechanics,* which remains a standard text on the subject. After becoming a naturalized citizen of the United States, von Neumann became a consultant to the Ballistics Research Laboratory of the Army Ordnance Department in 1937. After the attack on Pearl Harbor in 1941, he became more involved in defense research, serving as a consultant to the National Defense Research Council on the theory of detonation of explosives, and with the Navy Bureau of Ordnance on mine warfare and countermeasures to it. In 1943 he became a consultant on the development of the atomic bomb at the Los Alamos Scientific Laboratory in New Mexico.

At Los Alamos, von Neumann persuaded J. Robert Oppenheimer to pursue the possibility of using an implosion technique to detonate the atomic bomb. This technique was later used to detonate the bomb dropped on Nagasaki. Simulation of the technique at the Los Alamos lab required extensive numerical calculations which were performed by a staff of twenty people using desk calculators. Hoping to speed up the work, von Neumann investigated using computers for the calculations and studied the design and programming of IBM punch-card machines. In 1943 the Army sponsored work at the Moore School of Engineering at the University of Pennsylvania, under the direction of John William Mauchly and J. Presper Eckert, on a giant **calculator** for computing firing tables for guns. The machine, called ENIAC (Electronic Numerical Integrator and Computer), was brought to von Neumann's attention in 1944. He joined Mauchly and Eckert in planning an improved machine, EDVAC (Electronic Discrete **Variable** Automatic Computer). Von Neumann's 1945 report on the EDVAC presented the first written description of the stored-program concept, which makes it possible to load a computer program into computer memory from disk so that the computer can run the program without requiring manual reprogramming. All modern computers are based on this design.

Von Neumann's design for a computer for scientific research, built at the Princeton Institute for Advanced Study between 1946 and 1951, served as the **model** for virtually all subsequent computer applications. Those built at Los Alamos, the RAND Corporation, the University of Illinois and the IBM Corporation all incorporated, besides the stored program, the separate components of arithmetic function, central control (now commonly referred to as the central processing unit or CPU), random-access memory (or RAM) as represented by the hard drive, and the input and output devices operating in serial or parallel **mode**. These elements, present in virtually all personal and mainframe computers, were all pioneered under von Neumann's auspices.

In addition, von Neumann investigated the field of neurology, looking for ways for computers to imitate the operations of the human brain. In 1946 he became interested in the challenges of weather forecasting by computer; his Meteorology Project at Princeton succeeded in predicting the development of new storms. Because of his role in early computer design and programming techniques, von Neumann is considered one of the founders of the computer age.

While in Germany, von Neumann had analyzed strategies in the game of poker and wrote a paper presenting a mathematical model for games of strategy. He continued his work in this area while he was at Princeton, particularly considering applications of game theory to economics. When the Austrian economist **Oskar Morgenstern** came to Princeton, he and von Neumann started collaborating on applications of game theory to economic problems, such as the exchange of goods between parties, monopolies and oligopolies, and free trade. Their ambitious 641-page book, *Theory of Games and Economic Behavior,* was published in 1944. Von Neumann's work opened new channels of communication between mathematics and the social sciences.

Von Neumann and Morgenstern argued that the mathematics as developed for the physical sciences was inadequate for economics, since economics seeks to describe systems based not on immutable natural laws but on human action involving choice. Von Neumann proposed a different mathematical model to analyze strategies, taking into account the interdependent choices of "players." Game theory is based on an analogy between games and any complex decision-making process, and assumes that all participants act rationally to maximize the outcome of the "game" for themselves. It also assumes that participants are able to rank-order possible outcomes without error. Von Neumann's analysis enables players to calculate the consequences or probable outcomes of any given choice. It then becomes possible to opt for those strategies that have the highest probability of leading to a positive outcome. Game theory can be applied not only to economics and other social sciences but to politics, business organization and military strategy, to mention only a few areas of its usefulness.

After the war, von Neumann served as a scientific consultant for government policy committees and agencies such as the CIA and National Security Agency. He advised the RAND Corporation on its research on game theory and its military applications, and provided technical advice to companies such as IBM and Standard Oil. Following the detonation of an atomic bomb by the Soviets in 1949, von Neumann contributed to the development of the hydrogen bomb. He believed that a strong military capacity was more effective than a disarmament agreement. As chairman of the nuclear weapons panel of the Air Force scientific advisory board (known as the von Neumann committee), his recommendations led to the development of intercontinental missiles and submarine-launched missiles. Herbert York, the director of the Livermore Laboratory, said, "He was very powerful and productive in pure science and mathematics and at the same time had a remarkably strong streak of practicality [which] gave him a credibility with military officers, engineers, industrialists and scientists that nobody else could match."

In 1954 President Eisenhower appointed von Neumann to the Atomic Energy Commission. Von Neumann was hopeful that nuclear fusion technologies would provide cheap and plentiful energy. According to the chairman of the Commission, Admiral Lewis Strauss, "He had the invaluable faculty of being able to take the most difficult problem, separate it into its components, whereupon everything looked brilliantly simple, and all of us wondered why we had not been able to see through to the answer as clearly as it was possible for him to do." He received the Enrico Fermi Science Award in 1956, and in that same year the Medal of Freedom from President Eisenhower.

Von Neumann has been described as a genius, a practical joker, and a raconteur. Laura Fermi, wife of the associate director of the Los Alamos Laboratory Enrico Fermi, wrote that he was "one of the very few men about whom I have not heard a single critical remark. It is astonishing that so much equanimity and so much intelligence could be concentrated in a man of not extraordinary appearance."

Von Neumann married Mariette Kovesi, daughter of a Budapest physician, in 1929. Their daughter, Marina, was born in 1935. Mariette obtained a divorce in 1937. The following year, von Neumann married Klara Dan, from an affluent Budapest family. In 1955, von Neumann was diagnosed with bone cancer. Confined to a wheelchair, he continued to attend Atomic Energy Commission meetings and to work on his many projects. He died in 1957 at the age of 53.

VOTING PARADOX

Voting paradoxes can arise in any election involving three or more candidates; though they come in many different forms, they can all be summed up in a single statement: even if every voter is individually rational, society as a whole is not.

The most commonly used voting method in the United States is the plurality vote: each voter casts a ballot for one candidate, and the winner is the candidate who gets the most votes (even if this is less than 50 percent). Though this system is familiar, it is rife with inconsistencies. Consider, for example, a situation where nine people are deciding which one of three restaurants to eat dinner at. Each person ranks the three restaurants, and the rankings are as follows:

- 4 people rank A best, B second, and C third.
- 3 people rank B best, C second, and A third.
- 2 people rank C best, B second, and A third.

In a plurality vote, A would be the restaurant chosen—even though more than half of the voters rank it last. To make the irrationality of the plurality vote even more evident, suppose that restaurant B happens to be closed. The nine people recast their ballots, and now C, which used to be in last place, wins over A, five votes to four. (The three who voted for B would now cast their ballots for C, since it is their second choice). If an individual voter reacted this way, we would find it utterly illogical. (Imagine going into an ice cream parlor that has three flavors—chocolate, vanilla, and strawberry. You choose chocolate, and then the vendor tells you, "Oops, we don't have strawberry today." Would you then change your choice to vanilla?)

The paradox here is called dependence on irrelevant alternatives. That is, the outcome of the election between A and C depends on where people rank B. Even though no one changed their minds about the relative merits of A and C, the simple act of dropping B to last place on everyone's list (because the restaurant was closed) caused the social choice to change.

In view of this paradox, one might say that a "reasonable" voting system should be independent of irrelevant alternatives. One might also argue that a "reasonable" voting system should also satisfy the "Pareto condition": If every voter ranks A above B, then A should be ranked above B in the final outcome. According to a seminal **theorem** proved by Kenneth Arrow in 1950 (who later won a Nobel Prize for this work), if there are three or more candidates, then there is only one "reasonable" voting system: a system where one voter is a dictator, and no one else's vote counts. In essence, the irrationality of society can only be eliminated by giving all the power to one rational voter. Because this is not considered an option in a democratic society, this result is often called Arrow's impossibility theorem—it is impossible to find a "reasonable" voting system.

Ironically, Arrow's impossibility theorem—the master paradox of voting theory—had the effect of stimulating research into alternative voting systems, rather than discouraging it. Granted that no system is perfectly rational and perfectly fair, mathematicians and political scientists are still interested in determining what systems might be most rational and most fair under realistic conditions. Some other systems in common use are:

- runoff elections;
- approval voting (in which voters may cast a vote for all the candidates they approve of), used in several scientific societies;
- single transferable vote (in which the bottom candidate is dropped after each round and the voters who voted for that candidate have their votes "transferred" to their next-favorite), used in Ireland and Australia;
- Borda count (in which a voter's top-ranked candidate gets n points, the second-ranked candidate gets n-1, and so on), used in the ranking of sports teams.

See also Set theory

W

WALLIS, JOHN (1616-1703)

English algebraist

John Wallis was a founding member of the Royal Society, one of the oldest scientific organizations still in existence, and is considered by many the most influential British mathematician preceding **Isaac Newton**. He contributed the earliest forms, terms, and notations to nascent fields such as **calculus** and **analysis**. Wallis was the first to attempt to write a comprehensive history of British mathematics, striving to bring continuity to mathematical study and research. Among the many classical tracts Wallis translated and edited are two major works of **Archimedes**. He was also involved in government; as the Parliamentarians' cryptographer, Wallis was instrumental in deciphering enemy codes during the English Civil War. Not all of Wallis' discoveries were valid, as shown by the rare example of an attempt to analyze **Euclid**'s fifth **postulate** in 1663 that turned out to be a trivial **proof**.

Wallis was born on November 23, 1616, to John and Joanna (nee Chapman) Wallis, residents of Ashford in Kent. In 1622, Wallis' father, a rector, died, and three years later young Wallis left Ashford in order to escape an outbreak of the plague. In 1625, he attended boarding school in Ley Green, near Tenterden, and after five years there he transferred to Martin Holbeach in Essex county. During one Christmas holiday Wallis asked his brother to introduce him to **arithmetic** and he mastered it in two weeks. It turned out Wallis was a "calculating prodigy," who could solve problems like the **square** root of a 53–digit number to 17 places without notation.

Wallis' education continued at Emmanuel College in Cambridge, where he studied medicine and wrote one of the first papers on the circulatory system. He also took courses in physics and moral philosophy, for which he was awarded a B.A. in 1637 and a fellowship to Queen's College. During this time, civil unrest was building in England. Wallis became a minister of the Church of England in 1640, serving as private chaplain in Yorkshire and Essex. He would also eventually

become a royal chaplain to the court of Charles II and bishop of Winchester as well. Upon his marriage to Susanna Glyde in 1645 Wallis was forced out of his fellowship and moved to London. For his service in deciphering Royalist letters during the Civil War, Oliver Cromwell appointed Wallis Savilian Professor of Geometry at Oxford.

In 1655 Wallis produced his first major work, *Arithmetica Infinitorum*, which systematized the analytical methods of **René Descartes** and **Bonaventura Cavalieri's** method of indivisibles. It became a standard reference, influencing many mathematicians who thereafter wrote on the subject, and is still considered a monumental text in British mathematics. In 1658 Wallis was appointed an official archivist at Oxford, and in the following year his treatise on conics, *Tractatus de sectionibis conicis*, was published. In it, Wallis attempted to simplify Cartesian geometry—a notoriously recondite work—for consumption and use by his peers.

Among the vast range of mathematical subjects Wallis tackled included the quadrature of many curves, an infinite approximation for $4/\pi$, negative and fractional **exponents**, and calculating the center of gravity in **cycloids** by using indivisibles. He was the first to use the symbol ∞ to indicate **infinity** and the capital S for **sine**, and introduced the mantissa in **logarithms**. In his treatise on cycloids Wallis was the first to recast **conic sections** as curves of the second degree. His approximation for $4/\pi$, a formula now famous in mathematics, used "interpolation," a word peculiar to Wallis that became standardized.

Algebra: History and Practice, published in English in 1685, was significant not just as a historical document. Here, Wallis made the first recorded effort to graphically display the complex **roots** of a real quadratic equation. This intimated the geometric interpretations of **complex numbers** and **trigonometry** in use today in the Gaussian plane. He also prepared and published a second edition in 1693 as volume two of his *Opera Mathematica*. This enlarged version included the first systematic use of algebraic formulae—using numerical ratio to represent a given magnitude. That systematization had great impact on subsequent mathematical arguments about important unsolved problems in physics.

Wallis survived another political crisis, the Glorious Revolution (1688–1689), which led to the deposition of King James II. James's successor, William III, retained Wallis to decipher enemy communiques.

Wallis wrote on subjects other than mathematics, including physics, theology, etymology and linguistics, general philosophy, and pedagogy. He is recognized as the first hearing person to systematize a means of teaching deaf mutes. Wallis was well known as a partisan; he successfully lobbied against the adoption of the Gregorian **Calendar** and was opposed to crediting **Gottfried Wilhelm Leibnitz** as a co–founder of the calculus. Wallis launched into an argument with the philosopher Thomas Hobbes, ostensibly regarding the subject of geometry. The two men traded insults in pamphlets with such florid titles as "Due Correction for Mr. Hobbes, or School Discipline for not saying his Lessons." Wallis married and had one son and two daughters. His wife died in 1687.

Wallis kept his post as Savilian Professor until his death at Oxford on October 28, 1703. **Christopher Wren** and **Christiaan Huygens**, inspired by Wallis's use of analogy in his writings, extended their investigations to the next generation of European mathematicians.

WARING, EDWARD (1734-1798)
English algebraist and physician

Edward Waring was an established 18th–century mathematician and theorist who did groundbreaking work in the areas of **imaginary numbers** and their **roots**. He is best known for the Cauchy ratio testand Waring's **theorem**, which is also known as Waring's **problem**. In the former, Waring focused on the **convergence** and divergence of numerical series. In the theorem named after him, Waring postulated that any positive integer is the sum of not more than nine cubes, and is the sum of not more than 19 fourth **powers**. The result about cubes was not proved until 1910, and the result about fourth powers was not proved until 1986. Waring also posited that **integers** are composed of a specific combination of **prime numbers**. For example, an even integer, he asserted, was the sum of two prime numbers, whereas an odd number was either, in and of itself, a prime number or the sum of three primes. Waring was also known for his work with quartic curves and the approximation of imaginary roots.

Waring was born near Shrewsbury in Shropshire, a borough in western England, to John Waring, a wealthy farmer, and his wife, Elizabeth. Exact details about their oldest son's origins are sketchy: some sources say the mathematician was born in 1734, whereas others say 1736; Plealey has likewise been given as the more exact place of his birth, as has Old Heath.

Waring initially attended school in Shrewsbury before going on to Cambridge University, where he was enrolled as a sizar at Magdalene College in 1753. (A sizar, according to *Webster's Dictionary*, is a student who receives an academic stipend, often in return for performing work for other stu-

dents.) In 1757, having established himself and excelled in the field of mathematics, Waring graduated with a bachelor of arts and the title of senior wrangler. Around this time, he was also granted a fellowship, and this enabled him to continue with his studies. He was likewise honored with membership in Cambridge's recently founded Hyson Club.

The death of faculty member John Colson at Cambridge created a job opening for which Waring was nominated. Because of the complaints of some faculty members who faulted him for his age, work, and lack of teaching experience, Waring countered their opposition by showing some of his writings; he was ultimately offered the position, although he did not yet possess his master's degree, a requirement for this position. He eventually received the degree by royal mandate in 1760 and was awarded the title of "Lucasian professor of mathematics," (a position first held by **Isaac Newton**). Waring held this title for the remainder of his life and it would ultimately enable him to pursue other interests. Additional honors included Waring's induction into the Royal Society of London a few years after his M.A. was conferred upon him, as well as that organization's Copley Medal.

During his tenure at Cambridge, Waring published several nonfiction works and treatises about mathematics, the earliest of which were published in Latin. Portions of his first work, *Miscellanea analytica de aequationibus algebraicis, curvarum proprietatibus, fluxionibus et serierum summatione* (1759; "Miscellany of Analysis"), were used by Waring when he was attempting to strengthen his position against those critics who deemed him too young to be the Lucasian professor of mathematics and were protesting his nomination. In this work and its 1762 revised version, *Miscellanea analytica de aequationibus algebraicis et curvarum proprietatibus,* Waring focused in part on his theory of numbers and their composition (as a cube, a fourth power, or some combination of either cubes or fourth powers). Subsequent works published in Latin include the 1770 *Meditationes algebraicae* ("Thoughts on Algebra") and *Proprietates algebraicarum curvarum* ("The Properties of Algebraic Curves"), which was printed in 1772. His later works—notably *On the Principle of Translating Algebraic Quantities into Probable Relations and Annuities* (1792) and *An Essay on the Principles of Human Knowledge* (1794)—were first printed in his native language. In addition to his books, Waring wrote numerous other small articles and pieces. Despite the erudition of the ideas presented in his writings, Waring's books were often faulted by critics for their brevity, lack of explanation, and apparent lack, at times, of clarity and organization. For example, *Miscellanea analytica* was once described as "one of the most abstruse books written in the abstrusest parts of Algebra," a quotation that has been attributed to both Glieg and Charles Hutton, the latter being the English mathematician who had determined the Earth's **mean** density.

Perhaps because of the sometimes negative reception of his works upon their publication, it has since been noted that Waring's contributions stem from the originality of his work and research, and not necessarily as a writer or teacher of his ideas. Indeed, during his academic career, Waring spent little time teaching students (reports about his personality indicate that he would not have particularly suited to the job—he has been described as being somewhat arrogant, and it is even rumored that he once claimed that no one in his homeland was competent enough to read, let alone understand, his work) and actually devoted a portion of his life to medicine, eventually earning a medical degree from Cambridge in either the late 1760s or the early 1770s. (Although some sources disagree, it is often accepted that Waring participated in the dissection of corpses and was frequently found at work in the hospitals around London and Cambridge. For a time, it is believed that he actually practiced medicine at various English hospitals before retiring from the profession in the 1770s.)

Whatever the reception of Waring's work in their published format, his ideas about mathematics and science were groundbreaking but frequently misunderstood. Although some of his work still remains unproved and much of it was ignored or disparaged by his contemporaries, his ideas remain highly respected for their innovativeness. Waring died on August 15, 1798, in Pontesbury, Shropshire.

WEDDERBURN, JOSEPH HENRY MACLAGEN (1882-1948)

Scottish-American professor

Although mental illness cut his career somewhat short, Joseph Henry Maclagen Wedderburn made important contributions to **algebra** and wrote extensively on matrices. He taught at Princeton University for many years.

The son of a physician, Wedderburn was born the tenth of 14 children on February 26, 1882 in Forfar, Angus, Scotland. He began his academic career at the University of Edinburgh in 1898, earning a master's degree in mathematics from the school in 1903. During the 1903-1904 academic year, Wedderburn did postgraduate work at the University of Leipzig and the University of Berlin.

Making his first trip to the United States in 1904, Wedderburn used a Carnegie scholarship to study for a year at the University of Chicago. There he met and worked with **Oswald Veblen** for a short time before returning to Scotland and the University of Edinburgh. Wedderburn worked at the school until 1909, also serving as editor of the *Proceedings of the Edinburgh Mathematical Society* from 1906 to 1908.

Meanwhile, Wedderburn was doing his best work in mathematics. In 1905, he proved that a noncommutative finite **field** could not exist—a revelation that had important implications for projective **geometry** and number theory—and in 1907 he published his best-known paper on the classification of semisimple algebras. In this paper ("On Hypercomplex Numbers"), Wedderburn showed 1) that a semisimple algebra is a **matrix** algebra over a **division** ring and 2) that every semisimple algebra is a direct sum of simple algebras. These two theorems, which still bear his name, signaled the beginning of a new era in this field of research.

Wedderburn returned to the United States in 1909 to work as a preceptor in mathematics at Princeton University (New Jersey) and took on the additional role of editor of the *Annals of Mathematics* in 1912. He remained at Princeton until World War I began in 1914, when he volunteered for the British Army. He served in Britain and France until the war ended in 1918.

Returning to Princeton, Wedderburn obtained permanent tenure in 1921 and continued working as editor of the *Annals* until 1928. At about this point, however, Wedderburn apparently began suffering from emotional disturbances that gradually caused him to withdraw from life. Yet despite these problems, he managed to produce his most famous work, a textbook entitled *Lectures on Matrices,* in 1934.

Wedderburn continued to teach and do research at Princeton until 1945, when the school granted him early retirement due to his illness. He died there on October 9, 1948.

WEIERSTRASS, KARL THEODOR WILHELM (1815-1887)

German analyst and number theorist

Karl Wilhelm Theodor Weierstrass was considered one of the greatest mathematical analysts of 19th-century Europe. He is well known as a cofounder of the theory of analytic **functions** and their representation as **power series**. Weierstrass made crucial contributions to the arithematization of **analysis** and to the theory of **real numbers**. He showed the importance of uniform **convergence**, furthered the understanding of elliptic functions, and made contributions to the field of **differential equations**. Weierstrass's reputation for high standards of **proof** and **definition** is reflected in the modern development of calculusand analysis.

The eldest child of Wilhelm Weierstrass, a customs officer, and Theodora (nee Forst), Weierstrass was born in Ostenfeld, Germany, on October 31, 1815. The family soon moved to Westernkotten in Westphalia, where Weierstrass's three siblings, Peter, Klara, and Elise, were born. Shortly after Elise's birth in 1826, Weierstrass's mother died, and a year later his father remarried. His father, who had once been a teacher and had a reputation for righteousness and uncompromising authority, wished for his son to receive a solid education and pursue a bureaucratic career that might afford him a comfortable lifestyle. Since there was no school in the village of Westernkotten, Weierstrass was sent to study in nearby Muenster. At the age of 14 he entered the Gymnasium in Paderborn, from which he graduated in 1834. Earning awards in almost all of his subjects and finishing among the top in his class, Weierstrass appeared bound for the success his father had envisioned. It would not be long, however, before ambitions of a comfortable bureaucratic career were derailed. Weierstrass was directed by his father to attend the University of Bonn, where it was expected he would acquire an education in business and law. He enrolled at the university, but rather than taking his studies seriously, he embraced the sport of fencing and the habit of socializing in German pubs. After four years Weierstrass returned home without a degree. His father and siblings, reportedly shamed by his failure, decided he might salvage the family name by obtaining a teaching certificate. In 1839 he began studying toward this end at the Academy of Muenster.

Although Weierstrass had read *Celestial Mechanics* by **Pierre Simon Laplace** during his time in Bonn, he had, till 1839, shown no extraordinary interest in mathematics. At the Academy of Muenster, however, he came under the tutelage of Christof Gudermann, who taught a course in elliptic functions. Gudermann's special interest was in attempting to represent elliptic functions with power series, an approach to analysis that was to have a profound influence on Weierstrass' career. Weierstrass proved, in Muenster, to be an able and dedicated student. In 1841, when he took his examination for the teaching certificate, he asked Gudermann to provide a mathematical problem that might normally be presented to a doctoral candidate. Gudermann obliged, and proposed that Weierstrass find the power series developments of the elliptic functions. The examination results were so exemplary that Gudermann recommended Weierstrass be granted a university post, but the Academy was only able to grant him a teaching certificate.

Weierstrass began his teaching career at the Gymnasium in Muenster in 1841. In 1842 he took a position as a teacher of **mathematics and physics** in Deutsche–Krone, West Prussia. That same year his first mathematical work, "Remarks on Analytical Factorials," was printed in a school publication. The narrow circulation of the journal prevented Weierstrass' significant original contribution to the field from being immediately recognized. It would be more than a decade before the paper came to the attention of a wider, more scholarly audience. During this period Weierstrass was forced to confine his mathematical research to his spare hours, devoting most of his energies to the daily demands of his secondary school teaching. Occasionally, he would work until dawn on a mathematical problem of interest, unaware that the night had passed. Although he worked in isolation and had not yet established personal contact with the prominent European mathematicians of his day, he did manage to remain somewhat current in his readings. In 1848, Weierstrass moved to Braunsberg to teach at the Royal Catholic Gymnasium. By now he had taken an interest in the work of **Niels Henrik Abel**. Shortly after assuming his new teaching post, he published a paper in the gymnasium's journal entitled "Contributions to the Theory of Abelian Integrals." Once again, limited circulation meant his important discoveries would receive no immediate notice.

During the summer of 1853, Weierstrass spent his vacation in his father's home in Westernkotten, writing another paper on Abelian functions. This time, he chose to submit the work to the *Journal fur die reine and angewandte Mathematik*, a prestigious mathematical research publication begun in 1826 by **August Leopold Crelle**. The paper was accepted for publication and appeared in print in 1854, bringing Weierstrass instant acclaim. He had successfully solved a problem of hyper–elliptic integrals, and established himself as

one of the truly great mathematical analysts. The University of Königsberg bestowed on Weierstrass an honorary doctorate, and the gymnasium in Braunsberg granted him a leave so that he could pursue mathematics full time. In 1856, he was appointed professor at the Royal Polytechnic School in Berlin. He received a joint appointment as an assistant professor at the University of Berlin and was awarded membership in the Berlin Academy. That same year his first paper, "Remarks on Analytical Factorials," was reprinted in Crelle's journal, finally receiving an appropriate audience.

In Berlin, Weierstrass rapidly took on a heavy research and teaching schedule, and his lectures on Abelian functions and transcendents were widely attended. Although he wrote very few manuscripts, his ideas on analytic functions became widely known through the dissertations and other writings of his students. Eventually, in 1886, G.H. Halphen, a student of Weierstrass, published a comprehensive discussion of Weierstrass' theory of elliptic functions, entitled *Theorie Des Fonctions Elliptiques et des leurs Applications*.

At the core of Weierstrass's mathematical research was his work on the theory of analytic functions based on power seriesand the process of analytic continuation. Weierstrass demonstrated that the integral of an **infinite series** is equal to the sum of the integrals of the separate terms when the series converges uniformly within a given region. In 1861, Weierstrass demonstrated a function that is continuous over an **interval** but does not possess a derivative at any point on this interval. Before this, it had been assumed that a continuous function must have a derivative at most points. In 1863 he provided a proof of a Gaussian theoremthat **complex numbers** are the only commutative algebraic extensions of the real numbers. In the field of the **calculus** of variations, he brought clarity and rigor to necessary and sufficient conditions for elliptic functions. Toward the end of his career Weierstrass became interested in the astronomical problem of the stability of the solar system.

In addition to his numerous contributions to the field of mathematics, Weierstrass also earned a reputation as an excellent teacher and lecturer. He gathered about him a band of young students, many of whom themselves became successful mathematicians and propagators of Weierstrass's ideas. His students and the auditors in his seminars read like a who's who of 19th century mathematics. But undoubtedly, his most favorite pupil was the Russian mathematician **Sonya Kovalevskaya**. The two first met in 1870 when he was 55 and she was 20. Because the University of Berlin would not allow a woman to officially attend Weierstrass' lectures, he taught her privately for four years. Through his efforts, Kovalevskaya finally received her doctorate in absentia from Göttingen in 1874. They continued to correspond after her return to Russia and eventual appointment as professor of mathematics at the University of Stockholm until her untimely death in 1891.

Weierstrass remained at the University of Berlin for 30 years. His activities as a mathematician and lecturer in Berlin were interrupted from time to time because of chronic illness. He developed vertigo in the 1860s and later suffered from chronic bronchitis and phlebitis. In 1894 he became confined to a wheelchair. Weierstrass died of influenza, following a long illness, on February 19, 1897, in Berlin. A lifelong bachelor, he was buried in a Catholic cemetery alongside his two sisters, with whom he had lived for much of his adult life.

WEIL, ANDRÉ (1906-   )
French algebraist and number theorist

André Weil is responsible for important advances in **algebraic geometry**, **group theory**, and **number theory** and belonged to the group of French mathematicians who published many important works under the collective pseudonym of **Nicolas Bourbaki**. Many of his peers in the 1950s considered him the finest living mathematician in the world. In 1980, he was presented with the Barnard Medal by Columbia University; prior recipients of the medal, which is awarded every five years, include **Albert Einstein**, **Ernest Rutherford**, and Neils Bohr. The prize recognizes outstanding accomplishment in physical or astronomical science or a scientific application of great benefit to humanity.

Weil was born May 6, 1906, in Paris, France, to free-thinking Jewish parents. His father, Bernard, was a physician, and his mother, Selma Reinherz Weil, came from a cultured Russian family. His sister was the famous writer, social critic, and World War II French Resistance activist, Simone Weil. When he was eight years old, Weil happened upon a **geometry** book and began to read it for recreation. By the time he was nine, he was absorbed in mathematics and was solving difficult problems. In her biography of Weil's sister, Simone Pétrement quotes Weil's mother as saying that at nine years of age André "is so happy that he has given up all play and spends hours immersed in his calculations." Weil's father was drafted into the military in 1914, and the family accompanied him to various medical assignments around France during World War I. At age 16, Weil was accepted at the elite Ecole Normale Supérieure in Paris, where he received his doctorate in 1928. He also studied at the Sorbonne, the University of Göttingen, and the University of Rome. From 1930 to 1932, he taught at the Aligarh Muslim University in India. From 1933 to 1940, he was a professor of mathematics at the University of Strasbourg in France. In 1937, he married.

Weil was in Finland with his wife when France entered World War II. He believed that he could do France more good as a mathematician and refused to return to his home country and join the army. He was walking near an anti-aircraft gun emplacement when the Russians invaded Finland, and the Finns arrested him, thinking that he was a spy. The letters to Russian mathematicians in his room did not help his case, and for a while it appeared that he would be executed. The Finns, however, released Weil to the Swedes, who sent him to England, from where he returned to France to be imprisoned and tried for not reporting for military service. He was tried on May 3, 1940 and convicted, and he asked to be sent to the front. The court obliged, and he was to be sent to an infantry unit along the English Channel at Cherbourg. Weil's boat, however, wound up in a British port, and he made his way

back to France later in 1940. He soon rejoined his wife, Eveline, and they escaped the war to the United States. Their daughter, Sylvie, was born on September 12, 1942. Weil taught at Haverford and Swarthmore colleges in the United States in 1941 and 1942, and at the University of São Paulo in Brazil from 1945 to 1947.

In 1947, Weil was recruited to the mathematics department at the University of Chicago, where he taught until 1958. One of his colleagues at Chicago was Irving Kaplansky, who gives a sense of Weil's personality in *More Mathematical People:* "There we were at Chicago, lucky enough to have André Weil, one of the greatest mathematicians in the world. There were several times in my life that I've, one way or another, got that feeling, my gosh, here is a tremendous mathematician.... He was very impatient with what he regarded as incompetence." Kaplansky added, "Then there is his extraordinary quickness.... You can take an area of mathematics that he presumably never heard of before and just like that he'll have something to say about it." From 1958 until his retirement, Weil taught at the Institute for Advanced Study at Princeton.

Weil's mathematical innovations are highly technical and involve complex formulas. One of his discoveries was the concept of "uniform space," a kind of mathematical **space** that cannot be readily visualized like the three-dimensional space that we occupy in our daily lives. The *Science News Letter* pronounced Weil's discovery of uniform space one of the most important mathematical discoveries of 1939. In 1947, Weil developed some formulas in the field of algebraic geometry, which are known as the "Weil conjectures." Weil's conjectures, as Ian Stewart explains in *Scientific American,* "give formulas for the number of solutions to an algebraic equation in a finite field. In particular they allow one to deduce that a given equation does or does not have solutions; this information can be transferred to analogous **equations** involving **integers** or **algebraic numbers**.... [They] are of fundamental importance in algebraic geometry."

Weil's algebraic and geometrical innovations of the first half of the 20th century were especially important for the technological innovations of the second half of the 20th century. Complex computer software that models black holes for astronomers, scientific graphics for research physicists, and special effects visualizations for Hollywood filmmakers all rely in part on mathematical innovations in **algebra** and geometry. As the *Science News Letter* said in 1939, in the decades to come, mathematical innovations like Weil's may lead to "some concept that will illumine the universe as glimpsed by the 200-inch telescope or the atom as created or smashed by the powerful cyclotron."

In the mid–1930s, Weil and other important young French mathematicians—among them Jean Dieudonné, Claude Chevallier, and **Henri Cartan**—began to write a series of mathematical works under the pseudonym of Nicolas Bourbaki. As Paul Halmos said in *Scientific American,* one writer called Bourbaki a "polycephalic mathematician." The group has varied in number from ten to twenty and has been composed, predominantly, of those of French nationality.

Their purpose was quite serious: to write a series of books about such fundamental mathematical areas as **set theory**, algebra, and **topology**. The resulting series of books, which to date number over thirty, was called the *Elements of Mathematics*. As Halmos said, "The main features of the Bourbaki approach are a radical attitude about the right order for doing things, a dogmatic insistence on a privately invented terminology, a clean and economical organization of ideas, and a style of presentation which is so bent on saying everything that it leaves nothing to the imagination." Their work has been very thorough (for example, it took them two hundred pages to define the number "1") and influential. Among other things, they inspired the "new math" that was introduced into American schools in the 1960s.

While their purpose is serious, the Bourbakians cultivate an atmosphere of mystery about their identities: they attempt to keep their names secret, they like to make up stories about themselves, and they love pranks. One story about their origin, which could well be a hoax, is that they got the idea for their name from the annual visit of a character named Nicolas Bourbaki to the Ecole Normale Supérieure, where many of them were educated. This character was an actor who gave a mock-serious lecture on mathematics in double-talk. Some of their own double-talk consists of saying that the home institution of Nicolas Bourbaki is the "University of Nancago," a fusion of the Universities of Nancy and Chicago, where several members of the group teach. Another story reported is that the name was inspired by General Charles Denis Sauter Bourbaki, a colorful figure in the Franco-Prussian war. One of the group's pranks was to apply for a membership to the American Mathematical Society under the name of N. Bourbaki. They played another prank on Ralph P. Boas, the executive editor of *Mathematical Reviews*. Boas had said in one of the *Encyclopedia Britannica*'s annual *Book of the Year* volumes that Nicolas Bourbaki did not exist. The Bourbakians sent a letter to the editors of the *Britannica* complaining about Boas's charge. Later, as Paul Halmos said in *Scientific American,* the Bourbakians "circulated a rumor that Boas did not exist. Boas, said Bourbaki, is the collective pseudonym of a group of young American mathematicians who act jointly as the editors of *Mathematical Reviews*."

WESSEL, CASPAR (1745-1818)

Danish-born Norwegian number theorist and surveyor

Casper Wessel made a significant contribution to mathematics, but his legacy is to be a footnote in histories of other great mathematicians. **Karl Friedrich Gauss** and **Jean Robert Argand** are most often given credit for expressing complex numbers as geometric shapes, but it was an idea that Wessel, a surveyor and map–maker, was first to develop. Because he was not a professional mathematician and made no great effort to publish his one breakthrough paper, it went largely unknown for a century.

Wessel was born on June 8, 1745, in Jonsrud in Akershus county, Norway (then part of Denmark), to Jonas Wessel, a church vicar, and Maria Schumacher, his wife. He attended the

Christiania Cathedral School in Oslo from 1757 to 1763 and then spent a year at the University of Copenhagen. Wessel's career in cartography, or map–making, began in 1764, when he became an assistant to the Danish Survey Commission soon after leaving the University of Copenhagen. He earned a law degree in 1778 and rose to survey superintendent in 1798.

In 1797, Wessel published a little–noticed paper on the geometric representation of **complex numbers**, which he presented to the Royal Danish Academy of Sciences. Gauss reportedly had been working on the problem as early as 1799, but did not publish the idea until 1831. Argand, a bookkeeper, published the concept in 1806, and that work became very influential, hence many refer to a plane of complex numbers as the Argand plane or an Argand diagram. "And yet [Wessel's] exposition was, in some respects, superior to and more modern in spirit than Argand's," writes Phillip S. Jones in the *Dictionary of Scientific Biography*. Wessel's work lay forgotten for 98 years and was republished in French on the 100th anniversary of the original publication. "It is regrettable that it was not appreciated for nearly a century and hence did not have the influence it merited," writes Jones. Biographical information about Wessel is also lacking, as he had been deceased 79 years before his achievement was recognized.

Graphical representation of complex numbers is a key mathematical component of contemporary computer graphicsand was a major breakthrough for mathematicians grappling with understanding higher concepts. "The simple idea of considering the real and imaginary parts of a complex number $a + bi$ as the rectangular coordinates of a point in a plane made mathematicians feel much more at ease with imaginary numbers," Howard Eves wrote in *An Introduction to the History of Mathematics*. "Seeing is believing, and former ideas about the nonexistence or fictitiousness of **imaginary numbers** were generally abandoned."

Some scholars also credit Wessel with advancing the idea of adding vectors in a three–dimensional **space**, a basic concept in modern physics. The mapmaker realized that nonparallel vectors could be added together by laying the terminal end of one line at the beginning of a second, and then summing them by drawing a line from the beginning of the first vector to the end of the second.

Although he is now best known for the work on complex numbers, Wessel was actually quite famous in his day as a geographer. His work as a surveyor was so highly regarded that he was awarded a medal by the Royal Danish Academy of Sciences and was knighted in Danebrog in 1815. He had retired in 1805, but continued working until 1812, when his rheumatism became too severe. Wessel died on March 25, 1818, in Copenhagen, Denmark.

WEYL, HERMANN (1885-1955)

German-born American mathematical physicist

Hermann Weyl was one of the most wide-ranging mathematicians of his generation, following in the footsteps of his teacher **David Hilbert**. Weyl's interests in mathematics ran the gamut from foundations to physics, two areas in which he made profound contributions. He combined great technical virtuosity with imagination, and devoted attention to the explanation of mathematics to the general public. He managed to take a segment of mathematics developed in an abstract setting and apply it to certain branches of physics, such as relativity theory—a theory that holds that the velocity of **light** is the same for all observers, no matter how they are moving, that the laws of physics are the same in all inertial frames, and that all such frames are equivalent—and quantum mechanics—a theory that allows mathematical interpretation of elementary particles through wave properties. His distinctive ability was integrating nature and theory.

Claus Hugo Hermann Weyl was born on November 9, 1885, at Elmshorn, near Hamburg, Germany. The financial standing of his parents (his father, Ludwig, was a clerk in a bank and his mother, Anna Dieck, came from a wealthy family) enabled him to receive a quality education. From 1895 to 1904 he attended the Gymnasium at Altona, where his performance attracted the attention of his headmaster, a relative of an eminent mathematician of that time, **David Hilbert**. Weyl soon found himself at the University of Göttingen where Hilbert was an instructor. He remained there for the rest of his student days, with the exception of a semester at the University of Munich. He received his degree under Hilbert in 1908 and advanced to the ranks of *privatdocent* (an unpaid but licensed instructor) in 1910.

Weyl married Helene Joseph (known as Hella to the family) in 1913 and in the same year took a position as professor at the National Technical University (ETH) in Zurich, Switzerland. He declined the offer to be Felix Klein's successor at Göttingen, despite the university's central role in the mathematical world. It has been suggested that he wanted to free himself, somewhat, of the influence of Hilbert, especially in light of the fact that he had accepted an invitation to take a chair at Göttingen when Hilbert retired. In any case, he brought a great deal of mathematical distinction to the ETH in Zurich, where his sons Fritz Joachim and Michael grew up.

It is not surprising that Weyl's early work dealt with topics that which Hilbert held an interest. His *Habilitationsschrift* was devoted to boundary conditions of second-order linear **differential equations**. (The way the German educational system worked, it was necessary to do a substantial piece of original research beyond the doctoral dissertation in order to qualify to teach in the university. This "entitling document" was frequently the launching point of the mathematical career of its author.) In other words, he was looking into the way **functions** behaved on a given region when the behavior at the boundary was specified. His results were sufficient for the purpose of enabling him to earn a living, but he rapidly moved on to areas where his contributions were more innovative and have had a more lasting effect.

One of the principal areas of Weyl's research was the topic of Hilbert spaces. The problem was to understand something about the functions that operated on the points of **Hilbert space** in a way useful for analyzing the result of applying the functions. In particular, Weyl wanted to know where the func-

tions behaved more simply than on the **space** as a whole, since the behavior of the function on the rest of the space could be represented in terms of its behavior on the simpler regions. Different kinds of functions behaved in radically different ways on a Hilbert space, so Weyl had to restrict his attention to a subclass of functions small enough to be tractable (for example, the functions could not "blow up") but large enough to be useful. His choice of self-adjoint, compact operators was justified by their subsequent importance in the field of functional **analysis**.

Among the areas he brought together were **geometry** and analysis from the 19th century and **topology**, which was largely a creation of the 20th century. Topology sought to understand the behavior of space in ways that require a less-detailed understanding of how the elements of a structure fit together than geometry demanded. One of the basic ideas of topology is that of a "manifold," first introduced by **G. F. B. Riemann** in his *Habilitationsschrift* as a student of **Karl Gauss**. Riemann had little material with which to work, while Weyl was able to take advantage of the work of Hilbert and the Dutch mathematician **Luitzen Egbertus Jan Brouwer**. This effort culminated in his 1913 book on Riemann surfaces, an excellent exposition on how **complex analysis** and topology could be used together to analyze the behavior of complex functions.

Weyl served briefly in the German army at the outbreak of World War I, but before this military interlude, he did research that led to one of his most important papers. He looked at the way **irrational numbers** (those that cannot be expressed as a ratio of two **whole numbers**) were distributed. What he noticed was that the *fractional* parts of an irrational number and its integral multiples seemed to be evenly distributed in the **interval** between O and 1. He succeeded in proving this result, and it is known as the Kronecker-Weyl **theorem**, owing half of its name to an influential number theorist who had had an effect on Hilbert. Although the result may seem rather narrow, Weyl was able to generalize it to sequences of much broader application.

During his time in Zurich, Weyl spent a year in collaboration with **Albert Einstein** andpicked up a dose of enthusiasm for the relativity theory. Among the other results of this collaboration was Weyl's popular account of relativity theory, *Space, Time, Matter* (the original German edition appeared in 1918). In those early days of general relativity, which describes gravity in terms of how mass distorts **space-time**, the correct mathematical formulation of some of Einstein's ideas was not clear. He had been able to use ideas developed by differential geometers at the end of the nineteenth century that involved the notion of a tensor. A tensor can be thought of as a function on a number of vectors that takes a number as its value. Weyl used the tensor **calculus** that had been developed by the geometers to come up with neater formulations of general relativity than the original version proposed by Einstein. In later years, he took the evolution of tensors one step further while maintaining a strict mathematical level of rigor.

One of the most visible areas in which Weyl worked after World War I was in the foundations of mathematics. He had used some of the topological results of the Dutch mathematician

Brouwer in working on Riemann surfaces. In addition, he had looked at some of Brouwer's ideas about the philosophy of mathematics and was convinced that they had to be taken seriously. Although it was not always easy to understand what Brouwer was trying to say, it was clear that he was criticizing "classical" mathematics, that is, the mathematics that had prevailed at least since **Euclid**. One of the standard methods of **proof** in classical mathematics was the reductio ad absurdum, or proof by contradiction. If one wished to prove that P was true, one could assume that not-P was true and see if that led to a contradiction. If it did, then not-P must not be true, and P must be true instead. This method of proof depended on the principle that either P was true or not-P was true, which had seemed convincing to generations of mathematicians.

Brouwer, however, found this style of argument unacceptable. For reasons having to do with his understanding of mathematics as the creation of the human mind, he wanted to introduce a third category besides **truth** and falsity, a category we could call "unproven." In other words, there was more to truth than just the negation of falsity—to claim P or not-P, something had to be proven. This argument of Brouwer was especially directed against so-called nonconstructive existence proofs. These were proofs in which something was shown to exist, not by being constructed, but by arguing that if it didn't exist, a contradiction arose. For ordinary, finite mathematics it was usually easy to come up with a constructive proof, but for claims about infinite **sets** nonconstructive arguments were popular. If Brouwer's objections were to be sustained, a good part of mathematics even at the level of elementary calculus would have to be rewritten and some perhaps have to be abandoned.

This attitude aroused the ire of David Hilbert. He valued the progress that had been made in mathematics too highly to sacrifice it lightly for philosophical reasons. Although Hilbert had earlier expressed admiration for Brouwer's work, he felt obliged to negate Brouwer's philosophy of mathematics known as **intuitionism**. What especially disturbed Hilbert was Weyl's support of Brouwer's concepts, since Hilbert knew the mathematical strength of his former student. In the 1920s, while the argument was being considered, Hilbert was discouraged about the future of mathematics in the hands of the intuitionists.

Although Weyl never entirely abandoned his allegiance to Brouwer, he also recognized that **Hilbert's program** in the philosophy of mathematics was bound to appeal to the practicing mathematician, more than Brouwer's speculations. In 1927, responding to one of Hilbert's lectures concerned with the foundations of mathematics, Weyl commented on the extent to which Hilbert had been led to a reinterpretation of mathematics by the need to fight off Brouwer's criticisms. The tone of Weyl's remarks suggested that he would not have been unhappy if Hilbert's point of view was to prevail. This flexibility with regard to the foundations of mathematics indicates that Weyl was sensitive to the changes in attitude that others ignored, but also may explain why Weyl never founded a philosophical school: he was too ready to recognize the justice of others' points of view.

In general, Weyl took questions of literature and style seriously, which goes far to explain the success of his expository writings. His son recalls that when Weyl would read poetry aloud to the family, the intensity and volume of his voice would make the walls shake. He kept in touch with modern literature as well as the classics of his childhood. While he continued to enjoy the poetry of Friedrich Hölderlin and Johann Wolfgang von Goethe, he also read Friedrich Nietzsche's *Also Sprach Zarathustra* and Thomas Mann's *The Magic Mountain.* He could cite quotations from German poetry whenever he needed them. For those of a psychologizing bent, it has even been argued that his fondness for poetry may be in line with his preference for intuitionism as a philosophy of mathematics. The kind of poetry he preferred spoke to the heart, and he used quotations to add a human dimension to otherwise cold mathematical writing.

After he accepted a chair at Göttingen in 1930, Weyl did not have long to enjoy his return to familiar surroundings. In 1933 he decided that he could no longer remain in Nazi Germany, and he took up a permanent position at the Institute for Advanced Study, newly founded in Princeton, New Jersey. Although Weyl himself was of irreproachably Aryan ancestry, his wife was partly Jewish, and that would have been enough to attract the attention of the authorities. There may have been the additional attraction of the wealth of intellectual company available at the institute, between its visitors from all over the globe and permanent residents such as Einstein. Weyl took his official duties as a faculty member seriously, although his reputation could be terrifying to younger mathematicians unaware of the poet within.

Weyl's work continued to bridge the gap between physics and mathematics. As long ago as 1929, he developed a mathematical theory for the subatomic particle the neutrino. The theory was internally consistent but failed to preserve left-right **symmetry** and so was abandoned. Subsequent experimentation revealed that symmetry need not be conserved, with the result that Weyl's theory reentered the mathematical physics mainstream all the more forcefully. Another area for the interaction of **mathematics and physics** was the study of spinors, a kind of tensor that has proven to be of immense use in quantum **mechanics**. Although spinors had been known before Weyl, he was the first to give a full treatment of them. Perhaps it was this work that led **Roger Penrose**, one of the most insightful mathematical physicists of the second half of the twentieth century, to label Weyl "the greatest mathematician of this century."

One of the challenges of physical theories is to find quantities that do not change (are conserved) during other changes. **Felix Klein** in the nineteenth century had stressed the importance of **group theory**, then a new branch of mathematics, in describing what changed and what remained the same during processes. Weyl adapted Klein's ideas to the physics of the twentieth century by characterizing **invariant** quantities for relativity theory and for **quantum mechanics**. In a 1923 paper Weyl had come up with a suitable definition for **congruence** in relativistic space-time. Even more influential was his 1928 book on group theory and quantum mechanics, which imposed a **model** that would have been welcome to Klein due to the previously rather disjointed results assembled by quantum physicists. Weyl was an artist in the use of group theory and could accomplish wonders with modest mathematical structure.

After the end of World War II, Weyl divided his time between Zurich and Princeton. His first wife died in 1948, and two years later he married Ellen Bär. He took a serious view of the **history of mathematics** and arranged with Princeton to give a course on the subject. One of his magisterial works was a survey of the previous half-century of mathematics that appeared in the *American Mathematical Monthly* in 1951. Although he never became as fluent in English as he had in German, he retained a strong commitment to the public's right to be informed about scientific developments.

John Archibald Wheeler, an American physicist, called attention to Weyl's anticipation of the anthropic principle in cosmology. In a 1919 paper Weyl had speculated on the coincidence of the agreement of two enormous numbers of very different origin. In the 1930s this speculation had been given with the title "Weyl's hypothesis," although later authors referred more to its presence elsewhere. What cannot be denied is that the recent discussion of the anthropic principle concerning features necessary for human existence in the universe has, as Wheeler noted, taken up Weyl's point once again.

Weyl was unaware of the rules governing the length of time that a naturalized citizen could spend abroad at one time without losing citizenship. By inadvertence he exceeded the time limit and lost his American citizenship in the mid 1950s. To remedy the situation required an act of Congress, but there was no lack of help in securing it. In the meantime, Weyl celebrated his seventieth birthday in Zurich amid a flurry of congratulations. On December 8, 1955, as he was mailing some letters of thanks to well-wishers, he died of a heart attack. With his death passed one of the links with the great era of Göttingen as a mathematical center and one of the founders of contemporary mathematical physics, but even more, a mathematician who could convey the poetry in his discipline.

WHITEHEAD, ALFRED NORTH (1861-1947)

English American algebraist and logician

Albert North Whitehead began his career as a mathematician, but eventually became at least as famous as a philosopher. His first three books, *A Treatise on Universal Algebra, The Axioms of Projective Geometry,* and *The Axioms of Descriptive Geometry,* all dealt with traditional mathematical topics. In 1900, Whitehead first heard about the new system for expressing logical concepts in discrete symbols developed by the Italian mathematician **Giuseppe Peano**. Along with **Bertrand Russell**, his colleague and former student, Whitehead saw in Peano's symbolism a method for developing a rigorous, nonnumerical approach to logic. The work of these two men culminated in the publication of the three-volume *Principia Mathematica,* widely regarded as one of the most important books in mathematics ever written. In 1924 Whitehead became professor of philosophy at Harvard University, where

he devoted his time to the development of a comprehensive and complex system of philosophy.

Whitehead was born on February 15, 1861, at Ramsgate in the Isle of Thanet, Kent, England. Both his grandfather, Thomas Whitehead, and his father, Alfred Whitehead, had been headmasters of a private school in Ramsgate. Alfred Whitehead had later joined the clergy and become vicar of St. Peter's Parish, about two miles from Ramsgate. In his autobiography, Whitehead said of his father that he "was not intellectual, but he possessed personality." Whitehead's mother was the former Maria Sarah Buckmaster, daughter of a successful London businessman. As a young man, Whitehead often traveled to London to visit his maternal grandmother.

For the first fourteen years of his life, Whitehead was educated at home primarily by his father. Then, in 1875, he was sent to the public school at Sherborne in Dorsetshire. Whitehead described his education as traditional, with a strong emphasis on Latin and Greek. But, he continued, "we were not overworked," so that he had plenty of time for sports such as cricket and football, private reading, and a study of history. At Sherborne he also had his introduction to science and mathematics, at which he excelled. In fact, he was apparently excused from some Latin requirements in order to have more time for his mathematical studies.

In 1880 Whitehead received a scholarship to continue his studies at Trinity College, Cambridge. While at Trinity, all of Whitehead's formal education was in the field of mathematics, the British system not having yet accepted the concept of a broad liberal education for all students. Still, he later wrote, his mathematics courses "were only one side of the education" he experienced at Trinity. Another side consisted of regular evening meetings with other undergraduates at which virtually all subjects were discussed. Whitehead later referred to these meetings as "a daily Platonic dialogue." Through these dialogues, Whitehead rapidly expanded his knowledge of history, literature, philosophy, and politics.

Whitehead was awarded his bachelor of arts degree in 1884 for a thesis on James Clerk Maxwell's theory of **electromagnetism**. A few months later he was elected a fellow of Trinity College and appointed assistant lecturer in mathematics. Whitehead was awarded his M.A. in 1887, and in 1903, was named senior lecturer. Two years later he was granted his doctor of science degree.

While still at Trinity, Whitehead met Evelyn Willoughby Wade, described in the *Dictionary of American Biography* as the "daughter of impoverished Irish landed gentry" who was "witty, with passionate likes and dislikes, a great sense of drama, and... a keen aesthetic sense." Whitehead himself credits his wife with teaching him "that beauty, moral and aesthetic, is the aim of existence; and that kindness, and love, and artistic satisfaction are among its modes of attainment." The two were married on December 16, 1890. They later had four children, Thomas North in 1891, Jesse Marie in 1893, Eric Alfred in 1898, and an unnamed boy who died at birth in 1892. Eric later became a pilot with the Royal Flying Corps and was killed in March, 1918, during World War I.

Whitehead's first book, *A Treatise on Universal Algebra,* was begun in January, 1891, and published seven years later. The book was an attempt to expand on the works of three predecessors, **Hermann Grassmann**, **William Rowan Hamilton**, and **George Boole**, the founder of **symbolic logic**. Whitehead later wrote that Grassmann, in particular, had been "an original genius, never sufficiently recognized." All of Whitehead's future work on mathematical logic, he said, was derived from the contributions of these three men. Whitehead's book was a tour de force that earned him election to the Royal Society five years after its publication. In the book, Whitehead argues that algebraic concepts have an existence of their own, independent of any connection with real objects.

The year of Whitehead's marriage, 1890, also marked the beginning of another long and fruitful relationship, with Bertrand Russell. The two had met when Russell was still a freshman at Cambridge; Whitehead was one of his teachers. In later years their student-and-teacher relationship blossomed into a full-blown working relationship as professional colleagues. They were eventually to collaborate on a number of important mathematical works.

An important event in their association occurred in July, 1900, when they attended together the First International Congress of Philosophy in Paris. It was there that Whitehead and Russell were introduced to the techniques of symbolic logic developed by the Italian mathematician **Giuseppe Peano**. They immediately saw that Peano's symbolism could be used to clarify fundamental concepts of mathematics. When Russell returned to England, he began to incorporate Peano's approach into the book on which he was then working, *Principles of Mathematics.*

Before long, however, it occurred to Russell that he and his former teacher were both working on very similar topics, he on his *Principles of Mathematics* and Whitehead on a second volume of his *Universal Algebra.* The two agreed to start working together, with the result that the second volume of neither work ever appeared. Instead, they developed the three-volume masterpiece, *Principia Mathematica.* The fundamental concept behind the book was that the basic principles of mathematics can be derived in a strict way through the precise rules of symbolic logic. The *Principia Mathematica* has since been described by one of Whitehead's biographers, Victor Lowe, in the *Dictionary of American Biography* as "one of the great intellectual monuments of all time."

In 1910, the year in which the first volume of *Principia Mathematica* appeared, Whitehead ended a 25-year teaching career at Trinity College and moved to London. He remained without an academic appointment for one year, during which time he wrote his *Introduction to Mathematics,* which James R. Newman has called "a classic of popularization" of mathematics. Whitehead then accepted an appointment as lecturer in **applied mathematics** and **mechanics** at University College, London, and two years later, was made reader in **geometry** there. In 1914 he was named professor of applied mathematics at the Imperial College of Science and Technology in Kensington.

The London period was for Whitehead a particularly busy time in the political and administrative arenas. He served on a number of faculty and governmental committees and was outspoken in his concern about educational reform. Perhaps his best-known remarks on this subject came in a 1916 address to the Mathematical Association, "The Aims of Education: A Plan for Reform." In this address, Whitehead pointed out that the narrow view of education in which the classics are taught to a select number of upper-class men had become outmoded in a world of a "seething mass of artisans seeking intellectual enlightenment, of young people from every social grade craving for adequate knowledge."

During the latter years of his London period, Whitehead's interests shifted from mathematics to the philosophy of science. The fourth volume of *Principia Mathematica,* dealing with the foundations of geometry, was never completed. Instead, Whitehead began to write on the philosophical foundations of science in books such as *An Enquiry Concerning the Principles of Natural Knowledge* in 1919, *The Concept of Nature* in 1920, and *The Principle of Relativity, with Applications to Physical Science* in 1922.

The main theme of these books was that there exists a reality in the physical world that is distinct from the descriptions that scientists have invented for that reality. Scientific explanations certainly have their functions, according to Whitehead, but they should not be construed as being the reality of nature itself.

As early as 1920, Harvard University had been interested in offering Whitehead a position in its philosophy department. For financial reasons, a firm offer was not made until 1924, when Whitehead was sixty-three years old. He accepted the offer partly because he was nearing mandatory retirement age at Imperial College and partly because he looked forward to the opportunity of expanding his intellectual horizons. On September 1, 1924, Whitehead's appointment at Harvard became official.

Until the Harvard post became available, Whitehead had remained rather strictly within the areas of mathematics and natural science. After 1924, however, he extended the range of his writings to include far broader topics. His first work published in the United States, *Science and the Modern World,* discussed the significance of the scientific enterprise for other aspects of human culture. The book was an instant professional and commercial success and earned Whitehead an immediate reputation as a profound thinker and a writer of great clarity and persuasiveness.

His next book, *Religion in the Making,* was the first of a number that carried Whitehead far beyond the fields of mathematics and science. Its publication in 1926 was followed by *Symbolism, Its Meaning and Effect* in 1927, *The Aims of Education and Other Essays* in 1929, *The Function of Reason,* also in 1929, and a half dozen more books over the next two decades. In recognition of his work, Whitehead was elected a fellow of the British Academy in 1931, and he was awarded the Order of Merit, the highest honor that Great Britain can bestow on a man of letters, in 1945. Whitehead retired from active teaching at Harvard in 1937, at which time he was named emeritus professor of philosophy. He died at his home in Cambridge, Massachusetts, on December 30, 1947.

WHITEHEAD, JOHN HENRY (1904-1960)

English mathematician

John Henry Constantine Whitehead (known commonly as Henry Whitehead), had a large influence on the development of homotopy theory, which is based on a certain kind of **mapping** of topological spaces. With **Oswald Veblen**, he also wrote the classic *Foundations of Differential Geometry*. He was the nephew of mathematician and philosopher **Alfred North Whitehead**.

The son of a bishop, Whitehead was born in Madras, India on November 11, 1904. When he was about 18 months old, his parents took him to England and left him with his grandmother in Oxford before returning quickly to India. He would see virtually nothing of his parents until his father retired in 1920 or 1922 and moved back to England. Thus, he was raised primarily by his grandmother, who enrolled him in the local schools, where he did well both in his classes and at sports—especially boxing and cricket. He entered the prestigious school at Eton and began specializing in mathematics. Whitehead never seemed to be a brilliant student, perhaps because of the many other interests that distracted him and sadness over the long separation from his parents.

Nevertheless, he did sufficiently well at school to win a scholarship to Baliol College at Oxford University in 1923. There he also devoted most of his academic energy to mathematics, although he remained distracted by the wide range of nonacademic pursuits available to him, including sports and poker. Thus, when he graduated in 1927, Whitehead decided he was not fit for an academic career, despite earning high honors. He chose instead to join a stock brokerage. Meanwhile, he lived with his parents in Berkshire and commuted to the London firm.

After only a year, however, Whitehead realized he had made a mistake. He returned to Oxford University, where he soon attended a lecture by Veblen on differential **geometry**. He decided to dedicate himself to research on this topic, and quickly won a fellowship to work toward his doctorate at Princeton University in the United States. He arrived there in 1929. By the end of his three-year period there, Whitehead had become more interested in **topology**. He received his doctorate in 1932 after writing a thesis on representation of projective spaces, and the same year published *The Foundations of Differential Geometry*, which contains the first accepted definition of a differentiable **manifold**, with Veblen.

After returning to Oxford in 1933, Whitehead received a fellowship at Balliol College. Two years later, he published another pioneering work on differential geometry called *On the Covering of a Complete Space by the Geodesics through a Point*. He continued his study of manifolds and also established a school of topology at Oxford. When World War II began, however, Whitehead began dedicating much of his time and energy to helping Jews escape from Europe, even inviting some of them to stay in his home. He left Oxford in 1940 to do war work in London, including jobs at the foreign office and the Board of Trade, but returned to his home in Oxford when the war ended in 1945.

In 1947, Whitehead resumed his academic career when he accepted the chair of pure mathematics at Oxford University. During the 1950s, he wrote many papers with other mathematicians, indicating his eagerness to share his ideas and learn the results of others. When his mother died in 1953, he inherited a small farm and some cattle, so he and his wife (married in 1934) bought another farm nearby, moved in, and ran it as an active operation. He lived there until his death from a heart attack on May 8, 1960 while visiting Princeton University's Institute for Advanced Study. Many believed that he died at the height of his intellectual powers. Whitehead and his wife, a concert pianist whom he married in 1934, had two sons.

WHOLE NUMBERS

Whole numbers, counting numbers, natural numbers and integer numbers are all closely related to one another. Counting numbers, as their name implies, compose the set {1, 2, 3,...} (in this symbolism the brackets denote the elements of a set and the end dots denote a continuation of the sequence). Natural numbers are usually defined to be the set of counting numbers with **zero** added, that is the set {0, 1, 2, 3,...}. **Integers** numbers constitute the set {... -3, -2, -1, 0, 1, 2, 3,...}. Various authors have defined whole numbers differently: defining it as equivalent to the set of counting numbers, equivalent to the natural numbers, or to the set of integers. One reason for the various definitions of something as fundamental as whole numbers is most likely due to the fact that mathematics as a science has evolved over time. New number systems were created and previously defined systems and terms were changed or modified. In this article, whole numbers will be identified with the set of counting numbers, that is as the set {1, 2, 3,...}. Since integers can be partitioned into negative integers, zero (0), and positive integers, whole numbers can also be defined as the set of positive integers.

Before the modern concept of whole or counting number, people had not yet differentiated the "number" of a group from the objects of which it is composed. For example, people used different words to describe a pair of oxen and a pair of men. Whole numbers provided a way to compare groups on the basis of the quantity of elements they possess. Hence a group of 5 things has more elements than a group of 2 things. When used to describe quantity, whole numbers are called cardinal numbers. Now it is quite natural to arrange the elements of a set or group in a particular order. This is done all the time: the tallest to the shortest, the lightest to the heaviest, etc. Arranging the whole numbers according to their **cardinality** (the quantity of elements each number denotes), and proceeding from smaller to larger, results in the familiar sequence {1, 2, 3,...}. When whole numbers are used to place an ordering on some group, they are referred to as **ordinal numbers**. For example, customers arriving at a shop can be given a "number" to acknowledge who is number 1 in line, number 2 in line, and so forth. The arrival of a customer is, then, in the same order as the ordinal number they possess. Calling out the whole numbers in sequence yields the order in which the customers are served.

Historically, in 1430 the first known citation for the term whole number came from the *Art of Nombryng* and appears as the Middle English word "hoole," where its (now obsolete) definition is "a number composed of three prime factors." The term is found in its modern sense in a book anonymously published in 1537 titled *An Introduction for to leerne to reken with the Pen and with the Counters, after the true cast of arismetyke or awgrym in hole numbers, and also in broken.* In 1839 author J.R. Young refers to "a whole number or 0" in his book *Elements of the Integral Calculus,* and later referred to "a positive whole number."

The set of whole numbers is closed under the **arithmetic** operations of **addition** and **multiplication**, which means that the sum or product of any two whole numbers is another whole number. However, under **subtraction**, the inverse operation of addition, the whole numbers do not form a closed set. For instance, the difference of 5 - 2 is a whole number, but the difference of 2 - 5 is not a whole number. The stipulation that subtraction be defined for all values of the set of numbers it operates on leads directly from the whole numbers to the set of integers. In the set of integers the arithmetic expression "a - b" is closed regardless of whether a $\geq$ b or b $\geq$ a. A similar situation holds for **division**, the inverse operation of multiplication, where the quotient 4 / 2 is a whole number while the quotient of 5 / 2 is not. Once again, as for the case of subtraction leading to integers, the stipulation that division is defined for all values of the set of numbers it operates on leads from the whole numbers to the set of positive **rational numbers.** The synthesis of the positive rational numbers with the integers leads inevitably to the set of (positive and negative) rational numbers. All four arithmetic operations (addition, subtraction, multiplication and division, are valid and closed for all rational number values (except for division by zero).

The previous paragraph shows how the interaction between the whole numbers and the arithmetic operations leads to successive extensions of the number system (and the operations themselves) within the framework of arithmetic. The expansion and redefinition of the number system continued within the context of algebraic investigations so that ultimately mathematicians devised the real and complex number systems. Further developments within mathematics lead to the construction of mathematical objects such as vectors, matrices, **quaternions,** etc. The nineteenth century witnessed a concerted effort on the part of many scientists to reexamine the foundations of mathematics. Inasmuch as the more advanced number systems can trace their development back to the whole numbers, one consequence of this effort was a reexamination of the whole numbers. Italian mathematician **Giuseppe Peano** made a noteworthy accomplishment in that effort concerning the properties of the set of whole numbers. Peano's axioms (sometimes referred to as Peano's postulates) characterize the whole numbers by the use of five axioms and the principle of mathematical **induction.** (The axioms are sometimes stated with reference to the natural numbers.) Along with men like German mathematicians **Julius Wihelm Richard Dedekind** and

Georg Ferdinand Ludwig Philipp Cantor, Peano's axiomatic system helped show how the real number system, and hence most of mathematics, can be derived from a **postulate** set for the whole numbers.

See also Cardinal numbers; Complex numbers; Numbers and numerals; Peano axioms; Real numbers; Sets

WIENER, NORBERT (1894-1964)

American logician

Norbert Wiener was one of the most original mathematicians of his time. The field concerning the study of automatic control systems, called **cybernetics,** owes a great deal not only to his researches, but to his continuing efforts at publicity. He wrote for a variety of popular journals as well as for technical publications and was not reluctant to express political views even when they might be unpopular. Perhaps the most distinctive feature of Wiener's life as a student and a mathematician is how well documented it is, thanks to two volumes of autobiography published during his lifetime. They reveal some of the complexity of a man whose aspirations went well beyond the domain of mathematics.

Wiener was born in Columbia, Missouri, on November 26, 1894. His father, Leo Wiener, had been born in Bialystok, Poland (then Russia), and was an accomplished linguist. He arrived in New Orleans in 1880 with very little money but a great deal of determination, some of it visible in his relations with his son. He met his wife, Bertha Kahn, at a meeting of a Browning Club. As a result, when his son was born, he was given the name Norbert, from one of Browning's verse dramas. In light of the absence of Judaism from the Wiener home (Norbert was fifteen before he learned that he was Jewish), it is surprising that one of Leo Wiener's best-known works was a history of Yiddish literature.

As the title of the first volume of his autobiography *Ex-Prodigy* suggests, Wiener was a child prodigy. Whatever his natural talents, this was partly due to the efforts of his father. Leo Wiener was proud of his educational theories and pointed to the academic success of his son as evidence. Norbert was less enthusiastic and in his memoirs describes his recollections of his father's harsh disciplinary methods. He entered high school at the age of nine and graduated two years later. In 1906 he entered Tufts University, as the family had moved to the Boston area, and he graduated four years later.

Up until that point Wiener's education had clearly outrun that of most of his contemporaries, but he was now faced with the challenge of deciding what to do with his education. He enrolled at Harvard to study zoology, but the subject did not suit him. He tried studying philosophy at Cornell, but that was equally unavailing. Finally, Wiener came back to Harvard to work on philosophy and mathematics. The subject of his dissertation was a comparison of the system of logic developed by **Bertrand Russell** and **Alfred North Whitehead** in their *Principia Mathematica* with the earlier algebraic system created by Ernst Schröder. The relatively recent advances in mathematical research in the United States had partly occurred in the area of algebraic logic, so the topic was a reasonable one for a student hoping to bridge the still-existent gap between the European and American mathematical communities.

Although Wiener earned a Harvard travelling fellow-ship to enable him to study in Europe after taking his degree, his father still supervised his career by writing to Bertrand Russell on Norbert's behalf. Wiener was in England from June 1913 to April 1914 and attended two courses given by Russell, including a reading course on *Principia Mathematica*. Perhaps more influential in the long run for Wiener's mathematical development was a course he took from the British analyst **Godfrey H. Hardy**, whose lectures he greatly admired. In the same way, Wiener studied with some of the most eminent names in Göttingen, Germany, then the center of the international mathematical community.

Wiener returned to the United States in 1915, still unsure, despite his foreign travels, of the mathematical direction he wanted to pursue. He wrote articles for the *Encyclopedia Americana* and took a variety of teaching jobs until the entry of the United States into World War I. Wiener was a fervent patriot, and his enthusiasm led him to join the group of scientists and engineers at the Aberdeen Proving Ground in Maryland, where he encountered **Oswald Veblen**, already one of the leading mathematicians in the country. Although Wiener did not pursue Veblen's lines of research, Veblen's success in producing results useful to the military impressed Wiener more than mere academic success.

After the war two events decisively shaped Wiener's mathematical future. He obtained a position as instructor at the Massachusetts Institute of Technology (MIT) in mathematics, where he was to remain until his retirement. At that time mathematics was not particularly strong at MIT, but his position there assured him of continued contact with engineers and physicists. As a result, he displayed an ongoing concern for the applications of mathematics to problems that could be stated in physical terms. The question of which tools he would bring to bear on those problems was answered by the death of his sister's fiancé. That promising young mathematician left his collection of books to Wiener, who began to read avidly the standard texts in a way that he had not in his earlier studies.

The first problem Wiener addressed had to do with Brownian **motion**, the apparently random motion of particles in substances at rest. The phenomenon had earlier excited **Albert Einstein**'s interest, and he had dealt with it in one of his 1905 papers. Wiener took the existence of Brownian motion as a sign of randomness at the heart of nature. By idealizing the physical phenomenon, Wiener was able to produce a mathematical theory of Brownian motion that had wide influence among students of probability. It is possible to see in his work on Brownian motion, steps in the direction of the study of **fractals** (shapes whose detail repeats itself on any scale), although Wiener did not go far along that path.

The next subject Wiener addressed was the Dirichlet problem, which had been reintroduced into the mathematical mainstream by German **David Hilbert**. Much of the earliest work on the Dirichlet problem had been discredited as not

being sufficiently rigorous for the standards of the late 19th century. Wiener's work on the Dirichlet problem produced interesting results, some of which he delayed publishing for the sake of a couple of students finishing their theses at Harvard. Wiener felt subsequently that his forbearance was not recognized adequately. In particular, although Wiener progressed through the academic ranks at MIT from assistant professor in 1924 to associate professor in 1929 to full professor in 1932, he believed that more support from Harvard would have enabled him to advance more quickly.

Wiener had a high opinion of his own abilities, something of a change from colleagues whose public expressions of modesty were at odds with a deep-seated conviction of their own merits. Whatever his talents as a mathematician, Wiener's expository standards were at odds with those of most mathematicians of his time. While he was always exuberant, this was often at the cost of accuracy of detail. One of his main theorems depended on a series of lemmas, or auxiliary **propositions**, one of which was proven by assuming the **truth** of the main **theorem**. Students trying to learn from Wiener's papers and finding their efforts unrewarding discovered that this reaction was almost universal. As Hans Freudenthal remarked in the *Dictionary of Scientific Biography,* "After proving at length a fact that would be too easy if set as an exercise for an intelligent sophomore, he would assume without **proof** a profound theorem that was seemingly unrelated to the preceding text, then continue with a proof containing puzzling but irrelevant terms, next interrupt it with a totally unrelated historical exposition, meanwhile quote something from the 'last chapter' of the book that had actually been in the first, and so on."

In 1926 Wiener was married to Margaret Engemann, an assistant professor of modern languages at Juniata College. They had two daughters, Barbara (born 1928) and Peggy (born 1929). Wiener enjoyed his family's company and found there a relaxation from a mathematical community that did not always share his opinion of the merits of his work.

During the decade after his marriage, Wiener worked in a number of fields and wrote some of the papers with which he is most associated. In the field of harmonic **analysis**, he did a great deal with the decomposition of **functions** into series. Just as a polynomial is made up of terms like x, x^2, and x^3, and so forth, so functions in general could be broken up in various ways, depending on the questions to be answered. Somewhat surprisingly, Wiener also undertook putting the operational **calculus**, earlier developed by Oliver Heaviside, on a rigorous basis. There is even a hint in Wiener's work of the notion of a distribution, a kind of generalized function. It is not surprising that Wiener might start to move away from the kind of functions that had been most studied in mathematics toward those that could be useful in physics and engineering.

In 1926 Wiener returned to Europe, this time on a Guggenheim fellowship. He spent little time at Göttingen, due to disagreements with **Richard Courant**, perhaps the most active student of **David Hilbert** in mathematical organization. Courant's disparaging comments about Wiener cannot have helped the latter's standing in the mathematical community, but Wiener's brief visit introduced him to Tauberian theory, a fashionable area of analysis. Wiener came up with an imaginative new approach to Tauberian theorems and, perhaps more fortunately, with a coauthor for his longest paper on the subject. The quality of the exposition in the paper, combined with the originality of the results, make it Wiener's best exercise in communicating technical mathematics, although he did not pursue the subject as energetically as he did some of his other works.

In 1931 and 1932 Wiener gave lectures on analysis in Cambridge as a deputy for G. H. Hardy. While there, he made the acquaintance of a young British mathematician, R. E. A. C. Paley, with whom a collaboration soon flourished. He brought Paley to MIT the next academic year and their work progressed rapidly. Paley's death at the age of 26 in a skiing accident early in 1933 was a blow to Wiener, who received the Bôcher prize of the American Mathematical Society the same year and was named a fellow of the National Academy of Sciences the next. Among the other areas in which Wiener worked at MIT or Harvard were quantum **mechanics**, differential **geometry**, and statistical physics. His investigations in the last of these were wide ranging, but amounted more to the creation of a research program than a body of results.

The arrival of World War II occupied Wiener's attention in a number of ways. He was active on the Emergency Committee in Aid of Displaced German Scholars, which began operations well before the outbreak of fighting. He made proposals concerning the development of computers, although these were largely ignored. One of the problems to which he devoted time was antiaircraft fire, and his results were of great importance for engineering applications regarding filtering. Unfortunately, they were not of much use in the field because of the amount of time required for the calculations.

Weiner devoted the last decades of his life to the study of **statistics**, engineering, and biology. He had already worked on the general idea of **information theory**, which arose out of statistical mechanics. The idea of entropy had been around since the nineteenth century and enters into the second law of **thermodynamics**. It could be defined as an integral, but it was less clear what sort of quantity it was. Work of Ludwig Boltzmann suggested that entropy could be understood as a measure of the disorder of a system. Wiener pursued this notion and used it to get a physical definition of information related to entropy. Although information theory has not always followed the path laid down by Wiener, his work gave the subject a mathematical legitimacy.

An interdisciplinary seminar at the Harvard Medical School provided a push for Wiener in the direction of the interplay between biology and physics. He learned about the complexity of feedback in animals and studied current ideas about neurophysiology from a mathematical point of view. (Wiener left out the names of those who had most influenced him in this area in his autobiography as a result of an argument.) One area of particular interest was prosthetic limbs, perhaps as a result of breaking his arm in a fall. Wiener soon had the picture of a computer as a prosthesis for the brain. In 1947 he agreed to write a book on communication and control and was looking for a term for the theory of messages. The Greek word for messenger, *angelos,* had too many connec-

tions with angels to be useful, so he took the word for helmsman, *kubernes,* instead and came up with *cybernetics.* It turned out that the word had been used in the previous century, but Wiener gave it a new range of meaning and currency.

Cybernetics was treated by Wiener as a branch of mathematics with its own terms, like signal, noise, and information. One of his collaborators in this area was **John von Neumann**, whose work on computers had been followed up much more enthusiastically than Wiener's. The difference in reception could be explained by the difference in mathematical styles: von Neumann was meticulous, while Wiener tended to be less so. The new field of cybernetics prospered with two such distinct talents working in it. Von Neumann's major contribution to the field was only realized after his death. Wiener devoted most of his later years to the area. Among his more popular books were *The Human Use of Human Beings* in 1950 and *God and Golem, Inc.* in 1964.

In general, Wiener was happy writing for a wide variety of journals and audiences. He contributed to the *Atlantic, Nation,* the *New Republic,* and *Collier's,* among others. His two volumes of autobiography, *Ex-Prodigy* and *I Am a Mathematician,* came out in 1953 and 1956, respectively. Reviews pointed out the extent to which Wiener's memory operated selectively, but also admitted that he did bring the mathematical community to life in a way seldom seen. Although Wiener remarked that mathematics was a young man's game, he also indicated that he felt himself lucky in having selected subjects for investigation that he could pursue later in life. He received an honorary degree from Tufts in 1946 and in 1949 was Gibbs lecturer to the American Mathematical Society.

In 1964 Wiener received the National Medal of Science. On March 18, while travelling through Stockholm, he collapsed and died. A memorial service was held at MIT on the June 2, led by Swami Sarvagatananda of the Vedanta Society of Boston, along with Christian and Jewish clergy. This mixture of faiths was expressive of Wiener's lifelong unwillingness to be fit into a stereotype. He was a mathematician who talked about the theology of the Fall. He did not discover that he was Jewish until he was in graduate school but found great support in the poems of Heinrich Heine. Nevertheless, his intellectual originality led him down paths subsequent generations have come to follow.

WILES, ANDREW J. (1953-)
English-born American number theorist

Andrew J. Wiles conquered the most famous unsolved problem in mathematics—**Fermat's last theorem**, a conjecture that has frustrated professional and amateur mathematicians for more than 300 years. While the **proof** was a personal and professional victory of monumental **proportions**, the lasting value of Wiles's solution is the application of his innovative techniques to a wide range of other problems. Wiles was respected as a talented mathematician well before he announced his triumph on June 23, 1993, at the age of 40. John H. Coates, Wiles's former graduate studies advisor, wrote in *Notices of the AMS* that by 1986, Wiles already stood "amongst the select few over the last 150 years who have made profound contributions to algebraic number theory." Since proving Fermat's last **theorem**, Wiles has received many prestigious awards, including the Wolf, Schock, and Cole Prizes, and the National Academy of Sciences Award in Mathematics.

Andrew John Wiles was born in Cambridge, England, on April 11, 1953. His father, a professor of theology, taught at Oxford University. After earning a bachelor's degree from Oxford, Wiles pursued graduate studies at Cambridge University, earning his master's degree in 1977 and his Ph.D. in 1980. He came to the United States in 1982 to become a professor of mathematics at Princeton University. In 1988, Wiles received the Whitehead Prize from the London Mathematics Society, and he returned to England for two years as a research professor at Oxford. Wiles returned to Princeton in 1990, where he continues to work.

Fermat's last theorem (FLT) is compelling because it can be stated so simply that a child can understand it, yet it has stumped professional and amateur mathematicians for centuries. Wiles became intrigued with it when he was ten years old and continued to work on it throughout his teen years. During his university years, however, he realized it was more complicated than he had initially thought and pragmatically turned his attention to other interesting topics.

FLT concerns an equation similar to the well-known **Pythagorean theorem**: $a^2 + b^2 = c^2$. The **integers** 3, 4, and 5 satisfy this equation ($3^2+4^2=5^2$), as do many other **sets** of integers. In 1637, **Pierre de Fermat** jotted a note in the margin of his copy of Diophantus' *Arithmetic,* asserting that the analogous equation $a^n+b^n=c^n$ (where a, b, and c represent integers) can never be true if 2. In his note, Fermat asserted that he had found a marvelous proof, which the margin was too small to hold. In fact, Fermat later proved the theorem for n=4, as well as an extension of the form $a^4+b^4=c^2$. However, although he lived until 1665, he never again referred to a proof of the general theorem, leading to modern speculation that he had discovered a flaw in his original, supposed proof. Incidentally, FLT was not literally the last theorem proposed by Fermat, but its title reflects the fact that it was his last conjecture to remain unsolved.

More than 100 years after Fermat stated his famous theorem, **Leonhard Euler** proved it for n=3, demonstrating that there is no combination of integers that satisfy the equation $a^3+b^3=c^3$. Another century later, **P. G. Lejeune Dirichlet** proved the equation for n=5 and n=14. Other mathematicians whittled away at the problem, with Ernst E. Kummer proving in 1847, for example, that the theorem holds for all but three values of n less than 100. Even though a general proof was elusive, John Horgan wrote in *Scientific American,* "Techniques employed in these proofs have become standard tools in **number theory**, which has itself become vital to cryptography, error-protection codes and other applications." Modern computer-assisted calculations confirmed that the equation cannot be satisfied for any value of n less than four million, but this result was not established in the traditional sense of mathematical proof.

Fermat had originally proved his theorem for n=4 by developing the theory of elliptic curves. Although this topic played no role in intermediate advances on the problem, it played a key role in the final solution. In 1955, a Japanese mathematician named Yutaka Taniyama proposed a conjecture linking elliptical curves to modular forms of **equations**; the conjecture was refined by Goro Shimura of Princeton University and became known as the Taniyama-Shimura conjecture. During the 1980s, Gerhard Frey, a German mathematician, suggested that an elliptical curve might be used to represent all of the solutions to Fermat's equation; if so, proving the Taniyama-Shimura conjecture might indirectly prove FLT. In 1986, Kenneth Ribet, an American mathematician, accomplished the required formulation. This rekindled Wiles's hope that FLT was vulnerable, and he immediately dedicated himself to finding the solution. For the next seven years, other than fulfilling his teaching duties at Princeton, Wiles worked in solitude in an office in the attic of his home, pursuing his childhood quest.

An elliptic curve (whose graph is *not* an **ellipse**) is defined by an equation of the form $y^2=x(x-A)(x-B)$, where A and B are non-zero integers such that C=A+B 0. What Taniyama had done was treat elliptic equations, not as equalities, but as congruences modulo some prime number. (Two numbers are congruent modulo *p* if they each have the same remainder when divided by *p*.) Reducing an elliptic curve modulo *p* (i.e., dividing all values of x and y by *p* and retaining only the remainder) is useful because it results in a curve that has only a finite number of rational points. Furthermore, counting the number of solutions modulo *p* for many prime values of *p* can reveal other information about the elliptic curve. Some of this information is contained in a related mathematical expression called an L-series. A seemingly unrelated concept is that of modular forms, which are analytic **functions** defined on the upper half of the complex plane; they have nice properties such as being essentially **invariant** under certain transformations. What the Taniyama-Shimura conjecture asserted was that any elliptic curve can be associated with a modular form that has the same L-series. This provided a new way of analyzing elliptic curves.

Wiles believed he could prove that the elliptical curve representing the solutions to Fermat's equation could not exist and thereby prove FLT to be true. Wanting to develop the proof completely on his own, he told only his wife and one trusted colleague what he was working on. Although it took seven years to complete the solution, he did not get discouraged because he felt that he was making continual progress. By May of 1993 he had a proof that was complete except for one single but critical special case. While reading a paper by Harvard mathematician Barry Mazur, Wiles found a technique that would help him over this last hurdle.

Wiles announced his proof in an unusual manner. His withdrawal from most professional activities within the mathematics community had made his presentations at conferences rare. He asked to give three lectures at a small mathematics meeting in Cambridge, England, and rumors of an exciting announcement abounded, though the title of his lectures gave no hint about FLT. In his first lecture, Wiles remained secretive about the final outcome of his series of talks, but twice as many people attended his second lecture. He concluded the third lecture with a proof of a major portion of the Taniyama-Shimura conjecture. Almost as an afterthought, he added that this proved Fermat's last theorem. Word of Wiles's proof spread throughout the mathematical world at the speed of electronic mail, and both technical journals and the general press hounded him for interviews.

Only a few mathematicians in the world are qualified to analyze and confirm the entire 200-page proof. Reviewers did find some gaps, but Wiles corrected them quickly—except for one that took him 18 months and the help of a former graduate student, Richard Taylor. The proof is now considered complete.

During the seven years of his all-out assault on FLT, Wiles virtually withdrew from his professional activities to concentrate on his quest. He is quoted in the *MacTutor History of Mathematics* web site as saying, "my wife has only known me while I have been working on Fermat. I told her a few days after I got married. I decided that I really only had time for my problem and my family... I found with young children [he has two daughters] that it was the best possible way to relax. When you're talking to young children, they're simply not interested in Fermat."

When asked how he felt about completing his proof, Wiles told *People Weekly*, 'there is a sense of loss, actually." There are still many challenging avenues for Wiles (and other mathematicians) to pursue, however. In a series of talks Wiles gave in early 1996 to the joint meetings of the American Mathematical Society and the Mathematical Association of America, he proposed several significant questions generated by his famous proof that should be studied. As for his own future work, he told Horgan that "I have a preference for working on things that nobody else wants to or that nobody thinks they can solve. I prefer to compete with nature rather than be part of something fashionable.... I'm afraid I've made [elliptic curves] so fashionable that I may have to move onto something else."

WITCH OF AGNESI

Published by **Maria Gaëtana Agnesi** in her best-known work *Instituzioni analitiche ad uso della gioventù italiana* (translated: *Analytical Institutions for the Use of Italian Youth*, 1748) the Witch of Agnesi is a versed **sine** curve. (A versed sine, or versine curve, is an outdated expression for 1 - cos x, which was commonly used to describe curves in early mathematics.) It is alternatively known as Cubique d'Agnesi (Agnesi's cubic), Agnésienne, or the Agnesi curve, and was studied by both **Pierre de Fermat** and Guido Grandi as early as 1703. Termed "versiera" by Agnesi, the name of the curve was mistranslated "wife of the devil" or "witch" and hence the name "Witch of Agnesi" stuck. The curve is often most described algebraically, parametrically, or geometrically.

To construct the Witch of Agnesi geometrically, first draw a **circle** of **diameter** d centered at $(0, d/2)$—that is, the circle should be symmetric about the y-axis with a tangent line of $y = 0$ (the x-axis). Draw a second tangent line to the circle of $y = d$. To draw the Witch of Agnesi curve, draw a line segment from the origin $(0,0)$ to any point along the line $y = d$ (x,d); call this point Q. The line segment OQ will cross the circle at exactly two points—the origin $(0,0)$ and one other point (a,y)—call it point B—along the **circumference** of the circle. The intersection of all points (x,y) made by creating a vertical line (that is, parallel to the y-axis) through the point (x,d) and a horizontal line (that is, parallel to the x-axis) through the point (a,y) is Agnesi's curve. Agnesi's curve extends infinitely in both the positive x and negative x directions and is generally bell-shaped. The peak, or highest point, of Agnesi's curve will always occur at the point $(0,d)$, and its lower bound is the **asymptote** $y = 0$. The maximum height of the Agnesi curve is exactly the magnitude chosen for the diameter d.

An algebraic equation for Agnesi's curve, in Cartesian coordinates, is $y = d^3/(x^2 + d^2)$. The power of the numerator—3—is the reason that the curve is sometimes referred to as "Agensi's cubic." It has inflection points at the line $y = 3a/2$. Using Cartesian coordinates, one parametric form is $x = dt$, $y = d/(1 + t^2)$, while a parametric form of the curve in polar coordinates is $x = 2d\cot(, y = d(1 - \cos(2())$. Agnesi's curve can also be approximated using Taylor **polynomials**, with $x_0 = 0$. However, since the denominator $x^2 + d^2 = 0$ when $x = +di$ and $-di$, the Taylor approximation will hold only on the **interval** $(-d, d)$. Outside that interval, Taylor's polynomial will not converge to the curve.

Agnesi's curve and others that are similar were studied carefully by mathematicians such as Agnesi and Fermat in order to better explore the relationship between plane **geometry** and algebraic expressions, as well as the relationships between **plane geometry** and differential **calculus**. During this period, mathematicians and even educated elite were especially interested in simplifying mathematical expressions and concepts. Among the educated elite, even hosts would sometimes pose the simplification of a mathematical problem as a parlor game to their guests. Today, Agnesi's curve is used primarily as a **modeling** and statistical tool. Some computer models for weather and atmospheric conditions, for example, use Agnesi's curve to **model** mountains and other topographic peaks of terrain. In one application, for example, the mountain may be incorporated into the model using Agnesi's curve. Then scientists use the ideas of **slope** and velocity to determine wind speed and direction on the downslope. Although less common, the Agnesi curve is also used as a distribution model (akin to the **normal curve**) for **statistics**. Since the algebraic expression for the curve is relatively straight-forward to integrate over a specified interval, it holds advantages over using the standard normal curve in statistics.

See also Algebra; Cosine; Cubic equations; Equations; Geometry; Parametric equations; Plane geometry; Power functions; Statistics

WITTEN, EDWARD (1951-)

American mathematician and physisist

Often compared to **Albert Einstein** for the originality of his work, Edward Witten is a mathematical physicist whose primary field of research is string theory (introduced by physicist Gabriele Veneziano in 1968). His efforts to develop string theory into a unified "theory of everything" that would explain the fundamental workings of the universe have been among the most successful so far. Many scientists agree that Witten is leading the revival of the traditional symbiosis between **mathematics and physics**, which had become two virtually separate fields by the 1950s. However, he has many critics among physicists who believe that Witten's extensive use of mathematics in a traditional arena of physics is inappropriate.

Witten, born on August 26, 1951 in Baltimore, Maryland, is the son of a gravitational physicist and emeritus professor at the University of Cincinnati. As a preschooler, Witten reportedly enjoyed discussing physics with his father. He received his early education at Baltimore's Hebrew Park School, going on to earn a bachelor's degree in history from Brandeis University in 1971.

After graduating, Witten wrote for such publications as the *Nation* and the *New Republic,* also serving as an aide during George McGovern's 1972 run for president. Meanwhile, Witten thought about continuing his studies in graduate school, having decided that he lacked the right personality for a career in politics or journalism. He chose instead to study physics at Princeton University in New Jersey, where he completed a master's degree in 1974 and a doctorate in 1976.

After working as a junior fellow at Harvard University from 1977 to 1980, Witten returned to Princeton with an appointment as a full professor of physics at the unusually young age of 29. His research immediately centered on finding a simple, concise set of rules and **equations** that would finally explain the nature of all energy and matter in the universe. Just as the ancient Greeks once looked for the answer in the "elements" of earth, air, fire, and water, so modern scientists have looked to the atom for clues to the smallest components of matter.

At first, Witten dedicated much of his time at Princeton to quantum theory, winning a 1982 MacArthur Fellowship for his work. Yet although he soon gained a reputation as the world's most promising young physicist, he realized that his efforts to solve "the single biggest puzzle in physics-reconciling general relativity's gravity and **space** with quantum mechanics' nuclear-level events—were leading nowhere.

In 1984 Witten and another physicist published a paper about anomalies that occur during radioactive decay that could only be studied in terms of **topology** and only in 10 dimensions. Witten's hypothesis fit perfectly with a 1982 article by two other physicists that had suggested that string theory required the presence of a mind-bending 10 dimensions to explain the four known natural forces without any anomalies. They called their hypothesis "superstring theory." Suddenly, with this unlikely coincidence, physicists around the world became hopeful that science was on the verge of discovering what physicists have called the Grand Unified Theory (GUT), or the theory of everything.

String theory is the idea that instead of the different spherical particles (such as electrons, quarks, and photons) that modern physicists have agreed make up atomic structure, these particles are actually identical strands of one-dimensional string. The strings do not differ based on size, but rather according to the way they rotate and vibrate. Superstring theory is the same, but with addition of many more dimensions.

Witten has said that researching string theory is perfect for his abilities, since he believes it will "require a lot of new mathematics" and that "applying bizarre mathematics to physics is what [he's] good at." In 1985, he published 19 papers on string theory, also winning the Einstein Award, among others. That year he cowrote *Current Algebra and Anomalies.* In 1986, in addition to his teaching responsibilities, Witten collaborated with an astrophysicist to find out why galaxies tend to gather together at the perimeters of large voids. The result was their new theory of "cosmic strings," which they believe developed during the cooling process after the Big Bang that formed universe.

Witten, whom his students affectionately nicknamed "the Martian" because of his brilliance and soft voice, gave up his teaching duties in 1987 to concentrate on his research. Leaving the university's Physics Department, he transferred to the School of Natural Sciences at Princeton's Institute for Advanced Study (IAS). That same year, he published *Superstring Theory.* In 1989, Witten began concentrating on knot theory, a branch of topology, to look for further insights into string theory. The following year he won the Fields Medal, which is widely considered the equivalent of the Nobel Prize for mathematics. Witten is the first physicist ever to win the coveted award.

Witten had a major breakthrough in his string theory research in 1994, announcing that he and a colleague at Rutgers University had found a way to simplify the incredibly complex mathematics of quarks by assuming the existence of a particular kind of supersymmetry. Although physicists have not yet confirmed the existence of supersymmetry, Witten is confident that mathematics will eventually prove him correct. In the meantime, Witten and other physicists began lobbying Washington in 1993 to restore funding to a superconducting supercollider and other such facilities that they believe could soon lead to **proof** of supersymmetry.

Aside from his scientific interests, Witten also visits Israel frequently to help promote peace between Middle Eastern Jews and Arabs. He is married to a fellow IAS physicist and they have two daughters. His wife reports that Witten does calculations only in his head.

WOLFRAM, STEPHEN (1959-)

English physicist and mathematician

Stephen Wolfram shook up the scientific world from an early age, and has been recognized as a leading innovator in scientific computing since the first version of his computer program Mathematica was released in 1988.

Stephen Wolfram was born in London on August 29, 1959. His mother was a professor of philosophy at Oxford, and his father was a businessman and part-time novelist. Stephen attended the Dragon School at Oxford from 1967 to 1972, then entered Eton on a scholarship. He published his first scientific paper at the age of 15, which concerned particle physics. At 17, he entered Oxford, advancing rapidly through his studies. In 1978, he came to the United States to attend the California Institute of Technology; he received his Ph.D. in theoretical physics a year later, at age 20.

He received a "genius" grant from the MacArthur Foundation in 1981, becoming the youngest person ever to receive such an honor. The award was given in recognition of his work in physics and in the field of scientific computing, in which he was an early leader.

Besides his abiding interest in particle physics, he was also interested in cosmology, particularly the formation of the universe. During his research he developed a computer program that could handle symbols in algebraic statements as well as numbers, and named it the Symbolic Manipulation Program, or SMP.

He licensed the rights to produce SMP commercially to a California software company, and Caltech took him to court, claiming that because he had created SMP while working for the university, the university, not Wolfram, owned the program. Not long after the case was settled out of court, Wolfram left Caltech for Princeton, to join the Institute for Advanced Study. He was only 23.

Wolfram's interests now turned to discovering how complexity arises in nature. To do this, he practiced experimental mathematics: writing computer programs and running them to see what they would do, and then developing theories based on observations. The simple programs are known as cellular automata, and they can produce remarkably complex patterns—patterns that seem to have intelligence behind them. Wolfram's work with cellular automata became the foundation for a new field of science called complex systems research. Wolfram discovered ways to apply cellular automata for fluid dynamics and cryptography.

However, creating the programs was a laborious and time-consuming process, given the limits of the program being used (C) and the power of the computers the programs ran on. However, with new and more powerful computers being developed, Wolfram set out to create a program that would do that.

The result was Mathematica, a powerful computer program. It can produce three-dimensional graphics of each step in a calculation, no matter how complex, as the calculations are being done. Today, Mathematica has more than one million users around the world.

After leaving the Institute, he joined the faculty of the University at Illinois in 1986 as professor of physics, mathematics and computer science. In 1987 he founded his own company, Wolfram Research, Inc., based in Champaign, Ill. In 1988, he ended his active position at the University of Illinois to pursue other interests.

Today, Wolfram continues his scientific research even while serving as CEO of Wolfram Research. His long-term goals include finding the fundamental theory of physics, creating machines that think, and changing how science is organized.

More information is available on his web site, www.stephenwolfram.com.

WORK

Who is doing more work, a weight lifter holding up, but not moving, a 200-lb (91-kg) barbell, or an office worker lifting a pen? The weight lifter is certainly exerting more effort, and many people would say doing more work. To a physicist, however, the office worker is doing more work as long as the weight lifter does not actually move the barbell. The weight lifter does a considerable amount of work lifting the barbell in the first place, but not in holding it up.

The term work has a very specific meaning in physics that is different from the everyday use of the term. In physics, the amount of work is the **distance** an object is moved times the amount of **force** applied in the direction of the **motion**. If the force is not parallel to the direction of motion, the force must be multiplied by the **cosine** of the **angle** between the force and the direction of motion to get the component of the force parallel to the motion.

If the force applied in the direction of motion is **zero**, then the work done is zero regardless of the amount of motion. Likewise, if the distance moved is zero, then the work done is zero regardless of the force applied. Any number multiplied by zero is still zero. In the above example, the weight lifter is exerting a large force, but as long as he doesn't actually move the weight he is doing zero work, just exerting a lot of effort. The office worker does not need to exert much force to lift the pen, but the force is not zero. So lifting and moving the pen is more work than supporting but not moving the weight. Now think about the weight lifter actually lifting the weight. There is a large force required to lift the weight, and it moves several feet. The weight lifter is now doing quite a bit of work. To do work you must actually move something. Just exerting a force, no matter how large, is not enough.

WREN, CHRISTOPHER (1632-1723)

English architect and mathematician

Christopher Wren, best known as the architect who restored most of London's churches after a devastating fire, was also a skilled geometer, and did valuable work in the study of speed and the cycloid.

Wren was born on October 20, 1632, in East Knoyle, Wiltshire, England. His father, also named Christopher, was chaplain to Charles I. Even as a child young Christopher showed an aptitude for science, drawing, and building mechanical objects. He went to Westminster School and then to Wadham College, Oxford. He received his B.A. in 1651 and his M.A. in 1654. He excelled in his studies, including anatomy and astronomy. After graduation he was appointed a professor of astronomy in 1657 at Gresham College.

Wren's achievements were numerous. As a mathematician, he ranked with **Christiaan Huygens**. In 1658, he rectified the cycloid; that is, he found a straight line equal to an arc of that curve. He conducted experiments with pendulums and also devised a way in which to calculate the final speed of an object after impact, given its size and initial speed before impact.

His work with **geometry**, as well as his skill as a draftsman, made Wren an ideal architect, although he was not interested in architecture until he was over 30. His only journey out of England was in 1665, when he traveled to France to study building designs.

The following year a great fire destroyed much of London, including 87 churches. Wren was given the job of designing new churches, including the great Cathedral of St. Paul. He created a variety of ingenious designs for the new buildings, in which he used the baroque designs he had seen in France but adapted for English taste.

Wren was named surveyor general in 1669. That year he married Faith Coghill; together they had two sons, and after her death he remarried Jane Fitzwilliam, with whom he had another son and a daughter. Wren was knighted in 1673, and was among the founders of the Royal Society.

Wren died February 25, 1723. He is buried in St. Paul's Cathedral.

Y

YANG-MILLS THEORY

In the last half of the twentieth century, physicists succeeded in unifying three of the four fundamental forces of nature: **electromagnetism**, the "weak force" responsible for radioactive decay, and the "strong force" responsible for holding the nuclei of atoms together. The mathematical foundation underlying these advances is called Yang-Mills theory, after Chen Ning Yang and Robert Mills, the physicists who introduced it in a short paper in 1954. At present, it is the dominant approach to quantum **field** theory.

In quantum field theory, the classical distinction between particles and waves breaks down. A **light** wave, for example, can be thought of either as a wave or as a particle (a photon). A bowling ball, to take another example, is a superposition or "tensor product" of all the waves making up the neutrons, protons and electrons in its constituent atoms.

Instead of having a particular location in **space**, a particle/wave is described by its wave function psi(x, t). This function is sometimes described as giving the probability that the particle is observed at position x at time t. In fact, the story is a bit more complicated than that. For the purposes of electromagnetic theory, the value of psi(x, t) at each point and time is a complex number, and it is the magnitude squared, |psi(x, t)|^2, that actually represents the probability of observing the particle there. This subtlety takes on major importance below.

Having abandoned the idea that a particle is located at a single point in space, we also have to abandon the idea that it moves along a single path through spacetime. In fact, its evolution is a complicated "sum over histories," some of which are more probable than others. The probability of a particular history is given by evaluating a certain integral, called the "action"; in rough agreement with the principle of least action (Maupertuis' principle) from classical physics, histories with more action have a much smaller probability, and the history with the least action is the most likely.

For mathematicians, the question of how to define this "sum over histories" correctly is still a thorny and unsolved problem. For physicists, though, the most important question is what function to put inside the action integral. Different choices for this function, called the Lagrangian, lead to different physical theories.

Yang and Mills realized that the possible choices for a Lagrangian are strongly constrained by physical symmetries. In electromagnetic theory, all physical observables (such as a particle's electric charge) are unchanged when psi(x, t) is multiplied by a complex number of magnitude 1. This property of the electric field is called "gauge invariance." Invariance under **multiplication** by unit **complex numbers** is an especially simple kind of **symmetry**, because these numbers form an Abelian group, called $U(1)$. In their 1954 paper, Yang and Mills proposed that the weak and strong nuclear forces could be described by action functionals with more complicated, non-Abelian symmetry groups.

Yang and Mills's idea was not accepted immediately—in fact, Yang won his Nobel Prize for a completely different discovery—but in time, it turned out to be just what physicists needed. Steven Weinberg, Abdus Salam, and Sheldon Glashow used an $SU(2) \times U(1)$–**invariant** Lagrangian to unite electromagnetism and the weak **force**. Murray Gell-Mann found an $SU(3)$-invariant Lagrangian that described the strong nuclear force and predicted the existence of quarks. All of these physicists won Nobel Prizes for their work.

In many ways, the status of Yang-Mills theory today resembles the status of **calculus** in the 1700s. When Newton developed calculus in the late 1600s, his main motivation was not mathematical but physical: He wanted to understand the **motion** of planets subject to his inverse-square law of gravitation. Calculus was a powerful calculational device (thus its name) that suited this purpose, as well as many others. But its mathematical soundness was still very much open to debate for over a century. What were the infinitesimals or "fluxions" that Newton wrote of? How could you add up an infinite collection of infinitesimals and obtain a sensible, finite answer? These questions were finally answered by mathematicians like Cauchy and Riemann in the 1800s, and their work transformed calculus from a computational tool into the mathematical disciplines of real **analysis** and **complex analysis**.

Like calculus in the 1700s, Yang-Mills theory works, and works brilliantly. But, at present, it makes no logical sense. No one has yet proved the existence of a four-dimensional spacetime (such as our own) with well-defined solutions to the quantum Yang-Mills **equations**. At best, they have proved the existence of solutions in certain two- and three-dimensional "toy universes." For this reason, in 2000 the Clay Mathematics Institute named the **proof** of the existence of Yang-Mills fields as one of its seven Millennium Prize Problems. Anyone who solves the problem, and whose solution is generally accepted by the mathematical community, will receive a prize of one million dollars.

The Millennium Prize Problem comes with a second part, the "mass gap," which is also of great relevance to physicists. At present, there is no good explanation for the fact that the strong nuclear force is "local"—that is, we do not observe it at distances larger than the size of an atomic nucleus. One way to prove locality is to show that there is a minimum nonzero mass Delta for any solution to the quantum Yang-Mills equations. As a minimum requirement for a successful solution to the Yang-Mills existence problem, the Clay Mathematics Institute specified that the solution should explain this "mass gap" between the **zero** mass of the vacuum and the minimum mass Delta. There are other important physical phenomena that the solution should explain, such as "quark confinement" (the fact that quarks are never seen in isolation), but they are not required to win the million-dollar prize. According to mathematical physicists Arthur Jaffe and **Edward Witten**, "A solution of the existence and mass gap problem... would be a turning point in the mathematical understanding of quantum field theory."

See also Lie group; Lorentz transformations; Maxwell's equations

YOUNG, GRACE CHISHOLM (1868-1944)
English applied mathematician

A distinguished mathematician, Grace Chisholm Young is recognized as being the first woman to officially receive a Ph.D. in any field from a German university. Working closely with her husband, mathematician **William Henry Young**, she produced a large body of published work that made contributions to both pure and **applied mathematics**.

Grace Emily Chisholm Young was born on March 15, 1868, in Haslemere, Surrey, England, to Anna Louisa Bell and Henry William Chisholm. Her father was a British career civil servant who (following his own father) rose through the ranks to become the chief of Britain's weights and measures. Grace Emily Chisholm was the youngest of three surviving children. Her brother, Hugh Chisholm, enjoyed a distinguished career as editor of the eleventh edition of the *Encyclopaedia Britannica*.

As befitted a girl of her social class, Young received an education at home. Forbidden by her mother to study medicine—which the youngster wanted to do—she entered Girton College, Cambridge (one of two women's colleges there) in 1889. She was 21 years of age, and the institution's Sir Francis Goldschmid Scholar of mathematics. In 1892 she graduated with first-class honors, then sat informally for the final mathematics examinations at Oxford; there, she placed first. In 1893 she transferred to Göttingen University in Germany, where she attended lectures and produced a dissertation entitled "The Algebraic Groups of Spherical Trigonometry" under noted mathematician **Felix Klein**. In 1895 she became the first woman to receive a Göttingen doctorate in any subject. The degree bore the distinction magna cum laude.

She returned to London and married her former Girton tutor, William Henry Young, who had devoted years to coaching Cambridge students. After the birth of their first child, the Youngs moved to Göttingen. There, William Young began a distinguished research career in mathematics, which would be supported in large part by the work of his wife. Grace Chisholm Young studied anatomy at the university and raised their six children, while collaborating with her husband on mathematics in both co-authored papers and those published under his name alone. In 1905 the pair authored a widely regarded textbook on **set theory**. Grace Chisholm Young's most important work was achieved between 1914 and 1916, during which time she published several papers on derivates of real **functions**; in this work she contributed to what is known as the Denjoy-Saks-Young **theorem**.

The Young family lived modestly, and William Young traveled frequently to earn money by teaching. In 1908, with the birth of their sixth child, the Youngs moved from Göttingen to Geneva. William Young continually sought a well-paying professorship in England, but he failed to obtain such a position; in 1913 he obtained a lucrative professorship in Calcutta, which required his residence for only a few months per year, and after World War I he became professor at the University of Wales in Aberystwyth for several years. Switzerland, however, remained the family's permanent home.

With advancing years, Grace Chisholm Young's mathematical productivity waned; in 1929 she began an ambitious historical novel, which was never published. Writing fiction was but one of her many varied interests, which included music, languages, and medicine. She also wrote children's books, in which she introduced notions of science. Her children followed the path she had pioneered, becoming accomplished scholars of mathematics, chemistry, and medicine. Her son Frank, a British aviator, died during World War I.

Grace Chisholm Young had lived with her husband's extended absences for her entire married life, and the spring of 1940 found them separated again: she in England, and he in Switzerland. From that time onward, neither spouse was able to see the other again—both were prevented from doing so by the downfall of France during the war. William Young died in 1942, and Grace Chisholm Young died of a heart attack in 1944.

Z

ZENO OF ELEA (CA. 490 B.C.-CA. 430 B.C.)
Greek logician and philosopher

Zeno of Eleawas a Greek philosopher and logician whose development of paradoxical philosophical arguments about **motion** greatly influenced mathematical thought. One of the last major proponents of the Eleatic school of philosophy, he created his paradoxes to defend and confound this school of thought's detractors. His importance in early Greek philosophy and mathematics was noted by both **Plato** and **Aristotle**, who credited Zeno with the creation of dialectics.

Zeno was born in Elea, a southern Italian city, in the fifth century B.C. Although the exact date is unknown, historians place his birth at around 490 B.C., based largely on references by Plato in his book *Parmenides*. Plato describes Zeno as "tall and fair to look upon" and a favorite disciple of Parmenides, a Greek philosopher. He also reports that Zeno was about 40 years old when he accompanied his teacher to Athens in 449 B.C. In Athens, Zeno met and greatly impressed a young Socrates. Legend indicates that Zeno stayed for a number of years in Athens, where he made a good living by teaching philosophy. Athenian statesmen Pericles and Callias possibly studied under him.

According to Diogenes Laëritus, Zeno also created a unique cosmology in which he proposed the existence of several worlds. However, he was predominantly known for his paradoxes, which instigated centuries of philosophical debate. Little else is known about Zeno's life. He returned to live and work in Elea, where he eventually was caught up in a political intrigue that led to his death.

Zeno's teacher, Parmenides, was without question the greatest influence on his life. The founder of the Eleatic school of philosophy, Parmenides believed that there is only the "One" unchangeable being who constitutes the universe and argued that people's senses were mistaken in interpreting a world of motion and change. Parmenides was attacked and ridiculed by some, including the Pythagoreans. Zeno set out to defend his mentor's view of the world by creating a series of philosophical paradoxes.

At a very young age, Zeno wrote his famous and only known work, *Epicheiremata*. Only a few small fragments, amounting to about 200 words, survive. Fortunately, through Plato, Aristotle, Proclus, and other ancient authors, historians have been able to piece together the fundamentals of Zeno's work, which reportedly was published without his consent.

Zeno's unique approach to defending his master does not take the usual course of trying to prove Parmenides' position. Instead, he set out to create proofs that would reveal the absurd and self–contradictory nature of his opponents' views. The result was a series of paradoxesthat profoundly influenced both philosophical and mathematical thought.

Zeno's philosophical puzzles focus on the goal of proving that our sense of motion, time, and change are based on illusion. According to Plato and Proclus, Zeno developed 40 different paradoxes (of which only eight survive) designed to disprove, confound, and agitate Parmenides' detractors. Among the most famous is the paradox of "Achilles," who was a mythological Greek hero who was the "swiftest of mortals." This paradox states that Achilles could not catch a crawling tortoise with a head start because Achilles must always reach a point at which the tortoise has started from and already passed. The "Dichotomy" and "Arrow" paradoxes similarly focus on the illusion of motion as represented by points in **space** or time. The paradox of the arrow, for example, states that an arrow can never reach it target because an object can only reach one point at a time and to be someplace (or at some point) an object must be at rest. If the arrow as at rest at each point, then motion forward is impossible.

Simplistically, these paradoxes may appear to be mere philosophical "brain teasers," but they represent the foundation of a type of philosophical argument called dialectic. This method of **deduction** allows one to work out contradictory conclusions from a single **postulate** or proposition that is assumed to be true. The dialectic approach to philosophy influenced the work of Plato, Aristotle, Immanuel Kant, Georg Hegel, **Bertrand Russell**, and even **Albert Einstein**, who, like Zeno, was interested in the concepts of space and time.

Although developed primarily as a means of defending his teacher's philosophical views, Zeno's paradoxes also created difficulties for mathematicians. Some historians believe **Zeno's paradoxes** may have been directed toward the Pythagoreans (followers of the great Greek mathematician **Pythagoras**) and their handling of the **infinitesimal** in **geometry**. Eudoxius of Cnidus, who was a member of Plato's academy, may also have been influenced by Zeno in his development of a theory of **proportions** to accurately deal with the infinitesimal. However, since Greek mathematicians had not developed a strong concept of **convergence** or **infinity**, they mostly chose to ignore the mathematical dilemmas presented by the paradoxes. This early neglect was reinforced by Aristotle's declaration (without **proof**) that Zeno's paradoxes were merely fallacies.

Although Zeno's influence on early Greek mathematics probably was minimal, many of his paradoxes, like the "Achilles" paradox, relate to the concepts of infinity and the continuum, which would become major areas of interest in **calculus**. But it would take 19th– and 20th–century mathematicians to clearly understand the mathematics of Zeno's paradox, which is based on the distinction between measures of intervals and the number of points they contain. With the development of this concept, Zeno's paradoxes established an important influence on the work 19th–century German mathematician **Karl Weierstrass**, who helped start the "mathematical renaissance" and developed the notion of uniform convergence. The dilemmas presented by the paradoxes continued to exert a strong hold on mathematical thought through the work of **Bertrand Russell**, German mathematician **Georg Cantor**, who introduced the concept of infinite **sets**, and many others.

Despite being dubbed the first "great doubter" and "man of destiny" in mathematics, Zeno was primarily a philosopher and not a mathematician. It is a testament to Zeno's keen intellect that later mathematicians went on to form numerous mathematical theories and problems from his work. While Cantor's theories on infinite sets led to explaining some of Zeno's paradoxes, the philosophical problems he presented have remained of interest and debate.

As reported by Plato, Zeno admitted that his book developed from the "pugnacity" of a younger man defending his mentor. Zeno, who was reported to be an active participant in politics, exhibited a similar contentiousness later in life when he "defended" good government by joining a plot against the tyrant Nearchus. The plot failed, and Zeno was arrested. According to legend, Zeno was tortured to reveal the names of his co–conspirators. Instead, the unyielding Zeno named the tyrant's own friends. Another legend says that Zeno bit off his tongue and spit it at his interrogators rather than betray his friends. He was eventually executed around 440 B.C.

ZENO'S PARADOXES

A paradox (or antinomy) is a statement that appears self-contradictory or contrary to common sense. Scholars believe that the philosopher **Zeno** wrote his paradoxes around 465 B.C. He probably wrote about forty paradoxes, but only about two hundred words actually written by Zeno have survived; and these mention only two paradoxes. Thus, almost all of the information we have about Zeno's paradoxes comes from other authors, notably **Aristotle** and Simplicius. In particular, Aristotle dealt with Zeno's paradoxes of **motion** in his book *Physics;* he considered Zeno the father of dialectics, the method of argumentative reasoning in which a stated thesis is argued against by an interlocutor who tries to disprove it by showing that it produces a contradiction.

The point of Zeno's paradoxes is to prove that the idea of continuous motion is self-contradictory: that is, he wanted to first confute the ideas of Heraclitus, who thought that the world was in a state of perpetual change, and then to confirm the thesis of Parmenides, who considered the world to be a compact, homogeneous, motionless, immutable **sphere**.

Tradition ascribes to Zeno four classical paradoxes:

- Dichotomy paradox: before an object can travel a given **distance**, it must travel half that distance. In order to travel half that distance, it must first travel one quarter of it, and so on infinitely: the process of halving never reaches an end for there is always a distance to be halved, no matter how small. Despite the fact that a finite distance requires a finite amount of time to be traveled, Zeno inferred from this principle of "halving" that no distance could be traveled in a finite amount of time. Hence the paradox.

- Achilles and the tortoise paradox: in a race between the fleet-of-foot Achilles and the plodding tortoise, the former can never catch the later, if the latter is given a head start. This is because in the time it takes Achilles to arrive at some point formerly occupied by the tortoise, the tortoise has since moved ahead some distance; and while Achilles is closing the gap between this point and some farther one, the tortoise has again moved ahead. Thus, it would appear that, if the tortoise keeps moving, Achilles will never catch up, for he will have to travel an infinite number of finite distances.

- Arrow paradox: an arrow in flight is, at any single instant, indistinguishable from a motionless arrow in the same position. The question then arises: is the arrow moving or at rest in that instant? The paradox is that if you say the arrow is moving, how is it possible to be moving during one instant? And if you say the arrow isn't moving, it must therefore be at rest and thus cannot be in flight.

- Stadium paradox (sometimes called moving rows paradox): this is the most obscure of Zeno's paradoxes, little information about it survives. Its basic point appears to be that speed, usually considered an essential property of motion, is not an objective property but a relative one. The paradox involves parallel rows of seats (as in a stadium); it could also be visualized as three parallel trains, A, B, and C. A and C travel at the same speed in opposite directions; B, in the middle, is motionless. Zeno seems to conclude that A takes both the same amount of time and twice as much time to pass any part of B as C does.

The target of the first two paradoxes was probably the thesis that **space** and time can vary continuously, while the point of the other two was to confute the idea of discrete space or time. In order to overcome the apparent contradiction arising from the first two paradoxes, we have to consider that a **geometric series** converges. In other words, we have to accept the proposition that the sum of an infinite amount of finite quantities doesn't produces an infinite result!

For instance, if we assume that Achilles runs at a speed ten times faster than the tortoise's, and that the tortoise starts 100 meters in front of Achilles, Zeno's argument leads to the series: $100 + 10 + 1 + 1/10 + ... = 111 + 1/9$. Hence, after exactly $111 + 1/9$ meters, Achilles will reach the tortoise.

A similar argument based on the geometric series can deal with the dichotomy paradox.

The arguments needed to overcome the paradox of the arrow are probably more subtle, and depend on our assumptions on the nature of space and time—namely whether space (or time) is built by discrete irreducible atoms. A precise **definition** of the movement is also needed. A discussion about the problems arising in such definitions is in *The Principles of Mathematics*, by **Bertrand Russell**.

The stadium paradox, in the formulation classically ascribed to Zeno, is quite vague, and it can be probably overcome by the concept of relative speed. However, the same concept of relative speed presents some "paradoxes" in modern physics, since Einstein's relativity showed some inconsistency in the common sense definition of relative speed and relative time. Also, several philosophers, such as Russell and G. E. Owen, considered the stadium paradox in new ways, producing new logical paradoxes.

The paradoxes on movement were reconsidered by modern philosophers too. For instance, M. Black pointed out that the real logical problem should be that Achilles performed an **infinite series** of action in a finite time: this can lead to other paradoxes, such as the Thomson lamp. The Thomson lamp is not a real object, but a mental experiment. It deals with proving that machines that perform an infinite number of actions in a finite time can lead to logical problems and paradoxes. Thomson lamp is turned on for 1/2 minute, off for 1/4 minute, on for 1/8 minute, and so on. At the end of one minute, the lamp switch will be moved infinitely many times. The question (disregarding any objection based on the physical impossibility of such lamp) is whether the lamp is on or off after one minute.

Another paradox ascribed to Zeno deals with the relation between a whole object and its parts: if anything does exist, it must be composed of parts. Since these parts can be divided again and again, everything must be composed of infinite parts: such "elementary" parts should have no extension, otherwise they could be divided again. But if the situation is

like this, how is it possible that parts with no extension can produce anything extended?

Mathematically, such a paradox can be translated into the fact that points with no extension can produce a line. However, Zeno's argument was reconsidered by modern philosophers, such as William James and A. N. Whitehead, for metaphysical considerations about the discontinuity of the processes of change.

ZERMELO, ERNST FRIEDRICH FERDINAND (1871-1953)

German mathematician

Ernst Friedrich Ferdinand Zermelo made major contributions to mathematics in the area of **set theory**, especially by extending David Hilbert's work in solving the continuum hypothesis. The **Zermelo-Fraenkel set theory** is still in wide use.

Zermelo was born in Berlin, Germany, on July 27, 1871. His father was a college professor, which helped set the academic tone of Zermelo's childhood. He attended a gymnasium (a classical college-preparatory school) in Berlin, graduating in 1889. Then, as was customary at the time, Zermelo took classes in mathematics, physics, and philosophy at a variety of universities—namely those at Berlin, Freiburg, and Halle.

Earning his doctorate in 1894 from the University of Berlin, Zermelo began working as an assistant to physicist Max Planck. Soon he began studying for a teaching certificate, which he obtained in 1899 at the University of Göttingen with a dissertation on **hydrodynamics**. The school, which then was the world's leading center for mathematical research, appointed him as a lecturer based on the quality of his work.

At this point, Zermelo's research changed direction. In 1900, Hilbert suggested that mathematicians could solve Georg Ferdinand Ludwig Philipp Cantor's 1878 continuum hypothesis by first proving that any set can be well ordered. Zermelo accepted this challenge, and in 1902 published his first work on set theory, which centered on the **addition** of transfinite cardinals. Two years later, he met Hilbert's challenge by using the **axiom** of choice to prove that all **sets** can be well ordered. His achievement brought the mathematician immediate renown in the academic world and even led to a promotion to full professor at Göttingen in 1905.

However, Zermelo also endured much criticism for his work on set theory, since many mathematicians of the period did not accept the type of proofs that he had discovered. Thus, in 1908 he used another, more traditional **proof** to accomplish the same goal. Later that year, Zermelo published the results of his efforts to axiomize set theory despite his failure to prove that the seven axioms were consistent. Nevertheless, his system would turn out to be tremendously important to the development of mathematics. (It would eventually be improved by two other mathematicians in 1922, resulting in a 10-axiom system that is now commonly used for axiomatic set theory.)

Zermelo, who was beginning to suffer from poor health, left the University of Göttingen in 1910 to take up the chair of mathematics at the University of Zurich in Switzerland.

Despite efforts to rest and recover, by 1916 his health had not improved, so he resigned his post and moved to the Black Forest. He lived there until 1926, when he accepted an honorary chair at the University of Freiburg. By 1935 Zermelo had resigned the position to protest the regime of Adolf Hitler, but after the war, he accepted reinstatement.

Zermelo died in Breisgau, Germany, on May 21, 1953.

ZERMELO-FRAENKEL SET THEORY

The German mathematician **Georg Cantor** began a revolution in mathematics when invented the theory of **sets** in the 1870s. The parts of his theory dealing with infinite sets were the most controversial at that time. In an 1874 article, Cantor suggested that there were different kinds of infinite sets and different orders of **infinity**. This idea stirred up tremendous controversy among some editors of the journal and among mathematicians in general. Prior to this publication, all infinite sets had been considered alike. No one had ever suggested that there were different orders of infinity. By 1890 most of the furor over Cantor's ideas had settled down, and many mathematicians saw Cantor's **set theory** as an ideal foundation for the rest of mathematics. In particular, Gottlob Frege (1848-1925) used elements of Cantor's theory in producing a formal system of mathematics purportedly derived from logic. Unfortunately, during the years 1895-1903, paradoxes were discovered in Cantor's set theory, first by Cantor himself, then by the Italian mathematician Burali-Forti (1861-1931), and last, and most famously, by the British philosopher and mathematician, **Bertrand Russell**. Russell, while studying the work of Frege, discovered that Frege's unrestricted use of Cantor's theory of infinite sets allowed him to construct the rather bizarre "set of all sets that are not members of themselves." Then he asked the question, "Is this set a member of itself?" If he answered yes, then the set is not a member of itself; and if he answered no then the set is a member of itself. Hence the paradox. This set is a member of itself if and only if it is not a member of itself. This discovery crushed Frege and the rest of the mathematical community because Frege's system had been the best candidate, up to that time, for an adequate account of the foundations of mathematics. This sent everyone back to the drawing board to come up with a new account of set theory that would not be subject to the paradoxes. Enter **Ernst Zermelo**, a German mathematician, whose early career was in **applied mathematics** but who became interested in set theory by reading the works of Cantor.

While reading Cantor, Zermelo discovered his own paradox and decided to take on the task of axiomatizing Cantor's set theory in a way which would prevent the paradoxes from occurring. In Frege's system, the culprit that opened the door to the paradoxes was an **axiom** of abstraction, which said that for any property there is a set of all things having that property. This made Russell's set of all sets that are not members of themselves a legal formulation in Frege's program. Hence, Zermelo needed to come up with a list of axioms that would allow all the desirable results of Cantor's theory without allowing the paradoxes to be derived from them. In 1908, Zermelo introduced the first axiomatic set theory, containing eight axioms that put restrictions on the formation of sets. Zermelo took the terms "set" and "∈", as undefined, the latter of these being the symbol for "is an element of." His first two axioms established the meaning of the equality of two sets and the existence of the empty or "null" set which has no members. His third axiom was a restricted version of Frege's axiom of abstraction, which, unlike Frege's, prohibited asserting the existence of sets unconditionally. It required new sets to be constructed recursively in a finite number of steps using only pre-existing sets. This eliminated the possibility of Russell-like paradoxes. Zermelo needed additional axioms to allow his theory to include the Cantorian notions of union, intersection, Cartesian products, infinite sets, relations, **functions**, and a well-ordering principle. A well-ordering principle allows for the existence of a least element in each subset of a given set. The importance of this is that makes possible **proof** by a version of mathematical **induction** known as transfinite induction. The assertion of this well-ordering principle was the eighth and last of Zermelo's axioms presented in 1908. It was also the most controversial because it was equivalent to the so-called "axiom of choice", which was regarded as suspicious by a number of mathematicians of the day. The axiom of choice asserts that one may form a set by choosing exactly one element from any collection of sets. When the number of sets is finite there is no controversy; but when infinite sets are allowed, mathematicians known as "intuitionists" or "constructivists" regard this method of forming sets as unintuitive and non-constructive. Nevertheless, the vast majority of modern mathematicians regard the axiom of choice as indispensable to their work.

Zermelo's original eight axioms were later found to be inadequate for the development of the **arithmetic** of the natural numbers. The mathematicians Abraham Fraenkel (1891-1965), Thoralf Skolem (1887-1963), and **John von Neuman** added axioms to allow the natural numbers to be defined as sets. Thus 0 is associated with the empty set, 1 is the set containing just the empty set, 2 is the set containing the empty set and the set containing the empty set, and so on. Once defined in this way, a complete arithmetic of the natural numbers may be developed. From the natural numbers the **rational numbers** can be obtained, then the **real numbers** including the **irrational numbers** can be defined as sets in the manner of Dedekind. Due to the contributions of Fraenkel, the resulting ten axioms became known as the Zermelo-Fraenkel axioms. Zermelo-Fraenkel set theory is today regarded as the foundation upon which modern mathematics rests since every **theorem** of modern mathematics can be translated into the language of Zermelo-Fraenkel set theory and proved within the system.

ZERO

Zero is often equated with "nothing," but that is not a good analogy. Zero can be the absence of a quality, but it can also be a starting point, such as 0° on a temperature scale. In a mathematical system, zero is the *additive identity*. It is a number which can be added to any given number to yield a sum equal to the given number. Symbolically, it is a number 0, such that a + 0 = a for any number a.

In the Hindi-Arabic numeration system, zero is used as a placeholder as well as a number. The number 205 is distinguished from 25 by having a 0 in the tens place. This can be interpreted as no tens, but the early use of 0 in this way was more to show that 2 was in the hundreds place than to show no tens.

Zero is used in some ways that take it beyond ordinary **addition** and **multiplication**. One use is as an exponent. In an exponential function such as y = 10^x the exponent is not limited to the counting numbers. One of its possible values is 0. If 10^0 is to obey the rule for **exponents** then 10^m = 10^{m+0} = 10^m × 10^0 - 10^0 must equal 1. This is true not only for 10^0 but for any a^0, where a is any positive number. That is, a^0 = 1.

Another curious use of zero is the expression 0!. Ordinarily n! is the product 1 × 2 × 3 ×... × n, of all the **integers** from 1 to n. In a formula such as n!/r!(n - r)!, which represents the number of different combinations of things which can be chosen from n things r at a time, 0! can occur. If 0! is assigned the value 1, the formula works. This happens in other instances as well.

The symbol for zero does not appear before about 800 A. D., when it appears in connection with Hindu- Arabic base-10 numerals. In these numerals it functions as a place holder. The Mayans also used a zero in writing their base-20 numerals. It was a symbol which looked something like an eye, and it acted as a place holder.

The reason that the symbol appeared so late in history is that the number systems used by the Greeks, Romans, Chinese, Egyptians, and others didn't need it. For example, one can write the Roman numeral for 1056 as MLVI. No zero as a place holder is needed. The Babylonians did have a place-value system with their base-60 numerals, and a symbol for zero would have eliminated some of the ambiguity that shows up in their clay tablets, but was probably overlooked because, within each place, the numbers from 1 to 59 were represented with wedge-shaped tallies. In a tally system all that is required to represent zero is the absence of a tally. Sometimes Babylonians did use a dot or a **space** as a placeholder, but failed to see that this could be a number of its own.

The word zero appears to be a much metamorphosed translation of the Hindu word "sunya," meaning void or empty.

Zero also has the property a × 0 = 0 for any number a. This property is a consequence of zero's additive property.

In ordinary **arithmetic** the statement ab = 0 implies that a, b, or both are equal to 0; that is, the only way for a product to equal zero is for one or more of its factors to equal zero. This property is used when one solves **equations** such as (x - 2)(x + 3) = 0 by setting each **factor** equal to zero.

The multiplicative property of zero is also used in the argument for not allowing zero to be used as a divisor or a denominator. The law which defines a/b is (a/b)b = a. If one substitutes 0 for b, the result is (a/0)0 = a, which forces a to be 0. But even when a is 0, the law allows 0/0 to be any number, which is intolerable.

Another arithmetic is clock arithmetic. In this arithmetic 3 is three hours past twelve; 3 + 7 is ten hours past twelve; and 3 + 12 is fifteen hours past twelve. But on a clock, every twelve hours the hands return to their original position; so fifteen hours past twelve is the same as three hours past twelve. For any a, a + 12 = a. [In number-theory symbolism this would be written a + 12 intergral a (mod 12).] So in clock arithmetic, 12 behaves like 0 in ordinary arithmetic.

It also multiplies like 0. Twelve 3-hour periods equal 36 hours, which the hands show as 12. Twelve periods of a hours each leave the hands at 12 for any a (a is limited to **whole numbers** in clock arithmetic), so 12 × a = 12.

Thus, in clock arithmetic 12 doesn't look like zero, but it behaves like zero. It could be called 0, and on a digital 24-hour clock, where the number 24 behaves like 0, 24 is called 0. The next number after 23:59:59 is 0:0:0.

In this arithmetic, unlike ordinary arithmetic, the law "ab = 12 if and only if a, b, or both equal 12" does not hold. The "if" part does, but not the "only if." Six times 2 is 12, but neither 6 nor 2 is 12. Three times 8 is 12 (the hands go around twice, passing 12 once and ending at 12), but neither 3 nor 8 is 12. Thus in clock arithmetic there can be two numbers, neither of them zero, whose product is zero. Such numbers are called divisors of zero. This happens because we use 12-hour (or 24-hour) clocks. If we used 11-hour clocks, it would not.

ZETA-FUNCTION

Zeta-function is the name given to certain **functions** of the complex **variable** s = σ + it that play a fundamental role in analytic **number theory**. The most important example is the Riemann zeta-function ζ(s). In the right half plane {s ∈ **C**: 1 < σ} the Riemann zeta-function is defined by the **infinite series**

$$\zeta(s) = \sum_{n=1}^{\infty} \frac{1}{n^s}.$$

It can be shown that the infinite series defining the zeta-function converges absolutely and uniformly on all compact subsets of {s ∈ **C**: 1 < σ} and therefore the Riemann zeta-function is an analytic function in this **domain**. Series of this type are called Dirchlet series. More generally, if a(1), a(2), a(3),... is a sequence of **complex numbers** then

$$f(s) = \sum_{n=1}^{\infty} \frac{a(n)}{n^s}$$

is the associated Dirichlet series. The natural domain of **convergence** for such a series is always a right half plane, but the half plane may be empty or all of **C**. If the half plane is not empty then the series defines an analytic function f(s) in the interior of the half plane of convergence. The most important examples of Dirichlet series that occur in analytic number theory are those in which the coefficients $a(n)$ carry **arithmetic** information. For example, in the half plane $\{s \in \mathbf{C}: 1 < \sigma\}$ the **square** of the Riemann zeta-function is given by the Dirchlet series

$$\zeta(s)^2 = \sum_{n=1}^{\infty} \frac{d(n)}{n^s},$$

where for each positive integer n the coefficient $d(n)$ is the number of positive **integers** that divide n.

The Riemann zeta-function is important in analytic number theory because it has a second representation in the half plane $\{s \in \mathbf{C}: 1 < \sigma\}$ as a convergent infinite product over the sequence of **prime numbers**. More precisely, we have

$$\zeta(s) = \prod_{p} (1 - p^{-s})^{-1}$$

at each complex number s with $1 < \sigma$, where the product is over the sequence $p = 2, 3, 5, 7, 11,...$ of prime numbers. It can be shown that the partial products converge uniformly on compact subsets of the half plane $\{s \in \mathbf{C}: 1 < \sigma\}$ and converge to exactly the same value as the Dirichlet series for $\zeta(s)$. That is, we have

$$\sum_{n=1}^{\infty} \frac{1}{n^s} = \zeta(s) = \prod_{p} (1 - p^{-s})^{-1}$$

for all complex numbers $s = \sigma + it$ with $1 < \sigma$. The identity amounts to an analytic formulation of the fundamental **theorem** of arithmetic, which states that each integer $n \geq 2$ has a unique representation as a product of prime numbers. Because of this identity methods from complex **analysis** can be used to investigate problems about the distribution of prime numbers. For example, if $\pi(x)$ is defined to be the number of primes less than or equal to the positive real number x, then the **prime number theorem** asserts that

$$\lim_{x \to \infty} \frac{\pi(x) \log x}{x} = 1.$$

The first complete **proof** of the prime number theorem was given in 1896 by **Jacques Hadamard** and, independently, by C. J. de la Vallèe Poussin. The crucial step in both of their proofs was the discovery that $\zeta(s) \neq 0$ in a certain open subset of $\mathbf{C}$ that includes the closed half plane $\{s \in \mathbf{C}: 1 \leq \sigma\}$.

It can be shown that the Riemann zeta-function extends by analytic continuation to a function that is analytic on the complex plane $\mathbf{C}$ except for a pole of order 1 and residue 1 at the point $s = 1$. If we continue to write $\zeta(s)$ for the extended function then $\zeta(s)$ satisfies the functional equation

$$\pi^{-\frac{s}{2}} \Gamma\left(\frac{s}{2}\right) \zeta(s) = \pi^{-\frac{1-s}{2}} \Gamma\left(\frac{1-s}{2}\right) \zeta(1-s)$$

for all complex s. Here $\Gamma(s)$ is Euler's **gamma function**, defined by

$$\frac{1}{\Gamma(s)} = \lim_{N \to \infty} s e^{\gamma s} \prod_{n=1}^{N} \left(1 + \frac{s}{n}\right) e^{-\frac{s}{n}},$$

and γ is Euler's constant, defined by

$$\gamma = \lim_{N \to \infty} \left\{ \left(\sum_{n=1}^{N} \frac{1}{n}\right) - \log N \right\}.$$

The function $\Gamma(s/2)$ has poles at the nonpositive even integers. This fact together with the functional equation shows that $\zeta(s)$ has a **zero** at each negative even integer. It is also known that $\zeta(s)$ has infinitely many nonreal zeros $\rho = \beta + i\gamma$ that satisfy $0 < \beta < 1$. (The notation $\rho = \beta + i\gamma$ for a nonreal zero is generally used. Of course in this context γ does not refer to Euler's constant.) It was conjectured by **Bernhard Riemann** in 1859 that $\zeta(s) \neq 0$ in the open half plane $\{s \in \mathbf{C}: \frac{1}{2} < \sigma\}$, but this has never been proved. The conjecture is known as the **Riemann hypothesis**. An equivalent form of the Riemann hypothesis is the assertion that all of the nonreal zeros of $\zeta(s)$ occur on the line $\sigma = \frac{1}{2}$, that is, if $\rho = \beta + i\gamma$ is a nonreal zero of the zeta-function then $\beta = \frac{1}{2}$. The Riemann hypothesis is one of the most important unsolved problems in mathematics.

ZORN, MAX AUGUST (1906-1993)

Austrian-American mathematician

Max Zorn embarked on his academic career in Hamburg, Germany, studying under mathematician Emil Artin and receiving his mathematics doctorate in 1930. He then received an appointment to teach at the University of Halle, but after a short time, left Nazi Germany for the United States. Zorn worked at Yale in the early thirties, where he developed what became known as **Zorn's lemma**, an algebraic theory that states that if S is any nonempty partially ordered set in which every chain has an upper bound, then S has a maximal element.

Zorn left Yale in 1936 to move to a teaching position at the University of California, where he continued his work on algebraic structure theory. A decade later, he took a professorship at Indiana University, where he spent the remainder of his career until retiring in 1971. Zorn died in Bloomington, Indiana at the age of 87 of congestive heart failure.

ZORN'S LEMMA

Zorn's lemma, the well-ordering principle and the **axiom** of choice are three equivalent **propositions**. It has been said that from appearances, the axiom of choice has to be true, the well-ordering principle has to be false and Zorn's lemma is too confusing to figure out. Zermelo formulated the axiom of choice in 1904 in an attempt to solve the first problem on Hilbert's famous list, the continuum hypothesis. The axiom states that given any collection of mutually disjoint **sets** there is a set that contains one element from each of the given sets. Zermelo proved that this axiom is equivalent to the well-ordering principle. His axiom was controversial at the time. Many hoped that it was unnecessary. **Paul Cohen** dashed these hopes when he proved (in 1963) that the axiom of choice is independent of Zermelo-Fraenkel **set theory**.

To explain these notions, a little set theory is necessary. A partially ordered set is a set X with a relation ≤ which satisfies two properties: 1. If x, y and z are in X and x ≤ y and y ≤ x then x = y. 2. If x ≤ y and y ≤ z then x ≤ z. A total ordering is just a partial ordering such that every pair of elements is comparable, i.e. if x and y are in X then either x ≤ y or y ≤ x. The **real numbers** are totally ordered, for example. For any two pairs of real number (x, y) and (w, z), let's say that (x, y) is less than or equal to (w, z) if x is less than or equal to w and y is less than or equal to z. Under this order, the set of (ordered) pairs of real numbers is partially ordered by not totally ordered. A set is well-ordered if it is totally ordered and every subset has a smallest element. For example the smallest element of {1, 2, 3, 4, 5,...} is 1 but the smallest element of the **rational numbers** does not exist so neither the rationals nor the real numbers are well-ordered. The well-ordering principle states that any set can be well-ordered, i.e. a well-ordering of the set exists. Zermelo proved that if the well-ordering principle is true then given any two sets X and Y, either the **cardinality** of X is less than, equal to, or greater than that of Y. This is sometimes called the trichotomy law.

If S is a partially ordered set then a subset of S that is totally ordered under the given ordering is called a chain. An upper bound of a chain is an element which is greater than or equal to all elements in the chain. A maximal element of S in an element which is greater than or equal to all other elements of S. Zorn's lemma states if S is a partially ordered set with the property that every chain has an upper bound, then S has a maximal element. Zorn's lemma is sometimes stated as transfinite **induction**, a method of **proof** that is like finite induction except that the induction **variable** is assumed only to be contained in a well-ordered set.

Today, Zorn's lemma is used by mathematicians of all fields. However, many make a special note signifying the reliance of their methods on Zorn's lemma and then try to find a way to avoid its use.

See also Continuum hypothesis; Hilbert's problem; Zermelo-Fraenkel set theory

SOURCES CONSULTED

Abbott, David, ed. *The Biographical Dictionary of Scientists: Mathematicians.*, London: Blond Educational, 1985.

Bernardo, Allan, and Okagaki, Lynn. "Roles of Symbolic Knowledge and Problem-Information Context in Solving Word Problems." *Journal of Educational Psychology* 86, no. 2 (1994): 212-20.

Borowski, E. J., and J. M. Borwein. *The Dictionary of Mathematics.* London: Collins Reference, 1989.

Brown, Steven. "Subjective Behavior Analysis." Paper presented at the 25th Anniversary Annual Convention of the Association for Behavior Analysis in Chicago, IL. May 27, 1999. Available online at <http://facstaff.uww.edu/cottlec/QArchive/Aba99.htm>.

Brown, Stuart. "Mr. Babbage's Wonderful Calculating Machine." *Popular Science* 242, no. 3. (March 1993): 88-9.

Bueche, Frederick J. *Schaum's Outline Series: Theory and Problems of College Physics.* 7th ed. New York: McGraw-Hill, 1979.

Cajori, Florian. *A History of Mathematical Notations.* 2 vols. La Salle, IL: The Open Court Publishing Co., 1928.

Connors, Kenneth A. *Chemical Kinetics: The Study of Reaction Rates in Solution.* New York: VCH Publishers, Inc., 1990.

Considine, Douglas M., ed. *Van Nostrand's Scientific Encyclopedia.* 8th ed. New York: Van Nostrand Reinhold, 1995.

Courant, Richard, and Herbert Robbins. *What Is Mathematics: An Elementary Approach to Ideas and Methods.* 2nd ed. Revised by Ian Stewart. New York: Oxford University Press, 1996.

Daintith, John, *The Biographical Encyclopedia of Scientists.* 2nd ed. Philadelphia: Institute of Physics Publishing, 1994.

Dane, Abe. "Birth of an Old Machine." *Popular Mechanics* 169, no. 3 (March 1992): 99-100.

Ellis, Robert, and Denny Gulick. *Calculus with Analytic Geometry.* 3rd ed. San Diego: Harcourt Brace Jovanovich, Inc., 1986.

Ferguson, Cassie. "Howard Aiken: Makin' a Computer Wonder." *The Harvard University Gazette* (April 9, 1998). Available online at <http://www.hno.harvard.edu/gazette/1998/04.09/HowardAikenMaki.html>.

Gellert, Walter, ed. *The VNR Concise Encyclopedia of Mathematics.* 2nd ed. New York: Van Nostrand Reinhold, 1989.

Gillispie, Charles, ed. *The Dictionary of Scientific Biography.* New York: Scribners & Sons, 1973.

Hein, Morris, and Leo R. Best. *College Chemistry.* Encino, CA: Dickenson Publishing Company, 1976.

Henderson, Harry. *Modern Mathematicians.* New York: Facts On File, Inc., 1996.

Himmelblau, David. *Basic Principles and Calculations in Chemical Engineering.* 5th ed. Englewood Cliffs, NJ: Prentice-Hall, Inc., 1982.

Hwang, Ned. *Fundamentals of Hydraulic Engineering Systems.* Englewood Cliffs, NJ: Prentice-Hall, Inc., 1981.

Hyman, Anthony. *Charles Babbage: Pioneer of the Computer.* Princeton: Princeton University Press, 1982.

Kim, Eugene, and Betty Toole. "Ada and the First Computer." *Scientific American* 280, no. 5 (May 1999): 76-81.

•

Lapedes, Daniel N., ed. *McGraw-Hill Dictionary of Physics and Mathematics.* New York: McGraw-Hill, 1978.

Larsen, Richard J., and Morris L. Marx. *An Introduction to Mathematical Statistics and Its Applications.* 2nd ed. Edited by Maria McColligan. Englewood Cliffs, NJ: Prentice-Hall, 1986.

Leithold, Louis. *The Calculus with Analytic Geometry.* 3rd ed. New York: Harper & Row, Publishers, Inc., 1976.

Lindeburg, Michael. *Civil Engineering Reference Manual.* 6th ed. Belmont, CA: Professional Publications, Inc., 1992.

Lowry, Richard. *Concepts and Applications of Inferential Statistics.* Poughkeepsie, NY: Vassar College, 1999-2000. Available online at <http://departments.vassar.edu/~lowry/webtext.html>.

Lynch, Ransom V. *A First Course Calculus.* Edited by Donald R. Ostberg. Waltham, MS: George Springer, Xerox College Publishing, 1970.

The MacTutor History of Mathematics Archive. University of St Andrews, Scotland, School of Mathematics and Statistics. <http://www-groups.dcs.st-and.ac.uk/~history/index.html>.

McCabe, Warren, and Julian Smith. *Unit Operations of Chemical Engineering.* 3rd ed. New York: McGraw-Hill Book Company, 1976.

Meserve, Bruce and Max Sobel. *Introduction to Mathematics.* Englewood Cliffs, NJ: Prentice-Hall, Inc., 1964.

Nagle, R. Kent, and Edward B. Saff. *Fundamentals of Differential Equations.* 2nd ed. Edited by Sally Elliott. Redwood City, CA: The Benjamin/Cummings Publishing Company, Inc., 1989.

Nielsen, Kaj. *Calculus and Analytic Geometry: A Programmed Review.* Lincoln, NE: Cliff's Notes, Inc., 1970.

Olson, Reuben. *Essentials of Engineering Fluid Mechanics.* 2nd ed. Scranton, PA: International Textbook Co., 1966.

Pao, Richard. *Fluid Mechanics.* New York: John Wiley & Sons, Inc., 1961.

Perry, Robert, and Don Green. *Perry's Chemical Engineers' Handbook.* 6th ed. New York: McGraw-Hill Book Co., 1984.

Peters, Max, and Klaus Timmerhaus. *Plant Design and Economics for Chemical Engineers.* 3rd ed. New York: McGraw-Hill Book Co., 1980.

Rouse Ball, W. W. *A Short Account of the History of Mathematics.* 4th ed. 1908. A version is available online at <http://www.maths.tcd.ie/pub/HistMath/People/RBallHist.html>.

Suzuki, Mutsumi. "Magic Squares." <http://www.pse.che.tohoku.ac.jp/~msuzuki/MagicSquare.html>.

Trujillo, Karen, and Oakley Hadfield. "Tracing the Roots of Mathematics Anxiety through In-Depth Interviews with Preservice Elementary Teachers." *College Student Journal* 33, no. 2 (June 1999): 219-20.

Vennard, John. *Elementary Fluid Mechanics.* 4th ed. New York: John Wiley & Sons, Inc., 1961.

Weisstein, Eric W. *CRC Concise Encyclopedia of Mathematics.* New York: Chapman & Hall/CRC, 1999.

Wood, Arthur. *An Introduction to Engineering Principles and Practice.* Murfreesboro, TN: MTSU Press, 1964.

"The World's Latest Computer." *The Economist* 349, no. 8097 (December 5, 1998): 102.

Yam, Philip. "Intelligence Considered." *Scientific American* 9, no. 4 (Winter 1998). Available online at <http://www.sciam.com/specialissues/1198intelligence/1198yam.html>.

c. 35,000 B.C.

Marks engraved on bones found in Southern Africa are believed to be the earliest recorded form of counting and notation—the marks are thought to relate to the phases of the moon.

c. 7500 B.C.

During the Stone Age a vigesimal numeration system may have been in use—based on 20 (the fingers and toes)—some African and Australian tribes still use this today.

c. 6000 B.C.

Clay tablets found in the Middle East are believed to be counting tokens, different shapes have been found as well as some collected in hollow vessels.

c. 5000 B.C.

The Egyptians start using a number system based on ten.

c. 4700 B.C.

The probable start of Babylonian calendar.

c. 4241 B.C.

The Egyptian calendar is introduced.

c. 2700 B.C.

Huang Ti (The Yellow Emperor) begins his reign in China, during this period Ta Nao establishes the sexagesimal numeration system (counting using 60)—possibly taken from the Babylonians.

c. 2400 B.C.

The Babylonians start to use simple fractions in their calculations as recorded on clay tablets, the Egyptians are also using this system.

c. 2200 B.C.

A divinatory book of the lesser Taoists in China is written with details of the first recorded magic square (on the back of a mystical tortoise).

c. 2000 B.C.

Babylonian tablets are used to record quantities (e.g. grain, slaves etc.) with a symbolic representation of the quantity.

c. 1900–1600 B.C.

A Babylonian stone tablet from this period records a list of ratios equivalent to the modern secant, this is one of the first recorded examples of this aspect of plane trigonometry.

c. 18th century B.C.

During the reign of Hammurabi, a Babylonian lawgiver king, clay tablets are produced which show practical mathematics is already highly advanced in this society.

c. 1850 B.C.

The Babylonians are using a value of pi of 25 / 8.

c. 1700 B.C.

A Babylonian tablet describes a problem about a rectangle involving unknown numbers—this is a record of one of the first algebraic problems known.

c. 1650 B.C.

The Rhind papyrus (named after it's 1850's English purchaser A. Henry Rhind) is written by Ahmes (c. 1680–c. 1620 B.C.). This is one of the earliest surviving mathematical texts and is our primary source on the history of Egyptian mathematics. It is possible that Ahmes is not the author of this work, being only a scribe reproducing a document from c. 2000 B.C.

●

c. 1650 B.C.

The Egyptians are using a value of pi of 3 1/8.

1500 B.C.

Tablets found at Nippur in the 1930's by Otto Neugebauer (1899–1990) suggest that the Babylonians are using the abacus—some of the pictographs are reminiscent of the form of an abacus.

1500 B.C.

The basic rules of mensuration are known by the Babylonians as shown in clay tablets.

1500 B.C.

Indian mathematics is using a value of pi calculated from the square root of 10.

1122 B.C.

This is the start of known historical period of Chinese mathematics (as opposed to the astronomical observations of earlier periods)—the *Chou-Pei* is possibly written soon after this period (a Chinese math classic including work on the calendar).

c. 1100 B.C.

The *K'iu-ch'ang Suan-shu* ("Arithmetic in Nine Sections"), greatest of the Chinese mathematical classics, is written by an unknown author, it includes work on surveying, percentages, square and cube roots, volumes and what we now know as Pythagorean triangles.

670 B.C. This is the first known use of a monetary system (in China), knife money is initially followed rapidly by minted coins.

c. 650 B.C.

The first use of coins in the west by the Greeks, starting at the port city of Miletus. Coins made trade and commerce easier and more records of transactions and bookkeeping exist from now on.

650 B.C. Papyrus is introduced into Greece, allowing a greater spread of knowledge and a greater chance for it to be preserved.

c. 600 B.C.

Thales of Miletus (624 B.C.–546 B.C.), a Greek often regarded as the first western scientist and philosopher, discovers seven geometrical propositions including Euclidean theorems. It is possible that he gained his knowledge from Egyptian sources. He predicts the solar eclipse of 585 B.C.

c. 600 B.C.

The *Yi Jing* ("The Book of Changes") is published in China, it deals with combinations of numbers relating to different forms of the trigrams and hexagrams used in Chinese fortune telling.

594 B.C. Solon (c. 638 B.C.–559 B.C.) introduces a leap month into the Athenian calendar.

c. 550 B.C.

The Old Testament book of King's is written and it includes a value for pi of 3, much of the material it contains dates to 950 B.C..

c. 550 B.C.

Pythagoras (580 B.C.–520 B.C.) claims to have found the principles of all things in numbers, he also discovers irrational numbers, Pythagoras' theorem and the Pythagorean number theory. The original works are lost and now known only from later commentaries.

5th century B.C.

Followers of Pythagoras, the Pythagoreans, launch the ideas, and possibly the terms, of "arithmetic," "geometric" and "harmonic." They also start the study of amicable numbers at this time as well as recognizing figurate numbers.

480 B.C. Anaxagoras of Clazomenae (500–428 B.C.) introduces philosophy to the Athenians and when in prison for claiming the sun is not a god (450 B.C.) he tries to solve the problem of squaring the circle.

c. 460 B.C.

Zeno of Elea (490 B.C.–430 B.C.), a Greek philosopher, introduces Zeno's paradox (a race where the distance is constantly halving) no original texts of Zeno's are known to survive.

c. 440 B.C.

Hippocrates of Chios (470 B.C.–410 B.C.) becomes the first geometer to discover the area of a curvilinear figure (the lune—named after the crescent moon). Hippocrates also produces the first version of the elements of geometry as well as working on trigonometry and basic advances in algebra. The three classical problems of Greek mathematics—trisecting an angle, squaring the circle and duplication of the cube are also attributed to Hippocrates.

c. 430 B.C.

According to Plato (428 B.C.–347 B.C.) Theodorus of Cyrene (465 B.C.–398 B.C.) demonstrates that the roots of all non square numbers up to 17 are irrational.

c. 430 B.C.

Hippias of Elis (460 B.C.–400 B.C.) provides a description of quadratix which eventually paves the way for the solution of the trisection of an angle problem.

c. 430 B.C.

Antiphon (480 B.C.–411 B.C.) uses the method of exhaustion to solve calculations.

c. 420 B.C.

Democritus of Abdera (c. 460 B.C.–c. 370 B.C.) writes a number of mathematical books, "On numbers,"

"On geometry," "On tangencies," "On mappings," and "On irrationals." None of these survive but they influence many subsequent thinkers. He also holds an atomic view of the world.

c. 400 B.C.

Archytas of Tarentum (428 B.C.–350 B.C.) applies mathematics to practical problems and offers a solution to the problem of duplicating the cube—this uses a very complicated technique involving constructions in three dimensions—it is also called the Delian problem. Archytas also applies theories of proportion.

c. 390 B.C.

A lost work of Theaetetus (415 B.C.–369 B.C.) is the first to cover irrational numbers, and his discovery of the octahedron and icosahedron as regular solids.

c. 380 B.C.

Plato (428 B.C.–347 B.C.) believes that mathematics has a reality, independent of human thought. The Platonic solids (5 regular solids) are first described in *Timaeus*. Plato felt that mathematics should be an integral part of education and he laid down many of the foundations of mathematics.

c. 360 B.C.

Eudoxus of Cnidus (408 B.C.–355 B.C.) introduces the method of exhaustion to measure an area bounded by a curve—as such this can be seen as the forerunner of integral calculus. Euclid's theory of proportion is derived from the lost works of Eudoxus.

c. 350 B.C.

Aristotle (384 B.C.–322 B.C.), a member of Plato's Academy, is a commentator on the works of Zeno. At this time he forms his own Lyceum (School), where he carries out a great deal of work on mechanics, laying the foundations for the study of equilibrium and motion. Aristotle termed axioms the first things from which all demonstrative science must start, Aristotelian and Euclidean geometry were both formed using the axiomatic method.

c. 350 B.C.

Menaechmus (380 B.C.–320 B.C.) is the first to describe and study conic sections during this period.

c. 350 B.C.

Deinostratus is one of the first to employ quadratics which he uses in his attempt to square the circle.

c. 300 B.C.

The Alexandrian school (in Egypt) is founded—this brings together many mathematicians and philosophers and heralds a new golden age of mathematics.

c. 300 B.C.

Euclid (325 B.C.–265 B.C.) writes 13 volumes of *Stoicheion* ("The Elements"). They form the basis of (Euclidean) geometry of the triangle, circle, various quadrilaterals, as well as theory of proportion, elementary number theory, irrationals, solid geometry, parallel postulate, and the perfect numbers proof. Many of the proofs in these books are derived using the indirect method (*reductio ad absurdum*).

c. 300 B.C.

A symbol is used by the Babylonians to indicate a gap between numbers, this is an early form of positional notation and the use of a symbol for zero.

c. 270 B.C.

Aristarchus of Samos (310 B.C.–230 B.C.) claims that the sun lies at the center of the cosmos, not the Earth, in a lost work, he regards astronomy as a mathematical rather than descriptive science. He produces one of the first calculations of the sizes and distances of the sun and moon—incorrect assumptions lead to an inaccurate answer.

c. 250 B.C.

Archimedes of Syracuse (287 B.C.–212 B.C.) writes several books dealing with the measurement of a circle, quadrature of the parabola, plane geometry, solid geometry, hydrostatics, levers, specific gravity and centers of gravity. Archimedes also produces the Archimedes spiral—"On Spirals" shows how to create this figure and how to calculate the area. He is often called "the father of integral calculus" (see Eudoxus c. 360 B.C.) and he calculates a value of pi of 22 / 7.

c. 250 B.C.

Conon of Samosealry (280 B.C.–220 B.C.) is an early investigator in conics, and his work was is subsequently absorbed into that of Apollonius of Perga (262 B.C.–190 B.C.).

c. 240 B.C.

Chrysippus (280 B.C.–206 B.C.), using a form of logic based on propositions, introduces two rules of inference—the *Modus ponens* and the *Modus tollens*.

c. 240 B.C.

Eratosthenes of Cyrene (276 B.C.–197 B.C.) produces the sieve of Eratosthenes to find prime numbers, he also determines the circumference of the earth by comparing shadow lengths at two cities in Egypt.

c. 275 B.C.

Lysis, Archippus and Philolaus are the first to set down the teachings of Pythagoras (580 B.C.–520 B.C.) in writing, previously they had been passed down by word of mouth only.

c. 230 B.C.

Apollonius of Perga (262 B.C.–190 B.C.) produces 8 books of which 7 still survive ("Conics"). These books contained 400 propositions and define the parabola, hyperbola, ellipse and asymptote.

c. 200 B.C.

Diocles (240 B.C.–180 B.C.) writes on conics and mirrors, his works are known only from later translations.

c. 160 B.C.

Hipparchus (190 B.C.–120 B.C.) is the author of the first chord table (equivalent to modern sine tables) he also discovers precession of the equinoxes.

c. 100 B.C.

Roman numerals are derived from the Greek system of using letters to denote numbers.

c. 25 B.C.

Vitruvius Pollio (1st century B.C.) publishes *De architectura* ("The Architect") which considers construction and decoration of buildings, town planning, hydraulics, machine mechanics, sundials, water clocks and the Archimedes Eureka story. This is the principal Roman mathematical text of the Middle Ages.

c. 0–100 The Chinese mathematical compilation book *Jiuzhang Suanshu* ("The Nine Chapters of the Mathematical Art") is written during the Han dynasty, it considers 246 problems in arithmetic, geometry and a simple form of algebra along with appropriate solutions, this is the first known example of matrix methods.

c. 40 Hero of Alexandria (Heron) (10–75) writes many books on mensuration, the most important of which is *Metrica*. The books show how to calculate volumes of cones, prisms, pyramids, spherical segments, and the five regular polyhedra as well as a method of approximating square roots and the area of a triangle (Hero's formula).

c. 75 Writings by Pan Ku (c. 32–92) in China describe the workings of the abacus, made from bamboo, see also 1500 B.C.

c. 100 Menelaus of Alexandria (70–130) writes *Sphaerica* ("Spheres")—the earliest known examples of theorems of spherical trigonometry and a triangle theorem. He also writes "Chords in a Circle" and "Elements of Geometry"—neither of which survive. *Sphaerica* is now known only from a later Arabic translation. These books also contain Menelaus' theorem of colinearity.

c. 100 Nicomachos of Gerasa (c. 60–c. 120) writes *Arithmetike eisagoge* ("Introduction to Arithmetic") an early and influential book on number theory; it also contains the earliest known Greek multiplication table.

c. 140 Claudius Ptolemy (85–165) authors the 13 books of *Syntaxis Mathematica* ("Mathematical Collection") sometimes known as the *Almagest*. These contain corrected and extended versions of Hipparchus' (190 B.C.–120 B.C.) table of chords and details of how the table was constructed using Ptolemy's theorem, he also investigates trigonometric identities.

c. 150 Claudius Ptolemy (85–165) calculates a value of pi of 3.1416.

c. 240 Diophantus of Alexandria (200–284), authors *Arithmetica* (of which 10 of the 13 volumes still survive). These works consider 130 problems of what is now known as Diophantine equations. These include the first evidence of Greek use of algebra and indeterminate analysis.

4th and 5th century

Hindu mathematicians advance trigonometry by replacing the chords used by the Greeks with half chords of circles of known radii (equivalent to modern sine functions), these are found in a number of works called the *Siddhantas* ("Systems of Astronomy").

c. 320 Pappus of Alexandria (290–350) produces commentaries on Euclid (325 B.C.–265 B.C.), Appolonius (262 B.C.–190 B.C.) and Ptolemy (85–165). He is also the author of *Synagoge* ("Collections") of which 5 of 8 volumes survive. This work is considered an excellent guide to lost math and astronomy of late antiquities, it also includes Pappus' theorems, and gives definitions of the center of mass of solids of revolution (quoted but never defined in the work of Archimedes(287 B.C.–212 B.C.).

c. 400 Hypatia of Alexandria (370–415) is the first named female in math history, the daughter of Theon of Smyrna(c. 335–405), she produces commentaries on Diophantus (200–284) and Appolonius (262 B.C.–190 B.C.).

463 Chinese father and son team Tsu Ch'ung–Chih (430–501) and Tsu Keng–Chih obtain a value of pi which the West will not equal until the 17th century—355/113.

c. 440 Proclus (411–485) writes a commentary on Euclid's (325 B.C.–265 B.C.) *Elements*—this has survived and for some parts it is our only glimpse of the work of Euclid.

499 Aryabhata the Elder (476–550) writes *Aryabhatiya*, which is a summary of Hindu math up to that time (in verse). It contains much that is familiar now, including one of the first uses of algebra, and formulae for the area of circles and triangles and a very accurate value for pi (3.1416).

595 Marks the first known use of our number system, with base ten and positional notation, in India.

628   Brahmagupta (598–670) solves Diophantine equations.

c. 700   The Venerable Bede (673–735) in England defines a new way of calculating when Easter falls and his *Ecclesiastical History of the English Nation* did much to popularize the A.D. and B.C. calendar system we now use.

Early 9th century

A House of Wisdom is established in Baghdad—it's main purpose is to translate and circulate Greek texts and eventually to extend the knowledge in them. It is from these translations that much of our knowledge of Greek mathematics comes.

c. 800   Al-Khwarizmi (780–850) calculates a value of pi of 3.1416.

c. 820   Al-Khwarizmi (780–850) writes *Ilm al-jabr wa'l muqabalah* ("The Science of Transposition and Cancellation") which, when transliterated into Latin in c. 1140 by Robert of Chester, gives us the term algebra. His name also is the root of algorithm.

876   Indians are first to use the symbol for zero that we are now familiar with and it is rapidly taken up by the Arabs—the Hindu Arabic system.

960–1279

The Sung dynasty rules in China—during this period a complex counting board is developed to handle massive calculations, including simultaneous linear and quadratic equations, interest is also high in magic squares in China at this time.

11th century

Al-Karaji, al-Samawal and others discover the triangle of coefficients and a rudimentary form of induction.

c. 1050   Hermann the Lame (1013–1054), working in Germany, translates many scientific works, chiefly Arabic astronomy texts.

c. 1070   Omar Khayam (c. 1048–c. 1131) a multi talented mathematician, produces the first known complete solution of cubic equations.

12th century

Nasir al Tusi lays the foundations of non Euclidean geometry as well as providing important advances in optics.

12th century

William of Moerbeke is one of the most prolific translators of Greek math texts.

1110   al-Karaji develops various formulae involving surds.

1114   Bhaskara (1114–c. 1185) writes *Lilavati* ("The Beautiful") and *Bijaganita* ("Seed Arithmetic")

which use letters to represent unknowns and prove to be the forerunners of calculus.

1120   Adelard of Bath (c. 1075–c. 1160) starts to translate many important works including Eulcid's (325 B.C.–265 B.C.) *Elements* and the tables of al–Khwarizmi(c. 780–c. 850). Adelard also writes some works of his own on mathematics, most of which are strongly influenced by Arabic thought, this also establishes the Hindu–Arabic system of numerical notation throughout Europe.

c. 1136   The first Hebrew math encyclopaedia by is produced by Abraham bar Hiyya ha-Nasi (1090–1136)—it includes much on arithmetic, geometry, optics and music.

1140   Robert of Chester translates, into Latin, all of the works of al-Khwarizmi (780–850) including *Ilm al-jabr wa'l muqabalah* ("The Science of Transposition and Cancellation") which when transliterated into Latin gives us the term algebra.

c. 1144   Gerard of Cremona (c. 1114–1187) is the most prolific translator of Arabic mathematical texts at this time.

1202   Leonardo of Pissa (1170–1250)—better known as Fibonacci (son of Bonacci) publishes *Liber Abbaci* (which is subsequently updated and re-released in 1228). This is the work which first gives us Fibonacci numbers (largely ignored for the next three centuries), it also includes methods of calculation and multiplication.

c. 1240   Albertus Magnus (1200–1280) translates many mathematical works—chiefly relating to astronomy from Arabic and Greek.

1268   Roger Bacon (1214–1294) whilst lecturing at Oxford University, on mathematics and optics, starts to write *Communia mathematica* ("General Principles of Mathematical Science"), the parts which are written show remarkable insights into the topics mentioned as well as calendar reform.

1247   Qin Jiushao formulates the Chinese remainder theorem.

1261   Yang Hui (c. 1238–c. 1298) is using essentially modern decimal fractions as well as giving us an early version of Pascal's triangle.

14th century

Al Farisi determines a rule for calculating amicable numbers.

early 14th century

The Oxford Calculators of Merton College (a mainstay of British mathematical teaching and excellence) group is founded.

c. 1310   Levi ben Gerson (1288–1344) proves theorems on combinations and permutations by induction, the first person to do so.

c. 1303   Zhu Shijie writes *Siyuan yujian* ("The Jade Mirror of the Four Unknowns") it includes an early example of Pascal's triangle and its use in extracting roots, as well as techniques for solving polynomial equations.

Mid 14th century

Nicole d'Oresme (c. 1320–1382) of the Oxford Calculators publishes *De proportionibus proportionum* which considers the many applications of ratios, particularly within mechanics (he also speculates that the Earth rotates around its polar axis).

1350   Albert of Saxony (1316–1390) translates many earlier works and produces commentaries based on them.

Early 15th century

Leone Alberti (Italy) collaborates with Mariano di Jacomo Taccola (1381–1453?) and Fillippo Brunelleschi (1377–1446) to produce works on professional engineering.

c. 1400   Filippo Brunelleschi (1377–1446) is the first artist to make a serious study of perspective.

1427   Al-Kashi (1380–1429) produces an encyclopaedia of mathematics "The Key to Arithmetic," he is a member of the Samarkand astronomers led by Ulugh Beg, this group also compile trigonometric tables and calculate a value of pi correct to 16 decimal places.

1435   Leon Battista Alberti (1404–1472) produces *Della Pittura*, the first text on the mathematics of perspective in art.

1436   The invention of the printing press with moveable type by Johannes Gutenberg (c.1398–1468) allows greater access to texts and increased distribution.

1461–1476

The main work of Johannes Regiomontanus (1436–1476)—*De triangularis omni modis*—is complied in Italy during this period. It contains all the known theorems and techniques relating to triangles, along with new ideas including sine theorems. This work helped trigonometry become an important part of mathematics. It was not published until 1533.

1470's   Leonardo da Vinci (1452–1519) fills many (subsequently published) notebooks with work on ratios, gearings, machines and properties of flowing water.

1482   The first printed edition of Euclid's *Elements* is produced in Europe, it is based on a 1290 translation by Campanus of Novara.

1489   The first recorded printed use of the symbol "-" to denote subtraction, by Johann Widman in *Behenned und hupsche Rechnung* occurs in this year.

1494   Luca Pacioli publishes *Summa de Arithmetica*, which popularizes the work of Fibonacci (1170–1250) as well as dealing with general remarks on numbers, proto algebra and specifics of double entry book keeping.

1500's   The golden mean, although known to the Pythagoreans (as the section), gains prominence during the Renaissance as artists take it up as a divine proportion.

1525 and 1538

The years of publication of the original and updated versions of Albrecht Durer's (1471–1528) *Underweysung der Messung mit dem Circkel und Richtscheyt* ("Instructions on Measuring with the Circle and the Straightedge"). This work provides a mathematical view of perspective and how to achieve it in painting.

1535   Niccolo Tartaglia (1499–1557) is able to solve cubic equations although he proves reticent to publishing his method.

1543   *De Revolutionibus orbium coelstium* by Nicolaus Copernicus (1473–1543) is edited by Georg Joachim Rheticus (1514–1574) and published. It deals with celestial orbits and trigonometry.

1544   *The Grounde of Arts* written in English by Robert Recorde (1510–1558) is an important text in the instruction of practical math.

1545   Girolamo Cardano (1501–1576) publishes the cubic equation solutions of Niccolo Tartaglia (c. 1500–1557) in *Ars magna*.

1545   Lodovico Ferrari (1522–1565), writing in Cardano's *Ars magna*, provides the first method of finding an unknown in terms of the coefficients of a quartic equation.

1545   Michael Stifel (1487–1567) publishes *Deutsche Arithmetica* in German which popularizes the use of the plus (+) and minus (-) symbols as well as the square root symbol ($\sqrt{}$), his work shows he is very close to discovering logarithms, he is also the first to use the term "exponents."

Mid 16th century

The use of Hindu Arabic numerals overtakes the use of Roman numerals in the United Kingdom and Europe.

1557   Robert Recorde (1510–1558) writes *The Whetsone of Witte* which pioneers commercial mathematics and distance learning, he also includes many worked examples and gives the world the equals sign (=).

1570   The first English translation of Euclid (by Henry Billinglsey) appears in this year and it includes pop up diagrams.

1572   Raffele Bombelli (1526–1572) publishes *Algebra*, which in part draws on rediscovered works of Diophantus (c. 200–c. 284). This work focuses more on the properties of numbers as opposed to their numerical values and as such much time is spent on the discussion of complex numbers in algebra.

1579   positional notation as we know it is first used by Francois Viète (1540–1603) who also calculates some very accurate trigonometric tables.

1585   Francois Viète (1540–1603) calculates a value of pi which is correct to 9 places, this is a poorer approximation than that of al-Kashi (1380–1429) in the 1400's who achieved 14 places.

1582   Pope Gregory reforms the calendar by removing 10 days from this year to allow it to match with observations. This was based on work carried out by the monk Clavius and it gives us the Gregorian calendar that is still in use.

1583   Thomas Fincke (1561–1656) writes *Geometriae rotundi*, a mathematical text book which introduces such terms as tangent and secant.

1585   *De Thiende* ("The Tenth") and *La Pratique d'arithmetique* by Simon Stevin (1548–1620) are both published in this year. They introduce a form of decimal notation where the decimal point is indicated by a zero in a circle and units to the right hand side of the decimal separator are shown with corresponding locators for example 15.421 would be shown by 15(0)4(1)2(2)1(3), he also suggests the decimal system for measurement.

1591   Francois Viète (1540–1603) produces *In Artem analyticam isagoge* ("Introduction to the Analytical Arts"), one of the earliest western works on algebra, it includes Viète's formula and an updated method of analysis by proof.

1592   Galileo Galilei (1564–1642) is appointed professor of mathematics at the University of Padua (the University of the Republic of Venice). There his duties are mainly to teach Euclid's geometry and standard (geocentric) astronomy to medical students.

1593   Adrian van Roomen (1561–1615) calculates pi to 17 places.

1595   *Trigonometrica* by Bartholomaeus Pitiscus (1561–1613) introduces the term trigonometry and covers both spherical and planar trigonometry.

1596   The *Opus palatinum de triangulus* of Georg Joachim Rheticus (1514–1574) is posthumously published. It moves trigonometry away from the idea of chords to the functions of right angle triangles, many tables related to these functions are included, the work is ultimately finished by one of his students.

1599   Tycho Brahe (1546–1601) is appointed Imperial Mathematician to the Holy Roman Emperor, Rudolph II, his mathematics are concerned with gravity and celestial orbits and he attempts to prove that the Earth is at rest at the center of the universe.

17th century

The study of optics as a science starts here with Snell (c. 1580 –1626), Galileo (1564–1642), Grimaldi (1618–1663) and Newton (1643–1717)—prior to this there were only brief observations and oddities.

1600's   The cycloid is extensively studied during this century and their are many bitter arguments over priority, these include such people as Galileo (1564–1642), Roberval (1602–1675), Torricelli (1598–1647), and Huygens (1629–1695).

1600's   Paul Guldin (1577–1643) publishes Guldin's second rule in a four volume series of works covering such topics as gravity, cones and cylinders.

1609   Johannes Kepler (1571–1630) produces *Astronomia nova*—wherein he calculates the orbit of Mars and includes the first two of his three laws of planetary motion.

1610   Ludolph van Ceulen (1540–1610) calculates pi to 35 places, when he dies the value is engraved on his tombstone.

1614   John Napier (1550–1617) publishes his first table of logarithms (an abbreviation of logos arithmos—the number of the expression) in *Mirifici logarithmorum canonis descriptio* ("Descriptions of the Marvellous Rule of Logarithms"). These are also called natural logarithms and they are to base e. An early calculator called Napier's bones is used to calculate the values.

1615   Johannes Kepler (1571–1630) writes *Stereometria doliorum* ("Measurement of the Volume of Barrels") in which he finds the volumes of solids by imagining them to be composed of an infinite number of infinitesimally small elements—a major step forward in integral calculus. The third of Kepler's laws of planetary motion is published in this year in *Harmonice Mundi*.

1620   Edmund Gunter (1581–1626) invents a basic slide rule which is well received particularly for navigation at sea, whilst publishing mathematical tables he invents the word cosine.

1621   Proofs and causes (lacking in some of the previous works) of each of the three laws of planetary motion of Johannes Kepler (1571–1630) are included in *Epitome Astronomiae Copernicanae* which also lays down the principle of continuity.

•

1621 Willebrord van Roijen Snell (c. 1580–1626) publishes Snell's law of refraction as well as an improved method for the calculation of pi.

1623 Wilhelm Schickard (1592–1635) designs a calculating machine using rotating cylinders and linking mechanisms. His work is lost until the 1950's and credit for designing a mechanical calculator often goes to later workers such as Pascal (1642) or Leibniz (1671).

1624 Henry Briggs (1561–1630) in *Arithmetica logarithmica* ("The Arithmetic of Logarithms") publishes the first table of common (or Briggsian) Logarithms, these are to base ten as opposed to base e used by John Napier (1550–1617).

1629 Albert Girard (1595–1632) publishes an account of equation roots which leads him to formulate the fundamental theorem of algebra, although it was not given this name until Johann Carl Freidrich Gauss (1777–1855) did so in 1799.

1631 William Oughtred (1574–1660) gives the first recorded use of the symbol x to denote multiplication in mathematics in *Clavis Mathematicae*, the book includes a description of Hindu–Arabic notation and decimal fractions and a considerable section on algebra.

1631 The work of Thomas Harriot (c. 1560–1621) is published posthumously, *Artis analyticae praxis* considers some calculations using base 2 or binary. A few subsequent workers dabble with this oddity but it is not until the development of computers that binary comes into its own.

1634 Gilles Personne de Roberval (1602–1675) is appointed Ramus chair of mathematics at the Collège Royale, Paris, whilst in this position he carries out a number of ground breaking works on calculus and tangents.

1635 Marks the publication of *Geometria indivisibilibus* ("A New Geometry for Calculating Indivisibles") by Bonaventura Cavalieri (1598–1647). This introduce a new technique for calculating the area under curves—the method of indivisibles (integral calculus), it also includes Cavalieri's principle.

1636 Cartesian geometry and hence the Cartesian coordinate system is introduced in this year by René Descartes (1596–1650) and Pierre de Fermat (1601–1665), the system is also called analytic or coordinate geometry.

1636–1637
Marks the publication of *Harmonie Universelle* by Marin Mersenne (1588–1648), this covers some of Mersenne's work on acoustics and harmonics along with a consolidation of much of the trigonometric ideas of the time.

1637 René Descartes (1596–1650) publishes *Geometrie* in this year, it considers curves and loci and utilizes algebraic techniques in their solutions. It represents the first major advance from the formation of Euclidean geometry. In the same year Descartes publishes *Dioptrique* which looks at the geometrical and physical aspects of optics, he also formulates Descartes rule of signs for finding the maximum number of positive roots for a polynomial equation.

1638 Galileo Galilei (1564–1642) publishes *Discorsi e dimostrazione mathematiche intorno a due nuove scienze* ("Discourses and Mathematical Demonstrations Concerning Two New Sciences") which is largely a compilation of the bulk of his work till now. This provides a mathematical and experimental approach to the motion of objects, particularly with regard to their acceleration and velocity, and it includes the parallelogram rule.

1639 The most influential work of Girard Desargue (1591–1661), *Brouillon project d'une atteinte aux evenemens des rencontres du cone avec un plan*, is published in this year. This marks the founding of the discipline of projective geometry and gives the world what is now known as Desargue's theorem (largely ignored until the nineteenth century and Jean-Victor Poncelet (1788–1867)).

1640 Pierre de Fermat (1601–1665), in a letter to Marin Mersenne (1588–1648), outlines three propositions for discovering perfect numbers.

1640 Blaise Pascal (1623–1662) publishes *Essai pour les coniques* ("Essay on Conic Sections") which contains Pascal's theorem.

1642 Blaise Pascal (1623–1662) invents one of the first calculating machines (based on rotating discs).

1644 Evangelista Torricelli (1598–1647) shows that an object can have a finite measure in one dimension but an infinite measure in another.

1644 *Cogitata Physico Mathematica* is published by Marin Mersenne (1588–1648), and it introduces Mersenne numbers into number theory.

1653 Blaise Pascal (1623–1662) constructs Pascal's triangle.

1654 Correspondence between Blaise Pascal (1623–1662) and Pierre de Fermat (1601–1665) discusses games of chance, this is broadly recognized as the first mathematical foray into probability, they discussed a game called the problem of points.

1655 John Wallis (1616–1703) in *Arithmetica infinitorum* ("The Arithmetic of Infinitesimals") develops a new method of calculating the area of a circle and at the same time a new formula for pi (Wallis' formula) and introduces the infinity symbol, he also publishes extensively on algebra in 1680.

1658 Christopher Wren (1632–1723) gives us rectification of the cycloid and work on the hyperboloid.

1658 John Pell (1610–1685) gives Pell's equation to number theory.

1659 The first recorded use of the division symbol (occurs in this year in an algebra by Johann Heinrich Rahn (1622–1676).

1661 James Gregorie (1638–1675) is best known for his mathematical description of the Gregorian telescope in this year.

1662 Robert Boyle (1627–1691) publishes Boyle's Law, and is soon to be a founding member of the Royal Society of London, he also realizes matter is composed of primary particles (elements).

1665 Isaac Newton (1643–1717) developed his fluxions— what we now know as differential calculus.

1666 Gottfried Willhelm von Leibniz (1646–1716) publishes the first book on, and defines the term, combinatorics, *Dissertatio de arte combinatoria*.

1668 Nicolaus Mercator (1620–1687) publishes *Logarithmotechnica* ("Logarithmic Teachings") which includes the first appearance of the power series expansion for the logarithm.

1668 James Gregorie (1638–1675) in his second major work, *Geometricae Pars Universalis*, uses infinite convergent series to find the areas of the circle and hyperbola for the first time. Gregorie is also one of the first workers to distinguish between convergent and divergent series.

1670 Isaac Barrow (1630 –1677) in *Lectiones geometricae* ("Geometrical Lectures") publishes a method of finding tangents very similar to the technique now used in differential calculus. Isaac Newton (1643–1717) discovered this method and asked Barrow to include it in his book.

1670 The *Arithmetica* of Pierre de Fermat (1601–1665) is posthumously published by his son. It establishes many fundamentals of number theory, calculus and finding tangents to curves, it includes Fermat's last theorem, Fermat's spiral and Fermat numbers. Fermat he was a co founder, with René Descartes (1596–1650), of analytical geometry.

1671 Gottfried Willhelm von Leibniz (1646–1716) invents a calculating machine that can carry out addition and multiplication, the most complex calculating machine to this date.

1673 One of the first books by Christiaan Huygens (1629–1695), *Horologium oscillatorium* ("The Pendulum Clock"), deals with the problem of accelerated bodies falling freely. Huygens also publishes on probability theory and optics (Huygen's formula).

1674 Takakazu Seki Kowa (1642–1708) publishes *Hatsubi Sampo* which solves fifteen mathematical problems posed several years earlier, careful analysis of each is included. Kowa is recognised as the founder of Japanese mathematics.

1675 The first successful measure of the speed of light is made by O Roewer, prior to this the speed was assumed to be infinite.

1675 Gottfried Willhelm von Leibniz (1646–1716) carries out most of his initial work on differential calculus in Paris. The results are not published until 1684 with *Nova Methodus pro maximis et minimis* ("A New Method for Determining Maxima and Minima"). Subsequent work and publications revolve around integral calculus.

1676 Isaac Newton (1643–1717) gives, without proof, the binomial theorem, although it had been known about beforehand.

1678 Giovanni Ceva (1647–1734) rediscovers Menelaus' theorem (100) on colinearity in triangles, he also produces Ceva's theorem.

1679 Robert Hooke (1635–1703) talks about Hooke's law which discusses elasticity. Hooke has extensive correspondence with Isaac Newton (1643–1717) on gravity which leads him to accuse Newton of plagiarism.

1681 Christiaan Huygens (1629–1695) investigates the geometry of the catenary.

1683 Takakazu Seki Kowa (1642–1708) becomes the first person to study determinants. He also studies the Diophantine equations.

1687 The first edition of Isaac Newton's (1643–1717) *Philosophiae naturalis Principia mathematica* ("The Mathematical Principles of Natural Philosophy") is published in this year. The second revised edition comes out in 1713 and the third in 1726. This work covers Newton's laws of motion, law of universal gravitation and a system of mechanics capable of accurate descriptions of the motions of bodies. Newton also publishes several other books covering such topics as optics and algebra.

1691 Michel Rolle (1652–1719) publishes Rolle's theorem (calculus).

1692 Guillaume François Antoine Marquis de L'Hospital (1661–1704) publishes *Analyse des infiniment petits pour l'intelligence des lignes courbes* which is the first text book to be written on differential calculus, it includes L'Hospital's rule.

1692　Vincenzo Viviani (1622–1703), writing as D. Pio Lisci Pusillo Geometra, sets a mathematical problem which is solved by Gottfried Willhelm von Leibniz (1646–1716) using multiple integrals.

1694　Johann Bernoulli (1667–1748) publishes his L'Hopital's rule and subsequently goes on to found calculus of variations.

1699　Pierre Varignon (1654–1722) publishes on applications of differential calculus to fluid flow, and subsequently to clock springs.

1700　Antoine Parent (1666–1716) presents a paper to the French Academy which is the first systematic development of solid analytical geometry.

1706　The symbol for Pi, π, is introduced by William Jones (1675–1749); it is chosen because it is meant to represent periphery.

1707　Abraham de Moivre (1667–1754) discovers de Moivre's theorem in complex number theory.

1713　*Ars Conjectandi* ("The Art of Conjecture") by Jacob Bernoulli (1654–1705) is published, this concentrates on probability theory and introduces Bernoulli numbers and Bernoulli's theorem, it also includes a proof of the binomial theorem, the law of large numbers, the lemniscate of Bernoulli (calculus) and polar coordinates.

1714　Roger Cotes (1682–1716) publishes his only mathematical paper, *Logometrica*, which covers new methods of computing logarithms and for converting from one base to another. His collected notes are published posthumously in 1722 and they deal with integrated functions and multiple observations of astronomical data which he had tried to rationalize (an early application of statistics).

1715　Taylor's theorem, the expansion of which gives Taylor's and Maclaurins series, is first given in *Methodus incrementorum directa et inversa* ("Direct and Indirect Methods of Incrementation") by Brook Taylor (1685–1731).

1717　James Stirling (1692–1770) publishes a supplement to Isaac Newton's (1643–1717) enumeration of 72 forms of the cubic curve, entitled *Lineae Tertii Ordines Newtonianae*, ("Newtonian Third Order Curves"). A series expansion called Stirling's formula is named after him, though it was first derived by his countryman Abraham de Moivre (1667–1754).

1718　Abraham de Moivre (1667–1754) publishes one of the earliest works on probability theory (*The Doctrine of Chances*) in 1718 which includes the first appearance of the normal curve.

1720　Colin Maclaurin (1698–1746) publishes *Geometrica organica* ("Organic Geometry") which deals with the calculus devised by Isaac Newton (1643–1717).

1730　James Stirling (1692–1770) writes *Methodus Differentialis sive Tractatus de Summatione et Interpolatione Serierum Infinitarum* (Differential Method with a Tract on Summation and Interpolation of Infinite Series) on infinite series.

1731　Alexis-Claude Clauirat (1713–1765) is elected to the French Academy on the strength of his work on differential geometry.

1733　Girolamo Saccheri (1667–1733) narrowly misses discovering non-Euclidean geometry in *Euclides ab omni naevo vindicatus* ("Euclid Cleared From Every Strain").

1734　Bishop George Berkeley (1685–1753) publishes the first of two pamphlets, *The Analyst*, which tris to show that learning how to solve mathematical problems does not necessarily let us know what they are about, the second, *A Defence of Free-Thinking in Mathematics* is published in 1735.

1735　James Stirling (1692–1770) publishes *On the Figure of the Earth and On the Variation of the Force of Gravity at Its Surface* in this year.

1736　The Königsberg Bridge problem is solved by Leonhard Euler (1707–1783), this is one of the founding tenets of graph theory.

1736　Pierre-Louis-Moreau de Maupretuis (1698–1759) leads an expedition to measure the length of a degree along a meridian, and also formulates the principle of least action.

1738　Daniel Bernoulli (1700–1782) publishes *Hydrodynamica* which introduces the term dynamics and lays the foundations of hydrodynamics as a field of study, he also introduces Bernoulli's equation in this work.

c. 1739　Alexis-Claude Clairaut (1713–1765) finds Clairaut's equation, a differential equation, he also works on geodesy, celestial mechanics and cubic curves.

1740s　Leonhard Euler (1707–1783) and Jean Le Rond D'Alembert (1717–1783) use partial differential equations on the vibrating string problem, an integral forerunner of the wave equation.

1742　Christian Goldbach (1690–1764) lays out the Goldbach conjecture in a letter to Leonhard Euler (1707–1783).

1742　Colin Maclaurin (1698–1746) publishes *Treatise of Fluxions* which deals with the calculus devised by Isaac Newton (1643–1717). Maclaurin's theory deals with a resultant mathematical expansion.

1743 Thomas Simpson (1710–1761) publishes *Mathematical Dissertations on Physical and Analytical Subjects*, which introduces Simpson's rule on numerical integration.

1743 The *Treatise on Mathematics* by Jean Le Rond D'Alembert (1717–1783) introduces us to D'Alembert's principle, which states that the internal forces in a system of particles are in equilibrium.

1744 Leonhard Euler (1707–1783) provides the basic formula which is now know as the zeta function (this was in *Methodus inveniendi lineas curvas*), this was modified in 1859 by Georg Friedrich Bernhard Riemann (1826–1866) to produce the Riemann-Zeta function, Euler also introduced the gamma function.

1747 *On the General Cause of Winds* by Jean Le Rond D'Alembert (1717–1783) is written, a treatise on complex numbers.

1748 Colin Maclaurin (1698–1746) has published (posthumously) *A Treatise of Algebra in Three Parts*. Using a method now known as Cramer's rule Maclaurin calculates determinants.

1748 Maria Gaetana Agnesi (1718–1799) writes on differential calculus in *Istituzioni analitiche ad uso della giovent-italiana* ("Analytical Institutions of Italy") which has a discussion on the curve "the witch of Agnesi."

1748 Leonhard Euler (1707–1783) is the first to define continuity in a mathematical sense in this year.

1750 Gabriel Cramer (1704–1752) gives a classification of algebraic curves and Cramer's rule for the solution of systems of linear algebraic equations in *Introduction a l'analyse des lignes courbes algebriques* ("Introduction to the Analysis of Algebraic Curves").

1754 In a move ahead of its time Rudjer Boskovic (1711–1787) produces radical thoughts on differential calculus in *Elementa universae matheseos*. Many of these ideas are only taken up in the late twentieth century.

1759 The posthumous publication of Gabrielle Emilie Le Tonnelier de Breteuil Marquise du Chatelet's (1706–1749) translation of Isaac Newton's (1643–1717) *Principia Mathematica* into French. For many years this is the only French version available.

1760 Etienne Bezout (1730–1783) embarks on a multi volumed set entitled *Cours de mathemtique*. When finished it eventually covers trigonometry, roots of equations, analytic geometry, and Bezout's lemma and theorem.

1763 The work of Thomas Bayes (1702?–1761) is posthumously published including Bayes' Theorem which considers the likelihood of an event occurring which is conditional on another event occurring.

1768 Johann Heinrich Lambert (1728–1777) proves that pi is an irrational number.

1770 Johann Heinrich Lambert (1728–1777) provides the first systematic development of the theory of hyperbolic functions and a system of notation relating to them that we still use.

1770 Edward Waring's (c. 1734–1798) *Meditationes algebraicae* contains the first example of Waring's problem and also Wilson's theorem.

1770 Joseph-Louis Lagrange (1736– 1813) publishes a memoir on equation theory, followed by an extension in 1771. Lagrange's mean value theorem is first postulated in these papers.

1771 Alexandre Theophile Vandermonde (1735–1796) provides the final solution to solving quadratics.

1772 Johann Heinrich Lambert (1728–1777) lays down basic principles relating to the production of maps.

1776 Charles Augustin de Coulomb (1736–1806) wins the French Academy prize for a paper on stresses and forces under which beams are placed, this is the science of ergonomics.

1782 Adrien-Marie Legendre (1752–1833) wins the Berlin Academy projectiles prize with his *Recherches sur la trajectoire des projectiles dans les milieux résistants*, subsequent work on elliptical integrals leads to the Legendre symbol.

1782 The Societa Italiana is founded in Italy, its aim is to promote science and mathematics within Italy.

1785 The Voting Paradox is discovered by Marie Jean Antoine Nicolas de Caritat Condorcet (1743–1794) in *Essay on the Application of Analysis to the Probability of Majority Decisions*. Condorcet is a French mathematician, philosopher, economist, and social scientist. The voting paradox is largely ignored until Duncan Black explained its significance in a series of essays begun in the 1940s.

1785–1789

The *Magazin fur reine und angewandte Mathematik* is published in Germany for the first time by Carl Friedrich Hindenburg (1741–1808), this is one of the first mathematical periodicals.

1786 Simon Antoine Jean L'Huilier (1750–1840) wins an award from the Berlin Academy of Science with his essay on mathematical infinity entitled *Exposition elementaire des principes des calculs superieurs*. The standard concepts and notation for derivatives, and the elementary theorems on limits, used today, appear in a remarkably similar form in this publication.

1787 Caspar Wessel (1745–1818) gives a report on his surveying work. It includes new and improved techniques for surveying as well as the geometric interpretation of complex numbers. The Argand diagram is also included.

1788 Joseph-Louis Lagrange (1736– 1813) provides a rigorous text on post Newtonian mechanics with the publication of *Mecanique Analytique*, Lagrange also carries out pioneering work on gravity relationships.

1789–1793 The French Revolution is responsible for the deaths of a number of French mathematicians and many educational institutions and universities were abolished.

1795 French law adopts the metric system of measurement and gives legal definitions for each of the units.

1795 Playfair's axiom is an alternative version of Euclid's parallel postulate produced by John Playfair (1748–1819), it is first published in *Elements of Geometry*.

1799 Gaspard Monge (1746–1818), a pioneer in analytical geometry, publishes *Geometrie descriptive* which shows how three dimensional objects can be represented on a two dimensional plane.

1799 Johann Carl Freidrich Gauss (1777–1855) proves the fundamental theorem of algebra.

1799–1825 Sees the publication of the five volumes of Pierre-Simon Laplace's (1749–1827) *Traite de mecanique celeste* ("Celestial Mechanics") which attempts to update Newton's ideas on planetary motion involving newer data. During this period *Theorie analytique des probabilities* ("Analytic theory of Probability") is also published. These works include Laplace's equation and the Laplace transform.

1800s Marie-Sophie Germain (1776–1831) in correspondence, lays down the results which allow other mathematicians to show that Fermat's last theorem holds for all numbers where n < 100 (she assumed a masculine pseudonym).

1800 Jean Etienne Montucla (1725–1799) has a posthumously published history of mathematics in four volumes produced in this year.

1800 Louis-Francois-Antoine Arbogast (1759–1803) extends the algebraic approach to calculus by treating the objects as neither numbers nor geometrical magnitudes in *Calcul des derivations*.

1801 Johann Carl Freidrich Gauss (1777–1855) publishes *Disquisitiones arithmeticae* which introduces modular arithmetic, Gauss also publishes on astronomy and planetary motion as well as geometry (Gaussian curvature).

1804 Paolo Ruffini (1765–1822) discovers Horner's method some 15 years prior to William George Horner (1786–1837) but is largely ignored.

1805 Adrien-Marie Legendre (1752–1833) proposes the method of least squares—one of the first practical statistical tests.

1806 Jean-Robert Argand (1768–1822) thinks up imaginary numbers.

1806 Charles-Julien Brianchon (1783–1864) provides proof of the dual version of Pascal's theorem.

1808 Christian Kramp (1760–1826) introduces the symbol n! for the factorial of n.

1810 Bernard Bolzano (1781–1848) publishes a work on the foundations of elementary geometry in which he considers points, lines and planes as undefined elements and defines operations on them. This is an important step in the axiomatisation of geometry and an early move towards the necessary abstraction for the concept of a linear space to arise.

1810 Joseph-Diaz Gergonne (1771–1859) launches the journal *Annales de Mathematiques Pures et Appliquees*, (the first purely mathematical journal) in this year. It soon becomes a journal of unusual mathematics, with the majority of the contributions by Gergonne himself (publication ceases in 1832 when Gergonne's work load increased to such an extent he no longer has time to edit and write the journal).

1812 Jacques-Phillipe-Marie Binet (1786–1855) submits a paper on determinant product theorem to the Institute of France.

1813 The Analytical Society is founded by Charles Babbage (1791–1871), John Herschel (1792–1871) and George Peacock (1791–1858) at Cambridge University England. Their aim is to push forward the acceptance of Lagranges algebra.

1815 Simeon-Denis Poisson (1781–1840) and Augustin Louis Cauchy (1789–1857) discover the equations necessary to analyse deep fluid motion rather than the shallow motion that had previously been considered, this marks a great leap forward in the field of hydrodynamics.

1816 The Bolzano theorem is advanced by Bernard Bolzano (1781–1848) in *Der binomische Lehrsatz* ("The Binomial Theorem").

1817 Bernard Bolzano (1781–1848) gave the first account of a continuous function.

1818 Pierre-Simon Laplace (1749–1827) devises the central limit theorem which is subsequently modified in 1901 by Aleksandr Mikhailovich Lyapunov (1857–1918).

1819 William George Horner (1786–1837) submits his method for the solving of algebraic equations to the Royal Society of London, this is now known as Horner's Method.

1820–1871
Charles Babbage (1791–1871) works on the first mechanical computers, initially the difference engine and latterly the analytical engine. Due to circumstances beyond his control Babbage never completes working models of any of his machines; this is not carried out until 1991 by the Science Museum in London (the Babbage Engine).

1820 August Leopold Crelle (1780–1855) publishes factor tables.

1821 In Augustin Louis Cauchy's (1789–1857) *Cours d'analyse* ("A Course of Analysis") he introduces the notion of a limit which he then uses to define continuity, convergence and differentiability. This is a vital component of differential calculus. The Cauchy condition is included in this work, which also shows the uses of integrals with infinite limits, and he defines and produces a test for convergence.

1822 Jean Baptiste Joseph Fourier (1768–1830) publishes *Theorie analytique de la chaleur* ("Analytical Theory of Heat") which contains the technique known as Fourier analysis which utilizes the Fourier series to analyze waves.

1822 Jean-Victor Poncelet (1788–1867) privately publishes *Traite des proprietes projectives des figures* which revives the study of projective geometry and contains the principle of duality, it also extends the principle of continuity and limits formulated by Johannes Kepler (1571–1630).

1823 Augustin Louis Cauchy (1789–1857) develops the mean value theorem and studies derivatives and differentials, a set of results which is published in *Resume des Lecons donnes a L'Ecole Royale Polytechnique sur le Calcul Infinitesimal.*

1823 János Bolyai (1802–1860) demonstrates that it is possible to develop a consistent geometry in which the parallel postulate is rejected.

1824 Niels Henrik Abel (1802–1829) proves that the general quintic equation is unsolvable algebraically, he also provides the first solution of an integral equation.

1824 Bessel functions are originally published by Friedrich Willhelm Bessel (1784–1846), they were originally prompted by observations on planetary orbits.

1826 *Journal fur die reine und angewandte Mathematik* is founded in Berlin by August Leopold Crelle (1780–1855), this is one of the first journals devoted exclusively to mathematical research.

1826 Nikolay Lobachevsky (1792–1856) discovers hyperbolic geometry (independently of János Bolyai (1802–1860) in 1823), this is also called Lobachevskian geometry—a form of non-Euclidean geometry.

1827 Johann Carl Freidrich Gauss (1777–1855) develops differential geometry.

1827 Augustus Ferdinand Möbius (1790–1868) publishes a paper which outlines a one sided three dimensional shape—the Möbius strip.

1829 Évariste Galois (1811–1832) discovers what becomes the forerunner of group theory, it is not actually published until 1846. Called the Galois theory, the outlines of time field theory are explicit in this work.

1829 Peter Gustav Lejeune (1805–1859) works on Fermat's last theorem and the Fourier series, and he shows in this year an example of the convergence of series which can now be checked for by Dirichlet's test.

1829 Karl Gustav Jacob Jacobi (1804–1851) publishes *Fundamenta nova theoriae functionum ellipticarum* ("New Elements in the Theory of Elliptic Functions") which carries on and extends the work of Adrien-Marie Legendre (1752–1833) and introduces a powerful method of inverting integral functions (the Jacobi Inversion Problem).

1830 George Peacock (1791–1858) produces a treatise on algebra which looks at symbolic manipulation and its relation to mathematical truth—thus founding the study of symbolic logic.

1832 The Necker cube is first described, based on the observations of the crystallographer Louis Albert Necker.

1832 Jakob Steiner (1796–1863) attempts to create a grand theory of geometry utilising stereographic projection in *Systematische Entwickelung* ("Systematic Development").

1832 János Bolyai (1802–1860) produces his system of hyperbolic geometry which is one of the first clear accounts of non Euclidean geometry.

1833   William Rowan Hamilton (1805–1865) introduces quaternions and subsequently publishes several works on them, he also contributes Hamiltonian functions and Hamilton's principle to dynamics.

1835   *Sur l'homme et le développement de ses facultés, essai d'une physique sociale* is published by Adolphe Quetelet (1796–1874) which leads him to become the father of modern statistics, he invents the "average man" from his observations.

1835   Julius Plucker (1801–1868) publishes the first complete classification of plane cubic curves after making a number of contributions to analytical geometry between 1828 and 1831.

1836   Joseph Liouville (1809–1882) starts publication of the *Journal de Mathematiques Pures et Appliques* in France, from this period on mathematical journals became much more common.

1837   Pierre Wantzel (1814–1848) solves the classic Greek problem of trisecting an angle using Euclidean geometry but not with the conic techniques utilised by the Greeks and later workers.

1837   Simeon-Denis Poisson (1781–1840), by observing frequencies of occurrences and probabilities, discovers the Poisson distribution which he publishes in *Recherches sur la probabilite en matiere criminelle et en matiere civile*.

1840s   James Prescott Joule (1818–1889) after whom the unit of work, the joule, is named, studies the generation of heat from work—he is one of the founders of energy physics.

1844   Joseph Liouville (1809–1882) proves the existence of transcendental numbers and then provides an extensive range of transcendental numbers called Liouville's numbers.

1844   Herman Gunther Grassman (1809–1877) in *Die lineale Ausdehnungslehre* ("The Linear Doctrine of Extension") tries to use calculus to describe events in real space, this eventually leads to Grassman coordinates and is an early use of vector spaces.

1845   Leopold Kronecker (1823–1891) discovers algebraic numbers in *De Unitatibus complexis* ("On Complex Units"), his doctoral thesis.

1847   George Boole (1815–1864) shows how algebraic formulae can be used to express logical relations in *Mathematical Analysis of Logic*. This Boolean algebra is further developed in 1854 in his *The Laws of Thought*.

1847   Augustus De Morgan (1806–1871) proposes De Morgan's laws in calculus.

1849   The science of thermodynamics is essentially born and the term is first used in a paper by William Thomson (1824–1907).

c. 1850 Jules Antoine Lissajous (1822–1880) first produces the curves that are named after him—Lisajous figures.

1850   Bernard Bolzano (1781–1848) has a posthumous publication *Paradoxes of the Infinite* which is a precursor for some of the subsequent work of Georg Ferdinand Ludwig Phillip Cantor (1845–1918).

1850   James Joseph Sylvester (1814–1897) coins the word matrix and, along with the work of Arthur Cayley (1821–1895), Matrix theory is born.

1851   Enrico Betti (1823–1892) publishes widely on algebraic concepts of Galois theory and makes an important contribution in forming a bridge between classical and modern algebra.

1852   The four-color map problem is proposed by Francis Guthrie. It is not solved until 1976 by Wolfgang Haken (1928–) and Kenneth Appel (1932–).

1854   Arthur Cayley (1821–1895) publishes his work on algebraic geometry giving us Cayley's algebra.

1854   Georg Friedrich Bernhard Riemann (1826–1866) in a lecture entitled *Uber die Hypothesen welche der Geometrie zu Grunde liegen* ("On the Hypotheses that Lie at the Foundations of Geometry") puts forward his system of non Euclidean geometry—Riemannian (or elliptic) geometry..

1856   Karl Thodor Wilhelm Weierstrass (1815–1897) comes to the notice of the world with the publication of a paper on the inversion of hyperelliptic integrals.

1857   Arthur Cayley (1821–1895) introduces matrix algebras.

1858   Charles Hermite (1822–1901) using elliptic functions solves the general quintic equation for one variable.

1858   Arthur Cayley (1821–1895) notices that quaternions can be represented by matrices.

1859   Georg Friedrich Bernhard Riemann (1826–1866) gives the world the Riemann zeta function and the Riemann hypothesis, both of which are concerned with prime numbers. A modification of the work of Leonhard Euler (1707–1783) provides the basic formula which is now know as the zeta function (this was in *Methodus inveniendi lineas curvas*).

1859   Emile Leonard Mathieu (1835–1890) produces Mathieu's differential equation—a second order differential equation the results of which are Mathieus functions.

1860s Lazarus Fuchs (1833–1902) develops Niels Henrik Abel's (1802–1829) algebraic functions to give Fuchsian functions.

c. 1862 Emile Leonard Mathieu (1835–1890) studies a second order differential equation which is now know as Mathieus differential equation.

1865 James Clerk Maxwell (1831–1879) produces *A Dynamical theory of the Electromagnetic Field* which is the first work featuring Maxwell's equations, it is also responsible for uniting electricity and magnetism to give the science of electromagnetism, at this period he also studies the wave nature of light.

1868 Georg Friedrich Bernhard Riemann (1826–1866) has *Uber die Hypothesen, welche der Geometrie zu Grunde liegen* posthumously published—largely responsible for launching the topological interest in manifolds.

1870 Ernst Abbe (1840–1905) carries out studies in optical theory which greatly improve the microscope, he discovers the Abbe sine condition—a condition which a lens must meet to provide a good image.

1870 Heinrich Eduard Heine (1821–1881) uses Weierstrassian analysis to prove, for any finite number of points, the uniqueness of the Fourier series.

1870 Camille Jordan (1838–1822) in *Traite des substitutions et des equations algebriques* ("Treatise on Substitutions and Algebraic Equations") brings the work of Évariste Galois (1811–1832) back into the public notice and makes several amendments to algebra to give a new branch called Jordan algebra.

1871 Christian Felix Klein (1849–1925), amongst his works on geometry and topology, proposes a one sided closed surface object—the Klein bottle.

1871 Enrico Betti (1823–1892) publishes a work on topology which includes Betti numbers.

1872 Christian Felix Klein (1849–1925) delivers his inaugural address, to the Philosophical Faculty and Senate of the University of Erlangen, in group theory and advocates a programme of geometrical study now known as the Erlangen (Erlanger) program.

1872 Ludwig Eduard Boltzmann (1844–1906) whilst modelling behaviour of gases produces his H theorem.

1873 Charles Hermite (1822–1901) demonstrates the transcendence of e, and discovers Hermite polynomials, Hermite's differential equation, Hermite's formula of interpolation and Hermitian matrices..

1874 Pafnuty Lvovich Chebyshev (1821–1894) introduces into probability theory Chebyshev's inequality.

1874 Georg Ferdinand Ludwig Phillip Cantor (1845–1918) starts to develop the first clear account of transfinite sets and numbers (eventually leading to the foundation of topology as a subject). This also leads to the publication of the Cantor set, and the study of set theory, also the continuum hypothesis is evolved at the same time. Kurt Friedrich Gödel (1906–1978) in 1938 and Paul Cohen (1934–) in 1963 expand it,.

1875 Francis Galton (1822–1911) discovers regression in this year.

1875 Sonya Vasilyevna Kovalevskaya (1850–1891) improves and extends the use of one of Augustin Louis Cauchy's (1789–1857) differential equations, this is later immortalized as the Cauchy-Kovalevskaya theorem.

1876 Francois-Edouarde-Anatole Lucas (1841–1891) develops a test for primes and becomes the discoverer of the largest prime number without the aid of a computer. Subsequent work in number theory, particularly the Fibonacci series, provides Lucas numbers.

1877 The Rhind papyrus is discovered in Egypt and purchased by Henry Rhind.

1878 Arthur Sylvester (1814–1897) founds the first American mathematical research journal.

1879 William Thomson (1824–1907) is responsible for the name harmonic analysis as applied to an extension of the Fourier series—this is due to an appendix of his book *Treatise on Natural Philosophy*—co-authored with Peter Guthrie Tait (1831–1901)—which includes his harmonic analyzer.

1879 Lewis Carroll (Charles Lutwidge Dodgson) (1832–1898) whilst lecturing in math at Oxford University writes *Euclid and His Modern Rivals*, one of several maths books, this is most interesting for providing a portrait of the then current state in maths.

1879 Friedrich Ludwig Gottlob Frege (1848–1925) shows in *Begriffsschrift* ("Concept Writing") that arithmetic and parts of real variable analysis can be formulated entirely within logic.

1881 Josiah Willard Gibbs (1839–1903) in *Vector Analysis* (the origin of the name) introduces the mathematical tools into physics that eventually supplant the quaternions of Hamilton, included in these tools is the Gibbs phenomenon. Initially this publication is intended as notes for his students, a version for external distribution does not appear until 1901.

1881 John Venn's (1834–1923) publication *Symbolic Logic* uses diagrams of overlapping circles to represent relationships between sets, these are now known as Venn diagrams.

1882   Carl Louis Friedrich von Lindemann (1870–1946) proves that pi is transcendental and that it is thus impossible to square the circle using Euclidean methods.

1882   Moritz Pasch (1843–1930) adds a number of axioms to Euclidean geometry relating to ordering of points on a line and the topology of lines.

1883   Charles Sanders Pierce (1839–1914) develops the first formal theory of relations in logic.

1883   Edmund Husserl (1859–1938), a leader of the German phenomenological movement, gains his doctorate in the theory of calculus of variations and writes *Philosophie der Arithmetik* in this year and subsequently produces texts on the use of logic in mathematics.

1884   Charles Renard (1847–1905) a surveyor and topographic engineer also designs and builds the first practical airship.

1886   Gosta Mittag-Leffler (1846–1927) Popularizes Weierstrass math in his own journal started this year, *Acta Mathematica.*

1888   Julius Wilhelm Richard Dedekind (1831–1916) in *Was sind und was sollen die Zahlen?* ("The Nature and Meaning of Numbers") offers an account of natural numbers and defines irrational numbers in terms of the Dedekind cut.

1888–1893
Marius Sophus Lie (1842–1899) produces the Lie Group in *Die Transformationsgruppen* ("On Transformation Groups).

1890   Ernst Leonard Lindelof (1870–1946) publishes on the existence of solutions of differential equations and continues working with various types of analysis.

1890   Giuseppe Peano (1858–1932) discovers Peano's curve and develops a set of five axiom's (originally formulated by Julius Wilhelm Richard Dedekind (1831–1916)) for number theory which we now know as Peano axioms.

1890   Percy John Heawood (1861–1955) publishes his first paper on four color theorem (showing five colors can work), he subsequently publishes widely on geometry.

1892   Karl Pearson (1857–1936) writes *The Grammar of Science* which anticipates many of the ideas of relativity, he subsequently turns his mind to mathematics in the study of heredity.

1892   Aleksandr Mikhailovich Lyapunov (1857–1918) presents his doctoral thesis *On the General Problem of the Stability of Motion* to Kharkhov University. This work analyses different types of stability and pioneers a number of methods of solving non linear differential equations by linear approximations.

1892   Vilfredo Pareto (1848–1923) becomes professor of Mathematics at the University of Lausanne in Switzerland. He is interested in cycles of processes and he produces the principle of virtual work, published in 1902 and 1911.

1892   Jules-Henri Poincaré (1854–1912) in *Les Methodes nouvelles de la mechanique celeste* ("New Methods in Celestial Mechanics") defines the Poincaré conjecture—a topological assumption relating to manifold space.

1893   Camille Jordan (1838–1922) in the second edition of *Cours d'analyse de l'Ecole Polytechnique* describes the Jordan curve theorem and his research on analysis.

1893   Vilfredo Pareto (1848–1923), professor of political economy at University of Lausanne Switzerland, is known for his application of mathematics to economic analysis and for his theory of the "circulation of elites."

1894   Karl Thodor Wilhelm Weierstrass (1815–1897) develops Weierstrassian analysis and infinite power series amongst many other things and he soon becomes regarded as the "father of modern analysis." This year marks the publication of the first volume of his *Complete Works*, volume 2 appears in 1895 and the remainder are posthumously published.

1894   David Hilbert (1862–1943) introduces the world to Hilbert's matrix and subsequently several algebraic concepts as well as setting a number of problems for future mathematicians to solve, of which (by the end of the 20th century) only three quarters had been solved.

1894   Emile Borel (1871–1956) develops the Borel series in his doctoral thesis on set theory. The Heine-Borel property in the same document helps to lay the foundations of topology as a separate subject.

1895   Jules-Henri Poincaré (1854–1912) defines the fundamental group, this definition is extended by Eduard Cech (1893–1960) in 1932 and Witold Hurewicz (1904–1956) in 1935.

1895   Hendrik Antoon Lorentz (1853–1928) proposes that a body in motion contracts in the direction of movement and that its time alters, this is later explained by relativity (Lorentz transformations).

1896   Ferdinand Georg Frobenius (1849–1917), a pioneer of linear algebra, formulates Frobenius' theorem in this year.

1896   Jaques Hadamard (1865–1963) proves the prime number theorem using very complex techniques.

1896   Charles-Jean-Gustave Vallee-Poussin (1866–1962), independently of Jaques Hadamard (1865–1963), proves the prime number theorem.

1896   Vito Volterra (1860–1940) introduces Volterra's integral equation, part of the theory of infinite matrices.

1896   Tulio Levi-Civita (1873–1941) applies dynamics to absolute differential (or tensor) calculus, he also expresses a number of physical laws in both Euclidean and Riemannian curved space.

1897   The commencement, in Zurich, of the first International Congress of Mathematicians.

1897   Elie-Joseph Cartan (1869–1951) first applies hyper-complex numbers to group theory.

1898   Alfred North Whitehead (1861–1947) writes *A Treatise on Universal Algebra* which leads to subsequent collaboration with Bertrand Russell (1872–1970) when they attempt to derive all of mathematics from logical principles (the axiomatic method).

1898   After being appointed professor at the Sorbonne in Paris Charles-Emile Picard (1856–1941) introduces Picard's method for finding the numerical solution of differential equations.

1899   Ernest Rutherford (1871–1937) commences his identification of the alpha, beta and gamma radiations, and in 1902 formulates his spontaneous transformation radioactive decay theory.

1900s   Edward Vermilye Huntington (1874–1952), Eliakim Hastings Moore (1862–1932) and Oswald Veblen (1880–1960), inspired by Hilbert's geometry work develop model theory.

1900   Max Planck (1857–1957), by considering small units of energy, lays the foundations for the study of quantum mechanics.

1900   Erik Ivar Fredholm (1866–1927) publishes some of the earliest work on integral equations (previously largely ignored) in *Sur une nouvelle méthode pour la résolution du problème de Dirichlet*. He is also interested in mathematical physics, integral equations, and spectral theory. From here we have Fredholm's alternative.

1900   Ernst Freidrich Ferdinand Zermelo (1871–1953) founds axiomatic set theory, first publishing it in 1908 in *Untersuchungen uber die Grundlagen der Mengenlehre* ("Investigations on the Foundations of Set Theory"), and it is subsequently modified in 1922 by Abraham Halevi Fraenkel (1891–1965) to give the Zermelo-Fraenkel set theory.

1900   Karl Pearson (1857–1936), an English statistician introduces many key statistical concepts including Pearson's correlation coefficient, and the chi squared test.

1901   Among the earliest books devoted exclusively to combinatorics is the German mathematician Eugen Netto's (1848–1919) *Lehrbuch der Combinatorik* ("Textbook of Combinatorics") prior to this combinatorics had been used as a nineteenth century parlour game.

1901   The central limit theorem of Pierre-Simon Laplace (1749–1827) is modified by Aleksandr Mikhailovich Lyapunov (1857–1918).

1902   *Elementary Principles in Statistical Mechanics* by Josiah Willard Gibbs (1839–1903) puts the study of statistical mechanics on a firm foundation.

1902   The work of Guido Fubini (1879–1943) from his doctoral thesis on differential geometry is thrust into the public light by the publication of a book on this subject by his supervisor, this starts a wide and varied career which includes harmonic functions, differential equations, and group theory.

1902   Henri Lebesgue (1875–1941) works on measure theory in his doctoral thesis. This is expanded 2 years later into *Lecons sur l'integration* and also a general form of integration known as the Lebesgue integral based on the Lebesgue measure of a set.

1902   Bertrand Russell (1872–1970) whilst working on the foundations of mathematics discovers Russell's paradox in set theory.

1904   Nils Fabian Helge von Koch (1870–1924) introduces the Koch curve—an early example of the use of an equation to produce fractals (the direct ancestor of the snowflake curve).

1904   Waclaw Sierpinski (1882–1969) is awarded a University gold medal for an essay on number theory, the contents of which were later described as profound, many subsequent publications in number and set theory ensue.

1905   Jules Antoine Richard (1862–1956) discovers Richard's paradox involving the set of real numbers which can be defined in a finite number of words. The paradox first appears in a letter from Richard to Louis Olivier.

1905   Joseph Henry Maclagan Wedderburn (1882–1948) defines Wedderburn's structure theorem—a part of linear (associative) algebra.

1905 Albert Einstein (1879–1955) introduces his theory of special relativity whilst working as a patent clerk.

1906 G. G. Berry published his paradox—Berry's paradox.

1906 Grace Chisholm young (1868–1944) produces Theory of the Sets of Points—the first English language book on point set topology and measure.

1906 Maurice Frechet (1878–1973) names and formulates functional calculus in his doctoral thesis—the Frechet differential, he also inaugurates the study of abstract linear spaces.

1906 Andrei Andreyevich Markov (1856–1922) introduces the Markov chain into probability theory.

1906 William Young (1863–1942) produces his *Theory of Sets of Points* an influential work on set theory.

1907 Hermann Minkowski (1864–1909) publishes on number theory, and geometry of numbers in *Raum und Zeit* ("Space and Time") he subsequently works on four dimensional space-time continuum (Minkoswskian space time).

1908 Kurt Grelling (1886–c. 1942) proposes Grelling's paradox.

1908 Godfrey Harold Hardy (1877–1947) concurrently publishes what is now known as the Hardy Weinberg ratio for genetic analysis, much other work was carried out in collaboration with other mathematicians.

1909 Edmund Landau (1877–1938) provides the first systematic account of analytic number theory, including a much simplified proof of the prime number theorem.

1909 Agner Krarup Erlang (1878–1929) publishes the first of many papers on probability and telephone calls, he is also interested in mathematical tables.

1910s Nikolai Nikolayevich Lusin (1883–1950) works on function theory, developing Lusin's theorem. Lusin's main contributions are in the area of foundations of mathematics and measure theory. He also makes significant contributions to descriptive set topology..

1910 Eliakim Hastings Moore (1862–1932) in *Introduction to a Form of General Analysis* develops the underlying theory of general analysis.

1910 Herman Weyl (1885–1955) publishes widely on axiomatic set theory and Hilbert space.

1912 Egbertus Jan Luitzen Brouwer (1881–1966) formulates the idea of intuitionism (mathematical objects are mental in nature).

1912 George David Birkhoff (1884–1944) proves the three body problem and uses its properties in a 1927 publication, *Dynamical Systems*.

1913 Niels Henrik David Bohr (1885–1962) expresses the displacement of an electron in an atom in terms of the wave equation.

1914–1917
Srinivasa Ramanujan (1887–1920) publishes 21 papers in number theory whilst in Europe, many with Godfrey Harold Hardy (1877–1947).

1915 Egbertus Jan Luitzen Brouwer (1881–1966) produces Browuer's theorem which considers fixed point theorems in topology.

1915 Leopold Lowenheim (1878–c. 1940) proves the Lowenheim-Skolem theorem.

1915 Pavel Sergeevich Aleksandrov (1896–1982) proves that every non denumerable Borel set contains a perfect subset—the method he used now has an important place in set theory.

1916 Albert Einstein (1879–1955) introduced his theory of general theory of relativity, which is most well for the mass energy equation ($E = mc^2$).

1918 George Polya (1887–1985) publishes papers on series, number theory, combinatorics and voting systems, in the following years he adds astronomy and probability to the list.

1919 Felix Hausdorff (1868–1942) introduces the Hausdorff dimension which is used in the production of fractals.

1920s Linear programming is taking off as a subject after largely being ignored for a century. The foundations were essentially laid down by Jean Baptiste Joseph Fourier (1768–1830) and other workers.

1920s Emmy Noether (1882–1935) works on abstract algebra, including the structure of axiom systems, she is eventually appointed to Göttingen in 1922 after many difficulties due to her gender.

1921 Game theory is first formulated by Emile Borel (1871–1956) and expanded by John von Neumann (1903–1957).

1921 Kazimierz Ajdukiewicz (1890–1963), a Polish logician, writes *From the Methodology of Deductive Science* which contains formal (syntactical) definitions of such meta-logical concepts as consequence, logical consequence, logical theorem, and proof.

1922 Paul Levy (1886–1971) writes *Lecons d'analyse fonctionnelle* ("Lessons in Functional Analysis") the origin of the name of this branch of math (an extension of functional calculus).

1922 Oswald Veblen (1880–1960) writes *Analysis situs* which is instrumental in the early study of topology.

1922 Rudolf Carnap (1891–1970) publishes *Der Raum* ("Space"). His work deals with the theory of space from a philosophical point of view.

1922 Maurits Cornelius Escher (1898–1972) produces his first artwork based on regular divisions of the plane (*Eight Heads*), he subsequently studies plane symmetry groups and uses this knowledge in the production of his art works.

1922 Stefan Banach (1892–1945) commences work on a vector space more general than Hilbert space—Banach space—it is actually first mentioned in his dissertation of 1920, he subsequently develops the Banach-Tarski theorem.

1922 The axiomatic set theory, first published in 1908 in *Untersuchungen uber die Grundlagen der Mengenlehre* ("Investigations on the Foundations of Set Theory") by Ernst Freidrich Ferdinand Zermelo (1871–1953), is modified in this year by Adolf Abraham Halevi Fraenkel (1891–1965) to give the Zermelo-Fraenkel set theory.

1923 The work on the atomic theory of Danish physicist Niels Henrik David Bohr (1885–1962) leads him to formulate the correspondence principle, a philosophical guideline for the selection of new theories in physical science.

1924 Karl Menger (1902–1985) becomes interested in curves and publishes a definition of dimension, he subsequently continues working on this and other aspects of geometry.

1924 Louis Victor Pierre Raymond duc de Broglie (1892–1987) puts forward, in his doctoral thesis, the idea of electron waves underpinning wave mechanics and earning de Broglie the Nobel Prize in 1929, he is particularly interested in the statistical nature of physics.

1924 A comprehensive overview of mathematics in physics, *Methoden der mathematischen Physik*, is published in this year by Richard Courant (1888–1972).

1925 Werner Karl Heisenberg (1901–1976) takes the wave equation and shows that the position of an electron or its charge can be known, but not both at the same time (the Heisenberg Uncertainty Principle).

1925 Paul Adrien Maurice Dirac (1902–1984) extends the work of Heisenberg further and defines the properties p and q in algebra.

1925 Ronald Aylmer Fisher (1890–1962) introduces the student t test as well as variance in *Statistical Methods for Research Workers*.

1925 Pavel Sergeevich Aleksandrov (1896–1982) lays the boundaries of homology theory in this year.

1927 Stanislaw Lesniewski (1886–1939) starts his work on the foundations of maths, he publishes many papers on his theories of logic and maths.

1928 Frank Plumpton Ramsey (1903–1930) presents a paper on a *Problem of Formal Logic*, several theorems on combinatorics are included and a new area of study—Ramsey theory—is born.

1928 Jan Lukasiewicz (1878–1956) publishes *Elements of Mathematical Logic* which details three valued logic, a novel system of modal logic and a new logical notation (Polish notation).

1928 Rudolf Carnap (1891–1970) publishes *The Logical Structure of the World*, in which he develops a formal version of empiricism. He holds that all scientific terms are definable by means of a phenomenalistic language.

1930 Kurt Friedrich Gödel (1906–1978) proves the completeness of the first order functional calculus, in the following year he develops the first of his two incompleteness theorems (Gödel's theorem)—this leads to Gödel's number.

1930 Dame Mary Lucy Cartwright (1900–1998) works on complex functions and develops what is now known as Cartwright's theorem, she subsequently moves on to lay the foundations of the study of dynamical systems.

1931 Irmgard Flügge-Lotz (1903–1974) publishes the Lotz method in solving differential equations, subsequent work is on boundary layer problems in fluid dynamics and automatic control theory.

1932 John Henry Constantine Whitehead (1904–1960) writes *The Foundations of Differential Geometry*, and (in 1935) *On the Covering of a Complete Space by the Geodesics Through a Point*, these books contain pioneering contributions in homotopy theory.

1932 Robert Lee Moore (1882–1974) publishes *Foundations of Point Set Topology* (a phrase he invented).

1932–1934 Aleksandr Yakovlevich Khinchin (1894–1959) lays the foundations for the theory of stationary random processes, he studies number and probability theory.

1932 Nina Karlovna Bari (1901–1961) writes *Higher Algebra* and in 1936 *The Theory of Series*. This latter publication covers the theory of functions and the theory of trigonometric series.

1933 Erwin Schrödinger (1887–1961) receives the joint Nobel prize for his work on the wave equation and the formation of the study of Quantum mechanics.

1933 Andrei Nikolayevich Kolmogorov (1903–1987) provides the first general axiomatic treatment of probability theory—as well as the three series theorem and distribution test.

1934 Rudolf Carnap (1891–1970) publishes his most important contribution to logic—*The Logical Syntax of Language*.

1934 Oscar Reutersvard (1915–), an artist who specialises in drawing impossible figures, becomes the "father of impossible figures," with his first picture of a series of cubes.

1934–1936
Max August Zorn (1906–1993) establishes Zorn's lemma in infinite set theory, he also works on topology and algebra and proves the uniqueness of Cayley numbers.

1934–1939
Paul Isaac Bernays (1888–1977) publishes his two volume *Grundlagen der Mathematik* which attempts to build mathematics from symbolic logic, he spends the rest of his life devising axioms to unite all areas of mathematics.

1935 A logician who questioned the use of existence axioms in math and logic, Leon Chwistek (1884–1944), publishes his ideas in *Thoughts on Logic and Mathematics*.

1935 Alfred Tarski (1902–1983) writes *The Concept of Truth in Formalized Languages* which introduces ideas to avoid paradoxes within the foundations of set theory, Tarski also publishes on decision theory, model theory, and algebra in the study of formal systems.

1935 L. L. Thurstone develops with factor analysis.

1936 Alonzo Church (1903–1995) produces Church's theorem for calculus.

1936 Alan Mathison Turing (1912–1954) designs an abstract computer capable of complex serial calculations—the Turing machine, he also lays down the foundations of automata theory with this work.

1937 Jerzy Neyman (1894–1981) jointly publishes (with Egon Sharpe Pearson (1895–1980)) the Neyman-Pearson lemma, a theorem based on the likelihood ratio.

1937 George Polya (1887–1985) 1937 publishes his enumeration theorem for combinatorics.

1939 The first publication of Nicolas Bourbaki (*Elements de mathematiques*)—a collective pseudonym of French mathematicians who wished to produce a modern textbook—appears in this year.

1939 Leonid Vitalyevich Kantorovich (1912–1986) is one of the first to use linear programming as a tool in economics, he is awarded the Nobel Prize for economics.

1940s Warren S. McCulloch (1898–?) and Walter Pitts (1923–) advance automata theory by modelling it on the workings of animal nervous systems.

1941 Milutin Milankovic (1879–1958) writes *Kanon der Erdbestrahlung und seine Anwendung auf das Eiszeitenproblem* ("Canon of Insolation of the Earth and Its Application to the Problem of the Ice Ages")—a mathematical approach to weather and the ice ages.

1942 Richard Phillips Feynman (1918–1988) presents his doctoral thesis, which includes a new approach to quantum mechanics, for this, and his continued work, he is awarded the Nobel Prize in 1965.

1943 Howard Hathaway Aiken (1900–1973) constructs his automatic sequence controlled calculator—a 35 ton computer, subsequent work with IBM produces several more versions up to the mark IV.

1943 Shiing-Shen Chern (1911–) working on differential geometry discovers Chern characteristic classes in fibre spaces, and provides a proof of the Gauss-Bonnet formula.

1943 Stanislaw Marcin Ulam (1909–1984), working on the development of the atomic bomb at Los Alamos, uses his development of the Monte Carlo method of approximations.

1944 *The Theory of Games and Economic Behaviour* establishes games theory, it is written by John von Neumann (1903–1957). Along with Oskar Morgenstern (1902–1977) he applied games theory to economic competition. Von Neumann also carries out work on quantum mechanics.

1945 Samuel Eilenberg (1913–) and Saunders MacLane (1909–), in a paper of this year, introduce the idea of mapping categories within algebra.

1947 George Bernard Dantzig (1914–) whilst working as Mathematical Advisor for USAF Headquarters discovers the simplex method of optimization.

1947 Abraham Wald (1902–1950) formalizes sequential analysis based on the likelihood ration.

1948 Claude E. Shannon (1916–) proposes a general information theory—essentially a form of probability theory—thus founding the subject of information theory.

1948 Laurent Schwarz (1915–) researches and publishes widely on the theory of distributions.

1948   Norbert Wiener (1894–1964) founds cybernetics with his publication *Cybernetics, or Control and Communication in the Animal and the Machine*.

1949   Paul Erdös (1913–1996) found an elementary proof of the prime number theorem, with Atle Selberg (1917–). He also makes many contributions to number theory, graph theory, and combinatorics.

1949   Andre Weil (1906–1998)—one of the founders of the Bourbaki group—provides proof of the Riemann hypothesis for the congruence zeta functions of algebraic function fields—Weil conjectures.

1950   After a lifetime of studying statistics Gertrude Mary Cox (1900–1978) writes a classic text *Experimental Design*, the previous year she is the first woman elected to join the International Statistical Institute.

1951   Jean-Pierre Serre (1926–) applies spectral sequences to the study of the relations between the homology groups of fibre, total space and base space in a fibration. This enables him to discover fundamental connections between the homology groups and homotopy groups of a space and to prove important results on the homotopy groups of spheres.

1952   This year marks the first publication of Gerhard Ringel (1919–) on combinatorics and graph theory, the map color theorem, and embeddings of graphs into surfaces.

1952   Aleksandr Osipovich Gelfond (1906–1968) publishes two major books on transcendental numbers—*Transtendentnye i algebraicheskie chisla* and approximation and interpolation theory—*Ischislenie konechnykh raznostey* in this year.

1953   Lars Valerian Ahlfors (1907–1976) writes *Complex Analysis*, he also works on Reimann surfaces and Kleinian groups.

1954   Nelson Goodman (1906–) in *Fact, Fiction and Forecast* advances Goodman's Paradox on projection patterns—it is characterised by his statement "all emeralds are green."

1954   Thomson's lamp appears in a paper *Tasks and Super Tasks* by J. F. Thomson—it is a problem in analysis.

1956   Henri-Paul Cartan (1904–) publishes *Homological Algebra* and, in 1963, *Elementary Theory of Analytic Functions of One or Several Complex Variables*, he works on analytic functions, the theory of sheaves, homological theory, algebraic topology and potential theory and is one of the Bourbaki collective.

1957   Adrien Douady (1935–) publishes on dynamical systems.

1958   Alberto Calderón (1920–1998) publishes on uniqueness in the Cauchy problem for partial differential equations and then develops the theory of pseudodifferential operators in the 1960's.

1961   Texas instruments produce the first commercial planar integrated circuits utilising simple logic functions.

1963   Paul (Joseph) Cohen (1934–) solves the status of Cantor's continuum hypothesis, i.e. that it is insoluble.

1966   Lennart Axel Edvard Carleson (1928–) produces Carleson's theorem: the Fourier series of any square integrable function f converges to f almost everywhere.

1966   Michael Francis Atiyah (1929–) is awarded the Fields medal at the International Congress in Moscow for development of a technique in topology called K theory carried out in 1963, subsequently he works on elliptic differential equations.

1966   Alexander Grothendieck (1928–) is awarded the Fields medal for his work in algebraic geometry, and for unifying themes in geometry, number theory, topology and complex analysis.

1966–   Pierre-Rene Deligne (1944–) commences working on the fundamentals of algebraic geometry.

1968   J. W. T. Young writes on graph theory and chromatic numbers of all surfaces except plane or sphere.

1972   René Thom (1923–) creates the field of catastrophe theory with his publication *Stabilitie structurelle et morphogenese* ("Structural Stability and Morphogenesis").

1973   Lenore Blum (1943–) delivers *Inductive inference: A recursion theoretic approach* to the 14th Annual Symposium on Switching and Automata Theory.

1975   Tosio Kato (1917–1999) authors *Perturbation Theory for Linear Operators*, he is the father of modern mathematical physics.

1975   Chaos theory is born when T. Y. Li and James A Yorke (1941–) study oeriodic points for transformations of a real line.

1976   Wolfgang Haken (1928–) along with Kenneth Appel (1932–) proves the four color theorem (four colors are the minimum required).

1979   Apery's theory is put forward by Roger Apery (1916–1994), it is a constant used in non Euclidean geometry for the proof of the irrationality of the sum of the inverse of the cubes of integers.

1978   Stephen Wolfram (1959–) discovers early connections between cosmology and particle physics and invents Fox-Wolfram variables for analysis of event shapes in particle physics, he eventually moves to work on automata and writes the computer programme and book *Mathematica*.

1982   Simon Kirwan Donaldson (1957–) carries out work on 4 dimensional spaces using nonlinear partial differential equations, the paper is entitled *Self-dual Connections and the Topology of Smooth 4-Manifolds*.

1982   Benoit B Mandelbrot (1924–) writes *The Fractal Geometry of Nature* which introduces us to the Mandelbrot set—a group of numbers following a particular pattern which can give complex pictures, he also invents the word fractal.

1983   John Hubbard (1945–) begins teaching at the University of Richmond in computer science and mathematics, especially programming in Java and C++, data structures, algorithms, database systems, calculus, and numerical analysis.

1983   Subrahmanyan Chandrasekhar (1910–1995) is awarded the Nobel Prize for Physics on the mathematical study of collapsing stars, this work was first produced from 1930 onwards and a summary was published in his book *The Mathematical Theory of Black Holes*.

1986   Michael Freedman (1951–) is awarded the Fields medal for mathematics for his work on the Poincaré conjecture, providing a new area of study within topology.

1987   Krystyna M. Kuperberg (1944–) solves a problem in bi-homogeneity of dynamical systems.

1987   Edward Witten (1951–) publishes *Superstring Theory*, he also makes significant contributions to Morse theory, supersymmetry, and knot theory.

1988   The first graphing calculator is produced by Texas Instruments.

1988   Roger Penrose (1931–) receives the Wolf prize for physics, most of his work pertains to relativity theory and quantum physics, and a field of geometry known as tessellation, the covering of a surface with tiles of prescribed shapes. With his father he created the Penrose impossible staircase and the Penrose triangle, both of which were illustrated by Maurits Cornelius Escher (1898–1972).

1991   Alexandra Bellow (1935–) writes *Almost Everywhere Convergence: The Case for the Ergodic Viewpoint*.

1993   Linda Goldway Keen (1940–) writes and continues research on *Hyperbolic Geometry and Spaces of Riemann Surfaces*.

1994   Andrew Wiles (1953–) gives a lecture announcing his proof of Fermat's last theorem, it is published in 1995 (after an error in his 1993 proof was discovered).

1994   Ingrid Daubechies (1954–) writes *Ten Lectures on Wavelets* on time frequency analysis for which she is subsequently awarded the American Mathematical Society prize.

1995   Sun-Yung Alice Chang (1948–) is awarded the Ruth Lyttle Satter Prize in Mathematics for her work on partial differential equations, Riemannian manifolds and spectral geometry.

1995   Richard Lawrence Taylor (1962–) is appointed professor of mathematics at New College, University of Cambridge, where he works extensively on number theory.

A

Abacus, **1:1–2**, **1**:87, **1**:301
  duodecimal, **1**:184
  for negative numbers, **2**:443
  Roman numerals and, **1**:225
Abbe, Ernst, **1:2**, **1**:*3*
Abbe sine condition, **1**:2
Abel, Niels Henrik, **1:2–4**, **1**:*3*
  algebraic functions and, **1**:17
  Betti and, **1**:61
  elliptic functions and, **1**:360
  quintic equations and, **1**:2, **1**:3–4, **2**:527
Abelian functions, **1**:332, **2**:492, **2**:547, **2**:658
Abelian groups, **1**:253, **1**:276
  in electromagnetic theory, **2**:675
  infinite, **2**:438
  vector space built on, **1**:4, **1**:5
Abel–Ruffini theorem, **2**:556
Abel's theorem, **1**:4
Abscissa, **1**:21, **1**:105
Absolute differential calculus, **1**:364–365
Absolute extrema, **1**:215
Absolute geometry, **1**:208, **2**:468
Absolute motion, **2**:492
Absolute value equations, **1**:197
Absolute value function, **2**:476, **2**:536
Absolute value sign, **2**:444
Absolutely convergent series, **1**:20
  of analytic function, **1**:132
  rearrangement of, **1**:179
Absolutely normal numbers, **2**:455
Abstract algebra, **1**:7, **1**:14, **1:16**
  algebraic numbers and, **1**:18
  Betti's contributions to, **1**:61
  multiplication in, **2**:438, **2**:439
  Noether's contributions to, **2**:449–450
  See also Field(s); Group(s); Ring(s)
Abstract linear spaces, **1:4–5**. *See also* Vector spaces

Abstraction
  axiom of, **2**:680
  definition by, **1**:164
Abu Ma shar, **1**:7
Accelerating reference frames, **2**:539
  fictitious forces in, **1**:234–235
Acceleration, **2**:436
  as derivative of velocity, **1**:89, **1**:174
  force and, **1**:249, **2**:446–448
  integration of, **1**:324
  as vector, **2**:436, **2:644**
Accounting, **2**:466
Accumulation point, **1**:72–73
Accuracy, **1**:202–203
  of measurements, **2**:544–545
  See also Approximation
Achilles and the tortoise paradox, **2**:677, **2**:678
Action integral, **2**:675
Action-reaction forces, **2**:447–448
Actual infinite, **2**:628
Acute angle, **1**:25
Acute triangle, **2**:483, **2**:622, **2**:625
Acute-angle hypothesis, **2**:453
Adams, Frank, **1**:39
Adams, John Quincy, **1**:185
Adding machine, **1**:87
Addition, **1:5–6**
  of analytic functions, **1**:4
  axioms of, **1**:36
  in binary notation, **2**:404
  of complex numbers, **1**:6, **1**:7, **1**:133, **1**:282, **2**:469
  as first-order operation, **2**:502
  of forces, **2**:589
  of fractions, **1**:5–6, **1**:239, **1**:357
  of functions, **1**:6, **1**:245
  of imaginary numbers, **1**:313
  of infinite series, **1**:318–319
  of matrices, **1**:6, **1**:15–16, **2**:409
  of measurements, **2**:544
  of natural numbers, **1**:5, **1**:6, **1**:7

series expansions in, **1**:319, **2**:444
sieve of Eratosthenes, **1**:199, **2**:504, **2**:575
for tiling the plane, **2**:608
for Turing machine, **2**:632, **2**:633–634
Alhazen, **1**:13, **1**:27, **1**:369
Al-Kashi, **1**:160
Al-Khwarizmi, Abu Ja'far Muhammad ibn Musa, **1**:14, **1**:17, **1**:18,
1:**18–19**
quadratic equations and, **1**:19, **2**:519
translated by Adelard, **1**:8
translated by Gerard, **1**:262
''All,'' **2**:502, **2**:503, **2**:520, **2**:521
Al-Majriti, **1**:8
Almost everywhere convergence, **1**:99
Al-Salar, Husam al-Din, **2**:453
Alternating series, **1**:**19–20**
Alternating Series Test, **1**:19–20
Alternative hypothesis, **1**:122
Altitude of triangle, **1**:**20**, **1**:32, **2**:622, **2**:626
American Standard Code for Information Interchange (ASCII), **1**:64
Amicable numbers, **1**:**20–21**
Ampere, **2**:422, **2**:637
Ampere's law, **2**:414
Amplitude
of oscillation, **2**:464
of sine and cosine, **2**:627
Analysis, **1**:**21**
nonstandard, **1**:319, **1**:342, **2**:430
real, **1**:162, **1**:326
set theory and, **2**:570
See also Calculus; Complex analysis
Analysis of variance, **1**:232
Analysis situs. *See* Topology
Analytic continuation, **2**:658, **2**:682
Analytic functions
automorphic, **1**:297, **1**:344, **2**:485, **2**:492
Cauchy's integral theorem for, **1**:109
conformal mapping and, **2**:549–550
defined, **1**:109, **1**:132
Gelfond's research on, **1**:258
Liouville's theorem, **1**:377
meromorphic, **1**:9
modular forms, **2**:607, **2**:669
Picard's theorem on, **2**:485
power series of, **1**:132, **2**:657, **2**:658
vector space of, **1**:4–5
Weierstrass's work on, **2**:657, **2**:658
See also Complex functions; Zeta-function
Analytic geometry, **1**:**21–23**, **1**:170
Plücker's work on, **2**:489–490
Analytic number theory
Dirichlet's theorem, **1**:179
Hermite's work on, **1**:290
Landau's work on, **1**:354
zeta-function, **2**:**681–682**
Analytical Engine, **1**:41, **1**:42, **1**:43, **1**:88
Analytical Society, **1**:157
Anaxagoras of Clazomenae, **1**:**23–24**, **1**:*24*
Democritus and, **1**:165
squaring the circle and, **1**:23, **1**:24, **2**:583
''And,'' **1**:136, **1**:198, **1**:383, **2**:502, **2**:596. *See also* Conjunction
AND gate, **1**:383
Andrews, William, **2**:394

Angle(s), **1**:**24–25**, **1**:141
bisection of, **1**:25
complementary, **1**:25, **2**:624
congruent, **1**:25, **2**:625
defined, **2**:625
degree measure, **1**:24, **1**:25, **1**:141, **1**:299, **2**:529, **2**:571, **2**:625
in hyperbolic geometry, **1**:311, **1**:379
of incidence, **2**:578
on manifold, **1**:202
radian measure, **1**:24, **1**:25, **1**:141, **2**:**529**, **2**:625, **2**:637
of refraction, **2**:578
right, **1**:25, **2**:609, **2**:627
of rotation, **2**:487, **2**:599, **2**:620, **2**:625
sexagesimal numeration and, **1**:141, **2**:571
solid, **2**:488, **2**:529
supplementary, **1**:25
trisection of, **1**:25, **1**:207, **1**:208, **2**:**627–628**, **2**:647
between vectors, **2**:642
vertical, **1**:25, **2**:609
See also Trigonometric functions
Angle defect, **1**:257
Angular momentum
conservation of, **1**:339, **2**:599
of rigid body, **1**:212
as vector, **2**:642
Annuities, de Moivre's work on, **1**:157
Anselm, Saint, **2**:429, **2**:503, **2**:521
Antecedent, **1**:314, **1**:384, **2**:598
modus ponens and, **2**:432, **2**:509
modus tollens and, **2**:432
Anthropic principle, **2**:663
Anticlastic surface, **1**:257
Antiderivative, **1**:246, **1**:247, **1**:324
Anti-differentiation, **2**:421
Antilogarithm, **1**:382, **2**:554
Antinomies. *See* Paradoxes
Antiphon the Sophist, **2**:583
Aperiodic tilings, **2**:608
Apéry, Roger, **1**:26
Apéry's theorem, **1**:**25–26**
Apollonius of Perga, **1**:**26–27**
Euclid's solid geometry and, **2**:580
Gergonne and, **1**:262
Hypatia and, **1**:27, **1**:307
Pappus and, **2**:467
Regiomontanus' translation of, **2**:538
Snell's translation of, **2**:578
Apostles (at Cambridge), **1**:284, **2**:557
Appel, Kenneth I., **1**:**27**, **1**:236
Applied mathematics, **1**:**27–28**, **2**:421
Approximation, **1**:**202–203**
by Fourier series, **2**:420–421
by infinite series, **1**:319
by interpolation, **1**:91, **1**:**324–325**, **2**:420, **2**:655
by iteration, **1**:329, **2**:566
linear, **1**:90
in measurement, **2**:544–545
methods of, **2**:**420–421**
negligible terms in, **2**:444
Newton's method, **2**:554
numerical integration, **2**:421, **2**:437, **2**:**458–459**
order of, **2**:444

of polynomial division, **2**:556
by polynomials, **2**:420, **2**:498
by power series, **2**:501
by Riemann sums, **1**:247
for solving polynomial equations, **2**:528, **2**:554
of square roots, **1**:329, **2**:566
by Taylor series, **1**:260
Weyl's method, **2**:554
See also Numerical analysis
Arabic learning
Adelard's translations of, **1**:7–8
Gerard's translations of, **1**:261–262
Omar Khayyam, **1**:*341*, **1**:341–342, **2**:452–453, **2**:561
parallel postulate and, **2**:452–453
See also Al-Khwarizmi
Arabic system. *See* Decimal (Hindu-Arabic) system
Arago, Dominique-François, **1**:370
Arbogast, Louis François Antoine, **1**:28
Arc length, **2**:482, **2**:572–573
of cycloid, **1**:149, **2**:618, **2**:673
Arc of circle, **1**:125
angle subtended by, **1**:141
midpoint of, **2**:423
radian measure and, **2**:529
Arccos function, **1**:327
Arccosh function, **1**:327
Arc-degree, **2**:571
Archimedes' midpoint theorem, **2**:423
Archimedes of Syracuse, **1**:28–30, **1**:*29*, **2**:407
Apollonius and, **1**:26–27
area of circle and, **1**:29, **1**:30, **1**:32
area under curve and, **1**:90
Hero's formula and, **1**:291, **1**:292
method of exhaustion and, **2**:459
Pappus and, **2**:467
pi and, **1**:29, **1**:30, **1**:32, **2**:484
Regiomontanus' translation of, **2**:538
squaring the circle and, **2**:583
Stevin's mechanics and, **2**:589
Tartaglia's translation of, **2**:604
trisection of angle and, **2**:627
Wallis' translations of, **2**:655
Archimedes' Principle, **1**:29
Archimedes Screw, **1**:28, **1**:29
Archimedes' spiral, **1**:30, **2**:495, **2**:582
Archytas of Tarentum, **1**:30–31, **1**:*31*, **1**:185, **1**:208
Arc-minute, **2**:571
Arcsin function, **1**:327
Arcsinh function, **1**:327
Arctan function. *See* Inverse tangent function
Arctanh function, **1**:327
Area, **1**:31–32
of circle, **1**:29, **1**:30, **1**:32, **1**:125
of ellipse, **1**:196, **2**:580
as integral, **1**:32, **1**:90, **1**:163, **1**:246–247, **1**:324, **2**:592
of irregular figures, **1**:32
as limit of a series, **1**:7
of line segment, **1**:164
measure theory and, **1**:76
as multiple integral, **2**:437
of polygon, **1**:22, **1**:32
of rectangle, **1**:31–32

of snowflake curve, **2**:579
of square, **2**:501, **2**:580
of triangle, **1**:20, **1**:22, **1**:81, **1**:291, **1**:292–293, **2**:622, **2**:626
units of, **2**:422, **2**:637
Arens, R., **1**:91
Argand, Jean Robert, **1**:32–33
Argand diagram, **1**:32, **2**:660
Argument, **1**:314, **1**:318, **2**:509, **2**:594–595
Aristaeus, **2**:580
Aristarchus of Samos, **1**:33–34, **2**:625
Aristotelian logic, **1**:34, **1**:35, **1**:163, **1**:197, **2**:509
conjunction in, **1**:136, **1**:137
De Morgan's extension of, **1**:158
existence as predicate in, **2**:503, **2**:521
Lukasiewicz's interpretation of, **1**:386
syllogism in, **1**:34, **1**:163, **2**:509, **2**:513, **2**:594–595
Aristotelian physics
Galileo's refutation of, **1**:249, **1**:251
Stevin's refutation of, **2**:588, **2**:589
Aristotle, **1**:34–35, **2**:407
alchemy and, **1**:80
Archytas and, **1**:31
on infinity, **1**:326, **2**:628
Menaechmus and, **2**:417
on motion, **2**:446
on necessary truth, **2**:429
Nicholas of Oresme's translations of, **2**:448
on parallel postulate, **2**:452
Plato and, **2**:488, **2**:489
Ptolemy and, **2**:513
Tycho's observations and, **1**:80
Venn diagrams and, **2**:645
on Zeno's paradoxes, **2**:678
Aristoxenus, **2**:449
Arithmetic, **1**:35–37
algebra as generalization of, **1**:14
consistency of, **1**:123, **1**:297, **2**:651
fundamental theorem of, **1**:217, **2**:503, **2**:575, **2**:682
Gödel's incompleteness theorem for, **2**:500, **2**:630
Greek, **2**:449
nonstandard model of, **2**:430
Peano axioms for, **1**:163, **2**:474–475, **2**:500, **2**:510, **2**:666
symbols in, **2**:403
See also Addition; Division; Multiplication; Subtraction
Arithmetic mean, **1**:31, **2**:414
Arithmetic mean–geometric mean inequality, **1**:317
Arithmetic sequences, **1**:37, **1**:230
primes in, **1**:361
Arithmetic series, **1**:37, **1**:230, **2**:592
Arithmetization of real analysis, **1**:162
Armstrong, Neil, **1**:89
Arnold, V. I., **1**:346, **2**:442
Arrow, Kenneth, **2**:653
Arrow paradox, **2**:677, **2**:678
Arrow's impossibility theorem, **2**:653
Art and mathematics, **2**:403–404
Escher, **1**:203–204, **1**:311, **1**:379, **2**:481, **2**:545
perspective, **1**:12, **1**:168, **1**:186, **1**:187, **2**:404, **2**:483–484, **2**:606
proportion, **2**:403, **2**:404, **2**:512
Reutersvard's work, **2**:545
Artificial intelligence
Bayesian networks in, **1**:50
Leibniz's foreshadowing of, **1**:124

Artin, Emil, **1**:17, **1**:246, **2**:682
Aryabhata the Elder, **1:37–38**, **1**:247, **1**:261–262
ASCII (American Standard Code for Information Interchange), **1**:64
Asimov, Isaac, **2**:405
Associative algebras, **1**:102, **2**:479
Associative law of addition, **1**:5
    in a group, **1**:16
    for matrices, **1**:16
    for real numbers, **1**:36
    for vectors, **1**:376
Associative law of multiplication
    for matrices, **2**:409
    for real numbers, **1**:36, **2**:438
Associative magic square, **2**:393
Associative property, **1:38**
    division and, **1**:182
    of field, **1**:227
    of group, **1**:16, **1**:253, **1**:275
Assumptions. *See* Axioms; Postulates; Premises
Asterisk, **2**:507
Asteroids, **2**:462
Astrolabe, **1**:299, **1**:307, **2**:514
Astrology, **2**:406
    Arabic, **1**:7–8
    of Cardano, **1**:97, **1**:98
    of Hipparchus, **1**:298
    Kepler's tolerance for, **1**:338
    medicine and, **1**:140
    Nicholas of Oresme's opposition to, **2**:448, **2**:449
    Ptolemy's writings on, **2**:514
    of Regiomontanus, **2**:538
    of Simpson, **2**:576
Astronomical unit (AU), **1**:140, **1**:340
Astronomy
    Plato's theories, **2**:488
    quasars, **2**:540–541
    See also Planetary motion; Stars; Celestial mechanics
Asymptote, **1:38**
    of hyperbola, **1**:308–309
Atiyah, Michael Francis, **1:38–40**, **1**:165, **1**:183, **1**:184, **2**:568
Atiyah-Singer Index Theorem, **1**:39, **1**:92
Atmospheric pressure, **2**:617, **2**:618
Atomic clocks, **2**:544
Atomic theory
    Bohr model, **1**:69, **1**:70, **2**:562
    bonding, **2**:523
    of Democritus, **1**:165–166
    periodic table, **1**:324
    of Plato, **2**:489
    of Rutherford, **1**:68, **2**:558–559
    See also Quantum mechanics; Subatomic particles
Attractor, **1**:117, **1**:139
Augmented matrices, **1**:374, **2**:601
Augustine, Saint, **2**:482
Autological words, **1**:274
Automatic control theory, **1**:233–234
Automorphic functions, **1**:297
    Klein's work on, **1**:344
    Picard's work on, **2**:485
    Poincaré's work on, **2**:485, **2**:492
    Weil conjectures and, **1**:165
Automorphism, **1**:329

Average
    measures of central tendency, **2**:414, **2**:417, **2**:429, **2**:586
    Quetelet's interpretation of, **2**:527
    See also Mean
Avogadro's number, **2**:613
Axiom of abstraction, **2**:680
Axiom of choice, **1**:267, **2**:573
    Borel's rejection of, **1**:77, **1**:280
    Hadamard's acceptance of, **1**:280
    well-ordered sets and, **2**:619, **2**:679, **2**:680, **2**:683
    Zermelo's use of, **1**:95, **2**:619, **2**:679
    Zorn's lemma and, **2**:683
Axiom of reducibility, **2**:531
Axiomatic set theory, **1**:54, **2**:430, **2**:679. *See also* Zermelo–Fraenkel
   set theory
Axioms, **1:40**, **2**:610
    in abstract algebra, **1**:16
    of Aristotelian logic, **2**:594–595
    of arithmetic, **1**:36, **1**:163, **2:474–475**
    completeness theorem, **1**:266
    of deductive system, **1**:163, **2**:509
    formalism and, **1**:235, **1**:296–298
    for geometry, **2**:487 (*see also* Parallel postulate)
    of group, **1**:16
    in Hilbert's foundation for geometry, **1**:207, **1**:294, **1**:297, **1**:298
    metamathematics and, **2**:419, **2**:420
    of modal logic, **2**:428
    model theory and, **2**:430
    vs. postulates, **2**:499
    proof and, **2**:509
    of scientific theory, **1**:100
    See also Axiomatic set theory; Peano axioms; Proof
Axis (axes)
    in artistic composition, **2**:403–404
    of cone, **1**:134
    of ellipse, **1**:196
    of revolution, **2**:593*f*, **2**:594
    of rotational motion, **2**:436
    of sphere, **2**:582
    translation of, **2**:621, **2:622**
Axis of symmetry, **2**:598, **2**:599
    of circle or sphere, **1**:172
    of parabola, **1**:22, **2**:520
Azimuthal force, **1**:234
Aztec number system, **2**:648

B

Babbage, Charles, **1:41–43**, **1***:42*, **1**:88
Babbage, Henry P., **1**:88
Babbage's engine, **1**:41–42, **1:43–44**
Babylonian mathematics, **1**:17, **1**:19, **1:44**, **1**:301
    abacus, **1**:1
    angles, **1**:24, **1**:25
    duodecimal system in, **1**:185
    Pythagoras influenced by, **2**:514
    Pythagorean theorem in, **1**:44, **2**:517
    quadratic equations in, **2**:519
    rod numerals, **1**:87
    sexagesimal system, **1**:44, **2**:534, **2**:571
    zero and, **2**:571, **2**:681

C

Crystals
>> Penrose tilings and, **2**:481–482
>> space-filling patterns and, **2**:489
>> symmetries of, **2**:599

C-theory, **2**:567

Cubature, **2**:458

Cube(s), **2**:497
>> cross section of, **2**:580
>> magic, **2:390–391**
>> as Platonic solid, **2**:488, **2**:489
>> volume, **2**:439, **2**:501, **2**:650
>> Waring's theorem, **2**:656
>> *See also* Duplication of cube

Cube roots
>> Hero's approximation, **1**:292
>> *See also* Roots of a number

Cubic centimeter, **2**:649

Cubic curves, **1**:17
>> Stirling's work on, **2**:590
>> Witch of Agnesi, **1**:8

Cubic equations, **1:148, 1**:197
>> Bombelli's work on, **1**:73
>> Cardano's solution of, **1**:96, **1**:97–98
>> constructible numbers and, **1**:186, **2**:628
>> depressed, **1**:97, **2**:604–605
>> Diophantine, **1**:176
>> irreducible, **2**:628
>> Khayyam's work on, **1**:341–342
>> Riccati's solution of, **1**:310
>> Tartaglia's work on, **2**:604–605

Cubic lattice, **2**:608

Cubic spline interpolation, **1**:325

Cubit, **2**:422

Culture and mathematics, **2:405**

Cumulative distribution function, **2**:505

Curl of vector field, **2**:573, **2**:643

Curly brackets, **1**:16

Current density vector, **2**:414

Curvature
>> Gaussian, **1:257–258, 1**:261
>> geodesic, **1:259–260**
>> geometric mean, **1**:258
>> gravitation and, **1**:261, **2**:539
>> Levi–Civita's work on, **1**:364, **1**:365
>> of Lobachevsky plane, **1**:379
>> Monge's work on, **2**:434
>> Riemann's concept, **2**:546, **2**:550
>> of space-time, **2**:540

Curve(s)
>> algebraic, **1**:16–17, **1**:62, **1**:63
>> area under (*see* Definite integral)
>> equation for, **2**:489, **2**:490
>> family of, **1:218**
>> interpolated, **1**:324–325
>> Jordan curve, **1:332–333**, **2**:617
>> Menger's research on, **2**:417
>> motion on, **2**:644
>> as one-dimensional manifold, **2**:550
>> orthogonal, **2**:463
>> Peano, **2:475–476**
>> perimeter, **2**:483
>> smooth, **1**:314

>> on surface, **1**:259–260
>> vector integral on, **2**:643
>> *See also* Analytic geometry; Tangent line

Curves of indifference, **2**:470

Curvilinear motion, **2**:436

Curvilinear triangle, **2**:622

Cusp catastrophe, **1**:106, **1**:107

Cut, **1**:161, **1:162**

Cybernetics, **1:148, 2**:666, **2**:668

Cycle
>> of graph, **1**:129, **1**:130
>> of oscillation, **2**:464

Cyclic extension of field, **2**:527

Cycloid, **1:148–149, 1**:381
>> arc length, **1**:149, **2**:618, **2**:673
>> brachistochrone and, **1**:57, **1**:59, **1**:148
>> center of gravity, **2**:655
>> differentiability and, **2**:476
>> for gear wheels, **1**:168
>> isochronism of, **1**:148–149, **1**:304
>> parametric equations for, **2**:469
>> Pascal's work on, **2**:472
>> rectification of, **2**:673
>> Torricelli's work on, **2**:618

Cylinder, **1:149**
>> cross sections of, **2**:580
>> Gaussian curvature of, **1**:257
>> helix on surface of, **2**:582
>> topological properties of, **2**:616
>> volume, **1**:29, **2**:650

Cylinder functions. *See* Bessel functions

Cylindrical coordinates, **2**:463, **2**:495
>> for multiple integral, **2**:437

Cylindroid, **1**:149

D

Da Vinci, Leonardo, **1:152, 1**:153, **2**:404, **2**:422, **2**:484
>> Pacioli and, **2**:466
>> proof of Pythagorean theorem, **2**:517
>> squaring the circle and, **2**:583

Dal Ferro, Scipione, **1**:97, **2**:604

D'Alembert, Jean le Rond, **1:151–153, 1**:*152*
>> fundamental theorem of algebra and, **1**:246
>> Laplace and, **1**:355
>> vibration theory and, **1**:351

D'Alembert's paradox, **1**:305

D'Alembert's principle, **1**:151

Dali, Salvador, **2**:403

Dalton, John, **1**:324

Dantzig, George Bernard, **1:153–155**

Dantzig, Tobias, **1**:154

Darboux, Gaston, **1**:76

Data. *See* Statistics

Data compression, **1**:322

Daubechies, Ingrid, **1:155–156**

Davis, M., **1**:176

De Broglie. *See* Broglie, Louis Victor Pierre Raymond de

De Moivre, Abraham, **1:156–157**
>> Stirling's formula and, **1**:157, **2**:590

De Moivre's theorem, **1**:157

Descartes, René, **1:168–170**, 1:*169*
    algebraic curves and, 1:16
    analytic geometry of, 1:18, 1:21, 1:261, 2:655
    Arbogast and, 1:28
    Desargues and, 1:168
    existence of God and, 2:503, 2:521
    Fermat and, 1:219, 1:220
    function concept and, 1:244
    Huygens and, 1:303–304
    logarithmic spiral and, 2:582
    Mersenne and, 1:170, 2:418
    metaphysics of, 1:119–120
    Pascal and, 2:472
    Roberval's debate with, 2:551
    Snell's law and, 2:578
    symbols used by, 2:402, 2:439
    See also Cartesian coordinate system
Descartes' rule of signs, **1:170–171**
Descriptive geometry, 2:433, 2:434, 2:452
Descriptive statistics, 2:584, 2:585–586
Deshouillers, J.-M., 1:268
Deslarte, Jean, 1:78, 1:103
Destructive interference, 2:522
Detachment, rule of, 2:509, 2:598
Determinants, **1:171–172**, 1:171*f*–172*f*
    Bézout's use of, 1:62
    Cayley's work on, 1:110–111
    Cramer's work on, 1:147
    cross product represented by, 2:642
    Hadamard's inequality for, 1:317–318
    Jacobian, 1:332
    matrix inverse and, 2:410–411
    multiplication theorem for, 1:65, 2:411
    nonzero, 2:644
    Seki's discovery of, 2:565, 2:566
    Vandermonde's work on, 2:640
Devaney, Robert, 1:337
Diagonal
    of polygon, 2:497
    of polyhedron, 2:497
Diagonalization method of Cantor, 1:99
Dialectic, 2:677–678
Diameter, 1:125, 1:126, **1:172**, 2:482
    of sphere, 1:172, 2:582
    Thales' theorem, 2:609
Dichotomy paradox, 2:677, 2:678
Dictionary order, 2:619
Diderot's *Encyclopédie*, 1:151, 1:152
Dielectric constant, 1:71
Dieudonné, Jean, 1:78, 1:103, 1:274
Difference, 1:36, **2:592**
    of sets, 2:569–570, 2:570*f*, 2:592
    See also Subtraction
Difference Engine, 1:41–42, 1:43–44, 1:88
Difference quotient, 1:173
Difference rule, for derivatives, 1:174
Differentiable curves, 1:259
    Koch snowflake and, 2:579
Differentiable functions
    Cauchy and, 1:90
    continuity and, 1:167, 2:475–476, 2:658
    mean-value theorem for, 2:415
    See also Derivative(s)

Differentiable manifolds, 1:202, 2:550, 2:617
    defined by Veblen and Whitehead, 2:641, 2:665
    four-dimensional, 1:242
Differential, 1:59, **1:167**
Differential calculus, 1:90–91, **1:172–173**
    Arbogast's work on, 1:28
    chain rule in, **1:113–114**, 1:167, 1:314
    Fermat's work on, 1:219, 1:220
    integral calculus and, 1:247
    symbols in, 2:403
    See also Calculus; Derivative(s)
Differential equations, 1:89, **1:173–174**
    Bessel equation, 1:61
    Birkhoff's work on, 1:66, 1:67
    Boole's investigations of, 1:74
    Brouwer's fixed point theorem and, 1:86
    difference equations and, 2:537
    dynamical systems determined by, 2:591
    existence of solutions, 1:231, 2:485–486, 2:565
    family of solutions, 1:218
    first-order ordinary, **1:231**, 2:476
    group theory and, 1:368, 2:485
    higher order, 1:231
    integration of, 1:324
    inverse trigonometric function solutions, 1:328
    Lindelöf's work on, 1:371
    Maxwell's equations, 1:183, 1:195, **2:414**
    Monte Carlo solutions, 2:435
    nonlinear, 1:173
    numerical solution, 1:90, 1:375, 2:565
    power series solutions, 2:501
    second-order ordinary, **2:564–565**
    stability theory of, 1:388, 2:442
    systems of, 1:173
    uniqueness of solutions, 1:231, 2:565
    See also Partial differential equations
Differential forms, 1:102
Differential geometry, 1:59, 1:261
    Chern's work in, 1:121
    dimension concept in, 1:175
    Élie Cartan's work in, 1:102, 1:103
    Gauss's work in, 1:256–257, 2:489
    general relativity and, 2:661
    Henry Whitehead's work in, 2:665
    Levi–Civita's tensor calculus, 1:364–365
    on manifolds, 1:202
    Monge's work in, 2:433–434
    Riemannian, 2:468–469, 2:547, **2:550–551**, 2:641
Differential topology, 2:616, 2:617
Differentiation
    of algebraic functions, **1:174**
    covariant, 1:364
    of distance and velocity, **1:174**
    implicit, **1:314**
    partial, 1:167, 1:315, **2:471**, 2:642
    of power series, 2:501
    slope determined by, 2:577–578
    of transcendental functions, **1:175**
    See also Derivative(s)
Diffraction, 1:370
Digits, 1:161
    reliability of, 2:544–545

Dilation, **2**:620
    of parabola, **2**:467
Dimension, **1**:**175**, **1**:238
    fractal, **1**:175, **2**:396, **2**:476, **2**:501, **2**:579
    Hausdorff, **1**:238, **1**:**286–287**, **1**:288, **2**:574, **2**:579
    of Lie group, **1**:367
    of manifold, **1**:175, **2**:490–491
    Menger's work on, **2**:417
    of solutions to linear system, **2**:601
    topological, **1**:175, **1**:208, **1**:238, **2**:574
    of vector space, **1**:175, **2**:643–644
Dinostratus, **1**:372
Diophantine equations, **1**:**175–176**, **1**:197, **1**:323, **2**:455–456
    Chinese remainder theorem and, **1**:122
    Dirichlet's work on, **1**:178–179
    Hilbert's call for method, **1**:297
    in Hindu mathematics, **1**:37, **1**:63, **1**:176
    Rolle's work on, **2**:551, **2**:552
Diophantine geometry, **1**:176, **2**:455
Diophantus, **1**:17, **1**:19, **1**:**176–177**
    Bombelli's reintroduction of, **1**:73
    Hypatia's commentary on, **1**:306–307
    Latin translation of, **2**:552
    Regiomontanus and, **2**:538
Dirac, Paul Adrien Maurice, **1**:**177–178**, **1**:223
    magnetic monopoles and, **1**:184
Dirac delta function, **2**:562
Direct proof, **2**:508–509, **2**:512
Direct variation, **1**:130, **1**:**178**
Direction of translation, **2**:487
Directional derivative, **1**:167
    of potential energy, **1**:234
Directrix
    of cylinder, **1**:149
    of ellipse, **1**:196
    of hyperbola, **1**:308
    of parabola, **1**:22, **1**:135, **2**:467, **2**:520
Dirichlet, Johann Peter Gustav Lejeune, **1**:**178–180**, **1**:*179*
    Fermat's last theorem and, **1**:179, **2**:669
    on functions, **1**:179, **2**:533
    on prime numbers, **1**:361, **2**:456
    Riemann influenced by, **2**:547
Dirichlet principle, **2**:549
Dirichlet problem, **1**:179, **2**:667
Dirichlet series, **2**:681
Dirichlet's theorem, **1**:179
Dirigible, **2**:545
Discontinuous fractions, **1**:28
Discontinuous functions, **1**:138, **1**:138*f*
    catastrophe theory and, **2**:615
    Fourier series for, **1**:265–266
    integration of, **1**:357, **2**:416
    See also Continuous functions
Discrete dynamical systems, **2**:537
Discrete mathematics, Erdös and, **1**:200
Discrete variables, **2**:584
Discriminant, **1**:**180**
    of conic section, **1**:136, **1**:180, **1**:327
    Fisher's work on, **1**:232
    of polynomial, **1**:180
    of quadratic function, **1**:180, **2**:520
    Seki's use of, **2**:566
    Sylvester's introduction of term, **2**:595

Disjoint sets, **1**:16
    measure and, **2**:415
Disjunction, **1**:383–384, **2**:597, **2**:630
    tautology and, **2**:605
Disk
    hyperbolic, **1**:311
    unit disk, **2**:549
Disk method, **2**:572, **2**:594
Dispersion, **1**:**180**. *See also* Standard deviation
Displacement, **1**:181
Distance, **1**:**181**
    in Banach space, **1**:45
    in Cartesian coordinates, **2**:517
    derivative of, **1**:174, **1**:323
    in hyperbolic geometry, **1**:311, **1**:379
    in metric spaces, **1**:240
    motion and, **2**:436
    in n-dimensional Euclidean space, **1**:286
    from Newton's Second Law, **2**:446
    on number line, **1**:104
    from origin, **1**:21
    in polar coordinates, **2**:572
    preserved by isometry, **1**:329, **2**:608, **2**:620
    Pythagorean theorem and, **2**:517
    between real numbers, **2**:536
    of ship from shore, **2**:609
    of stars, **1**:60
    of translation, **2**:487
    trigonometry for measurement of, **2**:578
    between two points, **1**:21
    vector representation of, **2**:642
    Zeno's dichotomy paradox, **2**:678
    See also Length; Metric
Distance formula, **1**:21, **1**:312
Distance function, of metric space, **1**:240
Distribution curve, **2**:585
Distribution functions. *See* Probability density functions
Distributions, theory of, **2**:561, **2**:562, **2**:667
Distributive property, **1**:**181**, **2**:438
    of algebraic expressions, **1**:15
    in Boolean algebra, **1**:75
    division and, **1**:182
    of a field, **1**:227
    of scalar multiplication, **1**:4, **1**:5
    for subtraction of products, **2**:591
    in symbolic logic, **2**:597
Distributivity, axiom of, **1**:36
Divergence of vector field, **2**:573, **2**:643
Divergent sequence, **1**:139, **2**:566
Divergent series, **1**:319
    alternating, **1**:19–20
Dividend, **1**:181, **2**:508
Divine proportion, **1**:269
Division, **1**:36, **1**:**181–182**
    of complex numbers, **1**:133, **1**:182
    of fractions, **1**:182, **1**:239
    of functions, **1**:182
    of imaginary numbers, **1**:313
    as inverse of multiplication, **2**:508
    of measurements, **2**:544
    of polynomials, **2**:556
    of powers, **2**:502

of quaternions, **2**:479
quotient resulting from, **1**:36, **1**:181, **2:508**
significant figures in, **2**:544
symbol for, **2**:480
by zero, **1**:182, **2**:457, **2**:534, **2**:681
Division algebra, **1**:110, **1**:111, **2**:525
Divisors, **1**:181, **2**:508, **2**:575
prime, **2**:575
proper, **1**:20, **2**:482
See also Factors
Dodecahedron, **2**:488, **2**:489, **2**:497
golden mean and, **1**:269
Dodgson, Charles. *See* Carroll, Lewis
Domain, **1:182–183**, **1**:271, **2**:400
of parametric equations, **2**:470
range and, **2**:532–533
of relation, **1**:182–183
of transformation, **2**:620
of trigonometric functions, **2**:627
of truth function, **2**:630
Donaldson, Simon K., **1:183–184**, **1**:368
Dot product, **2**:439, **2**:507, **2**:642
Douady, Adrien, **1:184**
Double implication, **1**:198, **2**:596
Double integral, **2**:437
Double-entry accounting, **2**:466
Double-slit experiment, **2**:522, **2**:522*f*, **2**:523
Doubling the cube. *See* Duplication of cube
Drag forces, **1**:305, **1**:306
Dual vector space, **2**:644
Duality
of De Morgan's laws, **2**:631
in projective geometry, **1**:262, **2**:490, **2**:499, **2**:588
Dummy variable, in integral, **1**:163
Duodecimal system, **1:184–185**
Duplication of cube, **1**:30, **1**:31, **1:185–186**, **1**:199, **1**:207–208
Eratosthenes and, **1**:185, **1**:199
Hippocrates of Chios and, **1**:185, **1**:300
Menaechmus' work on, **2**:417
Viète's work on, **2**:647
Dürer, Albrecht, **1:186–187**, **1**:*187*
Dynamic symmetry, **1**:269
Dynamical systems theory
Birkhoff's contributions to, **1**:66–67
deterministic, **2**:591
discrete, **2**:537
KAM theory, **1**:346, **2**:442
Keen's contributions to, **1**:336, **1**:337
Markov chains, **2**:400–401, **2:401–402**, **2**:591
nonlinear, **1:117–118**, **1**:*118*
Seifert conjecture and, **1**:349
stochastic processes and, **2**:591
See also Chaos theory
Dynamics. *See* Mechanics

E

e, **1:189**
power series estimate of, **2**:607
proved transcendental, **1**:290, **1**:372
E=mc², **1**:192–193, **1**:194, **2**:402, **2**:543, **2**:544

Earth
circumference, **1**:24, **1**:199, **1**:301, **2**:514, **2**:625
climatic change, **2**:424
flat, **2**:609
magnetism of, **1**:195
meridian, **2**:578
orbit, **1**:33, **1**:94, **2**:424
Ptolemy's *Geography,* **2**:513, **2**:514
radius, **2**:578
rotation around axis, **1**:92, **1**:93, **2**:423
shape, **1**:126, **2**:412, **2**:461, **2**:493, **2**:515, **2**:578
See also Moon
Earth beads, **1**:1–2
Eccentricity, **1**:135, **1**:136, **1:189–190**, **1**:196
of hyperbola, **1**:136, **1**:308, **1**:309
Eccentrics, **1**:26
Eckert, J. Presper, **2**:652
Ecliptic, **1**:93
Economics
Nicholas of Oresme's writings, **2**:448
Pareto's mathematical analysis, **2**:470
Ramsey's papers on, **2**:531
See also Game theory; Linear programming
Eddington, Arthur, **1**:193
Chandrasekhar and, **1**:115
eclipse expedition, **2**:540
Edges
of graph, **1**:129, **1**:236
of polygon, **2**:482
of polyhedron, **1**:209, **2**:497, **2**:617
EDSAC (English Electronic Delay Storage Automatic Calculator), **1**:88–89
EDVAC (Electronic Discrete Variable Automatic Computer), **2**:652
Effective computability, **1**:124
Egan, Greg, **2**:405
Egyptian mathematics, **1:190–191**, **1**:301
fractions in, **2**:534
quadratic equations in, **2**:519
Rhind papyrus, **1**:10, **1**:190, **1**:301, **2:546**, **2**:583
Eiges, Aleksandr, **1**:13
Eightfold way, **1**:51, **1**:368
Eilenberg, Samuel, **1**:18
Einstein, Albert, **1**:*191,* **1:191–194**
Broglie and, **1**:84
Gauss praised by, **1**:323
Gödel and, **1**:266, **1**:268
Levi–Civita and, **1**:364, **1**:365
Lorentz and, **1**:385
Maxwell and, **2**:412, **2**:413
Minkowski and, **2**:424, **2**:425–426
Noether and, **2**:449
on photoelectric effect, **1**:191, **1**:192, **1**:193, **1**:370
photon and, **2**:524
quantum mechanics and, **2**:524
symmetry principle of, **2**:599
Weyl and, **2**:661
See also Relativity, general; Relativity, special
"Either. . .or," **1**:383–384
Ejection singularities, **2**:442–443
Elastic solids, equation of state, **2**:613
Elasticity, theory of, **2**:494

ENIAC (Electronic Numerical Integrator and Computer), **1**:11, **1**:88, **2**:633, **2**:652

Enigma machine, **2**:632

Enriques, F., **1**:17

Ensembles, Gibbsian, **1**:265

Enthalpy, **1**:265

Entire functions
  Ahlfors' research on, **1**:9
  Liouville's theorem, **1**:377

Entirely normal numbers, **2**:455

Entropy, **1**:264, **1**:265
  Boltzmann equation, **1**:70, **1**:71, **2**:668
  information and, **1**:321, **1**:322, **2**:668
  Kolmogorov's use of, **1**:346
  See also Second law of thermodynamics

Enumeration, **1**:128–129

Epicycles, **1**:26, **1**:299, **1**:307, **1**:339, **2**:513

Epsilon, in set notation, **1**:16

Equal algebraic expressions, **1**:14

Equal sets, **2**:568–569, **2**:680

Equal sign, **2**:402, **2**:536

Equality
  Euclid's "common notions" of, **2**:500
  Peano axiom for, **2**:474, **2**:500
  rules of, **2**:438

Equations, **1**:197
  functions and, **1**:244–245
  negligible terms in, **2**:444
  in scientific analysis, **1**:21
  solution by successive bisection, **1**:329
  See also specific categories of equations

Equations of motion, **2**:416

Equations of state, **2**:612–613

Equiangular triangle(s). *See* Equilateral triangle(s)

Equidistance Postulate, **2**:468

Equilateral triangle(s), **2**:483, **2**:622, **2**:625
  Sierpinski's fractal, **2**:**574**
  solids built from, **2**:488

Equilibrium
  stable, **2**:463–464
  in statics, **2**:640
  thermal, **2**:611, **2**:612
  unstable, **2**:463

Equilibrium vector, of Markov chain, **2**:402

Equinoxes, **1**:93, **1**:152, **1**:299, **2**:563

Equipartition of energy, **1**:70

Equipollent sets, **1**:95, **1**:98

Equivalence
  Einstein's principle of, **2**:539–540
  logical, **1**:158–159, **1**:**197–198**, **1**:384, **2**:521, **2**:597, **2**:598
  as one-to-one correspondence, **1**:141
  topological, **2**:**615–616**

Equivalence relations, **1**:128, **1**:129
  in field Q, **2**:535

Equivalent fractions, **1**:239

Equivalent system of equations, **2**:600

Eratosthenes of Cyrene, **1**:**198–200**
  Archimedes and, **1**:28, **1**:199
  circumference of earth and, **1**:199, **1**:301, **2**:625
  duplication of cube and, **1**:185, **1**:199
  Euclid and, **1**:205
  sieve of, **1**:199, **2**:504, **2**:**575**

Erdös, Paul, **1**:**200–201**
  four-color map theorem and, **1**:236
  prime number theorem proved by, **2**:503
  Pythagorean theorem and, **2**:517

Erdös number, **1**:200

Ergodic theorem, Birkhoff's proof of, **1**:67

Ergodic theory, Bellow's work on, **1**:52

Erlang, Agner Krarup, **1**:**201**

Erlanger program, **1**:**202**, **1**:344

Error analysis, **2**:544–545

Error bars, **2**:545

Errors, **1**:**202–203**, **1**:353
  in communicating information, **1**:321
  Lambert's theory of, **1**:353
  negligible terms and, **2**:444
  in numerical analysis, **2**:458
  See also Approximation

Escape velocity, **1**:340

Escher, Maurits Cornelis, **1**:**203–204**
  hyperbolic geometry and, **1**:311, **1**:379
  Penrose and, **1**:203, **2**:481
  Reutersvard and, **2**:545

Eskimo number system, **2**:648

Estimation
  reliability of, **2**:544–545
  See also Approximation

Étale cohomology, **1**:165

Ether theory, **1**:370
  Huygens and, **1**:304
  Lorentz's use of, **1**:385
  Michelson-Morley experiment and, **2**:539

Ethical issues, for mathematicians, **1**:336–337

Euclid of Alexandria, **1**:**204–205**, **1**:*205*
  algebraic problems and, **1**:17
  Apollonius and, **1**:26
  axiomatic approach of, **1**:40
  Eudoxus and, **1**:208, **1**:209
  Hippocrates of Chios and, **1**:300
  infinity of primes proved by, **2**:456, **2**:504
  irrational numbers and, **1**:328
  Pappus and, **2**:467
  perfect numbers and, **2**:482
  proof and, **2**:509
  proportions and, **1**:342
  Pythagorean theorem and, **2**:515
  quadratic equations and, **1**:341, **2**:519
  See also Elements (Euclid)

Euclidean algorithm, **1**:122, **1**:176, **1**:206

Euclidean construction, **1**:**207–208**
  Gauss's 17–sided polygon, **1**:256
  Poncelet–Steiner theorem and, **2**:588
  See also Duplication of cube; Squaring the circle; Trisecting an angle

Euclidean geometry, **1**:207, **1**:208, **2**:452
  "common notions" of, **2**:500
  Gauss's innovation in, **1**:256
  Klein's view of, **1**:344
  plane, **2**:**486–487**
  projective geometry and, **2**:484
  Russell's introduction to, **2**:557
  solid, **2**:**579–580**
  Tarski's algorithm, **2**:604

See also Elements (Euclid); Euclid's axioms; Non-Euclidean
    geometry
Euclidean group, **1**:367
Euclidean norm, **1**:46, **1**:316, **1**:317
Euclidean plane, **1**:208
    topology and, **1**:287
Euclidean space, **1**:208
    manifolds and, **2**:398–399
    n-dimensional, **1**:286
    transformations of, **1**:202
    as vector space, **1**:4, **1**:175, **1**:208
Euclidean structure, of manifold, **1**:202
Euclid's algorithm, **1:205–206**
Euclid's axioms, **1**:204–205, **1:206–207**, **1**:261, **2**:452, **2**:468, **2**:499
    Russell's objection to, **2**:557
    See also Parallel postulate
Eudoxus of Cnidus, **1**:32, **1:208–209**
    Apollonius and, **1**:26
    duplication of cube and, **1**:185
    Euclid's solid geometry and, **2**:580
    Menaechmus and, **2**:417
    Zeno's influence on, **2**:678
Euler, Leonhard, **1:210–211**
    algebraic numbers and, **2**:618
    amicable numbers found by, **1**:20
    Diophantine equations and, **2**:456
    Fermat's last theorem and, **1**:211, **1**:221, **2**:669
    function concept and, **1**:244
    gamma function of, **2**:682
    graph theory of, **1**:82–83, **1**:128
    infinity of primes and, **2**:456
    Königsberg bridges problem and, **1**:82–83, **1**:128, **2**:617
    Lagrange and, **1**:351
    on logarithms of rational numbers, **1**:258
    perfect numbers and, **2**:482
    pi and, **2**:484
    prime numbers and, **2**:504
    symbols used by, **1**:6, **2**:402–403, **2**:592
    tides and, **2**:389
    trigonometry and, **2**:625
    Venn diagrams and, **2**:645
    zeta-function and, **1**:26, **2**:548
Euler characteristic, **1:209–210**, **2**:617
Euler polyhedrons, **2**:497
Eulerian path, **1**:82, **1**:83
Euler-Lagrange equation, **1:211**
Euler-Mascheroni constant, **1**:255, **2**:403
Euler's constant, **2**:682
Euler's formula, **1**:133
Euler's laws of motion, **1:212**
Euler's method, **1**:90, **1**:375
Eutocius, **1**:26, **1**:27, **1**:291
Even integers, **2**:431
    as denumerable set, **2**:451
    as sum of two primes, **2**:656
Even powers, **2**:502
Event horizon, **2**:481
Events, in probability theory, **2**:505, **2**:506
Evolute, **1**:26
Exact first-order differential equation, **1**:231
Exact numbers, **2**:544
Exceptional Lie groups, **1**:368
Excluded contradiction, principle of, **1**:136, **1**:137, **1**:197

Excluded middle, law of, **2**:596
    Brouwer's fixed point theorem and, **1**:86
    indirect proof and, **2**:513
    for infinite sets, **1**:325–326
    intuitionism and, **2**:629
    Lukasiewicz on, **1**:386
    in propositional logic, **1**:136, **1**:137, **1**:197
    violated by m-valued logics, **1**:137
EXCLUSIVE OR gate, **1**:383
Exhaustion, method of, **1**:29, **1**:32, **1**:110, **1**:204, **2**:459
    in Euclid's solid geometry, **2**:580
    Hilbert's rejection of, **1**:298
    squaring the circle and, **2**:583
Exhaustion, proof by, **2**:509
Existence, **1:212–213**
    Chwistek's critique of, **1**:125
    of derivative, **1**:139
    intuitionism and, **1**:213, **1**:325, **2**:662
    as a predicate, **2**:503, **2**:521
    of sets, **2**:680
    of solution to differential equation, **1**:231, **2**:485–486, **2**:565
Existence axiom, for plane geometry, **2**:487
Existential quantifier, **2**:502, **2**:503, **2**:520–521
Expansion coefficients, **2**:611
Expectation
    of probability density function, **2**:504
    of probability experiment, **2**:507
Experiment
    controlled, **2**:587
    random, **2**:506, **2**:507, **2**:640
    scientific, **2**:563
Experimental design
    Cox's contributions to, **1**:145–146
    Fisher's contributions to, **1**:231, **1**:232
Experimental variable, **1**:315
Explicit function, **1**:314
Exponential decay, **1**:214
Exponential equations, **1**:197, **1**:213, **2:553–554**
    solution with logarithms, **1**:383
Exponential function(s), **1:213–214**, **1**:245
    base e in, **1**:189
    complex, **1**:132, **1**:285
    as elementary functions, **1**:195
    inverse function of, **1**:189, **1**:382
    Johann Bernoulli's work on, **1**:59
    MacLaurin series for, **1**:189
    power series for, **2**:501
    See also Power function(s)
Exponential growth, **1**:213–214
Exponential notation. *See* Scientific notation
Exponential probability density function, **2**:504
Exponential regression, **1**:375
Exponential sequences, **1**:230
Exponential-time algorithms, **2**:465
Exponents, **1:213**, **2**:439
    fractional, **1**:213
    irrational, **1**:297
    negative, **2**:502
    in power expression, **2**:501
    in scientific notation, **2**:564
    transfinite, **2**:619
    zero as, **2**:502, **2**:681
    See also Powers

L'Hôspital, Guillaume Francois Antoine, **1**:366
    Agnesi's memoir about, **1**:8
    Arbogast and, **1**:28
    Bernoulli and, **1**:58, **1**:59, **1**:366
Lhuilier, Simon Antoine Jean, **1**:209, **1:366–367**
Lie, Marius Sophus, **1:368–369**
Lie algebra, **1**:367
Lie groups, **1:367–368**, **1**:369, **2**:411
    Cartan's work on, **1**:102, **1**:103, **1**:367–368, **1**:369
Light, **1:369–370**
    black holes and, **2**:541
    Broglie's ideas on, **1**:84
    diffraction of, **1**:370
    dispersion of, **1**:180
    electromagnetic theory of, **1**:194, **2**:413
    geodesic path of, **2**:550
    Huygens' wave theory, **1**:303, **1**:304, **1**:369
    intensity, units of, **2**:637
    inverse square law, **2**:501
    Michelson-Morley experiment, **1**:385, **2**:539, **2**:541–542
    Newtonian particle theory, **1**:194–195, **1**:304, **1**:369, **1**:370, **2**:524
    photoelectric effect, **1**:191, **1**:192, **1**:193, **1**:370
    photon, **1**:192, **1**:370, **2**:524, **2**:675
    quantum theory of, **2**:523–524, **2**:675
    reflection of, **1**:369, **1**:370
    refraction of, **1**:369, **1**:370, **2**:578, **2**:606
    relativistic bending of, **1**:193, **2**:540–541
    velocity of, **1**:192, **1**:195, **1**:385, **2**:492, **2**:542–544, **2**:637
    See also Optics
Light years, **1**:60
Limit(s), **1:370–371**
    of alternating series, **1**:19–20
    in analysis, **1**:21
    Cauchy's theory, **1**:90, **1**:108
    continuity and, **1**:138–139
    defined, **1**:370
    improper integral defined by, **1**:315
    indeterminate, **1:315–316**
    of integration, **1**:163
    L'Hôpital's rule, **1:365–366**
    one-sided, **1**:371
    rate of approach to, **1**:296
    of sequence, **1**:139
    of sequence of functions, **1**:357
    of sequence of partial sums, **1**:319
    of series, **1**:7
Limit cycle, **1**:139
Limit point, **1**:72–73, **2**:570
Lindelöf, Ernst Leonard, **1**:9, **1:371**
Lindemann, Carl Louis Ferdinand von, **1**:172, **1:371–372**, **2**:484, **2**:619
Line(s)
    as algebraic curve, **1**:17
    in art, **2**:403–404
    as degenerate conic section, **1**:135
    distance and, **1**:181
    equations for, **2**:489
    intersecting, **1**:25, **1**:373, **1**:374
    orthogonal, **1**:25
    parallel, **1**:22, **1**:181, **2**:484
    perpendicular, **1**:25, **2**:463
    in Plücker's analytic geometry, **2**:490
    slope-intercept form, **2**:577

Line graphs, **2**:585
Line of best fit, **1**:374–375
Line of symmetry, **2**:598
    of circle, **1**:125
Line segment
    area of, **1**:164
    Euclid's postulate, **2**:468
    length of, **1**:21
    midpoint of, **1**:21
Linear algebra, **1**:15, **1:372–374**
    history of, **1**:18, **1**:372
    See also Matrix (matrices); Systems of linear equations
Linear combination, **1**:375
Linear differential equations, **1**:173
    first-order, **1**:231
    second-order, **2**:565
    Weyl's work on, **2**:661
Linear equation(s), **1**:22, **1**:197, **1:374**, **2**:599
    algebraic curves defined by, **1**:17
    Bézout's work on, **1**:62
    defined, **1**:15
    Diophantine, **1**:176
    ordered pair solutions, **1**:374, **2**:599, **2**:601
    standard form, **1**:22
    See also Systems of linear equations
Linear expansion coefficient, **2**:611
Linear functions, **1:374–375**
Linear interpolation, **1**:325
Linear map, on vector space, **2**:411
Linear model, **2**:429
Linear momentum, **1**:212. *See also* Momentum
Linear motion, **2**:436
Linear operators, **1**:336
Linear programming, **1:375–376**
    Dantzig's work on, **1**:153–155
    Kantorovich's work on, **1**:335
Linear regression, **1**:374–375
Linear spaces, **1:376**
    abstract, **1:4–5**
    See also Vector spaces
Linear systems. *See* Systems of linear equations
Linear transformations, **2**:620, **2**:644
    matrix of, **1**:112, **2**:620, **2**:644
    of quadratic forms, **2**:425
    of statistical distribution, **2**:584
Linearly independent solutions, **2**:564, **2**:565
Linearly independent vectors, **2**:642
Liouville, Joseph, **1:376–377**
    Galois and, **1**:252, **1**:253
    transcendental numbers and, **1**:290, **1**:376, **1**:377–378
Liouville numbers, **1:377–378**
Liouville's theorem, **1**:377
Liquid. *See* Fluid mechanics
Lissajous figures, **1:378–379**
Listing, Johann Benedict, **2**:427
Liter, **2**:423, **2**:636, **2**:637, **2**:649–650
Literature and mathematics, **2:405–406**
    Aristophanes, **2**:583
    Queneau, **2:525–526**
Littlewood, John Edensor, **1**:105, **1**:106, **1**:127
    Hardy and, **1**:282, **1**:283, **1**:284, **2**:530
Littlewood problem, **1**:127
Lituus, **1**:222

M

hyperbolic, **1**:340, **1**:341, **2**:462
of iterative function, **2**:398
parabolic, **1**:340
See also Planetary motion
Order
of differential equation, **1**:173
of group, **1**:253, **1**:352–353
of operation, **2**:502
Order axioms, of real numbers, **1**:36
Order of approximation, **2**:444
Order types, **2**:619
Ordered field
of rational numbers, **2**:535
of real numbers, **2**:535–536
Ordered pair(s), **2**:533
Cartesian product, **1**:182, **2**:439, **2**:508
complex numbers as, **1**:281–282
denumerable set of, **1**:166
of real numbers, **1**:21
relations consisting of, **1**:182, **2**:400
solution to linear equation, **1**:374, **2**:599, **2**:601
See also Function(s)
Ordered set, **2**:533, **2**:569, **2**:683
partially ordered, **2**:683
totally ordered, **2**:683
well-ordered, **1**:95, **2**:619, **2**:679, **2**:680, **2**:683
Ordered triples, **2**:642
Ordering relation, **2**:569
Ordinal numbers, **1**:98, **2**:462, **2**:665–666
infinite, **2**:619
well-ordered sets and, **2**:619
Ordinary differential equations
dynamical systems determined by, **2**:591
first-order, **1**:231, **2**:476
second-order, **2**:564–565
See also Differential equations
Ordinate, **1**:21, **1**:105
Orientable manifolds, **2**:399, **2**:491
Oriented surface, geodesic curvature of, **1**:260
Origin, of Cartesian coordinates, **1**:21, **1**:104
Orthocenter, **1**:20
Orthogonal complement, **2**:508
Orthogonal coordinate system, **2**:572–573
Orthogonal curves, **2**:463
Orthogonal functions
approximations with, **2**:420
of Hilbert space, **1**:295
Kronecker delta and, **1**:348
See also Fourier series
Orthogonal lines, **1**:25
Orthogonal planes, **2**:463
Orthogonal surfaces, **2**:463
Orthogonal vectors, **1**:316, **2**:642
Orthographic projection, **2**:433
Orthonormal basis, of Hilbert space, **2**:508
Orthonormal vectors, **2**:642
Oscillations, **2**:463–464. *See also* Vibration
Osculating circle, **1**:259
Osculating plane, **1**:59
Ostensive definition, **1**:164
Ostwald, Friedrich, **1**:71, **1**:265
Otto, Valentin, **2**:546
Oughtred, William, **2**:438, **2**:576–577

Outcomes, **2**:506
Outer measure, **2**:415
Outlier, **2**:414, **2**:429
Oval. *See* Ellipse(s)
Overtones, **1**:56
Owen, G. E., **2**:679
Ozanam, Jacques, **2**:552

P

P versus NP problem, **2**:465–466
Pacioli, Luca Barolomes, **2**:466
Dürer and, **1**:186, **1**:187
Packing, Minkowski's work on, **2**:425
Pade approximations, **2**:420
Paganini, Nicolo, **1**:20
Paley, R. E. A. C., **2**:668
Pandiagonal cube, **2**:391
Pandiagonal magic square, **2**:393, **2**:395
Pappus of Alexandria, **2**:466–467
Apollonius criticized by, **1**:26
Guldin's theorem and, **1**:276
torus studied by, **2**:618
Pappus's theorem, **1**:276
Parabola(s), **1**:22, **1**:134f–135f, **1**:134–136, **2**:467
Apollonius' investigations of, **1**:26, **1**:27
degenerate, **1**:180
discriminant and, **1**:180, **1**:327
family of, **1**:218
quadratic functions and, **2**:519–520
vertex form of equation, **1**:131, **2**:520
Parabolic geometry, **1**:208, **2**:453, **2**:486, **2**:579
Parabolic mirror, **1**:27, **2**:467, **2**:520
Parabolic orbit, **1**:340
Parabolic partial differential equations, **2**:471
Parabolic spiral, **1**:222, **2**:582
Parabolic umbilic catastrophe, **1**:107
Paradigms, **2**:563
Paradoxes
of Burali-Forti, **2**:619, **2**:680
in Cantor's set theory, **2**:680
of Cramer, **1**:147
of D'Alembert, **1**:305
defined, **2**:678
dichotomy paradox, **2**:677
garage paradox, **2**:543
of Goodman, **1**:270
of Grelling, **1**:274
of Hausdorff, **1**:288
of Hilbert, **1**:295–296
Petersburg paradox, **1**:55
of Russell, **1**:235, **2**:557–558, **2**:680
Thomson lamp paradox, **2**:679
Tristram Shandy paradox, **2**:628
twin paradox, **2**:543
in voting methods, **2**:653
of Zeno, **1**:137, **1**:260, **1**:318, **1**:319, **2**:677–678, **2**:678–679
Parallax, of star, **1**:60
Parallel displacement, **1**:365
Parallel lines
distance between, **1**:181
in projective geometry, **2**:484
slopes of, **1**:22

Parallel postulate, **1**:163, **1**:204–205, **1**:206–207, **1**:208, **1**:261, **2**:452–453, **2:468–469**
     Bolyai and, **1**:71, **1**:207
     hyperbolic geometry and, **1**:311, **1**:379, **1**:380
     Khayyam and, **1**:342, **2**:452–453, **2**:561
     Saccheri's work on, **2**:453, **2**:468, **2**:561
     Wallis' trivial analysis, **2**:655
     See also Non-Euclidean geometry
Parallelepiped, **2**:497
     volume of, **1**:318
Parallelogram rule, **2:469**, **2**:642
     for forces, **2**:589
Parametric equations, **2:469–470**
     of ellipse, **1**:196
     for motion, **1**:214, **2**:469
     of space-filling curve, **2**:477
Parentheses
     associative property and, **1**:38
     commutative property and, **1**:131
     with factorials, **1**:218
     in logical expressions, **2**:597
     in multiplication, **2**:403, **2**:438
Pareto, Vilfredo Frederico Damaso, **2:470–471**
Pareto condition, **2**:653
Parmenides, **2**:677, **2**:678
Partial differential equations, **1**:173, **2**:471
     Calderón's work on, **1**:90, **1**:91
     Cauchy–Kovalevskaya theorem, **1**:348
     d'Alembert's work on, **1**:152
     Dirichlet problem for, **1**:179
     elliptic, **2**:471, **2**:549
     Jacobi's work on, **1**:332
     Lie's work on, **1**:368
     Volterra's work on, **2**:649
     Yang-Mills equations, **1**:183, **1**:184
     See also Differential equations
Partial differentiation, **1**:167, **1**:315, **2:471**, **2**:642
Partial sums, **1**:319
     of alternating series, **1**:19, **1**:20
Partially ordered set, **2**:683
Particle accelerators, **2**:543
Particle-wave duality, **1**:69, **1**:85, **2**:675
Particular solution, of differential equation, **2**:564
Partitions, **2**:595
Pascal (unit of pressure), **2**:637
Pascal, Blaise, **2:471–473**
     Apollonius and, **1**:26
     Brianchon's theorem and, **1**:82
     calculating machine of, **1**:88
     Fermat and, **1**:219, **1**:220
     Huygens and, **1**:303
     hydraulics and, **1**:304
     Mersenne and, **2**:418
     probability theory and, **2**:505
     vacuum and, **2**:551
Pascal's law, **1**:304
Pascal's mystic hexagram, **2**:472
Pascal's principle, **2**:472
Pascal's triangle, **2**:472, **2:473–474**
     binomial theorem and, **1**:65, **1**:66, **1**:129, **2**:472, **2**:473
Pasch, Moritz, **2:474**
Pasch's axiom, **2**:474
Path, length of, **1**:181

Path integral, **1**:324
Pauli, Wolfgang, **1**:222, **1**:223
Peacock, George, **1**:18
Peano, Giuseppe, **1**:163, **2:476–477**
     Russell and, **2**:557
Peano axioms, **1**:163, **2:474–475**, **2**:500, **2**:510, **2**:666
     recursion and, **2**:537
     Russell and Whitehead influenced by, **2**:663, **2**:664
Peano curve, **2:475–476**
Pearson, E. S., **2**:448
Pearson, Karl, **1**:254, **2:477–479**
Pearson correlation coefficient, **2**:478, **2**:587
Peirce, Charles Sanders, **2:479–480**, **2**:631
Pell, John, **2:480**
Pell's equation, **1**:63, **1**:176, **2**:455, **2**:480
Pendulum, center of oscillation, **2**:551
Pendulum clock, of Huygens, **1**:304, **2**:418–419
Penrose, Roger, **2:480–481**
     Escher and, **1**:203, **2**:481
     Reutersvard and, **2**:545
Penrose staircase, **2**:480, **2**:481
Penrose tilings, **2**:480, **2:481–482**, **2**:608
     golden mean and, **1**:269, **2**:481
Pentagonal numbers, **1**:228–229, **1**:228*f*–229*f*
Pentagons, tilings of, **2**:608
Pentagrams, **1**:269
     magic, **2:392–393**
Percentage, **2**:512
Percentile rank, **2**:584
Perfect magic cube, **2**:390–391
Perfect numbers, **2:482**
     Mersenne primes and, **2**:419
Perfect square, in modular arithmetic, **1**:361
Perihelion, precession of, **1**:193, **2**:539
Perimeter, **2:482–483**
     of ellipse, **1**:196
     of rectangle, **2**:482
     of snowflake curve, **2**:579
     of triangle, **2**:622
     See also Circumference
Period of oscillation, **2**:464
Periodic functions
     elliptic, **1**:310
     trigonometric, **2**:576, **2**:627
     See also Fourier series; Harmonic analysis
Periodic table of elements, **1**:324
Periodic tiling, **2**:481
Permittivity of vacuum, **2**:414
Permutation groups, **1**:112, **1**:276
     model theory and, **2**:430
Permutations, **1**:128, **2**:505
Perpendicular lines, **1**:25, **2**:463
Perrin, Jean Baptiste, **1**:192
Perspective, **2**:404, **2:483–484**
     Alberti's laws of, **1**:12
     Desargues' ideas on, **1**:168
     Dürer's use of, **1**:186, **1**:187
     Taylor's work on, **2**:606
Perturbation theory
     Kato's work in, **1**:335–336
     of planetary orbits, **2**:461
Petersburg paradox, **1**:55
Phase change, **2**:612, **2**:613

Phase rule, **1**:264

Phase space, Hamiltonian, **2**:399

Phenomenology, of Husserl, **1**:302–303

Phi. *See* Golden mean

Phidias, **1**:28

Philosophy and mathematics, **2**:**407**
>   Menaechmus' theories, **2**:417
>   Tristram Shandy paradox, **2**:628
>   *See also* Formalism; Intuitionism; Logicism; Platonism

Phlogiston theory, **1**:119

Photoelectric effect
>   Einstein's paper on, **1**:191, **1**:192, **1**:193, **1**:370
>   Hertz's discovery of, **1**:370

Photon, **2**:524, **2**:675
>   Compton effect, **1**:370
>   photoelectric effect and, **1**:192

Physics and mathematics, **2**:**407–408**
>   symmetries, **2**:599, **2**:662, **2**:675
>   *See also* Electromagnetism; Mechanics; Optics; Quantum mechanics; Relativity; Thermodynamics

Pi, **1**:125, **1**:126, **1**:172, **2**:482, **2**:**484–485**
>   algebraic numbers and, **2**:583
>   Apollonius' calculation of, **1**:26
>   Archimedes' calculation of, **1**:29, **1**:30, **1**:32, **2**:484
>   as computable number, **2**:634
>   Euler's identity, **1**:51
>   Henry P. Babbage's calculation of, **1**:88
>   Hindu value of, **1**:37
>   Monte Carlo calculation of, **2**:435
>   in multiplicative notation, **2**:403
>   Pell's treatise on, **2**:480
>   power series estimate of, **2**:607
>   proved irrational, **1**:172, **1**:353, **2**:484, **2**:583
>   proved transcendental, **1**:371, **1**:372, **2**:484, **2**:618–619
>   Ptolemy's value of, **2**:513
>   randomness of digits, **2**:532
>   Seki's value of, **2**:566
>   squaring the circle and, **2**:583
>   from Taylor series expansion, **1**:319

Picard, Charles Emile, **1**:290, **2**:**485**

Picard group, **2**:485

Picard's method, **2**:**485–486**

Picard's theorem, **2**:485
>   Borel's proof, **1**:76

Pie chart, **2**:585–586

Pierce, B., **1**:18

Piero della Francesca, **2**:483

Pigeonhole principle, **2**:**486**

Pitiscus, Bartholomeo, **2**:**486**

Place-value numeral systems, **2**:456

Planck, Max
>   blackbody radiation and, **2**:523–524
>   photoelectric effect and, **1**:192, **1**:370
>   Second law of thermodynamics and, **2**:613–614

Planck's constant, **2**:524, **2**:637

Plane(s)
>   cross section and, **2**:580
>   Euclidean, **1**:208, **1**:287
>   Gaussian curvature of, **1**:257
>   orthogonal, **2**:463
>   points in, **1**:105
>   systems of equations and, **2**:600

Plane geometry, **1**:208, **2**:**486–487**
>   Archimedes' treatises on, **1**:28–29
>   *See also* Euclidean geometry

Plane of symmetry, **2**:598

Plane trigonometry. *See* Trigonometry

Plane-filling curves, **2**:475–476

Planetary motion, **1**:21
>   Apollonius on, **1**:27
>   Archimedes on, **1**:29
>   Aristarchus on, **1**:33
>   Aryabhata the Elder on, **1**:37
>   Brahmagupta on, **1**:81
>   calculus and, **1**:89
>   eccentricities of orbits, **1**:190
>   elliptical, **1**:21, **1**:338, **1**:339–340, **2**:461, **2**:500
>   history of theories, **2**:563–564
>   Kepler's laws, **1**:337, **1**:338, **1**:339–341**, **2**:500–501, **2**:564, **2**:606
>   Lagrange's calculations, **1**:351, **1**:352
>   Laplace's work on, **1**:355
>   Menaechmus' theory, **2**:417
>   *n*-body problem and, **2**:**442–443**
>   Neptune discovered, **1**:60
>   Newtonian gravitation and, **2**:447
>   Poisson's work on, **2**:493
>   Ptolemaic theory, **2**:513–514, **2**:563, **2**:598
>   Pythagorean theory, **2**:515, **2**:516
>   relativistic effects, **1**:193, **2**:539
>   sexagesimal numeration and, **2**:571
>   Tycho on, **1**:80
>   *See also* Celestial mechanics; Copernicus, Nicolaus; Orbit(s); Three-body problem

Plato, **2**:407, **2**:**487–488**, **2**:489
>   Archytas' friendship with, **1**:30
>   Aristotle and, **1**:34
>   duplication of cube and, **1**:185, **2**:417
>   Eratosthenes and, **1**:198
>   Eudoxus and, **1**:208
>   Zeno and, **2**:677, **2**:678

Platonic forms, **2**:487–488, **2**:629–630

Platonic solids, **2**:**488–489**
>   golden mean and, **1**:269

Platonism, **1**:163, **2**:629–630
>   Adelard of Bath, **1**:7
>   Bernays, **1**:54
>   Gödel, **1**:268
>   Hypatia, **1**:306
>   Nicomachus of Gerasa, **2**:449

Playfair, John, **2**:468

Playfair's Axiom, **1**:205, **2**:468

Plücker, Julius, **1**:17, **2**:**489–490**
>   Klein's association with, **1**:344
>   Lie influenced by, **1**:368

Plücker formulas, **1**:17

Plurality vote, **2**:653

Plus sign, **1**:5, **1**:6, **2**:402, **2**:592

P-norms, **1**:317

Poetry and mathematics, **2**:405

Poincaré, Henri, **2**:**491–492**
>   Ajdukiewicz's philosophy and, **1**:12
>   automorphic functions and, **2**:485, **2**:492
>   axiom of choice and, **1**:280

Postulates, **2**:499–500
    proof and, **2**:509
    See also Axioms
Potential energy, **1**:234, **2**:402
Potential infinite, **1**:326, **2**:628
Potential theory
    Cartan's work on, **1**:103
    Dirichlet's work on, **1**:179
    Laplace's creation of, **1**:356
Pound, **2**:637
Power function(s), **2**:500–501. *See also* Exponential function(s)
Power regression, **2**:500, **2**:501
Power rule, for derivatives, **1**:174
Power series, **2**:501
    of analytic functions, **1**:132, **2**:657, **2**:658
    of elliptic functions, **2**:658
    See also Taylor's series
Power set, **2**:452
Powers, **1**:213, **2**:439, **2**:501–502, **2**:507, **2**:553
    of an unknown, **1**:19
    of a set, **1**:95
    See also Exponents
Powers of ten, **2**:501–502
    scientific notation, **1**:382, **2**:502, **2**:544, **2**:564
    See also Decimal (Hindu-Arabic) system
Pragmatism, of Peirce, **2**:479
Prandtl, Ludwig, **1**:233, **1**:305
Precession
    of equinoxes, **1**:93, **1**:152, **1**:299, **2**:563
    of perihelion of Mercury, **1**:193, **2**:539
Precision. *See* Approximation; Reliability of digits and calculations
Predicate, **2**:502–503
Predicate calculus. *See* Predicate logic
Predicate logic, **2**:502–503
    quantifiers in, **2**:502, **2**:503, **2**:520–521
    See also Quantification theory
Predictor variable, **1**:140
Pre-images, **2**:486, **2**:620
Premises, **1**:40, **1**:318, **2**:508, **2**:509, **2**:598
    major, **1**:34, **2**:593
    minor, **1**:34, **2**:593
    self-evident, **2**:610
    of syllogism, **2**:594
    synonyms for, **2**:596
    See also Axioms
Pressure
    atmospheric, **2**:617, **2**:618
    in a fluid, **1**:304–305, **1**:306
    of gas, **2**:611, **2**:613
    unit of, **2**:637
    volume and, **2**:649
Prime divisors, **2**:575
Prime factors, **1**:217
Prime ideals, **1**:354
Prime magic pentagram, **2**:392–393
Prime magic square, **2**:394, **2**:395
Prime number theorem, **2**:456, **2**:503, **2**:504
    complex analysis and, **2**:503, **2**:682
    Goldbach's conjecture and, **1**:269
    Hadamard's proof, **1**:279, **2**:548, **2**:682
    Landau's proof, **1**:354
    Vallée–Poussin's proof, **2**:456, **2**:548, **2**:639, **2**:682
    von Koch's writings on, **2**:650

Prime numbers, **2**:503–504, **2**:575
    in arithmetic progressions, **1**:361
    Chebyshev's theorem, **1**:200
    Dirichlet's theorem, **1**:179
    distribution of, **1**:120, **1**:297, **1**:354, **2**:456
    Fermat's work on, **1**:219, **1**:220
    genus theorem, **2**:432
    Goldbach's conjecture, **1**:268–269
    infinite number of, **1**:204, **2**:456, **2**:504
    Legendre's work on, **1**:360
    Mersenne primes, **2**:419
    perfect numbers and, **2**:482
    quadratic reciprocity law, **1**:256, **1**:360, **1**:361, **2**:432
    Riemann hypothesis and, **1**:297, **2**:503, **2**:548, **2**:682
    sequence of, **2**:566
    sieve of Eratosthenes, **1**:199, **2**:504, **2**:575
    Waring's conjectures, **2**:656
Principal diagonal, of determinant, **1**:171
Principal root, **2**:554
Principia mathematica (Newton), **2**:445
    Châtelet's translation of, **1**:120
    Cotes' work on second edition, **1**:142
Principia Mathematica (Russell and Whitehead), **1**:235, **1**:266, **2**:531, **2**:558, **2**:630, **2**:663, **2**:664
    Wiener's analysis of, **2**:666–667
Principle of equivalence, **2**:539–540
Principle of excluded contradiction, **1**:136, **1**:137, **1**:197
Prior probabilities, **1**:49, **1**:50
Probability
    conditional, **1**:49, **2**:506
    defined, **2**:506
    in quantum mechanics, **2**:523, **2**:675
    stochastic processes and, **2**:591
Probability density functions, **2**:504–505
    chi square distribution, **1**:121–122
    linear transformations of, **2**:584
    in Monte Carlo method, **2**:435
    Poisson distribution, **1**:201, **2**:492–493
    range of, **2**:533
    Witch of Agnesi as, **2**:670
    See also Normal (Gaussian) distribution
Probability measure, **2**:455
Probability theory, **2**:505–507
    conditional probabilities, **1**:49, **2**:506
    in Fréchet's abstract spaces, **1**:240–241
    independent events, **1**:49, **2**:506–507
    Pascal's work on, **2**:472, **2**:473–474
Problem of Points, **2**:474
Problem solving, Pólya's book on, **2**:495, **2**:496
Proclus, **1**:206, **2**:452, **2**:468, **2**:609, **2**:677
Product, **1**:36, **2**:507–508
    infinite, **2**:682
    of powers, **2**:502
    See also Multiplication
Product topology, **2**:508
Projectile motion
    Cardano's work on, **1**:98
    Galileo's work on, **1**:250
    parabolic path, **2**:467, **2**:520
    parametric equations for, **1**:214
    Tartaglia's work on, **2**:605
    Torricelli's work on, **2**:617

Pythagoreans, **1**:51, **2**:515
    amicable numbers and, **1**:20
    Archytas of Tarentum, **1**:*30–31*, **1**:*31,* **1**:185, **1**:208
    golden mean and, **1**:269
    Hippocrates of Chios and, **1**:300
    incommensurable numbers and, **1**:162
    infinities and, **1**:26
    logic of, **2**:517
    musical theory, **1**:51, **1**:229, **1**:328, **2**:515
    neo–Pythagorean tradition, **2**:449
    number theory of, **2**:515, **2**:**516**
    Zeno's disagreements with, **2**:677, **2**:678

Q

Q methodology, **2**:408–409
Quadrants
    of coordinate plane, **1**:21, **1**:104
    trigonometric functions and, **2**:624–625
Quadratic equations, **1**:197, **2**:**519**
    Al-Khwarizmi's work on, **1**:19, **2**:519
    in analytic geometry, **2**:489
    completing the square, **1**:131, **2**:519
    Diophantine, **1**:176
    discriminant, **1**:180, **2**:520
    Hilbert's problem about, **1**:297
    imaginary number solutions, **1**:314
Quadratic extension, **2**:527
Quadratic forms
    Hermite's work on, **1**:290
    Minkowski's work on, **2**:424, **2**:425
Quadratic formula, **1**:131, **2**:519, **2**:554
Quadratic functions, **1**:245, **2**:467, **2**:**519–520**
Quadratic reciprocity law, **1**:256, **1**:360, **1**:361, **2**:432
Quadratic residue, **2**:432
Quadratic sieve, **1**:361
Quadrature, **2**:458
Quadrature of lune, **1**:299, **1**:300
Quanta, **2**:523–524
Quantification theory
    Peirce's work on, **2**:479
    undecidability of, **1**:123
    See also Predicate logic
Quantifiers, **2**:502–503, **2**:**520–521**
Quantitative predictions, **2**:563
Quantum black holes, **2**:541
Quantum chaos, **2**:548–549
Quantum computer, **2**:465
Quantum electrodynamics
    Dirac's contributions, **1**:177
    Feynman's contributions, **1**:222, **1**:223–224
    Yang-Mills equations and, **1**:183
Quantum field theory
    Morse theory and, **1**:211
    quantum electrodynamics, **1**:177, **1**:183, **1**:222, **1**:223–224
    Yang-Mills theory, **1**:183, **1**:184, **2**:**675–676**
Quantum mechanics, **2**:416, **2**:522*f,* **2**:**522–524**
    Bohr atom, **1**:69, **1**:70
    Broglie and, **1**:84, **2**:524
    Cauchy's integral theorem in, **1**:109
    Dirac theory, **1**:177–178, **1**:184
    Einstein's rejection of, **1**:194

    group theory and, **1**:368, **2**:662–663
    Hermite's quadratic forms and, **1**:290
    invariants in, **2**:662
    light and, **1**:195
    matrix mechanics, **1**:289, **1**:290, **2**:416
    measurement in, **2**:523, **2**:651
    spinors in, **2**:525, **2**:662
    topology and, **1**:184
    von Neumann's work on, **2**:651
    See also String theory; Subatomic particles
Quarks, **1**:51, **1**:224, **1**:368, **2**:671, **2**:675
Quart, **2**:636, **2**:649–650
Quartic curves, Waring's work on, **2**:656
Quartic equations, **1**:197, **2**:**524–525**
    Cardano's solution of, **1**:18, **1**:96, **1**:97–98, **2**:525
Quasar, **2**:540–541
Quasicrystals, **2**:482
Quasiperiodic orbit, **1**:139
Quaternions, **1**:110, **1**:111, **1**:282, **2**:**525**
    division of, **2**:479
    symplectic groups and, **1**:367
Queneau, Raymond, **2**:**525–526**
Quetelet, Adolphe, **2**:**526–527**
Queuing theory, **1**:201
Quintic equations, **2**:**527–528**
    Abel–Ruffini theorem, **2**:556
    Abel's contributions, **1**:2, **1**:3–4, **2**:527, **2**:554
    Galois theory and, **1**:253, **2**:527–528
    Hermite's solution, **1**:290, **2**:528
    symmetry of coefficients, **2**:599
Quotient, **1**:36, **1**:181, **2**:**508**. *See also* Division

R

Radar systems, **2**:495
Radians, **1**:24, **1**:25, **1**:141, **2**:**529,** **2**:625
    as SI unit, **2**:637
Radiation
    electromagnetic waves, **1**:195, **1**:370, **2**:541
    Rutherford's work on, **2**:558–559
    SI unit of, **2**:637
Radical conventionalism, **1**:12
Radical equations, **1**:197
Radicals
    in Galois theory, **1**:253, **2**:528
    polynomial equations and, **2**:527, **2**:528, **2**:556
    See also Roots of a number
Radioactivity, **2**:559
Radius
    of circle, **1**:22, **1**:125–126, **2**:482
    of circular motion, **2**:447
    of convergence, **2**:607
    of curvature, **1**:257
    of Earth, **2**:578
    radian measure and, **2**:529
    of sphere, **2**:582
Rainbow, **1**:180
Ramanujan, S. I., **2**:**529–530**
    Hardy and, **1**:282, **1**:283–284, **2**:529, **2**:530
    pi and, **1**:172
    prime number theorem and, **1**:279
    Weil conjectures and, **1**:165
Ramsey, Frank Plumpton, **2**:**531**

Scaling
in dynamical systems, **1**:117
fractal dimension and, **1**:175, **1**:238
Schechtman, Dan, **2**:482
Schemes, theory of, **1**:17, **1**:176
Scheutz, Edvard, **1**:88
Scheutz, George, **1**:88
Schneider, Theodor, **1**:259
Schreier, Otto, **1**:246
Schröder, Ernst, **2**:479, **2**:631, **2**:666
Schrödinger, Erwin
Broglie and, **1**:84
Dirac and, **1**:177
perturbation theory and, **1**:336
Schrödinger operators, **1**:336
Schrödinger wave equation, **2**:416, **2**:524
Dirac's extension of, **1**:177
Heisenberg theory and, **2**:651
Monte Carlo estimates, **2**:435
Schroeppel, Richard, **2**:394
Schubert's enumerative geometry, **1**:297
Schwartz, Laurent, **1**:274, **2**:561–562
Schwarz reflection principle, **2**:599
Schwarzschild, Karl, **2**:541
Schwarzschild radius, **1**:340–341
Schweitzer, P. A., **1**:349
Schwinger, Julian, **1**:223–224
Science fiction, **2**:405–406
Scientific method, **2**:562–564
Carnap and, **1**:100
Thales and, **2**:608
Scientific notation, **2**:502, **2**:564
in finding logarithms, **1**:382
significant figures in, **2**:544
S-circles, **1**:259–260
Screw, **1**:28, **1**:29, **1**:30
S-dimensional Hausdorff measure, **1**:378
Secant, etymology of, **1**:230
Secant function, **1**:247, **1**:248, **2**:623, **2**:626
hyperbolic, **1**:310
inverse, **1**:327
tables of, **2**:624–625
Secant method, **2**:554
Second category set, **1**:378
Second derivative, **1**:173
Second derivative test, **1**:180, **1**:215
Second fundamental tensor, **1**:257
Second law, Newton's, **1**:212, **2**:446–447
Second law of thermodynamics, **1**:264, **1**:265, **2**:613–614
Second-order approximation, **2**:501
Second-order ordinary differential equations, **2**:564–565
Seconds
of angle, **1**:25, **2**:571
of time, **2**:421, **2**:422, **2**:423, **2**:636, **2**:637
Seeley, Robert T., **1**:92
Segre, C., **1**:17
Seifert conjecture, **1**:349
Seki, Kowa, **2**:565–566
Selberg, A., **2**:503
Self-adjoint operators, **1**:336, **2**:661
Self-reference, **1**:267

Self-similarity, **2**:575
of logarithmic spiral, **1**:58
Mandelbrot's ideas on, **2**:396–397
of Sierpinski triangle, **2**:574
See also Fractals
Self-similarity dimension, **1**:286, **1**:287
Semantics
of Chwistek, **1**:125
Russell's theory of types and, **2**:531
of Tarski, **2**:603–604
Semicircle, angle inscribed in, **2**:609
Semiconductor, **1**:194
Semilogarithmic graph paper, **1**:375
Semimajor axis, of ellipse, **1**:189, **1**:196
Semiminor axis, of ellipse, **1**:196
Semi-perfect magic cube, **2**:390–391
Semisimple algebra, **2**:657
Semisimple Lie groups, **1**:102
Sentences. *See* Propositions (statements)
Sentential connectives. *See* Connectives
Separable differential equation, **1**:231
Sequence, **2**:566–567
arithmetic, **1**:37, **1**:230, **1**:361
Bolzano-Weierstrass Theorem and, **1**:73
of complex numbers, **2**:681
convergence of, **1**:109, **1**:139, **2**:566–567
defined, **1**:230, **2**:566
finite, **1**:230–231
as function, **2**:566
of functions, **1**:357
harmonic, **2**:567
negligible terms in, **2**:444
of partial sums, **1**:319
rate of growth, **1**:296
vs. series, **2**:567
Series, **1**:6–7, **2**:592
arithmetic, **1**:37, **1**:230, **2**:592
finite, **1**:230–231, **1**:260
harmonic, **1**:19–20
negligible terms in, **2**:444
See also Infinite series
Serre, Jean-Pierre, **1**:164, **2**:567–568
Set(s), **2**:570–571
algebra of, **1**:16 (*see also* Boolean algebra)
compact, **1**:77, **2**:415
difference of, **2**:569–570, **2**:570f, **2**:592
disjoint, **1**:16, **2**:415
elements of, **1**:16, **2**:568, **2**:680
equal, **2**:568–569, **2**:680
finite, **2**:451, **2**:569
inductive, **2**:511
intersection of, **1**:16, **1**:75, **1**:75f, **2**:569, **2**:569f
isomorphism of, **1**:329
Lebesgue measurable, **1**:358
notation for, **1**:16
open, **1**:45, **2**:415, **2**:573
ordered, **2**:533, **2**:569, **2**:683
partially ordered, **2**:683
power set, **2**:452
totally ordered, **2**:683
union of, **1**:7, **1**:16, **1**:75, **1**:75f, **2**:569, **2**:569f
Venn diagrams of, **2**:645, **2**:645f–646f, **2**:646

Solid geometry, 1:208, **2:579–580**
    Archimedes' writings on, 1:29
    Platonic solids, **2:488–489**
    See also Euclidean geometry
Solidification temperature, 2:612
Solids, volume of, 2:650
Solids of revolution, 2:593*f*, **2:593–594**
Solids with known cross sections, **2:580**
Solidus, 1:181, **2:508**
Solution
    of algebraic equation, 1:14, 1:15, 1:21–22
    of linear equation, 1:374, 2:599, 2:601
    of system of linear equations, 1:373, 2:410, 2:601
    See also Roots of an equation
Solution set, 1:374, 2:599
"Some," 2:502, 2:520
Soroban, 1:1, 1:87
Sosigenes, 1:92, 1:93
Sound waves, Bernoulli and, 1:56
Space, **2:581**
    geometric structure of, 1:202
    Zeno's paradoxes and, 2:679
Space shuttle
    Challenger investigation, 1:222, 1:224
    weightlessness in, 2:540
Space-filling curves, 2:475–477
Space-time, 2:539, 2:540, 2:542, **2:581–582**
    instantaneous events in, **1:322–323**
    Minkowski, 2:424, 2:425, **2:426**
    Noether's invariance theory and, 2:450
Space-time interval
    Lorentz transformations and, 1:385
    in Minkowski space-time, 2:426
Space-time singularity, 2:480–481
Special orthogonal group $SO(n)$, 1:367, 1:368
Special theory of relativity. *See* Relativity, special
Special unitary group $SU(n)$, 1:367, 1:368
    in quantum field theory, 2:675
Specific gravity, 1:29
Specific heats, 2:612
Specker, E., 1:54
Spectral lines, magnetic field and, 1:385
Spectral theory of operators, 1:336
Spectroscopy, Plücker's work on, 2:489, 2:490
Speed
    as rate, 2:535
    relative, 2:679
    See also Velocity
Sphere, **2:582**
    circumference, 1:125
    conformal geometry of, 2:549
    cross section, 2:580
    defined, 2:582
    diameter, 1:172, 2:582
    equation of, 1:23, 2:582
    Euler characteristic of, 1:209
    fundamental group of, 1:246
    Gaussian curvature of, 1:257
    geodesic on, 1:261
    Hausdorff paradox, 1:288
    as manifold, 1:175, 2:399
    as orientable manifold, 2:491

    Riemann curvature of, 2:550
    solid angle and, 2:529
    stereographic projection and, 2:588
    surface area, 2:582
    as surface of revolution, 2:593, 2:593*f*
    topological properties of, 2:616
    unit, 2:508, 2:642
    volume, 1:29, 2:501, 2:582
    See also Ball
Spherical coordinates, 2:463, 2:495
    for multiple integral, 2:437
Spherical functions, 1:289
Spherical geometry, 1:208, **2:452, 2:453, 2:469**
Spherical mirror, 1:27
Spherical triangles, 1:360, 2:550, 2:622
    Bolyai's work on, 2:453
    Hipparchus' use of, 1:299
    Napier's analogies, 2:442
Spherical trigonometry, 2:442, 2:576
    Pitiscus' writings on, 2:486
    Regiomontanus' writings on, 2:538
    Snell's writings on, 2:578
Spinors, 1:103, 2:525, 2:641, 2:662
Spiral, **2:582**
    of Archimedes, **1:30**, 2:495, 2:582
    of Fermat, **1:222**
    general equation for, 1:222
    logarithmic, 1:58, 1:269, 2:418, 2:582
    parabolic, 1:222, 2:582
    rotational symmetry of, 2:555
Spline interpolation, 1:325, 2:420
Square(s), **2:582–583**
    area of, 2:501, 2:580
    completing, 1:19, **1:131–132**, 2:519
    magic, **2:393–395**, 2:566
    tilings of, 2:481
    as unit of area, 1:31–32, 2:583
Square matrix, 1:15, 1:133
Square meters, 1:32, 2:583
Square numbers, 1:228, 1:228*f*, 1:229, 1:230, 2:473
    sequence of, 2:566
Square roots, 2:553–554, **2:583**
    approximation by iteration, 1:329, 2:566
    approximation of Hero, 1:291, 1:292
    computational algorithm, 1:228
    irrational, 1:162, 1:312
    of minus one, 1:313
    of two, 1:328, 2:457, 2:515, 2:516, 2:534, 2:618
Square wave, 1:138
Square wheel, 1:107
Square-integrable function, 1:99
Squaring the circle, 1:172, 1:207–208, **2:583–584**
    Anaxagoras' attempt, 1:23, 1:24, 2:583
    Euclid's Fifth Postulate and, 2:453
    Hippocrates' attempt, 1:300
    Lambert's investigation of, 1:353
    Lindemann's solution, 1:371, 1:372, 2:484, 2:619
Stability theory, 1:388, 2:442
Stable equilibrium, 2:463–464
Stadium paradox, 2:678–679
Standard deviation, **2:584**, 2:586
    of Gaussian distribution, 2:504, 2:584, 2:586

Subtraction, **1**:36, **2:591–592**
    complex numbers, **1**:133, **2**:469, **2**:591–592
    fractions, **1**:239
    functions, **1**:245, **2**:591
    imaginary numbers, **1**:313
    matrices, **2**:409, **2**:591
    measurements, **2**:544
    vectors, **2**:469, **2**:591
    whole numbers, **2**:666
Subtrahend, **2**:591
Successive approximations, method of, **2:485–486**
Successive bisection, method of, **1**:329
Successor element, **2**:619
Successor function, **1**:124
Successor relation, **2**:474, **2**:475, **2**:500, **2**:537
Successor statement, **2**:511
Sufficient condition, **1**:198, **2**:443, **2**:598
Sum, **1**:5, **1**:36, **2:592**
    of infinite series, **1**:318–319
    See also Addition
Sum over histories, **2**:675
Sum rule, for derivatives, **1**:174
Sumerian numeration system, **1**:301, **2**:571
Summands, **1**:5, **1**:6
Summation, **2**:592
Summation symbol, **1**:6–7, **2**:592
Sun
    bending of light by, **2**:540
    distance to, **1**:33, **2**:625
    spectrum of, **1**:370
    See also Planetary motion
Sundmann, Karl F., **1**:365
Superabundant numbers, **2**:482
Superfluidity, **1**:224
Superset, **1**:16
Superstrings. *See* String theory
Supersymmetry, **2**:671
Supplementary angles, **1**:25
Surface(s)
    Gaussian curvature, **1:257–258**
    geodesic curvature, **1:259–260**
    orthogonal, **2**:463
    particle constrained to, **2**:471
    of revolution, **2**:572
    Riemann curvature, **2**:546, **2**:550
    See also Differential geometry; Manifolds
Surface area, **1**:31, **1**:32
    as multiple integral, **2**:437
    of sphere, **2**:582
    of torus, **2**:618
    as vector integral, **2**:642
Surface integrals, **1**:324, **2**:437, **2**:642, **2**:643
Surface tension, **1**:305, **1**:306
Surfaces and solids of revolution, **2**:593*f,* **2:593–594**
Surgery, on manifold, **1**:242, **2**:491
Surjective function, **1**:99
Survey sampling, **2**:448
Surveying
    Bessel's work on, **1**:61
    Hero's formula in, **1**:292
    Tartaglia's work on, **2**:605
    trigonometry in, **2**:578
Swallowtail catastrophe, **1**:107

Switching systems, Erlang's work on, **1**:201
Syllogisms, **1**:34, **1**:163, **2**:509, **2**:513, **2:594–595**
    symbolic logic and, **2**:598
Sylow theorems, **1**:353
Sylvester, James Joseph, **2:595–596**
Symbolic logic, **2:596–598**
    Bernay's work in, **1**:54
    binary, **2**:630
    deductive reasoning in, **1**:163
    deontic, **2**:428
    equivalence in, **1**:158–159, **1:197–198**, **1**:384, **2**:521, **2**:597, **2**:598
    many-valued, **1**:386
    metamathematics and, **2:419–420**
    modal, **2:428–429**
    model theory, **1**:68, **2:430**, **2**:603, **2**:604
    Peano's work in, **2**:477
    symbols in, **1:383–384**, **2**:596
    temporal, **2**:428
    See also Aristotelian logic; Boolean algebra; Connectives;
        Predicate logic; Propositional logic; Rules of inference
Symbols, **2:402–403**
    additive notation, **1:6–7**
    in algebra, **1**:14, **2**:403
    Diophantus and, **1**:17
    formalism and, **1**:235, **2**:629
    Japanese notation, **2**:565–566
    logical, **1:383–384**, **2**:596
    for multiplication, **2**:507
    multiplicative notation, **2**:403, **2**:437, **2:438–439**, **2**:507, **2**:577
    numerals, **2:456–457**
Symmetric difference, of sets, **1**:16
Symmetric linear operators, **1**:336
Symmetric property, of equivalence relation, **1**:129
Symmetric Venn diagram, **2**:645
Symmetry, **2:598–599**
    axis of, **1**:22, **1**:172, **2**:520, **2**:598, **2**:599
    defined, **2**:599
    dynamic, **1**:269
    of extension field, **2**:527
    line of, **1**:125, **2**:598
    neutrino experiments and, **2**:662
    in plane geometry, **2**:487
    point of, **1**:125, **1**:172, **2**:598
    in quantum field theory, **2**:675
    rotational, **2**:481, **2**:555–556, **2**:599, **2**:621
    of tilings, **2**:481, **2**:608
    translational, **2**:599, **2**:621–622
Symmetry groups, **2**:608
Symplectic groups *Sp(n)*, **1**:367, **1**:368
Synchronous orbit, **2**:461–462
Synclastic surface, **1**:257
Synesius, **1**:306, **1**:307
Synodic month, **1**:92
Synodic periods, **1**:140
Synthesizers, audio, **2**:407
Synthetic geometry, **2**:489, **2**:490, **2**:587
System, thermodynamic, **2**:612
Système International (SI), **1**:32, **2:421–423**, **2**:636–637
    decimal comma in, **1**:160
    electric charge in, **1**:143
    See also Metric system
Systems of differential equations, **1**:173

Systems of linear equations, 1:15, 1:16, 1:374, **2:599–601**
  homogeneous, 1:172, 1:374
  in K-theory, 1:39
  linear algebra and, 1:372, 1:373
  in linear programming, 1:375–376
  matrix solution, 1:372, 1:373, 1:374, 2:410, 2:600, 2:601
  See also Cramer's rule
Szegö, Gabor, 2:495
Szilard, Leo, 1:194

T

Tables
  of data, 2:585
  as sequences, 2:567
  trigonometric, **2:624–625**
Tangent, etymology of, 1:230
Tangent function, 1:247–248, 2:623, 2:626, 2:627
  graph of, 2:627
  hyperbolic, 1:310
  inverse, 1:327
  Regiomontanus' introduction of, 2:538
  tables of, 2:624–625
Tangent line, 1:17, 2:463
  as best linear approximation, 1:375
  calculation of, 1:90
  derivative and, 1:172–173
  Roberval's methods, 2:551
  slope and, 2:577–578
  on surface, 1:259
Tangent plane, 1:260, 2:463
  of Lie group, 1:367
  normal vector and, 2:642
Tangents, law of, 2:623
Taniyama, Yutaka, 1:221, 2:669
Taniyama-Shimura conjecture, 1:221, 2:607, 2:669, 2:670
Tarski, Alfred, **2:603–604**
  Lesniewski and, 1:364
  Lukasiewicz and, 1:386
  on measurable sets, 2:415
  model theory and, 2:430
Tartaglia (Nicolò Fontana), 1:73, 1:97, **2:604–605**
Tauberian theory, 2:667–668
Tautochrone, 1:211
Tautologies, 1:198, 2:513, 2:598, **2:605**
Taylor, Brook, 1:319, **2:605–606**, 2:607
Taylor, Joseph, 2:541
Taylor, Richard Lawrence, 1:221, **2:607**, 2:670
Taylor's series, 1:260, 1:319, 2:420, 2:444, 2:606, **2:607**
  Hadamard's use of, 1:279
  Maclaurin series and, 2:390, 2:606, 2:607
  for Witch of Agnesi, 2:670
  See also Power series
Telephone switching systems, 1:201
Telescopes
  Copernicus' predictions and, 2:563
  of Galileo, 1:250–251
  of Huygens, 1:304
  Kepler and, 1:337, 1:339, 2:564
  parabolic, 2:467, 2:520
Teller, Edward, 1:223, 2:635
Temperature, 2:610–614
  volume and, 2:611, 2:614, 2:649

Temporal logic, 2:428
Tensor calculus
  of Levi–Civita, 1:364–365
  Weyl's use of, 2:661
Tensor product, 2:675
Tensors
  curvature, 2:550
  in general relativity, 2:661
  second fundamental, 1:257
Terminal side, of angle, 1:25
Terminal velocity, 2:446
Terminating decimals, 1:320
Terms
  negligible, **2:444**
  of series, 1:6, 2:592
Ternary Goldbach conjecture, 1:268
Tesselations (tilings), 2:480, 2:481, **2:607–608**
  decidability and, 1:159
  of Escher, 1:311
  of Penrose, 1:269, 2:480, **2:481–482**
Tesseract, magic, 2:391
Tetractys, 2:515, 2:516
Tetrahedron, 2:488, 2:497
  circumscribed sphere, 2:582
  Hilbert's problem about, 1:297
  in Plato's metaphysics, 2:489
  space-filling and, 2:489
  of triangular numbers, 1:229–230
Thabit ibn Qurra, 1:8, 1:20
Thales of Miletus, 1:247–248, 2:509, **2:608–609**, 2:610
  Pythagoras and, 2:514
Theology
  Anselm's ontological proof, 2:429, 2:503, 2:521
  Tristram Shandy paradox and, 2:628
  See also God
Theon, 1:306, 1:307
Theorems, **2:610**
  Aristotelian logic and, 2:595
  of deductive system, 1:163, 2:509
  postulates and, 2:499, 2:500
  as propositions, 2:512
  rules of inference and, 2:432
  See also Axioms; Proof
Theoretical probability, 2:505, 2:507
Theory of distributions, 2:561, 2:562
Theory of everything, 2:671
Theory of Types, 1:125, 2:531, 2:558
"There exists," 2:502, 2:503, 2:520–521
Thermal equilibrium, 2:611, 2:612
Thermodynamics, **2:610–614**
  Boltzmann's work on, 1:70–71
  first law, 2:611–613, 2:614
  Gibbs' contributions to, 1:264–265
  in hydraulics, 1:304
  information theory and, 1:321
  second law, 1:264, 1:265, 2:613–614
Thermometer, 2:610–611
Thermostat, 2:611
Theta functions, 2:528
Third law, Newton's, 1:151, 2:447–448
Thom, René Frédéric, 1:107, **2:614–615**
Thompson, Benjamin, 2:610
Thomson, J. J., 1:68, 2:490, 2:559

•

velocity, **2**:436, **2**:**644**
   zero, **1**:4, **1**:376
Vector algebra, **2**:**641–642**
Vector analysis, **2**:**642–643**
   Gibbs' development of, **1**:263, **1**:265
   Hamilton's ideas and, **1**:282
Vector derivative, **1**:167
Vector fields, **2**:642, **2**:643
   curl of, **2**:573, **2**:643
   divergence of, **2**:573, **2**:643
   gradient as, **2**:573, **2**:643
   Kuperberg's work on, **1**:349
Vector parametric equations, **2**:469
Vector spaces, **1**:376, **2**:**643–644**
   abstract, **1**:**4–5**
   of complex numbers, **1**:133
   dimension of, **1**:175, **2**:643–644
   dual, **2**:644
   Euclidean, **1**:208
   four-dimensional, **1**:110
   infinite dimensional, **2**:644
   of integrable functions, **1**:358, **2**:644
   of Lebesgue integrable functions, **1**:358
   linear maps on, **2**:411
   over finite field, **1**:5
   of solutions to linear system, **2**:601
   See also Banach space; Hilbert space
Velocity, **2**:436
   derivative of, **1**:89, **1**:174
   of escape, **1**:340
   of fluid, **1**:305–306
   instantaneous, **1**:322, **1**:**323**
   as integral of acceleration, **1**:324
   integration of, **1**:**324**
   Newton's laws and, **2**:446–447
   units of, **2**:422, **2**:637
   as vector, **2**:436, **2**:**644**
Venerable Bede, **1**:87
Venn, John, **2**:**646**
Venn diagrams, **2**:**645**, **2**:645*f*–646*f*, **2**:646
Venn graph, **2**:645, **2**:646*f*
Versed sine curve, **2**:670
Vertex (vertices)
   angle, **1**:24, **1**:25, **2**:625
   cone, **1**:134
   ellipse, **1**:22, **1**:196
   graph, **1**:129, **1**:236
   hyperbola, **1**:23, **1**:308, **1**:309
   parabola, **1**:22, **2**:467, **2**:520
   polygon, **1**:22, **2**:482, **2**:497
   polyhedron, **1**:209, **2**:497, **2**:617
   triangle, **2**:622, **2**:625
Vertical angles, **1**:25, **2**:609
Vertical asymptote, **1**:38
Vertical line, **1**:22
Vibration, **2**:**646**
   Lagrange's work on, **1**:351
   modes of, **1**:56
   Taylor's studies of, **2**:606
   See also Oscillations
Vienna Circle
   Carnap, **1**:100

Gödel, **1**:266
   Grelling, **1**:274
   Menger, **2**:417–418
Viète, François, **1**:14, **1**:18, **1**:247, **2**:**646–647**
Vigesimal numeration, **1**:185, **2**:**647–648**
Vinogradov, Ivan, **1**:268–269
Vis viva, **1**:56, **1**:59
Viscosity, **1**:305, **1**:306
Vogel, Wolfgang, **1**:27
Voltage, **2**:414
Voltaire
   Châtelet and, **1**:119
   d'Alembert and, **1**:152
   Maupertuis and, **2**:412
Volterra, Vito, **1**:279, **2**:**648–649**
Volume, **2**:**649–650**
   of cube, **2**:439, **2**:501, **2**:650
   of cylinder, **1**:29, **2**:650
   of gas, **2**:613
   as multiple integral, **2**:437
   in orthogonal coordinates, **2**:573
   of parallelepiped, **1**:318
   pressure and, **2**:649
   of solid of revolution, **2**:572, **2**:594
   of sphere, **1**:29, **2**:501, **2**:582
   thermal expansion of, **2**:611, **2**:614, **2**:649, **2**:650
   of torus, **2**:618
   units of, **2**:423, **2**:636, **2**:637, **2**:649–650
Volume element
   in cylindrical coordinates, **2**:437
   in spherical coordinates, **2**:437
Von Koch, Niels Fabian Helge, **2**:579, **2**:**650–651**
Von Neumann, John, **2**:**651–653**
   computers and, **1**:88, **1**:89
   Feynman and, **1**:222, **1**:223
   game theory and, **1**:254, **2**:436
   Monte Carlo method and, **2**:635
   set theory and, **1**:54, **2**:680
   Ulam and, **2**:635
   Wiener and, **2**:668
Von Neumann algebras, **2**:651
Von Neumann architecture, **1**:88
Von Neumann-Bernays system, **1**:54
Voting methods
   Carroll's work on, **1**:101
   paradoxes of, **2**:**653**

W

Wallis, John, **2**:484, **2**:**655–656**
Wang, Hao, **2**:481
Wang dominoes, **2**:481
Wang tiles, **1**:159
Wantzel, Pierre, **1**:185, **1**:207, **2**:627–628
Waring, Edward, **2**:**656–657**
Waring's problem, **2**:656
Waring's theorem, **2**:656
Wasan, **2**:565
Water
   steam engine, **2**:614
   thermal properties of, **2**:611, **2**:612, **2**:649, **2**:650
Waterwheel, Poncelet and, **2**:499
Watson, Thomas, Sr., **1**:11

For Reference

Not to be taken from this room